TELECOMMUNICATIONS FOR MANAGERS

Fifth Edition

Stanford H. Rowe II
Dow Corning Corporation
Midland, Michigan

page 465

Upper Saddle River, New Jersey
Columbus, Ohio

Library of Congress Cataloging-in-Publication Data
Rowe, Stanford H.
Telecommunications for managers / Stanford H. Rowe, II.—5th ed.
p. cm.
Includes bibliographical references and index.
ISBN 0-13-028486-6 (alk. paper)
1. Telecommunication. 2. Business—Communication systems.
I. Title.
HE7631.R69 2002
651.7—dc21 2001016305

Editor in Chief: Stephen Helba
Assistant Vice President and Publisher: Charles E. Stewart, Jr.
Production Editor: Alexandrina Benedicto Wolf
Design Coordinator: Robin G. Chukes
Cover Designer: Rod Harris
Cover Image: SuperStock
Production Manager: Matthew Ottenweller

This book was set in Palatino by Carlisle Communications, Ltd. It was printed and bound by R. R. Donnelley & Sons. The cover was printed by Phoenix Color Corp.

10 9 8 7 6 5 4 3 2 1
ISBN 0-13-028486-6

PREFACE

Telecommunications for Managers, Fifth Edition, presents a complete introduction to the fast-paced world of telecommunications. Designed for a first course in this field, it covers all facets of telecommunications, including both data and voice communications, as used by many types of organizations.

AUDIENCE

Telecommunications for Managers is aimed at the individual who has no background in telecommunications other than that which is obtained in the course of daily living. Some knowledge of information systems is assumed, but no more than a student would gain by working with a personal computer, by taking an introductory computing course, or by being a regular user of the Internet.

Students at all levels, from community colleges and technical schools through programs at universities, will be able to learn about telecommunications from this book. This has been clearly demonstrated by the success of the first four editions. In addition, *Telecommunications for Managers* will be helpful to people in industry who need to understand more about telecommunications concepts and terminology. The book's comprehensive coverage makes it a useful reference tool.

HIGHLIGHTS

It has been said that telecommunications is one of those subjects where you need to know everything before you can learn anything. *Telecommunications for Managers* reverses the cycle by presenting the material in a logical, building-block fashion. Words and terms are defined and explained when they are first used. Examples are used as a foundation and then expanded with more detail. By the end of the book, the student will have an excellent understanding of the subject.

Telecommunications for Managers takes an outside-in approach to its subject. Telecommunications applications familiar to everyone are discussed first as a way of easing the student into the material and explaining *what* the subject of telecommunications includes. The examples, references, and case studies all come from real-life settings and illustrate the uses of telecommunications. The student will learn *why* companies and corporations feel that telecommunications is vitally important as well as how the regulatory environment affects the telecommunications industry.

The International Organization for Standardization's model for communications is also introduced early in the book to give students a framework for the technical information they will be learning.

With that background, the text leads the reader into the technical details of telecommunications. The technology is explained in an easy to understand, yet thorough, manner. Current and emerging technologies are covered as well as traditional material. The student will gain an in-depth understanding of *how* telecommunications works and how networks are designed, constructed, and managed.

Equipped with an understanding of the applications and technical details of telecommunications and networks, the student is then introduced to the management of a telecommunications department. The student will learn *why* it is necessary to manage telecommunications. The functions of the telecommunications department are examined in detail, and alternative ways of organizing the department are shown. The book concludes with a quick look at several emerging applications and issues that telecommunications managers will be dealing with in the next few years.

CHANGES TO THE FIFTH EDITION

The fifth edition of *Telecommunications for Managers* has been updated with new technical material that is relevant in today's telecommunications environment.

Wireless Greatly expanded coverage of wireless communications has been added in several chapters of the book, including the use of cell phones to access the data networks and the Internet, and wireless LANs.

High-Speed Telecommunications Information about cable modems, DSLs, satellites, and ATMs has been updated and expanded. A comparison chart has been added that shows the pros and cons of the different high-speed ways to access the Internet.

Network Security Additional information about the importance of network security, security techniques, and public key encryption has been added, and the existing information was updated.

URLs Added The Internet addresses (URLs) of many telecommunications companies and resources have been added so that professors and students can get the most up-to-date information or do additional research on selected topics.

Sidebars A number of sidebars have been added throughout the book. Drawn from the trade and popular press, these examples bring telecommunications to life by describing real-world situations.

Voice Communications *Telecommunications for Managers* provides the most complete coverage of voice communications of any comparable beginning telecommunications text.

General Updating The entire book has been updated to include the most recent advances in the field and to make the examples relevant to today's students.

Improved Pedagogy The fifth edition has additional review questions at the end of all of the chapters, URLs for additional research, and an instructor's guide with accompanying CD-ROM that contains all of the artwork in the book.

ORGANIZATION

Telecommunications for Managers, Fifth Edition, has been reorganized to group topics even more logically than in earlier editions. The book is divided into six sections. Part One deals with the regulatory and architectural environment in which telecommunications operates.

- Chapter 1 introduces the subject matter and leads students to understand that they may know more about telecommunications than they realize, just by virtue of daily living experiences.
- Chapter 2 discusses the external environment. Deregulation and competition are covered, and the nature of the telecommunications industry is explained.
- Chapter 3 introduces and explains the ISO-OSI model for telecommunications and the TCP/IP protocol suite, which has been widely implemented. The importance of telecommunications standards is explained, and the need for both architectures and standards is discussed in detail. The advantages and disadvantages of both architectures and standards are explained. This chapter provides the student with a framework for the technical material in later sections.

Part Two examines the way that both data and voice applications use telecommunications technology. This material provides a reference set of applications for the student throughout the rest of the text.

- Chapter 4 examines several telecommunications applications in detail. It expands on applications that were introduced in Chapter 1 and introduces many others. Electronic mail receives considerable attention because of its importance in our society today.
- Chapter 5 explains voice telecommunications, with particular emphasis on the business setting. Wireless telephone systems are described and explained.

Part Three looks at the user interface to the telecommunications network.

- Chapter 6 describes various types of communications terminals, personal computers, and workstations and how they are used in a telecommunications environment. This material, familiar to many students, serves the purpose of setting a common level of knowledge.

Part Four delves into the technical details of telecommunications.

- Chapter 7 discusses how data is encoded for computing or telecommunications. The student learns the requirements for a good coding system, and why certain codes are appropriate and others not appropriate for telecommunications use. Data encryption is also explained.
- Chapter 8 explains how data is transmitted and how the terminal is interfaced to the communications circuit with a modem or other device. Four cases are studied: transmitting analog signals on analog circuits; analog signals on digital circuits; digital signals on analog circuits; and digital signals on digital circuits.
- Chapter 9 describes communications circuits. Various media are studied and their attributes are discussed. Multiplexing, error detection, and error correction are explained.
- Chapter 10 explains the data link control protocols, the "rules of the road" for communications circuits. Both wide area network and local area network protocols are emphasized.

Part Five examines the design and operation of all types of telecommunications networks.

- Chapter 11 presents the fundamental concepts of networks, showing how circuits can be combined into different topologies to form networks. Circuit switching, packet switching, frame relay, and asynchronous transfer mode are all explained. Internetworking is described and illustrated by referring to the Internet. Network connections to the computer and the role of computer software are discussed.
- Chapter 12 focuses on LANs: topologies, media access control, protocols, routing, standards, and real implementations such as Ethernet and token rings. The need for LAN software such as net-

work operating systems is explained, and selection criteria for choosing a LAN are provided. Managing the LAN is another topic covered in the chapter.

- Chapter 13 describes the management and operation of networks. Day-to-day operational procedures are explained, as well as the functions of problem management, performance management, configuration control, and change management. The alternative of outsourcing one or more of these activities is introduced. The critical role of the communications technical support staff is also covered.
- Chapter 14 explains the process of designing a network and implementing it. This process is broken into phases, each of which is discussed in detail. LAN and voice network design are covered.

Part Six deals with the management of the telecommunications department in the company.

- Chapter 15 focuses on the need for management, alternative ways to organize the telecommunications department, the functions of the department, and management's responsibilities.
- Chapter 16 provides a glimpse of future applications and issues that telecommunications managers and their staff will have to understand and deal with.

PEDAGOGICAL FEATURES

Telecommunications for Managers contains many pedagogical features designed to assist both the student and the instructor.

- A set of objectives appears at the beginning of each chapter, outlining what the student should learn.
- A running case study at the end of most chapters illustrates how the concepts and techniques have been applied by a real company. Questions have been added to stimulate students' thinking about the "real world" of telecommunications.
- Many Internet World Wide Web addresses (URLs) have been added throughout the text and in an appendix to direct the student to web sites where they may find the most current information or do additional research.
- An extensive word list at the end of each chapter serves as a checklist of important terms, concepts, and ideas.
- Review questions for each chapter give students an opportunity to test their knowledge of the material.
- Problems and Projects at the end of each chapter are designed to get the student "thinking" and "doing." The problems are challenging questions that will lead the student beyond the text. In

many cases, real-world situations are presented for the student's consideration. The projects often require the student to talk to telecommunications professionals and users, or to research information on the Internet.
- A comprehensive glossary of terms and a separate list of all of the acronyms used in the book appear at the end.
- A list of references to standard telecommunications books as well as current articles on topical subjects appear throughout the text.

Compared with the fourth edition, the fifth edition of *Telecommunications for Managers* contains more questions at the end of every chapter, additional problems and projects, and more illustrations and photographs to aid the student's understanding.

For the instructor there is a comprehensive Instructor's Manual that includes

- suggestions for several ways to organize the course, depending on the desired emphasis and focus;
- supplemental textual material that elaborates on some of the topics covered in the text;
- transparency masters of the chapter outlines;
- CD containing PowerPoint™ slides of all figures in the text;
- answers to the review questions in the text;
- suggested solutions to the problems in the text;
- hints for the presentation of material in the classroom; and
- test bank questions for examinations.

ACKNOWLEDGMENTS

Accurate and up-to-date coverage of a fast-changing field such as telecommunications requires the input of many people. I want to thank the following individuals for their excellent suggestions and assistance during the preparation of the text: Lori Andrews, Judy Burt, Marian Cimbalik, John DiCiaccio, Chet Floyd, Jim Marshall, Phil McCullough, Jeff Butcher, and Theresa Srebinski.

Warm thanks go also to the people who reviewed prior editions of *Telecommunications for Managers.* Prof. Tom Milham of DeVry Institute, Georgia, made many helpful suggestions that improved the content of this new edition.

Special thanks go to Majorie Leeson, who first encouraged me to write this book and who has offered significant help along the way.

I want to express my appreciation to my employer, Dow Corning Corporation, for permission to use the case study material and a number of photographs, and for their support in many other ways. In particular, my manager Barbara S. Carmichael, and dotted line manager Allen C. Ludgate, and my administrative assistant Denise A. Allen deserve special mention. Without their direct and indirect support it would not have been possible for me to do the fifth edition.

The team at Prentice Hall also deserves special mention. Charles Stewart provided encouragement and many helpful suggestions about changes that should be made to this edition. Alex Wolf provided excellent production coordination during the project. Others who made substantial contributions for which I am grateful are Rachel Besen, copy editor, and Melinda Anderson, photo researcher.

Finally, as always, I want to recognize the wonderful support of my wife, Pam, who put up with my schedule and me during the writing process. She was extremely flexible and tolerant of my using vacation time and disappearing in my office for hours on end while I worked on the fifth edition. Without her love and ongoing support, the fifth edition would not have become reality.

Stanford H. Rowe
Midland, Michigan

CONTENTS

To Steven Manning Rowe
September 26, 1978–March 28, 2000
My son who had so much to contribute but so little time.

TELECOMMUNICATIONS FOR MANAGERS

PART ONE

Introduction to the Business Telecommunications Environment

The purpose of Part One is to introduce you to telecommunications concepts, to give you background material about uses of telecommunications in business, and to explain the environment in which business telecommunications operates.

Chapter 1 introduces the subject of telecommunications, explains some basic vocabulary, and gives some simple examples of ways in which telecommunications is used. A brief history of telecommunications and an introduction to telecommunications regulation also are included.

Chapter 2 examines the external forces that shape the way telecommunications is used within business. The history of telecommunications regulation and the movement to a largely deregulated and competitive industry are described. An overview of today's telecommunications industry and major companies is given.

Chapter 3 discusses the architectures and standards that shape the way telecommunications systems are designed and operated. The need for these standards and architectures is discussed as well as their advantages and disadvantages. The International Organization for Standardization's model for Open Systems Interconnection (ISO-OSI) is introduced and explained. This model will give you a framework on which to build your telecommunications knowledge and will serve you well as you study the technical material later in the text.

After studying Part One, you will be familiar with basic telecommunications terms and concepts, understand a few of the ways that telecommunications can be used in business, and be aware of the forces that shape the way telecommunications technology and techniques can be applied. You will also have good knowledge of the ISO-OSI telecommunications model which will serve as a reference for your further study.

CHAPTER

1

Introduction to Telecommunications

OBJECTIVES

After you complete your study of this chapter, you should be able to

- define telecommunications and data communications;
- describe the basic elements of a telecommunications system and give examples of all of them;
- explain why telecommunications is an integral part of the contemporary business environment;
- explain why it is important to study and understand something about telecommunications;
- explain many of the requirements for voice and data communications systems;
- give several examples of the way telecommunications is used in business;
- summarize telecommunications history;
- explain why it can be difficult to keep your telecommunications knowledge up-to-date.

INTRODUCTION

We are members of the information age and the networked society, and the ability to share and communicate information and knowledge is more important than ever—and it is more possible. Every year brings an accelerating number of new discoveries about every facet of our world and the universe. Although information known only to one person can be very useful to that person for decision making, knowledge and information are most useful when they are shared with others. Sharing means communicating, and communicating can occur in many ways. A raised eyebrow, a shrug, a quizzical expression, a posture, a stance—all are effective ways of *body language* communication in certain situations. Communication can occur in other forms too. Music, art, and dance come immediately to mind as ways of expressing feelings and emotions that sometimes stir and inspire the soul.

Another example is *written* communication. Books, newspapers, magazines, and graffiti scrawled on a wall all exist to convey messages to readers. In some cases, the intent is to convey information that the reader wants to receive; in other cases, the writer wants to express his or her thoughts, ideas, or feelings on a topic of importance to him or her.

Radio and television are two forms of *broadcast* communication that cannot be ignored. Every day, most people around the world are bombarded by broadcasts designed to entertain, sell, or provide information.

Although one can ignore specific broadcasts, it is virtually impossible to deny the strong impact this type of communication has on our daily lives.

The amount of information that is available to us is accelerating rapidly. More books are being published and magazine articles written than ever before, to say nothing of the increasing number of research reports and all of the data generated by computers. Sharing this vast information bank is a staggering proposition, but all over the globe the *networked society* is taking form and growing at an equally staggering pace. At the start of the 1990s an estimated 1 million people were exchanging information through a primitive *Internet.* As joining the Internet became easier its population swelled to several hundred million at the beginning of the twenty-first century.

Internet

People log on to the Internet because it gives them value in the form of access to information that they never dreamed possible. Robert Metcalfe, who invented the Ethernet, a local area network that will be studied in Chapter 12, says that the power of a network—how much it can do—is measured by the square of the number of connected machines: $P(n) = n^2$. The implication is that the rapid growth of the Internet has given us an information tool that allows us to access unimaginable amounts of information. To say that the pace is breathtaking is probably an understatement!

DEFINITION OF COMMUNICATION

Communication is defined as "a process that allows information to pass between a sender and one or more receivers" or "the transfer of meaningful information from one location to a second location." *Webster's New World Dictionary* adds, "the art of expressing ideas especially in speech and writing" and "the science of transmitting information, especially in symbols." Each of these definitions has important elements. Communication is a process that is ongoing, and it obviously can occur between a sender and one or more receivers. The word *meaningful* in the second definition is significant. It is clear that communication is not effective if the information is not meaningful. One could also argue that if the information is not meaningful, no communication has occurred at all. There is a strong analogy here to the traditional question: If a tree falls deep in the forest and no one is within earshot, does the tree make a sound as it falls? The answer is that because sound requires a source and a receiver, without the receiver, there is no sound.

The last definition of communication relates most closely to the focus of this book. You will be studying the science of communication using electrical or electromagnetic techniques. The information being communicated will be in coded form so that it is compatible with the transmitting and receiving technologies. For our purposes, the practical applica-

The use of personal computer workstations has become so common throughout industries of all types that most workers use one or work with someone who does. Most of these workstations are connected to others using some form of telecommunications. (Courtesy of Terry Vine/Tony Stone Images)

tion of the communication science in the business environment is essential to the definition and study of communication.

DEFINITION OF TELECOMMUNICATIONS

Webster also tells us that the prefix *tele* means far off, distant, or remote. Practically speaking, the word *telecommunications* means communication by electrical or electromagnetic means, usually (but not necessarily) over a distance. Not long ago, telecommunications meant communication by wire. Although this is still accurate in many situations, it is not complete because telecommunication can also occur using optical fiber or radio waves.

DEFINITION OF DATA COMMUNICATIONS

Data communications is the movement of coded information from one place to another by electrical or electromagnetic means. Data communications is generally considered a subset of telecommunications that excludes voice communications. That distinction was accurate and useful years ago; however, today the notion of a subset is too narrow. In modern communication systems, voice transmissions are usually converted to digital signals. When this occurs, voice and data signals are indistinguishable from each other, and they appear and can be handled by the communications network in the same way. Voice communications can, therefore, be viewed as a subset of data communications, and the terms *data communications* and *telecommunications* can be used interchangeably. That is how they will be used in this book.

Data and Information

Perhaps you are not sure about the difference between *data* and *information*. According to the American National Standards Institute's *Dictionary for Information Systems,* the term *data* means "a representation of facts, concepts, or instructions in a formalized manner suitable for communication, interpretation, or processing by human beings or by automatic means." *Information* is "the meaning that is currently assigned to data by means of the conventions applied to those data." It's the meaning of the

This office worker is checking the status of her company's inventory while talking to a customer on the telephone. She will be able to tell the customer if the part he wants to order is available for immediate shipment. (Courtesy of Chris Marona/Photo Researchers, Inc.)

data that makes it information. The value of the information depends on its relevance to the individual receiving it.

Another way to look at the breadth of telecommunications is to look at four component parts that make up most of industry today.

Voice communications, also known as telephony, which is growing annually between 6 percent and 8 percent.

Data communications, which is the fastest part of the telecommunications industry at nearly 30 percent annually. Data communications is the foundation for computer networking, including local area networks and wide area networks, which you will study in Chapters 11 and 12.

Video communications, including video monitors for security, information displays, such as the flight information displays you see at the airport, and videoconferencing, which allows people in different locations to have a meeting and see each other.

You'll see, as we begin looking at the technology, the differences between these types of communications are rapidly disappearing. Inside networks, most voice, data and video is transmitted in digital form—a series of binary 1s and 0s. If you just look at the 1s and 0s and don't know how to interpret them, you can't tell whether they represent a voice conversation, a television picture, or a file being transferred between two computers.

Students often ask, "What's the difference between data transmission and text transmission or e-mail?" At the fundamental level of 1s and 0s, the answer is nothing—there is no difference. The difference is mainly in the way we think about what we're sending from one location to another. Usually, we think of text communications or e-mail as being relatively short transmissions, whereas when we say data transmission, we are usually thinking about a moderate or large quantity of data transferred between computers or other devices. The distinction is not solely based on size, however, because we could transfer a very small file between two personal computers, and call that data transmission, or we could send a very long e-mail. If you're feeling confused, don't worry. As you begin studying telecommunications, the distinctions, and the way the terms are used will become clear to you, and it won't be long before you're speaking the language of telecommunications like a pro!

BASIC ELEMENTS OF A TELECOMMUNICATIONS SYSTEM

A telecommunications system contains three basic elements: the *source, medium,* and *sink.* These are illustrated in Figure 1–1. More common terms would be *transmitter, medium,* and *receiver.* Examples of each element in

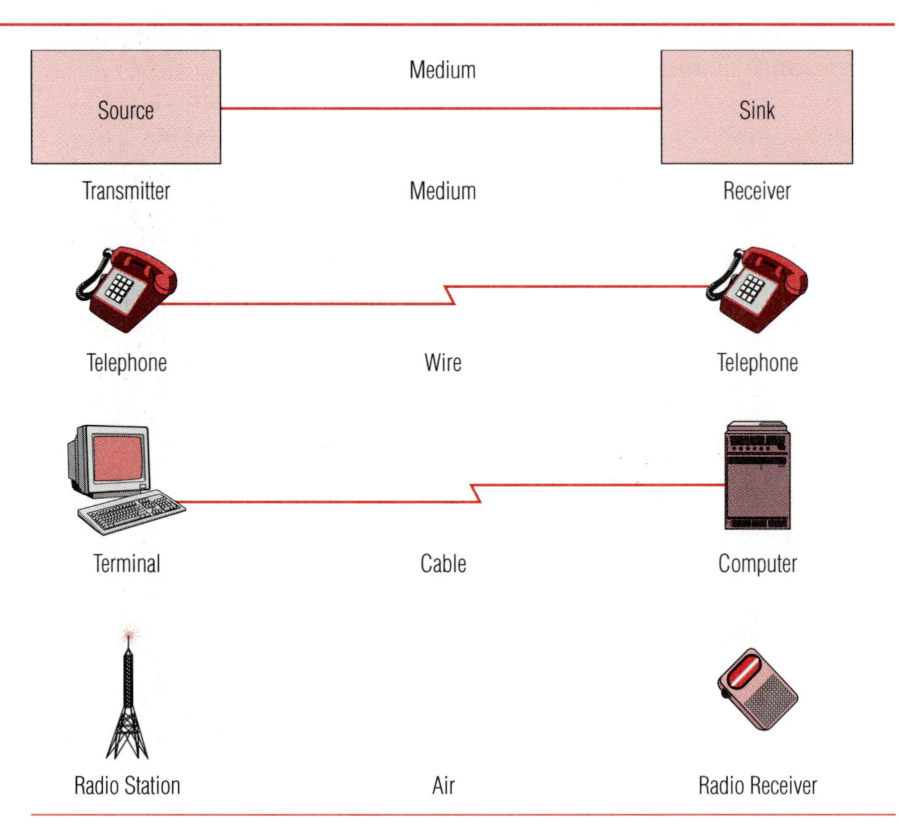

Figure 1–1
Basic elements of a telecommunications system.

the system are abundant: voicebox, air, ear; telephone, telephone lines, telephone; or terminal, data circuit, computer. There are many possible combinations of each of the basic elements.

We will immediately begin referring to the medium as the *communication line.* You can think of it as the telephone wire that comes into your house or office to connect your telephone to the telephone company office. The line can be implemented in many physical forms, such as copper wire or microwave radio, but visualizing it as a copper wire is convenient and not inaccurate.

communication networks

Communication lines are connected in many ways to build *communication networks.* It is easy to think of the telephone network as a series of lines connecting telephones to telephone company offices. As you undoubtedly realize, there are also many lines that connect the telephone company offices with each other. Together, all these lines make up the *telephone network.* In a similar way, businesses frequently build or lease their own networks, which connect all their locations. Lines and networks are precisely defined and more thoroughly discussed in Chapters 9, 11, and 12.

To ensure that communications will be successful and effective, rules must guide its progress. Although these rules are not technically a basic

SIDEBAR 1–1

BLOCK DIAGRAMS

Note that in this context, the word *system* means the complete process of communicating. Figure 1–1 is a very simple and high-level type of *block diagram* that is designed to illustrate the communications system. Block diagrams can be used to illustrate any process or flow of work. Perhaps you have seen or used a block diagram showing the process of registering for a class at school. From the diagram you can see what steps have to be performed to complete the process. Block diagrams are usually read left to right and/or top to bottom, and each block represents a step or stage in the overall process. What is sometimes confusing when you are first learning to read block diagrams is that they can exist at any level of detail. Figure 1–1 is called a high-level diagram because it shows very little detail. A low-level block diagram would show much more detail about what actually goes on when a telephone call is made or when a terminal is connected to a computer. You'll see diagrams of that type later in the book.

element of telecommunications, they are absolutely necessary to prevent chaos. For two entities to communicate successfully, they must speak the same language. If you are a person who speaks only Spanish and you try to communicate with someone who speaks only German, you won't be successful. Perhaps, through trial and error, you'll be able to negotiate a compromise, such as both of you speaking English, in which case you'll be at least partially successful in your communication, depending on how well both of you speak your second language.

communication rules

At a different level, there are unwritten rules that guide our use of the telephone. Most of us learned them when we were very young. When receiving a telephone call, the answering party traditionally initiates the conversation by saying "hello" if he is American. The caller and answerer then go through a brief dialogue to identify each other before they launch into the purpose of the call. This is illustrated in Figure 1–2. The process is often shortcut by combining some of the exchanges or when one of the parties recognizes the other's voice. Some examples of this process are shown in Figure 1–3.

In some aspects of a telephone communication, well-defined rules don't exist. For example, what happens if the conversation is cut off due to some fault in the telephone equipment? Who calls whom to reinitiate the conversation? When you have been cut off, have you ever received a busy signal when you redialed the call because the person you were talking to also tried to redial? Eventually one person waits, and the problem is solved.

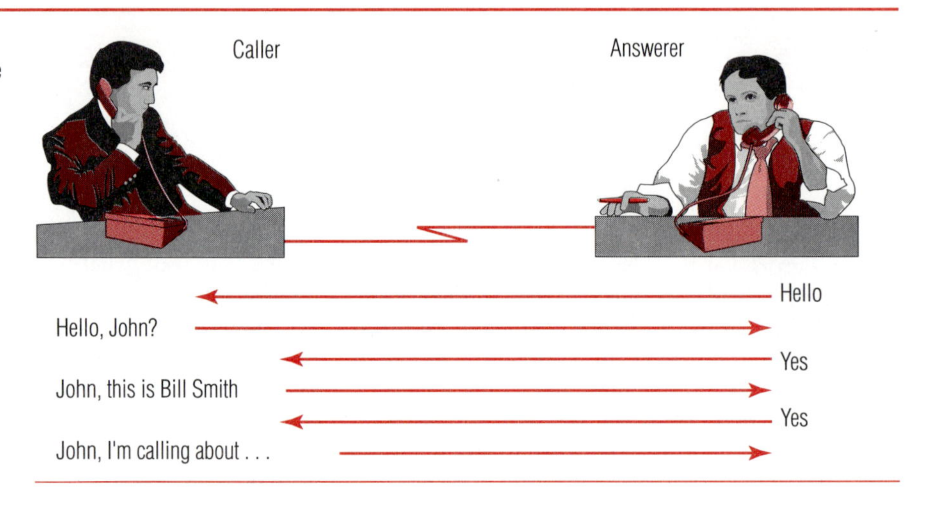

Figure 1–2
Unwritten rules of telephone communication: the initiation of the conversation.

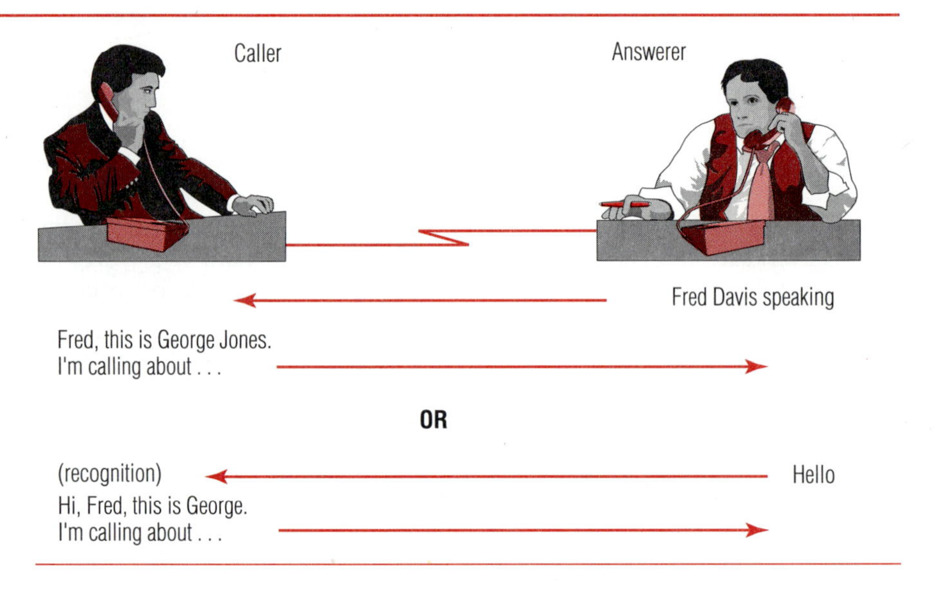

Figure 1–3
Shortcutting the unwritten rules of telephone call initiation.

In a similar way, when two pieces of equipment such as a terminal and a computer communicate, rules are needed to determine which device will transmit first, how the terminal and computer will be identified to each other, what happens if the communication gets cut off, and so forth. Unlike the voice communication example, however, when equipment is communicating automatically, all of the rules must be defined precisely, and they must cover all situations—usual and unusual—that can occur. The rules of communication are called *protocol.*

protocol

SCOPE OF THIS BOOK

For the purposes of this book, telecommunications includes the transmission and reception of information using electrical or electromagnetic means from a transmitter to a receiver over a medium. Various types of transmitters, receivers, and media, and alternative sets of rules that can be used for those communications, will be introduced and described in some detail. The primary focus of this book is the communication of information in commerce or business taken in its broadest sense. Although most of the examples will come from business and industry, the points being made will in most cases apply equally well to governmental, educational, medical, or other not-for-profit institutions. The types of communication that will be discussed involve the transmission and reception of voice, data, text, graphics or images, or combinations of these forms. You will see that when they are reduced to their most basic electrical form, all of these communications look alike. There are no fundamental differences among voice, television and data transmissions. This book does not specifically describe or analyze commercial radio or television; however, most of the principles and techniques described here apply equally well in those settings.

It is important for you to realize that this is an introductory book. Most of the chapters in this book are themselves the subject of other books that delve into far more detail than is appropriate for an introductory course in telecommunications. Furthermore, communications engineering is highly technical and mathematical, and is beyond the scope of this book. You will find, however, that *Telecommunications for Managers* gives you an excellent introduction to this complex field and provides a solid foundation for further study if you decide to expand your knowledge of the subject.

IMPORTANCE OF TELECOMMUNICATIONS TO BUSINESS

Melding of Information Systems and Telecommunications

Telecommunications is an important part of the information systems environment in many companies. Certainly, if an organization has more than one location, there is a high potential for data communications. But even if the company has only a single location, it is likely to have multiple personal computers that are candidates to be interconnected on a small, local network so that data and information can be shared. Or, if the company has a mainframe computer, it is likely that data communications is used to connect terminals located more than a few hundred feet from the computer room. Data communications with or between computers is one of the fastest-growing segments in the communication marketplace and the foundation of the Internet.

Furthermore, modern telephone switching equipment is a specialized type of computer, designed to handle voice conversations instead of data. It has many of the same capabilities and physical requirements as the traditional data processing computer. Many companies have realized that they can improve their communication efficiency by managing all of their voice *and* data communication activities together. Communication lines can be shared, costs can be managed, and new technology can be assimilated in ways that provide more effective communication service to the employees of the company. This topic is discussed in greater detail in Chapter 15.

Having Information Available in the Right Place at the Right Time

Businesses in the information age generate more information faster than ever before. At the same time, most companies realize that information is an increasingly valuable asset that must be managed with the same care and attention as the company's finances, buildings, machines, and people. Having the right information in the right place at the right time can mean the difference between the profitability and unprofitability of the business—and ultimately can determine its success or failure. Having information available to the right people isn't a new requirement of business success. What is new is the amount of information that can be delivered near or far at speeds that were unthinkable a few years ago. For years, companies have had telephones for voice communication, and reports and analyses were hand-generated and mailed to arrive days later. Now, with the marriage of communication and computer techniques, you can make the detail report or summary analysis available anywhere virtually instantaneously. People in widely different locations can look at the data and discuss it soon after it becomes available.

In the world at large, but especially in today's competitive business world, information is power. Organizations that understand what their competitors are doing, and can quickly and efficiently distribute that information to their marketing and sales people in the field, will have a leg up on their competition. Organizations whose employees keep in touch with one another using electronic mail find that their productivity increases dramatically. Transportation companies that can track the location of their vehicles and their customers' shipments have a significant advantage over competitors that don't have similar capabilities. News organizations with a large global presence and sophisticated telecommunications networks dominate their industry. Startup companies have grown to powerhouses in their industry by focusing on doing business entirely on the Internet. The more information you or your company has about a particular subject of interest, the better positioned you are to lead, control, or teach others.

Capturing Basic Data About Business Operations as They Occur

Another reason that telecommunications is increasingly important is that, combined with the computer, it is being used as an input mechanism for capturing data about the basic operations of the business as soon as they occur. Online computer applications, in which a terminal is connected directly to a computer with virtually instantaneous communication, are being used by businesses to enter customer orders, record customer payments, give notification of product shipments, and track inventory. These operations are the *basic business transactions* that are fundamental units of business operations. Once data about the business transaction is captured and stored in a computer, it is available, via telecommunications, to others.

basic business transactions

A classic example is the airline reservation system, in which the traveler's airline representative or travel agent makes a flight reservation. Information about this reservation is recorded in a computer database and has the effect of reducing the number of seats that are available for the particular flight. This reduction of seat inventory is a basic business transaction of the airline. As soon as the database is updated, other airline reservation agents or travel agents in other locations around the country or around the world can check to see if seats are still available on the flight for their customers. Imagine how this was done before computers and telecommunications.

In another case, a customer who wishes to order a product from a company calls the company and talks to an order processing clerk. Through a computer terminal, the clerk can check to see if the required product is available or, if not, when it will be. If the product is available and the customer places the order, shipping instructions can be processed in the warehouse nearest the customer to ensure the most rapid delivery. Later, the reports detailing this and all other sales transactions can be generated and transmitted to analysts who can spot trends or detect inventory problems quickly. These reports often generate follow-up questions that analysts and management confer about over the telephone or through terminals.

In an increasing number of cases, customers might do most of the ordering directly through the Internet. Online shopping for both consumer and industrial products is increasingly widely used, and while the customers do more of the work themselves, they are also in control of the transaction and can seek as much or as little information as they need from the vendor's web site, assuming of course that the web site is properly designed and rich in content.

With online systems, businesses are increasingly dependent on the computing and telecommunications technologies on which those systems are based. In the transportation, insurance, and finance industries, the dependence on telecommunications and computer systems has become critical. If a massive failure occurred, business could not be conducted, and financial failure would follow. As a result, some companies have gone to

Satellite antennas, such as this dish, can transmit many different kinds of telecommunication signals, such as voice, data, and television, at the same time. (Courtesy of Mark Gibson/Corbis)

great lengths to ensure that they have complete backup facilities in place so that major computer and telecommunications "downtimes" cannot occur. Most companies find that whether or not the computer and telecommunications systems are vital to their operation, it is prudent to make contingency plans for how business operations would be conducted if a prolonged outage did occur.

Allowing Geographic Dispersion of Facilities and People

Telecommunications allows people in diverse locations to work together as if they were in close proximity. Branch banking clerks, car rental agents, and insurance agents can all share common information and have most of the same capabilities they would have if they were located in the home office. In some industries, it is much less expensive to do manufacturing offshore in foreign countries such as Malaysia or the People's Republic of China. Companies in these industries need telecommunications to connect their far-flung operations and eliminate the barriers that distance would otherwise impose.

At the same time, companies in industries that are not *required* to have operations in widely separated geographic regions have a relatively inexpensive opportunity and the flexibility to do so, thanks to telecommunications. While corporate headquarters remain in a major metropolitan area, such as New York City, other facilities, such as sales and marketing offices, can be located close to customers, and manufacturing plants can be located close to sources of raw materials or natural resources used in the manufacturing process. With telecommunications, the company can still operate as a single, coordinated entity and, for most purposes, ignore the geographic dispersion.

Internet Marketing

Using the Internet, many companies are making information about their products available to potential customers and selling products directly. This capability, which has only been available on a practical basis since the mid-1990s, has expanded potential markets dramatically for many companies. Small companies especially find that they are able to reach potential customers globally whom they never would have been able to contact in the past. Product descriptions and photographs can be displayed, and sometimes demonstrations can be given online. Software companies can distribute trial copies of their products for prospective customers to try. Payments can be collected by asking customers to submit credit card numbers, which can then be automatically verified. Telecommunications is providing significant new capabilities that were unthought of 10 years ago.

In summary, telecommunications is becoming an integral part of the way companies conduct business because its capabilities provide efficient, effective ways of conducting business. In some cases, the combination of telecommunications and computing allows business to be performed in ways that are impossible using manual techniques. Business can be conducted faster and more accurately, and decisions can be made with more timely information than was previously possible. It is important to point out, however, that making decisions faster and with better information does not necessarily lead to making "better" decisions, although it should contribute to their overall quality.

REASONS FOR STUDYING TELECOMMUNICATIONS

There are several important reasons for learning about telecommunications, its terminology, and its applications.

Telecommunications Is Shrinking the World

In these days of worldwide news organizations, instant financial information from stock exchanges around the world, national paging systems, global electronic mail, and countless other examples, it is apparent that telecommunications, in all forms, is becoming a significant part of all our lives. Only a few years ago, calling grandma across the country on Christmas Day was a big event—and moderately expensive. Now people pick up the phone on a whim and think little about calling or sending a fax halfway around the world to place an order with a mail-order clothing company or to make a travel reservation. People log on to information networks such as America Online to exchange messages, see information, or register complaints with vendors. There is much talk about the "information highway" which will link all of our homes and offices with incredibly fast communications of high capacity and capability. We're inundated with information about new services that are available now. Telecommunications is affecting all of us and, for the most part, improving the quality of our lives.

Direct Use on the Job

In businesses of all types, telecommunications is an integral part of the way work is done. Whether it be the automatic teller machine at a bank, the supermarket checkout scanner that reads bar codes on food products, the wand and computer terminal that the librarian uses to check books out of the public library, or the terminal that the clerk at the Internal Revenue Service uses to enter the income tax information from a tax form into the computer, telecommunications often is involved. More and more people are using computer terminals connected to computers via telecommunications lines.

It has been estimated that as far back as 1989 there was a computer terminal for *every* person in the U.S. workforce. This won't surprise you if you work in the information intense industries, or companies that rely on telecommunications to survive. In fact, today many workers have more than one terminal—perhaps a PC on their desk and a laptop to take when they work at home or when they travel. Scientists often have a PC in their office and one or more in their laboratory. Stockbrokers usually have multiple terminals so that they can keep track of several markets of interest. It is literally true that most workers in the industrialized countries use a terminal on their job in one way or another.

Indirect Use on the Job

Even if your job doesn't deal directly with telecommunications, you will most likely work with people who do. Knowledge of the subject and its vocabulary will help you communicate with "telecommunications workers." In business, it could help you request new information or services and understand problems. In some cases, people find that because of a job change or promotion, they are suddenly thrust into a new job where they work directly with telecommunications terminals or equipment. They are thankful for any telecommunications background they have obtained in the past.

SIDEBAR 1–2

A BIG 10-4 FOR THE INTERNET

E-Mail and Satellite Tracking Revamp Trucking

By Justin Bachman, *International Herald Tribune*. Reprinted with permission.

Atlanta—Ah, the lonely, tedious life of a long-haul trucker. Rest stops and road-stop coffee. Lonely phone calls home and a CB radio on the road.

Times change. Citizens-band radios are giving way to Internet sites. Truckers now punch up an electronic mail rather than wait by the phone. And global positioning systems are used to give traffic advice in a matter of seconds.

"As far as all these computers are concerned, I get directions to everywhere I go," Don Buchta said as he fueled his rig at an Atlanta truck stop on his way to Massachusetts.

Before his company installed a satellite tracking system to manage trucks and communicate with drivers, Mr. Buchta used to spend much of his time sitting by a pay phone—waiting for load orders, directions and "calling back, calling back, calling back," he said.

Now, trucks are wired for e-mail, dispatchers send route changes to a driver's personal video display and an engine diagnosis is as likely to take place while driving on the highway as it is in a maintenance shop.

One company, Park 'n View, provides access to cable, Internet, pay-per-view and telephone services at 125 truck stops nationwide. For $30 per month, subscribers receive a card that allows them to plug in.

Freight companies say that global positioning satellites, which can locate a truck to within a 50-foot (15-meter) radius, have reduced the practice of drivers' visiting girlfriends or hangouts when they should be on the road.

But most importantly to the industry, the technology cuts delays and costs.

"The pace of technology coming to this industry is just unbelievable," said Seth Skydel, editor of Trucking Technology Magazine.

A truck's maintenance schedule used to be posted on a chalkboard in the repair depot, he said. Now, systems that monitor engine performance allow technicians to receive data while the truck is on the road. Technicians can determine whether a quirky noise is cause for immediate repair or can be ignored until the driver returns.

"Computers were unheard of in this industry 10, 12 years ago," Mr. Skydel said. "The accounting people may have had them, but you never saw them in dispatch or maintenance."

Ten years ago, the largest U.S. trucking company, Wisconsin-based Schneider National Inc., became the first of the industry's major players to adopt two-way satellite communications on its 14,000-truck fleet. "It paid for itself years ago," said John Lanigan, president of Schneider's transportation division.

Despite drivers' initial suspicions that satellite tracking meant Big Brother-style surveillance, managers report that the systems have become a recruitment tool in the fierce competition for truckers.

Drivers, and especially their families, appreciate knowing that help is a push of a distress button away, regardless of location, said Andy Dougherty, operations manager at Georgia-based Ready Trucking, which spent $300,000 to wire its 135 trucks.

"They want to know that someone is looking out for them," he said. "Some employees say, 'I don't want to be watched,' but those are probably the ones that need to be."

The satellite also tracks the truck's progress, with an updated arrival time sent directly to the customer. Some haulers post the information on the World Wide Web.

Wide Use at Home

Telecommunications techniques and products are becoming more widely used every day. Nearly everyone has or will have some contact with them. At home, using the telephone is perfectly natural, but we have to make some decisions about the type of telephone service we want. We have the option of purchasing our own telephones rather than renting them from the telephone company. We have to select which company we want to carry our long distance telephone calls.

Being able to use the Internet and the World Wide Web is the reason that many people made their first use of data communications. Whether sending electronic mail, getting assistance with homework by searching for information, tracking the prices of stocks or mutual funds, or for countless other reasons, millions of people are using the Internet every day. Getting started has become so simple that people with no computer or data communications background successfully log on with no help. Making optimal, efficient, and cost-effective use of the myriad of capabilities and tapping the full potential requires some knowledge, however.

Knowledge of telecommunications is also desirable to help us make intelligent decisions about the communications services we want to purchase. Many people have cable television at home, and this is another form of telecommunications. Subscribers can select from a wide variety of available programming, some of which is included in the basic fee paid each month and some of which is available only at extra cost. In most sections of the country, it is possible to do banking using the telephone or a personal computer in the home. In the future, we will be faced with ever-increasing choices, and a knowledge of telecommunications principles and terminology will be even more useful.

A Possible Career in Telecommunications

Another reason for learning about telecommunications is to help you assess whether you might want to consider making telecommunications your career. A partial list of telecommunications jobs is shown in Figure 1–4. As the use of telecommunications grows, so does the need for knowledgeable people to design, install, repair, maintain, and operate telecommunications systems. We can all relate to the job of telephone operator, computer terminal operator, or repairperson because we have seen these people at work.

related careers

Not so obvious, however, are the thousands of people who work "behind the scenes" to sell, design, and install telecommunications systems and keep them operating reliably and properly. Network analysts and designers are people who are knowledgeable about the types of telecommunications hardware and services that are available. They identify the communication requirements for a company and then design a telecommunications solution that will provide the required capabilities. Communication programmers write special computer software that en-

Telephone operator
Terminal operator
Network control operator
Telecommunications administrative support specialist
Communication repair technician
Network analyst
Network designer
Telecommunications programmer
Telecommunications technical support specialist
Network and services sales
Telecommunications management

Figure 1–4
A partial list of telecommunications career opportunities.

ables computers to communicate with one another or with terminals. Network operators monitor the day-to-day operation of a communication network, solve problems when they occur, capture statistics about the network's performance, and assist the users of the network to take advantage of its capabilities.

All of these jobs and others are described in detail in Part Six of the text. Suffice it to say now that good, well-prepared telecommunications people—who creatively apply technology and techniques to take advantage of new business opportunities or to help solve a company's communications problems—are always in demand. They are especially valuable if they can communicate well with others in the company and have a solid business background with knowledge of finance, accounting, and marketing in addition to their specialized telecommunications knowledge. Several universities offer degrees in the technical and management aspects of telecommunications to help people get the best possible preparation for a career in this exciting field. Well-prepared people can earn a very good income because the value of their skills and the services they provide are recognized by the companies and organizations they serve.

NEW TERMINOLOGY

Like any other subject, telecommunications has its own vocabulary to master. When you first learned about computers, you were probably overwhelmed and mystified by such words as *disk, CPU,* and *file.* After some study, however, these words became a comfortable part of your vocabulary.

It is the same with telecommunications. There are many new words to learn, but your computer background will stand you in good stead and provide a foundation on which to build. Unfortunately, many people assume that since they know the language of computers, they also know and understand the language of telecommunications. This is usually not the case. There are even some words, such as *dataset,* that mean one thing in the computer sense and something entirely different in a telecommunications context. There are also quite a few acronyms to be mastered. Some are directly related to specific vendors' products, but many are used more generally. A list of commonly used telecommunications acronyms can be found in Appendix G.

There are three words used rather generically throughout the book that need to be explained up front. *Business* generically includes the business, education, government, and medical fields. A sentence that begins, "Telecommunications is becoming more widely used in," could end with business, education, government, or medicine and be equally correct.

Company is used generically to mean organization, bureau, hospital, college, or farm. A company is a specific organizational unit within the business field, just as a hospital is a specific organization within the medical field.

Virtually anyplace the pronoun *he* is used, *she* could be substituted. Telecommunications is truly an equal opportunity career field. Both men and women are commonly terminal operators, network designers, repair technicians, salespeople, and telecommunications managers.

COMMON EXAMPLES OF TELECOMMUNICATIONS

You have been exposed to telecommunications in many forms—probably more than you realize. Let's look at some common activities that rely on telecommunications technology and techniques.

Telephone Call

Figure 1–5 is a simplified diagram of the components involved when you make a telephone call. The line that looks like a small lightning bolt con-

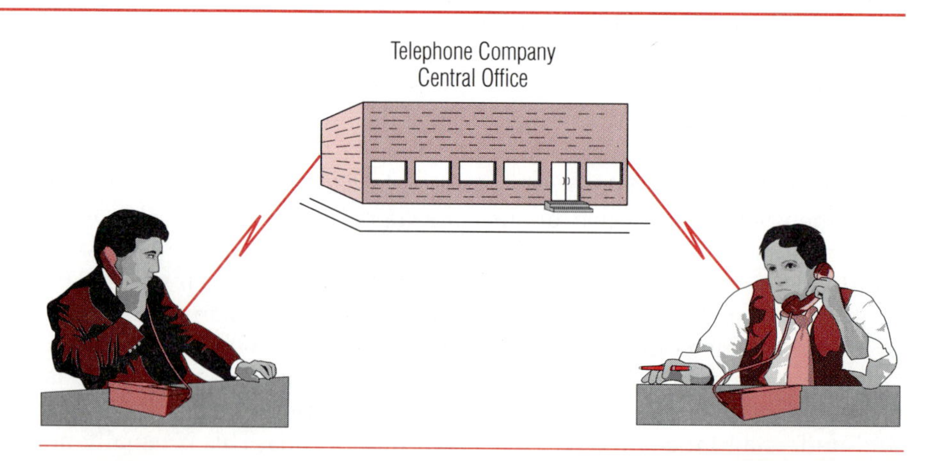

Figure 1–5
Simplified diagram of the components of a standard telephone call.

SIDEBAR 1–3

The symbol in Figure 1–5 that looks like a lightning bolt and connects the people to the telephone company central office is commonly used in telecommunications block diagrams to indicate a communications connection. This connection can be over long or short distances—no distance is implied by the symbol itself. Contrast this with the solid line connecting the computer and modem in Figure 1–6. The solid line is used to indicate a cable connection, usually of short distance, and no telecommunications. Later in the text you'll learn the difference between sending electrical signals over a simple, short cable and telecommunications signals over a longer distance.

necting the telephone to the telephone company is the common symbol used to indicate a telecommunications connection. This connection can take many forms, as you'll see in Chapter 9. As an experienced telephone user, you probably have a number of perceptions about what is involved when a telephone call is made. Probably most of your perceptions are correct. You know that when you touch the buttons on a push-button phone or dial a rotary telephone, a signal is transmitted to the telephone company that causes a connection to be made to the party to whom you wish to talk (assuming that you dialed the number correctly). Normally, you will soon hear a buzzing sound that tells you that the other telephone is ringing. With luck, the person you want to talk to will answer the telephone, and the conversation will begin.

You probably have the perception that no people (telephone operators) are involved at the telephone company in making the connection or otherwise completing the call, and you are correct. Today's telephone switching equipment in the telephone company central office is completely automatic, and the vast majority of calls are completed without human intervention. You probably also feel that the telephone system works fine most of the time, except maybe when the weather is very bad or on Christmas or Mother's Day. Again you are correct, for the vast majority of calls are completed accurately and with high quality. Occasionally, the telephone system gets overloaded, but such occurrences are rare.

The "telephone company" in Figure 1–5 is probably a bit more nebulous. You know that you have seen the telephone company building in your town, telephone poles with wires strung on them, repair trucks, telephone stores, and certainly the monthly bill, but you are probably unclear about how these pieces fit together to provide the telephone service you have come to expect. Suffice it to say that in addition to the parts of the telephone company and its operations that you know about, there are at least as many parts that you probably have never heard about or been exposed to. The details are described in Chapter 5.

Figure 1–6
The telecommunications between a home computer and an Internet access provider.

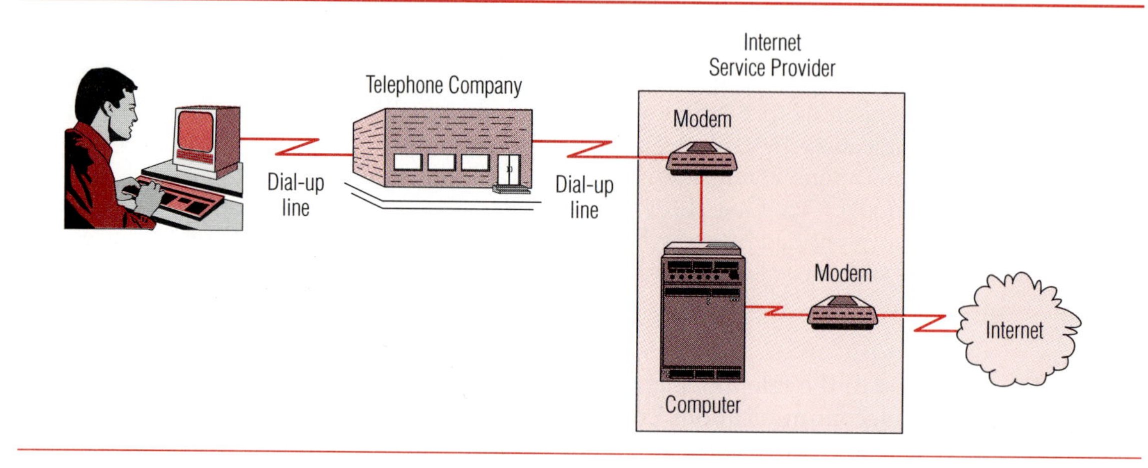

To put this example in terms defined earlier, the calling party in the example is the *source;* the connection to the telephone company, the telephone company itself, and the connection to the called party are the *medium;* and the called party is the *sink.* This is an example of person-to-person communication. The caller and receiver are the direct users of the telecommunications system, which in this case is the telephone system. They operate the terminal (telephone), and that operation is so routine and familiar that they don't think twice about it. Most people learned how to operate this telecommunications system when they were very young.

Using the Internet from Home

Many companies have direct connections to the Internet that allow instant access from any personal computer as easily as accessing software stored on the hard disk of the PC. Accessing the Internet from home typically requires a few more steps. Although there are various options for home connection, a typical one is for the home user to make a dial-up connection from a home PC to an *Internet service provider* (ISP), a company that provides a network connection to the Internet, as shown in Figure 1–6. Instructions are given to the PC's modem to dial the telephone number of the service provider's modem and computer. When the call is answered, the service provider's computer usually performs some logging and accounting operations, so that it can bill the customer, and then it automatically completes the connection to the Internet. Depending on the type of Internet service to be used, the customer may activate a *web browser,* such as Microsoft Internet Explorer or Netscape Navigator, or other software on the home PC to simplify the online activities. When the activity is complete, the user instructs the software to disconnect the call and the modems break the connection.

Internet service provider

web browser

Airline Reservation System

Figures 1–7 and 1–8 illustrate a slightly more complicated use of telecommunications: making an airline reservation. There are several variations on how the reservation can be made; this example illustrates the main points that occur when you call the airline directly. The first part of the figures are the same as Figure 1–5: You make a telephone call through the telephone company to a person—the airline reservation agent. The reservation agent

Figure 1–7
The telecommunications connections between a traveler and an airline reservation computer through a local area network.

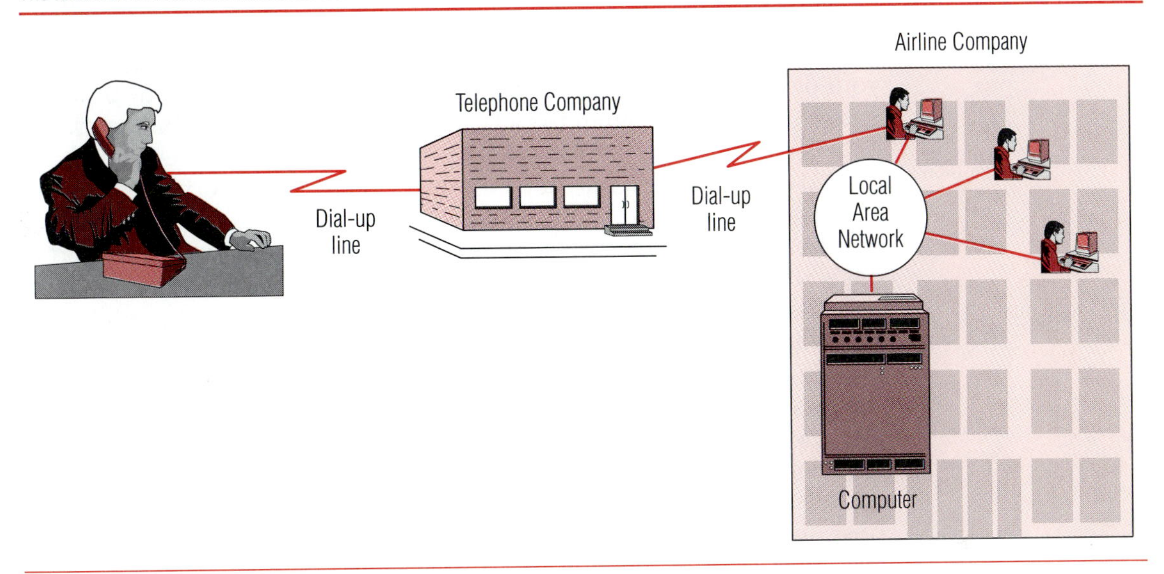

Figure 1–8
The telecommunications connection between a traveler and an airline reservation computer through a leased line.

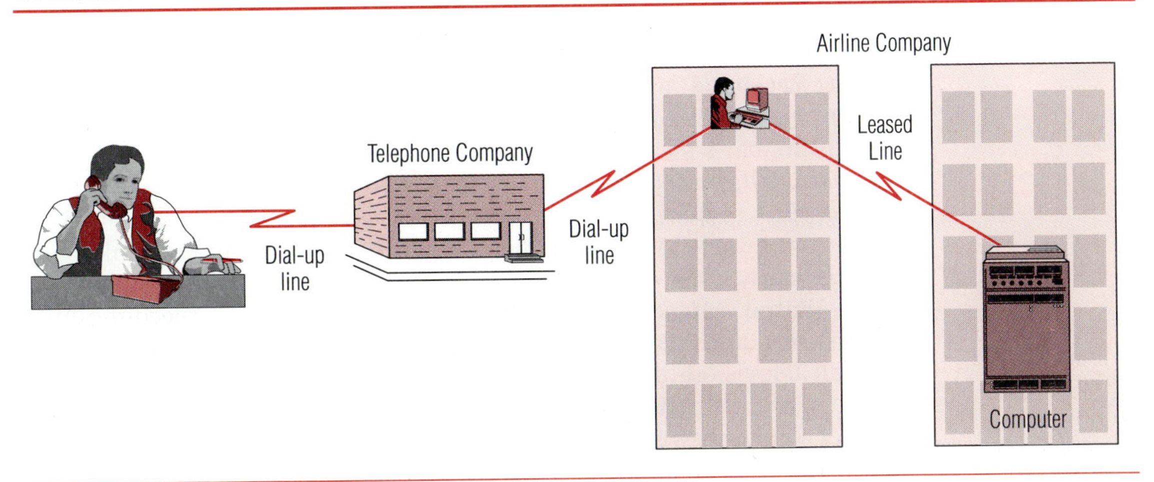

operates a personal computer or *video display terminal (VDT)*, a terminal with a screen that looks like the monitor on a PC and a keyboard. The PC or VDT is connected via a small *local area network* or a telecommunications line to the airline's central computer. You describe the reservation you would like to make, and the agent requests information about flight schedules, alternative connections, and ticket prices from the computer using the terminal. The information displayed on the screen of the PC or terminal is in a highly coded form, which the reservation agent interprets and relays to you on the telephone. Two dialogues are taking place: one between you and the reservation agent and one between the reservation agent and the computer.

Ultimately, you decide on the flight, and the reservation agent enters your personal information, along with the flight number and the time and date of the flight, through the terminal into the computer's memory. The data is copied from there onto disk storage. A ticket may be printed and mailed to you, or you may pick up the ticket at the airport when you arrive for the flight.

The telecommunications line between the reservation agent's terminal is used so much that it is usually advantageous for the airline to install special communications wiring and a local area network (which will be studied in Chapter 12) if the terminal and computer are in the same building, as shown in Figure 1–7. If the computer is in a separate building or some distance away from the reservation agents, the airline may rent the telecommunications line full-time as a *private* or *leased line* from the telephone company, as illustrated in Figure 1–8. The airline pays the telephone company a flat monthly fee for its exclusive use. By way of contrast, the connection between your telephone and the airline is called a *dial-up* or *switched line*. You pay a basic charge to make telephone calls, plus a usage charge based on the number, distance, and duration of the calls you make. Dial-up lines are discussed in more detail in Chapter 5.

leased line

Banking with an Automatic Teller Machine (ATM)

Figure 1–9 illustrates the modern way of banking: interacting directly with the bank's computer through a computer terminal called an *automatic teller machine (ATM)*. The ATM may be located at the bank, shopping center, airport, grocery store—in fact, just about anywhere. Telecommunications allows the ATM to be connected to the bank's computer many miles away.

With ATMs, consumers with little knowledge about telecommunications or computers operate computer terminals. The ATM contains a small television-like screen similar to the airline agent's VDT. On the screen are instructions or questions, such as "Insert your card," "Please enter your password," and "Would you like to make a deposit, a withdrawal, or pay a bill?" The machine "walks users through" the process of

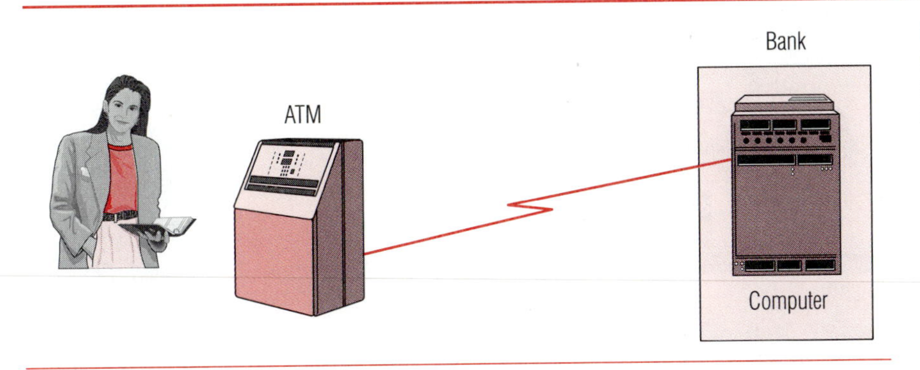

Figure 1–9
An ATM connected to a computer by a telecommunications line.

performing banking transactions. The ATM interacts with customers and with the computer, getting information about the transaction the customer wants to perform and relaying the information to the computer. For withdrawals, the ATM verifies that the customer has enough money in the account by requesting that the computer check the balance.

ATMs may contain a small computer that can be programmed to provide the user with instructions. A programmable ATM is an example of an *intelligent terminal.* The intelligent ATM handles all interactions with the user and only communicates with a central computer to check account balances or to relay the results of the ATM's processing. Alternately, the ATM may receive all of its capability from a computer to which it is connected via a telecommunications line. In that case, the ATM is considered to be a *dumb terminal.* The instructions to the user come from the computer and all input from the keyboard of the ATM is passed to the computer for processing.

Initially, many people are somewhat afraid to use ATMs. Some people never become comfortable using an ATM and prefer the human contact that a live teller provides. The bank prefers to have customers use ATMs because it reduces the number of human tellers, which in turn reduces the bank's costs. ATMs can also improve the bank's service by allowing customers to bank 24 hours a day, 7 days a week.

Automatic Remote Water Meter Reading

Figure 1–10 illustrates another example of the use of telecommunications: water meters with telecommunications capability in residences and commercial buildings. In one city, a computer at the water company office dials the telephone line connected to the water meter in the residence. An encoder on the meter is activated by an interrogation signal. The encoder senses the meter reading and sends back meter data to start the billing process.

The cost advantage to the water company is having no meter readers who walk to all of the houses and buildings in the city to read the water

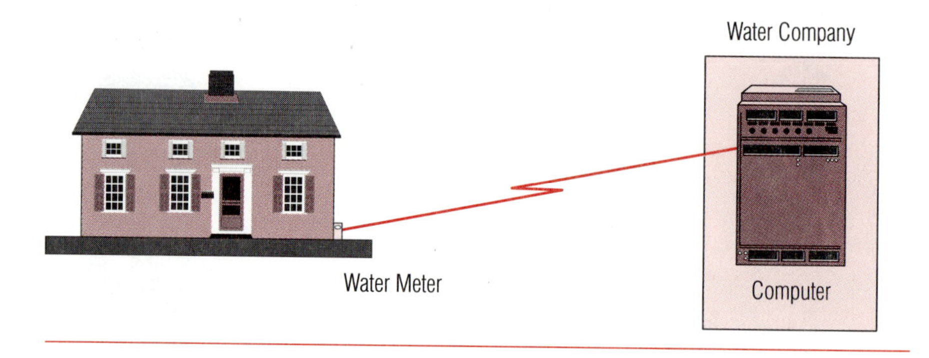

Figure 1–10
Remote reading of a water meter using telecommunications.

meters every month. Although the initial cost to install these automatic, communicating water meters is high, the payback comes rather quickly when fewer people are needed to read meters. In addition, the automatically generated meter readings are more accurate than those read manually.

In some cases, there are practical alternatives to the telecommunications technologies that are described and used as examples in this book. For example, a water company may be able to substantially increase the accuracy of meter readings by equipping its human meter readers with electronic data recorders. Although the meter reader still walks to all of the locations on the route, the time it takes to record the meter reading and the data entry cost are significantly reduced. The key to finding the best solution for any business problem is first to define the problem and then to explore several alternative solutions before selecting one to implement. This process is discussed more fully in Chapter 14.

REQUIREMENTS FOR TELECOMMUNICATIONS SYSTEMS

Users of telecommunications systems have certain expectations about the capabilities the system will provide and the system operations. In this section, you will examine typical user requirements for voice and data communications systems.

Requirements for Telephone Systems

Put yourself in the position of defining requirements to the telephone company for your telephone service. You might want the system to be

- available when you need it;
- trouble-free and reliable;
- easy to use and easy to learn;
- a universal service so that you can call to and from any location;
- fast (doesn't take a long time to get a call completed);
- inexpensive.

Some people might have special requirements, such as

- amplifiers on the telephone handset if they are hard of hearing;
- easily accessible public phones if they travel;
- telephones with built-in directories and the capability to dial calls automatically if they use the telephone frequently;
- a second line so that they can connect to the Internet without tying up their main telephone line.

If you look at the list of "typical" requirements shown, you would probably conclude that the telephone system in the United States has done a pretty good job of meeting your needs. In fact, this is the general opinion of the American public. In many respects, the telephone companies have been sensitive to user requirements and have provided new services and capabilities as needed.

In a similar way, the telecommunications department in a company must be sensitive to the requirements of the telecommunications users in the organization. The requirements for telephone service in a company are, in general, similar to those for public telephone service. However, there are usually some special requirements, such as

- multiple-line telephones so that secretaries can answer their bosses' calls;
- hold buttons so that one call can be held while a conversation is conducted with a second party;
- call-transferring ability to move a call to someone else in the organization.

There may also be some other more specialized requirements, such as an automatic call distribution (ACD) service as described in the airline reservation example in Chapter 4.

Some of these capabilities have been available for a long time. Others are just becoming available and/or economical because of advances in electronic technology. The important thing, however, is to understand what is required by the users of the telecommunications system.

Requirements for Data Communications Systems

From a user's perspective, many of the characteristics that are desired from the telephone system are also desired from a data communications system:

- availability—the system is ready and operating when it is needed;
- reliability—the system is trouble-free and does not introduce errors into the communication process;
- online and realtime—users are able to operate the system through interactive terminals;

- responsive—the system is quick enough so that it helps the user do the job and does not hinder him by imposing delays on the communication;
- ease of use—it is easy to accomplish the needed communication and dialogues allow the user to interact with the computer;
- ergonomics—the users' workstations and terminals must provide for long periods of comfortable use;
- flexibility—the system must be easy to change.

As in the case of voice systems, some data communications users may have special requirements, such as

- the ability to get hard copy output easily and conveniently;
- multiple terminals or personal computers (e.g., for scientists and stockbrokers);
- terminals that are also high-powered computer workstations;
- large monitors or display screens.

Let's now examine each of these requirements in detail.

Availability *Availability* is having the system or service operating when the user wants or needs to use it. Take, for example, the telephone system. We expect telephone service to be available anytime, 24 hours a day, 7 days a week, 365 days a year. What would it be like if the telephones did not operate during certain hours, say from 10 P.M. to 7 A.M., or during the lunch hour, or on holidays? In many cases, we could get used to these reduced hours of operation. We would complain, and we would certainly have to adjust some of our habits, but in most cases the reduced hours of operation would not be critical.

There are some situations, however, when not having the telephone system available could be disastrous. What if we needed to call the fire department or the police? What if somebody desperately needed to reach us? For these cases, other communication methods would need to be developed.

variable requirements

The real requirement for availability varies by application. Many data applications in business only need to be available during business hours. For example, the order entry application discussed in Chapter 4 may only need to be available during business hours because customers won't be at work to place orders at other times. On the other hand, if we advertise on television at all times of the day or night and our advertisement says "Call 1-800-555-XXXX. Operators are standing by to take your order," our order entry application may need to be available around the clock.

If our company does business nationwide, the hours of availability will undoubtedly have to be longer than from 8 A.M. to 5 P.M. local time.

Figure 1–11
The business days in many major cities of the world have little overlap with one another. Telecommunications systems that serve international locations must have extended operating hours in order to be available during the business day in foreign cities.

BUSINESS DAY – MONDAY | TUESDAY
Sydney 8am 9 10 11 NOON 1 2 3 4 5pm 6 7 8 9 10 11 MID 1am 2 3 4 5 6 7 am

BUSINESS DAY – MONDAY
Paris 11pm MID 1 2 3am 4 5 6 7 8am 9 10 11 NOON 1 2 3 4 5pm 6 7 8 9 10pm

SUNDAY | BUSINESS DAY – MONDAY
New York 5pm 6 7 8 9 10 11 MID 1am 2 3 4 5 6 7 8am 9 10 11 NOON 1 2 3 4pm

If we are located on the East Coast, we will probably want to keep our systems available until 8 P.M., which is 5 P.M., the close of the business day, on the West Coast. Conversely, if we are located on the West Coast, we may need to start our system operating at 5 A.M. because that is 8 A.M. in the east, and our customers there are ready to do business with us.

If the communication network serves an international business, the window of availability will be longer. Figure 1–11 illustrates the situation. Allowing for an hour or two variation caused by daylight saving time, when it is 8 A.M. in Paris, it is 2 A.M. on the East Coast of the United States. Therefore, the telecommunications system may have to be available from 2 A.M. until 8 P.M. eastern standard time (if it is also serving the West Coast of the United States). Again, allowing for 2 or 3 hours' variation due to daylight saving time, Japan and Australia are approximately 15 hours ahead of the U.S. East Coast. That means when it is 8 A.M. Monday in Sydney, it is 5 P.M. Sunday in New York. When the Australians go home at the end of their business day at 5 P.M. Monday, it is 2 A.M. Monday in the U.S. eastern time zone. If a single network served the United States, Europe, and Asia, it would have to be available nearly 24 hours a day, 7 days a week.

Holidays are another consideration that must be studied when determining the availability requirements of a network. Even within the United States, we celebrate certain regional holidays in some parts of the country and not in others. Different customs for holidays and lunch hours also prevail. For an international network, the situation gets even more complicated. For example, Thanksgiving is celebrated in Canada but on a different day from the United States, and whereas Americans close businesses on July 4, no other country celebrates America's Independence Day.

Determining the real requirement for network availability is key. Whereas the public telephone system and some businesses, such as hospitals, are required to be operational 24 hours a day, 7 days a week, 365 days a year, most other business organizations do not have to operate on that schedule.

The use of cellular telephones has grown rapidly all over the world in the last several years. (Courtesy of Will & Deni McIntyre/Photo Researchers, Inc.)

Reliability *Reliability* in telecommunications is trouble-free operation. When someone uses the system, it must work. Just as we don't want our car to break down when we are on a trip, we don't want our telephone connection to break in the middle of a call. Similarly, we don't want the terminal and computer to go down when we are interacting with it.

One of the most frustrating situations for users of communication and computer systems is unpredictability. Users generally understand that systems occasionally fail, no matter how well they are designed. Once a system fails, users would generally rather have it stay out of service until it is fixed than have it come up but fail again within a short period of time.

MTBF and MTTR

A classic measure of reliability on a system is *mean time between failures (MTBF)*. The MTBF is the average time between the failures of a system. A related measure is the *mean time to repair (MTTR)*. MTTR is a measure of how long, on average, it takes to fix the problem and get the system back up after a failure has occurred. Reliability is often measured

Component	Reliability
Computer	.98
Circuit	.97
Terminal	.99

System Reliability = .98 × .97 × .99 = .941 = 94.1%

Figure 1–12
The overall reliability of a three-component, serial telecommunications system.

9 hour day = 540 minutes

At a reliability of	The system would be down
.999	32 seconds
.995	2.7 minutes
.990	5.4 minutes
.98	10.8 minutes
.97	16.2 minutes
.96	21.6 minutes
.95	27.0 minutes
.90	54.0 minutes

Figure 1–13
The maximum number of minutes a system can be down in a day and still achieve a given level of reliability.

in terms of probability. If a system is 98 percent reliable, that means it is working 98 percent of the time and out of service 2 percent of the time.

Most telecommunications systems are made up of a number of components—such as the terminal, line, and computer—connected in series. This is known as a *serial system*. Each component has a certain MTBF and MTTR. To get the overall reliability of a serial system, the reliability of each of the components is multiplied together. Figure 1–12 shows an example of a three-component telecommunications system in which each component has a certain reliability. The overall resultant reliability of the system is of course *lower* than any of the components individually. You can see from the example that in order to achieve a reliability of even 95 percent, which is a low-typical requirement of a telecommunications system, each of the components must be considerably more reliable than .95.

serial system

The fact that the overall reliability is less than any of its components is an important concept of combinatorial probability. It is particularly relevant, since most communication systems are serial systems and have many components connected in series. To put it in other terms, if a telecommunications system is scheduled to be operational from 8 A.M. to 5 P.M. each day (9 hours), the chart in Figure 1–13 shows how many minutes the system would be down, on average, each day at different levels of system reliability. For many business applications, 20 or 30 minutes of outage each day is simply not good enough. For those applications, reliability must be greater than 95 or 96 percent.

Because the public telephone system is expected to be operational 24 hours a day, 7 days a week, with a very high reliability, the central office telephone equipment is designed with many redundant components. When a component failure occurs, the backup component is automatically switched into operation. With this type of fail-safe design, the expected failure rate of an entire central office is approximately once in 40 years!

Most business systems don't need that kind of reliability. But again, the important thing is for the business to determine the real reliability requirements for each of its applications or systems and to design its systems to meet these requirements. High reliability can be achieved by designing redundancy into any system. Redundant telecommunications lines, computers, or terminals can be put in a telecommunications system to make it more fail-safe. The costs of the redundant equipment must be assessed, however, and an economic analysis must be performed to determine whether the benefits of the increased reliability equal or outweigh the costs.

Online and Realtime For our purposes, *online* simply means "connected to the computer." Most computers are designed for online operation through terminals. With microcomputers, the computer is built into the terminal, and with most microcomputers today, there is no capability to attach additional terminals. In any case, the microcomputer is designed to be used by one person sitting at its VDT working interactively.

If the computer can support multiple terminals, they may be directly cabled to the computer, in which case they do not use telecommunications facilities at all. Alternately, the terminals may be connected via telecommunications lines.

Realtime is a rate of response or operation fast enough to affect a course of action or a decision. The response usually is measured from the time the terminal operator presses the ENTER key, or its equivalent, on the terminal to signal the computer to perform some processing to the time the computer delivers the first part of the output back through the network to the user's terminal. We will discuss more about response time later in the chapter.

Real Enough Time The definition of realtime is "a rate of response or operation fast enough. . . ." The question is, "What is fast enough?" The answer depends on the needs of the particular application. We can therefore think about a concept of *real enough time.* This notion suggests that the response time that is good enough for one application may not be good enough for another. For example, traditional wisdom says that a 2-second response time is good enough in most applications. It isn't good enough, however, when you pick up the telephone handset on your telephone and expect to hear a dial tone by the time you get the handset to your ear. You get the telephone to your ear in less than 2 seconds, and if you don't hear a dial tone, you are annoyed and immediately begin to wonder what is wrong. Two-second response time also is not good enough for the flight controller at Cape Canaveral who is trying to destroy a rocket that is off course and headed for a populated area. The controller wants to give the command to the computer and have the rocket blown up within milliseconds to avoid a disaster. Similarly, 2-second response time is not good enough in many industrial applications in which computers are controlling machine tools or chemical processes. In those applications, real enough time usually means something less than 1 second.

On the other hand, after you complete the dialing of your telephone call, you do not expect (nor do you probably require) instantaneous completion of the call. In that situation, real enough time is more like 10 or 15 seconds. Similarly, in the airline reservation or customer order entry application, multisecond response time is normally acceptable at the completion of a reservation or order. While the computer is busy updating files and completing the transaction, the operator has time to put papers back in a file folder, take a sip of coffee, and get ready to handle the next call.

One must look at the real needs of each application to determine what is real enough time. Indeed, one must look within the application at the various transactions or interactions that occur, since some of them may require a faster response than others.

Response Time *Response time* is traditionally defined as the time between pressing the ENTER key on the terminal signaling the computer that processing is needed until the first character of output is received at the terminal. Many people argue, however, that since the operator can do little when the first character is received, the time ought to be extended until the last character is received at the terminal and the operator is able to begin work again.

The overall response time that the user sees is made up of several components. In most cases, when the operator keys data into a terminal, it is stored in the terminal's memory until the ENTER key is pressed. At that time the data is transmitted over the telecommunications line to the computer. After the computer receives the data, it must process it and formulate a response. Then the response is transmitted back over the telecommunications line to the terminal, where it is displayed or printed. The time it takes to transmit the data on the telecommunications line is a function of the number of characters to be transmitted and the speed of the line.

The transmission time may be extended by delays encountered when a line is shared among several terminals. Sometimes a transaction may have to wait for another transaction to finish using the line. This wait is called *queuing*, and it can have a significant effect on the overall response time.

queuing

The processing time on the computer is a direct function of the number of instructions that must be executed to interpret the input message, process it, and formulate a message for transmission back to the terminal. This time often is extended when a computer is dealing with several terminal users concurrently. Because a computer can only process one transaction at a time, arriving transactions may encounter a *queue*, or waiting line, of other transactions waiting to be processed. Thus, if an input message from terminal A arrives while a message from terminal B is being processed, it will be placed in a queue and delayed until the processing for terminal B's transaction is complete or interrupted.

processimg time

Processing time at the computer is most often calculated on an average and probabilistic basis. The usual way of stating processing time is in the form, "computer processing takes *X* seconds *Y* percent of the time." In addition to the average processing time, one is also interested in the variability. It is one thing to know that processing is completed for 95 percent of the transactions in 1 second, but what about the other 5 percent? Do those transactions take 5 seconds to process? Or 10 seconds? Or 90 seconds? Determining the averages, variances, and probabilities for a given computer system is a complicated task. It depends on a knowledge of the mathematics involved and also a knowledge of the characteristics of the workload on the computer.

Figure 1–14 shows a typical but simplified response time calculation. No queuing has been considered. It is evident that the telecommunications network plays a very significant role in determining the overall response time that the user sees. The people who configure the computer and those who design the telecommunications network must work together to ensure that the response time requirements of the user can be met.

IBM response time study

Two studies conducted in different parts of the IBM corporation have attempted to quantify the value and economic benefit of rapid response

Figure 1–14
A simplified response time calculation. Queuing time is not considered.

Action	**Seconds**	**Cumulative Seconds**
Operator types transaction—presses ENTER key		
100-character transaction transmitted to the computer	.10	.10
Computer receives transaction—processes it	.40	.50
1000-character response transmitted to the terminal and displayed	1.04	1.54 total response time

Telecommunication Time = 1.14 seconds = 74% of total response time

time to users doing different kinds of work with the computer. Walter Doherty, who worked at the IBM research laboratories in Yorktown Heights, New York, said that if the response time is fast enough, the computer becomes an extension of the human brain. In reality, the person at the terminal is usually thinking several steps ahead of the work she is doing at any point in time and remembering those steps in the short-term memory of the brain. If response time is slow, however, the short-term memory contents are replaced as the mind wanders. When the computer responds, the person must refocus attention on the work being done with the computer.

Another type of response time is *user response time.* User response time, also called *think time,* is the time that it takes the user to see what the computer displayed, interpret it, type the next transaction, and press the ENTER key. User response time is a significant part of the overall productivity of the human–machine system. Doherty says that the faster the machine responds, the faster the person will respond with the next transaction. His studies showed that the overall productivity kept climbing as the computer response time decreased to less than .5 seconds, which was the fastest computer response time that could be obtained in that particular situation.

think time

The telecommunications network designer must look at the needs of the particular application to determine what is adequate response time or real enough response time for each application. In practice, the response time is a function of the network design and cost. With a good design, and by spending enough money, the response can be reduced to a fraction of a second if that is what is needed, however without a proper design, the response time objectives will never be achieved no matter how much money is spent.

Usable Dialogues As the use of terminals becomes widespread, many people who have had no previous computer experience are becoming telecommunications and computer users. The need to make the interaction with the computer easy is greater than ever before. The interaction or series of interactions between the user at a terminal and the computer is called a *dialogue.* A typical requirement of the user is that the dialogue with the computer be *user friendly.* User friendliness is an attribute of computer interactions that is difficult to describe and in many cases is strictly relative. A dialogue that may be easy for one person to use may not appear to be easy, or friendly, to another person.

dialogue

It helps to break the concept of user friendliness into two subconcepts: "easy to learn" and "easy to use." Easy-to-learn interactions enable a person with little instruction or guidance to figure out how to operate the terminal or system and get the desired results. Easy-to-use interactions are ones that may take some training to learn but, once the education has been completed, are simple to perform in the desired way. The two attributes are not mutually exclusive or always mutually desired.

For example, if the dialogue between an airline reservation agent and the computer is being designed, the major concern should be for ease of use because the same set of transactions will be completed thousands of times as reservations are made. Shortcuts and flexibility are very important so that the reservation agent can jump around within the reservation process when necessary. Usually the input data is highly coded so that a minimal number of characters needs to be typed. Ease of learning the system is of secondary importance, since the operators receive some training and usually work in an environment where other, more experienced operators can provide assistance when necessary.

By contrast, the design of the dialogue for use with an automatic teller machine needs to be easy to learn, since many of the users are completely inexperienced with terminals and may use the ATM infrequently. Furthermore, there is no teacher or coach standing by ready to assist, particularly in off hours when the bank is closed. Ease of use is a secondary concern. This implies that the transactions through an ATM may be somewhat rigid in sequence, content, and format, with little variation allowed. There should be an option to abort the transaction or go back to the beginning, but beyond that the user should be taken through a straightforward series of inputs and actions to accomplish the desired results.

Ergonomics Ergonomics is another user requirement that the telecommunications designer must consider and understand. As with other requirements, the goal is to ensure that the network and all of its components meet user requirements. *Ergonomics* is, according to Webster, "the study of the problems of people in adjusting to their environment, especially the science that seeks to adapt work or working conditions to suit the worker." In recent years, as the use of computer terminals has become more widespread, ergonomic issues have received considerable attention.

The telecommunications managers or designers need to understand the ergonomic issues and deal with them, but they do not have total responsibility for resolving them. The design of the individual office or workspace and the selection of furniture is not usually within the telecommunications manager's responsibility. Often this is done by office designers or consultants who lay out the office and select the furniture to go in it. Hardware manufacturers determine the physical characteristics of their terminals, and the telecommunications department has little direct influence over the decisions made. However, equipment manufacturers have typically been sensitive to the ergonomic issues and have designed their equipment with the flexibility to meet personal preferences or legislative requirements. The specific details of workstation ergonomics are discussed in Chapter 6.

Flexibility and Growth One thing is certain about telecommunications systems and networks: They change and grow. It is important when you plan and design a network to strive for flexibility and growth possibilities. As companies grow, the network must handle increasing traffic volumes by adding capacity. As companies reorganize, the network may have to be reconfigured to handle new traffic patterns. As technology changes, the user community will demand the newest terminals and capabilities.

Users will expect that changes can be made to the telecommunications system and will have little empathy with technical arguments as to why changes are difficult to make. However, users will not directly ask for telecommunications flexibility, so the network designer must ask probing questions to draw the users into thinking about the future. Although not all changes can be foreseen, the network designer should work with and anticipate as many of the users' needs as possible.

HISTORY OF TELECOMMUNICATIONS

What do you think of first when you think of telecommunications history? Alexander Graham Bell? Samuel F. B. Morse? Telegraphers sitting with telegraph keys to send messages across the continent? The old-style, telephone your grandmother used to have? All of these images are certainly a part of telecommunications history. The timetable in Figure 1–15 lists many of the significant events of recent telecommunications history (since communication was first performed using wire as the medium).

Let's go back further in time. Long distance or *tele* communications was used in ancient times, when people used fires to communicate simple messages over long distances. Similarly, drums, smoke signals, and the printing press were used by cultures long before electricity or magnetism was known. Humans have always needed to communicate over long distances and have found ways to do so without sophisticated tools.

Invention of the Telegraph

The history of modern telecommunications as we think of it really begins with Samuel F. B. Morse, who invented the telegraph and first demonstrated it on September 2, 1837. In 1845, Morse formed a company with private money to exploit the telegraph, and his idea caught on. By 1851, there were 50 telegraph companies in the United States, and the invention was beginning to be used by railroads, newspapers, and the government. In 1856, the Western Union Telegraph Company was formed. By 1866, it was the largest communication company in the United States and had absorbed all of the other telegraph companies.

Invention of the Telephone

On March 7, 1876, a patent entitled "Improvements in Telegraphy" was issued to Alexander Graham Bell. This patent, which did not mention the word *telephone,* discussed only a method for the electrical transmission of

1837	Samuel F. B. Morse invents the telegraph.
1838	Telegraph demonstrated to government. Government declines to use.
1845	Morse forms a company with private money to exploit the telegraph.
1851	Fifty telegraph companies in operation.
1856	Western Union Telegraph Company formed.
1866	Western Union Telegraph Company was the largest communications company in the United States.
1876	Patent issued to Alexander Graham Bell for telephone.
1876	Bell offers to sell telephone patents to Western Union for $100,000. Western Union declines.
1877	Bell Telephone Company formed.
1878	First telephone exchange with operator installed.
	Western Union Telegraph Company sets up its own phone company, sued by Bell for patent infringement, gets out of phone business, and sells network to Bell.
1885	American Telephone and Telegraph Company (AT&T) formed to build and operate long distance lines interconnecting regional telephone companies.
1893–94	Original Bell patents expire; independent telephone companies enter market.
1911	Bell Telephone franchise companies reorganize into larger organizations known as the Bell Associated Companies. Beginning of the Bell System.
1913	Invention of the vacuum tube.
1941	First marriage of computer and communication technology.
1943	Development of submersible amplifier/repeaters.
1947	Invention of the transistor.
1948	Hush-a-Phone case.
1956	First trans-Atlantic telephone cable installed.
1957	First satellite launched.
1968	Carterfone decision.
1971	Computer Inquiry I.
1981	Computer Inquiry II.
1982	Modified Final Judgment.
1984	Divestiture.
1986	Computer Inquiry III.
1996	Telecommunications Act of 1996.

(Note: These events are discussed in Chapter 2.) [applies to 1957–1996]

Figure 1–15
Significant events in telecommunications history.

"vocal or other sounds." At the time Bell's patent was issued, he did not have a working model of the telephone—only plans and drawings—but within a week of the patent's being issued, he and his assistant, Thomas Watson, got the telephone to work in their laboratory. In July of 1877, one of Bell's financial backers, Gardinar Hubbard, created the Bell Telephone Company. By the fall of that year, there were approximately 600 telephone subscribers.

first telephone company

This early telephone "system" was a little different from what we are used to today. Telephone switching equipment, which could connect any telephone to any other telephone, had not yet been invented. Telephone subscribers in 1877 had one pair of wires coming into their home or business for each telephone they wanted to connect to! Since there were 600 subscribers, it is possible, though not likely, that one individual could

A re-creation of Alexander Graham Bell's original laboratory. (Property of AT&T Archives. Reprinted with permission of AT&T)

have had 600 pairs of wires coming into the home and then would have to find the right wires and connect them to the telephone before making the call! There were also no ringers or bells on the telephone. If the other party didn't just happen to pick up the telephone when the caller wanted to talk, there was no conversation!

The technology advanced quickly. By January of 1878, the first telephone exchange with an operator was in place. With that innovation, it was possible to have just one pair of wires connecting a telephone to the central switchboard. The operator, with a series of plugs and jacks, could make the connection from one telephone to any other telephone that was connected to that switchboard.

Later in 1878, Western Union Telegraph Company set up its own telephone company and took advantage of its network of telegraph wires, which was beginning to blanket the country. Bell sued Western Union for patent infringement. After studying the situation, the Western Union attorney became convinced that Bell would win the suit. Western Union settled out of court, got out of the telephone business, and sold its network to the Bell Telephone Company.

In 1885, American Telephone and Telegraph Company (AT&T) was formed to build and operate long distance telephone lines. These lines in-

The first successful telephone call in 1876. The first complete sentence to be transmitted electrically was, "Mr. Watson, come here, I want to see you!" (Property of AT&T Archives. Reprinted with permission of AT&T).

terconnected the regional telephone companies that had established franchises with Bell to provide telephone service in various parts of the country.

competition begins

In 1893 and 1894, the original Bell patents for the telephone expired. Many independent telephone companies entered the market and started providing telephone service. For the first time there was substantial competition in the telephone business as these independent telephone companies competed with the regional telephone companies that were franchised by Bell.

Bell System

In 1911, the regional companies with Bell franchises were reorganized into larger organizations, which became known as the Bell associated companies. This marked the beginning of the *Bell System* as the collection of companies, headed by AT&T, came to be known.

computers, communications, cables, and satellites

Since the structure of the telecommunications industry and its basic capabilities were established in the early 1900s, technological advances have allowed new capability, new services, and reduced cost almost continually. The marriage of computers and communications in the early 1940s was a major milestone that had synergistic effects on both technologies as they developed. The development of the submersible amplifier/repeater made undersea telephone cables possible and greatly expanded long distance calling capabilities. The launch of the first satellite in 1957 opened up a totally new type of communication capability.

The invention of the transistor in 1947, and subsequently the development of large-scale integrated circuits and microprocessors, have undoubtedly had the most profound impact of all of the technological advances to date. These devices made it possible to develop miniaturized devices with low power requirements, such as amplifiers and microprocessors, without which the networked world we know today would not exist. We would long ago have reached the limits of physical size, required power, and reliability which are absolutely essential to the network capability we have available today and have come to rely on.

transistors, integrated circuits, and microprocessors

Regulatory actions have been another major area of activity in the communications industry during this century. The regulatory posture was first oriented to protecting the fledgling telecommunications industry and allowing a nationwide compatible network to get on its feet. Recently, the movement has been toward deregulation to encourage competition, innovation, and lowering of costs. Since we will be studying the events and impact of regulation extensively in Chapter 2, they are not included in the present discussion.

Telecommunications and the Computer

It is interesting to note that the connection between telecommunications and computers first occurred in 1941, before computers had even emerged from development laboratories. In that year, a message recorded in telegraph code on punched paper tape was converted to a code used to represent the message data on punched cards to be read into a computer. Although it was certainly unsophisticated, this process demonstrated a way that these two technologies could work together.

THE CHALLENGE OF STAYING CURRENT

One of the things that make telecommunications such an interesting field is that change is occurring rapidly. The rapid technological change is similar to the type of change that has occurred in the computer field. The basic building blocks, the electronic chips and circuits, are very similar. However, in the telecommunications field, we also have experienced the advent of satellites and fiber-optic technology, both of which present new alternatives for connecting pieces of telecommunications equipment to one another.

Another dimension of change affecting telecommunications is the legislative and regulatory process. The movement from a monopolistic to a competitive structure has had a profound impact on the companies within the industry.

The third dimension that continues to change rapidly is the business environment. Some strong competitors in the telecommunications industry didn't even exist five years ago. Some of the older companies are not the powerhouses they once were. Everyday you can read about startups, mergers, acquisitions and dissolutions of telecommunications companies. It's hard to keep up with who owns what!

While all of the change can be frustrating at times, it also makes it possible, practical, and economical to use telecommunications in new and innovative ways. The good news is that this keeps work interesting and exciting for people in the field. The bad news is that it is very difficult to keep current with all that is happening and to understand the significance or applicability for your organization or your personal life.

Serious telecommunications students and professionals must constantly educate and reeducate themselves about products, technology, and trends that are appropriate and relevant for their companies. Fortunately, there is no limit to the number of opportunities to do so. Seminars are taught frequently in major cities, numerous telecommunications trade magazines exist, and specialized books on a wide variety of telecommunications topics are available.

This book distinguishes between the basic telecommunications principles and concepts that are relatively constant and stable, and current products or implementations that change rapidly. Understand, however, that if you are going to stay current in the telecommunications field, you must be prepared to continue your study.

SUMMARY

The focus of this book is telecommunications in a business environment. Telecommunications is important to study because it is used widely and is an important part of our lives at work and in the home. The technologies on which telecommunications is based are changing rapidly. Advances in microelectronics and transmission media, such as fiber optics and satellite communication, exemplify this change. Telecommunications is changing in another dimension—the regulatory one. For many years the movement in most countries of the world has been toward deregulating the telecommunications industry. Today the United States has a mixed environment with part of the telecommunications industry being regulated and part being unregulated. The business environment is in constant flux too, as companies enter and leave the market. Like any discipline, telecommunications has its own terminology to be learned. You have begun building your telecommunications vocabulary in this chapter, and will continue to do so throughout the book.

CASE STUDY

Telecommunications at Work in the Home

Virtually everyone, even those who don't work in telecommunications is aware of the lightning-speed pace by which the industry is changing. This progress is not only limited to the corporate world; perhaps, even more remarkable is the impact that telecommunications advancements have made in the home. For some, it has simply made keeping in touch with far-away family and friends easier, while for others it has dramatically altered day-to-day life.

Lori and John Andrews live in Connecticut, are in their mid-thirties and are the parents of a two-year-old. They are a good example of how advancements in telecommunications have affected the home user. "We 'surf' the net, send electronic mail to friends around the world and pay our bills on-line," said Lori. Until I started talking with the Andrews, they hadn't really thought about their use of telecommunications or the role it plays in their lives. In addition to a standard phone line, they have a second line, which is shared with the fax machine and their personal computer. They each have a cellular phone, and John has a pager for work.

When asked how their lives have changed because of developments in home telecommunications, John was quick to respond. "Life is so much easier, especially when it comes to banking. We hardly ever go to the bank anymore. If we need a statement we go on-line. If we want to transfer money from one account to another, we can do so on-line. It's remarkably convenient. I log on to PC Banking and within seconds I'm looking at all the activity in our accounts, paying bills or e-mailing customer service with a question."

Most banks now offer free PC banking, but just 2 years ago that wasn't the case. The concept of free on-line banking is becoming widely available, and this makes good business sense for financial institutions. Every transaction conducted with an actual teller costs a bank approximately \$1–\$2; every transaction processed through an automated teller machine costs around 40 cents, but an on-line transaction costs an institution only a fraction of a penny. In addition to PC banking from traditional brick and mortar banks, there are now virtual Internet banks as well as an on-line bill paying services that e-mail you when a bill is received and with one click of your authorization, the bill is paid.

Another telecommunications innovation, the cellular phone, has made the Andrews' lives less stressful. Lori was the first to get a cell phone, primarily for safety reasons. At one point, she traveled an 80-mile round-trip route twice a week, not arriving home those evenings until 11:00 P.M. "I just felt safer knowing I wouldn't be stranded on Interstate 95 with a flat tire or a dead engine. The cell phone was my safety net." Lori continued, "But now with cell rates becoming more and more competitive, I find myself using the cell phone for other reasons." John added, "I use my cell phone as a business tool. I commute 2 hours a day on the train, so I can get caught up with work and listen to my voice mail. It saves a fair amount of time and probably even lets me leave work earlier than I could have before the advent of the cell phone."

Although Lori and John have had home access to the Internet for more than 6 years, I asked them if they still think about the changes it has brought to their day-to-day lives. "Definitely!" said Lori immediately. "Being a stay-at-home mom has brought me a more appreciative view of the net.

I probably do 60 percent of my shopping on-line—baby items, books, gifts for others, etc. My son Alec goes down for a nap, I log on and within 30 minutes, I've accomplished what it would have taken me 3 hours to do in the car." Lori laughed, "I even find myself not buying from stores that aren't on-line shopping-ready yet. I'll find another store with the same item and buy it from there instead. But in all seriousness, the whole business to customer aspect of the Internet is life-changing and for the better, I believe."

Lori told how she was amazed at the number of friends who did not bank or shop on-line. "They have personal computers, but for some reason are not taking advantage of it. They all e-mail and do research, but that's it." A friend once referred to Lori as an "Internet maniac." Lori laughed. "What I do on the net is a mere fraction of what's available to me, yet these people think I'm at the height of computer savvy. I'm not sure why they aren't taking advantage of the net and its uses. Maybe it's their age; after all, those of us in our mid to late thirties didn't grow up with computers in the classroom. We typed college papers on electric typewriters and used Whiteout. It's not that these people are anti-computer; that's not it at all. It's more like it's overwhelming." John laughed, "And here we are considering buying a second computer, and I can easily see that evolving into a network of computers in our home within a few years."

John added another Internet story. "A few months ago we were concerned that Lori had contracted Lyme disease. We used the search engine Yahoo!, typed in Lyme disease and within seconds we had located a list of the symptoms, what to be aware of, plus a color picture of what to look for on the skin. Because of the information we had found on the Internet, we were able to immediately rule out Lyme disease." He concluded, "Not only was it a big time-saver, but it put our minds at ease in a very short period of time."

When I asked about their phone service, I learned that although the Andrews primarily rely on electronic mail to keep in touch with their overseas friends, they still make a number of international calls every month. "After our first few bills from AT&T, I knew something had to be done," said Lori. She went on to explain that she had been bombarded with phone calls and mailings from MCI, Sprint and other long distance companies, all claiming to offer better rates than AT&T. I always ignored them, but after realizing our high phone bills were becoming a problem, I began to pay attention to their offers." Lori received an offer from MCI for calls to Japan at a lower rate than what she was currently paying. "I liked AT&T's service and really didn't want to get into the phone company wars, but I did phone Japan a lot, so I decided to call AT&T and see what they could do." Lori explained the situation to the customer representative at AT&T and within a short time the representative called Lori back and offered her a rate lower than the MCI offer. "I realize we probably make more international calls than the average residential phone customer, but I was still impressed that AT&T was willing to negotiate a deal, so to speak."

In finishing our interview, I asked the Andrews if they felt they were a techno-savvy family; in other words, were they up to date with the latest in home telecommunications offerings. Lori said, "If you would have asked me that question 3 years ago, I wouldn't have hesitated to answer 'yes, definitely.' But now, I have to admit that not only are there internet services available that I don't understand, but I don't think we've changed much in how we use telecommunications in our home over the past 3 years. Yes, we shop more on-line or send more e-mail, but those are both things that were available 3 years ago. The reason we shop more is because there are more businesses on-line. The reason we e-mail more is simple—more people have Internet access."

John went on, "Broadband access to the Internet is becoming available in our area, yet we haven't taken advantage of that." Lori added, "It's something that I want to understand more. It's probably naive, but I have a problem with being connected to the net 24 hours a day, even if my computer is turned off. I'm uncomfortable with that."

"MP-3 is something we haven't taken advantage of," Lori said. "I know it's a way to download music from the net, but that's about it. I definitely feel out of it on that one. Technology is speeding by us," added Lori. "Wireless Internet is another one," said John. "I have a Palm Pilot, but I know people who use their Palm Pilots for wireless Internet access."

In closing the Andrews said, "The bottom line is that telecommunications advancements, for us, are convenient and time saving which has made life noticeably less stressful."

QUESTIONS

1. How has your life changed because of the advancement of telecommunications? Think of the telecommunication devices you use regularly. Is your life easier or more complicated?
2. What advancements in telecommunications do you see in the future? How will they benefit the home?
3. The high demand for home telecommunication devices has brought about the need for people with a skill set particular to this industry. Explain how this has affected the economy.

CASE STUDY

Dow Corning Corporation—Background

This case study is intended to describe how a company is using telecommunications. The Dow Corning case will continue throughout the book, with a different aspect of the company's telecommunications network and management being discussed at the end of most chapters. It is hoped that the student will see how many of the ideas and concepts discussed in the text apply in a real life situation.

INTRODUCTION TO THE COMPANY

Dow Corning Corporation was founded in 1943 as a 50-50 joint venture of The Dow Chemical Company and Corning, Inc. The company's business is to develop, manufacture, and market silicones and related specialty materials. Dow Corning's corporate headquarters is in Midland, Michigan, located about 120 miles north of Detroit. Midland is a town of approximately 38,000 people and is also the corporate headquarters for The Dow Chemical Company. From Midland, Dow Corning oversees a worldwide operation with 1999 annual sales of about $2.6 billion. The company has sales offices in major cities throughout the world and manufacturing plants in many industrialized countries.

Since silicone products can take many forms, Dow Corning's product line is extensive. In various forms, including fluids, rubbers, and resins, silicone products function as additives to other products, as ingredients in consumer products, as processing aids, as maintenance materials, and as final products.

Dow Corning Corporation's corporate headquarters in Midland, Michigan. (Courtesy of Dow Corning Corporation)

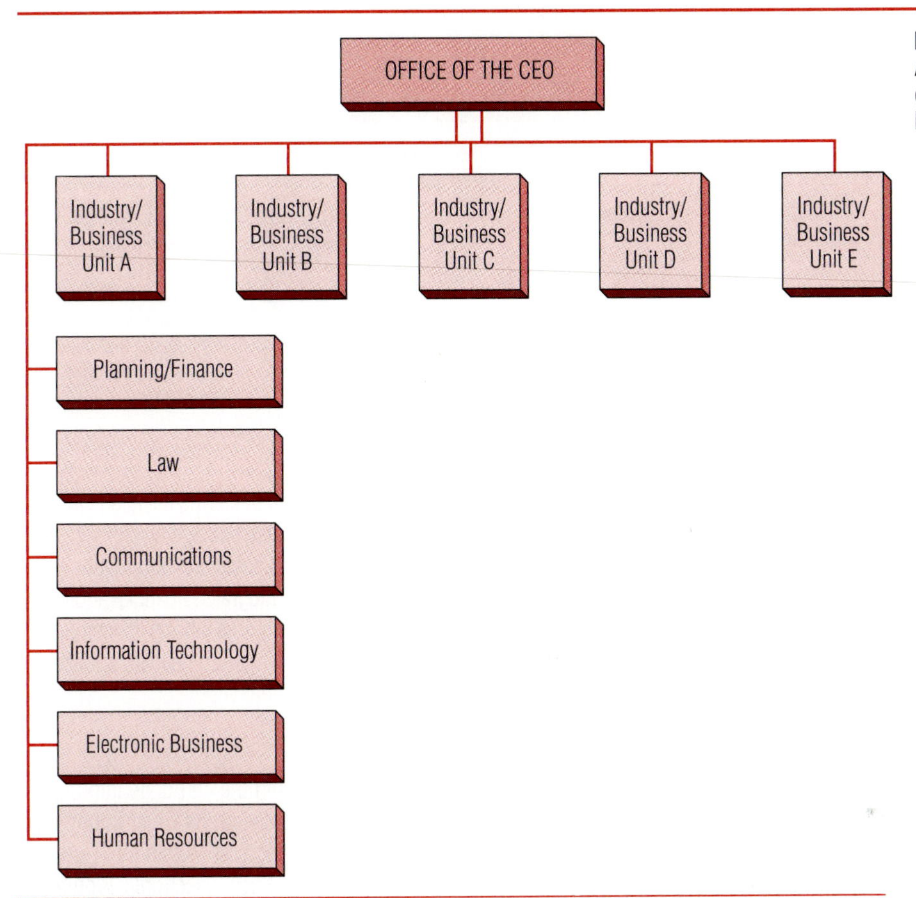

Figure 1–16
A diagram of Dow Corning's organization. (Courtesy of Dow Corning Corporation)

Dow Corning's manufacturing activities are highly integrated. Products produced by one plant may be sold to customers directly or used as chemical intermediates in other manufacturing processes. For this reason, there must be close coordination and communication between marketing and manufacturing to determine which products will be sold and which will be shipped to other plants for further processing. Furthermore, the plants must coordinate their activities to ensure that they have a steady supply of materials from one another.

Dow Corning's organization is divided into five major product groups called Industry Business Units (IBUs) as shown in Figure 1–16. IBU general managers are responsible for setting strategy for the products for which their groups are responsible, and for the worldwide profitability of those products. The industry managers who report to IBU general managers are responsible for the worldwide sales of Dow Corning products to the industry for which they are responsible. Dow Corning also loosely divides the world into geographic units called regions. People located in the regions focus on cultural, marketing or product differences that are required in the countries they serve. The focus of operation for North and South America is in Midland, for Europe in Brussels, and for Asia in Tokyo.

Historically, Dow Corning's management style and methods have been quite centralized. Most of the major decisions about company operations have been made at the Midland

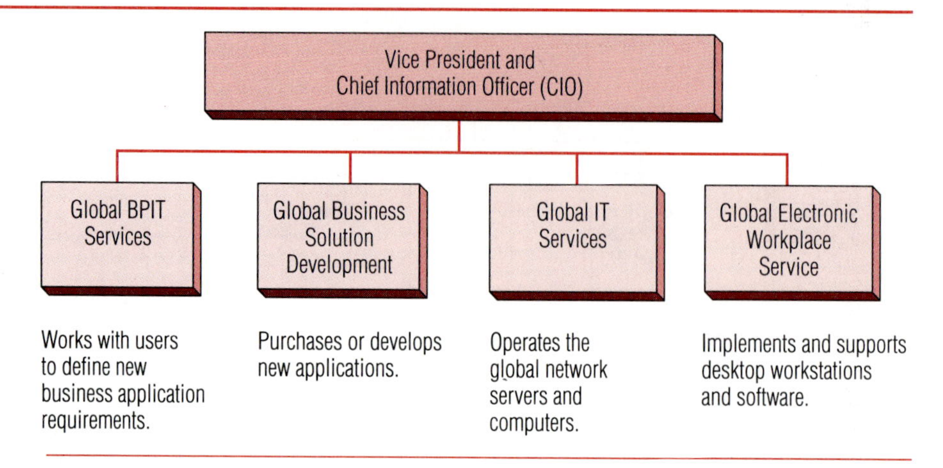

Figure 1–17
Dow Corning's Global IT Organization

headquarters, with some freedom given to the foreign regions to make operating decisions that are appropriate for their countries. At the same time, the company's executive management expects that the strategic direction for finance, human resources, communications, legal and information technology will continue to be centralized. There is a strong feeling that integrated financial and information systems are essential to maintaining the close coordination required among the company's operating units. There is also a belief that these systems provide information needed to make the appropriate decisions that enable the company to maintain its profitability.

TELECOMMUNICATIONS AT DOW CORNING

The responsibility for telecommunications in Dow Corning lies within the corporate Information Technology (IT) department. The Vice President in charge of IT is called the Chief Information Officer (CIO), and reports to the Executive Vice President of the corporation. Within the IT department, the directors of Global Information Technology Services, Global Electronic Workplace Services, Global Business Solutions Delivery, and Global Business Process/IT Services report to the Vice President and CIO, as shown in Figure 1–17. Telecommunications responsibility falls within the Global Information Technology Services group.

The telecommunications staff has reported to different parts of the organization at various stages in its evolution. Before 1982, data communication was managed by the IT department, but voice communication reported to a different part of the company. In 1982, the data and voice responsibilities were integrated in the IT organization. In 1985, a separate telecommunications department was established within IT, and in 1994, telecommunications was incorporated into the Global IT Services organization. These changes reflect differing needs and emphases placed on telecommunications over the years.

Beginning in 1991, a much stronger emphasis was placed on ensuring that the telecommunications staff in the IT department and the people in the manufacturing function who had telecommunication expertise and requirements worked more closely together. It was recognized that the two groups were gradually drifting apart in philosophy and technical implementations of networks, and there was strong recognition of the need to ensure the ongoing maintenance and further development of the integrated data network that Dow Corning had always enjoyed.

As in the rest of the company, the staff size of the telecommunications staff is kept lean. There are 15 people, not including the people who work on the help desk, most of whom are contractors. The department develops and operates a worldwide data communications network that connects all major Dow Corning locations with one another. The staff is also responsible for the company's voice network. There are more than 11,000 personal computers on the data network, about 5,800 of them in the United States. One goal of the data network is to provide very rapid response time for most transactions. Although this objective is being met in many cases, constant monitoring is required to keep the performance optimized and to try to anticipate impending bottlenecks before they occur.

QUESTIONS

1. What are the pros and cons of the telecommunications organization being a separate department versus reporting to another department?
2. What are the important attributes the senior executive should possess to successfully oversee the systems and telecommunications departments?
3. Why do you suppose the telecommunications and manufacturing departments in Dow Corning began to drift apart in philosophy and technical details of network implementations?
4. Why would a network such as Dow Corning's experience almost constant change?

REVIEW QUESTIONS

H.W

1. Define telecommunications. What are the advantages of using a telecommunications system?
2. Define data communications.
3. Distinguish among communication, telecommunications, and data communications.
4. Why is it important to know something about telecommunications?
5. Why was the invention of the transistor important to the progress of telecommunications technology?
6. Explain the terms:

source	*private line*	*ATM*
medium	*switched line*	
sink	*VDT*	

7. Describe the human protocol of making a telephone call.
8. Discuss reasons why telecommunications is becoming increasingly important to business.
9. In what ways can telecommunications education be obtained?
10. List some of the careers that are available in the telecommunications industry. Which ones do you think might be of interest to you? Why?
11. What are the factors that determine the hours during which a telecommunications network must be available?
12. Compare and contrast the terms *reliability* and *availability* as they apply to a telecommunications network.
13. Explain how the reliability of a serial system is calculated.
14. Explain the difference between an *online* system and a *realtime* system.
15. Explain the concept of *real enough time.*
16. How is *response time* measured?
17. What are the factors that determine what a system's response time will be?

18. Why is *queuing* a factor in calculating response time?
19. Why does fast response time help the productivity of a terminal user?
20. Why is response time normally calculated on a probabilistic basis?
21. What is *think time?*
22. Compare and contrast the concepts *easy to use* and *easy to learn.*
23. Why does the telecommunications manager need to understand ergonomic issues?
24. Why is it important for a telecommunications network to be flexible?
25. Explain why it is important for a telecommunications department to be proactive and forward thinking.
26. Explain why senior management in a company is interested in cost-effective telecommunications.
27. Briefly describe the history of the telephone industry from its inception through the early 1900s.
28. What company passed up the opportunity to own the basic patents on the telephone?
29. In what situations is it necessary to stay current with developments in telecommunications? Why is it difficult to stay current?

PROBLEMS AND PROJECTS

1. Data communications is one way of making information available at the right place at the right time. Identify other ways that information can be transported rapidly without using data communications.
2. List the basic business transactions that are fundamental to the operation of a retail store. Which ones are candidates for data communications?
3. If an ATM is 98 percent reliable, the telecommunications line to which it is connected is 96 percent reliable, and the computer to which they are attached is 99 percent reliable, what is the overall reliability of the three-component system?
4. If you were designing an "ideal" telephone system for the United States, what characteristics would it have that the present system lacks?
5. Make a chart that shows the time differences between Chicago, Brussels, Hong Kong, and Tokyo at different times of the year. (You may have to do some research to investigate the impact of daylight saving time in the various countries.)
6. Calculate the average reliability that each component of a three-component communication system would need for the reliability of the total system to be 99 percent.
7. You have just been named the manager of a full-service department store, one of several in a statewide chain. The company will soon be installing its first point-of-sale terminals, which will have hand-held scanners to read merchandise tickets. You have heard that the terminals in your store will be connected to the central computer at company headquarters, but you have no other details. Next week two analysts from the corporate telecommunications department will be coming to talk to you about your requirements for the new system. In preparation for their visit, make a list of the attributes and operating characteristics that you hope the new terminals and communication network will have, and the questions you will want to ask the analysts.
8. The text gives several examples of how telecommunications is shrinking the world. List several examples of the way you use data communications today that were not possible or would have been considered extravagant 15 years ago.
9. Describe how airline reservations were made before the advent of computers.
10. Do you feel that the Internet and CD-ROM-based encyclopedias will eliminate the need for printed reference books and libraries' reference sections? Why or why not?

Vocabulary

networked society
the Internet
communication
telecommunications
data communications
data
information
source
medium
sink
protocol
basic business transactions
Internet service provider (ISP)
web browser
video display terminal (VDT)
private line
leased line
dial-up line
switched line
automatic teller machine (ATM)
intelligent terminal
dumb terminal
availability
reliability
mean time between failures (MTBF)
mean time to repair (MTTR)
serial system
online
realtime
real enough time
response time
queuing
queue
user response time
think time
dialogue
user friendly
ergonomics
Bell System

References

American National Standards Institute. *Dictionary for Information Systems.* X3. 172–1990.

Doherty, W. J., and R. P. Kelisky. "Managing VM/CMS Systems for User Effectiveness." *IBM Systems Journal* 18, no. 1 (1979).

"The Economic Value of Rapid Response Time." IBM brochure, no. GE20-0752-0 (November 1982).

Levy, Steven. *Hackers: Heroes of the Computer Revolution.* New York: Anchor Press/Doubleday, 1984.

Martin, James. *Design of Man-Computer Dialogues.* Englewood Cliffs, NJ: Prentice-Hall, 1973.

Ramo, Joshua Cooper. "Welcome to the Wired World." *Time* (February 3, 1997): 36.

Stallings, William. *Data and Computer Communications,* 5th ed. Upper Saddle River, NJ: Prentice-Hall, 1997.

CHAPTER

2

External Influences on Telecommunications in the Enterprise

OBJECTIVES

After studying the material in this chapter, you should be able to

- describe the major forces that are changing the telecommunications industry;
- tell how the United States progressed from no telecommunications regulation to a highly regulated environment and then back to deregulation;
- distinguish between the regulated and unregulated portions of the telecommunications industry;
- describe the regulatory status in several other countries;
- describe the structure of the U.S. telecommunications industry.

INTRODUCTION

A company's telecommunications organization is subject to many external influences. Not only is technology changing rapidly (as in the computer arena), but the telecommunications regulatory environment is changing throughout the world. Furthermore, the emergence of voice and data communications standards is affecting the telecommunications environment.

The purposes of this chapter are to look at the background, current situation, and trends of the regulatory environment in the world, and to look at the telecommunications industry in the United States in more detail.

THE REGULATORY ENVIRONMENT

During most of the twentieth century, the telecommunications industry throughout the world was regulated. Companies in the industry were subject to legislation affecting the way they operate, the prices they charge, and the profits they make. In the United States, the telecommunications industry has consisted of private companies that were subject to regulation, whereas in most other countries the telecommunications industry is operated by a government organization or bureau, such as the one that operates the post office and mail system.

In the United States the communications regulatory situation has changed dramatically since 1984 and it continues to evolve. The communications industry has moved from being almost totally regulated in 1983 to being largely deregulated and competitive today. Many other countries, such as the United Kingdom and Japan, are following the same trend although, in some cases, at a slower pace.

Why Regulation?

Why has the world's telecommunications industry been subject to heavy regulation? As the fledgling communications industry caught hold in the years around the turn of the century, it was recognized that widespread, general purpose communications could have significant national benefit if and only if they developed in a uniform, compatible way. Localized networks were not good enough. What was needed was universality, the ability of a person at any telephone to communicate with a person at any other telephone.

ensure compatibility

The only way to ensure this universality was to control what was designed and implemented and to insist that all parts of the telephone system worked in the same way. It was also felt that it was important to provide uniform communications capability to all parts of the country. People believed that the same communications services ought to be available in Billings, Montana, as in New York City. It was also important to guarantee the communications companies a fair profit for their investment and for the risk they were taking, and to motivate them to stay in the communications business.

The telecommunications industry is not unique in this regard. Other industries where universality is desired or that have come to be viewed as a national resource have been regulated too, such as transportation, banking, insurance, and utilities.

protect companies

Regulation can also protect new companies in a fledgling industry by preventing competition and allowing them time to get their feet on the ground. This is particularly important in industries that must make heavy capital investments to build an infrastructure with which they can operate. For example, railroads spent millions of dollars to gain right-of-ways and to lay track before they could run a single train. Similarly, communications companies not only needed equipment, such as telephones, but also networks of wires and central office switching equipment.

prevent interference

In the communications industry, regulation also serves the purpose of preventing interference. What if all of the radio stations in a metropolitan area decided to broadcast on the same frequency? None of the programs would be intelligible to the listeners because they would all interfere with one another. Communications regulations allocate certain frequencies to each radio or television station and separate them so that they don't interfere with one another.

common carrier

In certain industries, the regulatory concept of a *common carrier* emerged. This concept limits the number of companies that can provide key public services, such as transportation or communications. The idea is to prevent the duplication of services or the expensive, capital-intensive infrastructure, such as railroad tracks, communications transmission facilities, or pipelines necessary to provide the services. In the communications industry, the companies that provide the telecommunications networks are the common carriers, sometimes just called *carriers*. Even today their activities are regulated, in return for which they are given monopoly power within the territory or region they serve.

carriers

Milestones of Telecommunications Regulation in the United States

Figure 2–1 summarizes the important milestones of telecommunications regulation in the United States. It is important that you have some understanding of the series of events that led the country into regulation and, more recently, back to a largely deregulated status. This information is a useful aid to understanding the seemingly complex telecommunications industry.

Although AT&T had been founded to provide the nationwide network and interconnection service between Bell franchised telephone companies, many other companies sprang up to provide telephone service in limited markets when the original Bell patents expired in the early 1890s. As companies entered the telephone industry, competition was rampant. At one time, in the early 1890s, more than 125 companies were in operation providing duplicate telephone services. These companies naturally tended to focus their efforts in major population centers where the market was the largest, while ignoring less populated areas of the country. Big cities had telephone service long before it reached small towns.

early competition

1907	First state regulatory agencies established in New York and Wisconsin.
1910	Mann-Elkins Act
1913	Kingsbury Commitment
1921	Graham Act
1934	Communications Act of 1934—FCC established
1948	Hush-a-Phone case
1956	AT&T Consent Decree
1968	Carterfone Decision
1969	MCI Decision
1971	Computer Inquiry I
1971	Open Skies Policy
1981	Computer Inquiry II
1982	Modified Final Judgment
1984	Divestiture
1986	Computer Inquiry III
1996	Telecommunications Act of 1996

Figure 2–1
Milestones of United States telecommunications regulation

Not only was their coverage not uniform, but there was no interconnection between the various networks. Customers could place telephone calls only to other customers who were on the same telephone company's network. In many cases there was no way to call a telephone subscriber connected to a different telephone company.

Naturally, these independent companies wanted to connect to AT&T's nationwide network, but AT&T refused to make the connection. As a further defensive measure and in an attempt to maintain its unique position in the market, AT&T bought out as many of these competitors as it could.

Mann-Elkins Act Congress finally reacted to the heavy competition and, in 1910, passed the Mann-Elkins Act, which placed the activities of all telephone companies doing interstate business under the control of the Interstate Commerce Commission (ICC). The main focus of this action was to install a uniform system of accounting for all telegraph and telephone common carriers. It was the first step toward regulation of the telecommunications industry.

Kingsbury Commitment The competitive pressure exerted by AT&T became burdensome to the independent companies, and in 1912 they complained to the federal government about AT&T's practices. Their major concern was the fact that AT&T was buying out the independent telephone companies and stifling competition. In early 1913, the U.S. Department of Justice and the ICC both informed AT&T that some of its activities appeared to be thwarting competition and were in violation of the Sherman Antitrust Act. The ICC began an investigation of AT&T's activities.

AT&T was concerned that an antitrust suit by the government could cause irreparable damage to its company and business, and so, at the end of 1913, the company agreed to

1. dispose of its stock in Western Union;
2. get approval from the ICC before acquiring any additional independent telephone companies;
3. allow the independent telephone companies to interconnect to its facilities so that they could offer nationwide telephone services to their customers.

This agreement is known as the Kingsbury Commitment after the AT&T vice president, Nathan C. Kingsbury, who authored it for AT&T. The Kingsbury Commitment is significant because it set the telecommunications system on a course toward *universal service.* For the first time it would be possible for all of the communications companies to be interconnected into one nationwide network.

universal service

Graham Act In 1921, Congress passed the Graham Act, which recognized and legitimized AT&T's monopoly. The Graham Act specifically exempted the telecommunications industry from the Sherman Antitrust Act in terms of consolidating competing companies. It also allowed the independent telephone companies to become monopolies in their own geographic territories. This ensured the establishment and building of a nationwide and nonoverlapping telephone network covering the entire country. The view of the day was that allowing these monopolies to exist was in the public's best interest.

The Communications Act of 1934 and the Federal Communications Commission (FCC) The Interstate Commerce Commission had been empowered to regulate interstate telephone services in 1910 but proved to be more interested in regulating the transportation industry than in dealing with the emerging telecommunications issues. As a result, Congress passed the Communications Act of 1934, which created a new agency, the *Federal Communications Commission (FCC),* whose sole purpose was to regulate the telecommunications industry. The FCC, as stated in the opening paragraph of the act, was created

FCC

> For the purpose of regulating interstate and foreign commerce in communication by wire and radio so as to make available, so far as possible, to all the people of the United States a rapid, efficient nationwide and worldwide wire and radio and communication service with adequate facilities at reasonable charges. . . .

The formation of the FCC was the beginning of serious regulation in the communications field. The FCC is a board of commissioners appointed by the president of the United States. The commission, supported by a technical staff, has the power to regulate all interstate telecommunications facilities and services as well as international traffic within the United States. The FCC also controls the radio and television broadcasting industry and issues individual station and operator licenses. You may want to access the FCC's web site at www.fcc.gov for further information.

The descriptions of all regulated telecommunications services and the rates to be charged are called *tariffs.* The common carriers file tariffs with the FCC whenever they want to offer a new interstate or international service or change the price of an existing service. The four major rate categories for tariffs are

tariffs

1. the charge for the time the communication service is used, including possible variations for different times of the day; example: long distance telephone charges;
2. a flat rate for full-time use of a service; example: a leased communications line;

3. monthly minimum charge for a basic amount of usage with additional charges when the usage exceeds the basic amount; example: 800 service (discussed in detail in Chapter 5);
4. a charge for the amount of data sent; example: packet data transmission.

All tariffs are a matter of public record. The FCC reviews tariffs as they are submitted and ultimately approves or disapproves them.

Hush-a-Phone Case In 1948 the Hush-a-Phone Company developed a device that could be placed over the transmitting part (microphone) of a telephone handset to block out background noise. Even though the device was passive and had no electrical or magnetic connection to the telephone set or telephone wires, AT&T did not approve of the device and threatened to suspend telephone service to users who put the device on their telephone. Hush-a-Phone took the case to the FCC, which, after several hearings, ruled in favor of AT&T. Hush-a-Phone appealed, and in 1956 the appeals court overturned the FCC's decision and stated that the device would not harm the AT&T network. The decision had a significant impact on the telecommunications industry, as it opened the door for other companies to attach devices to the telephone network. This was the first chink in the monopolistic structure of the U.S. telecommunications industry.

AT&T Consent Decree In 1949, the U.S. Department of Justice sued AT&T for violation of the Sherman Antitrust Act. The specific charge was that the absence of competition had tended to defeat effective public regulation of the rates charged to subscribers. Of specific concern was Western Electric Company, the Bell System's manufacturing arm, which the Department of Justice wanted separated from the rest of AT&T.

The suit was settled in 1956 by a consent decree that permitted AT&T to keep Western Electric but specifically limited the Bell System companies to the telephone business. Western Electric was prohibited from manufacturing noncommunications equipment and was ordered to limit its sales to Bell System companies. Since computing was a very young science at the time, the significance of this decision, which kept AT&T and the Bell System out of data processing activities, may not have been fully realized.

Carterfone Decision In 1968, the FCC decided that Carter Electronics Corporation, a small Dallas-based company, could attach its Carterfone product to the public telephone network. The Carterfone was designed to allow the interconnection of private radio systems to the public telephone network, and Carter Electronics Corporation sought permission to sell its product. When AT&T refused to allow the Carterfone to be attached to its network, Carter Electronics sued AT&T and won.

That decision widened the door for the attachment of nontelephone company devices to the telephone network and spawned a new segment of the communications industry. After the Carterfone decision, many companies saw an opportunity to make equipment for attachment to the telephone network. This equipment provided customers with new alternatives, such as decorative telephones, private telephone systems for business, and the option to buy rather than rent telephone equipment. The new segment of the telecommunications industry that emerged has become known as the *interconnect industry.*

interconnect industry

MCI Decision The next step toward deregulation and competition was a decision the FCC made in 1969. This decision allowed Microwave Communications Incorporated (MCI) to provide intercity telecommunications links to organizations that wanted to lease them on a full-time basis for their exclusive use. This type of private line service had been offered by AT&T and the Bell System for many years so the product wasn't new, but allowing a private company to provide network facilities was. The MCI decision required the telephone companies to interconnect the MCI lines into the public telephone network, giving their customers nationwide access.

The MCI decision opened the door to network services, and several companies quickly jumped in to provide comparable network offerings. Since they were not required to provide uniform capabilities throughout the country, the new companies tended to offer their services only on high-density routes between major cities, such as Chicago to New York or New York to Washington. Southern Pacific Communication Company, a subsidiary of Southern Pacific Railroad, was one of the first companies in the business and offered network services by taking advantage of the railroad right-of-ways it already owned. It was a relatively simple matter for Southern Pacific to run telecommunications wires alongside the tracks, and it became the first company outside of the Bell System to offer a coast-to-coast alternative service. These companies became known as *other common carriers (OCCs)* or *specialized common carriers (SCCs),* although those terms are rarely used anymore.

Computer Inquiry I (CI–I) In 1971, the FCC concluded a study known as *Computer Inquiry I (CI–I)* in which it examined the relationship between the telecommunications and data processing industries. At issue was the fact that although the two industries were rapidly growing closer, the computer industry was unregulated and the telecommunications industry was regulated. The FCC attempted to determine which aspects of both industries should be regulated over the long term. The outcome of CI–I was that the FCC said that the computer industry was not subject to its control.

Open Skies Policy Also in 1971, the FCC reversed a previous decision regarding communication by satellite. The resulting open skies policy declared that, with few restrictions, anyone could enter the communication satellite business. This decision opened the door to the formation of satellite transmission companies. Several existing companies, such as Western Union and RCA, soon jumped into the market, and they were quickly joined by several newly formed companies that wanted to take advantage of the open skies policy and offer satellite communication services. Today the satellite business is a very active subindustry of the overall telecommunications marketplace. Major players include Scientific Atlanta, AT&T, and Verizon.

Computer Inquiry II (CI–II) The year 1981 marked the conclusion of another study by the FCC to look at the relationship between computer and telecommunications services. This study resulted in a decision known as *Computer Inquiry II (CI–II)*, which stated that

1. computer companies could transmit data on an unregulated basis;
2. the Bell System was allowed to participate in the data processing market;
3. customer premise equipment and enhanced services would be deregulated and provided by fully separate subsidiaries of the carriers;
4. basic communications services would remain regulated.

CI–II provided structural safeguards against cross-subsidization by requiring the carriers to have different subsidiary companies handling the enhanced and basic services.

enhanced service

In *enhanced service*, some processing of the information being transmitted occurs. The processing may be simply a conversion of the speed or the coding system, but it can be more extensive. Other enhanced services that you may be familiar with include call forwarding, call waiting, and caller identification. Enhanced services are not regulated by the FCC and can be provided by the communications carriers at little additional cost. Hence they result in high profit margins, which explains why the carriers try hard to sell them.

basic service

In contrast with enhanced services are the *basic services*, which are tariffed and regulated. Basic service provides only the transportation of information. No processing or other change to the information is allowed. Basic services are provided by the traditional common carriers.

Because of CI–II, the *Bell Operating Companies (BOCs)*, the 22 telephone companies that were members of the Bell System, and AT&T organized new subsidiary companies to handle the unregulated products and services. AT&T chose the name American Bell for its unregulated subsidiary; however, this company name was to be short-lived.

Modified Final Judgment A landmark year in the U.S. telecommunications industry was 1982. It marked the completion of an eight-year antitrust suit by the federal government against AT&T. The suit, which had originally been filed in 1974, was settled by a consent decree.

Many in the computer industry thought that the monopolistic nature of AT&T prior to 1982 was having a serious and detrimental impact on the U.S. telecommunications system. Problems included the lack of willingness to meet customer—particularly business customer—requirements and slow incorporation of technological change into offered products. These were the primary reasons the government brought suit against AT&T.

The original consent decree issued by the court was modified slightly by Judge Harold Green before it received final approval, and it became known as the *modified final judgment (MFJ)*. The judgment stipulated that on January 1, 1984, AT&T would divest itself of all 22 of its associated operating companies in the Bell System, such as Michigan Bell Telephone Company, Pacific Bell Telephone Company, and Southwest Bell Telephone Company. These companies became known as the Bell Operating Companies (BOC). The MFJ also stated that the Bell name was reserved for the use of the divested operating companies.

The restructuring that followed the modified final judgment was traumatic for AT&T, the BOCs, their employees, and customers. It meant that the operating companies were, for the first time, independent. There was a great deal of confusion about which companies would supply specific services and how they would be provided. Employees were transferred from AT&T to the operating companies and vice versa. (Later in the chapter, we will look at the current state of the telecommunications industry in detail.)

The changing cost structure of the local telephone companies and AT&T caused prices for products and services to be adjusted, in some cases radically. For years, the Bell System had subsidized less profitable services with revenue gained from services that were more profitable. For example, the long distance telephone rates for calls on a route between two large cities with a high volume of traffic were set high enough to cover the costs of calls between cities in remote locations or with a low volume of calls. Callers from New York to Chicago helped pay for the cost of calls between Sioux City and Fargo. The modified final judgment specifically prohibited this type of cross-subsidization and essentially stated that each service must pay for itself.

Computer Inquiry III (CI–III) *Computer Inquiry III (CI–III)* was a study conducted by the FCC to determine how and to what extent the carriers could offer enhanced services. In a decision announced in mid-1986 and reaffirmed in early 1987, the FCC ordered that the BOCs and AT&T can offer unregulated, enhanced services if the companies agree to a complex set of provisions called *open network architecture (ONA)*. The independent telephone companies, such as GTE however, were exempted from the FCC order.

Telecommunications Act of 1996 On February 1, 1996, Congress passed the *Telecommunications Act of 1996.* The act, the first comprehensive rewrite of the Communications Act of 1934, dramatically changed the ground rules for competition and regulation in virtually all sectors of the communications industry. The 22 Bell companies that until 1984 had been part of AT&T had been organized into seven regional corporations. The 1996 act allowed these regional corporations to provide long distance service outside of their regions immediately, and inside their regions after certain steps were taken to ensure competition. They were also allowed, for the first time, to manufacture telecommunications equipment. The long distance carriers, such as AT&T, MCI, and US Sprint, were allowed to offer local telephone service. Telephone companies were permitted to offer cable television services. The act also relaxed the rules regarding the ownership of radio and television broadcasting stations, allowing a higher concentration of ownership. A notable subpart of the act created criminal penalties for knowingly transmitting material considered to be indecent to minors over the Internet. These provisions also made it a crime to make any computer network transmission with the intent of harassing the recipient.

Deregulation

In the 118 years since the telephone was invented, there has been a transition from unrestricted competition to complete monopoly powers vested in AT&T and the Bell System and now back to virtually unrestricted competition. In the meantime, computing and data processing have come on the scene, and the nationwide telecommunications network—which was originally designed strictly for voice communication—has become heavily used for data transmission.

The FCC remains as the national regulatory agency that approves all rates and services offered on an interstate basis. Each state has a *public utility commission (PUC),* sometimes called the *public service commission (PSC),* which has jurisdiction over intrastate rates and services. The first PUCs were established in New York and Wisconsin in 1907.

PUC

The implication of having state-level PUCs is that it's possible for all 50 states to have different rates for the same service. Fortunately, this is usually not the case; the state PUCs do talk to one another and to some extent coordinate their activities.

The common carriers file tariffs with the appropriate PUC whenever they want to offer a new intrastate service or change the price of an existing service. The process is essentially the same one that is used for filing interstate tariffs with the FCC. Having to file 51 different tariffs (one for each state plus one to the FCC for the interstate tariff) for a nationwide service places a large burden on the common carriers.

Implications of Deregulation

Thanks to the Telecommunications Act of 1996, the year 1997 marked the first time that every sector of the telecommunications industry became truly competitive. Long distance carriers and local telephone companies entered one another's markets. The Telecommunications Act of 1996 made provision for both *interconnect* and *wholesaling*. The wholesaling provision requires owners of a telephone network or equipment to resell network capacity or equipment usage to other companies. The interconnect provision prohibits the seller from charging unreasonable rates. Thus, any company can get into the communications business. If it does not own a network or switching equipment, it can lease capacity from a company that has it, and resell the service. For example, a long distance company such as WorldCom, which did not have a network in a certain city, could lease capacity from the local telephone company and resell it. In this way, WorldCom would be able to provide a total package of services to its customers, including local and long distance services. Similarly, a local telephone company that wanted to offer long distance service but didn't have a long distance network could lease long distance capacity from WorldCom or AT&T and resell to its local customers, thus also providing a total package.

The intent of deregulating the U.S. telecommunications industry was to provide better, more economical service and new, more flexible products to telecommunications customers. It was felt that competition would stimulate development of innovative products. Furthermore, it was expected that prices for communication services would drop as competition increased. Indeed, 5 years after the Act was passed, its goals are largely being fulfilled. The competitive environment has forced telecommunications companies to be more aggressive in developing and pricing new products and services. You can find more information about the Telecommunications Act of 1996 on the FCC's web site at www.fcc.gov/telecom.html.

For the telephone companies, this type of competition was a culture shock. In the past, being monopolies, they simply took orders for products and services. The products they chose to offer were the only products available; marketing was unnecessary.

competition

In today's competitive environment, all telecommunications companies, including AT&T and the BOCs, must actively market their products and services or risk losing business. Thousands of new companies have entered the marketplace, some with broadly based product lines that they market nationally or even globally, and others with specialized products sold to a niche market. AT&T and the BOCs are also allowed to market their products and services essentially anywhere they choose, and some of the companies offer their products outside of the United States.

When deregulation first began, there was some concern that the long-term results might be unfavorable for the industry and its customers.

Some people felt that the integrated telephone network might start to crumble and that the traditionally excellent products of the Bell System might be replaced by inferior products from other companies. Fortunately, the reverse has been the case. Competition has forced all of the participants in the industry to sharpen their product development and marketing skills. Consumers and business customers have a range of products and services that is much larger than in the past when the industry was tightly regulated. Price competition has also worked to the customer's benefit.

While heavy regulation was desirable in the formative years of the industry when it simplified the building of an integrated nationwide telecommunications network, the tight controls are no longer needed and a less regulated industry has behaved like an unshackled giant to everyone's benefit.

TELECOMMUNICATIONS IN OTHER COUNTRIES

Historically, telecommunications in most countries has been heavily regulated. This is partly a function of the state of development, with less developed countries tending to have more regulation, and more developed countries moving toward deregulation. If you consider that two-thirds of the people in the world have never made a phone call, you get a perspective that is quite different from the modern, wired society in which we live. China and India, with a third of the global population, have fewer than two phone lines for every 100 people. Indonesia has one; Brazil, seven. By contrast, the United States has 60.

The trend in developed and developing countries, however, is very clearly toward deregulating telecommunications. The European Union made the big step and deregulated on January 1, 1998. The decision to deregulate on that date was actually made in the early 1990s, so the regulatory agencies and telecommunications companies in Europe had many years to prepare. Now competition is beginning to flourish as it has in other countries where deregulation has already taken place.

In February 1997, 69 nations negotiating under the auspices of the World Trade Organization (WTO) agreed to open 95 percent of the $600 billion annual world telecommunications market, long constrained by government-run phone monopolies, to foreign trade. This change allowed companies to access foreign telecommunications markets that were previously closed to them and was especially beneficial for U.S. telecommunications companies, which previously had access to only 17 percent of the top 20 global markets. Increasingly, countries now buy communications equipment (phones, switching centers, fiber-optic cable) and services (the transmission of signals) from the cheapest sources. Even a few years ago this was barely imaginable. Telecommunications was seen as a natural monopoly and, some countries today still have monopolies or near-monopolies that buy equipment from a few captive suppliers.

For international callers the WTO agreement is expected to reduce long distance rates, in some cases by up to 80 percent. In the U.S., where communications competition is most advanced, long distance rates have dropped 70 percent since competition began in 1984 and today average 10 to 15 cents a minute. In Japan, comparable rates are 95 cents a minute; in Brazil, they're about 65 cents. Overseas calls are also high-priced. From the U.S. they now average about $1 per minute, because they blend the lower American rates with higher rates abroad. There are no obvious technical reasons for these differences. Satellite or fiber-optic transmission costs about the same between New York and Los Angeles as between New York and Tokyo. The gaps mainly reflect monopolistic pricing and high overhead costs. Competition will erode these prices and international rates may ultimately drop 80 percent from $1 per minute to 20 cents.

Generically, the regulatory agencies in most other countries are called the *PTT,* which stands for *post, telephone, and telegraph.* The name comes from the fact that in most countries the same regulatory agency has the responsibility for the postal service as well as telecommunications. Generally, the government owns and operates the common carriers and, in some cases, the equipment manufacturers. With deregulation occurring in the United States, the more advanced countries are watching U.S. developments closely, and some countries have made deregulatory moves of their own. A list of the regulatory agencies in selected countries is shown in Figure 2–2. The regulatory status of the telecommunications industry in several countries is described in the next few paragraphs.

PTT

Australia

Australia finished its move to full deregulation and competition in 1997 after a 5-year transition period. There are three major domestic carriers,

Country	Regulatory Agency
Australia	Department of Communications (DOC)
Belgium	Belgian Institute for Postal and Telecommunications Services (BIPT)
Brazil	Ministry of Communications
Canada	Canadian Radio-television and Telecommunications Commission (CRTC), and provincial governments
France	Direction Generales des Postes et Telecommunications
Germany	Budesministerium fur Post und Telekommunikation (BMPT)
Japan	Ministry of Posts and Telecommunications (MPT)
Switzerland	Federal Office of Communications (OFCOM)
United Kingdom	Office of Telecommunications (OFTEL)

Figure 2–2
Regulatory agencies in selected countries.

with Telstra being the largest, and a host of smaller companies. There are also three main carriers for cellular telephone systems. In many ways, the telecommunications environment in Australia parallels that of the United States.

Belgium

Although it is a small country, Belgium is important because it has encouraged foreign companies to establish offices and plants in the country through liberal tax laws and other incentives. As a result, many international corporations have established their European headquarters in Belgium.

Telecommunications in Belgium was largely deregulated in late 1992. The former Regie des Telegraphes et des Telephones (RTT) in French, or—because Belgium is officially a bilingual country—the Regie van Telefoon en Telegraaf in Flemish, was divided into two organizations. Belgacom, which is the commercial and operational company, is the unregulated portion, and the Belgian Institute for Postal and Telecommunications Services is the regulated part. Belgacom's mission is to be a leading international supplier of telecommunications services. The company is more customer- and market-oriented than the RTT, which was primarily engineering driven. Belgacom operates the public telephone system and telex network, and provides leased lines for voice and data use.

The Belgian telephone system has for years been among the world's most advanced. For example, it provided direct dialing to many foreign countries in the mid-1960s, long before that service was generally available elsewhere. The quality of telephone connections is usually excellent, yet there are some interesting quirks. In order to obtain telephone service in a home or business, it is necessary to deposit a substantial sum of money with Belgacom (several hundred dollars for a residence telephone) in case the bill isn't paid. A wait of several weeks for telephone service to be installed is not uncommon.

Data transmission capabilities to and from Belgium are of good quality and readily available. Many of the international companies located in Belgium have extensive data networks radiating to other European countries.

Brazil

Historically, Brazil has had a very tightly regulated telecommunications environment. Although telephone service was relatively open, there were many years when data communications outside the country was prohibited, ostensibly for national defense reasons. Today the state-owned telephone company, Embratel, reviews and approves all applications for telecommunications services, and while some justification is required for new services, the environment is much less restrictive than in the past.

Canada

The Canadian Radio-television and Telecommunications Commission (CRTC) performs functions similar to the FCC in the United States. Only transmission facilities are regulated. Service and rates are defined by tariffs, much as in the United States. The nationwide voice and data network is provided by a number of companies, the largest of which is Bell Canada, which also has several operating subsidiaries. There are also a number of Canadian competitors, such as Telus, Corp., and many U.S.-based companies, such as AT&T, WorldCom, and Ameritech, are also active in the Canadian market. Canada also has a nationwide network called Dataroute, which was specifically designed to carry data. Since its introduction in 1972, Dataroute has been continually upgraded with new capabilities and service offerings.

Japan

Japan's telecommunications regulation is governed by the Ministry of Posts and Telecommunications (MPT), and the agency has made moves toward deregulating the Japanese telecommunications industry. In 1985, the major domestic telephone company, Nippon Telephone & Telegraph Company (NTT), shifted from being fully government owned to semi-public status. Taking advantage of deregulation in the United States, NTT began offering telecommunications services to U.S. customers in late 1998. In 1999, NTT was divided in three companies, a long distance carrier and two regional domestic carriers. NTT's long distance company competes with the other major Japanese long distance carrier, Kokusai Denshin Denwa (KDD), whose primary business until recently has been international long distance. KDD itself merged with two smaller carriers in late 1999 in an attempt to create a full service rival to AT&T. DDI, as the new company is known, is able to offer its customers everything that NTT offers them except local telephone service, which is increasingly under pressure from the proliferation of cellular telephones and the services available on them.

Until recently, Japan has been closed to the importation of telecommunications equipment made in other countries. Japan participated in the 1997 WTO agreement, however, so the market in Japan has opened to foreign suppliers. This move will relieve some international tension because, while Japan's domestic market has been closed to foreign companies, Japan's own telecommunications equipment manufacturers such as Nippon Electric Company (NEC) and Toshiba have steadily increased the amount of telecommunications equipment they have exported. Naturally this has angered telecommunications manufacturers in other countries.

Mexico

Mexico has deregulated the national telephone company, Telefonos de Mexico (Telemex), and the primary effect has been to attract millions of dollars of foreign capital to help upgrade the telecommunications network and infrastructure. Nine new carriers of long distance service entered the market shortly after deregulation, and telephone customers were given a choice of which one they wanted to use. Like NTT from Japan, Telemex is selling its services in the United States, but some Mexican customers think it should keep its focus on Mexico because it still takes a long time to get a telephone.

Russia

Just a few years ago Russia's 85 regional telephone companies were part of a mammoth state monopoly. Now partly privatized, they are boosting revenues by increasing residential rates, until now among the world's lowest. The companies are using the extra cash for much-needed improvements in the network infrastructure. In some areas the system is so decrepit that calls between cities are impossible. Fixing the system will mean laying thousands of kilometers of cable and installing modern switching equipment, and the overhaul is expected to take at least 20 years.

United Kingdom

In the United Kingdom, the Office of Telecommunications (OFTEL), a part of the Department of Trade and Industry, regulates telecommunications. The first steps toward deregulating telecommunications were taken when the government-owned British Telecommunications company became publicly owned in 1984. A competitor, Mercury Communications, was formed, and the ensuing years have seen many new services offered and a general upgrading of the technology of the British public telecommunications network.

In November 1991, the British government announced that it would deregulate most of the nation's telecommunications industry, including eliminating most of the legal restrictions separating the telephone and cable television industries. Access to the local, long distance, and international service markets is now largely unrestricted. Although British Telecommunications still holds a vast share of the communications market in the United Kingdom, the presence of competition has forced it to quickly become more flexible and competitive than it might otherwise have.

Other Countries

The regulatory and competitive status in other countries, particularly in Asia, is evolving rapidly. Additional information can be gleaned almost daily from newspapers and general business publications. Prices of everything from international calls to cellular handsets are plunging, and many companies are finding it increasingly difficult to remain profitable. The pressure will accelerate as the full impact of the 1997 WTO agreement becomes evident and competition becomes even more intense.

The General Post Office Tower in London is the support structure for many types of radio antennas which are used to transmit radio signals for data, voice, and other types of transmission. (Courtesy of Adam Woolfitt/Corbis)

One interesting development is coming from poorer third world countries that do not have an extensive telecommunications infrastructure (wires, cables, switches, etc.) installed. Some countries are considering skipping the cost of building the wired infrastructure and jumping right to wireless technology. Instead of installing thousands of miles of cables, a country can build a radio transmission and relay system that allows cellular phones and data terminals to communicate effectively. In most cases, building the radio-based transmission system is much less expensive than laying cables.

TRANSNATIONAL DATA FLOW

A concern in the international telecommunications arena is the subject of *transnational data flow (TNDF)*. Several countries, especially in Europe, have enacted legislation to prohibit or restrict the transmission of data and information across national borders. One of their concerns has to do with national defense and the fear that national secrets will be transmitted to unfriendly countries. Another concern involves personal privacy and the fear that private data about citizens will be transmitted to foreign countries. A third concern is the export of information processing, resulting in the loss of jobs. Canada, for example, has been very concerned that because of the close proximity and the common language with the United States, U.S. companies with operations in Canada would choose to do all of their data processing in the United States, thereby limiting the growth of the Canadian information processing industry.

Multinational companies are big generators of transnational data flows. Data is transmitted across borders for processing customer-related transactions, for financial management, and for the coordination of production and inventory management. The question being raised by multinational companies is whether restrictions imposed by countries on the installation and use of international telecommunications and computer processing will create unworkable limits on the management of worldwide activities. Some examples of the types of regulation that could be damaging would be

- a tax or tariff on information transfer;
- monitoring the content of international communications;
- restricting the availability of private leased lines;
- privacy legislation mandating that no personal information be transmitted across international borders.

Several international agencies have been active in transnational data flow issues, such as the Council of Europe (COE), the Organization for Economic Cooperation and Development (OECD), and the United Nations. The OECD has developed guidelines about the protection of personal privacy when information flows across borders. These guidelines have been adopted by most of the countries of Europe and by Japan.

Individual countries have taken widely differing positions on transnational data flow issues. The U.S. position is that a free flow of information should be encouraged, and data processing and telecommunications uses should be unregulated. By contrast, the Scandinavian countries have been leaders in establishing laws to severely restrict the international transmission of personal data. The proactive countries passed laws related to the transmission of information from their countries in the early 1980s and have been getting experience with their impact since then.

THE TELECOMMUNICATIONS INDUSTRY IN THE UNITED STATES

Since January 1, 1984, when divestiture took effect, the structure and participants in the telecommunications industry in the United States have changed dramatically. Now we will look at the current status of the industry and some of the key companies that participate in it.

BOCs and RBOCs

The 22 Bell Operating Companies, which were divested from AT&T, were grouped into seven *Regional Bell Operating Companies (RBOCs).*

1. NYNEX Corporation;
2. Bell Atlantic;
3. Bell South;
4. American Information Technology (AMERITECH);
5. Southwestern Bell Corporation;
6. U.S. West;
7. Pacific Telesis (PACTEL).

Figure 2–3 shows how the 22 BOCs were originally regrouped to form the seven RBOCs. Each RBOC is a publicly held company. When they were originally established, the RBOCs among them had about 80 percent of the former Bell System's assets—about $17 billion each. Each RBOC started with about 90,000 employees, making them roughly the same size as GTE, and each company was in the top 50 of *Forbes* magazine's ranking of the top 500 companies, as measured by their assets.

Since that time, a series of mergers and acquisitions has occurred that has reduced the number of RBOCs as shown in Figure 2–4. Bell Atlantic purchased NYNEX and then merged with independent telephone

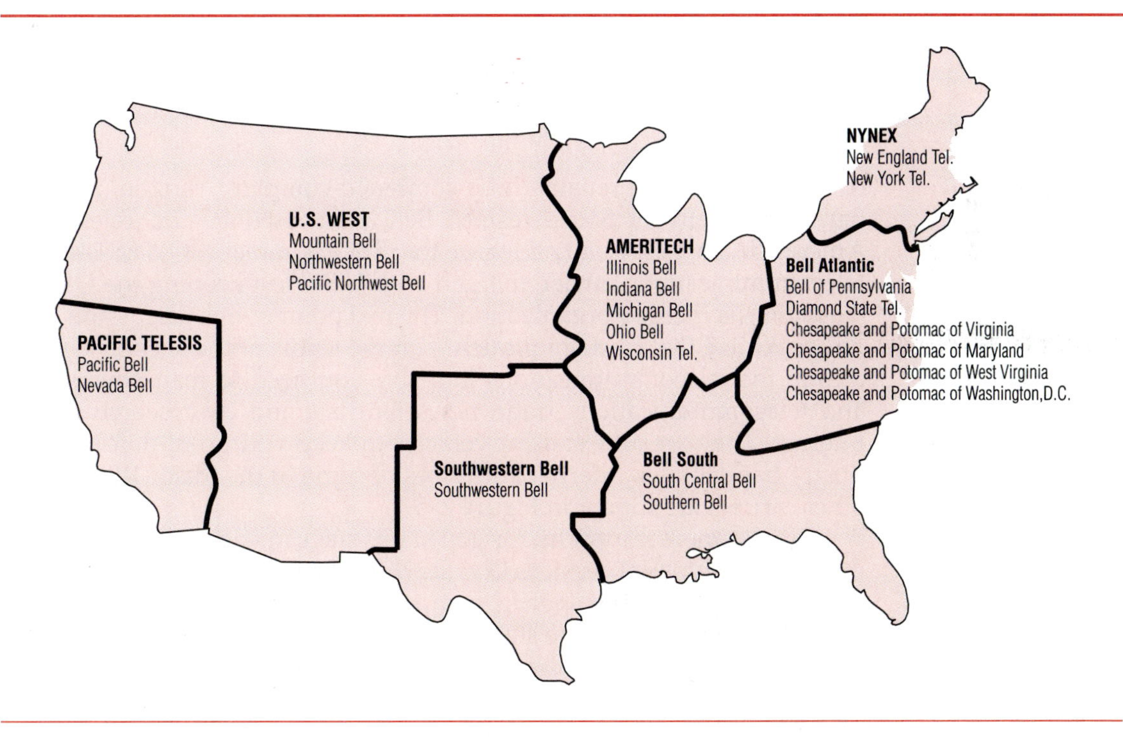

Figure 2–3
The territories served by the original Regional Bell Operating Companies.

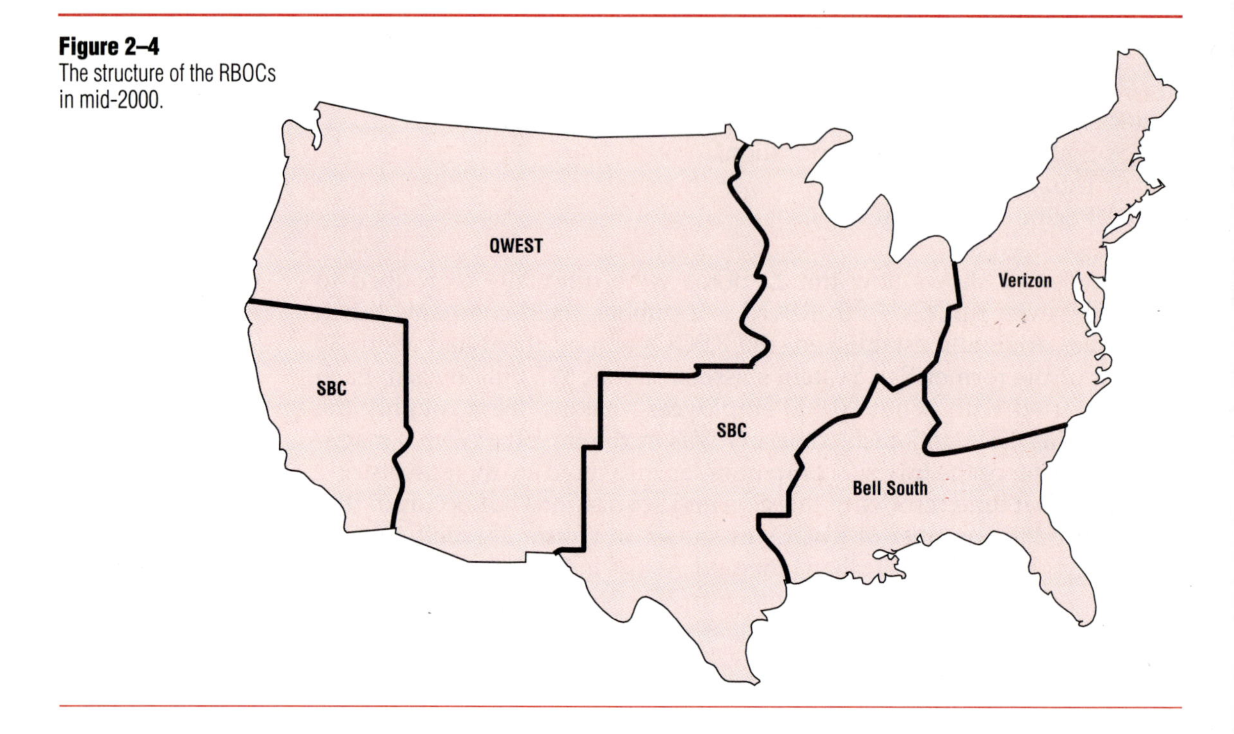

Figure 2–4
The structure of the RBOCs in mid-2000.

company, GTE, and renamed the combined company Verizon. SBC, the parent company of Southwestern Bell, purchased Pacific Telesis and Ameritech, so today only four of the RBOCs remain. QWEST, a new telecommunications carrier, merged with U.S. West keeping the QWEST name for the merged organization. The real point of this discussion is for you to realize that to stay in touch with the status of the specific companies in the telecommunications industry you need to read newspapers and telecommunications trade news journals and do research on the Internet. They are all excellent vehicles to keep you-up-to-date with industry happenings. Internet addresses of some of the major RBOCs and other carriers are shown in Figure 2–5.

What is more important for you at this stage in your learning is to understand the types of services that the different types of companies in the industry can offer. The specific companies may change quickly, but the service categories are more stable since the law governs some of them.

The RBOCs provide local telephone service within the areas they serve, using the network that was in place before they were divested by AT&T, supplemented by the network upgrades and additions they have made since that time. The regional companies in most cases conduct their

Company	Internet WWW Address
AT&T	www.att.com
Bell South	www.bell-south.com
SBC	www.sbc.com
QWEST	www.qwest.com
Verizon	www.verizon.com
Worldcom	www.worldcom.com

Figure 2–5
Internet addresses of some of the RBOCs and other major U.S. carriers.

day-to-day business through the BOCs, many of which are organized along state boundaries, such as Wisconsin Bell Telephone Company and Illinois Bell Telephone Company. The BOCs know the regulatory issues within the states they serve and can work more effectively with their respective public utility commissions than the RBOCs could.

Many of the RBOCs have actively pursued other business opportunities besides traditional telephone service. For example, Ameritech and Bell Atlantic invested in New Zealand's telephone company. U.S. West invested $2.5 billion in Time Warner to help the company rebuild its cable television systems. Bell South is a major provider of paging services nationwide. The companies have been most successful when they have stayed within the communications business. NYNEX bought a chain of computer retail stores; Southwestern Bell started an interior design service; U.S. West got into financial services. None of these ventures did well.

In addition to the RBOCs and BOCs, there are many independent telephone companies that provide telephone service but were never a part of the Bell System. Since they were not part of the consent decree breaking up the old Bell System, these companies can offer local and long distance service if they choose. Before it merged with Bell Atlantic, GTE Corporation was the second largest telephone company in the United States. Other companies include Continental Telephone, United Telephone, and approximately 1,375 others. Many of these are very small "mom and pop" companies that provide telephone service in a very limited geographic area. With modern equipment, the telephone operation is totally automatic and virtually trouble free. The BOCs and independents together are called *local exchange carriers* or *incumbent local exchange carriers (LECs or ILECs).*

local exchange carriers (LEC)

Competitive Local Exchange Carriers (CLECs)

A new group of companies called *competitive local exchange carriers (CLECs)* has sprung up in the last few years. CLECs, which have no ties to the old world of the Bell system, are taking advantage of the Telecommunications Act of 1996. These companies are of three types:

- CLECs that already own their own network,
- CLECs that lease their network infrastructure from someone else,
- CLECs that build an entirely new network.

When a CLEC builds a new network, it is typically faster and cheaper than the old telephone networks and designed from the ground up to handle a mixture of voice, data, and video, rather than having that capability added, as the older (legacy) networks that were originally designed for voice have had to do.

While these companies could represent the beginning of the next-generation telephone companies, the future is not entirely certain. The large companies in the industry, such as SBC, AT&T, and WorldCom could decide to build new digital networks themselves, or could buy up the CLECs. However, even if the CLECs should falter, their presence in the market is forcing the established companies to change their ways. The initial impact is strong competition, which is forcing prices down for many products and services. U.S. long distance rates, which decreased nearly one third between 1998 and 2000, are one example where the impact can clearly be seen.

More sophisticated telecommunication features are likely to be the most noticeable and long lasting impact, however. Resort operators might be able to talk to prospective customers while simultaneously showing them pictures of rooms and the grounds over the same telephone line. If you need more telephone or data capacity at home, you may be able to request a capacity increase to give you the ability to handle, for example, two simultaneous phone calls and Internet access, without ordering a separate line. In effect, you would order a fatter line then you have today!

The old-line phone companies are of course not sitting idle. They all have aggressive programs to upgrade their facilities and offer new products and services, and they all have more financial resources to draw on than the CLECs. The question will be whether they can break out of their traditional operating styles in order to fully participate in the new, fast moving telecommunications world. To learn more about the CLECs, visit the web site www.clec.com.

As part of divestiture, 165 local calling areas called *local access and transport areas (LATAs)* were defined within the United States. A diagram of the LATAs in the state of Michigan is shown in Figure 2–6. The LECs provide telephone service within LATAs, but inter-LATA telephone traffic was given to the long distance carriers such as AT&T, WorldCom, and Sprint. These long distance carriers are also known as *interexchange carriers (IXCs)*. The Telecommunications Act of 1996 effectively eliminated the differences in the areas and type of traffic the LECs and IXCs could pro-

interexchange carrier (IXC)

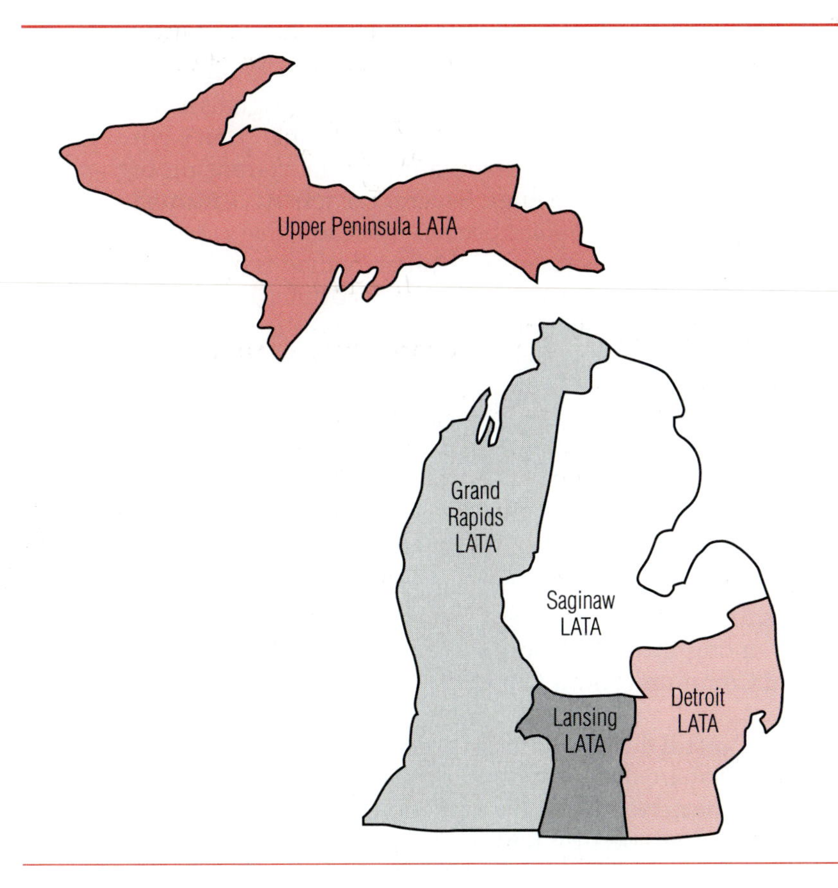

Figure 2–6
The LATAs in the state of Michigan.

vide, and beginning late that year and in 1997 the IXCs began offering local telephone service while the LECs began offering long distance service.

The responsibility of telephone companies for lines and equipment extends up to the residential premises but not within them. Inside wiring and equipment, such as telephones, are the customer's responsibility. That is, if there is a problem with the wiring or telephone in the house, the telephone company will charge for repairing it; if the problem occurs outside the house, the telephone company will fix it at its expense. The point where the telephone company responsibility ends and customer responsibility begins is called the *demarcation point.*

demarcation point

The situation is more complicated with businesses. For small business locations, the rules are the same as for residences. For large business campuses or multistory buildings, the telephone company's responsibilities frequently extend into the building to a central equipment room or even to an equipment room on each floor. Frequently the location of the demarcation point is negotiated between the customer and the telephone company.

A significant challenge that all of the carriers face today is getting high-speed communications service into the home. Standard telephone service does not have the capacity to carry all of the signals a family might want to access at a speed they would find acceptable. That's why carriers are working on new technologies to allow them to transmit high-speed data over existing telephone lines. Several carriers have formed alliances with cable television companies because the cable that carries a few television channels today has far greater capacity than is being used. On the other hand, the cable television network was primarily designed and engineered to carry signals in one direction—from the source to the home. When used, for example, to access the Internet, some cable television systems have difficulty carrying signals from the keyboard of the computer back through the cable to be fed to the Internet access provider. High-speed communication services will be discussed in more detail in Chapter 9.

Long Distance Carriers

There are now many long distance carriers in the United States but three companies are the market leaders, holding nearly 90 percent of the market share among them. They are AT&T, WorldCom Corporation (MCI), and Sprint Communications Company (Sprint).

AT&T AT&T is still the largest long distance carrier, holding about 55 percent of market share in the U.S. and providing nationwide long distance service using the network that it installed before divestiture, considerably enhanced and upgraded by new technology such as optical fibers and electronic switching equipment. In 1996 the company split itself into three pieces: AT&T, providing domestic and international transmission and online services to consumers and businesses; Lucent Technologies, which manufactures communications equipment; and NCR Corporation, a company AT&T had purchased several years earlier and which manufactures computers. In late 1996 AT&T began offering local telephone service, taking advantage of the provisions of the Telecommunications Act of 1996. AT&T's major thrusts for the immediate future are long distance service, wireless products, cable, and international service. The company has long-time experience offering long distance, and to some extent, international service, but wireless and cable television (also to be used for data and voice transmissions) are newer areas where the company has less expertise.

WorldCom WorldCom started out as a CLEC but really came into its own when it purchased MCI Communications Corporation in 1997. MCI was founded in 1963 in Illinois as a small common carrier providing niche services, mostly in major cities. MCI was a key player in the breakup of the Bell System in 1984, and grew to offer a full range of domestic and international long distance services to more than 180 countries. The company began offering local telephone service in the U.S. in 1997. In November 1997 MCI was purchased by WorldCom and the name was changed to MCI WorldCom, but the MCI portion of the name disappeared in 2000.

Sprint Communications Company (Sprint) Sprint is the third largest long distance carrier. Although its roots go back to 1899, the company started operating under its current name in 1984. Sprint provides services that are similar to those of AT&T and WorldCom. Sprint is also involved in cable television, with several partners, and with wireless communication (which will be described in Chapter 5). Sprint has several telecommunications "firsts" to its credit, including being the first to offer residential 800 service, the first to build a nationwide all-digital fiber-optic network in the U.S., and the first to offer prepaid calling cards.

International Carriers

International carriers provide communications services between countries. AT&T, World Com, and Sprint are all international carriers in the United States. Teleglobe Canada (Canada), FTC Communications (France), British Telecommunications (United Kingdom), Kokusai Denshin Denwa Co. (Japan), and Telstra (Australia) are some of the international carriers in other countries. In general, international carriers provide international voice and data services, including dial-up and full-time connections, as well as a variety of special services for customers with high transmission volumes.

The costs for international voice and data lines are divided into two parts. Each country charges for one-half of the line; unless special arrangements are made, the customer receives a bill from the company providing each end of the line. Conceptually, on a trans-Atlantic line, it's as though each carrier billed for the line from the carrier's country to the middle of the ocean!

Equipment Manufacturers and Providers

There are literally thousands of companies operating in the telecommunications industry today. Telephone sets, computer terminals, private telephone systems, wiring and cabling, and satellite equipment are just a small sample of the types of products that these companies supply in a totally unregulated way in most countries. The RBOCs and long distance carriers have marketing units or subsidiary companies to sell these types of products and they are often in competition with one another. Trade journals abound filled with advertisements for equipment and services. The telecommunications department in a company has a wide choice of products from which to choose.

Several multinational companies with headquarters outside the United States also compete in the global telecommunications marketplace.

Figure 2–7
Some of the largest companies in the global data communications marketplace.

Company	Home Country	Internet Web Address
Nortel Networks	Canada	www.nortelnetworks.com
Nokia	Finland	www.nokia.com
Alcatel	France	www.alcatel.com
Siemens	Germany	www.siemens.com
Canon	Japan	www.canon.com
Fujitsu	Japan	www.fujitsu.com
Matsushita	Japan	www.matsushita.co.jp
Nippon Electric Company (NEC)	Japan	www.nec.com
Ricoh	Japan	www.ricoh.com
Toshiba	Japan	www.toshiba.com
N.V. Philips	Netherlands	www.philips.com
Ericsson	Sweden	www.ericsson.com
AT&T	United States	www.att.com
Cisco Systems	United States	www.cisco.com
Hewlett Packard	United States	www.hp.com
IBM	United States	www.ibm.com
Lucent Technologies	United States	www.lucent.com
Motorola	United States	www.motorola.com

Although naturally strong in their home country, these companies all work hard to expand their territory beyond national boundaries. The United States is fertile territory because of its size and the relative sophistication of its telecommunications applications. Many of these companies are actively selling their products in the U.S. A list of major telecommunications equipment companies and their web addresses is shown in Figure 2–7.

■ SUMMARY

Regulation was a very important factor in ensuring that the telecommunications industry developed in a unified, compatible way providing users with universal service. Now that the industry is more mature, deregulation is allowing market forces to shape the services and products that are developed and made available. The resulting competition and innovation has benefited businesses and consumers alike.

CASE STUDY

Impact of the Telecommunications Industry and Deregulation on Dow Corning's Telecommunications

Like most other companies, Dow Corning Corporation weathered the change that divestiture and deregulation imposed on the vendors and the telecommunications industry in the 1980s. Because its location (Midland, Michigan) is not in a large metropolitan area, not as many of the small telecommunications vendors call on the company as a company would experience in a larger city. Historically, the primary communications vendors in Midland have been Ameritech, WorldCom and AT&T, IBM, Amdahl, Sun, Hewlett-Packard, and DEC have been the major vendors of computers and PC workstations, but many other vendors also compete in this market. Several of the vendors have marketing teams in or near Midland. These teams are typically responsible for all of the marketing by their companies to Dow Corning throughout the United States, and in some cases, globally.

Most of the people in the telecommunications department have grown up in a world of fully competitive telecommunications. They are very comfortable working in the environment and take advantage of the competitive situation by frequently negotiating rates and services with vendors. The result is that, compared to previous decades, Dow Corning gets far more value for the money it spends on telecommunications than it did in the past.

QUESTIONS

1. Do you think that telecommunications managers must be more skilled today than in the past when there was no competition in the industry? Give some examples.
2. Do companies such as Dow Corning generally prefer the competitive conditions of the telecommunications industry today, or prefer the regulated, more monopolistic environment of past years?
3. Based on your life experience, can you think of some telecommunications products and services that might not be available today if competition had not been allowed in the telecommunications industry?
4. What telecommunications products and services can you think of that have come into the marketplace in the past 12 months?

REVIEW QUESTIONS

1. Explain why the telecommunications industry was regulated in its early years.
2. Explain the terms:

BOC	carrier	ILEC
RBOC	tariff	CLEC
divestiture	international carrier	

3. Describe the roles of the FCC and state PUCs.
4. Explain the four major rate categories for tariffs.
5. What was the significance of Computer Inquiry II?
6. What was the significance of the Carterfone decision?
7. What was the modified final judgment? What significance did it have for AT&T?
8. Explain the difference between deregulation and divestiture.
9. What happened on January 1, 1984? How was the Bell System reorganized?
10. What is the function of a PTT?
11. What is the concern about transnational data flow?
12. What is the U.S. position on transnational data flow issues?
13. Explain the significance of the Telecommunications Act of 1996.
14. Compare and contrast the functions of an LEC and an IXC.
15. What did the WTO agree to in early 1997 and why was the agreement important?
16. Are other countries deregulating their telecommunications as the United States is? Are they encouraging competition?
17. How are CLECs shaking up the communications industry? Why are they a worry to "old line" telecommunications companies?

PROBLEMS AND PROJECTS

1. Find out what percentage of the market for long distance telephone service is held by each of the companies in the industry. How has this changed in the last year? Do you think the smaller companies will be able to stay in business and remain profitable? Why or why not?
2. Why do you think the United States is leading the world in deregulating its telecommunications industry? Are deregulation and competition really better than a regulated monopoly for the communications industry? Is deregulation leading to lower prices for users? Defend your position.
3. At divestiture, the RBOCs were all about the same size, as measured by their revenue and number of employees. Do some research and find out what has happened to their size in the years since divestiture occurred. Which companies have grown? Which have shrunk? Which are the most profitable today?
4. Compare your telephone bill at home for a period of time (quarter or year) with the same period last year. Separate the changes in the number of calls that were made and determine whether your rates went up, down, or stayed the same. Find out what the telephone company claims happened to the rates and compare its claim to your own experience.
5. Do some research to determine the current status of the "information highway" in the United States.
6. Investigate all of the possibilities available to you for obtaining telecommunications service in your home. Some of the questions you may want to investigate are: Can you get telephone service from your cable TV provider? What about access to the Internet? Are satellite services available for TV or Internet access? Does your telephone company provide any other high speed Internet access capability such as DSL? Do any vendors offer package deals if you purchase a range of services from them?

Vocabulary

common carrier
carrier
universal service
Federal Communications Commission (FCC)
tariff
interconnect industry
other common carrier (OCC)
specialized common carrier (SCC)
Computer Inquiry I (CI–I)
Computer Inquiry II (CI–II)
enhanced service
basic services
Bell Operating Companies (BOCs)
modified final judgment (MFJ)
Computer Inquiry III (CI–III)
open network architecture (ONA)
Telecommunications Act of 1996
public utility commission (PUC)
public service commission (PSC)
post, telephone, and telegraph (PTT)
transnational data flow (TNDF)
Regional Bell Operating Company (RBOC)
local exchange carrier (LEC)
incumbent local exchange carrier (ILEC)
competitive local exchange carrier (CLEC)
local access and transport area (LATA)
interexchange carrier (IXC)
demarcation point
international carrier
value added carrier

References

Allen, Doug. "The Year of Living Dangerously: When Is a CLEC Not a CLEC?" *Telecommunications* (November 1999): 26–32.

Allen, Doug. "Everything Old is New Again." *Telecommunications* (October 1999): C1–C16.

Arnst, Catherine. "Telecommunications." *Business Week* (January 13, 1997): 62.

Clifford, Mark. "Asia's Furious Phone Derby." *Business Week* (February 17, 1997): 20–24.

Elstrom, Peter, Andy Reinhardt, Susan Jackson, and Catherine Yang. "The New Trailblazers." *Business Week* (April 8, 1998): 46–56.

Jackson, Susan, and Catherine Arnst. "Trench Warfare in Long Distance." *Business Week* (February 11, 1997): 43.

Kibati, Mugo, and Donyaprueth Krairit. "The Wireless Local Loop in Developing Regions." *Communications of the ACM* (June 1999): 60–72.

Kupfer, Andrew. "Mike Armstrong's AT&T: Will the Pieces Come Together?" *Fortune* (April 26, 1999): 82–89.

Kupfer, Andrew. "The Telecom Ward." *Fortune*" (March 3, 1997): 136–142.

Masud, Sam, and Ben McClure. "AT&T/BT: Doing the Right Thing?" *Telecommunications* (May 1999): 31–34.

Matlack, Carol. "All Circuits Are Busy." *Business Week* (March 3, 1997): 31.

McGinty, Meg. "Telecom Act, Scene One." *Communications of the ACM* (July 1999): 15–18.

Development

CHAPTER

3

Telecommunications Architectures and Standards

OBJECTIVES

After studying the material in this chapter, you should be able to

- explain the difference between communications architectures and communications standards;
- explain the need for communications standards;
- explain the need for a communications architecture;
- describe the seven layers of the ISO-OSI model architecture;
- describe the five-layer TCP/IP architecture;
- discuss the advantages and disadvantages of layered architectures.

INTRODUCTION

This chapter examines telecommunications architectures and standards. The first telecommunications architectures were developed in the 1970s to tie many of the individual pieces of telecommunications into a unified whole. Standards in the telecommunications industry are much older and have been developed by many national and international organizations. We will first look at the differences between architectures and standards and the need to have both. Then we will look at some specific standards and architectures in more detail.

DEFINITION OF ARCHITECTURES AND STANDARDS

In general terms, an *architecture* is a plan or direction that is oriented toward the needs of the user. It describes "what" will be built but does not deal with "how." A traditional architect must consult with the eventual occupants of a home to ensure that it matches the family's lifestyle or special requirements. If a family is active in sports, the architect may design special closets or racks to store sports equipment. There might also be shelves in the family room to hold sports trophies.

In the same way, a telecommunications system architect must be aware of the needs of communications users in order to design an archi-

network architecture

tecture to meet their needs. A *network architecture* is a set of design principles used as the basis for the design and implementation of a communications network. It includes the organization of functions and the description of data formats and procedures. An architecture may or may not conform to standards.

Since most architectures (plans) are made for the long term, the architect must not only consider today's requirements but also ensure that the architecture is flexible enough to meet new requirements and support new capabilities that will arise in the future. This is particularly challenging in a field such as communications, where technology is advancing rapidly and visions become reality in only a few years.

communications standards

Communications standards are the rules that are established to ensure compatibility among similar communications services. Communications standards are the flesh on the architectural skeleton. They specify "how" a particular communications service or interface will operate.

With the success of early telecommunications systems and with the reduction of communications costs based on improving technology, other online applications became justifiable. Because there had not been an overall plan or architecture for telecommunications, standard ways of implementing the early systems, or a vision for the future, the existing networks were not flexible enough to support new requirements. As a result, separate networks had to be built for each application. This meant that one location of a company might have had multiple communications lines running to it, each one attached to a different set of terminals. Lines and terminals could not be shared between applications or networks because the rules for using the lines were different and because the communications software was often built into the application on the host computer. Each application contained specialized (nonstandard) communications programming designed to meet the specific needs of that application. Little consideration was given by designers to sharing programs, terminals, or networks with other applications. As each new application was justified, its designers and programmers started over. They made design decisions to meet what they perceived to be the unique requirements of their application with little regard for what had been done before or what might follow.

As telecommunications networks and systems evolved, it became obvious to network designers and users that there had to be a better way. It didn't make good business sense to continually start over and build new, unique communications networks for each new application. There was a growing recognition of the need for an overall plan or architecture to guide future network and application developments. What gave rise to this realization?

First, users recognized that systems were becoming too complex. Because of the lack of an overall plan and the desire to optimize communications systems to particular applications, a proliferation of transmission techniques, programs, and communications services had evolved. Most of them were incompatible with one another. It became very difficult for a company to manage the diversity of hardware and software in its communications networks. Making changes to a network was difficult and risky because of the complexity and the possibility that the implications of a change would not be fully realized and ensuing problems would crash the system.

Second, users wanted to be isolated from the complexities of the network. Communications networks do change. There is a need to add lines and terminals almost constantly and a strong economic incentive to add new applications as they are justified and developed. Terminal users should be isolated from these changes. They should not have to worry about modifications to the network infrastructure. A person using the network should not see changes in how the network reacts from day to day. On the one hand, consistency, from the user's perspective, is a virtue! On the other hand, the network designer must have the ability to make changes in the topology of the network, the services it provides, or other characteristics without affecting current users. As user requirements change, traffic volume increases or decreases, and new terminals or other products come to the market, the designer should be free to incorporate them into the network.

Third, network users wanted to connect different types of devices to the network. Networks had to be able to service different types of terminals. Both interactive devices, such as video display terminals, and batch terminals that contain high-speed line printers were needed by most companies. Furthermore, it is desirable for a company to be able to acquire these terminals from different vendors and to take advantage of new technology improvements as they become available.

Fourth, distributed processing became a network issue. With the rapid increase in the number of minicomputers and personal computers, the ability to do data processing on a distributed basis exists and is often desirable. However, these distributed processors do not exist in a vacuum. There must be communication between them in order for them to share and refer to common data.

Finally, users developed a need for integrated network management. It is very desirable to have an integrated set of tools with which to manage the communications network. Diagnostic and performance measurement capability can help ensure that the communications network is delivering the service for which it is designed efficiently and effectively.

Clearly, to be most effective and to meet user requirements for network capabilities today, a telecommunications architecture supported by appropriate standards is required to mask the physical configuration and capabilities of the network from the logical requirements of the user. Discipline is required to get the full benefits of networking and distributed processing. The chaos of multiple terminals on multiple lines using multiple transmission rules can be avoided only if an overall plan (architecture) supported by appropriate standards is in place.

Fortunately, several vendors and the *International Organization for Standardization (ISO)*, a group made up of representatives from the standards organization in each member country, recognized the need for a telecommunications architecture and for standards to support it.

STANDARDS AND STANDARDS-MAKING ORGANIZATIONS

Before discussing specific architectures, we will look at the communications standards-making organizations and some of the standards they have developed.

The communications industry, probably because it is much older, recognized long before the computer industry that standards are required to ensure that communications equipment will work together. By the nature of their industry, communications vendors recognized that their equipment would have to connect to the equipment of other vendors, hence the need to develop standard ways of handling these connections and interfaces.

By way of contrast, computer vendors have been slower to recognize the need for national or international standards and traditionally have tried to develop their own proprietary standards in order to force customers to use one brand of equipment. With the growth in interconnected, computer-based networks, customers insisted that they be allowed to buy equipment from whichever vendor provided the best solution to their business problem, and demanded that computer vendors provide ways to connect equipment. The only solution was standardization, and the result is that standards now exist throughout both the computer and communications industries.

A number of organizations in the world establish communications standards. Many of these organizations, and the main focus of their standardization efforts, are shown in Figure 3–1. Because of the recognized need for common international standards, there has been a great deal of cooperation among these organizations, especially in recent years.

Normally the way that a standard is developed is through consensus of the members of the standards organization. Remember that depending on the standards organization, the members may be companies, countries, or both, so adoption of a new standard is usually very far reaching, and a major event. While all of the standards organizations vary in their exact standards-making methodology, they all have specific processes in place. Basically, a proposal for a standard is usually assigned to a committee for study, investigation, discussion of the approach, and production of the first draft of the proposed standard. The draft is circulated to other members of the standards organization and their comments, both technical and non-technical, are invited. With that feedback, the stan-

Standards Organization	Main Telecommunications Focus	Internet Web Address
International:		
International Telecommunications Union—Telecommunications Standardization Section (ITU-T)	Telephone and data communications	www.itu.ch
International Organization for Standardization (ISO)	Communications standards of all types (coordinates with the ITU-T)	www.iso.ch
Internet Engineering Task Force (IETF)	Sets standards for how the Internet will operate	www.ietf.org
United States:		
American National Standards Institute (ANSI)	Data communications in general	www.ansi.org
Electrical Industries Association (EIA)	Interfaces, connectors, media, facsimile	www.eia.org
Institute of Electrical and Electronic Engineers (IEEE)	802 LAN standards	www.ieee.org
National Institute of Standards and Technology (NIST)	Standards of all types	www.nist.gov
National Exchange Carriers Association (NECA)	North American wide area network standards	www.neca.org
User/Vendor Forums: (These groups feed information to the standards-setting bodies listed above.)		
European Computer Manufacturers Association (ECMA)	Computer and data communication standards (feeds input to ISO)	www.ecma.ch
European Telecommunications Standards Institute (ETSI)	European telecommunications standards	www.etsi.org
Corporation for Open Systems (COS)	Promotes the use of equipment that meets ISO standards	

Figure 3–1
Some of the organizations involved in setting telecommunications standards or in passing input to the standards-setting bodies.

dards committee revises the initial draft, sometimes in the form of another draft, and sometimes as the final proposal. There may be several iterations of the process, but eventually the final draft standard is submitted to the members for voting, and if the vote is positive, the standard becomes official and is adopted by the members. The process is not fast, usually being measured in months or years.

In many cases, the United States has developed standards for domestic use ahead of the rest of the world. More recently, when the need arose in other countries, they improved on the U.S. standards, and those improvements became the international standards. It is not uncommon to find one communications standard in use in the United States and another in the rest of the world. For example, the international telex system, an early form of electronic mail that is now used mainly in a few Third

World countries, was an international standard. Telex uses the 5-bit Baudot code for individual characters and the transmission speed is 66 words per minute. In the United States, AT&T developed a similar system, the teletypewriter exchange system (TWX), which used an 8-bit code and transmitted at 100 words per minute. The two systems operated according to different sets of standards and were incompatible. It took the advent of computers to allow messages from one system to be translated and exchanged with the other system.

telex and TWX

In the telephone industry, there were similar problems. Early standards were mostly electrical in nature, such as the voltage required to cause the telephone bell to ring, the electrical resistance or impedance of telephone lines, and the drop in signal strength permitted over various distances. Not all countries adopted the same standards, of course, which was fine as long as telephone calls stayed within national boundaries. When calls were made overseas, converters were inserted in the line to adjust to the standards of the called country so that the call could be completed. International standards organizations eventually got the differences ironed out and countries modified their telephone systems to meet the standards so that now it is not only possible but also easy to make calls virtually anywhere in the world.

In the data communications world, standards evolution has occurred more slowly, but today many international standards are in place. The situation is more complicated than with voice communication. In addition to electrical standards, protocols, the "language" of data communications must be standardized so that terminals and computers can understand each other. The issue is further complicated by the fact that the data processing industry has been unregulated, and each company originally set its own communications standards. IBM's data transmission techniques were different from and incompatible with Hewlett-Packard's, which were different from Digital Equipment Corporation's. Although they all followed the same *electrical* standards for connecting their equipment, that was not enough to allow the equipment to communicate. Fortunately this problem has been largely resolved, as you will see.

It's one thing to make a connection and another to communicate. For example, in the voice world, it is possible to make a "connection" between a telephone in the United States and a telephone in Japan, but if the American only speaks English and the Japanese person only speaks Japanese, "communication" will not occur.

THE V. AND X. STANDARDS

The ITU-T has created two sets of standards for the electrical connection of terminals to a communications network. *Note that these are only electri-*

cal standards for connection. Many other standards and rules are required before communication occurs. Attention is drawn to the V. (pronounced V-dot) and X. (pronounced X-dot) standards here because they will be referred to frequently throughout the book.

V. standards define the connection of digital equipment (terminals and computers) to the public telephone network's analog lines. (Analog lines will be discussed in Chapter 9, but for now suffice it to say that they are the type of lines you use when you make a normal telephone call from home.) *X. standards define the connection of digital equipment to the public telephone network's digital lines.* (Digital lines will also be discussed in Chapter 9.)

The wording of the standards themselves is highly technical, and even a brief summary uses many terms that you are not familiar with yet. The names and meanings of these standards will be more clear as you get into the technical details of data communications in subsequent chapters.

THE ITU-T X.400 AND X.500 STANDARDS

The ITU-T *X.400* and *X.500 standards* for electronic mail systems deserve special mention because they are more completely defined than some other standards. Systems based on X.400 are in widespread use, and systems based on the X.500 standard are beginning to be deployed.

The X.400 standard was developed to allow users on incompatible public electronic mail systems to communicate with each other. Since adoption of X.400, its support has been announced by every major computer and communications vendor. X.400-based electronic mail systems are provided by AT&T, WorldCom, and many other carriers worldwide. What makes X.400 so attractive is the ability to connect public and proprietary electronic mail systems that operate according to different protocols and standards. The connection is transparent to the users—they may not even know that their correspondent is on a different network.

X.500 is a standard that specifies how to create and maintain a directory of e-mail users and their network addresses. When an X.500 directory is in place, any user can access the directory, even if it is located on another computer, and search the directory to find the e-mail address of another user. The directory itself may be decentralized—that is, spread among several computers, but when that's the case, the directory management software communicates to keep all of the parts of the directory up-to-date, so that to the user it looks like a single telephone book. Eventually, as more and more X.500 directories are built and linked with one another, looking up e-mail addresses, no matter where they are located, will become much easier.

ADVANTAGES AND DISADVANTAGES OF STANDARDS

So far, the discussion of standards has probably left you with the impression that standards are entirely beneficial and yield positive results. While the advantages almost certainly outweigh the disadvantages, there are two sides of the coin.

On the positive side, standards allow products from multiple vendors to be connected to one another and to communicate. We all appreciate this flexibility when we are purchasing equipment. Furthermore, having one or only a few sets of standards allows vendors to concentrate on fewer products and sell more items of the same type, leading to economies of scale. The world today has only three sets of standards for sending television signals. Vendors produce television sets to receive pictures sent by one or more of these standards, but the maximum is three. Imagine how it would be if each country had established its own standards for television picture transmission.

On the negative side, the standards process is slow. Writing and getting standards approved frequently takes years. In our fast-paced world, technology has sometimes moved on to better approaches before standards for the old approach get written. Standards can be obsolete before they are approved. Once they are approved, standards tend to freeze the technology, discouraging innovation, because the new way is "nonstandard." Furthermore, because there are a number of national and international standards-making bodies, there are sometimes multiple conflicting standards for the same thing. Fortunately, most of these organizations have been cooperating with each other in recent years, so the number of instances of multiple standards is getting smaller.

For internal use, it is very important that a company adopt standards—not necessarily the international or national ones—but a set of standard products, protocols, and technologies for use within the organization. Without these "internal standards"—if every department can implement whatever telecommunications technologies it wants—keeping the internal telecommunications systems working together can become difficult or impossible. Furthermore, without standards, when changes are needed, or new capabilities must be added to meet changing business conditions, the effort and cost are multiplied several times. Telecommunications management must insist on a set of internal standards.

COMMUNICATIONS ARCHITECTURES

Working with the ITU-T, the International Organization for Standardization (ISO) developed a communications architecture called the *open systems interconnection (OSI) reference model* in 1978. When the model was developed, it was expected that it would be followed by the definition of a set of standards that would allow products to be developed to implement the model. However, the standards development process took so long that the OSI model never came into widespread use. It is, however, the standard reference point by which data communications networks are measured.

Understanding the model will provide you with a framework into which you will be able to fit the technical information that is presented in subsequent chapters. With such a framework, you will be able to apply more quickly the information on communications hardware, networks, and protocols, and make sense of what you learn.

THE ISO-OSI MODEL

Objective

The architects of the OSI model had as their primary objective to provide a basis for interconnecting dissimilar systems for the purpose of information exchange. For their purposes, a system is viewed as consisting of one or more computers, associated software, and terminals capable of performing information processing. The intent was to define communications rules that, if followed, would allow otherwise incompatible systems made by different manufacturers to communicate with each other.

The OSI model uses a layered approach in which each *layer* represents a component of the total process of communicating. A diagram of the seven OSI layers is shown in Figure 3–2. In reality, each end of the communication (a computer or a terminal) must have an implementation of the seven-layer architecture because each layer in one end system

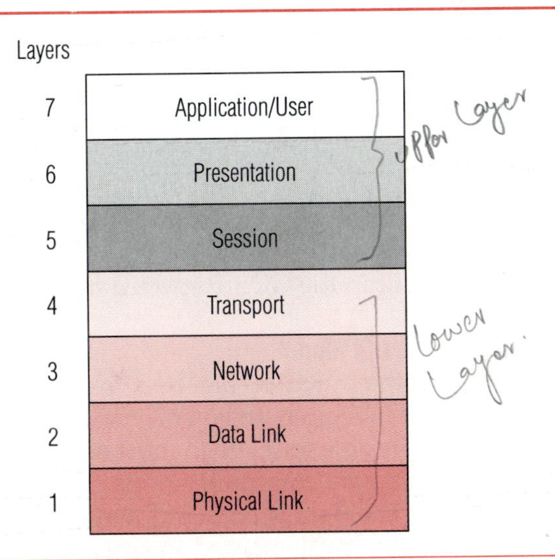

Figure 3–2
Layers of the ISO-OSI reference model.

communicates with its peer on the other end system. Layer 7 in a remote terminal talks to layer 7 in a host computer, layer 6 to layer 6, and so forth.

A helpful way to visualize the layers is to picture an onion and think of the way a computer system is layered. At the heart is the computer hardware. The hardware is surrounded by operating system software. Outside the operating system is other specialized software, perhaps a communications program or a database system. Finally, the outermost layer is the application program. The application program is shielded from the complexities of the hardware by the operating system and other software. At the same time, the operating system and other specialized software provide useful services to the application program.

Another way to view the interaction of the layers of the OSI model is that each one requests services from the layer below it. The lower layer honors the request using its own capability or, if necessary, requesting services from layers below it. For example, the network control layer, layer 3, assumes that the data link control layer, layer 2, has passed good data to it, but it is not concerned with the details of how that data was transmitted or how any errors were corrected. In the OSI model, the highest layer, layer 7, is the end user or a program that makes the original request for service. The lowest layer, layer 1, communicates across the transmission line to the lowest layer in the other system.

X.25 standard

In the OSI model, the first three layers are well defined. Standards have been written and agreed upon, and these layers have become more widely known and understood than the other layers. The combination of these three layers is the *X.25 standard for data transmission* used in packet switching networks.

GOSIP

The United States government has defined a subset of the OSI model that vendors must support if they want to sell network services or equipment to the government. This subset is called the *Government Open Systems Interconnection Protocol (GOSIP)*. GOSIP is not a new protocol or architecture, but rather a more precise specification of what parts of the OSI model the government will use. Included in the GOSIP specification are the standard local area network architectures and the Internet protocol portion of TCP/IP (discussed below). GOSIP is important because the government represents a huge customer for communications suppliers, and in order to do business with that customer, they need to ensure that their products fall within the GOSIP specification.

The Seven Layers of the OSI Model

We will now look at the functions and responsibilities of the seven layers of the OSI model. We'll begin with layer 7, the application layer, since that is the layer you will be able to relate to most easily. Remember that when a message is sent on a data communications network, it begins at layer 7 on the sending end. The data is passed down through the layers, which, by the way, are mostly implemented in software, to layer 1. Layer 1 is where the actual physical transmission across the telecommunications line takes place. At the receiving end, the data passes upward through the layers, each one performing its specific role, until it reaches layer 7, the application at the receiving end.

In many cases, the layers at the sending end surround the original data with control characters or other information for message routing or error checking. At the receiving end, these control characters are checked and removed by the corresponding layer.

The analogy most frequently used is one of putting a letter (the original data) in an envelope and writing some addressing characters on the envelope. In big companies, the envelope is sent to the mail room where employees may insert the original envelope in a larger envelope that only contains mail (data) for a certain location. Other addressing characters are written on the second envelope. The large envelope is then sent through the postal system to the destination location (and, by the way, most of the time we don't care exactly how it gets from one location to another). At the receiving location, the large envelope goes to the mail room where the address characters are checked to be sure the mail has arrived at the right place. Then the envelope is opened, the smaller envelopes inside are removed, and the large envelope is discarded. The addresses on the small envelopes inside are checked, and each envelope is sent to its addressee. Each recipient checks to be sure the mail has been delivered correctly and then opens the small envelope, discards it, and looks at the original message that was sent.

In data communications there are other steps, such as breaking the message into smaller pieces before it is sent, and more extensive error checking, but the process is very similar to the mail analogy.

The OSI model divides the tasks of moving information between networked computers into seven smaller task groups. A group of tasks is then assigned to each of the layers. Each layer is reasonably self-contained so that the tasks assigned to it can be implemented independently. This enables the implementation of one layer to be updated without affecting the other layers.

The seven layers of the OSI model can be divided into two categories: upper layers and lower layers. The three upper layers, the session, presentation, and application layers deal with application issues and are usually implemented in software. The four lower layers, physical, data link, network, and transport handle the data transport on the network and are implemented with a combination of hardware and software. The physical layer is closest to the network medium, such as the network cabling, and is responsible for actually placing data on the medium.

Now we'll look at the layers in more detail.

Application Layer The application layer is represented by the user at a terminal or a computer program. The application layer determines what data is to be sent on the network and, at the receiving end, processes data that is received. This is the starting and ending point for a data transmission. Some examples are a user (the application layer at the sending end) sending a message to an application program in a mainframe computer (the application layer at the receiving end); and an application program in one personal computer transferring a file to an application program in another personal computer.

Presentation Layer The presentation layer handles the changes of data formats that are required between the user or application program and the network. For example, an application may be programmed to send data to terminals with a certain screen size. If a user has a terminal with a different screen size, the presentation layer is responsible for converting the data format to fit. This layer also handles the compression/expansion and encryption/decryption of data if required.

Session Layer The session layer is responsible for establishing the communications rules between specific machines or applications. If users of an application use video display terminals with differing screen sizes, for example, the session layer ensures that the application program knows what size blocks of data it must send to each type of terminal to fill the screen. The session layer also is responsible for pacing the rate of transmission so data is not sent faster than the receiver can handle it. It is also responsible for certain accounting functions.

Transport Layer The transport layer identifies the actual address of the terminal or computer that will receive the message. The transport layer knows the final destination but does not know the detailed routing that will actually be used when the message is sent. (In large networks, there may be more than one route between some heavily used locations on the network.) This layer also calculates a checksum for the entire message, which is recalculated by layer 4 at the receiving end and compared. Using this technique, layer 4 can determine if the entire message was received correctly. This layer also plays a role when messages are sent between two separate but interconnected networks.

Network Layer The network layer is responsible for routing a message all the way through the network from the transmitter to the receiver. It normally maintains a routing table that tells how to send messages to different destinations. The network layer may break the message into pieces, called transmission units, and route each transmission unit through the network on a different route. When this happens, the network layer at the receiving end is responsible for assembling the entire message before delivering it to the transport layer.

Data Link Layer The data link layer is responsible for establishing a link between two points on a network and ensuring that data is transferred between the two points successfully. This layer must include means for detecting errors in the transmission and correcting errors when they do occur. The data link layer may also further divide the data into smaller units for transmission. It adds header and trailer information to each unit of data it sends to assist with error control. Assuming that several terminals are sharing the line, this layer also determines which terminal can use the line at any moment.

Physical Link Layer The physical link layer specifies the electrical characteristics between the communications line and the terminal or computer system. It specifies how the signals are carried on the wires, what types of connectors are used, their physical shape, and which pins in the connectors are used for which signals. It tells which way data is allowed to flow on the line. This is the only layer at which actual bits pass, and it is also the only layer that is implemented strictly in hardware. All other layers are implemented in software or a combination of hardware and software. All messages must be passed down to layer 1 to be communicated across the medium to layer 1 in the receiving system. However, layer 1 only knows about a stream of bits—it does not know about the meaning of the bits or how they might be grouped.

In the following chapters, reference will be made to the layers of the OSI model to help you fit that information into the OSI framework. Coupled with this book's overall outside-in approach that leads you from familiar telecommunications applications to unfamiliar technical details, you have two excellent bases on which to build your knowledge. A more technical explanation of the seven layers of the OSI model can be found in Appendix A.

THE TCP/IP PROTOCOL ARCHITECTURE

When the OSI architecture model was developed in 1978, it was felt that it would come to replace the vendor-specific architectures, such as IBM's Systems Network Architecture (SNA), that were in widespread use in the late 1970s and 1980s. Also in the 1970s the U.S. Department of Defense wanted to interconnect computers and networks it had acquired from different vendors. Its Advanced Research Project Agency developed a set of protocols called the *Transmission Control Protocol/Internet Protocol (TCP/IP)* to allow the interconnection. The original use of these protocols was in the ARPANET, a network that connected various government and university research laboratories. Eventually the Department of Defense mandated that TCP/IP be used in all of its computers and networks, which

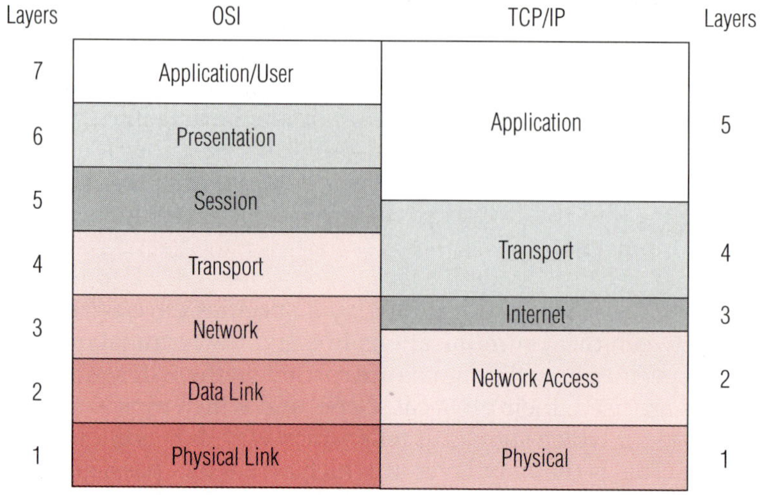

Figure 3–3
A comparison of the layers of the OSI reference model and the implementation of TCP/IP.

Layers	OSI	TCP/IP	Layers
7	Application/User	Application	5
6	Presentation		
5	Session		
4	Transport	Transport	4
3	Network	Internet	3
2	Data Link	Network Access	2
1	Physical Link	Physical	1

automatically provided a huge base of equipment with the protocols installed, and a large market opportunity for software and equipment vendors. As a result of this and the slowness of OSI standards development on which products could be based, and the fact that the Internet uses TCP/IP, it has become the de facto standard architecture for a growing majority of networks.

Although there is no official TCP/IP model like the OSI model, the TCP/IP structure is also a layered organization with five layers:

- application layer;
- transport layer;
- Internet layer;
- network access layer;
- physical layer.

These layers can be roughly aligned with the OSI layers as shown in Figure 3–3.

The *application layer* contains the programming required to support the user's application. Different modules are required for each application, such as electronic mail or file transfer. The *transport layer* is responsible for providing reliable communication, including error-checking procedures. This work is handled by the TCP protocol, which will be discussed in detail in Chapter 11. In many situations today, the two parties of a communication, which could be two users at personal computers, or a user at terminal communicating with a server at a remote location, are on different networks. The *Internet layer* uses the IP protocol to route data between networks when necessary.

The *network access* layer handles the connection between the end system and the network to which it is attached. The specific implementation depends on the type of network because there are different standards and requirements for local and wide area networks. This will be discussed in more detail in Chapters 11 and 12. The *physical layer* specifies the physical (e.g., connectors, plugs, adapters) and electrical (e.g., voltages and currents) interface between the data communication device and the network, and again this varies by the type of network. Interfaces are discussed in Chapter 8.

TCP/IP can reliably send data across disparate networks with excellent assurance that transmission errors will be detected and the data will arrive at the receiving end error free. Its universality has propelled it to widespread use throughout the world.

MANUFACTURERS' ARCHITECTURES

Even before the International Organization for Standardization began its work developing the OSI model, the major computer manufacturers, such as IBM, Digital Equipment Corporation, Burroughs, and Univac, were working to develop architectures on which to base their communications products. In the early 1970s, many of these companies found themselves supporting a wide variety of communications terminals and protocols. Most companies felt they could not afford to continue proliferating incompatible communications hardware and software, so they began working to develop telecommunications architectures on which to standardize their future products. In 1974, IBM announced the *Systems Network Architecture (SNA)*. Soon after, Digital Equipment Corporation announced Digital Network Architecture (DNA), Burroughs announced Burroughs Network Architecture (BNA), and so on.

Note, however, that the original focus of the manufacturers was different from that of the International Organization for Standardization. *The manufacturers were primarily interested in developing proprietary architectures on which to base their future products. They were not especially interested in having an open architecture that would allow easy interconnectivity of each other's equipment.* The computer vendors' motivations were more along the lines of controlling their customers and locking them in to one vendor's products. In most cases, their work was very practical, oriented toward solving immediate problems, and driven by economics. ISO's primary goal was to connect dissimilar networks and systems. The OSI model was intended to define standards for connecting the proprietary architectures of the manufacturers as well as to connect to public packet switching networks. But because of customer/market pressures, the situation has now changed, and the computer manufacturers are now fully supportive of the international communications standards. They have, for example, implemented TCP/IP on their personal computers, workstations, and mainframes so that these machines can fully participate in the Internet and interconnect with the equipment of other manufacturers.

We will look at IBM Corporation's Systems Network Architecture (SNA) and Digital Equipment Corporation's Digital Network Architecture (DNA) in Chapter 11 after you have studied more of the technical details of data communications.

A CAVEAT ABOUT ARCHITECTURES AND STANDARDS

It is important to understand that writing the specifications for a telecommunications architecture and the related standards is difficult work. Precision is needed but difficult to achieve. The architecture and standards are always subject to interpretation. Therefore, it is entirely possible for two companies to implement networks that they believe correspond to a set of standards but then have the resulting networks unable to communicate with each other. True compatibility in telecommunications must be confirmed by extensive testing of the components thought to be compatible.

ADVANTAGES OF LAYERED ARCHITECTURES

Layering forces a modularization of function that simplifies the structure of all of the aspects of data communications. The thinking process that is required to define the layers in the first place forces clarification of ideas and resolution of troublesome areas. The output of the layering definition process is standards that help ensure that implementers who subscribe to the standards will produce products that can communicate with one another. The combination of the layering and the standards is an aid to understanding the entire telecommunications process.

To the extent that the interfaces between layers are clearly defined, the implementation of one layer can be changed or modified without affecting the other layers as long as the interface specifications are met. This means, for example, that certain layers of the architecture could be implemented in software or hardware, depending on the cost and the requirements for performance and flexibility. Furthermore, different implementations can be substituted relatively transparently. For example, by changing OSI layer 3, one type of line could be substituted for another type of line and the rest of the communications process would not have to be changed or even aware. This again provides a way to meet changing telecommunications requirements in a modular, nondisruptive manner.

DISADVANTAGES OF LAYERED ARCHITECTURES

The implementation of a layered architecture requires reasonably sophisticated intelligence at each end of the connection. When layered architectures were originally conceived in the 1970s, in the days before personal computers, providing the intelligence at the terminal end of a connection was a problem because the necessary logic could not be put into each terminal at a reasonable cost. So initial implementations of layered architectures were frequently between mainframe computers, which had dedicated front-end communications processors to provide the processing for the layers of the architecture. The logic for terminals was often put in a device called a terminal control unit, which could typically support 8 to 32 attached terminals, and thus the cost of the logic chips could be shared.

Today, this obstacle, which was originally a real stumbling block for companies considering moving to a layered architecture, is largely mitigated because virtually any personal computer has enough power to perform the processing necessary to support a layered architecture such as TCP/IP. Frequently, part of the processing, especially at the lower layers, is implemented in hardware in the form of circuit cards that plug into a slot in a PC and provide an interface to the network.

Another minor disadvantage of layered architectures is that the communication software and rules are likely to be more complex than a nonlayered approach. However, the vendors have made it relatively straightforward to buy hardware and/or software to implement a communications architecture such as TCP/IP and hence this "disadvantage" is really of little concern.

For virtually all applications, the advantages of using a layered architecture greatly outweigh the disadvantages. Most communications applications today are implemented using a layered approach.

THIS CHAPTER IS A REFERENCE

The material presented in this chapter provides a framework for your further study of telecommunications. It is important that concepts such as "layered architecture" are introduced to you early, and yet many of the details, such as the purposes of the layers, will probably not stick with you at this time because they are so new. The layers of the OSI and TCP/IP architectures will be mentioned frequently throughout the rest of the book, and I encourage you to refer back to the material in this chapter and in Appendix A, to reinforce your understanding of the models, and the functions of their layers.

SUMMARY

Telecommunications is a highly architected field with a high degree of standardization. Telecommunications architectures provide the framework for the design and implementation of communications networks and systems. Layered architectures, when supplemented with standards, provide a modularity of implementation that allows the flexibility to meet current and future requirements.

The International Organization for Standardization developed a model for Open Systems Interconnection, which still serves as the basis for describing communications systems. However, the model was never broadly converted into implementable standards and products, so the TCP/IP model and protocols have become the de facto standard by which most of the world's data communications networks are implemented.

Layered architectures require computer intelligence at both ends of the connection; however, with the advent of powerful microprocessors, the necessary intelligence is widely available and affordable. Hence layered architectures have wide acceptance throughout the telecommunications community and the advantages of an architected, layered approach far outweigh the disadvantages.

CASE STUDY

Dow Corning's Architecture Decisions

Dow Corning was an early user of IBM's SNA communications architecture. The company made the decision to switch from older communications technology to SNA in 1975 for one primary reason: SNA allowed the consolidation of two separate communications networks that were serving many of the same locations. One of the old networks was providing remote job entry (RJE) capability and the other was serving the video display terminals (VDTs). The consolidation of the two networks eliminated several redundant telecommunications lines, and the company realized significant savings. It was also simpler to manage one network than two. When the decision was made, the company recognized that it was making the decision to go with a proprietary architecture, which would limit its choices in the future. However, at the time, more than 25 years ago, there were no "open" architectures available, and the advantages of the IBM architecture far outweighed the disadvantages.

Despite the fact that Dow Corning was a heavy user of IBM computer and communications equipment for many years in the 1970s and 1980s, the company came to the realization in the early 1990s that one vendor was not going to be able to meet its future telecommunications requirements. At the same time, the telecommunications industry began moving quickly toward open standards and architectures that allowed the equipment of many vendors to be connected together. Since Dow Corning was very interested in maintaining its flexibility and wanted to have the ability to use equipment from any vendor that provided a new capability or was more cost effective, the company began to evolve its network from one based on the proprietary SNA standards of IBM to a network based on international, public standards. One of the telecommunications managers stated that adhering to standards was absolutely critical because the company implements telecommunications solutions in multiples. For example, there are multiple plants and multiple sales offices, and if each were allowed to implement its own networks or workstations in its own way or followed different standards, maintenance of the network and connectivity would be a nightmare.

Furthermore, while Dow Corning connects computers from various manufacturers into a single network, the company has established the requirement that the network must appear as a single entity to the users. That is, regardless of location or type of equipment, the user must be able to access any authorized computer or data. This requirement presents a challenge to the telecommunications staff, since, even with the growing number of international communications standards, it is difficult to provide universal connections that are easy to use and economically effective.

To assist in defining network requirements and parameters, a Network Architecture Board (NAB) was established in 1991. The NAB focused on both short-range and long-range requirements and paid particular attention to local area networks (LANs) and communications in the manufacturing plants. The intent of the group was to ensure that Dow Corning's telecommunications architecture met evolving telecommunications needs in the early 1990s and provided a foundation for the network that would be required in the future.

In 1992, the NAB approved a communications architecture for Dow Corning's data network evolution. This architecture was based on the use of three protocols (which you will study in Chapters 9 and 10), SNA-SDLC from IBM, DECNET-DDCMP from Digital Equipment Corporation, and TCP/IP, a protocol used for linking other networks together. In addition, token ring and Ethernet

LANs and their protocols were supported. This architecture was a significant step in combining the IBM-based networks of the business community and the DEC-based networks of the manufacturing and research functions in the company.

After seeing the company through the transition to an international standards-based network focused on local area networks interconnected by a wide area network, the NAB was dissolved in 1995, and telecommunications management and direction-setting responsibility reverted to the line organization in IT.

QUESTIONS

1. The decision to use SNA was a very wise move for Dow Corning at the time, as it allowed them to build a single, unified network, albeit based on one vendor's (IBM's) products. Do you think Dow Corning has suffered by moving away from a single vendor's architecture and products in the 1990s?
2. What justification would an organization have today for staying with a single vendor's architecture?
3. Dow Corning officially ended its NAB in the mid-1990s. Do you think this was a wise move for the long term?

REVIEW QUESTIONS

1. Explain what a telecommunications architecture is.
2. Why were telecommunications architectures not devised when data communications systems first began to be used?
3. What were some of the problems that companies experienced when they installed early data communications networks? How did an architecture help to solve those problems?
4. What role does the ISO perform for communications architectures?
5. Why is the establishment of standards more difficult in the data communications environment than in the voice environment?
6. What is the ISO-OSI reference model? Why is it important for you to learn about it?
7. What are the functions of the physical link control, data link control, and network control layers of the OSI model?
8. Explain the difference between *connection* and *communication.*
9. The ______ layer defines the electrical standards and signaling required to make and break a connection on a physical circuit.
10. The OSI data-link protocol is called ______ . It is defined in the ______ layer of the OSI architecture.
11. Describe X.25 in terms of the layers of the OSI model.
12. Layers 4 through 1 together provide the ______ for the user's message.
13. When data is received, the presentation layer is responsible for ______ before presenting it to the user.
14. Why might two telecommunications systems, both implemented according to the OSI model, not be able to communicate with each other?
15. ISO and computer vendors had different reasons for developing communications architectures. What were they?

16. Explain why TCP/IP has become the de facto standard for a growing majority of communications networks.
17. What is the function of the Internet layer and the IP protocol in a TCP/IP-based network?
18. Discuss the relative advantages and disadvantages of using a layered architecture.
19. What is the X.400 standard and why is it important?
20. What are the advantages and disadvantages of standards and the standards-making process?
21. What is the difference between the OSI model and TCP/IP?

PROBLEMS AND PROJECTS

1. The text used the mail system as an analogy for the seven layers of the OSI model. Make up another analogy to test your understanding of the OSI concepts.
2. Think of some areas of telecommunications in your life where there are multiple, conflicting standards.

Vocabulary

architecture
network architecture
communications standards
International Organization for Standardization (ISO)
X.400 standard for electronic mail
X.500 standard for directories
open systems interconnection (OSI) reference model
layer
X.25 standard for data transmission
Government Open Systems Interconnection Protocol (GOSIP)
Transmission Control Protocol/Internet Protocol (TCP/IP)
Systems Network Architecture (SNA)

References

Aschenbrenner, John R. "Open Systems Interconnection." *IBM Systems Journal* 25, nos. 3–4, p. 369.

Backman, Frank F. *Advanced Function for Communications—IBM's Implementation of Systems Network Architecture.* Armonk, NY: IBM Corporation, 1976.

Coy, Peter, and Neil Gross. "Cowboys vs. Committees." *Business Week* (April 10, 1995): 56–57.

Czubek, Donald. "What Are the Differences Between SNA and DECNET?" *Communications Week* (February 9, 1987): 54.

Digital's Networks: An Architecture with a Future. Maynard, MA: Digital Equipment Corporation.

Randesi, Stephen J. "IBM Communications Are a Lot More Than Just SNA." *Information Week* (January 19, 1987): 17.

Systems Network Architecture—General Information. Armonk, NY: IBM Corporation, 1975.

PART TWO

Telecommunications Applications—How the Enterprise Uses Telecommunications

Part Two examines the uses of telecommunications. You will look in more detail at the applications that were introduced in Chapter 1, and several new applications will be introduced.

Chapter 4 categorizes applications in several ways and gives examples of each type. The focus is on data applications, but you will see that data and voice applications cannot be clearly separated or considered in isolation.

Chapter 5 delves into the detail of telephone systems, first looking at the telephone handset, then discussing the role of the central office, and finally discussing the public telephone network. Modulation and frequency division multiplexing are explained, and the technical discussion provides a good introduction to many of the data communications topics.

After studying Part Two, you will be very familiar with the large number of ways in which telecommunications are used by businesses and other organizations. You will understand the operation of telephone systems and will have a good idea about how several telecommunications applications actually operate.

CHAPTER

4

Data Communications Applications

OBJECTIVES

After you complete your study of this chapter, you should be able to

- categorize telecommunications applications in several ways;
- describe in nontechnical terms how a telephone call is completed;
- explain how data applications evolved from early message switching applications;
- compare and contrast electronic mail to older messaging systems;
- describe several telecommunications applications, such as an airline reservation system and a supermarket checkout system;
- describe the fundamental operation of the Internet;
- discuss the special factors to consider when telecommunications applications are planned;
- explain the importance of planning for telecommunications system outages.

INTRODUCTION

This chapter looks at some practical uses (applications) of telecommunications in the business environment. The purposes of this chapter are

1. to give you a better feeling for the importance of telecommunications to business;
2. to give you some specific knowledge of telecommunications applications for reference as you learn the technical details later in the book;
3. to introduce some additional words and concepts and further build your telecommunications vocabulary;
4. to lay a practical foundation for the technical material in later chapters.

The applications described in this chapter and the next are all representative of layer 7, the application layer of the OSI model.

CATEGORIES OF APPLICATIONS

Human-Machine Interaction

Telecommunications applications can be categorized in many ways, as shown in Figure 4–1. One way is according to how people and machines

Figure 4–1
Several ways to categorize data communications applications.

Category	Example
Human-Machine Interaction	
Person-to-Person	Telephone call
	Electronic mail
Person-to-Machine	Database inquiry
	Transaction processing system
Machine-to-Person	Automated distribution of information from a computer system
Machine-to-Machine	Process monitoring and control systems
	Automated purchasing application to automated selling application
	File transfer between servers
Type of Information	
Voice	Telephone call
	Voice mail
Structured Data	Transaction processing system
Unstructured Data	Word processing
	Text manipulation and retrieval
Image	Video conferencing
	Security monitoring
	Distribution of files containing pictures or movies
Timeliness	
Online, Realtime	Transaction processing system
	Interactive messaging system
Store-and-Forward	Voice mail
	Some e-mail systems
	Some fax systems

person-to-person

interact with each other in the process of communicating. A "person-to-person" communication or application occurs when one person communicates directly with another. Machines may be used in the middle of the conversation, but their presence is transparent to the people. This type of communication is typified by the standard telephone conversation or e-mail. The machine in this example is the telephone company's central office computer or the computer that manages the e-mail system. Its role is transparent to the two people conversing.

person-to-machine
machine-to-person

Person-to-machine and machine-to-person applications usually go hand in hand. They are typified by the user of a data terminal who carries on a dialogue with a computer. First the person, through the terminal, sends a message or command to the computer. Then the computer sends a response to the user. This interchange normally continues until the user gets the result. Examples of this application are the airline reservation agent checking seat inventory on the computer, an ATM terminal user making a deposit or withdrawal, and a warehouse employee determining the location of certain merchandise in the warehouse.

machine-to-machine

A third type of communication, the machine-to-machine interaction, occurs when one machine automatically communicates with another, without human intervention. It is typified by the automated instrumentation found in a modern chemical or manufacturing plant. Intelligent, microprocessor-based instruments gather data about manufacturing processes and relay the data via a telecommunications line to a control computer. Here the data is analyzed, and any appropriate action is taken. No person is directly involved in the communication between the instrument and the control computer, although results of data collected over time or from several instruments may be analyzed and displayed for operator interpretation. Companies such as Monsanto and Ford have thousands of these communicating instruments installed in their plants around the world.

Another type of machine-to-machine application occurs when an automated purchasing application running on one company's server communicates, typically through the Internet, to an automated selling application running on another company's server. The purchasing application might search on the Internet for a product the company wants to buy, checking several sources, looking for information about specifications, quality, and of course price. In a fully automated scenario, the purchasing application would have the authority to select the product and initiate the buying process without human intervention. A less sophisticated application, but one that is more typical of the applications that exist today, would require human intervention before the purchase and sale are consummated. You can envision the day in the not too distant future, however, when the purchase of certain products will occur totally under computer control without human intervention.

One other type of machine-to-machine application is the file transfer process that takes place frequently between computers connected to a network. The file may contain information, a picture, or a computer program that will be useful to the recipient. File transfer is a very common application on the Internet.

Type of Information

Another way to categorize telecommunications applications is by the type of information carried. The telephone call is the simplest example of a "voice" application. "Data" applications are often grouped according to whether the data is "structured" into records and fields in typical data processing fashion or "unstructured" text such as is found in word processing or text manipulation. Applications that transmit "images" can be divided into those that transmit static images, such as facsimile and freeze-frame television, or dynamic images, such as normal, full-motion television.

Timeliness

A third way to categorize telecommunications applications is by the timeliness of the transmission and reception. *Online, realtime* applications, such as interactive messaging or inquiry-response, require that the data

online, realtime

or information be delivered virtually instantaneously. Airline reservation systems certainly fit in this category.

store-and-forward

In *store-and-forward* applications the input is transmitted, usually to a computer, where it is stored, and then later delivered to the recipient. Voice mail applications work this way, as do many electronic mail applications. There are also some fax services that work on a store-and-forward basis. Faxes are sent to a computer, which stores them until the evening or nighttime hours when telephone rates drop. Then the faxes are sent, taking advantage of the lower rates for the call. Of course this approach eliminates the realtime element of fax transmission, but for many types of faxes, such as those being sent overseas, the store-and-forward approach is perfectly acceptable and more cost effective.

batch

Batch applications are usually thought of as computer applications that don't require telecommunications. In some cases this is true, but in many batch applications, the data is collected via a telecommunications system before it is processed by the batch computer programs. One method of collecting the data is to have it entered through VDTs or other workstations, one record at a time. Another method is to have the data collected into a batch and then transmitted to the computer as a unit. This method is called *remote batch* or *remote job entry (RJE).* The name comes from the fact that frequently the data is prefaced by control statements that instruct the computer to execute a certain program or job as soon as the data is received. In this type of application, the results of the computer processing are frequently returned to the terminal that submitted the data. When the terminal is located off-site from the computer, it is probably connected to the computer by a communication line. Hence, even batch applications may use telecommunications.

With virtually any method of categorization, there is overlap. Most applications have elements from several of the categories. For example, the airline reservation system discussed in Chapter 1 has an element of person-to-person communication between the person wanting to travel and the airline reservation agent. There is also an element of person-to-machine application between the reservation agent and the computer. Both of those elements occur in realtime. Since the data terminal is connected to the computer, it is *online.* If the reservation agent instructs the computer to print a ticket for the traveler at the airport on the day of departure, the computer *stores* the data until the day of departure, when it is *forwarded* to the airport and printed on a ticket form.

DATA APPLICATION EVOLUTION

Before we take a more detailed look at typical data applications that use telecommunications, let's look at how these applications have evolved. The earliest type of telecommunications equipment (other than smoke signals and drums) that did not use voice was the telegraph. Telegraphs sent a message from one person, group, or location to another. Business use of the telegraph evolved quickly. Many companies required rapid communication between their staff at widely separated locations. Originally, companies employed telegraph operators to send the messages using Morse code. Then *teletypewriters* came into widespread use. Remembering the definition of *tele* as distant, we can think of the teletypewriter as a device on which a person types a message; it prints on a similar machine at a distant location.

teletypewriter

Of course, teletypewriters were connected to each other with telecommunications lines, and this application became known as *message switching*. It was one of the first applications of data communications, and conceptually it is still widely used today, though it is now usually called electronic mail.

Message Switching

Figure 4–2 shows the way a company might have set up its telecommunications network for message switching during the 1950s. The company headquarters are in Chicago, and branch offices are in Detroit, New York, and Washington. A telecommunications line connects each of the outlying locations with company headquarters. This type of connection is called a *point-to-point line*. The line runs from one point to another. In the Chicago office, there would be three teletypewriter machines—one connected to each of the three lines. If a person in the Chicago office wanted

point-to-point line

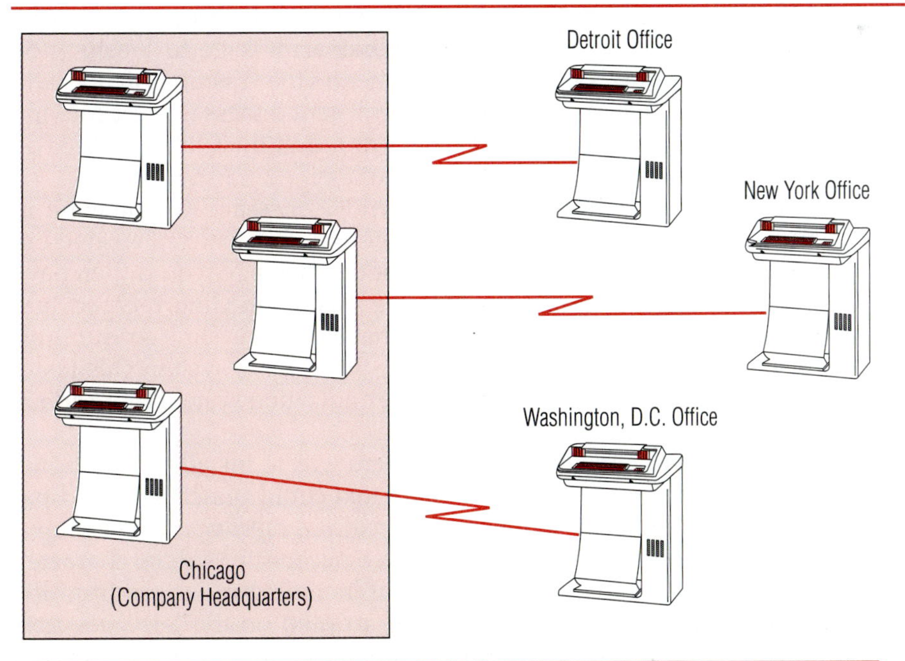

Figure 4–2
A simple network with point-to-point lines connecting offices in Detroit, New York, and Washington, D.C., with company headquarters in Chicago.

An older style teletypewriter. The paper tape punching mechanism is at the center of the machine inside the glass door. The paper tape reader is on the flat surface at the left side of the machine below the two black buttons. (Courtesy of Post Street Archives, The Dow Chemical Company)

to send a message to Detroit, the message would have been typed on the teletypewriter connected to the Detroit line, a message to Washington would have been typed on the teletypewriter on the Washington line, and so on. Conversely, when a person in Detroit sent a message to company headquarters, the message would have printed in the Chicago office on the teletypewriter connected to the Detroit line.

If someone in Detroit wanted to send a message to someone in Washington, the message would have been typed in Detroit, received in Chicago, and printed. The printed copy of the message would then have been taken to the machine connected to the Washington line, retyped, and then sent to Washington where it would have printed. Time consuming? Yes. Labor intensive? Yes, yet at the time, this type of system was faster and more effective than any message communication system that had existed before.

punched paper tape

The first improvement to such a system was to install an additional piece of hardware on the teletypewriters that could punch and read paper tape. Figure 4–3 shows a typical early *punched paper tape.* Each column of holes across the tape represents one letter, number, or special character. The advantage of the paper tape was that incoming messages from the outlying offices could be simultaneously printed on the teletypewriter and punched into the paper tape. If the message was to be forwarded to

Figure 4–3
Five-level, Baudot coded paper tape from a teletypewriter.

Figure 4–4
The message switching network with offices in New York and Boston sharing a multipoint line.

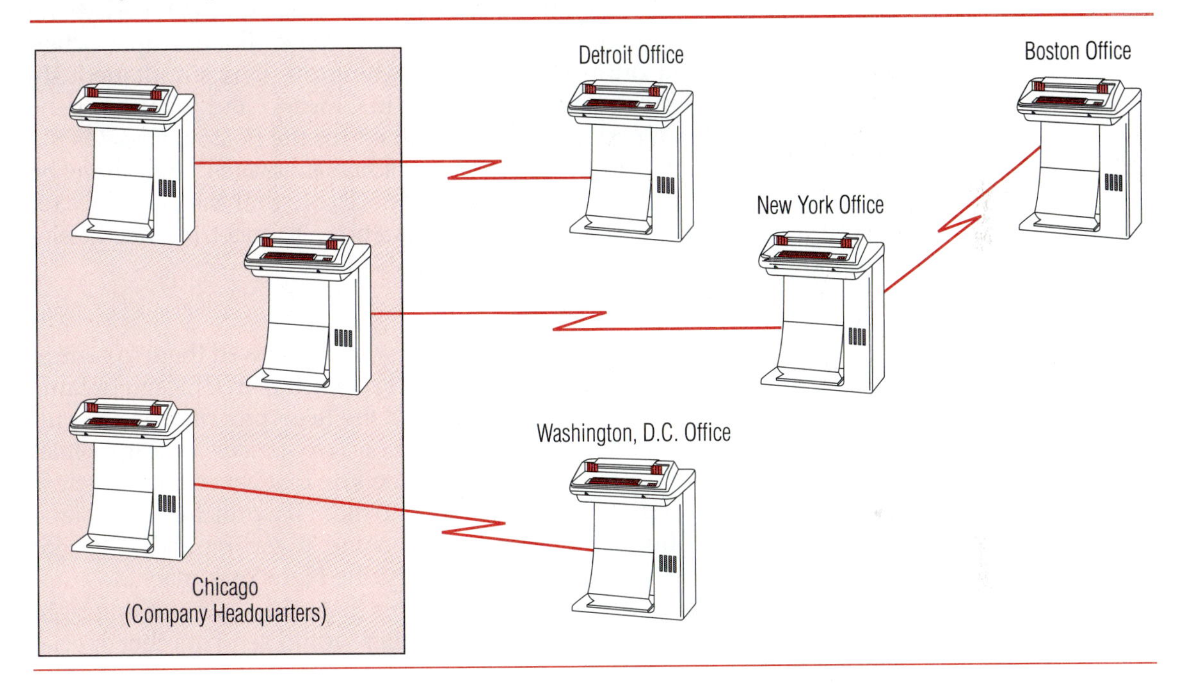

another location, the paper tape could be torn off one machine and read into another machine to be sent to the destination. This eliminated the need to retype the message. This type of operation became known as a *torn tape message system.* In large companies, the *message center* became a large noisy room filled with dozens of machines and many people. Racks were often needed to hold the tapes containing incoming messages that had not yet been forwarded to their destinations. The whole system took on a store-and-forward connotation. Although outmoded by today's standards, torn tape message switching systems vastly improved the efficiency of the message center by eliminating the need to retype all of the messages that were to be relayed.

Getting back to our example, Figure 4–4 shows that the company has grown and opened a new office in Boston. If the network were to expand as it had in the past, an additional point-to-point telecommunications line would be added between Chicago and Boston. However, since Boston is close to New York, it is reasonable to consider running a new line from New York to Boston and having the two locations share the line between New York and Chicago. Indeed, this was the economical thing to do, particularly if the line between Chicago and New York was not always busy and could handle the additional messages to and from the Boston office. This new type of line, which is shared between several terminals, is called a *multipoint line.*

multipoint line

Complications arise, however, when multipoint lines are used. On a point-to-point line using teletypewriters, when a message is typed at one end, it is printed at the other end as it is being typed. For example, when a message is typed on a machine in Washington, it is simultaneously printed and/or punched on the machine in Chicago. But what happens when a message is typed in Chicago destined for the Boston office? Is the message also printed in New York? For some messages, this might be okay, but clearly for many messages it is desirable that they are only received and printed at the intended destination. To meet the need, teletypewriters had to have additional sophistication.

Polling Additional components in teletypewriters allowed them to control the use of the line. In our example, the teletypewriters in Chicago became the *control* or *master station* on each line, and the teletypewriters in the outlying offices assumed the role of *subordinate* or *slave stations.* Chicago's teletypewriter sent out special characters on the line that asked the question, "New York, do you have a message to send to me?" If a punched paper tape was in the transmitter of the New York machine, it was immediately sent on the line to Chicago. If no tape was ready, the New York machine automatically responded with a special character that said, "No, nothing right now." In that case, the Chicago teletypewriter would send another special character sequence that said, "Boston, do you have a message to send to me?" If Boston had a message to send, it would send it; if not, it would respond with the special character saying, "No, nothing right now."

polling

This technique, in which a control terminal asks each slave terminal if it has a message to send, is called *polling.* Having a control terminal poll terminals on the line ensures that two terminals do not try to transmit a message at the same time. This occurrence, which some types of telecommunications systems allow, is called a *collision.* It invariably causes both messages to be garbled and unintelligible at the receiving end. When a collision or garbling is detected, the message must be retransmitted. Detecting a garbled message is usually not too difficult when textual messages are sent between two people. The person looking at a garbled message can usually figure out what a garbled message like this one says:

> I'm cxming tw Lrs Anceles on Moy 17sh. Arriving Nztqwest, flijt 762. P-eabe pick me up ut thx azrport.

An early message switching center in a large industrial company. Short sections of punched paper tape can be seen in the rack in the foreground. Storage bins of punched tape are on the back wall. (Courtesy of Post Street Archives, The Dow Chemical Company)

Obviously, the problem is more difficult when the message contains numerical data. One hopes that the digits in the date and flight number in the above message were received correctly, but asking the sender to retransmit the message would be prudent.

The primary disadvantage of a system that uses polling is the additional number of characters that have to be transmitted (the polling characters) on the line. Another disadvantage is that the slave terminals must wait until they are polled before they send data. If there are a large number of terminals on the line, the delay can be long.

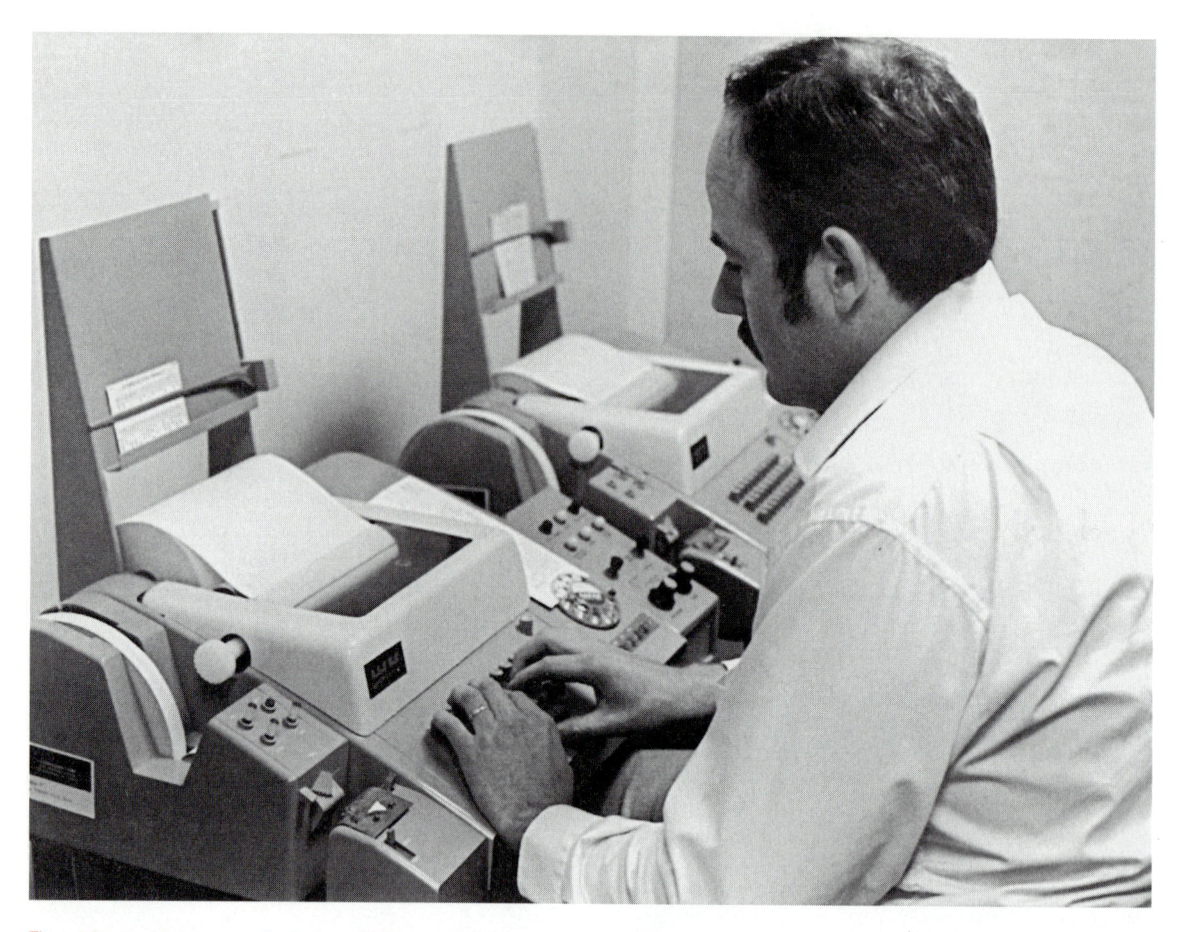

These telex machines are a particular type of teletypewriter, and are connected to the public telex network for the exchange of typed messages. The telex network is being rapidly replaced by more up-to-date electronic mail systems in most countries. (Courtesy of Dow Corning Corporation)

addressing

Addressing A complementary line control technique that evolved at the same time as polling is called *addressing.* When the control terminal has a message to send to one of the subordinate terminals, it first sends special control characters, called *addressing characters,* on the line. The addressing characters are recognized only by the terminal that has the specified address. To that terminal, the addressing characters say in effect, "Slave terminal, I have a message for you. Are you ready to receive it?" The addressing characters also tell all other terminals on the line that the following message is not for them and that they should not print it. If the slave terminal is ready, it responds with a character that says, "Ready to receive," and the control terminal immediately sends the message. After the message is sent, the control terminal either addresses another terminal (if it has more messages to send) or resumes polling the slave terminals to solicit messages from them.

Addressing has the added benefit of preventing the transmission of messages to a terminal that is not turned on, is out of paper, or is otherwise not ready. If a message is sent to a terminal that is not ready to receive, the message usually is lost. Assuming there is a way of detecting when messages have been lost and that the originals have been saved by the sending terminal, they would have to be retransmitted later. If the slave terminal is not ready when it is addressed, it simply does not respond, and the control terminal knows not to send the message.

> The control station *polls* the *subordinate* stations, asking them if they have messages to *send*.
> The control station *addresses* a *subordinate* station, asking if it is ready to *receive* a message.

The complementary techniques of polling and addressing are a simple form of line *protocol,* rules under which the line operates. Most telecommunications systems today use some sort of protocol to control the lines; various protocols are examined in detail in Chapter 10.

protocol

The Evolution to Computers

Soon after businesses started using computers, hardware was developed that allowed telecommunications lines to be connected to the computer. With special programming, data could be read from the line or sent out on the line. It became obvious that message switching was an ideal application for the computer. It could perform all of the functions of the control terminal and, because of its speed, could handle many lines simultaneously. All that was needed was to connect the telecommunications lines to the computer, as illustrated in Figure 4–5.

With proper programming, the computer could read a message from one line and send it out on another line without human intervention. In addition, as computer programming grew in sophistication, computer-controlled message switching systems provided added features, such as adding the time and date to each message, collecting statistics about the number of messages sent and received, logging all messages, and storing messages for later retrieval or retransmission.

Inquiry-Response

Soon there came the realization that a message could be sent to the computer itself (or, more specifically, to a program running in the computer). A very simple example is a coded message sent to the computer asking it to send back the current time. A more useful example is a coded message asking the computer program to look for certain data in a computer file, format it, and send it back to the requestor. This was the beginning of computerized inquiry systems. Figures 4–6 and 4–7 show two examples of simple inquiries to the computer and the responses the computer might generate.

Figure 4–5
The control terminals on each line of the message switching network have been replaced by a computer.

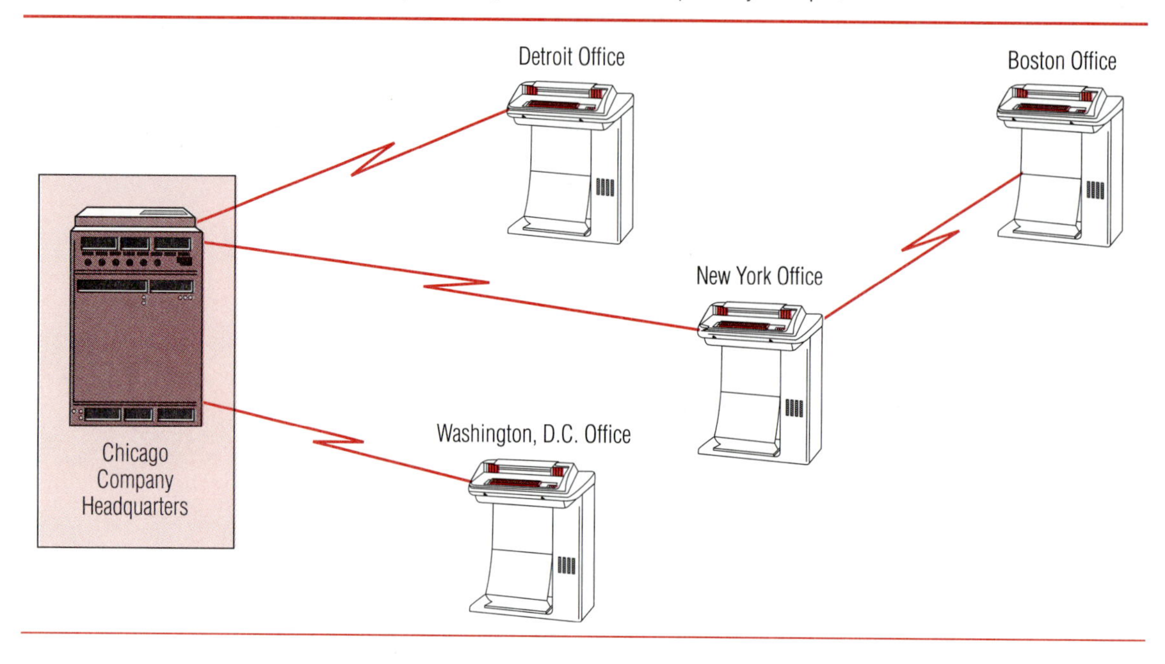

Figure 4–6
An example of a simple inquiry to a computer system and the response from the computer.

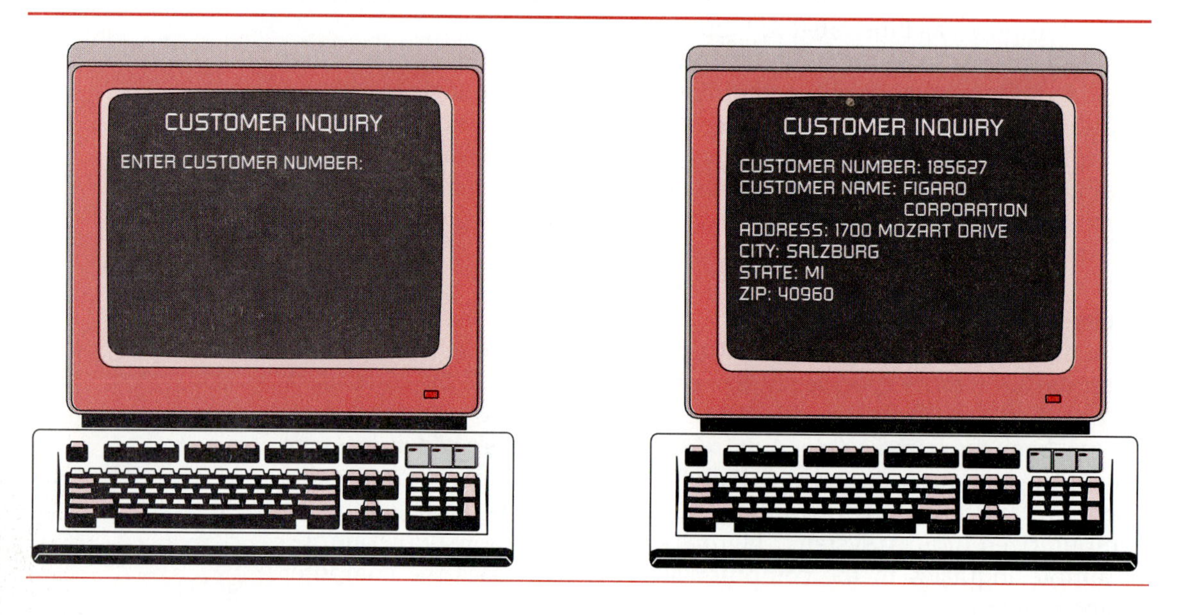

File Updating

From simple inquiries and responses, it is easy to imagine the evolution to more elaborate computer programming in which data is sent to the computer and checked, and ultimately a computer file is updated. Finally, combining a series of inquiries, responses, and file updates gives us the foundation for many modern online computer-based processing systems, such as airline reservation, customer order entry, and inventory applications. Applications of this type, marrying telecommunications and computer technology, were first developed in the late 1950s but did not really come into widespread use until the late 1960s. Today they are common in almost every industry.

Timesharing

Parallel to the development of online applications for business transaction processing, such as those mentioned here, computer software (control programs) was developed to control teletypewriter terminals used for relatively unstructured activities, such as writing programs, executing existing programs, or playing games. This was handled by a process commonly called *timesharing* that first came into its own in colleges and universities for educational purposes, where there were large computers that could handle multiple applications for multiple uses. One student might

Figure 4–7
Another simple inquiry and the computer's response.

run a program for an economics class while another student wrote a program for a FORTRAN class while a third student wrote a program to solve a mathematics problem.

Transaction Processing Systems

By way of contrast, *transaction processing systems* are generally characterized as those in which the users use prewritten programs to perform business transactions, generally of a somewhat repetitive nature. Typical examples are the airline reservation process, an order entry application, or an online inventory application.

similarities between processing systems

In point of fact, timesharing and transaction processing systems had more similarities than differences. Both share the time of a computer among many users. Furthermore, it is entirely possible that a university system designed primarily for student use might also be used by administrative staff to update student grade records or tuition payments—that is, for transaction processing as well. Similarly, the computer in a business that primarily handles the company's accounting, inventory, and other business transactions might also be used by the scientists in the research department or programmers in the data processing department for data analysis or programming—typical timesharing tasks.

From the standpoint of the computer and telecommunications requirements, the two types of systems had many similar elements and a number of different requirements as well. The workload on a timesharing system was relatively less structured, less defined, and less predictable than the workload on a commercial business transaction processing system. Generally, the workload on transaction processing systems is somehow related to the business volume of the company. As airline customers buy more tickets, more reservation transactions are generated. As bank customers use ATMs more extensively, more financial transactions are generated.

Whereas the distinction between the types of workloads led to the distinction between transaction processing and timesharing systems in the early days of telecommunications and computing, the distinction has largely disappeared, and the term *timesharing* is not often used today.

Distributed Processing and Client-Server Computing

For many years there has been a desire to be able to have multiple computers in an organization that can work together as one when appropriate. For some companies, it makes more sense to have departmental or divisional computers with programs that focus on meeting the specific needs of the people in that organizational unit. Minicomputers gained wide popularity in the 1970s to satisfy this need, and applications were built to meet users' requirements.

The problem has always been that when it was time to add up the total results of the company and get a composite picture of sales, financial,

A bank customer uses an ATM. Some ATMs are designed to be installed inside a building, while others are relatively weatherproof and can be installed outside in a less protected environment. (Courtesy of NCR Corporation)

or production operations, it was very difficult to pull the data together because the applications running on the minicomputers were usually designed and programmed differently, and the computers themselves may have even been incompatible. Trying to link the computers with telecommunications was difficult and, even if they were linked, adding up the data was sometimes even harder or impossible.

distributed processing

In the late 1970s and early 1980s, the concept of *distributed processing* was introduced. The idea was that applications would be designed as an integrated whole, but in such a way that they could execute on multiple computers, which could of course be located in different parts of the company. Vendors introduced new minicomputers that were designed to be connected with a host computer and to share the workload with it. It was much easier to connect the computers and get them to communicate. The problem was that the software capabilities of the day just didn't measure up to the complexity of the work.

Designing applications that execute in a coordinated way on multiple computers and share data is a complicated task. There were no guidelines for how it should be done, little experience among the people who were trying to do it, and the software tools with which to implement the distributed systems were weak. Few distributed applications were successful. The problem was not with the concept of distributed processing, but with the lack of experience of the applications designers, the lack of adequate software tools, and sometimes the lack of adequate computer power on the remote (distributed) computers. It may have been a good idea, but its time had just not arrived.

In the 1980s, technology kept marching ahead: Computer hardware got faster, personal computers became widely used, software tools improved, communications capabilities became easier to implement and use, and applications and software designers became more sophisticated.

client-server computing

In the early 1990s, a new term began to be heard in the computer-communications industry—*client-server computing.* The concept of client-server is that some computers, typically personal computers, are programmed to meet the exact needs of their users, whereas other computers are programmed primarily to store data and perform general functions for a wide group of people. The specialized computers and their users are the *clients.* The generalized, database computers are the *servers.* The two are linked together with telecommunications in a way that makes telecommunications essentially invisible.

client server

As a very simple example, imagine your personal computer being online to a larger computer that is a larger PC, a server, or a mainframe—you don't really care, and you may not even know. You need two years of sales history to prepare a special sales analysis for your boss. You have a statistics software package on your PC that can do the analysis, but the sales data is on the larger computer. With appropriate client-server software, you would simply write your analysis program—using the statistics package—and give a command that in effect says, "Use the last 2 years of sales history. I don't know where it is stored. Please find it." The client-server software would consult tables and directories, and locate the data, which in this case happens to be on the server computer to which you are attached. It gives your analysis program access to the data as though it were on the hard disk of your personal computer.

That's a simple example of how client-server computing works. Obviously it requires very sophisticated database management software, including a directory telling where data is located, and telecommunications software that is transparent to the user (client). Much more complex uses are in use today. Suppose the data is stored on several computers and

Figure 4–8
A small local area network (LAN).

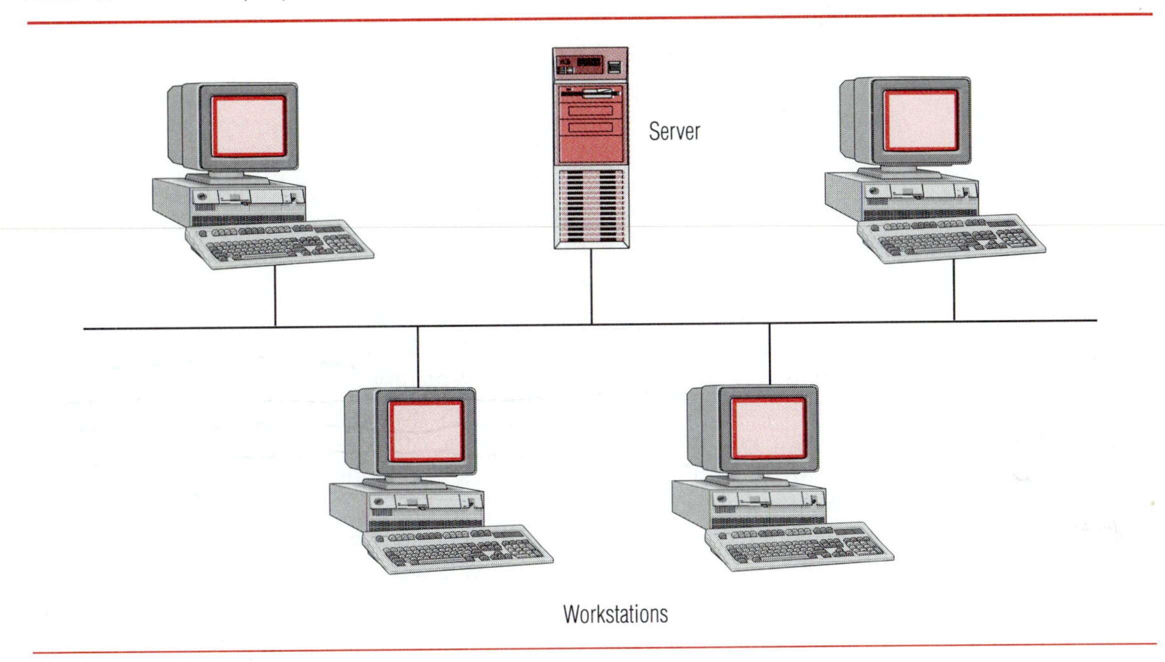

has to be brought together; suppose there are thousands of clients accessing the data on hundreds of servers, all at the same time; suppose the data includes sound and video as well as traditional structured data.

Today client-server systems are most commonly found in companies that have several PC workstations connected together on a *local area network (LAN)*. A LAN frequently has one or more servers that handle the majority of the processing and database access. LANs, which will be studied in detail in Chapter 12, use cable or a wireless technology to connect the workstations and servers, but all of them must be within a relatively small area, usually no more than 1,000 feet apart. That distance, however, is sufficient for most offices, laboratories and other workplaces, so LANs are very widely used and popular. Figure 4–8 is an illustration of small LAN with a single server.

LAN

Client-server software is in widespread use today and software vendors and individual companies have built many client-server applications. The technology continues to evolve in several directions. Tools are becoming available to let client-server applications support multimedia applications that incorporate audio and video images. Also, the development tools used to write client-server programs and test them have become very sophisticated, allowing programmers to develop applications more quickly and with fewer bugs. As the applications become more sophisticated, and especially if they include audio or video components, they need increasingly high-speed telecommunications lines to provide adequate response time.

Types of Computers

Several types of computers have been mentioned so far, and it is appropriate to clarify the distinctions between them. You are probably most familiar with *personal computers (PCs)* sometimes called *microcomputers.* PCs normally have between one and four microprocessors, including the main processor, such as an Intel Pentium chip, and several specialized processors to handle functions such as graphics processing. *Minicomputers* are higher-end computers that have dozens of processors to manage larger databases and more intense processing than a PC could accommodate. An example of a minicomputer is IBM's AS-400. A *mainframe computer* is a very high-end, expensive machine that typically has hundreds of processors. They are used by large companies or government agencies that have a very high volume of transactions or other work to process. IBM and Compaq, through its Digital Equipment Corporation (DEC) division make this type of computer.

MAIN frame

scalability

Any of the above computers can be used as a server depending on the volume of work the computer is expected to process. Frequently organizations will install a small computer as a server and then upgrade it to a larger size as the workload grows. The term *scalability* is used to describe the attribute of being able to upgrade in a compatible way without having to change software or reorganize data. Organizations want to ensure that whatever server they install is scalable and can be easily upgraded when necessary.

TYPICAL APPLICATIONS OF DATA COMMUNICATIONS

Now that you've seen the differences between processing methods and read a bit of history about how telecommunications and computer processing have merged, it's time to explore several applications that make extensive use of data communications. The applications are grouped according to the primary type of information they process.

Structured Data Applications

automatic call distribution

Airline Reservation System With your new knowledge of the way in which data applications have evolved, look again at the airline reservation and ATM applications that were presented in Chapter 1. Figure 4–9 shows an expanded view of the airline reservation application. Multiple travelers can call multiple reservation agents, each using a terminal connected to the central computer. Between the callers and reservation agents is a new piece of hardware, an *automatic call distribution (ACD) unit.* The ACD unit

Figure 4–9
The telecommunications of an airline reservation system.

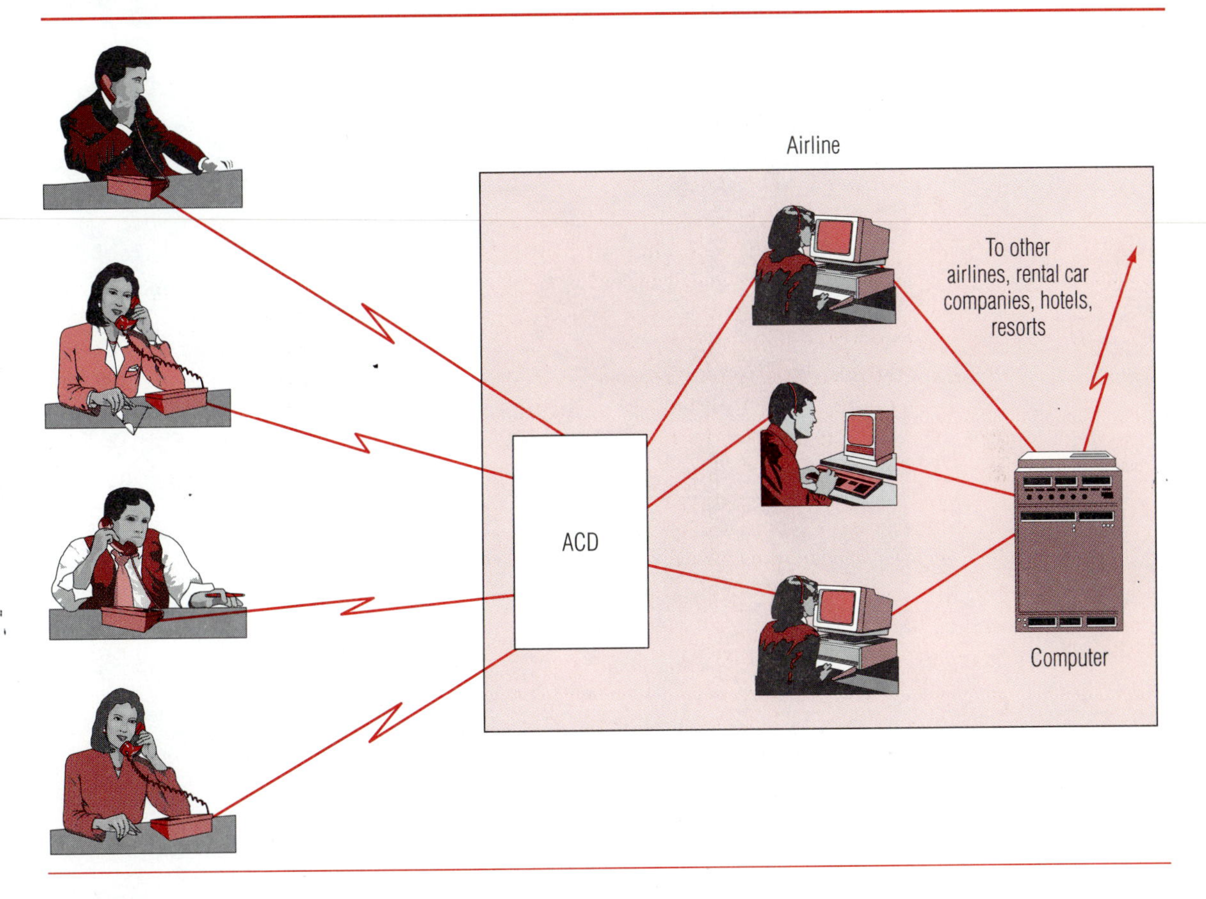

is an adjunct to the telephone system in the airline office. Its purpose is to route the next incoming call to the next available reservation agent based on various criteria, such as the longest wait time, location of caller, or, in slack times, to the agent who has been idle the longest. If all reservation agents are busy, the ACD unit can play a voice recording to the caller stating that the next available reservation agent will serve him or her as quickly as possible. Most of us have encountered ACD units and their "all agents are busy" messages when we have called reservation lines, mail order companies, or software technical support desks!

ACD units work well in any situation where multiple callers must be routed to multiple employees who handle the calls sequentially. Examples include other types of reservation systems, such as rental car companies; customer service lines at utility or manufacturing companies; and companies that have a large number of customers calling to place orders for merchandise, such as catalog sales or toll-free order centers.

These airline reservation agents are using a wide array of telecommunications and computer equipment to service their customers. (Courtesy of American Airlines.)

The data portion of this application has multiple terminals connected to the central computer of the airline. Also, the airline's computer is connected via a telecommunications line to the computers of other airlines. In this way, reservation agents from United Airlines can make reservations on the flights of American Airlines or any other airline whose computer is connected to the network. All of the airlines in the world, except the very smallest, have interconnected their telecommunications net-

A TWA 747 passenger plane. (Courtesy of AP/Wide World Photos.)

works and computers so that travelers can call one airline and make all of the reservations for a trip, even though several airlines may provide the transportation. Furthermore, the computers of the rental car companies and many hotels and resorts also are connected to the same network so that all of the reservations for an entire trip can be made with one telephone call.

American Airlines's Sabre system and United Airlines's Apollo reservation system have been significantly enhanced with additional computer programming. Both airlines have actively sold the right to access the systems to travel agencies and large corporations whose employees travel extensively. The services these systems offer are extensive and go far beyond making airline reservations. Hotel accommodations and theater tickets can be ordered in addition to transportation. The revenue generated by the reservation systems has become a substantial part of the total income of the airlines, sometimes exceeding the amount of revenue generated by the airplane flights themselves.

An airline reservation system is a high-volume transaction processing system that must provide fast response time to the reservation agents. Reservation systems are critical to airline operations, and the airlines spend a great deal of money to ensure that their computer system will always be available by having backup computers ready to take over instantly if a primary computer fails.

In 1985, Delta Airlines had over 4,000 voice communication lines and 18 reservation centers throughout the United States. Delta had 66 data communications lines and handled 340 to 350 transactions per second at peak times. Over 23,000 terminals were connected to the Delta data network—many of these located at the 2,800 travel agencies that were online to Delta. The response time goal was 3 seconds or less 95 percent of the time. While this is an old example, it is still valid for illustrating the high volumes that an active online system needs to be prepared to deal with. And imagine what the transaction volumes must be today, if Delta airlines was experiencing these volumes 15 years ago!

Cathay Pacific Airlines, the flagship carrier of Hong Kong, made a decision to centralize all of its reservation agents in Australia. After looking at all of the factors, the airline concluded that it could provide better customer service at a lower cost by handling reservations from one location, and Australia won the competition. The traditional airline approach of having potential passengers dial a local telephone number regardless of where they are located is complicated by the huge geography of Asia. Long distance telephone lines from each country in Asia tie to the hub in Australia—and the distances are very long! Another complication was language, since customers expected to be able to make their reservations in their native language, and Asia has at least eight to ten major languages. The consolidation of their reservation system was an ambitious undertaking for Cathay Pacific, but one that had significant benefits to the company.

In addition to reservation systems, airlines also operate other online systems for sending administrative messages, tracing baggage, scheduling airplane crews, and scheduling airplane maintenance. A typical 747 flight may involve more than 29,000 transactions, of which 27,000 concern passengers. With such systems, airlines are some of the most advanced, sophisticated users of voice and data telecommunications.

Automatic Teller Machine Figure 4–10 shows an ATM network in which multiple ATMs are connected to computers and several computers are connected together. Networks interconnecting the ATM systems of several banks are common throughout the world. Typically, each branch of a bank has at least one ATM, but in addition, banks may place ATMs in local shopping centers, train stations, or even large business establishments. Banks that have historically been competitors in a local area or region are now cooperating with one another, at least to the extent of interconnecting their ATM networks.

Figure 4–10 shows that any ATM can make a transaction with any bank on the network. It is possible to walk up to a single ATM and withdraw money from an account at one bank and then put the cash back in the machine for deposit in an account at another bank.

Since ATM networks are expensive to operate, service charges are often levied for certain types of transactions to help defray the costs of the

Figure 4–10
A network of ATMs from several banks.

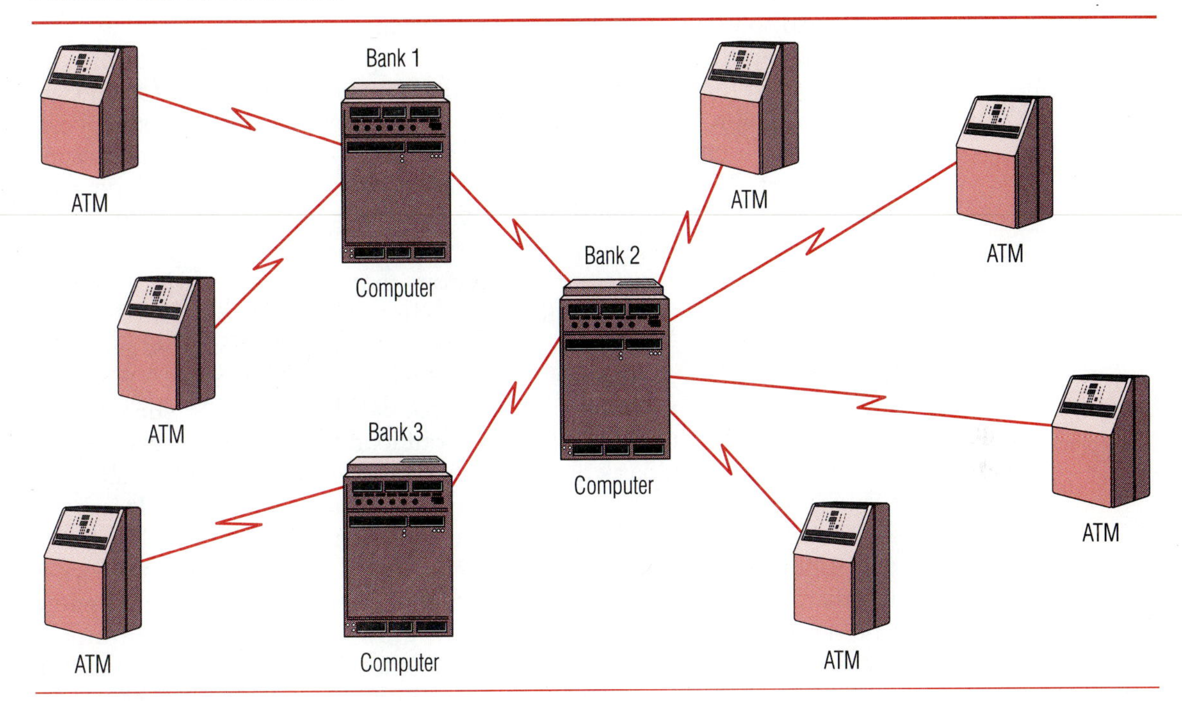

network. Customers whose transactions require access to another bank's computer or network are especially likely to be charged. In Japan, service charges also are assessed for transactions that occur outside of the bank's normal operating hours. That is, ATM transactions are free to the customer if they occur while the bank branches are open. After the branches close, the customer pays a fee for each transaction. It would seem that market competition might change this situation within a few years.

Security in the ATM application is obviously very important. One concern is the physical security of the machine, since it contains hundreds of dollars in cash and checks. This problem is generally solved by keeping the money in a small vault within the machine and having the machine itself mounted in the wall of a building.

Another aspect of security is the identification of the user. Bank customers are identified by issuing each user a plastic card much like a credit card. The back of the card contains a magnetic stripe on which the user's name and account number are coded. When the card is inserted in the ATM, the machine reads the characters on the magnetic stripe. If the account number is valid, the machine asks the person for his or her unique *personal identification number (PIN)*. A computer compares the PIN the user enters with the PIN for that account, which is stored on the computer's disk. If the two PINs match, the user is permitted to proceed with the banking transaction.

personal identification number (PIN)

ATM applications are realtime transaction processing systems that handle a high volume of transactions. Although ATM operation is not yet as critical to banking as the airline reservation system is to airlines, it is obviously in the bank's best interest to keep the ATMs and network operating with high reliability, especially after hours when users can't get assistance with banking elsewhere. In addition, banks save staff costs when people use the ATM rather than a human teller. One nationwide network of ATMs experiences uptime that is better than 97 percent. *Uptime* is defined as the probability that the system is being "up" and available for use whenever someone wants to use it.

The number of different transactions that ATMs can perform has increased dramatically. Originally, the machines did little more than dispense cash. Now, consumers can make deposits, check their balances, get credit advances, pay bills, and transfer funds between accounts; and it is expected that many of these systems will be expanded to new tasks, such as customer orders for checks or inquiries about loan or investment services.

Internationally, bank networks are connected by the SWIFT network. SWIFT, an acronym that stands for Society for Worldwide Interbank Financial Telecommunications, connects most major banks and financial institutions worldwide. Most of the traffic sent on the SWIFT network is interbank financial transactions or confirming messages for verification. The network experienced a 12 percent growth in the message traffic it handled for many years and had to make a complete technical overhaul of its network to cope with the message volumes in the early 1990s.

Both the airline reservation and ATM applications have revolutionized the way business is done in their respective industries. In both cases, the application requires the marriage of telecommunications and computing technology. In addition, communication and cooperation between competing companies must occur. Furthermore, all of the companies must subscribe to the same standards for how transactions are processed. In addition, standards are required to allow computers in different banks, which may be from different computer vendors, to be interconnected with telecommunications. Without telecommunications, these applications would not be possible. These examples clearly illustrate how telecommunications is changing the way business is done and the structure of business itself.

Sales Order Entry Look at another application of telecommunications in the business environment, particularly one found in a manufacturing company. Some of the products the company makes are produced and stored in a warehouse in anticipation of customer demand. Other products are produced only to fill a specific customer order.

Figure 4–11 shows a conceptual view of a sales order entry system that a manufacturing company might have. Order entry operators sitting at terminals may receive orders by mail or telephone (this example as-

Figure 4–11
The telecommunications in an industrial sales order entry system.

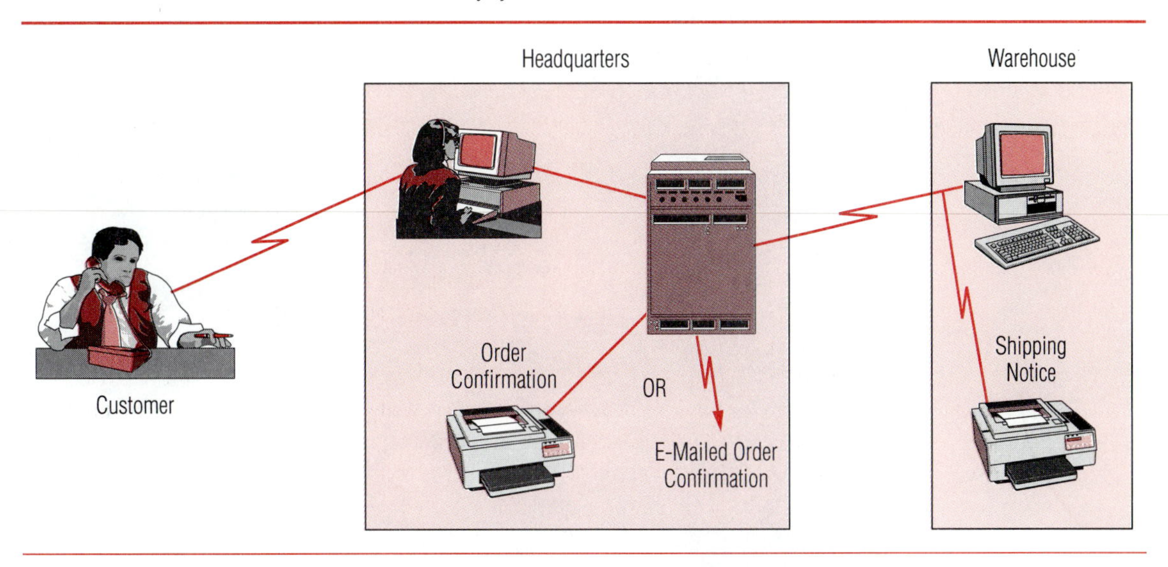

sumes a telephone order). The customer call may be handled as in the airline reservation system with an automatic call distributor routing the call to the next available order entry operator. The order entry operator keys the data required for the order into the computer. When all of the data has been entered, the operator tells the computer to process the order. The computer subtracts the quantity ordered from inventory, and either prints a confirmation notice to be mailed or sends an electronic mail message directly to the customer. The system would also send a notice to the warehouse to ship the order immediately if appropriate.

Notice that in this example there are several uses of telecommunications. The operator uses a terminal connected to the computer via a telecommunications line. The computer is also connected to a local printer that prints the order confirmation notice and a remote printer in the warehouse that prints the shipping document. Electronic mail may be used to contact the customer. There is also a terminal in the warehouse that shares the telecommunications line with the printer in a multipoint configuration. The warehouse employee uses the terminal to notify the computer when the order has been shipped. Obviously, with additional programming, the equipment used primarily for the order entry application could also send messages between the order entry clerk and the warehouse (or vice versa). The warehouse employee could also use it to request the computer balances for any inventory item.

Now imagine adding an Internet-based front end to the application. In that case, the customer might search an online product catalog, select the merchandise to be ordered, fill out the order form, supply shipping

A sales order entry clerk takes an order from a customer and enters the information into the computer through the VDT. (Courtesy of Dow Corning Corporation)

and payment information, and confirm that the order is correct. An application designed in this way would eliminate, or at least significantly change, the nature of the order entry operators. They would now become customer/product consultants, standing by to assist customers who have questions or have trouble using the Internet-based order entry process. The hope would be, of course, that, for a high percentage of the orders, customers could complete order entry transactions themselves.

Point of Sale Systems in a Retail Store or Supermarket *Point of sale (POS) terminals* are used widely in large retail stores and supermarket chains. Although the technology of the terminals used in retail stores differs from that used in supermarkets, the use of telecommunications and the basic application are quite similar.

Many years ago, the grocery industry standardized the Universal Product Code, a bar code that can be used to identify virtually every item stocked in a supermarket. This bar code is quickly and accurately read by a laser scanner built into the supermarket checkout counter. Good design enables the products to be read when they are passed through the laser beam at virtually any angle or speed. The supermarket checkout clerk ensures that the bar code on the item is facing in the general direction of the laser and passes the product through the beam. The laser reads the bar code and sends the product number to the store computer, which looks up the product description and price in a database. The computer trans-

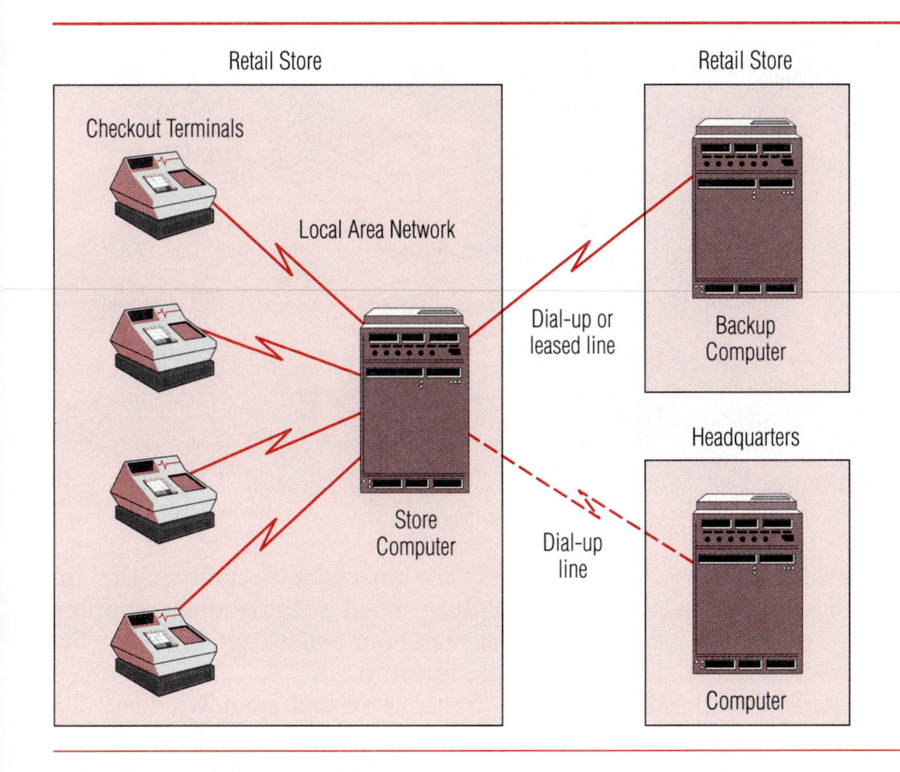

Figure 4–12
The telecommunications of a retail checkout system.

mits that information back to the terminal, which then prints the cash register tape. Occasionally, a product cannot be read or does not have a bar code. In that case, the checkout clerk can key the price into the terminal. It is printed on the tape and added to the running total.

The retail store has a similar type of system. Unfortunately, products in retail stores are not universally coded in machine-readable form. Some stores have merchandise tags with a magnetic stripe or punched holes that can be read by the retail terminal, but in most cases the clerk must manually enter the information.

The most frequent design of supermarket and retail store systems is for the store's computer to contain product and price information, which is completely self-sufficient for normal operations during store hours. Often a telecommunications link is provided to a nearby store for backup purposes. If the store's primary computer fails, the checkout terminals can continue to operate off the backup computer. These telecommunications links are illustrated in Figure 4–12.

At night a central computer at the company headquarters dials the computer at each store to collect data about the day's sales and to check inventory levels. Since the connection between the headquarters computer and the store computer is only needed for a few minutes, it would not be cost effective to have a full-time leased line connecting the two

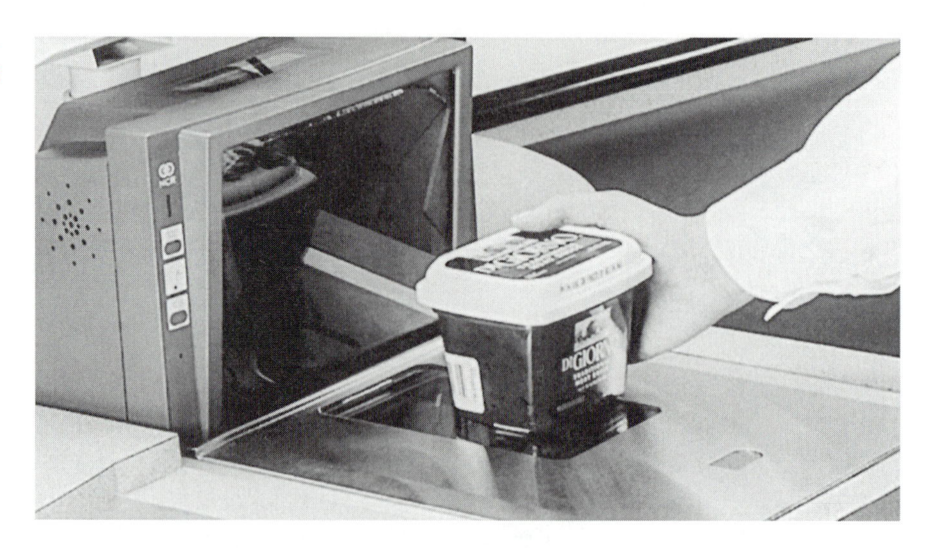

An automated checkout terminal scans the bar code on a can of raspberries. (Courtesy of NCR Corporation)

machines. In this case, the computers are connected by a standard dial-up telephone line. The call is automatically initiated by the central computer. This call is exactly like a normal voice telephone call except that when the connection is made, data, not voice, is transmitted.

When all of the data has been transmitted and acknowledged, the connection is broken by electronically hanging up. In this way, only the time actually used for transmitting data must be paid for. After the central computer collects the data from one store, it immediately dials the next store in the chain and collects similar data and so on until all stores have been contacted. Once all of the data has been collected on the central computer, it can be processed to generate daily sales reports of various types. Also, since inventory levels were checked at each store, the computer can generate automatic restocking information.

For the customer, scanners reduce the wait in the checkout line by 25 percent, reduce cashier errors, and produce receipts showing an exact description of the products as well as the prices. Although they were largely introduced as labor-saving devices, store scanners have changed the way business is done by revolutionizing inventory control. Indeed, retailing is turning into more of a science than an art because stores can see patterns of sales that were invisible a few years ago. Grocers and customers alike recognize the value and benefits that scanning has brought to the grocery business.

Obviously, the role of telecommunications and the computer is critical in point of sale retail store systems. If a telecommunications line or the computer is not operating, customers cannot buy merchandise. Since competition is heavy in this industry and there are many supermarkets and retail stores in a given area, a failure of the computer or telecommunications system usually means that customers go elsewhere and business is lost.

A retail point of sale (POS) terminal that could be used in many different types of stores. (Courtesy of NCR Corporation)

Unstructured Data Applications

Electronic Mail *Electronic mail* applications, or *e-mail* as they are often called, are some of the most heavily used applications in distributed systems. In 1996, for the first time, more e-mail messages were sent in the U.S. than letters! E-mail is similar to the message switching application described earlier, but is designed so that the user can create, format, and send the message from a computer or terminal. By way of contrast, in older systems the message sender usually wrote the message by hand, gave it to a secretary who typed it, and then delivered it to the communications room where it was rekeyed onto a communications terminal for transmission. Clearly the newer systems are more productive because they eliminate a lot of duplicate work.

As shown in Figure 4–13, an e-mail application requires a terminal or personal computer from which the user can access the e-mail system software. Although e-mail is conceptually a simple application, the software is quite sophisticated. The software provides a directory and translation function so that people can address each other by name, nickname, or at least a mnemonic code. These must then be translated to an address for the user that the network can understand if the destination is on another computer, or simply a mailbox address if the recipient uses the same computer as the sender.

Figure 4–13
A simple electronic mail configuration with users on two computers.

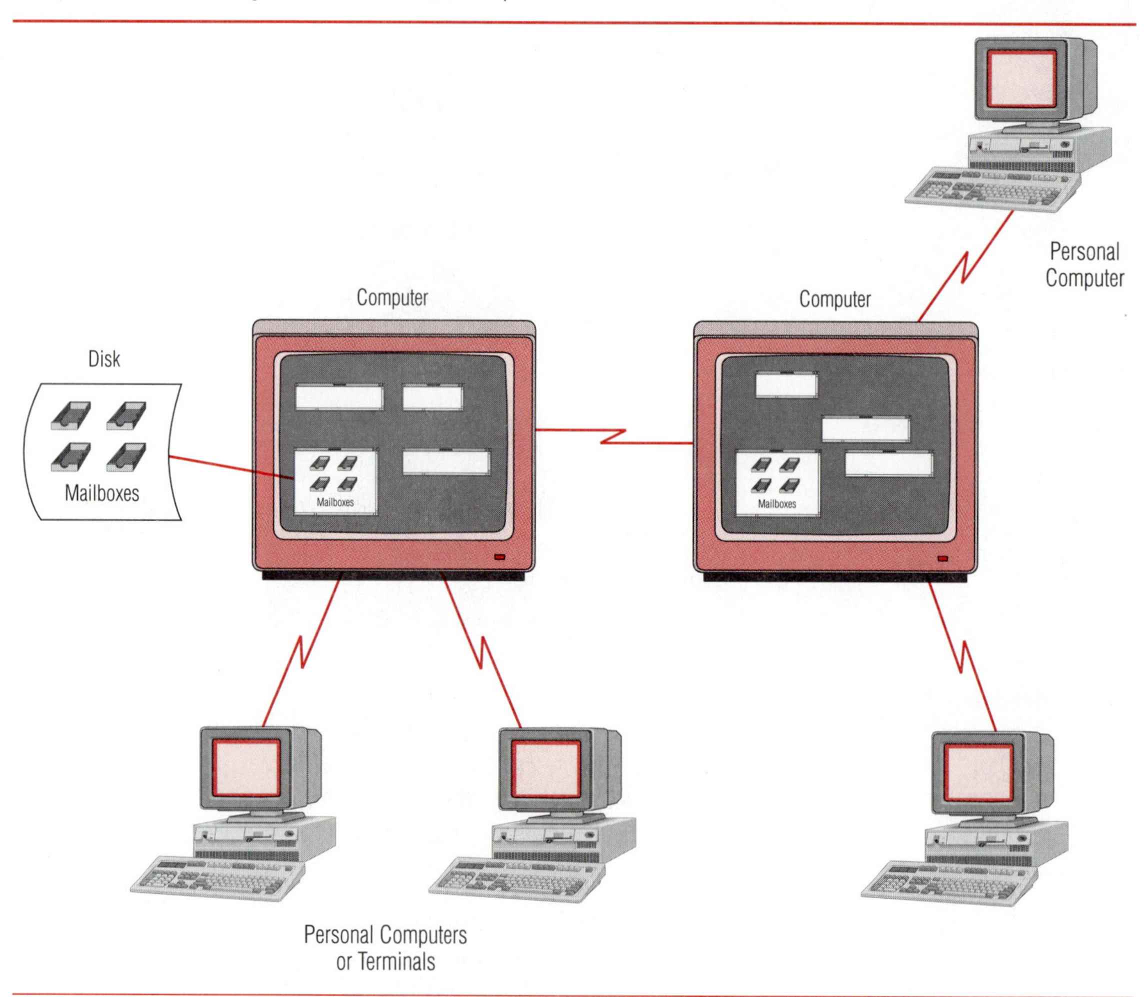

electronic mailbox

The host computer must have adequate disk storage to store incoming messages. Users are each assigned a space on the disk known as the *electronic mailbox.* Messages are stored in the electronic mailbox until the user deletes them, preferably after he or she has read them! Although most users may check their mailboxes and delete messages they have read on a daily (or more frequent) basis, others may leave messages on the system for days or weeks, so the disk capacity must be large enough to accommodate these differences.

The e-mail system provides the user with exclusive access to his or her stored messages. But in a company setting the question is whether the messages are really private. Court rulings have held that if your computer belongs to the company, so does its content. The law lets the company

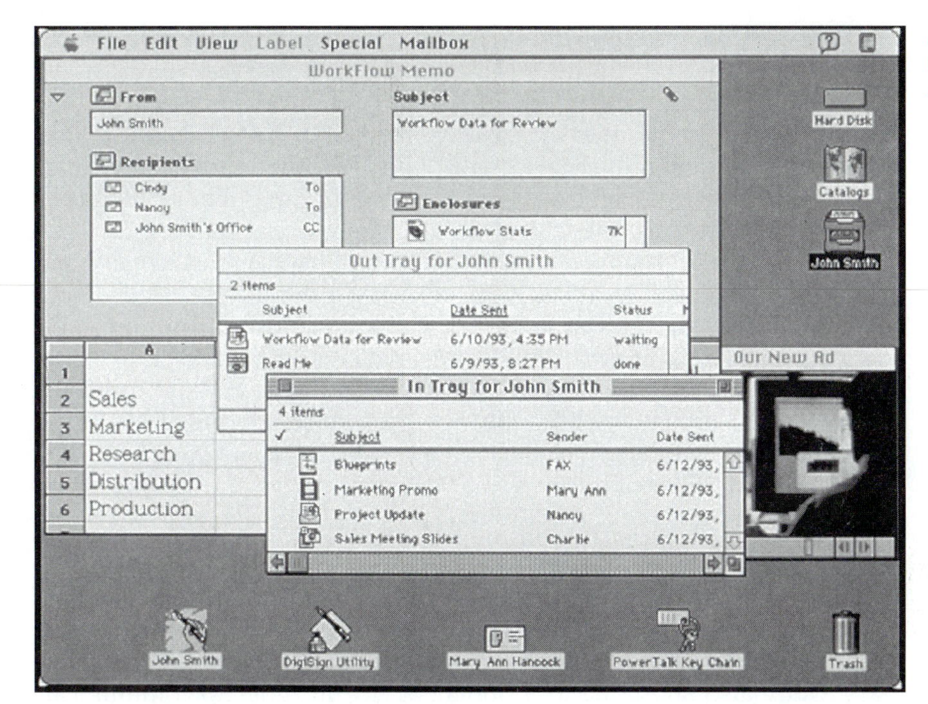

A screen from an electronic mail system. (Courtesy of Apple Computer, Inc.)

read what you put on the computer, and because of the large memory capacities of today's machines, there's little that can't be recalled, even if the user thinks it has been deleted. You shouldn't be fooled into thinking a message or file is really gone just because you have deleted it. Many e-mail systems routinely store copies of all mail that passes through them.

Some TCP/IP-based e-mail systems use the *simple mail transfer protocol (SMTP),* one of several application layer protocols, to provide a way to transfer messages between computers. Messages are created by a program and passed to SMTP for transmission. Although SMTP is generally not concerned with the format or content of the messages, there is normally a header that tells the address of the recipient and a body containing the message to be sent. The SMTP software on the sending computer takes the message and transmits it, using SMTP transactions, over a TCP connection to the destination computer. SMTP software on the receiving computer accepts the message and either places it in a user's mailbox, or puts it in an outgoing queue if forwarding is required.

SMTP has some limitations that are important in today's environment:

- it cannot transmit text that contains characters from foreign character sets, such as German characters with an umlaut;
- it cannot transmit executable files, such as programs, which a user might want to attach to a message;
- it cannot handle mail over a certain size.

To overcome these limitations, a more sophisticated protocol, which is an extension of SMPT, was defined. The *multipurpose Internet mail extensions (MIME)* protocol eliminates the restrictions of SMTP in a compatible way, so that SMPT encoded messages can be passed to a MIME-based system and will be handled properly. The MIME protocol allows messages to contain foreign languages, binary files, and essentially an unlimited file size.

Several companies are in the business of providing public e-mail services. For a small fee, anyone can subscribe and be assigned an electronic mailbox. Subsequent charges vary, but they are usually based on the number of messages sent or received. Leading vendors of public e-mail are MCI and AT&T. E-mail service is also available from online services and Internet access providers such as America Online and CompuServe. Public e-mail services are normally accessible through one of the public networks, via the Internet, or through toll-free telephone numbers.

Almost all types of businesses or business people find public e-mail services very attractive and beneficial:

- independent business people who need to communicate with their vendors and customers who may be located anywhere in the country;
- small companies that cannot afford their own private e-mail service;
- large companies, which may have their own private e-mail system, may also subscribe to a public service in order to communicate with customers, vendors, or employees;
- salespeople and other travelers, who are frequently away from their office, find public e-mail an ideal way to keep in touch with the home office and to conduct business almost as though they were at their desk.

E-mail systems such as Microsoft's Outlook can also be installed on a company's servers for its private use. Companies that install their own private e-mail system may use their regular data communication network and simply treat e-mail as another application. Usually the private systems have a link to the Internet so that employees can send e-mail outside of the company as well as to other employees.

Unified Messaging System

Some e-mail software allows e-mail, voice mail, and fax messages to be brought together into a *unified messaging* system. These services will let you send voice mail, e-mail, and faxes to a single digital phone number or mailbox. End users need only tap into one mailbox to access all of their messages, greatly simplifying communications and boosting convenience for consumers by reducing the need to carry multiple devices such as pagers, cell phones, and palmtop computers. They can choose between a computer or telephone to receive their messages, because the system will "read" textual e-mail messages to them on the telephone, access voice messages from the PC, or even have e-mail messages faxed. Automatic media conversion provides universal access to messages—e-mail, voice, or fax—no matter where the user is or what type of device is available.

SIDEB

MESSAGE CENTER

Home Is Where the Hub Is

Today's always-on economy means we're always on call, even far beyond power centers of the network economy. The communications devices strewn around a typical American household generate an average of 115 unique outgoing and incoming messages each week. Our daily dose of interruption, er, connection, comes via telephone, postal service, e-mail, voice mail, cell phone, pager, and fax, according to a household messaging study by Pitney Bowes. Of those communications, more than 40 percent are for work-related matters. Are we approaching the saturation point? Not likely. The number of continuous Net connections (DSL lines and cable modems) in US homes is expected to top 25 million by 2003. Add to that a total of nearly 200 million cell phones in the US by year-end—up 100 million from 1999—and it becomes clear that the only way to escape the all-encompassing datasphere is to . . . please hold.

Weekly communications sent/received	Average household	High-volume household
Phone	54	102
Mail	35	50
Email	16	59
Voicemail	4	12
Cell phone	3	8
Pager	2	7
Fax	1	2
Total	**115**	**240**

Reprinted with permission from *Wired Magazine*

Clearly this is the type of mail system that people will be using in the near future since the software is already available.

Image Applications

Facsimile A *facsimile (fax) machine* scans a sheet of paper electronically and converts the light and dark areas to electrical signals, which can then be transmitted over telephone lines. At the other end, a similar machine reverses the process and produces the original image on a sheet of paper. Individual characters are not sent as such; only as contrast between light and dark. As a result, the facsimile is ideal for sending preprinted business documents or forms, as well as letters, contracts, and even photographs. Each facsimile machine, therefore, has two parts: a reader-transmitter and a receiver-printer.

The Sharp FO-6600 facsimile uses plain paper and can send a page in six seconds. (Reproduced with the permission of Sharp Electronics Corporation)

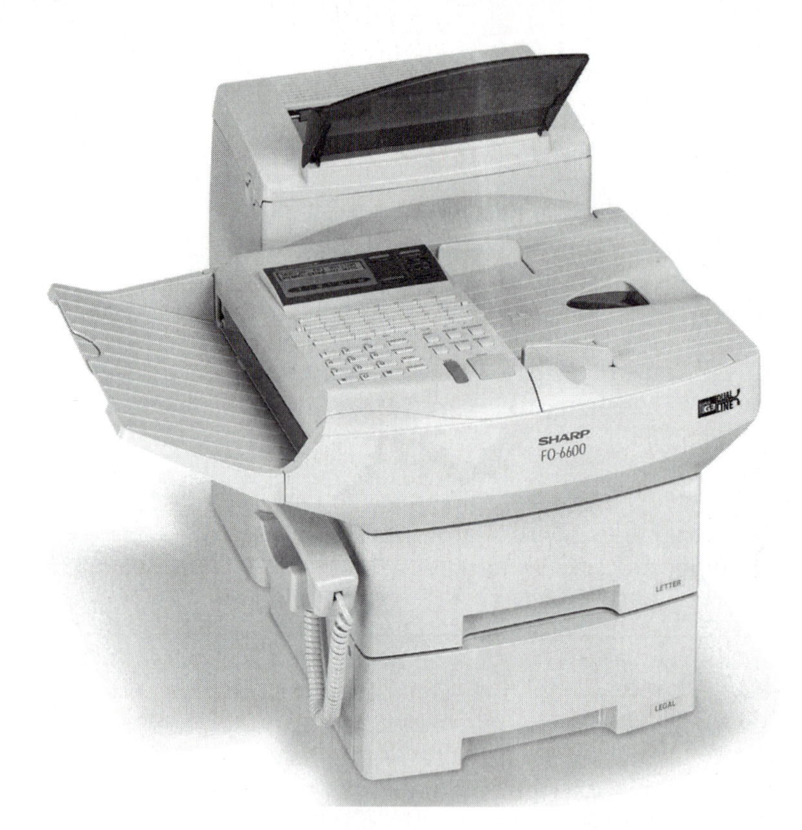

Since the machine deals with light and dark areas (not letters, words, or numbers), any image, photograph, drawing, or graph can be transmitted. The facsimile machine scans back and forth across the original document in a series of scan lines. The most sophisticated facsimile machines can detect several shades of light or dark. The more shades detected, the more faithful the reproduction is at the receiving end. The speed to transmit and print an 8 1/2- × 11-inch document varies from 6 minutes to approximately 2 seconds, depending on the techniques used for transmitting.

Facsimile transmission has several advantages over other data communication methods in certain applications:

- if a document is already printed, it does not need to be rekeyed in order to be transmitted;
- operation of facsimile machines is very simple and requires little training;
- since the recipient receives an exact duplicate of the original image, graphs, charts, and handwritten notes can be sent as easily as typed documents.

An industrial security guard works at his VDT. The television monitors in the background are connected to cameras at various locations throughout the plant and allow the guard to watch many locations simultaneously. Other equipment monitors the use of badge readers at entrance doors. (Courtesy of Dow Corning Corporation)

In business, facsimile is used to transmit any document that must arrive quickly. Examples include contracts going to a law firm or a prospective client, shipping documents sent from the corporate headquarters to a warehouse, and simple engineering drawings going to a contractor or outside engineering firm. Radio stations are now accepting requests by facsimile, and some individuals have facsimile machines at home. In Japan, where addresses are complicated and streets usually are not named, facsimile is regularly used to send a map explaining how to get to one's office or home.

Television Television is used in business in several ways. Among them are

- to monitor doors, parking lots, or other facilities;
- to provide information to employees;
- for video conferencing—conducting a meeting where the participants are in different locations but can see each other on television.

Security Monitoring In the security monitoring application, the television camera can be on top of a building, in a doorway or corridor, or in the corner of a room viewing the room or area being monitored. The camera is unattended, and its video signal is fed to a central monitoring point such as a security desk. A video tape recorder may be connected so that any unusual activity can be recorded and then later replayed for analysis.

A video conference in progress. Cameras are mounted just above the television sets to take pictures of the participants and send them to the other end.

For security monitoring, black-and-white television pictures are most often used, and in many cases the picture is updated only every 30 to 90 seconds. This periodic refreshing of the picture is referred to as *freeze-frame television.* It requires a much slower, less expensive telecommunications line than one that must transmit *full-motion television* pictures. Full-motion pictures are like those seen on commercial television stations; to achieve the motion, 30 frames are sent every second.

Providing Information to Employees Providing information to employees is another use for television in business. The information may be textual information that announces promotions, job changes, or upcoming events. An announcement might be displayed on the screen for 15 or 20 seconds and then automatically replaced with the next one. Information could also be conveyed through regular full-motion, full-color television broadcasts. The programming might include interviews with company officers, company news broadcasts, or training films.

Whether textual or full-motion, in-house television is most successful if television sets to receive the broadcasts are located in convenient locations around the office or plant. Having the sets in coffee break areas, the cafeteria, and other gathering places helps ensure that the messages are conveyed to the maximum number of people.

Video Conferencing A third use of television in business is *video conferencing.* In this application, meetings are conducted in rooms equipped with television cameras and receivers. The participants in one room can view those in the other rooms via the screen. The television signals are transmitted between the rooms on telecommunications lines.

There are actually two types of video conferencing: one-way and two-way. In one-way conferencing, a company broadcasts a program from a central location, which is received simultaneously at numerous receiving stations equipped with relatively inexpensive receivers. Some audio audience interaction is possible using telephone lines, but the picture communication is one-way. This is in contrast to two-way teleconferencing, which is more like a conventional meeting and requires cameras and transmission equipment at each end of the connection.

Another way to categorize video conferencing is whether it is full-motion or freeze-frame. In this application, the trend has been toward simple operation using freeze-frame television. In most meetings, full-motion pictures are not necessary. Once participants are in the room and seated, a picture can be transmitted to the other locations so that the participants can see who they are talking to. After that, having an updated picture every 30 to 60 seconds is usually adequate. The use of freeze-frame television greatly reduces the amount of information that has to be transmitted and keeps equipment and transmission costs down.

Television equipment costs and complexity vary widely. Factors that affect them are

- black-and-white versus color;
- full-motion versus freeze-frame;
- lighting sensitivity;
- special equipment for graphics or text transmission.

The more the video conferencing capability becomes like commercial television, the more elaborate and costly the equipment. Sending 30 frames per second obviously requires a much higher transmission speed than sending 1 frame every 30 seconds. Studios and control rooms with editing equipment are sometimes found in the corporate environment, but in most cases they are unnecessary and difficult to justify. They are used where the television broadcast has the elements of a television production.

Most video conferencing today is done from a conference room setting, not in a studio. One camera is usually focused on the meeting participants, and a second camera is used to transmit pictures of transparencies, slides, or drawings that illustrate points during the meeting. Ideally, the video conferencing equipment is simple enough to be operated by the meeting participants.

public video conferencing centers

Several companies have set up public centers in major cities for video conferencing. The centers have fully equipped video conferencing rooms that are rented by the hour. The rental fee includes the use of the room, any technicians required to operate the equipment, and transmission of the video conference by satellite to another, similarly equipped room. Using public centers is ideal for a company that occasionally needs to use video conferencing but cannot justify installing its own equipment. In addition to using the rooms for conducting meetings, some companies regularly use them to announce new products, conduct sales meetings, or communicate with customers. The conference might feature the vice president of marketing in a room in New York City and salespeople gathered in rooms in other major cities. A new product could be announced to the entire sales force simultaneously, and the two-way nature of the media would allow the salespeople to ask questions and interact with the vice president.

A number of companies have established video conferencing networks. Ford Motor Company established its network in 1985 and has since expanded it to plants, district offices, and other facilities in North America. General Motors has a network that links its worldwide facilities as do General Electric and IBM. These links enable management to reach people they were unable or unwilling to reach by traveling. IBM has used one-way video conferencing for several years for employee and customer education. The instructor works in a studio with complete graphic aids available. Students sit in classrooms at IBM facilities throughout the country and watch the instructor on large screen televisions. They can ask questions and respond to the instructor using audio communication.

New video conferencing units are smaller and don't require dedicated facilities. Several companies are now making adapter cards for personal computers that allow the PC, with a small camera mounted on top, to be used for a small video conference between two people or a very small group of people at each end.

Compression technology continues to improve so that pictures can be sent over slower telecommunications lines than in the past and with acceptable quality. Simultaneously, the cost of higher speed lines is dropping so the purchaser of a video conferencing system has more choices, tradeoffs, and possibilities for using this technology than a few years ago.

Delay and Loss Sensitivity It is important to point out that some kinds of telecommunications transmission are more sensitive to delay than others. On the one hand, if your voice is delayed when you are talking on the telephone, it is very noticeable and irritating. Similarly, you may have noticed times when the voice and picture on a television are not synchronized so that the words you hear from the speaker don't match the picture on the screen. This can be caused by several technical problems, but one of them is that either the voice or picture is delayed in transmission relative to the other. The term we apply to this attribute of voice and television transmissions is that they are *time* or *delay sensitive.* On the other hand, if we are sending an e-mail message to a friend, we are normally not concerned whether the message arrives in 1 second, 1 minute, or sometimes even 1 hour. However, we don't want the message to be unreadable as it could be if some of the characters in the message were garbled or lost during the

delay sensitive

transmission. We say that such a signal, a data signal, is highly *loss sensitive*, but not time sensitive.

loss sensitive

Returning to the first example, if a voice or television signal suffers some loss of data during the transmission, you might hear a missing syllable or see white specks on the TV picture, but it would not normally cause you to miss the message or impair your understanding of what was being said or shown. So voice and television signals are not highly loss sensitive, unless of course the loss is carried to an extreme.

The Internet

The Internet has become so widely used in the past several years that special coverage may not be necessary. Nonetheless, while extensive coverage of the multifaceted capabilities of the Internet is beyond the scope of this book, a few words are in order to put the Internet into context with the rest of the material in this book. The Internet and its uses are a microcosm (or maybe a macrocosm!) of all of the applications discussed so far. Whatever application one can imagine seems to be in use on the Internet. Structured, unstructured, voice, data, text, image, video, interactive, batch—they're all there!

The Internet and *World Wide Web* are so widely used that I assume students reading this book have some experience using them. In the mid-1990s the arcane letters *http://www.royal.gov.uk* wouldn't have meant anything to anyone, but now many people know that they're used by the Queen of England to identify the royal family's site on "the web." Similar letters appear routinely in virtually every newspaper, magazine, and advertisement, and on news shows to direct people to sites on the Internet that may be of interest. I did most of the research for this edition of this book using worldwide resources available on the WWW while sitting at my personal computer in my home. Information that was previously hard to come by is now easily and quickly available. The rapid rise in the use of the Internet is a phenomenon unparalleled in history.

World Wide Web

Some recent statistics serve to illustrate how pervasive the WWW has become. It is estimated that there are over 130 million users of the Internet, and in a given month, over 82 million people actually use it. One survey showed that those users are online from home an average of almost 10 hours per month, and for those who use it at work, about 22 hours. Another survey showed that the average overall use was approximately 12 hours per week, far more than is spent reading magazines.

The Internet is a huge interconnected network of networks using telecommunication lines running under the TCP/IP architecture/protocol. Originating in 1969 with the ARPANET, which was sponsored by the U.S. Department of Defense, the connections spread, first to universities and research institutions, then to other government agencies, to private corporations, and finally to individuals. Several studies have been performed attempting to address the subject of who is using the Internet today and what they are doing with it. A summary of many of the studies can be found, on the Internet of course, at *www.cyberatlas.com.* Some typical uses are:

- communication (e-mail);
- downloading software;
- interactive discussions;
- using another computer;
- realtime audio and video;
- looking up information/research.

Access to the Internet occurs in various ways. Some companies and institutions have full-time leased lines operating at very high speed, such as 1.54 million bits per second. This option is on the expensive side, however if the cost is spread over a large number of users who are making productive use of the Internet, it may actually be very economical on a per-user basis.

Access to the Internet through cable television systems is spreading rapidly as more cable companies upgrade their equipment to make Internet access possible. Cable connections are extremely fast, and have the added advantage that the user is always connected to the Internet—there is no logon process. If the computer workstation is on, Internet access is enabled. The downside, however, is that the permanent connection leaves the user more exposed to certain security issues.

ISP

The majority of home users still have a dial-in connection from the modem on their PC to an online service such as America Online or to an Internet service provider (ISP). These connections typically operate at 28.8 or 33.6 thousand bits per second. This option comes at an affordable cost, typically much less than what one would pay for a cable television service on a monthly basis. There are several types of ISPs to which one might subscribe.

- Tier 1 ISP—An organization that has its own, typically national, network that supports very high speeds (45 Mbps) at its core. These ISPs often sell wholesale services to local telephone companies and local ISPs.
- Tier 2 ISP—Organizations that do not have their own network but lease capacity from other service providers. Tier 2 ISPs also typically support more than 100 thousand customers. America Online (AOL) is an example of a company in this category.
- Tier 3 ISP—These ISPs only support regional or local customers through a limited local network. There are currently about 4,000 Tier 3 ISPs in the United States.

At first, the Internet was used primarily for e-mail. Soon after, *user groups,* groups of people who get together on the Internet to exchange ideas or discuss a particular topic, began to develop. File transfer was another early application that is still widely used today. Files contain anything from pictures, to programs, or information on a topic of interest. When people started to discover that there was a lot of potentially interesting information "out there" on the Internet, they needed ways to locate information of interest. The University of Minnesota was one of the early organizations to respond with a program appropriately called *Gopher* (later, a new version was called Gopher+), which searches for information based on parameters specified by the user. Other programs in the same genre are Archie and Wide Area Information Server (WAIS).

By far the most common, and probably easiest, way to use the Internet, however, is by using a *browser program* to access the World Wide Web (WWW). A browser is a program designed to read the information stored on WWW sites and to handle the hypertext that these sites contain. The two most popular commercially available browser programs are Netscape's Navigator and Microsoft's Internet Explorer. The WWW contains textual and non-textual information stored in the form of *hypertext*. Hypertext allows the user to click on a word, which then connects to related information. For example, one might enter the web searching for information about Colorado. Reading a description of the state there would likely be a reference to skiing. Clicking on the hypertext word "skiing" with the mouse would take you to information about skiing, which could lead you off to read about skis, ski equipment, ski resorts, etc.

browser

Two concepts that are important to understand about the way the WWW works are *universal resource locators (URL)* and *hypertext transfer protocol (HTTP)*. URLs are essentially the addresses of objects located anywhere on the Internet and today there are over five million registered names. The URL consists of the name of the access method for the object, followed by a colon, and then by an identifier of a resource. The major access methods are:

URL

- http (hypertext transfer protocol);
- ftp (file transfer protocol);
- gopher (the Gopher protocol);
- telnet (reference to interactive sessions);
- mailto (electronic mail address);
- wais (wide area information servers);
- news (USENET news).

The http URL scheme designates Internet resources that use the hypertext transfer protocol. In particular, these are sites on the WWW. Other Internet resources are accessed using the other protocols shown. An http URL is of the form: http://(host):(port)/path, however, the port normally defaults to a value of 80, and the path is optional, so a typical http URL looks like this: http://www.netscape.com. The URL www.netscape.com

is the name of the host, and since no path was specified, this URL points to the home page of that WWW *site.* A WWW site is a computer (server) containing a collection of information stored in http format and connected to the Internet. Organizations of all types and many individuals have WWW sites providing information on every imaginable topic. Furthermore, many of these same organizations are using the same networking and software technologies to set up private, internal versions of the WWW for their employees. These internal "webs" are called *intranets.*

intranet

The Internet is changing the way that applications are designed as companies look to the web as the entry point for electronic commerce. Internet access provides customers an additional way to find product information and to place orders for products of interest. If customers or prospects prefer to talk in person with a sales representative, that option is of course still available, but if customers prefer to research product options themselves, having an online catalog, product data sheets, brochures, material safety data sheets, and other information available through the Internet is convenient for customers and provides a new channel to the market for the seller.

The Internet and the WWW can serve as an excellent reference point for you as you study the rest of the material in this book. Most of the principals and concepts that will be discussed in later chapters can be illustrated with Internet-based examples.

SPECIAL CONSIDERATIONS OF DATA COMMUNICATIONS APPLICATIONS

Most data communications applications involve people. True machine-to-machine communication with no human involvement is less common. Since people are involved in the majority of data communications applications, a number of human factor elements must be considered in the planning or design of a data communications application or network.

Response Time

Matching the response time of a telecommunications system or network to the user's expectations, or vice versa, is an important human factor consideration. Of equal importance is response time consistency so that the user has the feeling that the system is always performing in the same way. We looked into this issue in detail in Chapter 1.

Security

System security is another important consideration in telecommunications systems, particularly where computer data is involved. Businesses are becoming extremely sensitive about the protection of their data and are insisting that ever more stringent security measures be put in place to protect it. When a telecommunications system is being designed, security must be carefully evaluated, and security techniques appropriate for the application must be implemented. We discuss these techniques in detail in Chapter 15.

Planning for Failures

A third consideration of terminal-based systems is the special procedures required when the computer or network fails. Companies using telecommunications networks tend to become extremely dependent on the network to conduct normal business operations, yet network or computer failure does occur. The telecommunications system designer must plan for how the business will operate when the computer or network is down. Telecommunications adds an element of complexity to computer applications, particularly if long distance lines are involved, since they are subject to problems caused by severe weather, electrical interference, and misguided bulldozers or backhoes.

failures will occur

Therefore, it can be safely assumed that someday when a user picks up the telephone or sits down at the terminal, it won't work. The outage may or may not be a problem. Some telephone calls are more important than others; some can wait or were optional in the first place. The same is true of computer applications. In the sales order entry system described earlier in this chapter, the company may not need to enter a customer order or a material today if the product is not to be shipped for several weeks or months. In contrast, if the supermarket checkout system is not operating, customers will go elsewhere to buy their groceries.

The key is planning and having thought through what will happen when (not "if") the computer or network fails. Several options are available, including

- wait and do nothing except try to determine the reason for the failure and the likely amount of downtime;
- fall back to manual procedures, the types that were used before a computer and telecommunications network were in place;
- manually switch to a backup computer or telecommunications line;
- have a standby computer or telecommunications line in place ready to take over automatically in case of failure. This is called a *hot standby* system.

hot standby

Each of these alternatives must be evaluated in light of the business situation. In all but the most critical applications, the usual approach is to try to bring the system back up. At the other extreme, the airlines with their reservation systems have standby computers and duplicated networks to make sure that when a failure occurs, redundant facilities are available to take over immediately without missing a reservation. Few companies would actually wait and literally "do nothing" when the computer or communication system is down.

The problem with falling back to manual procedures is that once staff becomes used to automated systems, almost nobody remembers how to do things manually. Most companies find that manual procedures are not a very acceptable backup alternative. In addition, the volume of transactions often cannot be handled manually.

The difference between having backup computers and lines and a hot standby system is that the backup computers and lines usually are used for other types of work until they are needed. When a failure of the critical computer or network occurs, work is switched off the backup computer and the application is restarted on it. Careful planning will ensure that the time required to switch to the backup computer is not excessive. With a hot standby system, the switchover takes place rapidly and automatically. Again, the speed required is dependent on the application. Hot standby systems are expensive, and the benefits must be weighed against the costs.

planning is key

The key to handling failures is the planning itself. System outages need not be catastrophic if someone has thought through the implications of an outage and actions to be taken when the outage occurs. Many companies conclude that although it would not be convenient, the best alternative for many of their applications is to simply try to bring the system back up as quickly as possible and then catch up on the processing by having people work extra hours. Of course, in some applications, this is just not possible. More elaborate procedures are needed.

Disaster Recovery Planning Disaster recovery planning is an extension of the discussion about planning for normal system outages. For our purposes, a *disaster* is defined as a long-term outage that cannot be quickly remedied. A fire, flood, or earthquake may be the cause, for example. No immediate repair is possible, and the computer facility and equipment are unusable. Figure 4–14 shows a checklist for disaster recovery planning.

Again, planning is the key. Having a computer in an alternate site and the ability to switch the telecommunications lines to the alternate computer is a viable alternative in many situations. In lieu of that, quickly obtaining a new computer from the vendor may be possible. Most vendors state that they will "take the next computer off the manufacturing line" to replace a computer damaged in a disaster. Given the thousands of servers that are produced each day, this sounds like it should guarantee quick delivery of a replacement machine. However, in practice, getting a server with the right configuration, memory size, disk capacity, and other needed features sometimes takes weeks rather than days. One small Internet access company recently lost its e-mail service and a lot of customers when it was unable to get a replacement e-mail server from the

Considerations for Disaster Recovery Planning

1. What level of service should be maintained during a disaster?
2. How will the organization communicate internally?
3. Where will help desks and command centers be located?
4. Are policies and procedures in place to handle customer and other incoming calls in a professional manner and to provide timely information?
5. Are computers and their data sufficiently backed up?
6. What happens in the event of an evacuation?
7. What is the sequence for recovery? Which departments must be put back in operation first, second, etc.?
8. Are procedures in place for regular testing of the disaster recovery plan?

Figure 4–14
A checklist for disaster recovery planning.

manufacturer for over 4 weeks. The customers were without e-mail service for all that time, and the fact that they weren't billed was not enough to keep them from moving to other access providers who could give reliable service.

Even if a replacement computer can be obtained, it is of no use if a suitable facility cannot be found to house it. Of equal importance is the ability to switch the telecommunications network to the new computer site, again something that may be relatively easy if it has been planned for ahead of time. Telephone companies have developed techniques and facilities to allow networks to be switched to alternate sites, a capability that is demanded by companies that are putting disaster recovery plans in place.

mutual aid pact

Sometimes it is possible to work out a mutual aid pact with another company. Both organizations agree to back each other up in case of a disaster. They may agree to provide computer time on the second or third shift in case of a disaster, even if it means not running some of the low-priority processing. Computer centers are easier to back up with mutual aid pacts than are telecommunications networks. With proper planning, however, extra lines could be installed between the companies providing at least some backup transmission capability. If the two companies are in close physical proximity, they must be concerned about a disaster that would hit both of them simultaneously. An earthquake or tornado could easily hit companies in a several-mile radius and effectively neutralize any backup plans.

Several companies exist for the sole purpose of providing disaster backup recovery facilities. These companies, such as Comdisco Continuity Services (www.Comdisco.com), have one or more large computer centers filled with hardware ready for use in an emergency. In order to use the facilities, a business must subscribe to the service by paying a membership fee and annual dues. Then, if the use of the emergency facility is necessary, a one-time activation fee must be paid as well as usage charges for the actual time the disaster site is occupied. Subscribing to one of these services is much like buying a form of insurance.

Whatever plan is developed for disaster recovery, it must be specific for different kinds of disasters. Corning Incorporated, located in Corning, New York, had a disaster plan detailing how the company would recover from a fire. In 1972, however, the town of Corning was hit by a massive flood, which literally put the Corning computer center under water. Although some of the procedures previously developed for a fire were appropriate, many were totally useless or inappropriate for problems caused by water damage.

testing is very important

Disaster recovery plans must be tested. Rarely will all of the problems of a real disaster be covered when a plan is written, and although testing does not recreate a real disaster, it does identify weaknesses in the plan. If a contract has been signed with a disaster recovery firm, that firm will assist in testing the disaster plan. The contract normally provides for a specified number of hours of test time at the disaster recovery site. Companies use this time to ensure that they can transfer their software from their mainframe computers to the backup computer and to test the telecommunications links that connect the disaster site to parts of the company not affected by the disaster.

Tests of the disaster plan can be conducted in other ways. One company's data processing manager worked with the computer vendor to develop a disaster test. Late one night, the vendor's service manager removed a critical but obscure part of the mainframe computer. When the computer operations staff could not bring the computer up the next morning, they called the vendor for service. The vendor's technicians tried for several hours but were unable to precisely diagnose the problem; therefore, they called for help from the national support center. In the meantime, the prolonged computer outage had caused the company to activate its disaster recovery plan and because only the data processing manager, the company president, and the vendor service manager knew the real nature of the outage, the "test" was extremely realistic. It was allowed to run for over 24 hours before it was revealed to be a test. Although there was some grumbling, there was also general consensus among the computer users and management that much valuable information had been gained.

As a company becomes increasingly reliant on telecommunications and computer networks, it must constantly reassess how long it can afford to be without the systems. Getting the users involved in assessing the impact of an extended outage caused by a disaster is one way to build a case for management to support spending time and money on developing and maintaining disaster recovery procedures.

SUMMARY

This chapter has looked at a number of business applications that involve the use of telecommunications and the computer. Telecommunications and computing go hand in hand in providing modern business systems that help companies to be more effective and efficient. Simple uses of telecommunications for message switching and online data collection have given way to sophisticated, interactive applications that enable a company to do business in new ways. The rise in the availability and usage of the Internet and the World Wide Web have introduced a new set of data communication–based applications for business and the general public's use.

The integration of these applications into the business brings with it new responsibilities. Provisions must be made for appropriate security to ensure that company data is not lost or misappropriated. Contingency planning must be done to determine how the company will operate when short or long outages occur. By now, you should have a good understanding of the many ways in which companies use telecommunications and the breadth and depth at which it can penetrate an organization's activities.

CASE STUDY

Telecommunications at Work in a Small Business

F/S Associates is a small computer software development firm located in Hermosa Beach, California. The company specializes in the development of software for telephone companies to aid in the formatting and printing of telephone directories, especially the Yellow Pages. Chet Floyd, president of F/S Associates, incorporated the company and is also the chief software developer. He uses two IBM personal computers for the software development work.

Several years ago, Chet established a business relationship with another company, Strategic Management Systems (SMS), a software development firm located in New Jersey. At the present time, F/S Associates is under contract to Strategic Management Systems to develop and test software for several large customers.

F/S Associates is using telecommunications in its business activities in four different ways. Chet has a small, relatively standard telephone system for his office, which he purchased at a local discount store and installed himself. It provides an intercom capability for easy communication with his administrative assistant, and allows him to use either of the two telephone lines in his office to make telephone calls. Normally, he uses one of the lines for voice calls and the other for data calls from his personal computers. In his business, it is of course very useful for him to be able to talk on the telephone to a client while simultaneously accessing information or sharing data from his personal computer on the other line. A headset is used in these situations to allow full use of the hands to call up support material from the local computers. He says that the capability is extremely useful when he is debugging programs. It allows him to test complex programs as if he were at the customer's site. He adds, "And my availability to a project is increased while costs and lead times are reduced."

Strategic Management Systems uses other software developers besides F/S Associates, and sometimes it is necessary for several of the developers to talk with each other simultaneously. When necessary, the companies establish a conference call through the AT&T audio conferencing service. Because these calls sometimes last more than an hour, Chet bought a high quality speakerphone as a part of his telephone system. Using it saves him from having to hold the telephone handset up to his ear during the calls. "The speakerphone is essential when more than one party is at the local site. Frequently, the other programmer is required in these conferences, and occasionally we might also have a vendor or a contractor on site as well," Chet says. "Sometimes five or six of us will be on a teleconference at the same time," "It's amazing how fast we can exchange information and decide who is going to work on what part of a program or solve a problem. Of course we all have to remember not to get too excited and speak at the same time. Common courtesy is a necessary part of the calls to help make them productive," he adds.

Because F/S Associates is located on the west coast and Strategic Management Systems on the east coast, electronic mail is the third frequently used communications technique. Both companies use the Internet to exchange e-mail. Chet says, "Years ago we had to write our own software just to automate the dialing process, but now, with Windows 98, I can connect to my Internet service provider effortlessly using Windows 98's built-in dialing capability. It takes just a few seconds and I'm online ready to send e-mail or look up other information I may need." When Chet has long or complicated documents to send on e-mail, he prepares them offline using the Microsoft Word word processing program. "Word gives me great power to create and edit messages or complicated documents," he says, "and I'm not tying up the phone or paying to be online while I'm preparing the document. When I'm ready to send the document, I prepare a simple e-mail and attach the document to it using the e-mail program's 'Attach' command."

The fourth way F/S Associates uses telecommunications is for remote testing and support of software. Chet bought a package called ControlIT, and with it running in an F/S Associates customer's computer and one of Chet's machines, he makes a standard dial-up telephone connection between the two computers. As the developer, Chet can then run or test his program remotely on the customer's computer. This is a big help for demonstrating the programs capabilities or in finding and correcting bugs. Chet says that the ControlIT software also has an excellent file transfer program he uses to send new copies of the program or other data files to his customers over the telephone. The capability is also useful for exchanging updated versions of programs with Strategic Management Systems.

"Without telecommunications capability, it would be impossible for me to conduct business the way I am doing it," Chet says. "We're doing business in a way that lets us get the job done but still be located where we want to live. There would be no practical way to codevelop software with a company across the country if we couldn't send programs, data files, and messages to each other electronically. Nor could we support our customers who are widely distributed, geographically. Internet access is very important as well. We use an economical Internet provider to maintain a customer support site that is linked in from SMS. This provides first level support for frequently asked questions, posting of software updates downloadable by authorized customers, and it has an e-mail submission capability for more detailed inquiries. In addition, we have our own web site for the solicitation of new business not connected with SMS. The Internet service provider hosts all inquiries, so we don't require a dedicated line nor a 24 × 7 server (or staff!) to support it."

QUESTIONS

1. Could F/S Associates survive as a business if it only had telephone capability but not e-mail or the ability to exchange programs and data files?
2. Do you think it would be economically advantageous for a small firm like F/S Associates to consider renting office space in an e-business office park where the costs of the telecommunications capabilities and other office facilities, such as copy machines, fax machines, and secretarial support are shared between all of the small business owners? What would be the downside of such an arrangement?

CASE STUDY

Dow Corning's Applications That Use Telecommunications

Dow Corning is dependent on its telecommunications network to do business. Like most companies, it relies on an effective, cost-efficient voice network, but the real payoff for Dow Corning comes in the way it uses data communications. For a number of years, the company's philosophy was to minimize the number of mainframe computers it used for business applications processing and concentrate on a few large computer centers that were accessible by users through a telecommunications network. Experience showed that by having a few large computer centers the overall unit cost of computing was lower, even though the telecommunications expense was a higher percent of the total expense.

In 1995, the company began moving from the use of mainframe computers to a LAN-based computer-processing network using multiple application servers. Dow Corning acquired the SAP enterprise management software and began the long, expensive transition from applications developed by its own analysts and programmers to the use of the purchased SAP software. Since SAP operates in client-server mode, the computer processing and telecommunications networks needed to be completely overhauled, a process that took about 3 years. Two of Dow Corning's mainframe computers were removed by the end of 1999, and the third mainframe will be removed at the end of 2001.

The situation is different within Dow Corning's manufacturing plants, where computers (typically DEC VAXs) are installed because of unique requirements for sharing information stored on process controllers. However, the major business applications, including plant management applications, for Dow Corning are handled globally on server computers in Midland, Michigan. The major business applications that are processed on these servers include:

- customer order processing;
- inventory management and control;
- manufacturing planning and scheduling;
- finance and accounting;
- personnel and human resource management;
- planning and budgeting.

Users around the world sign on to these applications through the telecommunications network and process transactions in realtime, updating databases, generating routine reports, and finding answers to unique questions.

In most of the plants, local area networks connect sensors and other automation equipment to process computers, which are in turn connected to plant computers. These computers do some plant-specific processing and also pass much of the data on to corporate applications and databases. A data network is also used to connect to customers and suppliers for electronic document interchange (EDI). Routine business transactions are sent from one company to another, saving the delay of mail, and in most cases, eliminating paperwork. Some customers have specifically requested that Dow Corning send documents via EDI instead of in paper form.

The company is making its first forays into the world of electronic business through its Internet web site, www.dowcorning.com. Customers and prospects are able to obtain information about many of the company's products, request additional information through e-mail, and, with appropriate passwords, order some of the company's products in certain countries. This web-based capability will continue to expand, and Dow Corning believes that it must be on the forefront of electronic business in order to remain competitive. The company has also found that there are savings associated with doing business electronically, which help its profitability.

In 1983, an internal study concluded that Dow Corning should provide a comprehensive electronic mail and office automation application on the mainframe. A mainframe-based solution from IBM was selected, and it was used by virtually all employees in the company until it was replaced in 1999 by a significantly upgraded product, Microsoft's Outlook. Having a single e-mail and office system has been a very successful application of communications and computing technology for the company. Over 9,000 employees, contractors, and employees of other companies that Dow Corning works with use the system. The only Dow Corning employees who do not use the e-mail system are certain manufacturing people who have little need to communicate with others outside of their immediate work group.

The company also makes extensive use of voice mail (which will be discussed in Chapter 5) in the United States and other locations around the world. The company has over 4,000 voice mailboxes that employees use to exchange voice messages and leave voice reminders and other information for each other.

Another use of telecommunications is audio and video conferencing. Virtually all locations have audio conferencing capability in many of their conference rooms, which use high quality speakerphones. Video conference centers have been established in Midland, Brussels, Wales, Tokyo, Sydney, Hong Kong, Singapore, Seoul, and at several plants in the U.S. Employees and management routinely "meet on television" to discuss their progress on projects and a wide variety of other topics. Although they have not proliferated widely, a few PC-based video conferencing units are also used at small locations.

From the above examples, you can see that Dow Corning is truly dependent on its voice and data communications systems to conduct its business and to be competitive.

QUESTIONS

1. Suppose that Dow Corning had decided to manage and operate its computer systems and data networks in a less centralized way and to allow each Industry Business Unit (IBU) to implement its own systems and networks. What would be the impact on the company's telecommunications network? Which approach do you think is more cost effective?

2. Given the rapid growth in the power of personal computers, do you think Dow Corning would follow the same centralized philosophy if it were starting over today?

3. Do you think the centralized approach to systems and networks will be an appropriate model for the future, considering the rapid growth of client-server computing? (If you're not familiar with the term "client-server computing" it will be explained in Chapter 12.)

4. How does Dow Corning's centralized approach to communications and computing heighten the need for good disaster recovery planning of its network and computer centers?

REVIEW QUESTIONS

1. Discuss the three ways in which telecommunications applications can be categorized.
2. Explain the following terms:

   central office
   multipoint line
   polling
   transaction processing system
   freeze-frame
   store-and-forward
   master or control station
   collision
   video conferencing
   World Wide Web

3. Why are airline reservation systems so vital to airlines?
4. Describe how a supermarket checkout system works.
5. Describe some situations where e-mail is inadequate for communicating a message and a facsimile is required.
6. What types of businesses are not candidates to use public e-mail systems?
7. Would you rent an apartment or buy a home based on a picture you had seen on the WWW? What other information would you require so that you could make a decision without visiting the house or apartment?
8. Identify and explain three different uses for television in business.
9. Explain the advantages and disadvantages of using video teleconferencing to conduct a business meeting.
10. Discuss the reasons why a company should have a disaster recovery plan for its telecommunications and computer systems.
11. Identify five types of "disaster" that could disable a computer-communications network.
12. Categorize the applications shown on the left into one or more of the categories shown on the right.

   car rental reservation system
   data collection for machines at a paper mill
   computer-aided drafting
   telemetry from a spacecraft
   a facsimile transmission

   voice
   realtime
   store-and-forward
   structured data
   unstructured data
   person-to-person
   person-to-machine
   machine-to-machine
   image

13. What is the function of a browser program?
14. Describe the purpose of the http protocol.
15. Identify ten types of information you might find on the Internet that would be useful to you in your everyday life.
16. What is a unified mail system?
17. Describe the purpose of the following Internet tools: FTP, Gopher, Archie.
18. What is the World Wide Web? What tool do you need to use it?
19. Identify some ways that the Internet can be used for business.
20. Compare and contrast the terms personal computer, minicomputer, and server.
21. Explain the meaning of delay sensitivity and loss sensitivity and their implications if you were sending a file to a friend over the Internet.

PROBLEMS AND PROJECTS

1. In order to install a video conferencing system, a company must invest money in cameras, monitors, and other equipment. What costs, if any, will be offset by the video conferencing? Is video conferencing an alternative to some other type of communication?
2. As the dean of your school, you have been asked by the president to develop a disaster recovery plan for the school's computers. Describe such a plan, focusing on the recovery procedures in case of fire. If classroom space could be found so that classes could continue,

how would the computing for the students, faculty, and administration be handled?

3. Visit a supermarket that has an automatic checkout system with scanners. Observe several of the checkers for 5 minutes each and keep a count of the number of items whose bar code cannot be read and for which the data must be entered through the keyboard. What is the "read failure" rate? Do you notice any significant difference between the checkers?

4. Visit a travel agency that is connected to one of the airline reservation systems and find out how they conduct business when their computer terminal is down. How often do they experience downtime? What has happened to their telephone use since they installed the reservation terminals?

5. For your school or your organization (if you are working), conceptualize a new multimedia application that will take advantage of new technologies that were not available two to five years ago. For this exercise, don't worry about the cost justification, just be creative and use your imagination to invent a new way to use computers and communications to improve the productivity or quality of life for students, faculty, administrators, employees, or managers.

6. Talk to someone who works in a business that uses e-mail and find out:
   a. to what extent e-mail messages have replaced formal written correspondence for communication within the company and for communication outside the company with vendors and customers;
   b. to what extent correspondence is being faxed instead of being sent through the mail;
   c. to what extent they are using the Internet in their work.

7. Use the Internet to find information about your school or a nearby university.

Vocabulary

online realtime
store-and-forward
batch
remote batch
remote job entry (RJE)
teletypewriter
message switching
point-to-point line
punched paper tape
torn tape message system
message center
multipoint line
control station
master station
subordinate station
slave station
polling
collision
addressing
addressing characters
protocol
timesharing
transaction processing system
distributed processing
client-server computing
client
server
local area network (LAN)
personal computer
microcomputer
minicomputer
mainframe computer
scalability
delay sensitivity
time sensitivity
loss sensitivity
automatic call distribution (ACD) unit
personal identification number (PIN)
uptime
point of sale (POS) terminal
electronic mail (e-mail)
electronic mailbox
simple mail transfer protocol (SMTP)
multipurpose Internet mail extensions (MIME)
unified messaging
facsimile (fax) machine
freeze-frame television
full-motion television
video conferencing
World Wide Web (WWW)
user groups
file transfer
browser program
hypertext
Uniform Resource Locator (URL)
hypertext transfer protocol (HTTP)
site
intranet
hot standby
disaster

References

Behar, Richard. "Who's Reading Your E-Mail?" *Fortune* (February 3, 1997): 57–70.

Boccadoro, Diane. "It's The Next Best Thing To Being There." *Teleconnect* (December 1999): 68–78.

Case, Carol. "Do You Have a Disaster Recovery Plan?" *Teleconnect* (March 1997): 156.

Cortese, Amy. "A Way Out of the Web Maze." *Business Week* (February 24, 1997): 40–45.

Fowler, Thomas B. "Internet Access and Pricing: Sorting Out the Options." *Telecommunications* (February 1997): 41–70.

Hoffman, Dona L., William D. Kalsbeek, and Thomas P. Novak. "Internet and Web Use in the U.S." *Communications of the ACM* (December 1996): 36–46.

Jainschigg, John. "Video for Everyone." *Teleconnect* (September 1995): 126–135.

Jerome, Marty, Jason Toates, and Ron White. "25 Top Technologies That Will Transform Your Business." *SMARTBUSINESSMAG.COM* (May 2000): 98–135.

Levine, John R., and Carol Baroudi. *The Internet for Dummies.* San Mateo, CA: IDG Books Worldwide, 1993.

Mueller, Milton. "Emerging Internet Infrastructures Worldwide." *Communications of the ACM* (June 1999): 29–36.

Nanneman, Don. "Unified Messaging: A Progress Report." *Telecommunications* (March 1997): 41–42.

O'Brien, Ross. "Telecom: Building a Better Asian Network." *Asia, Inc.* (December 1993): 26–32.

Ramo, Joshua Cooper. "Welcome to the Wired World." *Time* (February 3, 1997): 36–48.

Robinson, Teri. "The Revolution Is Here." *Information Week* (November 18, 1996): 106–108.

Rosenbush, Steve. "Telecom Market Prepares for Unified Messaging." *USA Today* (December 7, 1998).

Wallace, G. David. "The Electronic Tutor Is In." *Business Week* (January 20, 1997): 8.

———. "How the Internet Works: All You Need to Know." *Business Week* (July 20, 1998): 58–60.

CHAPTER

5

Voice Communications

OBJECTIVES

After you complete your study of this chapter, you should be able to

- describe the purpose and functions of the various parts of the telephone system;
- describe the evolution of telephone switching equipment;
- differentiate between public and private telephone switching equipment;
- describe the characteristics of analog signals;
- describe several types of telephone lines;
- explain several ways to acquire telephone service at a discount;
- describe several types of special telephone service;
- describe the way wireless voice communications systems work.

INTRODUCTION

Perhaps you think it is a bit strange to be studying voice communications in a book that is primarily devoted to data communications. Fifteen or 20 years ago it would have been strange! But since that time, there has been a merger of the computer and communications industries that has markedly changed both and brought them closer together until, today, they operate as one.

By the time you finish this chapter, you will understand that a study of data communications today wouldn't be complete without understanding voice communications. Furthermore, if you are working, or planning to work, in an organization that uses data communications, you'll find that voice communications is an important part of their total telecommunications picture. Eighty percent of business telecommunications expense in most companies is for voice and only 20 percent is for data, and because it is so large, revenue from voice communications is extremely important to telecommunications carriers. Most probably, in your company, voice and data operations have come together. Perhaps they share the same telecommunications lines. Or perhaps both are managed by the same person. Hence, it is important for you to learn about voice communications and telephone systems. Your knowledge of telecommunications would not be complete without a good understanding of voice communications concepts, capabilities, and products.

This chapter describes and discusses the four basic elements of voice communication systems:

- the telephone handset;
- telephone company switching equipment;
- telephone lines;
- telephone signals.

Through your study of this chapter, you will gain a detailed understanding of the way these four components work together to provide the reliable telephone service that is ubiquitous through most of the world and which we have come to expect. You'll also see how voice signals are sent on telephone lines, which is a basic step in understanding how data is sent. Finally we'll look at other aspects of telephone communications, such as the special capabilities and services provided by the telephone companies.

Telephone communications is one very significant application of telecommunications. It is an application we are familiar with because we use it every day. As such, it serves as an excellent jumping-off point and provides a solid foundation for your study of data communications in the rest of this book. The majority of data transmissions in the world today occur on lines and networks that were originally designed to carry voice. In those cases, the data transmissions are adjusted to fit the parameters of the voice network.

THE TELEPHONE SET

customer premise equipment

The telephone is a primary example of a class of equipment known as *customer premise equipment (CPE).* CPE includes any communications equipment that is located on the customer's premises and includes telephones, personal computers, fax machines, and telephone key systems, and PBXs (both of which you'll study later in this chapter). The term CPE is used frequently in the communications industry and is one with which you need to be familiar.

Primary Functions

The primary functions of the telephone set are to

- convert voice sound to electrical signals for transmission;
- convert electrical signals to sound (reception);
- provide a means to signal the telephone company that a call is to be made or a call is complete (off-hook and on-hook);
- provide a means to tell the telephone company the number the caller wants to call;
- provide a means for the telephone company to indicate that a call is coming in (ringing).

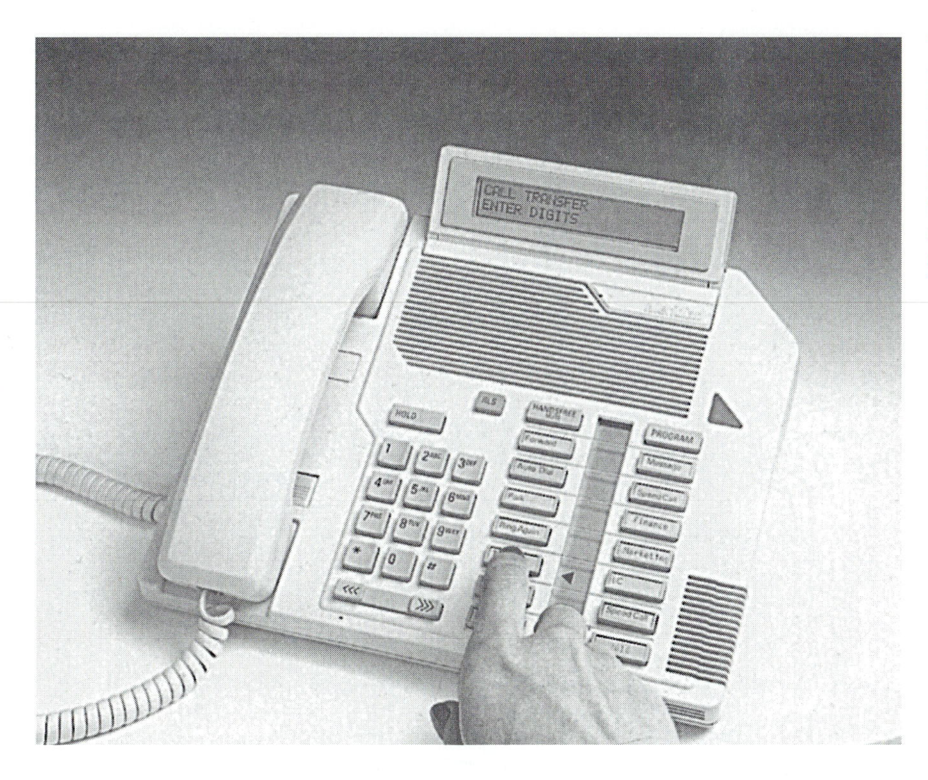

This modern telephone handset has several extra buttons that can be preprogrammed to automatically dial frequently called numbers, or to activate special features of the telephone system such as call transfer. (Courtesy of Northern Telecom, Inc.)

Telephone sets come in many sizes, shapes, styles, and colors. Years ago, there was just one type—the traditional black desk phone—and it was available only from the telephone company. Now phones to fit any decor or color scheme are available from a variety of vendors.

Telephones may be equipped with amplifiers to make the incoming voice signal louder for the hard of hearing. They may have loud bells or lights to indicate ringing in noisy areas. Explosion-proof telephones are available for use in chemical plants or other places where a spark might set off an explosion. Also, telephones with clocks, calendars, and elapsed timers are available.

Telephone Transmitter

Built into the handset of the telephone is the transmitter. It converts sound waves into electrical signals that can be transmitted on telephone lines. When you speak, your voice generates sound waves or vibrations that move the air in ever-widening circles, as shown in Figure 5–1. Some of these sound waves (which are really variations in air pressure) flow into the telephone mouthpiece, causing a thin diaphragm in the microphone to move back and forth. The vibrations put alternately more and less pressure on carbon granules in the microphone. As more pressure is applied, the granules are packed more tightly against one another and conduct

Figure 5–1
In the telephone transmitter, the sound waves of the voice cause the diaphragm to vibrate. The vibration puts varying amounts of pressure on the carbon granules, causing more and less electrical current to flow.

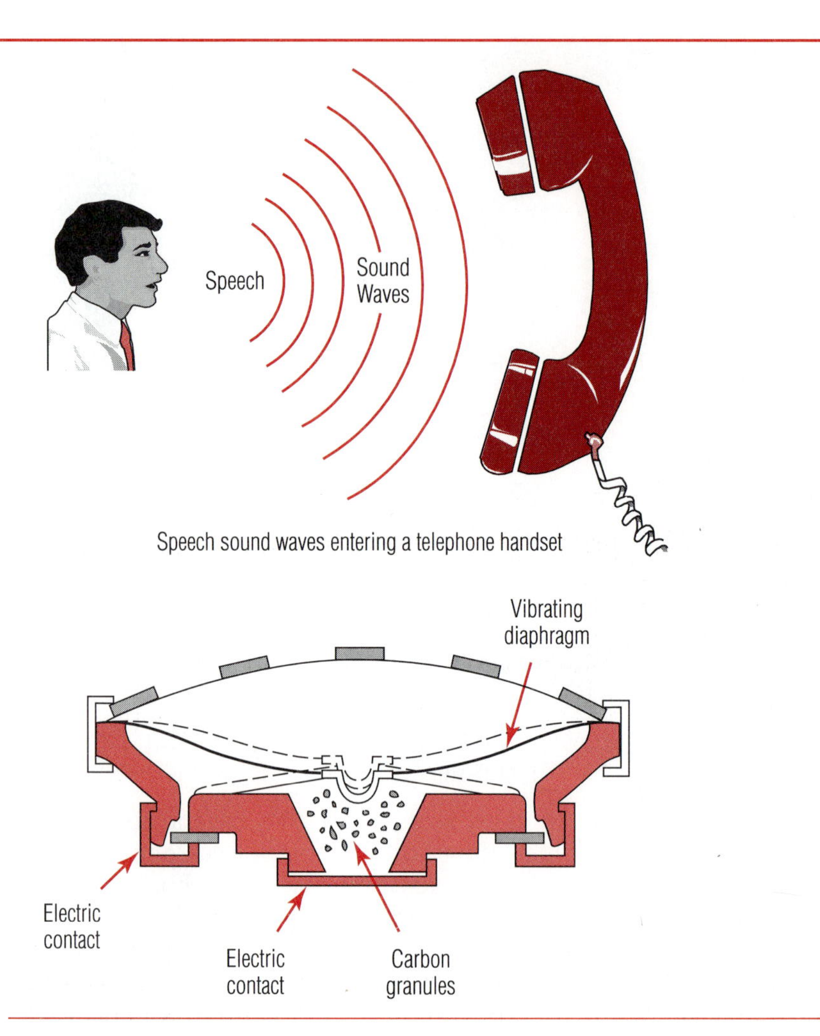

electricity better. Conversely, between the pressure waves, the granules move apart and do not conduct electricity as well. With voltage applied across the electrical contacts, this varying resistance of the carbon granules in the microphone causes a varying amount of current to flow. The varying current is an electrical representation of the sound waves generated by our voice.

analog signal

Since the electrical signal is analogous to the sound waves, we call it an *analog signal.* If we graphed a typical voice signal or watched it on an oscilloscope, it would look like the wave shown in Figure 5–2. To be more complete, an analog signal is one that is in the form of a continuously variable physical quantity, such as voltage.

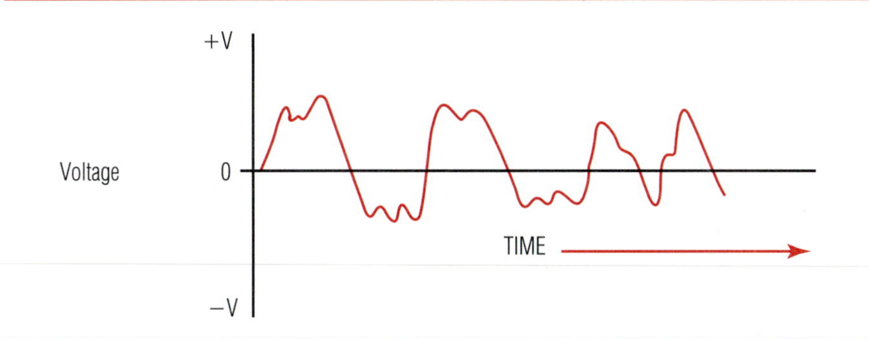

Figure 5–2
Wave of a typical voice signal.

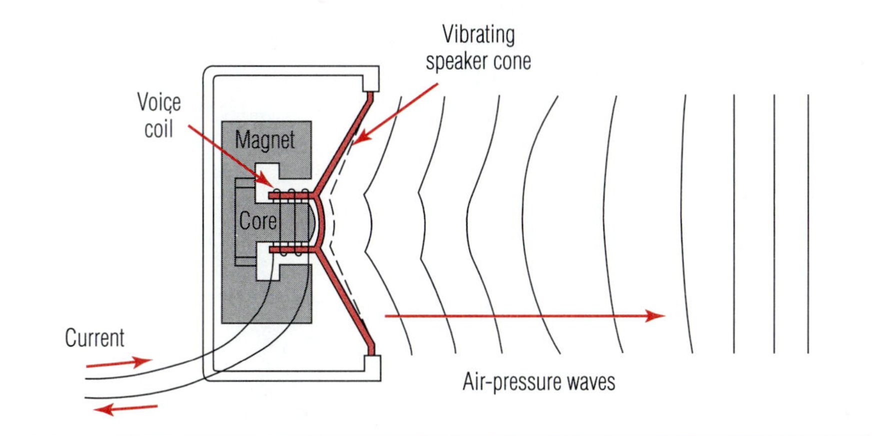

Figure 5–3
In the telephone receiver, the electrical current activates the voice coil, causing the speaker cone to move. The vibrating speaker cone creates air pressure waves that are heard as sound by the ear.

Telephone Receiver

At the other end of the telephone handset, the receiver's job is just the opposite. The receiver converts incoming electrical signals to sound waves that can be heard by the listener. Figure 5–3 is a diagram of this process. The incoming electrical signal flows through the voice coil in the receiver, and the interaction of the current in the voice coil with the magnet causes the coil to move. The coil is attached to a large diaphragm called the *speaker cone,* which vibrates back and forth causing sound waves that flow outward from it and into the ear of the person listening.

sidetone

An important aspect of the telephone and its handset is that when you talk into the transmitter, some of the electrical current generated by the microphone in the transmitter is fed into your own receiver so that you hear yourself talk. This small amount of signal is called *sidetone.* Although sidetone is not technically required for the telephone to operate properly, it is generated for human factors purposes. Without sidetone, the telephone seems dead, and people tend to talk too loudly. With too much sidetone, people talk too softly. Sidetone is a form of feedback to the

The Merlin cordless telephone can connect to up to five lines through a key system. (Property of AT&T Archives. Reprinted with permission of AT&T)

person who is speaking and gives the person a reference point for determining whether he or she is speaking too loudly or too softly.

Telephone Switchhook

The telephone *switchhook* derives its name from old-style telephones that had a hook on the side on which the handset was hung. In many telephones, the switchhook consists of the buttons that are depressed when the handset is placed in the cradle of the telephone. These buttons might better be called just "the switch."

When a user lifts the handset, the switch is closed, and electrical current flows to the telephone company's central office. This current signals the central office that a call is about to be made. This condition is called *off-hook.* The central office equipment responds with a *dial tone,* which lets the caller know that the central office is ready to accept the call. At this point, the line is said to be *seized.*

off-hook

At the end of a call, replacing the handset in its cradle depresses the switchhook buttons, opens the circuit, and signals the central office that the call is complete. This condition is called *on-hook.* When the caller hangs up the phone, the on-hook signal causes the creation of a data record in the central office computer that is used for billing the customer at the end of the month.

on-hook

MAKING A TELEPHONE CALL

Now we will look at the actions that take place when we make a telephone call. Some are obvious and visible to the caller, but many occur behind the scenes in the telephone company central offices.

"Dialing" a Number

Many years ago, when someone lifted the handset on a telephone, the response from the telephone company central office was not a dial tone but instead the voice of a human operator asking "Number, please?" In those days, the caller spoke the number he or she wished to call and the operator made the connection manually. As the volume of telephone calls grew, telephone company forecasts showed that many additional operators would be required. In fact, the forecasts showed that all young women leaving high school would need to be employed as operators just to handle the projected volume of telephone calls! It was evident that the process of completing the call had to be automated.

automating call connections

Rotary dials were added to telephones, which transferred the manual aspect of making the call to the caller. In more recent years, rotary dials have largely been replaced by numerical pads also known as dual-tone-multifrequency tone generators. Both the *dial pulsing* and dual-tone-multifrequency methods of signaling the central office the number of the telephone to be called will be discussed.

Dial Pulsing The *rotary dial* shown in Figure 5–4 generates pulses on the telephone line by opening and closing an electrical circuit when the dial is turned and released. The number of pulses is determined by how far the dial is turned. When the dial is released, the pulses are generated as a spring rotates the dial back to the resting position. The pulses are generated at the rate of 10 pulses per second. Each pulse is 1/20 of a second long, with a 1/20 of a second pause between pulses. This process is called *out-pulsing.*

out-pulsing

The rotary dial mechanism has largely been replaced by an integrated electronic circuit. This allows the telephone to have a push-button keypad, but the signaling is still done by electrical pulses that are generated

Figure 5–4
A diagram of a telephone's rotary dial.

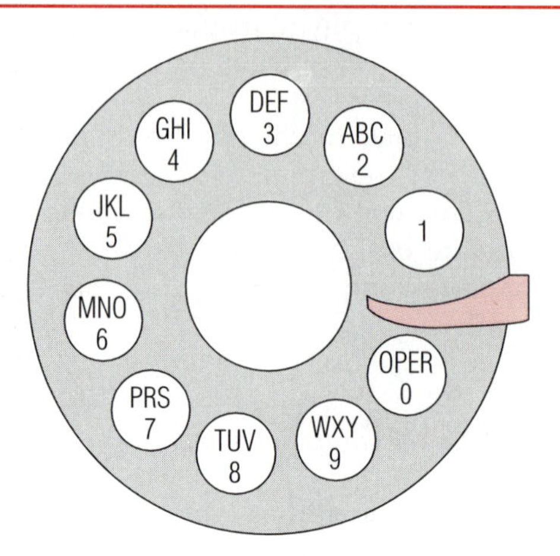

by the electronic circuitry. Since people can push the buttons faster than the pulses can be sent on the telephone lines, this type of telephone must be equipped with an electronic buffer to store digits that have been keyed but not yet out-pulsed.

Worldwide, many telephones still use the rotary dial and dial pulse technique for signaling the desired number, but it has largely been replaced in the United States and many other countries by the dual-tone-multifrequency method.

Dual-Tone-Multifrequency (DTMF) The newer technique for signaling the desired number to be called is called *dual-tone-multifrequency (DTMF)*. It is accomplished by sending tones on the telephone line. In its most common implementation, the telephone is equipped with a 12-button keypad. When a button is pressed, electrical contacts are closed that cause two oscillators to generate two tones at specified frequencies, much like pressing two keys on a piano at the same time. The combined tones are the signal for one of the digits. Figure 5–5 shows the combinations of tones generated for each key on the telephone keypad. The frequencies were carefully selected to be different from other tones or signals on the telephone line. To be accepted by the central office, the dial tones must have a duration of at least 50 milliseconds.

Tone dialing is considerably faster than pulse dialing. For example, dialing the number zero with tones takes 50 milliseconds for the tones plus 50 milliseconds between tones for a total of 100 milliseconds or 1/10 of a second. Using the dial pulse technique at 10 pulses per second, it takes 1 second to dial the digit zero. The digit 1 takes the same length of time for either pulse or tone dialing. The average telephone number can be dialed 10 to 15 times faster with tone dialing than with rotary pulses. Tone dialing is replacing pulse dialing because it is faster and because it

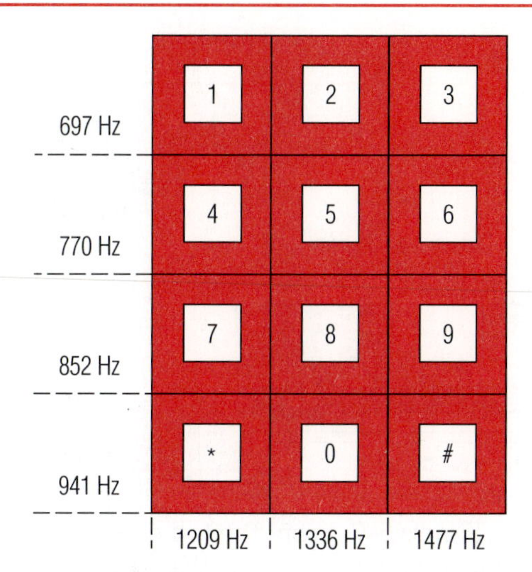

Figure 5–5
DTMF (Touchtone) pad. When the digit 6 is pressed, for example, tones of frequencies 770 Hz and 1477 Hz are sent on the telephone line.

is generated electronically rather than mechanically and is therefore more reliable. The tradename for DTMF dialing is *Touchtone.*

Making the Connection

Figure 5–6 shows the equipment and the facilities involved when a call is made. The connection between the telephone in a residence or business and the local telephone company's central office is called the *local loop.* The building that houses the telephone equipment for a specific geographical area is called the *central office.* When a caller lifts the telephone handset from the cradle, an electrical signal is sent to the central office, signaling it that the person is placing a call. The central office responds by sending the dial tone back on the local loop to the telephone. Assuming things are working normally, the person hears the tone by the time the telephone handset reaches his or her ear.

When the caller dials the number, the number is stored in the telephone switching equipment in the central office. The switching equipment is usually a computer, but in older offices it might still be an electromechanical device. In either case, it is called the *central office switch.* The first three digits of the telephone number determine whether the call is local or long distance. If it is local, the switch determines whether it can complete the call by itself or it needs to forward the call to another nearby central office that handles that telephone number. If the switch can complete the call (because the number being called is also handled through the same central office), it does so, as described below.

If the caller's central office determines from the first three digits of the number that this is a long distance call, it passes the call on to a communication line called a toll trunk to another central office called the toll

Figure 5–6
Diagram of the components of a telephone call.

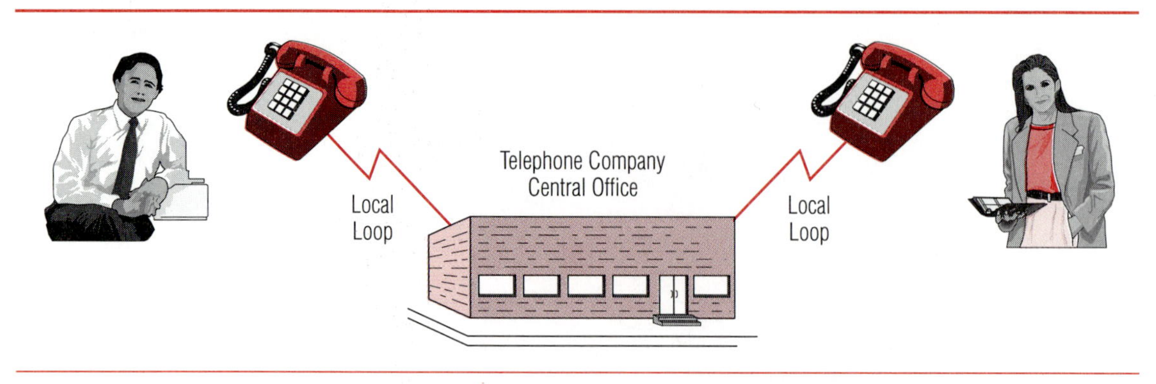

office, as shown in Figure 5–7. From the toll office the call may be passed directly to the central office that handles the receiving telephone or to another toll office. Ultimately a toll office passes the call to the central office that handles the receiving telephone. It is impressive to realize that all of these steps take place in a matter of seconds.

Ringing

When a call has been dialed and a connection through the central office of a telephone network has been made, the telephone company's central office that handles the receiver's telephone indicates the incoming call to the receiving telephone by sending a ringing signal (voltage) to it. This voltage may activate a bell, an electronic ringer, a light, or other device. In the United States and Europe, the ringing signal is sent for 2 seconds with a 4-second pause between rings. Other countries have different timings.

At the same time the ringing signal is being sent to the called telephone, an audio ringing signal is sent to the calling telephone. This signal lets the caller know that the telephone company has completed the call connection process and that the called telephone is ringing. The two ringing signals are generated independently, however, and occasionally the called telephone will be answered before the ringing signal back to the caller has been generated. You may have been surprised by this situation yourself when someone you were calling seemingly answered the call "before it rang."

When the called party picks up the telephone (goes off-hook), a signal is sent to the central office that tells it to stop sending the ringing signal to both the caller and receiver.

tracking call costs

In most cases, it also signals the telephone company to begin charging for the call, although some carriers use other methods to determine when the call begins. Among the long distance carriers, AT&T uses this signal to time all of the long distance calls it handles. By doing so, it mea-

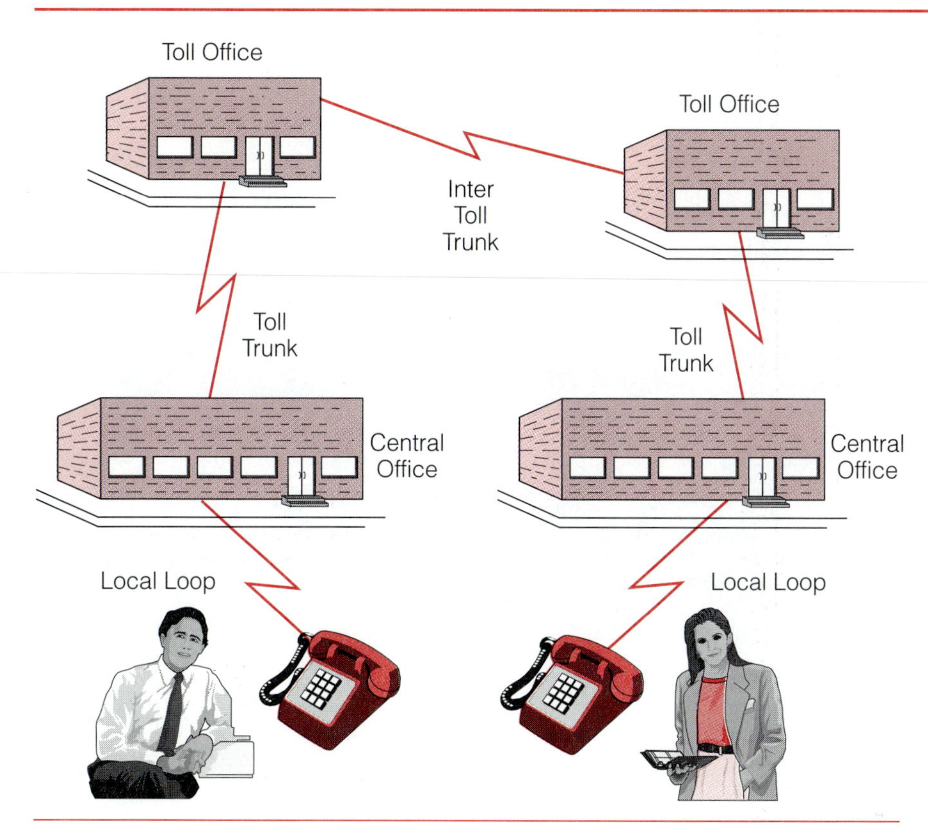

Figure 5–7
Diagram of a telephone call involving toll trunks and toll offices.

sures and charges only for the duration of the conversation. Some of the other carriers include the time to connect the call, which is called *call setup time,* in their charges. Still other carriers subtract a predetermined number of seconds to compensate for an average call setup time, and others measure noise on the line and begin charging for the call when talking begins.

call setup time

CENTRAL OFFICE EQUIPMENT

For a short time in the early days of telephones, there was no central office or switching equipment. As described earlier, all telephones had to be connected directly to each other or they could not communicate. Clearly this situation was cumbersome and neither economical nor manageable. Switching (connecting) equipment quickly evolved. The concept is quite simple: All telephones are connected to a central office switch, and the central office switches are connected with one another as illustrated in Figure 5–8.

The early manual telephone switchboard was installed in New York City in 1888. (Property of AT&T Archives. Reprinted with permission of AT&T)

Figure 5–8
It is more efficient to connect telephones to a central office switch and then connect the central office switches together than to connect all telephones to each other.

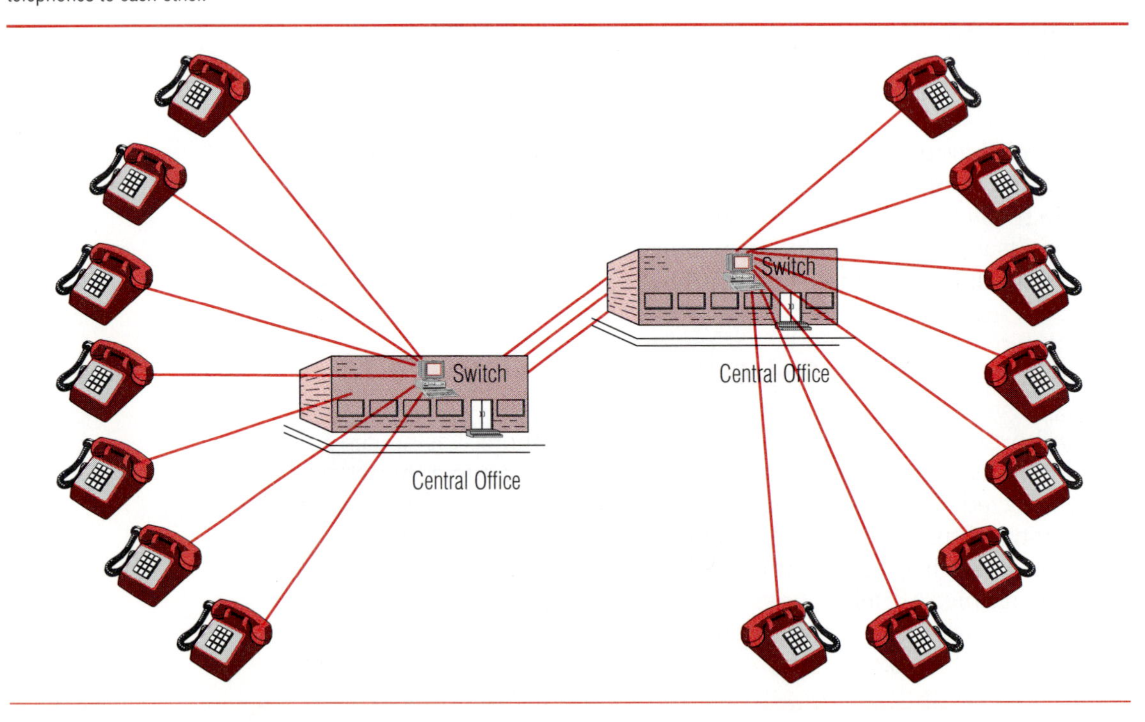

Manual Switching

The earliest type of switching equipment had a central console where all telephone lines terminated in jacks similar to the headphone jack on a stereo receiver. When a caller went off-hook, a light on the console lit, and the operator plugged her headset into the jack and talked to the caller. She got the number to be called and plugged a *patch cord,* a cable with plugs on both ends, into the caller's and receiver's jacks to connect the two. Special jacks were available so that the operator could connect her switchboard to other operator's switchboards and to special lines that went to other cities for long distance calls.

patch cord

Automatic Switching

From the 1920s (when telephones with dials began to replace older style telephones) until today, the connection of telephones at the central office has been increasingly automated. The equipment used has progressed through a series of technologies from *step-by-step switches* (also known as *Strowger switches* after their inventor, Almon B. Strowger), to *crossbar switches* and *reed relays.* Each of these switches was electromechanical and, therefore, more prone to failure than today's electronic switching equipment.

Electronic Switching

With the development of the transistor and the evolution of the integrated circuit, telephone switching equipment has moved rapidly from the electromechanical to the electronic age. Modern central office switches are entirely electronic, and the connection of calls is completed much more rapidly than before. Since the switches are electronic, they are much less prone to hardware failure than their electromechanical predecessors. However, since they are really programmable computers, they are susceptible to program bugs. Central office switches are designed to be highly redundant so that the failure of any portion of the equipment can be circumvented. The most sophisticated central offices are designed to have no more than one total failure in 40 years.

rare central office failures

Electronic central office switches handle thousands of lines. Compared to switches implemented with older technologies, such as step-by-step switches, they are physically much smaller by thousands of square feet. They are also significantly quieter because there are no relay contacts opening and closing or crossbars moving in frames.

As central offices have evolved to electronic switches, the staffing requirement in the central office has dropped dramatically. The main reason is that the newer equipment is more reliable, and fewer maintenance people are required to keep it operating. Another reason is that much of the electronic equipment can be tested remotely and spare equipment switched into the system if necessary. At the same time, the education and skill levels of the central office work force have increased because the newer equipment is more sophisticated. Diagnosing the problems that do

staff costs reduced

A modern electronic central office switch is a specialized digital computer and is considerably more compact and reliable than earlier mechanical switches. (Property of AT&T Archives. Reprinted with permission of AT&T)

occur requires sensitive electronic test equipment and a knowledge of how to use it to isolate the failure.

Design Considerations

Blocking When telephone facilities are designed to provide good service, designers must make a trade-off between the amount and cost of central office switching equipment and the level of service it provides. It is not necessary to include enough equipment or lines to handle simultaneous calls from every telephone connected to a central office. In normal circumstances, only a small percentage of the telephones are in use at any one time. Therefore, the central office can be designed to handle a fraction of the maximum theoretical number of simultaneous calls. If, however, all of the central office equipment is in use and one more person tries to make a call, that call is said to be blocked because there are no central office facilities available to handle it. The call cannot be completed even though the called number's telephone is not in use. *Blocking* can also occur when all of the lines connecting the central offices are in use. In that case, the central office handling the caller's telephone may be able to handle the call, but it is unable to make a connection to the central office handling the receiver's telephone. Locations where blocking can occur are shown in Figure 5–9.

blocking

In the U.S. telephone system, blocking is rare. When it does occur, the condition is signaled to the caller by either a fast busy signal or a voice recording. The most common occurrences of blocking are when all lines are busy on major holidays, such as Thanksgiving or Mother's Day.

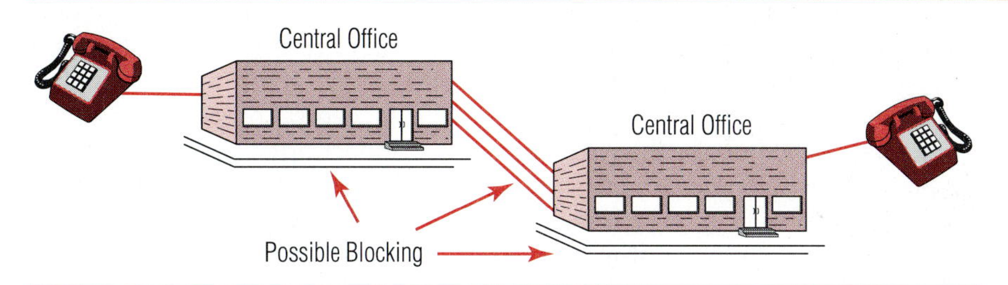

Figure 5–9
Blocking can occur at any central office or on the lines connecting them.

Busy Hour When deciding how much equipment and how many lines to install, the usual procedure is to study or forecast how many calls will occur during the busiest hour of the day. In businesses in the United States, the *busy hour* usually occurs between 10 and 11 A.M., and sometimes another peak occurs between 2 and 3 P.M. By way of contrast, in Japan, the busy hour is between 9 and 10 A.M. Most offices in Japan open at 9 A.M., and Japanese people typically get on the phone as soon as they come to work. Obviously, if enough equipment and lines are installed so that all calls can be handled during the busy hour, there are sufficient facilities available to handle calls made during other hours of the day.

A study of the number of calls to be handled during the busy hour must also take into account their durations. Clearly, eight 3-minute calls tie up more equipment and lines than do eight 1-minute calls. Three 8-minute calls tie up a different mix of equipment than either of the other situations.

For cost reasons, the equipment and lines in a telephone system frequently are designed to handle less traffic than occurs during the busiest hour. When a detailed analysis is performed, the results often show that there is a significant savings in equipment costs by designing the facilities to handle 99, 98, or even 95 percent of the busy hour traffic. This means that 1, 2, or 5 percent of the calls will be blocked, and the caller will have to try again. For most telephone systems, some level of blockage is tolerable.

The study of telephone traffic patterns involves probability and statistics and has grown to be a very rigorous mathematical discipline, although many of the critical calculations have been reduced to tables. The *grade of service* is the proportion of blocked calls to attempted calls expressed as a percentage. If 5 calls of 100 are blocked, the grade of service would be 5/100 or 5 percent. This is designated as a P.05 grade of service. Most public telephone facilities are designed to provide at least a P.01 grade of service, meaning that only 1 call out of 100 would be blocked.

grade of service

CENTRAL OFFICE ORGANIZATION

Each central office serves all of the telephones within a specific geographical area. The size of the area served depends on the density of the telephones. In rural areas, the central office might serve many square miles, whereas in New York City many central offices are needed to handle all of the office buildings and residences.

In the United States, there are about 25,000 central offices. The central office acts as the hub for the wires and cable connecting all of the telephones it serves. Wiring comes together in a rack called a *main distribution frame* in the central office from which it connects to the switching equipment.

Central offices are optimized to perform specific functions. Some are primarily designed to switch local telephone calls from and to businesses and residences. This type of central office is commonly known as an *end office.* The central office to which a specific telephone is connected is known as that telephone's *serving central office.*

toll offices

Other central offices, called *toll offices* or *switching offices,* are designed primarily for forwarding long distance calls to other parts of the country. Offices are connected in a weblike pattern that provides a high level of redundancy and alternate routing, as shown in Figure 5–10. Any central office may be connected to any other central office by direct lines, but a major consideration as to whether to connect two offices is the amount of traffic (number of telephone calls) that flows between them. In some cases, several lines are needed to carry the traffic between offices; in other cases, a direct connection is not necessary, and traffic is routed in a less direct fashion.

A call is handled at the lowest level office that can complete it or provide the required service. If a call cannot be handled by the serving central office that first received it, the call is forwarded to the most appropriate central office. The routing is determined by tables stored in the memory of the central office switch. For example, a call from a business in Dallas going to New York City might be routed from the serving central office in Dallas to a toll center serving part of Dallas. The toll center's tables might tell it that calls for New York should be routed to a certain toll office in Atlanta or, if Atlanta is busy, to Saint Louis. The objective of the network is to route the calls to the destination central office in the shortest, fastest way possible.

THE PUBLIC TELEPHONE NETWORK

The *public telephone network,* sometimes called the *public switched telephone network (PSTN),* consists of many distinct pieces. Telephones in a home or business are most commonly connected to their serving central office by a pair of copper wires called the *local loop.* In telephone company jargon, these two wires are often referred to as *tip and ring.* In a residential neighborhood, the wire coming from the house, called the *drop wire,* runs to a pole (or underground equivalent) where it joins other similar wires to form a *distribution cable.* Eventually, several distribution cables join to-

local loop

Figure 5–10
The connection of the central offices in the nationwide telephone network.

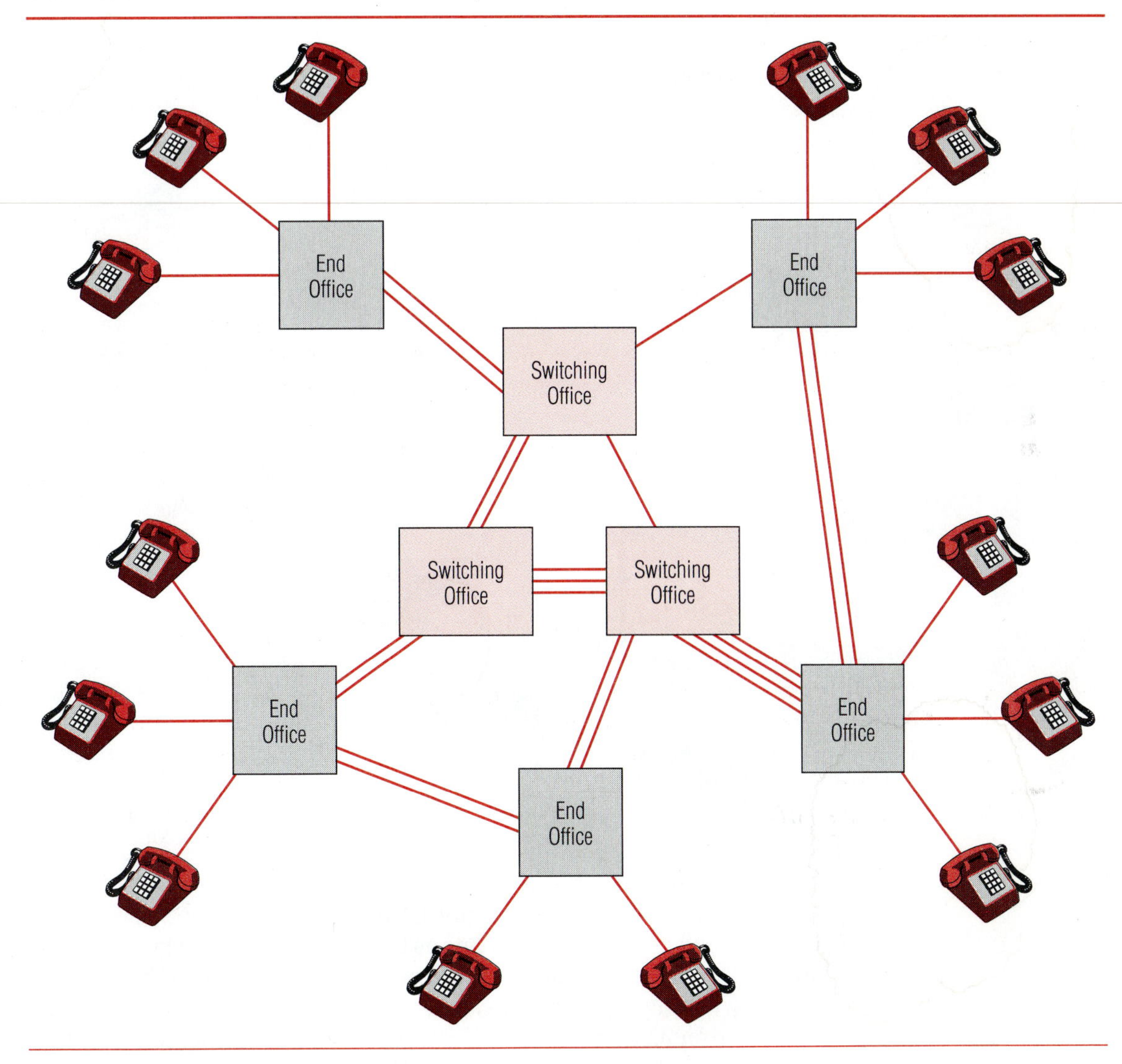

gether to form a *feeder cable* that terminates at the central office, as shown in Figure 5–11. Clearly, the cable gets physically larger the closer it gets to the central office. Cables may run above ground on poles or underground, but most new installations being made today are underground where they are better protected and out of sight.

trunk

Central offices are connected to each other with multiple lines called *trunks*. A trunk is defined as a circuit connecting telephone switches or

Figure 5–11
Residential telephone cabling.

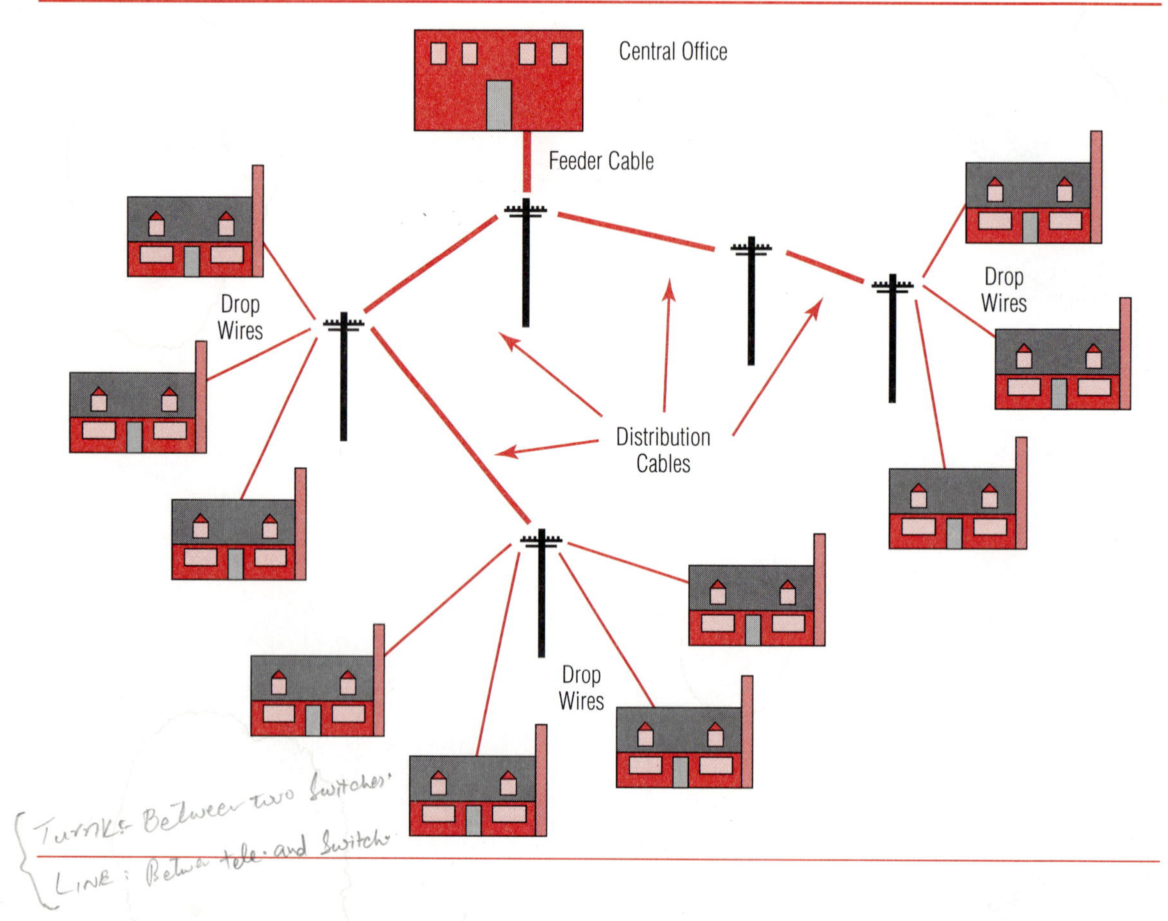

switching locations. (The term *circuit* is defined in Chapter 9, but in the meantime, think of a circuit and a line as being synonymous.) Functionally, a trunk and a line do the same thing—carry communications—but a trunk connects switching equipment, whereas a line connects to a telephone, computer terminal, or other device. Trunks may be implemented with regular copper wire but are more often implemented with coaxial cable, microwave radio, or fiber-optic cable.

The PSTN, with its millions of miles of lines, handles virtually all of the voice and data communications in the world. It is diverse and highly redundant; therefore, it is impervious to massive outages. It is resilient when failures do occur. Most countries consider their public communications network to be a national asset and take steps, including regulation, to protect it.

In the United States, the PSTN is facing a new challenge. Local telephone networks have been engineered under the assumption that the average telephone call lasts about 3 minutes. When users of the Internet or other online services dial up from their computers, they stay on the line an average of 20 minutes, and sometimes for hours. With those local lines tied up, the network and switching equipment can become congested, making it impossible for other people to make calls. Telephone companies and others are studying alternative solutions to this problem, including charging data users on a per-minute basis. One way or another, the problem will be solved, and over time the solution is likely to cost the data user more money.

ANALOG SIGNALS

When voice is converted to an electrical signal through a microphone, it provides a continuously varying electrical wave like the one shown in Figure 5–2. This electrical wave matches the pressure pattern of the sound that created it. The wave is called an *analog signal* because it is analogous to the continually varying sound waves created when sound is generated by speech or other means. (Another type of signal frequently used to transmit voice or data signals is a digital signal, which you will learn about in Chapter 8.) In order to understand the characteristics of the telephone network, you need to understand something about the characteristics of analog signals.

Signal Frequency

The sound waves we generate when we speak and the electrical waves that result after the sound has been converted for transmission have many common characteristics. One attribute is their *frequency,* which for sound waves is the number of vibrations per second that cause the particular sound. If you strike the A key above middle C on a piano keyboard, you generate a very pure tone that is created by the A string on the piano vibrating back and forth 440 times per second. If you held your telephone up to the piano and struck the A key, the microphone in the telephone handset would convert the 440 vibrations per second to an electrical signal on the telephone line that also changes 440 times per second. This signal is commonly diagrammed as a *sine wave,* as shown in Figure 5–12. Each complete wave is called a *cycle,* and the frequency of the signal is the number of cycles that occur in 1 second. The unit of measure for frequency is the *Hertz,* abbreviated *Hz.* We say that the A key we struck on the piano generated a tone with a frequency of 440 Hz. The corresponding (analogous) electrical signal also has a frequency of 440 Hz. By way of comparison, a higher tone on the piano has a higher frequency. The A key that is an octave above A 440 has a frequency of 880 Hz. The lowest note on the normal piano keyboard has a frequency of 27.5 Hz, and the highest note has a frequency of 4,186 Hz.

sine wave

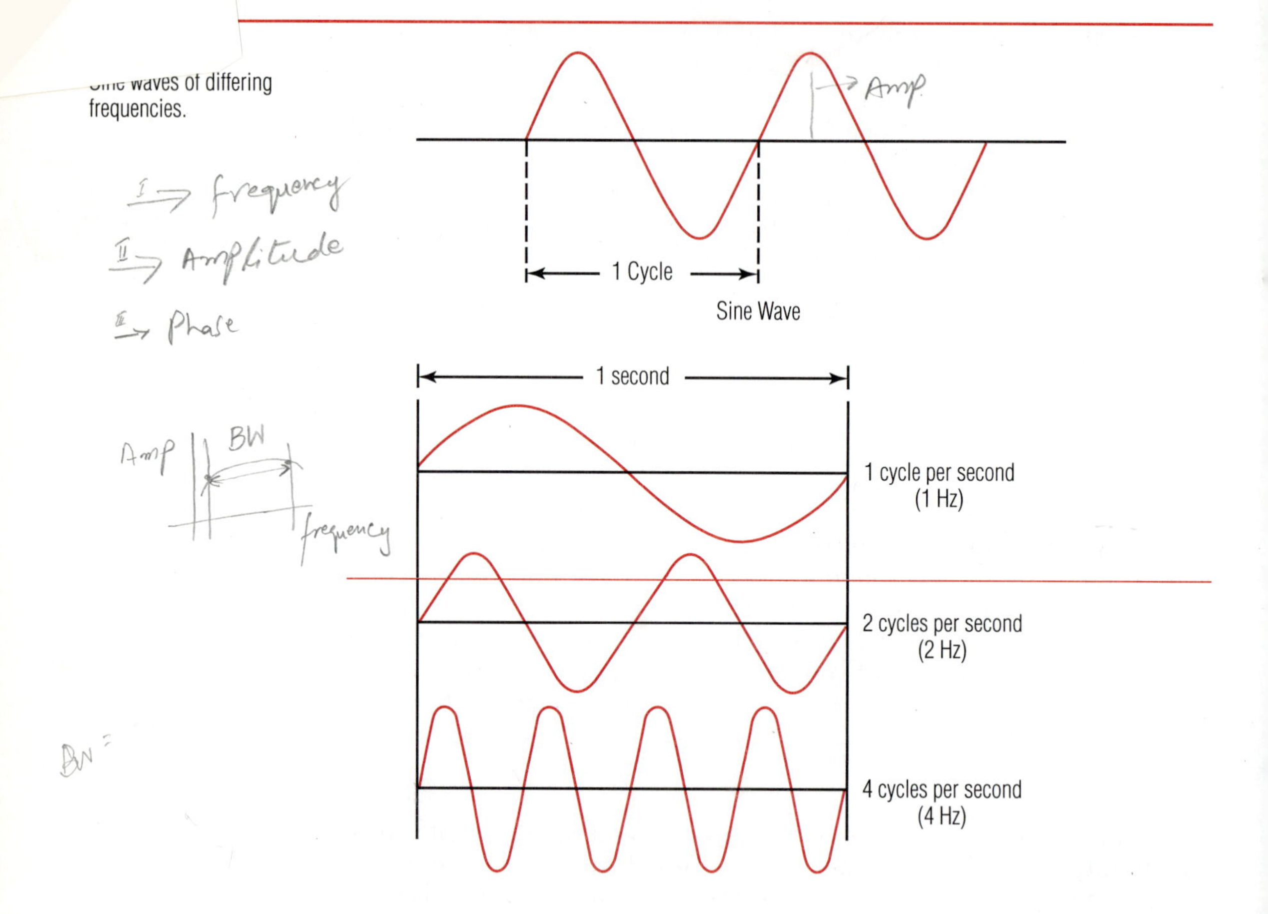

...ine waves of differing frequencies.

The human ear can hear sound with a range of frequencies from about 20 Hz to approximately 15,000 Hz. Between 15,000 and 20,000 Hz, most people can sense the sound but not actually hear it. Good stereo systems reproduce sounds up to approximately 20,000 Hz and have better fidelity (sound better) than systems that do not reproduce frequencies that high. Figure 5–13 illustrates some of these frequency ranges.

Sound waves, electrical waves traveling in a wire, and electromagnetic waves traveling through space, such as radio waves, have essentially the same characteristics. All are represented as sine waves whose frequency is measured in Hertz. Figures 5–14 and 5–15 show the frequency spectrum, the full range of frequencies from zero Hertz to several hundred thousand million Hertz. When referring to very high frequencies, we commonly use the designations *kilohertz (kHz)*, *megahertz (MHz)*, and *gigahertz (gHz)* to more easily describe the frequencies. Figure 5–16 shows the full range of abbreviations used for very large and small units of measure in the scientific world.

Figure 5–13
The frequency ranges of some common sounds.

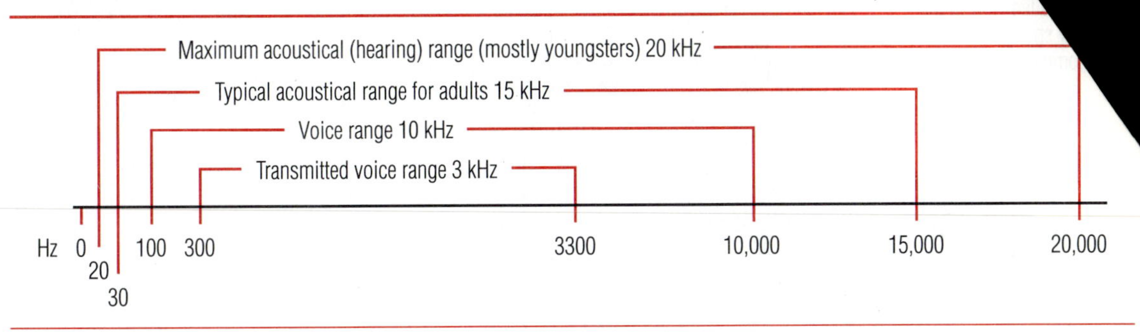

Figure 5–14
The frequency spectrum showing the common names applied to certain frequency ranges.

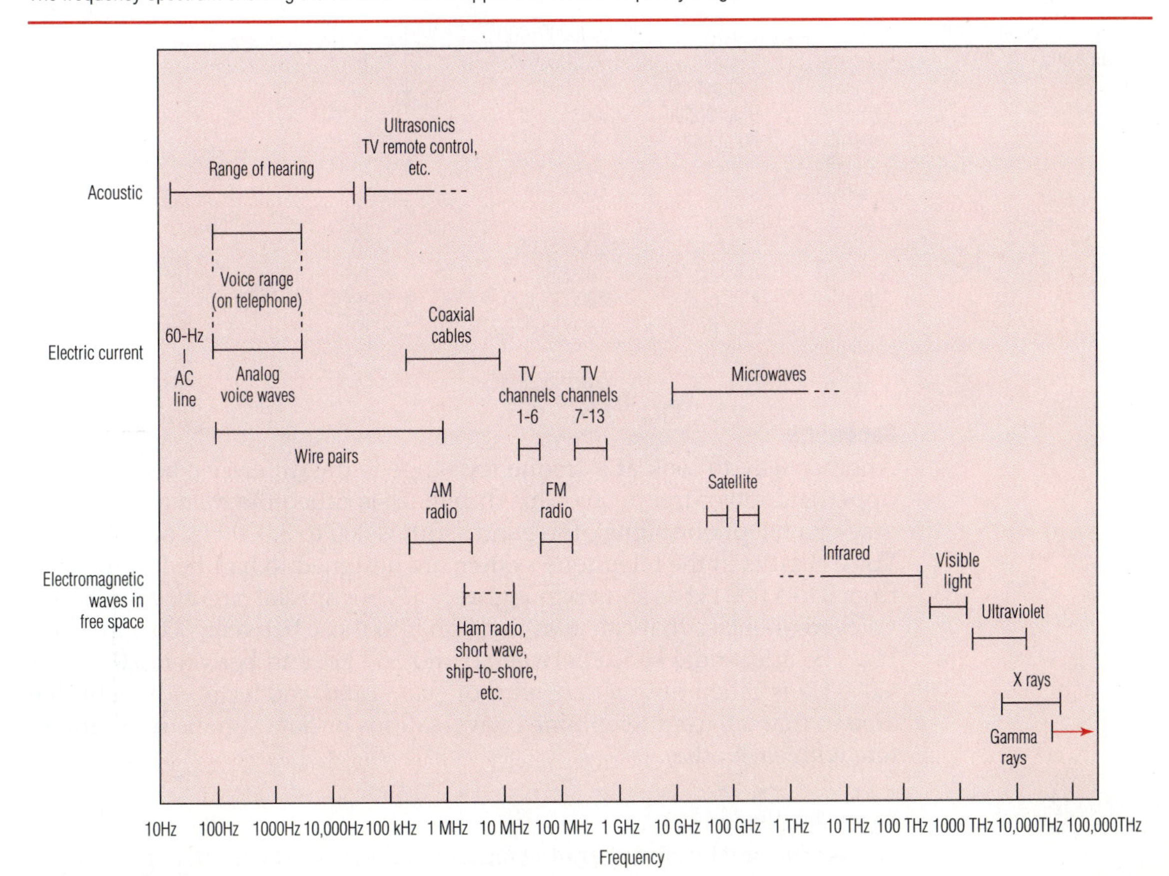

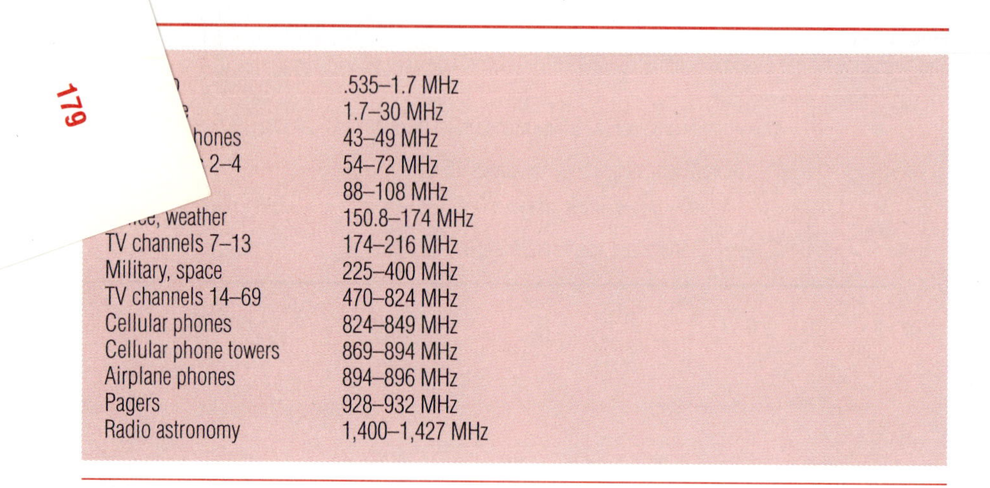

[illegible]	.535–1.7 MHz
[illegible]	1.7–30 MHz
[illegible]hones	43–49 MHz
[illegible] 2–4	54–72 MHz
[illegible]	88–108 MHz
[illegible], weather	150.8–174 MHz
TV channels 7–13	174–216 MHz
Military, space	225–400 MHz
TV channels 14–69	470–824 MHz
Cellular phones	824–849 MHz
Cellular phone towers	869–894 MHz
Airplane phones	894–896 MHz
Pagers	928–932 MHz
Radio astronomy	1,400–1,427 MHz

Figure 5–16
Common abbreviations for very large and very small quantities.

pico	trillionth	.000000000001	1×10^{-12}
nano	billionth	.000000001	1×10^{-9}
micro	millionth	.000001	1×10^{-6}
milli	thousandth	.001	1×10^{-3}
centi	hundredth	.01	1×10^{-2}
deci	tenth	.1	1×10^{-1}
deca	ten	10	1×10^{1}
centa	hundred	100	1×10^{2}
kilo	thousand	1000	1×10^{3}
mega	million	1000000	1×10^{6}
giga	billion	1000000000	1×10^{9}
tera	trillion	1000000000000	1×10^{12}

Bandwidth

Another way to look at a frequency range is the difference between the upper and lower frequency. This difference is called the *bandwidth.* In the case of a telephone signal, the bandwidth is 300 to 3,000 Hz, or 2,700 Hz. Voice circuits in the telephone system are designed to handle frequencies from 0 to 4,000 Hz as shown in Figure 5–17, but special circuitry limits the voice frequencies that can pass through it to those between 300 and 3,000 Hz. The additional space between 0 and 300 Hz and between 3,000 and 4,000 Hz is called the *guard channel* or *guard band,* and it provides a buffer area so that adjacent telephone conversations or data signals don't interfere with each other.

guard band

Signal Amplitude

Another characteristic of analog signals is their loudness, or *amplitude.* As you speak more loudly or softly into the telephone, the sound waves cre-

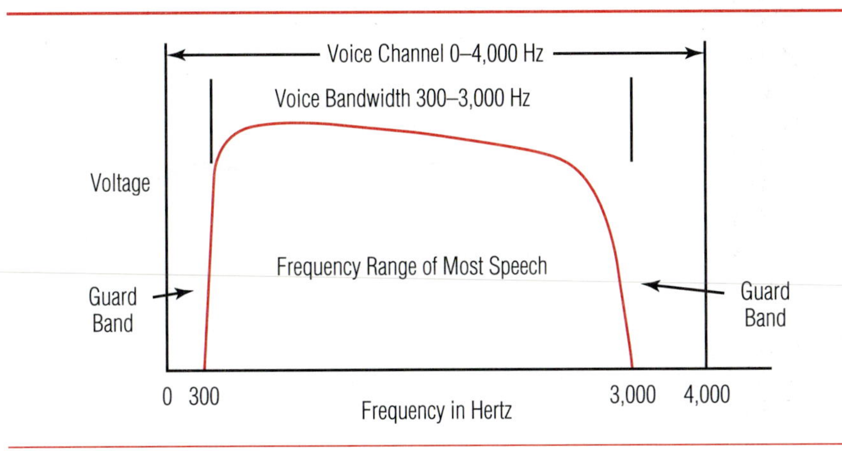

Figure 5–17
Bandwidth of a voice channel.

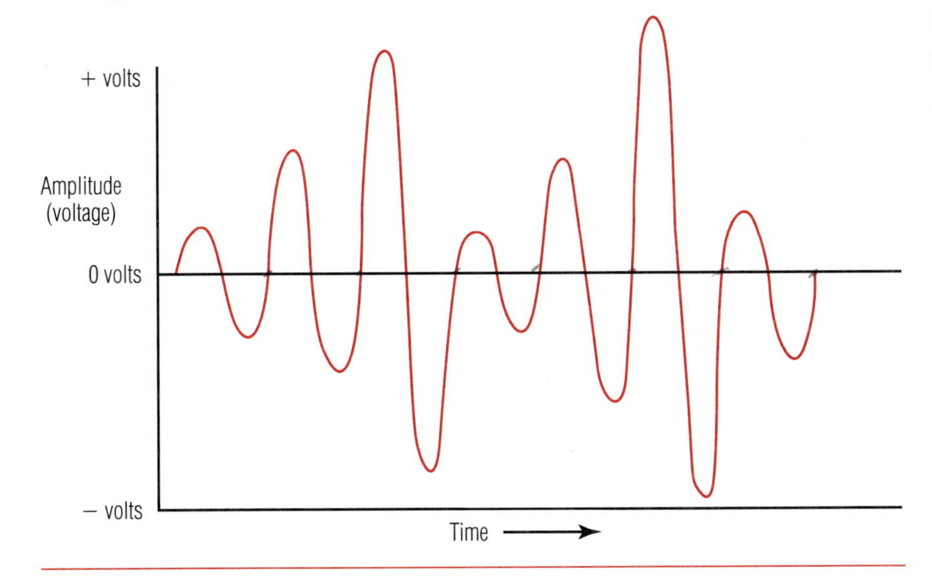

Figure 5–18
Analog wave with constant frequency and varying amplitude.

ate larger and smaller electrical waves that are represented by higher peaks and valleys of the signal's voltage (shown in Figure 5–18). The amplitude of the signal is also called its *level,* and whereas with sound the amplitude relates to loudness, in an electrical signal the amplitude is the difference between its most negative voltage (the lowest point in the sine wave) and its most positive voltage (the highest point in the sine wave).

Analog signal level is measured in *decibels (dB),* which is a logarithmic ratio of signal input and output power. Because the dB is a logarithmic measure, doubling the strength of a signal increases its level by 3 dB.

decibels

Figure 5–19
The relative power of a signal measured in decibels.

Decibels	Relative Power
+30 dB	1000
+20	100
+10	10
+3	2
0 dB	1
−3 dB	1/2
−10	1/10
−20	1/100
−30	1/1000

This is true regardless of the signal's original strength. If we say that a signal increased by 3 dB, we mean it doubled in strength, without knowing what the original or new signal strengths are. In the same way, increasing its strength by a factor of 10 raises its level by 10 dB, by a factor of 100, 20 decibels, and so on. Working in the other direction, we find that reducing the signal to 1/2 of its former level causes the strength to be measured as −3 dB; 1/10 of the power is −10 dB; 1/100 of the power is −20 dB, and so on. Figure 5–19 shows these values.

For electrical telecommunications signals, 0 dB is defined as 1 milliwatt of power. An increase of the power to 2 milliwatts is a doubling of the power, and the signal would therefore have a relative strength of +3 dB. Doubling the power again to 4 milliwatts would yield a signal with a strength of +6 dB. The mathematical formula for the relationship between power and signal strength is

$$\text{dB} = 10\log(\text{Power out}/\text{Power in})$$

The quotient of power out to power in is a mathematical way to show the number of times the power was increased. If the power out is 30 watts and the power in was 10 watts, then the power was increased three times and the multiplier used in the formula would be 3. So the formula may be expressed in words as: The decibels = ten times the logarithm of the power increase. By the way, the formula works if the power is decreased too. If the power is reduced from 5 milliwatts to 2 milliwatts, then the multiplier used in the formula would be .4. The reason we need to use logarithms in this equation is that signals traveling through a medium weaken *exponentially* fast. Logarithms relate to exponents and therefore accurately reflect the signal's behavior through a transmission medium.

Decibels and the strength of a signal are of considerable interest in telecommunications. If too much power is put on a line, a particular type of interference called *crosstalk* (which is discussed in Chapter 9) can occur. The loss of signal strength is also of interest because if a signal does not

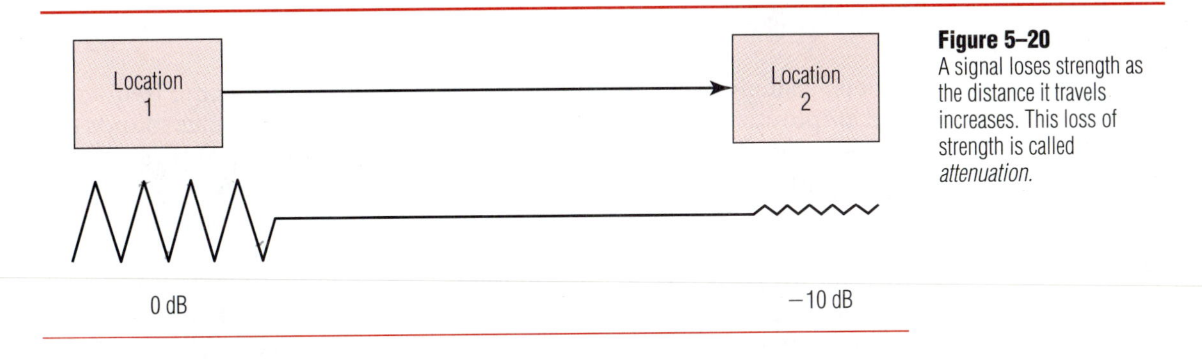

Figure 5–20
A signal loses strength as the distance it travels increases. This loss of strength is called *attenuation.*

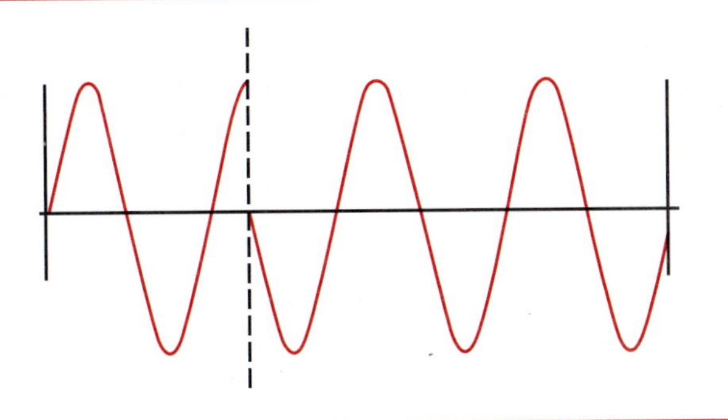

Figure 5–21
Example of a phase shift.

have enough strength at the receiving end of a communication path, it will be unusable. This loss, which is called *attenuation,* is measured between two points on a line as shown in Figure 5–20. At the point where the signal is injected on the line, it has a certain strength. As the signal moves away, its strength is reduced due to the attenuation of the line. The reduction is measured in decibels.

attenuation

Signal Phase

A third attribute of an analog signal is called its *phase.* In contrast to frequency and amplitude, phase is harder to relate to the physical world and, therefore, is somewhat harder to understand. Sine waves can be measured in degrees, where 360° is one complete cycle of the wave. A signal's phase is the relative position of the sine wave measured in degrees. Figure 5–21 shows a sine wave that appears to break and start again, skipping a portion of the wave. This is a phase shift. Since 1/4 of the wave has been skipped, it is called a 90° phase shift. Phase shifts are created and detected by electronic circuitry.

While amplitude and frequency changes can be detected by the human ear, phase changes cannot, and they therefore are of little importance in voice transmission. They are very important in data transmission, however, and Chapter 8 discusses phase shifting in more detail.

ATTRIBUTES OF A VOICE SIGNAL

Whereas single tones produce clean sine waves of a specific frequency and amplitude, the human voice, music, noise, and most other sounds are made up of a large range of frequencies and amplitudes. As a result, the wave pattern is far more complex than the simple sine waves we have looked at thus far. Normal speech is made up of sounds with frequencies in the range of 100 to 6,000 Hz, but most of the speech "energy" falls in the 300 to 3,000 Hz range. Although some people with high-pitched voices emit occasional sounds above 6,000 Hz, the majority of the sound still falls in the range of 300 to 3,000 Hz. That is why the public telephone system is designed so that all of the lines, handsets, and other components will pass voice frequencies in that range. Frequencies outside that range are filtered out by electronic circuitry and are not allowed to pass.

FREQUENCY DIVISION MULTIPLEXING (FDM)

While the individual telephone circuit has a bandwidth of 4,000 Hz, the pair of wires or other media carrying it has a much higher bandwidth capacity. Twisted-pair wires have a bandwidth of approximately 1 million Hz. Dividing 4,000 Hz into 1 million Hz shows us that, at least theoretically, a standard pair of telephone wires should be able to carry approximately 250 telephone conversations. This is a very theoretical number; in practice 12 or 24 voice signals normally are carried. Naturally, the telephone company would like to take advantage of the ability to carry multiple conversations on one line, especially on trunk lines between central offices in which hundreds of telephone calls are handled simultaneously.

The technique of packing several analog signals (phone calls in this case) onto a single wire (or other media) is called *frequency division multiplexing (FDM).* It is accomplished by translating each voice channel to a different part of the frequency spectrum that the media can carry. Using the telephone wire pair as an example, if a second voice signal could be relocated from its natural frequency of 0 to 4,000 Hz to, say, 4,000 to 8,000 Hz, and a third voice signal relocated to 8,000 to 12,000 Hz (as shown in Figure 5–22), many telephone conversations could be packed on one pair of wires.

MODULATION

Frequency division multiplexing is accomplished by transmitting a sine wave signal in the new frequency range in which the original signal is to be relocated. The new sine wave is called a *carrier wave,* not to be confused with a *common carrier.* The carrier wave in itself contains no information, but its attributes are changed corresponding to the information in the

carrier wave

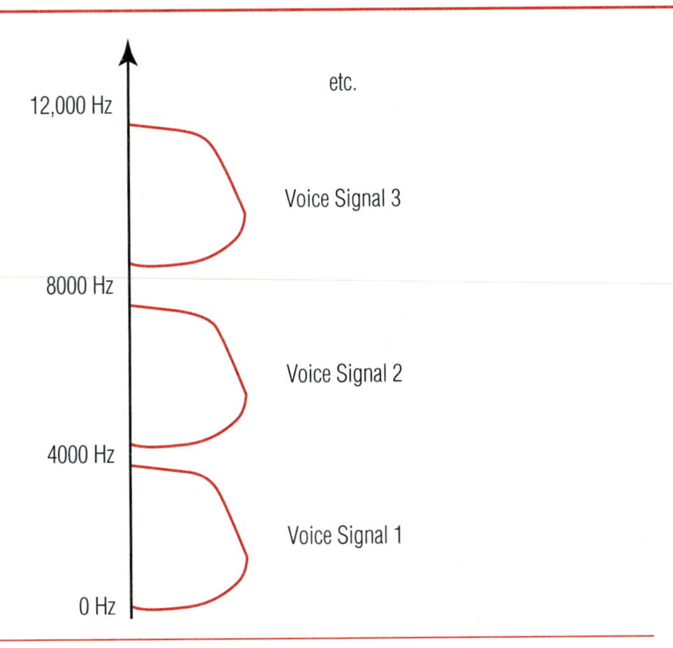

Figure 5–22
Frequency multiplexed voice signals.

original signal. This change to the carrier wave is called *modulation.* Modulation converts a communication signal from one form to another, more appropriate form for transmission over a particular medium between two locations.

AM, FM, and PM

You have learned that there are three attributes of the sine waves that can be changed. If the amplitude is changed, it is called *amplitude modulation (AM);* changing the frequency is called *frequency modulation (FM);* and changing the phase is called *phase modulation (PM).* Amplitude and frequency modulation are shown in Figure 5–23; phase modulation is discussed in Chapter 8. Combinations of these modulation techniques are also possible, for example, phase amplitude modulation (PAM).

Shifting the frequency of a signal to a different frequency range is one important use of modulation (we will look at the other use in Chapter 8), and the result is that the original signal is relocated to a different set of frequencies. At the receiving end, an electronic circuit called a *detector* must be able to unscramble the modulated signal and relocate it back to the original frequencies—a process called *demodulation.* In the telephone example, the modulation of the original 0 to 4,000 Hz voice signal occurs at the central office serving the person who is speaking, and the demodulation occurs at the serving central office near the listener.

multiplexers

Multiplexing equipment (multiplexers) at the central office pack groups of twelve 4 kHz voice signals into 48 kHz signals called *base groups, channel groups,* or just *groups.* Other multiplexers then pack five 48 kHz groups into *supergroups,* which have a bandwidth of 240 kHz and contain 60 voice

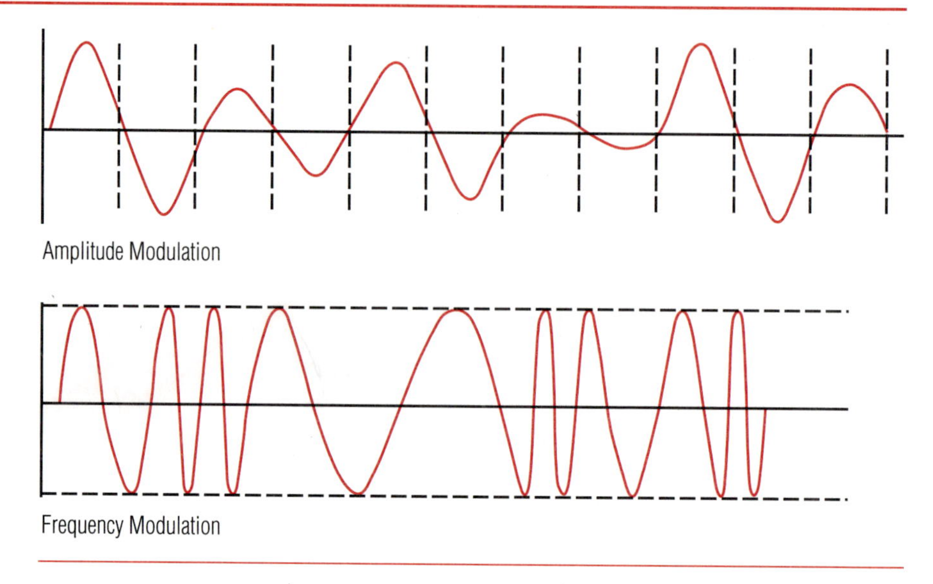

Figure 5–23
Amplitude and frequency modulation.

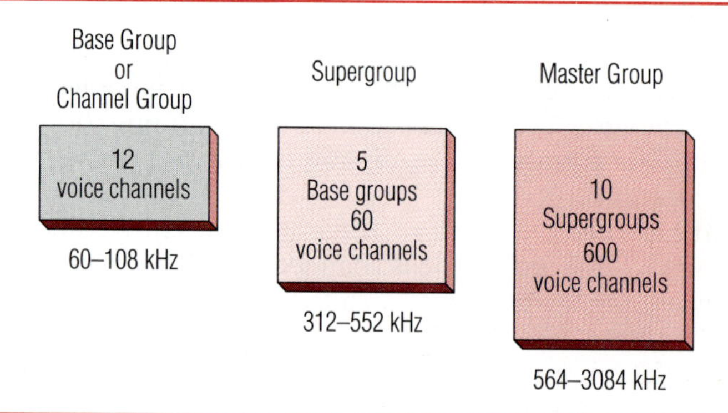

Figure 5–24
The hierarchy of voice channels as they are multiplexed together.

signals. Ten supergroups are multiplexed into *master groups,* or even larger *jumbo groups* for long distance transmission. You can see from Figure 5–24 that as it travels from transmitter to receiver, the individual 4 kHz voice signal is modulated to different frequencies several times by the FDM equipment of the telephone company. As a result, lines and trunks are used efficiently between telephone company offices where traffic volume is high.

Although the above discussion shows how the telephone companies can achieve efficiency through FDM techniques, it must be pointed out that they have changed virtually all of their high-speed trunks and interoffice communications to digital transmission, which yields even higher efficiencies. Digital techniques are of course applicable to data transmission as well, and so they will be covered in Chapter 8 with a specific focus on data. Suffice it to say for now that when a voice signal is digitized, it becomes a stream of bits that is indistinguishable from digitized data. As a result, digital voice and data signals can be multiplexed together on the same line, and the efficiencies are higher than with the analog techniques that have been discussed so far.

TASI VOICE TRANSMISSION

A technique that has long been used on undersea cables is called *time assignment speech interpolation (TASI)*. TASI is another way of packing multiple voice conversations onto a single telephone line. TASI takes advantage of the fact that there are pauses in speech and that approximately 10 percent of the time in a normal voice conversation, no one is speaking. TASI equipment detects when a person starts speaking and, within a few milliseconds, assigns a communication circuit to the speaker. Although a very small amount of the first syllable may be lost, it is almost undetectable by the listener. When the person stops talking, the communications circuit is taken away and assigned to another speaker. When the person starts talking again, a new circuit is assigned.

Using TASI, 100 talkers require only about 45 circuits to carry their conversations. Therefore, when the volume of calls is high, TASI provides a way to economize on the number of circuits required.

INTEROFFICE SIGNALING

Direct Current (DC) Signaling

In addition to carrying voice signals, the telephone lines must also carry various other kinds of signals used to set up a telephone call and indicate its status. One type of signaling occurs simply by opening or closing the electrical circuit between the telephone handset and the central office. This is called *direct current (DC) signaling*. DC signaling is used primarily between the serving central office and the customer. It is analogous to turning on a light switch that allows electrical current to flow to a light bulb. The electrical current required for DC signaling is generated by a power supply at the serving central office.

As used on the local loop, DC signaling works as follows. When the telephone handset is on-hook, the circuit is open, and no current can flow. When the handset is lifted, the circuit is closed, current flows to the central office, and it sends a dial tone (a tone signal) to the handset.

Another type of DC signaling is pulse dialing. When a digit is pulsed, the flow of current is interrupted by the pulsing mechanism (a rotary dial or its electronic equivalent), which opens and closes the circuit a certain number of times depending on which digit is being dialed.

Tone Signaling

Another type of signaling used in the telephone system is *tone signaling.* When you lift the handset, you hear a *dial tone* (assuming everything is working) that is a combination of a 350 Hz tone and a 440 Hz tone. On telephones with DTMF dialing, each button pressed creates a tone also made up of a combination of two frequencies. As the call is set up, you either hear a *ringing signal* that is a combination of 440 Hz and 480 Hz tones, a *busy signal* that is a combination of 480 Hz and 620 Hz tones, or a congestion signal, which means that toll trunks between central offices are busy. This is sometimes called the *fast busy* and is made up of tones with a frequency of 480 plus 620 Hz that are sent more rapidly than a normal busy signal. If you leave your telephone off-hook, you get an *off-hook* signal that combines tones at 1,400 Hz, 2,060 Hz, 2,450 Hz, and 2,600 Hz. This signal is much louder than the others in order to get your attention. All of these tone signals are collectively called *progress tones* because they indicate the progress of your call.

fast busy

Other signal tones are used between central offices. For example, the DC signal generated by a pulse dial telephone cannot be transmitted between central offices, so it is converted to tone signals at the originating central office. This conversion is an example of *E&M signaling*—a special type of signaling that takes place between switching equipment.

Notice that all of the tones mentioned so far fall in the 300 to 3,000 Hz frequency range allowed for the voice signal. These are called *in-band signals.* Most of the tones that the central offices use for signaling each other also use in-band signals, but frequencies between the 3,000 Hz cutoff for the voice signal and the 4,000 Hz boundary of the telephone circuit are sometimes used. These are called *out-of-band signals,* and the most commonly used frequency is 3,700 Hz.

in-band signals

out-of-band signals

Common Channel Signaling

The most important signaling between central offices occurs on a special network of lines that are reserved exclusively for signaling information. This network, called the *common channel interoffice signaling* (*CCIS*) system, uses a set of signals called *Signaling System No. 7* or *SS7.* SS7 was first proposed by the ITU-T in 1980 and updated in 1984 and 1987, and the implementation is essentially complete in the United States and in most other countries.

SS7

SS7 uses separate lines to set up telephone calls from those used for the actual voice or data transmission. The advantage of using a separate signaling system is that you don't have to tie up a regular telephone line until the call is actually established. Since up to 40 percent of calls that are attempted result in busy signals or no answer, SS7 saves a great deal of time on the actual voice lines.

SS7 is optimized for use in digital telecommunications networks in conjunction with intelligent, computerized switches in the central offices.

It allows for database access as a part of the call setup, and that allows the telephone companies to provide certain enhanced telephone services, such as automatic callback and calling number identification. Implementing SS7 has been a big job for the telephone companies in the last few years, but the benefits to the companies and telephone users far outweigh the costs.

TELEPHONE NUMBERING

Under the guidance of the international standards group, the ITU-T, a reasonably consistent numbering plan exists for telephones around the world. This ensures that every telephone number, in its fully expanded form, is unique. According to the ITU-T plan, the world is divided into 9 geographic zones:

1. North America;
2. Africa;
3. Europe (part);
4. Europe (part);
5. South and Central America;
6. South Pacific;
7. Russia and Eastern Europe;
8. Far East;
9. Middle East and Southeast Asia.

Countries within each zone are assigned country codes beginning with the zone's digit. The countries that have, or are projected to have, the most telephones are assigned one-digit country codes; countries with the fewest number of phones are assigned three-digit country codes. In general, the form of a telephone number is as shown in Figure 5–25. The area code is sometimes shortened to one or two digits. In some countries it is called a *routing code* or *city code,* but the results are the same.

The North American Numbering Plan (NANP) (www.nanpa.com) covers the United States, Canada, and some Caribbean countries. This territory has been divided into areas, each with a unique three-digit area code. Areas in close geographical proximity to one another have area codes that are quite different to avoid confusion and accidental misdialing. Approximately 800 area codes are available for assignment.

The first three digits of a seven-digit telephone number are called the *exchange code.* Within each area code, the exchange codes are unique. Most central offices handle more than one exchange code, although some of the smaller offices only handle one. For example, a central office might handle exchange codes beginning with 631, 839, and 832.

exchange code

Figure 5–25
The general form of a telephone number and sample country and area/city codes of several countries.

Country Code		Area/City Code		Exchange Code		Subscriber Code
XXX	–	NXX	–	NXX	–	XXXX

where X = 0 – 9 (any digit)
N = 2 – 9
0/1 = 0 or 1

Sample Countries and Area/City Codes

Country	Country Code	Area/City Code	Area
United States	1	212	New York City
		616	Western Michigan
Australia	61	2	Sydney
Brazil	55	11	Sao Paulo
Ireland	353	1	Dublin
		91	Galway
Japan	81	3	Tokyo
United Kingdom	44	1	London
		222	Cardiff
Germany	49	069	Frankfurt
		89	Munich

LOCAL CALLING

Local calling is defined as telephone service within a designated *local service area.* The local service area includes telephones served by the central office and usually several other central offices nearby. Calls within a local service area are *local calls.* Local calls are charged in one of two ways. *Flat rate service* gives the user a specified and sometimes unlimited number of local calls for a flat monthly rate. *Measured rate service* bases the charges for local calls on the number of calls, their duration, or the distance.

Local service areas frequently overlap as shown in Figure 5–26. Mayville's local service area includes Middleburg, Freeland, and Wanigas. Bayport's is Hopedale, Freeland, and Wanigas. Freeland's local service area includes all of the communities shown.

LONG DISTANCE CALLING

There are two types of *long distance calls.* Both are sometimes referred to as *direct distance dialing (DDD). Toll calls* are calls outside of the local service area but within the LATA. Toll calls are handled by the local exchange carrier (telephone company), which also does the billing. The other type

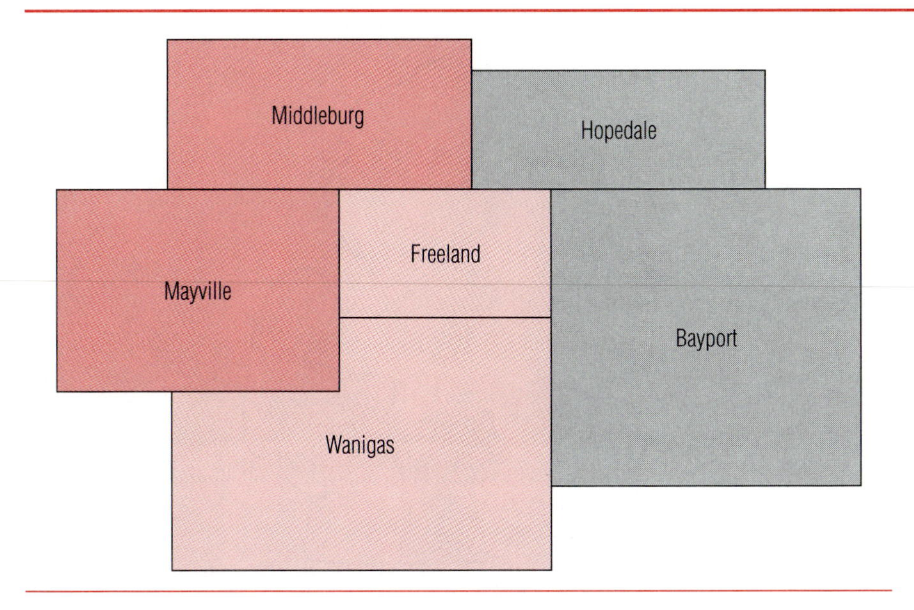

Figure 5–26
A local service area usually includes the territory covered by several central offices.

of long distance call is one that crosses LATA boundaries and must be handled by an interexchange carrier (IXC).

Recall from Chapter 2 that LECs provide telephone service within a service area consisting of one or more LATAs. Until the advent of the Telecommunications Act of 1996, LECs could not provide service across LATA boundaries; that type of service was classified as long distance and had to be provided by the long distance carriers (IXCs). Looking at Figure 5–27, you can see that because the caller and receiver are in different LATAs, Ameritech had to pass a call between Lansing and Detroit to one of the IXCs, even though Ameritech serves both cities. Because the call crossed LATA boundaries it was a long distance call. Now that restriction is lifted, and Ameritech is allowed to carry the call.

When someone calls out of a LATA, a long distance carrier's facilities are accessed by appending an access code to the front of the telephone number. Each long distance carrier has its own three-digit code. To reach a long distance carrier, the access code 10 must first be dialed, to let the central office know that a carrier's code—and not a telephone number—follows. The total sequence is

Access Code	Carrier's Code	Area Code	Telephone Number
10	XXX	NXX	NXX-XXXX

To reduce the number of digits that have to be dialed, each telephone customer designates one long distance carrier as its primary carrier, and

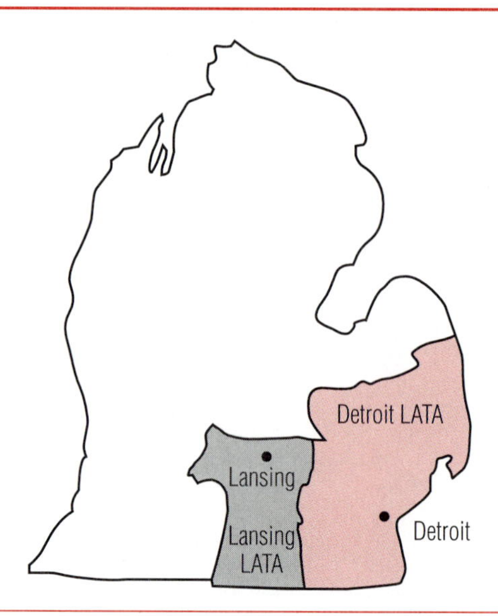

Figure 5–27
An inter-LATA call from Lansing to Detroit is considered to be a long distance call even though Ameritech serves both cities.

that company can be reached with the single-digit access code 1. The shortened form of the long distance number is

Prefix	Area Code	Telephone Number
1	NXX	NXX-XXXX

To reach a carrier other than the designated primary carrier, the caller must dial that company's prefix plus the carrier's normal three-digit code.

Telephone calls to most foreign countries can be completed automatically by dialing an international access code, which is 011 in the United States, then the country code, the city/area code, and the foreign telephone number. Some foreign telephone numbers are seven digits long, as in the United States, and some are eight digits, as in Japan or Australia. Dialing an international telephone is called *international direct distance dialing.*

international direct distance dialing

operator service

Although the vast majority of long distance calls are self-dialed by the caller, operator services still exist. When a telephone operator places a call, a premium is paid for the operator assistance, but the per-minute rates for the call are the same as for self-dialed calls. Certain types of calls, such as collect calls, calls billed to a third number, and person-to-person calls, must be set up by an operator. However, calling card calls are rapidly being automated. The burden is placed back on the caller to enter the calling card number through a DTMF telephone.

As an additional way of automating the process of placing calls, many of the telephone companies have implemented speech recognition systems. Callers are instructed to "touch or say" a digit or series of digits to obtain the service they need. Callers with rotary dial telephones may speak the digit to select the service they want. Of course, callers with a DTMF telephone may either touch the digit on the telephone's keypad or speak the digit to obtain the service.

SPECIAL TYPES OF TELEPHONE SERVICES

Everything that has been discussed in this chapter so far can loosely be called *plain old telephone service (POTS)*, and while the term is somewhat humorous, in fact it is used industry wide and is well understood among telecommunications professions. POTS is basic telephone service with no extra features or frills.

plain old telephone service

Many people and businesses have special situations, however. They may make a large number of calls, need to have the telephone answered when they aren't at home, want to know who is calling before they answer, or have any number of other special requirements. To meet these needs, telephone companies offer a number of special calling services, including extra service features available on residential and business phones, and discounted prices for high-volume telephone use. Many of these services are specially oriented to the customer who has particular calling patterns, such as a large number of calls from a certain part of the country or the desire for the receiver to pay for the call instead of the calling party. Plans that are aimed at long distance calling are offered by the interexchange carriers (IXCs), while those aimed at residential or business telephone use are offered by the local exchange carriers (LECs).

Telephone Service Features

LECs offer a number of enhanced features that can simplify or improve telephone service beyond what is available with POTS. The specific services each company offers will vary, and customers must evaluate each service on its own merits to determine whether it provides useful value.

Caller identification provides the number of the calling party to the called party as the telephone is ringing. The number is displayed on the telephone, if it has the ability, or on a small box that sits alongside the telephone. Telephone subscribers can specify that they don't want their number displayed when they make a call.

Call forwarding automatically transfers incoming calls to a different number. This feature is often useful to people who have a home office. When they are away, they can still receive calls, and the caller may not be any the wiser that the person they are calling is not "in the office."

The call waiting feature automatically sounds a tone that alerts you of another incoming call. You can put one call on hold to accept another by pressing the receiver button and continue to switch back and forth between calls if necessary.

Three-way conference calling allows you to speak to two other people at the same time. It can often eliminate a lot of back-and-forth calling.

Discounted Calling

Customers who make a large number of long distance calls are eligible for discounts from the normal charges. One category of discounting is based on call volume, measured by the amount of money spent for long distance calls each month. For example, if an individual or business has more than $100 per month in long distance calls, it may receive a 10 percent discount. If the call volume is greater than $500 per month, a discount of 20 percent may apply. Call volumes over $3,000 per month might receive a 35 percent discount, and so on. Each long distance carrier has its own plan, but all are similar.

Very large customers, those spending more than $10,000 per month, qualify for even higher discounts of 45 or 50 percent. Usually, however, they are required to have a direct connection to the long distance carrier's nearest point of presence (POP). This full-time line is capable of carrying at least 20 or 25 simultaneous voice calls. Long distance calls are routed through the dedicated line to the POP and from the POP on the public telephone network. For these very large customers, the cost of the special access line is usually well justified because they can obtain the larger discounts.

Regardless of the size of the customer and whether it is residential or business, the telephone companies have a wide variety of discount plans that may save money. The plans change quickly so it is worth inquiring and then staying in touch with the telephone companies. This is not to suggest that it makes sense to switch residential services every time one of the telephone companies calls, but rather that if you or your company has established calling patterns or specific calling needs, a discussion with a telephone company representative about your requirements may yield cost savings.

800 Service

The *800 service* provides numbers that allow toll-free calls; the cost of the call is paid by the sponsor of the 800 service. There is a widespread perception that 800 calls are "free," but indeed they are not! Telephone numbers that are part of an 800 service plan begin with the familiar 800 or the newer 888, 877, 866, or 855 area codes. Companies provide 800 numbers so that their customers or prospective customers can call for information, to place an order, or for technical service.

Each of the long distance carriers offers a variety of 800 service plans based on the coverage needed, the number of hours a month the service will be used, and in some plans, the distance of the calls.

All 800 numbers are stored in a nationwide database. When an 800 number is dialed, the LEC queries the database, using SS7, to find out which IXC provides service for that number, and passes the call to that company. Because of the central database many special features are available to 800 service users. Incoming calls may be routed to different locations depending on the time of day, the day of the week, or the location from which the call originated. The caller may be prompted to enter a digit to indicate which of several parties he or she wishes to speak to. Incoming calls can be automatically rerouted in case of an emergency or a temporary shutdown of a call center, for example, for training.

Outbound 800 service provides discounted calling with the cost depending on the duration, distance, and volume of the calls. Companies use this service when employees must make many calls to various parts of the country. The price would be less, for example, if all of the calls were to adjoining states rather than spread throughout the country.

Many countries offer toll-free plans similar to 800 service. There is also a coordinated global service called *Universal International Freephone Numbering (UIFN)*, which, for countries that choose to participate, allows the same toll-free number to be used globally. Billing for UIFN calls goes back to the sponsor of the toll-free number, regardless of where the call originates.

The 800 service offerings are very profitable for the telephone companies because, in most cases, no special equipment is required and standard telephone lines are used. The main difference in the service is the way the calls are billed. As a result, the 800 service offerings tend to be quite dynamic because the telephone companies can essentially offer a new service by changing the computerized billing program.

900 Service

The *900 service* is a sponsored service for businesses that have a message they wish to convey to the public. The message may be prerecorded, such as an advertisement, or live, such as listening to the astronauts on the space shuttle. The message (recorded or live) is controlled by the sponsor and can be changed as often as needed.

Normally the caller pays for the call, although the sponsor may elect to pay. The charges for the call vary, but after a certain volume of calls or call-minutes is reached, the sponsor begins earning income on each additional call-minute. There are other charges to the sponsor that depend on the number of calls made to the 900 number.

Another use for 900 service is for taking public opinion surveys. A question may be asked of the public in newspapers, on television, or through other mass media. Callers call one 900 number to express a positive vote and another 900 number for a negative vote.

Because of controversy surrounding the costs of 900 calls—which can be quite high—and the messages that some of the sponsors play, the telephone companies offer a service to block outgoing 900 calls. Any telephone subscriber can tell the telephone company not to allow 900 calls to be made from his or her telephone, and any attempted calls will be stopped (blocked) in the telephone company office.

Software Defined Network

The *software defined network (SDN)* is another bulk pricing offering designed for large companies. A business accesses SDN on either dedicated or dial lines, after which the calls are carried on the carrier's normal long distance network. In addition to a discounted price, SDN provides some additional services, such as seven-digit dialing to all company locations that are connected to the network and special billing. Additionally, the network can be defined so that special authorization codes are required by individuals to make certain types of calls.

Foreign Exchange (FX) Lines

If a telephone customer makes or receives many telephone calls from a particular city, a *foreign exchange (FX) line* can be installed. An FX line provides access to a remote telephone company central office so that it appears as though the subscriber has a telephone in that city. If a company in Dallas had a foreign exchange line to Houston, employees in Dallas could make Houston telephone calls at local Houston rates. Also, the company would have a Houston telephone number that, when called, would ring at the company switchboard in Dallas. Companies located outside a major city often have FX lines to the heart of the city if they make many telephone calls or want to provide a local telephone number for their customers who are located downtown.

An FX line can handle calls in either direction but, of course, only one call at a time. The subscriber pays a flat monthly rate for the line, plus a per-call charge from the local carrier.

In order to decide which of the many discount pricing plans is best for a given company, a network analyst needs a knowledge of the number and duration of the telephone calls that will use the service, calling patterns by location and time of day, and information about the carrier's discount pricing plans. The carriers will help the customer analyze a firm's requirements for these services by performing a traffic study of telephone call frequency, length, and distribution patterns, hoping, of course, to influence the customer to buy more services.

Integrated Services Digital Network (ISDN)

The *Integrated Services Digital Network (ISDN)* is an offering of the telephone companies that is applicable in some businesses and homes. ISDN service, which will be discussed in detail in Chapter 9, provides a high-speed line that may be used for one voice circuit and one 64 kbps data circuit, or one 128 kbps data circuit. For homes or small businesses that need simultaneous voice and dial-up data capability (for example, to access the Internet), ISDN may be a good fit. In the United States the ISDN offerings of the telephone companies all vary slightly, and have been notoriously hard to install and make operational. However, progress has been made, and the service is becoming more widely used. In other countries, such as Japan and Australia, ISDN has been standardized at a national level and is widely available and is used frequently by businesses and residential customers.

Telephone Calls on the Internet

Voice over IP

Making telephone calls on the Internet uses a technology called *voice over IP (VoIP)*. Internet telephone calls are only a subset of the uses of this technology because IP is primarily a data communications protocol that is used much more widely than on the Internet. But anywhere the VoIP technology is employed, voice calls can be made. The IP protocol will be discussed in Chapter 11, but suffice it to say here that to use VoIP, telephone calls must be converted to digital signals and divided into small pieces called *packets*, which are sent through the network independently. Each packet may travel by a different route to its destination, and each packet risks the possibility of being delayed. The result of these conversions, path differences, and timing issues is that it is possible for voice calls that travel on the Internet (or any other IP-based network) to be of variable and sometimes poor quality compared to what we expect from the normal voice network. Nonetheless, the technical problems are being worked out, and there is an intense amount of interest in developing the VoIP technology to the point where the quality of the calls is at least as good as what we are used to on the public switched network today.

The reason that is usually stated for wanting to place telephone calls on the Internet is that the calls are free. While this may be true in the short term, if the number of telephone calls grows, the capacity of the Internet will have to be expanded, and that cost will eventually get passed back to the users. So while users might not pay for Internet telephone calls directly, they may find that the cost of Internet access goes up faster than it would if the Internet wasn't used for telephone calls. While this is difficult to measure, except in a macroeconomic sense, one must rely on the basic principle that nothing is free, not even the Internet.

So the real reasons that companies are so interested in developing VoIP capabilities are that it allows voice signals to be transmitted using the latest digital technologies, and to avoid the constraints of the current voice network, which was originally designed for analog transmission and then later converted to be able to carry digital signals. Using VoIP, voice transmission can be sent using much less bandwidth than is required on the PSTN, yielding more efficient use of wire circuits, fiber optic cables, and the other components that make up a digital network. You'll learn more about the specific technology improvements in Chapter 8, when you study digital signals, and in Chapter 11, when you examine the IP protocol.

PRIVATE TELEPHONE SYSTEMS

When a business requires more than two or three telephones, it usually acquires some type of private telephone system to provide special services and help manage the telephone traffic. As a company grows, a large number of the telephone calls that are made are intraoffice calls from one department to another or one building to another. Without a private telephone system, each telephone would require a local loop connection to the central office, and each call would have to go through the public telephone network—even if it were destined for an office just down the hall. With a private telephone system connected as shown in Figure 5–28, the intraoffice calls can be handled internally, and only external calls must be sent to the telephone company's central office.

Earlier in the chapter, a *trunk* was defined as a circuit connecting telephone switches or switching locations. Our previous use of the word has been in the context of trunk lines connecting telephone company central office switches. A private telephone system is another type of telephone switch. Therefore, the lines connecting it with the switch in the telephone

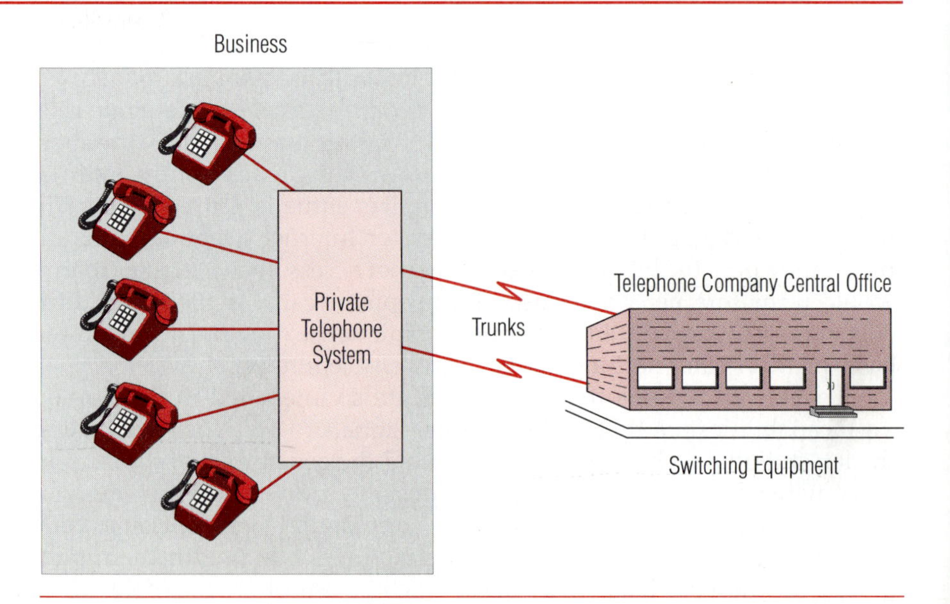

Figure 5–28
The private telephone system could be a key system or a PBX.

company central office are also trunks. Since a high percentage of the calls in a business are internal, the number of trunks connecting the private telephone system to the central office can be substantially less than the number of telephones in the office.

Key Systems

Small private telephone systems are called *key systems.* This name is a holdover from earlier days when a telephone connected to this type of system had push buttons or keys that allowed a line to be selected. In a typical key system, each telephone can access two or more lines, and lamps on the telephone indicate whether each line is busy. The caller selects a line by pushing a button on the telephone to seize the line, then dials the call. The telephone usually also has a hold button so that a call on one line can be held while a second call is made or answered.

Key systems typically handle from 3 to 50 telephones. This segment of the telephone equipment market is the largest and fastest growing because of the thousands of small businesses that can use this size phone system. In the past, key systems were available only from telephone companies, and the capabilities they offered were closely tied to the capabilities of the central office to which they were connected. Since the Carterfone decision and particularly since deregulation in the early 1980s, many vendors of key systems have entered the market, and the equipment has become significantly more sophisticated and better able to provide capabilities independent of the telephone company's central office equipment.

Key systems are available with a wide range of capabilities. The price of the system is partly related to the number of features that the system offers. The types of features available are similar to those found on PBXs.

Private Branch Exchange/ Private Automatic Branch Exchange/ Computer Branch Exchange

Private branch exchange (PBX), private automatic branch exchange (PABX), and *computer branch exchange (CBX)* are terms often used interchangeably to describe private telephone switching systems that are larger and usually more sophisticated than key systems. In this book, the term *PBX* is used to refer to any of these telephone systems.

A PBX is a private telephone system designed to handle the needs of a large organization. It is the next step up from a key system in capacity and the number of special features it supports. Private branch exchanges (PBXs) are typically designed to handle from 50 to more than 10,000 telephones. Each telephone on a PBX (or key system) is called a *station* or an *extension.* Larger sizes of PBXs are similar in capacity and capability to the switching equipment used in telephone company central offices. Since PBXs are designed for the exclusive use of one company, however, they are located on the company's premises.

These two PBX systems can handle different numbers of lines and handsets depending on the size and needs of the customer. (Courtesy of Northern Telecom)

PBXs, like the switching equipment in central offices, are computers especially designed for handling and switching voice telephone calls. Older PBXs, like older central office switches, had mainly mechanical components, and although some of these older PBXs still exist, the new units are all electronic and programmable. That is, they require an environment suited to computers—clean and air conditioned; they can be upgraded to add more capacity; they are physically smaller than their predecessors; and they require software to operate. PBXs, like key systems, are available from telephone companies and many other firms.

purchasing a PBX

The decision to acquire a PBX is not one to be taken lightly. The prospective purchaser needs to be sure he or she understands the firm's needs as well as the types of maintenance, training, and other support required by the firm as opposed to support provided by the PBX vendor. The level of support the vendor will provide is usually somewhat flexible. Support, along with price, are two of the major points of negotiation between the PBX vendor and a prospective customer.

A PBX gives its owner more control over its telephone system and usage than telephone systems provided by the telephone company. In addition, the PBX may provide features to improve the capability and efficiency

Station Features	System Features
Automatic redial	Automatic call distribution (ACD)
Automatic reminder	Class of service
Call forwarding	Data communication
Call park	Direct inward dialing
Call transfer	Hunt group
Call waiting	Least cost routing
Camp on	Paging
Conference calls	Pickup group
Distinctive ringing	Ring-down station
Do not disturb	Station message detail recording/Call detail recording (SMDR/CDR)
Speed dialing	Voice messaging

Figure 5–29
Features of a PBX. Many of these same features are also found in key systems.

of the telephone service that are otherwise not available. Features are usually divided into two categories: system features and station features. System features are capabilities that operate for all users of the PBX and that are in many cases transparent or unnoticeable to the user. Station features are customized to each user or telephone to provide the separate capabilities most useful to the individual. A list of system and station features is shown in Figure 5–29. The most common features are described next.

System Features

PBX *system features* are available to all users. In reality, all of the features may not be activated for all users because some are designed for particular needs and may be applicable only in certain departments or parts of the company. We will look at some of these features in more detail.

Data Communications Most PBXs have the capability to handle data communications as a standard part of their hardware. With this feature, a user with a computer terminal who needs to access several computers can use the PBX as a switch. First, the user dials one computer and connects the terminal to it through the PBX, and then, when the communication to it is finished, the user can dial the number of another computer and repeat the process. The data communications feature can be used to connect the terminals or personal computers in a company to other computers within the company, or through outside telephone lines the PBX can make the connection to computers in outside service bureaus or other organizations.

Direct Inward Dialing (DID) This feature gives outside callers the ability to call directly to an extension number so calls don't pass through an operator. Outside callers dial the normal seven-digit telephone number, and the call passes from the telephone company central office through the PBX and directly to an individual's telephone. Without the DID feature, all incoming calls pass through an operator who makes the connection to the desired extension.

Hunt Group The hunt group is another method of distributing calls to one of several individuals in a predetermined sequence. Hunt groups are often set up for departments in which several individuals can handle incoming calls but where there is a definite preferred sequence. When a call comes in, it is passed to the first extension in the hunt group. If that extension is busy or is not answered, the call is passed to the second extension and so on throughout the entire group.

Least Cost Routing The least cost routing feature attempts to place outgoing long distance calls on the line over which the call can be completed at the least cost. For example, if a PBX had foreign exchange lines and 800 service lines connected to it, the least cost routing feature would use a table stored in the memory of the PBX to determine which line should be used to place a particular call. If the preferred type of line is not available, the second choice is used and so on. When all alternatives have been tried, the call is sent out on the standard long distance facilities as a regular DDD call (the most expensive alternative). The least cost routing facility usually provides statistics showing the number of calls that went out on each type of line as well as the number of calls that had to overflow to a more expensive alternative. Using these statistics, the network analyst can determine whether more lines of a particular type are required.

Pickup Groups Pickup groups allow any member of the group to answer an incoming call. For example, in a group of marketing people that are connected on a pickup group, if a ringing telephone is not answered, any member of the group may pick up his own phone, key in the pickup code (usually an asterisk), and have the call transferred to his telephone so that it can be answered. The idea is that incoming calls will be answered by someone who does a similar kind of work and who can potentially help the caller.

Station Message Detail Recording (SMDR) Station message detail recording (SMDR), sometimes called call detail recording (CDR), is the feature of the PBX that records statistics about all calls placed through the system. The data recorded includes at least the calling and called extension numbers, the time of the call, and its duration. It may also include other statistics, such as the user's class of service or the line numbers used. The statistics usually are accumulated on a magnetic tape or disk attached to the PBX. Once the data is captured, it can be used for recording and billing purposes. Usually the data is taken off the PBX and transferred to another computer that performs the reporting and billing functions. Some PBXs can transmit the call detail data over a communications line to another location for processing.

The SMDR capability is very important to network management and analysts. In addition to providing the source data for telephone billing, the SMDR data can:

- show the busy hour thereby providing data as input to network staffing decisions,
- provide input for analysts monitoring for telephone abuse or toll fraud,
- be used to analyze whether the mix of toll free lines and other discount calling services is correct,
- show the cost of every call,
- show the effectiveness of an advertising campaign,
- be used to analyze the effectiveness of operators in a call center, and identify training needs,
- show business trends over time when they are related to call volume or duration.

In most organizations, SMDR data is regularly analyzed as a way of ensuring that telephone service is optimized for cost and capability.

Station Features

Station features are activated by a PBX system user. Whereas the features may be made available to individuals or groups of people, it takes some action by the individual to use the feature.

Automatic Reminder The automatic reminder feature lets a user tell the PBX to call back at a specified time. Its most common use is in a hotel for wake-up calls. You tell the hotel operator when you want to be called, and the operator instructs the PBX to ring your telephone at the specified time. Often the PBX can play a prerecorded message when it makes the automatic reminder call, such as "Good morning, it is 7:00 A.M."

Call Forwarding Call forwarding lets calls for one extension ring at another extension. If a person is going to be out of the office, she can activate call forwarding to have all of the calls ring at the secretary's desk or at whatever extension she will be. Call forwarding can also be set to forward the call if the extension is busy. A person might have calls forwarded to different extensions depending on which condition (such as a ring with no answer or a busy signal) occurs.

Call Transfer The call transfer feature allows calls to be transferred to another extension. In the blind transfer, a transfer code and the extension

This model of Northern Telecom's Meridian 1 is very compact. Line cards and additional cabinets of various sizes can be added or removed for smaller or larger configurations. (Courtesy of Northern Telecom)

number are keyed in. When the person who was originally called hangs up the telephone, the call is transferred. In the consultation transfer, keying in the transfer code places the caller on hold. The new extension is dialed, and when it is answered the person originally called and the person to whom the call is being transferred can converse. When the original answerer hangs up, the transfer is completed.

Call Waiting The call waiting feature indicates an incoming call while a call is in progress. The second call is indicated by a tone or lamp, and the parties in conversation can decide whether to take the second call.

Camp On When you place a call and the number you called is busy, the camp on feature lets you tell the system to call you back when the number is free. The PBX tests the extension that was called and, when it is free, calls you back. If you answer your telephone, it automatically redials your call for you. This keeps you from having to continually redial a busy number yourself.

Distinctive Ringing With the distinctive ringing feature, a telephone may have different ringing signals for calls from within the company and outside calls. Other distinctive rings may be available for emergency calls or trouble calls.

Do Not Disturb The do not disturb feature may be implemented in several ways. In its simplest form, it gives the caller a busy signal even though the extension called is not in use. Another implementation causes the caller to receive a distinctive busy signal that indicates the person being called has his or her telephone in do-not-disturb mode. A third implementation signals the PBX to automatically forward calls to another extension, such as to a voice mail system.

Speed Dialing Speed dialing allows frequently called numbers to be stored in the PBX's memory and then accessed with a shorter set of digits. For example, a PBX may allow every user to store 100 ten-digit telephone numbers. Each one can then be accessed by dialing a speed dialing code followed by a two-digit number, 00–99. A common use for speed dialing is to store the numbers of other company locations or frequently called customers or suppliers.

Several of the features just described are becoming available on the public telephone system. These features become available when the serving central office is upgraded to an electronic switch, which is a big brother of the PBX. Call waiting, call forwarding, and conference calling are available from most telephone companies today. Since there is an extra charge for these features, you must notify your telephone company if you wish to have them activated for your telephone. In addition, automatic redial is a feature available on many home telephones today.

PBX Security

During the 1980s, businesses became aware of *hackers* on telephone and computer networks and the millions of dollars in damages and expenses they can cause. Until several years ago, telephone fraud was mostly limited to employees placing long or expensive personal calls during business hours.

With the increased sophistication of PBXs and the granting of external access to the PBX by legitimate users, telephone hackers have found a lucrative niche by breaking into organizations' telephone systems and placing or selling long distance calls. Phone hackers use computers with auto-dialing modems to break security passwords and gain access to telephone systems. Once inside, they use or sell long distance calls, leaving the PBX owner holding the bill.

toll fraud

AT&T reports that toll fraud cost U.S. businesses more than $2 billion in 1993. Companies should take preventive measures including blocking outgoing international calls during off-hours, blocking remote access features of PBXs, and changing access codes and passwords frequently.

HYBRID SYSTEMS AND COMMUNICATIONS SERVERS

hybrid systems

It used to be that the distinction between key systems and PBXs were quite clear. Then more manufacturers declared their products to be *hybrid systems,* which have characteristics of both a key system and a PBX, such as the ability to be programmed to act like a key system (select the line you wish to use to make a call) or like a PBX (pick up a handset and the PBX selects an outbound trunk for you when you dial "9"). Whereas key systems typically use digital or analog phones, hybrid systems can use both.

communications servers

Key systems and PBXs are increasingly using personal computers as their main hardware platform. With appropriate software, they are sometimes categorized as *communications servers.* The term is very loosely defined, so the capabilities and features that a communications server from one vendor provides may be somewhat different from the offerings from another vendor. Typically communications servers are connected to both the outgoing telephone lines and a company's local area data network or LAN. The server acts as the office's phone system and has many of the standard features of a PBX, but may add other capabilities such as the ability to dial the telephone from the PC screen or from a program. The server can also answer incoming calls, handle incoming faxes and e-mails, page you if you're not at your desk, or transfer the caller to voicemail. You can expect to hear a lot more about communications servers in the very near future as their capabilities evolve and they become more widely used.

CENTREX SERVICE

Centrex service, which some telephone companies call by a different name for marketing purposes, provides a telephone service much like that of a PBX, but the telephone companies use equipment located at their central office, not at the customer site. The central office equipment may be dedicated to the customer, but more likely it is the same equipment used to provide normal public telephone switching functions. It is important to understand that Centrex service is not precisely defined or standardized even though the name is used nationwide. Since each telephone company can offer Centrex service and several different types of central office equipment are used, the Centrex offering is a combination of the technical capabilities of the central office equipment and the features that the telephone company decides to make available to its customers. Centrex service is regulated by the state public utilities commission; a PBX is a private system that is unregulated.

Centrex capabilities have advanced rapidly in the past few years. The switches used in telephone company central offices are very powerful computers that allow many unique features to be programmed. Some

PBX	Centrex
Ultimate control	Control shared with telephone company
Not regulated	Regulated by state PUC
Total ability to manage	Management shared with telephone company
Total responsibility to manage	Telephone company is primarily responsible
Requires capital to buy the system	No capital required
User/vendor must provide service	Telephone company provides all service
Usually less redundancy is built in	More redundant hardware may mean higher reliability
Growth capability depends on the inherent design of the PBX	Essentially unlimited growth capability

Figure 5–30
Factors to be considered in selecting a PBX versus a Centrex system.

people feel that Centrex features and technical capabilities are now better than what is offered by PBXs.

You might wonder why a business would consider acquiring a PBX if telephone company-provided Centrex would give the same or better service. There are definitely pros and cons to both types of systems, as shown in Figure 5–30. On the one hand, a PBX gives the company ultimate control of its telephone system and the total ability to manage it. On the other hand, it requires the company to make the capital investment in the PBX equipment and to provide space, power, and air conditioning, as well as skilled technical people to operate it. By way of contrast, when a Centrex system is used, the telephone company manages and services the equipment, and the customer pays the telephone company a service fee.

PRIVATE VOICE NETWORKS

When a company has several locations, each with its own PBX or key system, it is often desirable to tie the locations together with telephone lines. By renting lines to connect the PBXs or key systems, a private network can be built that saves money compared to the cost of making standard long distance calls. In addition, it simplifies the dialing and, in many cases, can appear to the telephone user as a single, integrated private telephone system.

Tie Lines

Leased lines that connect the private PBXs or key systems are called *tie lines* or, more properly (because they connect switching equipment), *tie trunks*. They are acquired from the common carrier or can be installed privately if the locations are in close proximity. If leased from the telephone company, a fixed price is paid for full-time, 7-day, 24-hour use.

tie lines
tie trunks

An audio teleconference in progress. The flat microphone on the table is omnidirectional and can easily pick up voices from anywhere in the room. (Courtesy of Dow Corning Corporation)

With tie trunks in place, a telephone user at one location dials an access code to access the trunk and PBX at the other end. Usually, a second dial tone is heard from the remote PBX, and then the extension number of the called party is dialed. If a speed dialing feature is installed, this process may be simplified.

Private Networks for Large Organizations

When an organization has a very large volume of calls it may be economical to build a private network. In past years companies leased a variety of full-time lines from the telephone companies to build such networks; however, today most of the networks that are built are virtual. That is, they use standard telephone company facilities rather than dedicated lines. *Virtual networks* and discounted telephone calling go hand-in-hand, and a company must work with telephone company sales representatives and engineers to define the best solution. Service offerings from the telephone companies change frequently, and new capabilities become available regularly because of the onward march of technology.

AUDIO TELECONFERENCING

Another type of telephone service that some businesses find useful is audio teleconferencing. Usually used in a conference room, an audio teleconferencing setup has an omnidirectional microphone in the center of

The Octel 200 voice messaging system includes many advanced features. It is modular and can be expanded to support several thousand users. (Courtesy of Lucent Technologies Octel Messaging Division)

the table and a speaker, both of which are connected to the telephone. Participants sit around the table, and one participant makes a standard telephone call to an individual or to another similarly equipped conference room. All the people in the room can hear what is said at the other end of the connection on the speaker, and the microphone picks up everything that is said and transmits it over the telephone lines. In some audio teleconferencing systems, multiple rooms can be connected so that several groups of people can converse. About the only special requirement—in addition to the microphones and speakers—is the human factor consideration of taking turns and being polite.

Businesses have found audio teleconferencing useful for allowing a group of people to converse with a specialist or expert at a remote location, for having status meetings between groups of people at diverse locations, and for holding meetings when the participants are in many locations. In some cases, it can be a substitute for travel. In one recent audio teleconference, a company conducted a sales meeting. Eleven locations, each with approximately 20 participants, were connected in a large audio teleconference.

VOICE MAIL

Voice mail provides an electronic voice mailbox where callers can leave messages for other people. Voice mail is the voice equivalent of the e-mail systems discussed in Chapter 4. Voice mail systems may be a system feature built into a PBX or provided as a separate piece of hardware. Some companies make voice mail units that may be attached to a wide variety of PBXs or that work with Centrex systems.

Whether it is a feature on a PBX or a stand-alone system, the voice mail system can be used by callers inside a company or those calling in from the outside. In addition to simply providing a voice mailbox capability, a voice mail system provides other functions as well. It allows voice messages to be sent to several people at the same time. For example, a manager could communicate to everyone in a department and be sure that they all heard the same message said the same way. Comments may be added to voice messages, and the original message with the comments can be forwarded to another party. Voice messages may be put in the system with instructions to place them in voice mailboxes at a later time.

Using this capability, a department manager could, before leaving on a trip on Tuesday afternoon, put a voice message in the system for all of the department's employees, announcing a job change or new product. The manager could tell the system not to put the message in the employees' mailboxes until Thursday morning if it were important that the announcement not be made until a certain time.

A voice mail system is not just a simple tape recorder. It converts an analog voice signal to a digital signal and stores it on a magnetic disk for later recall. When a person retrieves the calls in a mailbox, the digital signal is converted back to an analog signal, and the voice message is spoken to the recipient.

Voice mail capability is useful in a variety of situations. Research has shown that up to 83 percent of all business telephone calls do not reach the called party on the first attempt. An important advantage of voice mail is that the calling and called parties do not need to be present at the same time. Voice mail allows the caller to easily leave a detailed message. If the caller can go into detail and ask a question or explain a situation, it is possible to avoid the need for a callback in 50 percent of the cases. For these situations, telephone tag can be minimized or avoided altogether. Some people expect that electronic conversing, where the two people never actually talk to one another, will become an accepted form of intracompany communication.

Another use for voice mail is in minimizing the problems caused by time zone differences. Callers on the east coast can leave messages for business associates on the west coast before they get to work. Conversely, west coast callers can leave messages after east coast workers have gone home for the day.

Voice mail is also useful for salespeople and marketing people who do a lot of traveling. They can call into the voice mail system from any tone-dialing telephone and have the system play back all of the messages stored for them. Then they can respond to the messages by sending a reply, forwarding the message to another person for handling, or sending new messages.

While special hardware is required to set up a voice mail system, technology has shrunk the circuitry so that small systems may now be established by installing a few special-purpose circuit cards in a personal computer. A key attribute to be considered is the size of the computer's hard disk, because digitized voice messages take a lot of disk space and rapidly fill a personal computer's hard disk.

Automated Voice Response

The automated processing of incoming calls has made rapid advances in recent years. There are three concepts to understand, and they are often confused: *automated attendant*, *audiotex*, and *interactive voice response*. They can be implemented separately but are frequently used in combination. All require the use of a Touchtone telephone by the caller.

Automated Attendant

The automated attendant capability is at work when you call a telephone number and an automated voice asks you for some information in order to route your call to a person. Substitute the word *operator* for *attendant*, and it is easy to imagine how an operator might answer you, ask you a few questions, and then pass on your phone call to an individual who could serve you. The automated attendant brings the power of the computer to bear by eliminating the person who routes the call. You provide information to the computer by pressing keys on your telephone rather than by speaking to a person.

Audiotex

Audiotex is the service that provides fixed information when you call a certain telephone number. The information is fixed in the sense that the caller cannot select information. Of course the information itself can be changed as often as necessary. If you call a number to get the current time or the weather report, you are being served by an audiotex system.

Interactive Voice Processing

When you call a telephone number and are answered by an automated voice that asks you for information, and then you are routed to another part of the automated system, you are using interactive voice response. Visualize the information as being stored in a tree structure. Depending on the response you give to the question—using your Touchtone telephone—you proceed down a branch of the tree, perhaps answering more questions as you go. Finally you get to the information you need, and the system reads it to you using the automated voice. Note that the fixed information itself is a form of audiotex.

Of course, one option in the tree structure may be to route your call to a person if the system does not have the information you want prestored. In that case, the system is behaving like an automated attendant system.

It is important to point out that the features of automated voice response systems must be carefully designed and programmed to avoid setting up tedious and time-consuming messages, loops, or other situations that may frustrate the caller. If prospective customers get caught in the tangle of a poorly programmed automated voice response system they may well take their business elsewhere—to a company where a real person answers the phone! Automated voice response systems should not be viewed as a cheap substitute for people!

WIRELESS COMMUNICATIONS

One of the fastest-growing segments of voice communications is *wireless communications.* First we'll look at two applications of voice wireless technology that you are probably already familiar with, cordless and cellular telephones. Then we'll explore the rapidly expanding world of wireless beyond these two.

Cordless Telephones

Cordless telephones represent a subset of wireless technology that has become very familiar to most of us. The base of the telephone, which is connected to a normal telephone line, contains a small radio transmitter and receiver that broadcast a signal to the handset. The handset, operated by rechargeable batteries, also contains a small transmitter and receiver for communicating with the base. The normal radius for clear, static-free operation of a cordless handset is 50 to 800 feet from the base.

Cordless telephones are usually designed so that they operate in standby mode, conserving battery power but able to detect incoming calls and signal the user with a buzz whenever the base unit sends a ringing signal. When users want to answer or place a call, they switch the transmitter on and then continue with the call. Most units are designed so that the handset will recharge its batteries when placed in a cradle on the base unit.

broadcast

An important word in the above description is *broadcast,* for indeed the base and the handset broadcast their signal like a small radio station.

That means that anyone listening on the right frequency and located fairly close can pick up the telephone conversation and listen in. Telephone conversations conducted using cordless telephones or any other wireless technique are not private! In the United States, a federal law makes eavesdropping a crime, but it is virtually impossible to enforce.

Digital cordless phones are also available, and they address the security issue and some other problems that users often experience. Specifically, digital cordless phones offer improved transmission quality (less noise), greater range, and security against eavesdroppers. Rather than transmit the phone conversation on a single channel, most of the new digital cordless phones use a sophisticated technique called *spread spectrum.* Although there are several different spread spectrum techniques, it is easiest to visualize how it works by thinking of the transmission jumping between several frequencies. With the digitized signal dispersed among multiple frequencies, the chances of successful eavesdropping are significantly reduced.

spread spectrum

Most vendors claim that digital cordless phones have an operating range two to four times greater than conventional analog cordless phones, allowing the users greater freedom to move around while talking.

Cellular Telephone Service

Cellular telephone service was originally designed to provide mobile telephone service as the caller moves around within a relatively small geographic area—typically a city. In the past, cities had a single powerful radio transmitter and receiver for mobile telephone service. Each telephone call required a separate frequency, and although a city like Chicago had approximately 2,000 frequencies available, there was a large demand for more.

The problem of insufficient frequencies became so acute in large cities that the whole concept of mobile telephone service was reworked in the early 1980s to take advantage of modern computer technology. Cities were divided into cells (shown in Figure 5–31), and low-powered radio equipment controlled by a central computer was installed in each cell. Because the transmitters use low power, the same frequencies can be reused in nearby cells without interference. Thus if a city has 10 cells, 5 to 8 times as many channels are available for telephone conversations as were available in the previous system.

This type of system is called a *cellular telephone system;* its basic technology was developed at the Bell Laboratories in the early 1960s. Access to the system is made from a telephone that also contains a radio transmitter and receiver. When a call is initiated from a cellular phone it is picked up by the nearest cellular antenna and forwarded, usually on wire, to a *mobile telephone switching office (MTSO),* the equivalent of a central office in the regular telephone system. MTSOs are connected to the regular telephone network so that calls from cell phones can be made to other cell phones or to regular phones, and vice versa as shown in Figure 5–32.

mobile telephone switching office

Figure 5–31
The layout of a cellular telephone system.

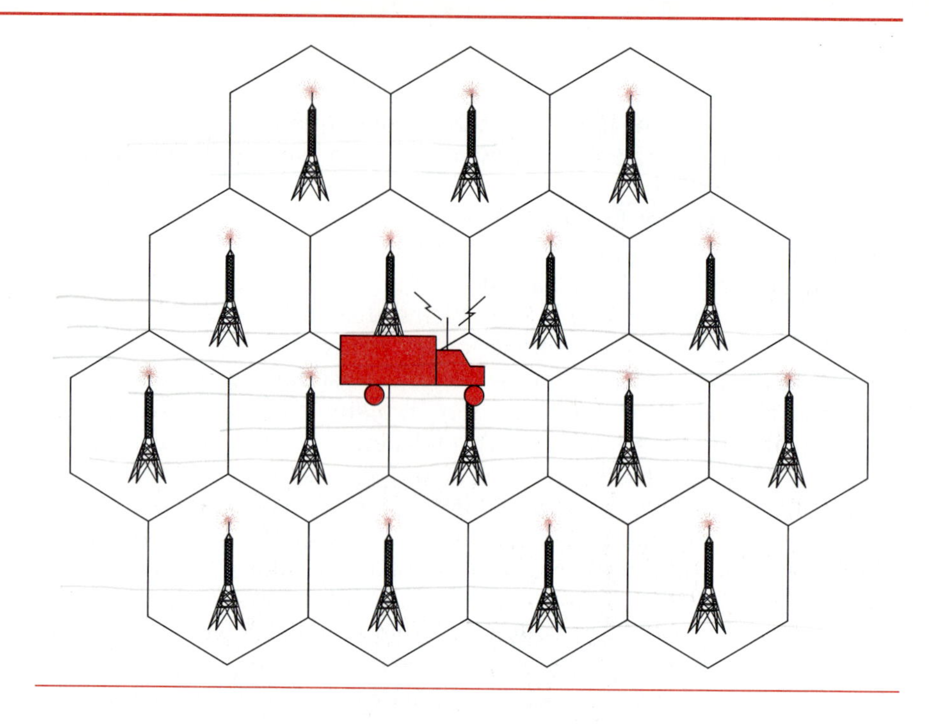

Figure 5–32
Cellular telephone users can make calls to other cellular users or to telephones on the public (wire based) switched network.

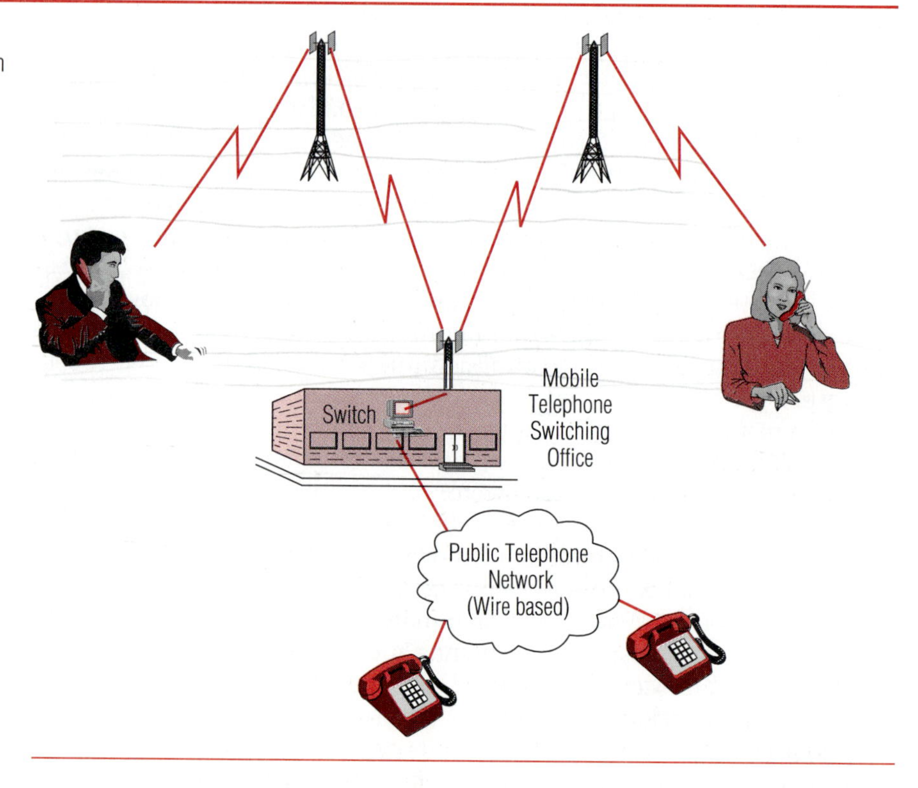

Cellular telephones can be useful in cases of emergencies as well as for normal business or personal conversations. (Courtesy of AirTouch Cellular)

Cellular telephone calls are monitored every few seconds by a central computer in the MTSO. When the signal from the cell phone starts getting weak, the computer determines which cell to switch the call to and instantaneously switches it to the new cell's radio equipment. The switch occurs so fast that it is unnoticeable by the individuals who are conversing. (It can present problems if data is being transmitted on the cellular connection, however.) One of the key attributes of cellular telephone systems is that they operate completely automatically. No operator is required to set up, monitor, or switch calls from one cell to another.

A growing number of people are using cellular telephone service for facsimile and data transmissions. Data transmission can be nearly 100

This Motorola cellular telephone is extremely small and compact, yet contains a full set of features for cellular operation. (Courtesy of Motorola, Inc.)

percent reliable if the proper error-checking circuitry is built into the equipment. One ambulance company is using a cellular telephone system to transmit data about patients to the hospital while the ambulance is en route. Contractors are sending facsimiles of drawings from the office to the job site by cellular phone. And the American Chemistry Council maintains a database of information about hazardous chemicals that can be transmitted via a cellular telephone system to the site of a chemical spill.

An interesting sidelight to cellular phone usage: A study conducted by the *New England Journal of Medicine* showed that motorists are four times as likely to have an accident when they are using cellular phones. Inattention to their driving rather than dexterity in using the phone is the major reason why, the study showed. Among other findings:

- Phones that allowed the driver's hands to be free did not appear to be safer than hand-held phones.
- Younger drivers were at a somewhat higher risk.
- More drivers tended to get in a crash during the work week than during the weekend. Wednesdays were the worst day; Saturdays the best.

Since cellular telephones also transmit calls by radio, there has been some concern about eavesdropping. Cellular calls can be picked up by some radio scanners as far as 5 to 10 miles away from the transmitter. Users need to be aware that calls on cellular telephones are not as secure as calls placed on conventional telephones.

Cellular telephone service has been a big commercial success. Although cellular service is more expensive than a regular telephone line, it can be a boon to any company that makes its living from customer service. It also is widely used for business and pleasure by people who spend large amounts of time commuting or tied up in traffic.

Today's cellular systems are so-called second-generation systems and they are not all globally compatible with one another. People who use a cellular phone system that uses one technology cannot communicate with people on a system that uses a different technology. Currently there are three ways that digital cell phones transmit their signals: TDMA, GSM, and CDMA.

Time division multiple access (TDMA) is the oldest technology. TDMA divides radio frequencies into separate channels, assigns a specific channel to your call, and then broadcasts the call with precision timing. Since its capacity is not high, TDMA technology is not well suited for applications such as e-mail or web browsing, and call quality is often not very good. Nonetheless, TDMA is the most frequently used technology for cellular phone calls in the United States.

Global system for mobile communications (GSM) is based on TDMA technology but has a higher capacity and therefore better call quality. It supports both voice and data transmission and has become the standard for cellular phone service in Europe and Asia. GSM service is available in the United States but only in limited areas.

Code division multiple access (CDMA) is the newest of the technologies and broadcasts calls within the same channel. Conversations are distinguished from one another by special codes that are assigned to them. A CDMA phone actually receives all the calls being transmitted, but only those with your call's special code are assembled back into voice for you to hear. Like GSM, CDMA supports both voice and data transmissions, and the quality of the voice is the best of the three technologies.

So today, a European traveler can use a GSM cell phone in Europe and Australia, but is likely to need a different phone and a subscription to a U.S. cell phone plan when traveling in the U.S. This problem is gradually getting resolved with newer cell phones that can operate on any of the technologies, but it is still a major inconvenience.

Third generation cellular systems are scheduled for operational start-up in Japan and Europe in 2001–2002. The hope for these systems, which are generically called 3G (third generation), is to create a universal system that will be used worldwide. It is planned for 3G phones to be able to transmit data at a very high rate, especially compared to today's cell phones.

Formerly equated only with cellular phone technology, wireless now includes mobile access to the Internet and other kinds of data networks as well. People are constantly on the go, and there are times when they want to be able to initiate communications of various types—to check the stock market on the Internet, to make a dinner reservation while traveling home on the train or in a car, or to exchange e-mail when they aren't connected to their company's network. Convenience seems to always be high on people's list of priorities as it relates to their ability to communicate.

As most Europeans and Japanese know, the Internet is entering a new phase in its short but dramatic existence. Engineers of every major computer, telecommunications, and software company in the world are working hard to give customers anywhere, anytime access to the Internet. Millions of Japanese and Europeans (especially Scandinavians) can exchange e-mail and surf certain web sites from their mobile phones, though they use proprietary standards that are interim to 3G. The basic Japanese system is called i-mode and is offered by NTT's cellular services division, DoCoMo. Over seven million Japanese surf the Internet and exchange e-mail on wireless devices. Today's phones have a tiny screen, but despite its small size, i-mode users see a menu of choices such as news, games, chat, and search. Roughly 10,000 web sites have been specially formatted so that pages will download quickly to i-mode phone screens.

Europe is further behind but is investing heavily in the *wireless application protocol (WAP)* technology, which downsizes fat, graphics-rich web pages so that they are usable on small cell phone displays. In the U.S., only a small number of cell phone and Palm Pilot customers can access such services because they are not widely available. Basically, the rest of the world has bypassed the United States in cell phone technology, and the U.S. is playing catch-up.

In addition to the new WAP and 3G standards for wireless communications, a third standard called *Bluetooth* is emerging. Bluetooth is a standard for wireless connectivity within a 33-foot radius. It allows wireless communication between mobile phones, laptop computers, personal digital assistants, and other portable devices. Being a radio-based link, Bluetooth doesn't require a line-of-sight connection in order to establish communication. A person could, for example, wear a wireless headset that communicates to a cell phone being carried in an attaché case, as long as all three devices are in range of each other. Likewise, a PDA could communicate with a laptop computer, automatically exchanging data when the two are within range of each other.

You can see that the cellular and wireless technologies are merging, and will give us many useful capabilities in the coming years. However, since the concept is relatively new, there are still several questions that vendors are struggling to answer, such as the following:

- What kinds of services do customers really want?
- How much are customers willing to pay for such services?
- What is the likely mix of consumers and business customers?
- Will this be a profitable market?

We can expect major evolution in the coming few years.

SIDEBAR 5–1

WHAT IS I-MODE?

Think of i-mode as a mini-version of Internet sites delivered to 1.5-inch phone screens.

DoCoMo, like Yahoo, runs a megasite that offers everything from news to airline tickets to weather reports. When i-mode is clicked on, users see a menu. By pushing buttons on the phone, they navigate through DoCoMo's site, which links to 500 content providers. An additional 15,000 Net sites are i-mode compatible, but users have to enter the sites' addresses.

In a matter of seconds, and for less than 25 cents per item, i-mode users can do such things as bank via phone, check airline schedules, check the weather, send e-mail, read the news, play simple video games and even make phone calls.

I-mode transmits data at 9.6 kilobits per second, much slower than most dial-up modems in U.S. homes. But most graphics and banners are eliminated so text flows faster.

DoCoMo already sends a monthly bill to its customers for their cellular phone use. It tacks on $1 to $3 for subscriptions to some i-mode sites, such as Nikkei news. DoCoMo kicks back 91% of the revenue to Nikkei and keeps 9% for itself.

DoCoMo also makes money when users send e-mail or flip through Internet pages. Fees run about 4 cents for a 250-character e-mail or about 3 cents per page view.

By Julie Schmit, *USA Today,* July 7, 2000. Reprinted with permission.

Marine and Aeronautical Telephone Services

Yet another variation of mobile telephone service are calls made from ships or airplanes. These are called *marine telephone service* and *aeronautical telephone service.* In both marine and aeronautical telephone services, radio communication is established between a transceiver on a boat or airplane and a land-based transmitter and receiver that are connected to the public telephone system. The land-based equipment is usually voice-activated so that it switches automatically from transmit to receive as the parties at either end talk. In most cases, an operator must initially establish the call and monitor its progress to ensure that the radio transmitter and receiver are working properly.

Aeronautical telephone service has the special problem that because the airplane is moving so fast, it quickly leaves the range of the transmitter. Unless a means exists to pass a call from one transmitter to another as the plane moves, the call is terminated as the plane goes out of range. A small system of coordinated ground-based equipment has been established along some major air corridors. The transmitters and receivers are close enough together so that if a plane goes out of range of one, it is within range of another, and the call can be passed with no interruption.

Many commercial airlines offer telephone service on their flights. In the most common implementation, telephones are built into the backs of one or more seats in each row. A passenger inserts a credit card, which allows the handset to be removed from the back of the seat. The telephone number is dialed, and the call is broadcast from the airplane to the nearest ground receiver. When the connection is made, the call progresses like any other telephone call. Upon completion of the call, the passenger returns the handset to the telephone and retrieves the credit card. Billing is done through the credit card. There is a basic flat rate for the call plus a per-minute charge.

THE VOICE NETWORK USED FOR DATA TRANSMISSION

Much of the ubiquitous nationwide public telephone network that we take for granted is, to all appearances, an analog network that was originally designed and optimized to handle analog voice communications. Advances in technology and computer circuitry have made digital transmission possible, and, as you will see in succeeding chapters, there are certain advantages to digital transmission that make it preferable. Because of these advantages, the telephone network has for the most part been changed so that today most of it operates in digital mode.

Furthermore, of the over 300 million telephone calls that take place in a day, several million are connections between terminals and computers or between two computers, sending data that originated in digital form. These transmissions have certain unique attributes and different requirements than analog voice data, but the data transmissions are usually adapted to fit the parameters of the analog telephone network. You will see how this is done in the following chapters.

SUMMARY

This chapter presented the various components of the voice telephone system. It looked at the telephone instrument itself, the switching equipment located in the central office, and the hierarchy of central offices that compose the public telephone network. You have also studied how voice signals are converted to electrical signals, the characteristics of those signals, and the way in which they are modified (modulated) for transmission on the telephone network. In addition, you studied private telephone systems and their features, systems like those found in most businesses today. Finally, you were introduced to a number of the services offered by telephone companies.

Many of the concepts and facilities presented in this chapter will come up in later chapters that discuss data communications. By understanding the vocabulary and subject material presented in this chapter, you will have a good foundation for the chapters that follow.

CASE STUDY

Telecommunications at Work in a Brokerage Office

John Duncan, a vice president of a major brokerage firm, says that he couldn't do his job without the telecommunications and personal computer technology in his office. John manages a team of five people who handle the investment needs and manage the portfolios of over two hundred clients. The brokerage firm that they work for provides a mainframe and personal computer-based information system with workstations on each desk and linked via a wide area network to the firm's mainframe computers at headquarters. This system provides the basic capability that allows John and his staff to buy and sell securities for their clients. Also provided is an AT&T telephone system, which includes call forwarding and voice mail, and is vital to maintaining contact with clients.

What makes John's team unique, however, is the private PC-based system that he has installed to supplement the system provided by his employer. Each member of John's team has a second PC on his or her desk that runs the Goldmine contact management software. "With Goldmine we can keep a complete set of information about each of our clients and the many other people with whom we deal," John says. "Of course we keep basic demographic data, such as names, addresses, and phone numbers, but also information about their securities portfolio, trades, financial goals and the progress they are making toward their goals. We mail a unique birthday greeting to each client on their birthday showing them what happened in history on the day they were born. We use a program called Special Days, and clients love the fact that we remember their birthdays."

Regardless of where he works, John uses the contact management software to help schedule his day. "We preschedule monthly and quarterly phone calls with clients, and record daily meetings and appointments. At the beginning of each day, my secretary prints the schedule for the day, which shows me what I have to do and also indicates the priority of contacts," John says. On a typical day he calls 15 to 25 people in addition to holding meetings with clients, his staff, and associates at his firm.

Perhaps the most distinguishing feature of the operation, however, is that John has connected the telephone and computer systems in his office to a two-line telephone and a PC at his vacation home in another state, which allows him to do his job as effectively there as in the office. For example, when the phone rings in the office, John's secretary can just as easily transfer the call to his vacation home as to his office. When John answers the call, the client doesn't even know that he may be relaxing by the pool or consulting his computerized database from his computer at his vacation home. John says, "I can be just as effective for my clients at our second home as I can in the office. Through the use of telecommunications I have access to all of the same data, and my staff can support me in the same way as when I'm in my office." And John's firm agrees, for they fully support his operation from his second home.

"The biggest improvement in technology in the last three years," John says, "is the full implementation of the Palm Pilot into our daily lives. We now have virtually all of the Goldmine data on our Palm Pilots. We are currently using model Vx, which has 8 megabytes of memory and fits in my shirt pocket. All of the financial data as well as the last five history comments about each client is contained in the Pilot. Naturally we have our calendar and phone numbers for each client as well. We use the Pilot for world news and financial information when we travel. A quick update to

Avantgo.com by a modem that we connect to the Pilot keeps us up to date. Frequently when we travel, we get an update from an airport lounge by dialing a toll-free number, and in a matter of 2 to 3 minutes the update is complete. We also have a list of stocks that gets updated along with the news stories for each company. The publications we download are the *Wall Street Journal, Bloomberg, Financial Times, USA Today,* and Hollywood.com. The last one tells us what movies are playing and the times for each movie theater in our particular zip code. Map Quest is also on the Pilot and we can input a request for directions and it shows us a map with driving instructions, just like on the computer. I also have The Weather Channel with Manhattan programmed in so that when I travel to New York each quarter, I get an accurate forecast, so I know what to pack. On the Palm, we also have Vindigo, a restaurant, nightlife, and movie guide for several cities. I input my exact location and I choose what type of food I would like to eat, and it gives me a *Zagat* review as well as the address, phone number, and walking instructions and distance from where I am. I can sort the results by distance, price, or review. When traveling to a different city it has proved invaluable. A modem connection is not needed to get the results. A periodic update through the regular synching process is all that is needed. We synchronize the Palm with our desktop computer daily when we are in the office. It is a 5-minute process. In addition to synching with Goldmine, it goes to the Internet and updates all of the aforementioned services and deletes the outdated information. Any changes we make in the Pilot, say an address change, gets changed in Goldmine when we synch. When we are out of the office, we synch with the Internet only for our daily information. We could synch with Goldmine over the telephone line, but it isn't something we have needed to do. I read my Pilot sometimes when I am at lunch or even at home to keep updated. I have canceled most of my paper subscriptions."

"Contained within the Pilot is a clock that keeps track of different time zones and has an excellent alarm feature. Recently I stayed in a very upscale New York City hotel and asked for a 6:00 A.M. wakeup call. I never did get the call, but I had programmed the Pilot to wake me at 6:15 A.M., which it did."

"We use a program to store all of our passwords, logons, and scripts for the many electronic gatekeepers we need to pass through. The data is securely encrypted on the Palm for security.

"Recently while in a business meeting, I took out the Palm and the pocket keyboard for note taking. The keyboard opens to full size. The Pilot sits atop on a connector. I have replaced the laptop as a tool I take to meetings with the Pilot and the keyboard. Both the Pilot and keyboard fit in my pocket. The power in the Palm lasts me about 2 to 3 weeks before it needs to recharge. When the Pilot sits in the cradle near my desktop for synching, it also recharges so the need for a full charge is only needed when I return from a business trip."

"While on the road, I send and receive all of my e-mail using the Pilot and the modem. My cell phone can also receive short e-mails, so that if I am traveling and cannot receive phone coverage, I can see the e-mail when I get back into range and can take appropriate action."

"Three years ago, I said that what I was looking for in the future was a portable device that would give me most of the PC capabilities while I'm on the move and not at either my office or second home. Everything I asked for then is available to me in my Palm Pilot now! So what do I want going forward from here? The integration of the telephone, the Palm Pilot, and the computer communicating wirelessly from anyplace on the globe is technically possible. I envision information being pushed to me in an intelligent manner that learns my preferences. Home, office, and handheld devices will communicate easily and effectively. Technology such as Bluetooth is paving the way. Voice recognition, I believe, is still several years away from being practical. Technology is leveraging my time both personally and business-wise."

QUESTIONS

1. Is it unusual today for a company to allow employees to work from a second home the way John does? How does the firm measure and monitor an employee's productivity and effectiveness when he or she isn't working in the office?
2. Is it unusual for an employee to supplement the employer's information and communications system with a private system as John has done with his team? What factors would influence or motivate an employee to invest his or her money in this way?

CASE STUDY

Dow Corning's Telephone System

Dow Corning's main telephone system in Midland is a digital Centrex system provided by Ameritech. In actuality, because of the way its locations are spread in relation to the telephone company central offices, there are three different Centrex systems that are loosely connected together. These three systems between them have over 5,000 lines connected. The systems have direct inward dialing (DID) and provide four-digit dialing between stations on the same Centrex, but seven-digit dialing is required between systems. The DID feature allows the company to provide good telephone service with only two central telephone operators on duty at one time, and two others trained as backups. These operators also serve as receptionists for the Dow Corning Center headquarters complex.

In the past, each department in the company paid for the local and long distance calls of its employees. The company produced an internal telephone bill each month from magnetic data tapes provided by Ameritech and AT&T and some internal tables of information that Dow Corning maintained. However, in 1999 a decision was made by the finance department to eliminate all charge backs within the company, so now the telephone costs are all kept in the IT department.

Users in the Midland area make long distance telephone calls by dialing the standard ten-digit telephone number (three-digit area code plus seven-digit telephone number). Calls are routed through one of the Centrex systems and out through AT&T's software defined network (SDN), which was installed in 1992. The result is that Dow Corning receives a very low per-minute rate for long distance calls—much lower than it would if the SDN arrangement had not been implemented.

Voice calls to one plant are not routed through SDN. The plant is in Hemlock, a town fairly near Midland, and Dow Corning has leased lines to the plant location so voice calls are routed on the leased line. Users dial a three-digit access code and then the four-digit extension number of the person they are calling at the plant. Of course, if a caller who forgets about the special line and access code may dial the Hemlock plant directly, which costs the company more money; but the manager of voice communications says that this has not been a big problem.

Dow Corning's plants and sales offices in the United States have a variety of equipment suited to their unique requirements. Some of the plants have small PBXs and others have key systems. Many of the sales offices are quite small, so key systems serve their needs adequately. If a plant needs an 800 line or other special service, the requirement is rolled into the overall voice network arrangement with AT&T and managed by Dow Corning's telecommunications staff.

The company's central customer service department, which enters and processes all orders for Dow Corning products from customers throughout the United States and Canada, has its own PBX. It was installed to handle the special needs of that department, including handling a high volume of incoming 800 calls from customers and distributors of Dow Corning's products and the need for special telephone usage reporting. The department has its own 800-service network for customer calls. Calls pass through an ACD system that allows customers to enter a customer service representative's extension number, if it is known, and directs callers who want to order product literature to another department. Callers are then transferred to a customer service operator who screens out a few other calls that have been misdialed and then routes the call to the appropriate customer service representative. The customer service PBX is connected to one of the Midland

Dow Corning's telephone operators also serve as receptionists for the corporate headquarters. (Courtesy of Dow Corning Corporation)

A Dow Corning customer service representative takes an order. She is using a headset connected to her Rolm telephone and a flat panel video display terminal that allows her to log on to four different applications simultaneously. (Courtesy of Dow Corning Corporation)

area Centrex systems and all of the telephones on the PBX have four-digit telephone numbers that can be accessed by Centrex users the same way they access any other telephone.

Connected to the main Centrex system is a voice mail system provided by Lucent Technologies. Voice mail was first investigated in 1982 and seemed to have applicability, particularly for sales people who travel a great deal. In 1983, Dow Corning began using voice mail on a public service bureau system and then installed its own voice mail computer in 1984. Today the Lucent system has more than 4,000 mailboxes for voice mail users throughout the company. Like the overall telephone costs, the costs of the voice mail system are borne by the IT department. The Lucent system has proven to be very reliable, and the users are extremely happy with the capability it provides.

The staff that manages the system has noticed that some people prefer to use voice messaging and others prefer to use e-mail when communicating with other employees. In general, people from marketing, sales, and the corporate communications departments are the biggest voice mail users, perhaps because much of their job content involves verbal communication. Research and IT people tend to prefer e-mail. Of course people in all of these departments have access to and use both the e-mail and voice mail systems, but the relative preferences are an interesting observation.

QUESTIONS

1. The use of a Centrex telephone system is a good example of "outsourcing." What benefits would Dow Corning have accrued by using a Centrex system compared to having its own large, in-house, PBX-based telephone system? What benefits or trade-offs would the company realize if it replaced the Centrex system with a PBX?
2. Assuming it is true that a PBX requires more in-house staff to manage than a Centrex system, do you suppose there could be a conflict between lower-level managers who want to expand their staff to increase their influence, and senior managers who want to reduce staff size?
3. Because standards exist, is tight control needed to build a company-wide voice communications network, or are products and services standardized enough so that each location in the company can be given autonomy, knowing that the voice network can be tied together after the local units decide what voice communication equipment they are going to buy?
4. Employees at Dow Corning and most large companies have a variety of ways to communicate with each other. e-mail, voice mail, fax, and the standard telephone connections are just some of the examples. How does an employee decide which is the most effective method for a particular communication?

REVIEW QUESTIONS

1. List the five functions of the telephone set.
2. Compare and contrast dial pulsing with the DTMF technique for dialing a telephone number.
3. If the telephone system is designed for a P.03 grade of service, how many calls would you expect to be blocked for every 500 calls placed?
4. Distinguish between a telephone line and a trunk.
5. In normal speech, some sounds above 3,000 Hz are generated. What happens to these frequencies when they are sent through the public telephone network?
6. Explain the term *modulation*. For what is it used?

7. Is the electrical representation of the voice signal that is transmitted between the home and the telephone company central office modulated? Explain your answer.
8. What are the three attributes of a sine wave?
9. Explain what a foreign exchange line does.
10. Compare and contrast key systems and PBXs.
11. Distinguish among drop wires, local loops, distribution cables, and feeder cables.
12. What is the bandwidth of the AM radio broadcasting band of frequencies in the United States? The FM radio band? (If necessary, do some research at the library.)
13. An analog signal is fed into one end of a twisted pair of wires. At the other end, the relative strength of the signal is measured as −10 dB. How much has the power of the signal dropped as it traveled through the wire?
14. List an example of an electrical wave with a frequency of 60 Hz, a radio wave with a frequency of 640 kHz, a radio wave with a frequency of 102 mHz, and a sound wave with a frequency of 880 Hz.
15. What are the attributes of a tone used for tone signaling in the telephone system?
16. What is the importance of Signaling System 7?
17. Explain the difference between LATAs and area codes. Can an LEC carry a telephone call between two telephones whose telephone numbers have different area codes? If so, under what conditions?
18. Explain 800 service.
19. What factors must the network analyst consider when selecting the type of long distance service that is best for his or her company? How can he or she get help in analyzing the data and making the decision?
20. Compare and contrast a voice mail system with an ordinary telephone answering machine.
21. Explain how a cellular telephone system works.
22. Explain toll fraud and some of the steps that can be taken to prevent it.
23. What are the security concerns surrounding cordless and cellular telephones?
24. Why do dial-up circuits have more errors than leased circuits?
25. Why is the world's telephone system being changed to digital circuits?
26. Describe how TASI works.
27. Describe some situations in which a business might want to have 800 service for its employees to use.
28. Why are voice signals multiplexed?
29. What are the trade-offs between a Centrex telephone system and a PBX?
30. Compare and contrast cordless phones and cellular systems.
31. What is the difference between e-mail and voice mail?
32. What are the reasons companies might like to use the Universal International Freephone Numbering service?
33. Why do some telephone customers not want to use the automatic number identification feature that is available on most public telephone systems? Why wouldn't they want to use the call waiting feature? 193,
34. Why is the United States behind in cellular telephone technology? What capabilities do other countries have with their cell phones that are not available in the U.S.?

PROBLEMS AND PROJECTS

1. A company with 300 employees in Texas wants to acquire a private telephone system. Would you suggest a key system or a PBX? Why?
2. Visit a local hotel/motel and find out what type of telephone system it uses. Does the system have the automatic reminder feature for wake-up calls? If not, how are wake-up calls handled? Are there any special features on

the hotel's telephone system that weren't described in the text? If so, what do they do?

3. Visit a local small business that has a key system installed. How many telephones and outside lines will the system handle? How much expansion capability does it have? Did the company purchase the system, or is the firm leasing/renting it? What features does the key system have? (Use the features described for PBXs as a checklist.)

4. Pick one of the following services to investigate: cellular telephone system; voice mail system; audio teleconferencing system. Find a company in your area or do research at the library, and write a two-page report describing the capabilities, features, and shortcomings of the system you have chosen.

5. Investigate the current status of Internet telephone service and report to your instructor or class.

Vocabulary

customer premise equipment (CPE)
analog signal
speaker cone
sidetone
switchhook
off-hook
dial tone
on-hook
dial pulsing
rotary dial
out-pulsing
dual-tone-multifrequency (DTMF)
Touchtone
central office
central office switch
call setup time
patch cord
step-by-step switch
Strowger switch
crossbar switch
reed relay
blocking
busy hour
grade of service
main distribution frame
end office
serving central office
toll office
switching office
public telephone network
public switched telephone network (PSTN)
local loop
tip and ring
drop wire
distribution cable
feeder cable
trunk
frequency
sine wave
cycle
Hertz (Hz)
kilohertz (kHz)
megahertz (MHz)
gigahertz (gHz)
bandwidth
guard channel
guard band
amplitude
level
decibel (dB)
attenuation
phase
frequency division multiplexing (FDM)
carrier wave
common carrier
modulation
amplitude modulation (AM)
frequency modulation (FM)
phase modulation (PM)
detector
demodulation
multiplexer
base group
channel group
group
supergroup
master group
jumbo group
time assignment speech interpolation (TASI)
direct current (DC) signaling
tone signaling
fast busy
E&M signaling
in-band signals
out-of-band signals
common channel interoffice signaling (CCIS)
Signaling System No. 7 (SS7)
routing code
city code
exchange code
local calling
local service area
local calls
flat rate service
measured rate service
long distance calls
direct distance dialing (DDD)
toll calls
international direct distance dialing (IDDD)

plain old telephone service (POTS)

800 service

Universal International Freephone Numbering (UIFN)

900 service

software defined network (SDN)

foreign exchange (FX) line

Integrated Services Digital Network (ISDN)

voice over IP (VoIP)

key system

private branch exchange (PBX)

private automatic branch exchange (PABX)

computer branch exchange (CBX)

station

extension

system features

station message detail recording (SMDR)

call detail recording (CDR)

station features

hacker

hybrid systems

communications servers

Centrex

tie line

tie trunk

virtual network

voice mail

automated attendant

audiotex

interactive voice response

wireless communications

cordless telephones

spread spectrum

cellular telephone service

cellular telephone system

mobile telephone switching office (MTSO)

time division multiple access (TDMA)

global system for mobile communications (GSM)

code division multiple access (CDMA)

wireless application protocol (WAP)

Bluetooth

marine telephone service

aeronautical telephone service

References

Arnst, Catherine. "Uncle Sam Please Pick a Cell-Phone Standard." *Business Week* (February 24, 1997): 34.

Baker, Stephen, Neil Gross, Irene M. Kunii, and Roger O. Crockett. "The Wireless Internet." *Business Week* (May 29, 2000): 136–144.

Barrett, Amy. "Before We Get a Worldwide Busy Signal." *Business Week* (December 16, 1996): 72.

Barrett, Amy, Peter Elstrom, and Catherine Arnst. "Vaulting the Walls with Wireless." *Business Week* (January 20, 1997): 45–46.

Bass, Steve. "Nightmare on ISDN Street." *PC World* (January 1997): 236.

Castaneda, Carol J. "Drivers on Phone Proven 4 Times More Likely to Crash." *USA Today* (February 13, 1997).

Creswell, Julie. "The Battle to Control Your Cell Phone." *Fortune* (March 29, 2000): 170–174.

Maney, Kevin. "Megahertz Remains a Mega-Mystery to Most." *USA Today* (February 13, 1997).

Mitchell, Dan. "Waiting for Wireless." *PC Computing* (January 2000): 72–77.

Ousey, Alison. "Simplify Your Life—Centrex Gives You Easier Access to Phone Features." *Teleconnect* (November 1996): 68–71.

Pollack, Andrew. "Cheaper Mobile Phones Capture Japan's Ear." *International Herald Tribune* (January 21, 1997).

Shannon, Victoria. "The Wireless Way: Are We Obsessed?" *International Herald Tribune* (February 24, 2000).

Stripp, David. "The Idol of the Geeks." *Fortune* (March 3, 1997): 40.

Varshney, Upkar, and Ron Vetter. "Emerging Mobile and Wireless Networks." *Communications of the ACM* (June 2000): 73–81.

PART THREE

The User Interface to the Network

Part Three describes the user interface to the communications network—the terminal or personal computer. This is a subject that you may be familiar with, but the discussion will serve as a level-setting exercise for the material in subsequent chapters.

Chapter 6 discusses data terminals. Various types of terminals are examined, and video display terminals and personal computers are discussed in depth. The functions of a cluster controller are introduced, and the chapter examines various amounts of terminal intelligence and the uses to which the intelligence is put.

After studying Part Three, you will have updated your knowledge about communications terminals and personal computers, and you will be ready to learn about the more technical details of telecommunications systems.

CHAPTER 6

Data Terminals and Personal Computers

OBJECTIVES

After you complete your study of this chapter, you should be able to

- classify terminals in several different ways;
- explain the capabilities and characteristics of several types of terminals;
- describe uses for several specialized types of terminals;
- distinguish among intelligent terminals, smart terminals, and dumb terminals and describe the characteristics of each;
- explain the criteria used when terminals are selected for a particular application;
- explain the concept of the "Total Cost of Ownership."

INTRODUCTION

This chapter begins your look at the details of data communications systems by studying the parts with which you are most likely to be familiar—the data terminal and personal computer. The study of data communications in this book takes an outside-in approach. The discussion begins at the terminal and, through the next six chapters, works its way into the computer to which the terminal is attached. For most students, this approach will take them from familiar territory—the terminal or personal computer—through the unfamiliar and more technical aspects of data communications, and then back to more familiar ground at the computer. Coupled with the structure provided by the ISO-OSI model that was introduced in Chapter 3, you will have two ways in which to categorize and classify your new knowledge.

This chapter relates to layer 6 of the ISO-OSI model, the presentation layer. Terminals, especially those with intelligence such as personal computers, have a great deal of control over the way that data is formatted to users, so hence the connection with layer 6.

Although you may have studied the material about terminals in a data processing class, the information in this chapter will serve as a quick review of the different types of terminals in common use. The chapter also highlights the telecommunications implications of the various types of terminals. It may also serve to fill in the gaps in your knowledge of data communications terminals and how they work.

DEFINITIONS

A *terminal* is an input/output device that may be attached to a computer via direct cable connection or via a communications line. The terminal may be dependent on the computer for computational power and/or for data. If the terminal is directly connected to the computer by a cable, telecommunications may not be involved. Such terminals are said to be *hardwired,* and they may use signaling methods other than those commonly used in telecommunications.

A VDT or other device does not need to be connected to the computer full time to qualify as a terminal. A personal computer is an example of a device that has its own intelligence and may operate independently of a host computer a good deal of the time. At other times, however, a connection may be made to allow the personal computer to be operated as a terminal attached to a host computer.

data terminal equipment

data circuit-terminating equipment

In communications terminology, both the terminal and the computer to which it is attached are properly known as *data terminal equipment (DTE).* Data terminal equipment operates internally in digital format. Its output signal is a series of digital pulses. *Data circuit-terminating equipment (DCE),* which you will study in Chapter 8, provides the interface between data terminal equipment and the communications line. As shown in Figure 6–1, when analog lines are used, DCE provides the translation between the digital format of the DTE and the analog format of the transmission line. Even when digital communications lines are used, DCE is required between the DTE and the line.

TERMINAL CLASSIFICATION

Data terminals can be classified into categories, but the classification scheme may not be definitive because the categories have wide overlap. Because of the overlap, any classification of terminals is questionable. Nonetheless, there are several categories that will serve us well for descriptive and discussion purposes. They are:

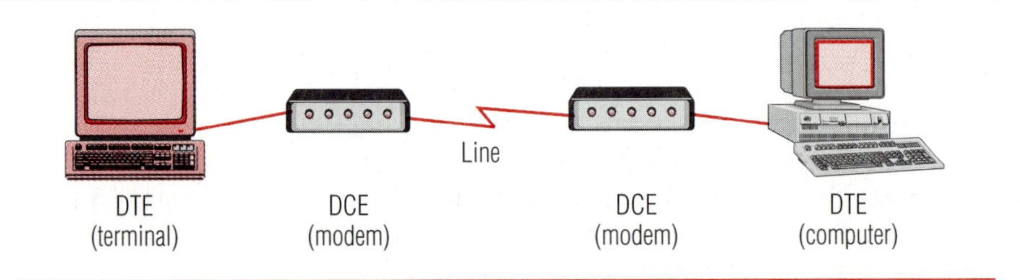

Figure 6–1
The location of data terminal equipment (DTE) and data circuit-terminating equipment (DCE) in a communications system.

- teletypewriters;
- video display terminals (VDTs);
- industry-oriented transaction terminals;
- intelligent terminals;
- remote job entry terminals;
- specialized terminals.

You can see the overlap already. Personal computers are intelligent devices, but they certainly use video display terminal technology; video display terminals are used in many transaction-oriented terminals such as bank ATMs.

Teletypewriter Terminal

The teletypewriter was the primary terminal in use before 1970. The simplest way to think of a teletypewriter terminal is as a typewriter with additional electronics and other features added for communications purposes. Indeed, some early terminals were just that—a typewriter equipped for communication. Most teletypewriter terminals in use today were originally designed for communications and will not operate as ordinary typewriters.

The teletypewriter is equipped with a keyboard and a mechanism for continuous feeding of paper. The keyboard is similar to that of a typewriter, although the keys are usually laid out a bit differently. The printing mechanism varies in how it forms the characters on the paper. Most use a matrix of wires that push the ribbon against the paper and form characters by a series of closely spaced dots, known as a *dot matrix* printing mechanism. Others employ heat as well as a matrix of wires to cause a chemical reaction in specially treated paper to form characters. This type of printing mechanism does not require a ribbon.

dot matrix

Teletypewriters and other terminals operate in one of two ways: *unbuffered* or *buffered*. In an unbuffered terminal, a character is transmitted to the computer as soon as a key on the keyboard is pressed. In a buffered terminal the keyed characters are stored in internal memory, called a *buffer*, until a special key such as the RETURN or ENTER key is pressed. Then all of the characters stored in the buffer are transmitted to the computer in one operation. At a buffered terminal users can correct typing mistakes before data is sent to the computer. Buffered terminals are the most common type today. Buffering is not unique to teletypewriters; it is found in most other types of terminals as well.

buffer

Some other characteristics of teletypewriter terminals are:

- slow speed—seldom more than 15 characters per second;
- very mechanical—therefore not as reliable as all-electronic devices;
- may be appropriate where hard copy is required;
- often connected to a computer only by dialing up.

Video Display Terminal (VDT)

The names *video display terminal (VDT), video display unit (VDU), cathode ray tube terminal (CRT),* and, on a PC, *monitor* are often used interchangeably. Not all video display terminals use cathode ray tube technology, however, so *video display terminal* is more general and appropriate. Other technologies used for VDTs are liquid crystal displays (LCDs), light emitting diodes (LEDs), gas plasma, and electroluminescent displays. These technologies yield a display that is much flatter and takes up less space on a desk, hence the generic name, *flat panel displays.*

flat panel displays

Other characteristics of VDTs are:

- all electronic—highly reliable;
- buffered (usually)—data is stored until the user presses the ENTER key;
- capable of very high-speed display of data;
- various types range in capability from simple, unintelligent "glass teletypewriter" to very intelligent, programmable ones.

The Screen The VDT frequently contains a cathode ray tube on which an electron beam causes phosphors to glow, forming the desired letters, numbers, special characters, or other patterns. Three electron beams and different types of phosphorus are used to create a color image in the same way that a color television picture is formed. The most common size of VDT screens is 15 inches, measured diagonally like television screens. However, 17-inch screens are becoming increasingly popular, and 19-, 20-, and 21-inch screens are available, though quite a bit more expensive. In applications that require high resolution, have detailed graphics, or require many windows to be open simultaneously, the larger screen sizes are definitely a good investment and easier on the eyes.

In a VDT that uses a cathode ray tube, characters are formed when an electron beam energizes selected dots of phosphorus within a matrix. Typical matrix sizes are 5×7, 7×9, and 8×10 dots. The characters are formed so that there is at least one unused row of dots around them to provide spacing between the characters. This concept is illustrated in Figure 6–2.

Screens may be either alphanumeric, which allow screen positions to be addressed at the character level, or *all-points-addressable (APA),* which allow individual dots on the screen to be controlled. These dots are called *picture elements,* commonly known as *pixels* or *pels,* and they can be turned on or off or set to a specific color under a program's control. APA displays are the norm for new video displays today; they are required for windows, graphics, and pictures to be displayed.

pixels

VDTs that use liquid crystal display (LCD) technology have two sheets of polarizing material with a liquid crystal solution between them. An electric current passed through the liquid causes the crystals to align

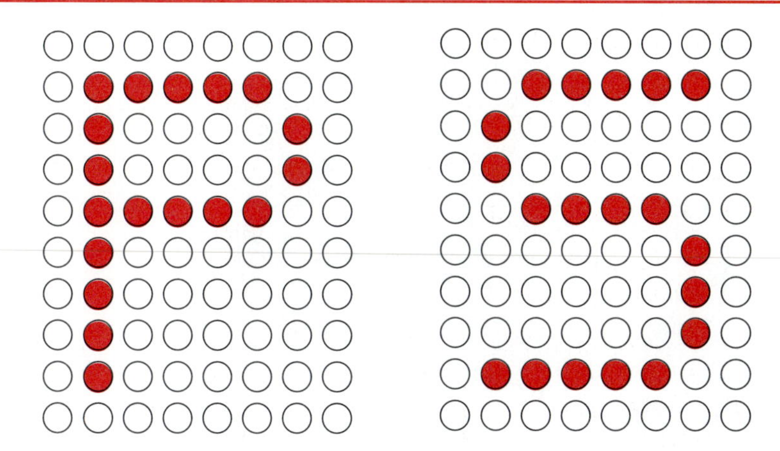

Figure 6–2
Characters formed by a dot matrix.

so that light cannot pass through them. Each crystal, therefore, is like a shutter, either allowing light to pass through or blocking the light. Both monochrome and color LCD screens are available but color screens are more widely used. LCD-based devices have a very slim profile, and a bright distortion-free image on a flat panel, making them especially attractive where space is limited. The problem has been that they were very expensive; however, the cost is coming down, so we can expect to see more of these devices in our offices and homes in coming years.

The screen of a VDT also displays a special place-marking character called a *cursor,* which indicates where the next character from the computer or keyboard will be displayed. In some cases, the cursor can be made to blink or display at a higher intensity than other characters on the screen so it can be seen easily. The keyboard of a VDT has special keys, called *cursor control keys,* for moving the cursor. These are often designated by arrows that point up, down, left, and right and that move the cursor in the indicated direction. Moving the cursor with the cursor control keys is not the same as moving it with the SPACE BAR or BACKSPACE key. The SPACE BAR inserts a space character, which will be transmitted to the computer, in the data stream. The BACKSPACE key removes a character; using a cursor control key has no effect on the data stream.

VDTs have several methods of highlighting characters for easy identification by the user. One technique is called *intensifying,* in which a character (or any collection of dots) on the screen is made brighter than the other characters around it. Some VDTs have several brightness levels. Another technique is to cause a character to *blink* by varying the intensity at which it is displayed several times each second. A third technique is called *reverse video,* which reverses the character and background colors. If, for example, the normal display shows green characters on a black

This Toshiba personal computer is typical of many found in the market today and is designed for multimedia use. (Courtesy of Toshiba America Information Systems, Inc.)

background, a reverse video character would be black on a green background. A fourth technique for highlighting characters is with the use of *color,* for example, displaying most of the characters in green and certain characters, such as error messages, in red.

When teletypewriter terminals receive output from a computer, they print it one line at a time. There is no inherent concept of a page, except as a collection of lines. This is known as *line-by-line mode.* VDTs can operate in a similar way, receiving and displaying one or a few lines at a time. When a VDT operates in line-by-line mode, the new lines normally appear at the bottom of the screen. All other lines are moved up, and the top line disappears off the top of the screen. This is much like what the user sees as printed teletypewriter output moves up and eventually disappears over the back of the terminal.

VDTs normally operate in *page* or *formatted mode.* There are several ways this is implemented. It is simplest to think of the screen of the VDT as a "page," consisting of a specific number of lines, each with a set number of characters. Programmatically, output can be placed anywhere on the screen, and if the communications lines are fast enough, the entire screen is displayed to the user at the same time. Page mode allows the screen to be laid out like a paper form with headings, field identifiers, and fields to be filled in as shown in Figure 6–3. The operator may be required

Quality Assurance System **10/11/01**
Lot Number Test Results **10:57**

Lot Number: FA128639

Item ID	Lot Number	QA Status	Status Date	Sample Type	Test Results
1959042	FA128639	Approved	20SEP94	1A	1:56.2 2: OK 3:7.1cm 4:HIGH AVERAGE
2876691	BQ212598	Approved	29SEP94	2	1:41.0 2:OK 3:4.5cm 4:OKAY
1464552	RM213356	Rejected	30SEP94	1A	1:55.1 2:LOW 3:1.1cm 4:HIGH

Figure 6–3
Page mode allows the VDT to be laid out like a paper form.

This keyboard for an IBM personal computer has a numeric keypad at the right side. Twelve program function keys are at the top. (Courtesy of IBM Corporation)

to fill in certain fields, whereas others are optional. After the operator types certain information on the form, for example, a material's lot number, the computer may respond by filling in other fields, such as the product's name and status.

The Keyboard VDT and personal computer keyboards, like teletypewriter keyboards, are similar to standard typewriter keyboards. The extra keys for cursor control already have been mentioned. In addition, there are usually special *function keys,* sometimes called *program function keys,* that direct the computer to perform actions predetermined by the software being run. Some keyboards have a *numeric keypad,* which is like a 10-key calculator keyboard. The numeric keypad is useful if a lot of numerical data must be entered, and its keys are in addition to the regular numerical keys on the keyboard.

Two other useful keys on a typical VDT keyboard are the INSERT and DELETE keys that the operator uses to insert or delete characters. When the cursor is moved under a character and DELETE is pressed, the character disappears from the screen, and all characters to the right of it on the line are shifted one position to the left to fill the gap. The INSERT key works just the opposite; it allows characters to be inserted on a line with all following characters shifted to the right.

Some terminals have specialized keyboards that usually are designed to help the operator perform his or her job more quickly and efficiently. The terminals in McDonald's restaurants are one example. They have keys for the clerk to press to indicate the item the customer is ordering and the quantity. The keys are actually labeled "Hamburger," "Cheeseburger," "Large Fries," "Shake," and so on. Other specialized keyboards are used on terminals in laboratories, plants, banks, and so on.

Another type of specialized keyboard is one used in countries that use alphabets made up of characters other than Roman letters. In Japan, China, Korea, and the Arab countries, for example, the alphabets and characters are much different from the ones we use in Western countries, and keyboards must be designed to handle those characters. Minor modifications are required in almost every country so that the keyboard has keys for national characters, such as those with an umlaut in German-speaking countries and accented characters where French and Spanish are spoken.

Other Input Mechanisms In addition to keyboards there are other ways to control the movement of the cursor. The most common of these is the *mouse* with which every user of an Apple Macintosh or Windows-based computer is familiar. Similar mechanisms are the *trackball,* often found on laptop computers, and the *joystick,* most commonly used with computer games.

The screens of some VDTs are *touch sensitive.* If the computer displays a question on the screen with a list of possible answers, the user can indicate a choice by touching an appropriate place on the screen. Touch screens have been successful in applications where people unfamiliar with computers must use the terminal because the screens are easy to use and require no training. One example is a VDT at an information booth at a fair or exposition. The general public may use the VDT to obtain information about the location of exhibits or other facilities. Obviously there is no opportunity to train the users, many of whom may never have used a computer terminal before.

VDT Selection Criteria When you select a VDT, consider the following factors:

- Does the face of the screen have a nonglare surface to prevent reflection from nearby light?
- Can the screen be tilted or swiveled to a comfortable viewing position?

- On a monochrome VDT, is the character display in one of the two standard colors (green on a black background or amber on a black background)?
- On a color VDT, can the default background and character colors be selected by the user?
- Is the screen image flickerfree? (Flickering screen images cause eyestrain.)
- Is the size of the screen appropriate for the application?
- Is the resolution of the characters fine enough to minimize eyestrain?
- Is the VDT programmable?
- Does it have graphics capability?

Personal Computers Used as Terminals

Personal computers (PCs), sometimes called *microcomputers,* are now the "norm" when one thinks of terminals on a communications network. In most applications, personal computers are the terminal of choice because of their ability to participate in a distributed processing or client-server system. Software that is commonly found in the personal computer environment, such as word processing and spreadsheet programs, may be used to manipulate data before it is sent to a server or mainframe for further processing, or data from the mainframe may be sent to the personal computer for analysis or formatting before it is presented to the user.

microcomputers

The wide adoption of Microsoft's Windows, by personal computer users, following the lead of the Apple Macintosh computers, has raised the use of personal computers to new levels of sophistication. The *graphical user interface (GUI)* that these operating systems provide is easier to use and more intuitive than previous, command-driven operating systems and is preferred by most people. Of course, users who are very familiar with the commands of DOS must make some adjustment to the GUI because it makes extensive use of the mouse and less use of the keyboard.

graphical user interface

These operating systems simulate or provide a true *multitasking* capability that allows multiple programs to be, or appear to be, operating at the same time. Thus, a user can be communicating with a host computer in one window of his VDT screen, while performing word processing, spreadsheet, or other processing in another window—and of course the number of windows is not limited to two.

Personal computers may be connected to networks and mainframe computers at several levels of sophistication. Sometimes they are connected so that they act like dumb terminals because this is the simplest, least expensive type of connection that can be made. In this case, they emulate a terminal that the host computer recognizes. Terminal emulation is accomplished by special hardware in the personal computer and a program called a *terminal emulation program.* The host thinks it is working with a standard terminal because the hardware or software in the PC

terminal emulation program

This personal computer, manufactured by Dell, has a CD-ROM player, sound card, and speakers, and is designed for multimedia use. (Courtesy of Dell Computer Corporation.)

responds just as the terminal would. The most common emulated terminals are the Digital Equipment Corporation (DEC) VT-100 and VT-220, and the IBM 3270.

uploading and downloading

Terminal emulation is a simple, inexpensive way to use a personal computer to communicate; however, when a personal computer emulates a dumb terminal, the power of the personal computer is not used to its best advantage. The host doesn't know it is communicating with another computer so it cannot tap the intelligence that resides in the PC. More sophisticated programming is required to allow the two computers to communicate in a distributed processing fashion or as peers. One simple enhancement is to add software that allows the personal computer and the mainframe to transfer files back and forth. Either computer can create a file using its own software, and then the operator can have it sent to the other computer using the SEND or RECEIVE command of the file transfer software. This type of file transfer is called *uploading* if the file is sent from the personal computer to the mainframe or other computer, and *downloading* if the file is sent from the mainframe to the personal computer. In a TCP/IP-based network the *file transfer protocol (FTP)* is most commonly used to accomplish these transfers.

network computer

One special type of computer is the so-called *network computer.* Network computers are designed to allow full connection to the Internet and other networks, but do not have all of the features of a traditional PC. For example, the size of the disk drive may be limited, or they may be required to download software from the network. The primary motivation for producing these machines is to reduce the cost of the computer required to attach to networks, especially the Internet, so that more people will be able to afford to connect.

More sophisticated software for distributed processing allows the host and personal computers to interact through a communications network in a more realtime fashion. Called *client-server computing*, such software allows a user or program on the personal computer, called the *client*, to request information or service from the host computer, called the *server*. Similarly, a user of a personal computer (client) might store or view data on another computer (server) as a simple extension of the disk capacity of the PC. This level of sophistication requires considerably more elaborate software and more powerful computers than are required for terminal emulation.

client

server

When personal computers are connected to a network or other computers, users must be concerned and vigilant about protecting the machines from *computer viruses*. Viruses may be inadvertently downloaded with a file or program, and can cause severe disruption and data destruction if not detected and eliminated. Fortunately virus detection and correction software is available, and it does a good job if it is regularly used.

viruses

Portable, laptop, or notebook computers are often used by sales representatives or other travelers in two modes. They use the machines as stand-alone computers during the day to perform processing independent of the host. At night, when the representative returns to the hotel room, the machines are used as terminals, connected via ordinary telephone lines to a host computer at the company or division headquarters. The results of the day's activities are transmitted to the host, and messages or electronic mail are received from it.

Sometimes the terms *personal computer* (or microcomputer) and *workstation* are used interchangeably, and their use can be confusing because there is little consistency in the industry about the use of these terms. However, generally speaking, when the term *personal computer* is used, people generally think of the hardware, operating system, and perhaps some basic applications, such as word processing, spreadsheets, or business graphics packages. When the term *workstation* is used, people think either of

workstation

- more powerful hardware and operating system software and a more complete, more specialized set of applications, such as would be used by a stockbroker or an engineer doing computer-aided design work; or
- the place where a person sits or stands to do work.

You can see why confusion exists! In this book, the term *workstation* is used both ways, but it will be clear to you which meaning is intended.

This mouse can be moved around on the desk top to control the location of the cursor on the screen of a VDT. This mouse is especially designed for children. (Courtesy of Logitech, Inc.)

Engineering Workstations

computer-aided design (CAD)

Engineering workstations are large, all-points-addressable VDTs with very high resolution. Engineering graphics VDTs are used for engineering design and drafting, the application being called *computer-aided design* or *computer-assisted drafting (CAD)*. The terminals usually contain their own microcomputers to perform specialized calculations to enlarge or reduce a drawing or to rotate the viewpoint of a drawing. Telecommunications for this type of terminal is used for downloading a drawing from the host computer to the terminal and uploading it back to the computer for storage after it has been modified.

Engineering workstations frequently place a heavy load on a communications line. The APA characteristic of the terminals coupled with the large screen size and the complexity of the drawings dictate that many bits or characters must be exchanged with a host computer. The amount of communication is inversely related to the amount of intelligence in the terminal. If the terminal is intelligent, it may only need to communicate with the host to retrieve and store drawings. All other work can be carried out on the terminal. If the terminal is unintelligent, it relies on the host to assist with every change made to the drawing, and the communication is very frequent and lengthy. Most CAD terminals fall somewhere

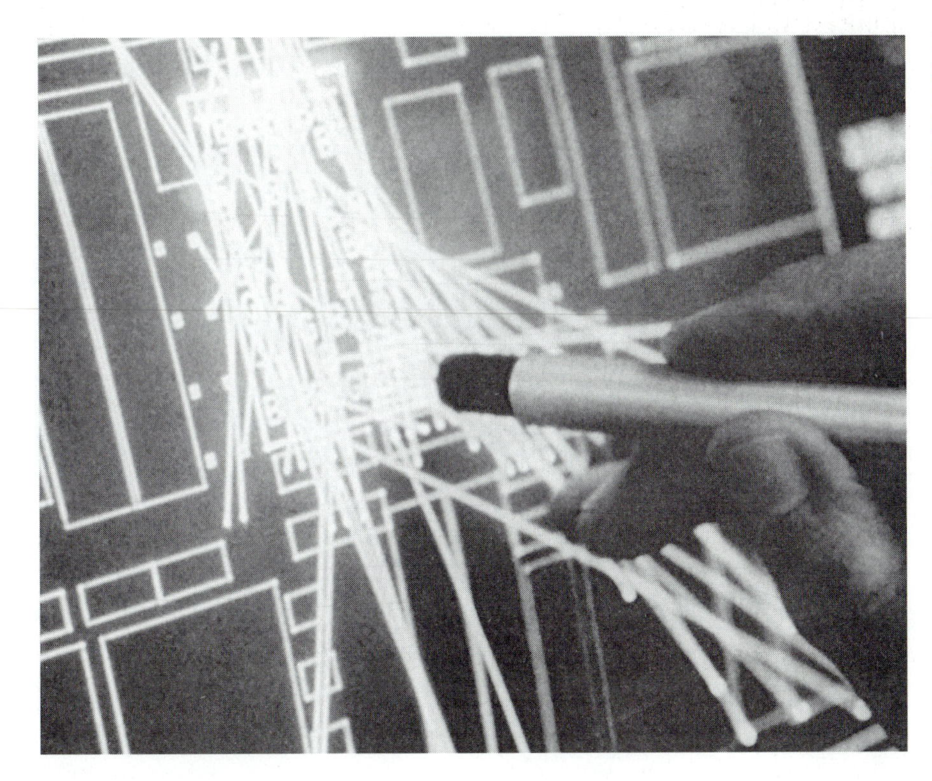

This user is holding a light pen, which is used for selecting items displayed on the screen of the VDT. (Courtesy of International Business Machines Corporation. Unauthorized use not permitted.)

between the two extremes, but the direction is clearly toward more intelligence in the terminal.

Industry-Oriented Transaction Terminals

Industry-oriented transaction terminals are specifically designed for the efficient processing of online transactions in a certain industry. Common examples of these types of terminals are the automatic teller machines and supermarket checkout terminals mentioned in previous chapters. Others include the terminals used by bank tellers. Many of these have special printers for recording changes in passbooks. Point of sale terminals read the data on merchandise tags in retail stores. Badge readers are used for time and attendance reporting in factories. These terminals are all designed around the requirements of a particular type of business transaction. They are designed to be easy to use, even by untrained operators or laypeople. Since the market for these specialized terminals usually is relatively small, the price is often higher than that for a more general terminal with the same electronic components and similar packaging.

Remote Job Entry (RJE) Terminals

Remote job entry (RJE) terminals, as the name implies, were originally used for the remote submission of batch programs or jobs to run on a host

This high-powered workstation looks similar to a standard personal computer but is designed for calculation-intensive engineering applications. (Courtesy of Digital Equipment Corporation)

computer and for receiving and printing the output of those jobs. The classic RJE terminal consisted of a reader for decks of punched cards containing the control statements or programs to be run on the computer and a medium- to high-speed printer for printing the results of the jobs' execution. To the host computer, the job appears to have been submitted from a card reader directly attached to the computer in the computer room, and the output appears to be printed on a computer room printer. The fact that telecommunications is involved is masked from the computer and from much of its operating system software.

With the obsolescence and decline in the use of punched cards and the widespread availability of interactive VDTs, work for the computer is submitted online with a few commands. As a result, the classic RJE terminal is seldom seen, although the high-speed remote printer is still in wide use. Printers of 100 to 1,000 lines per minute are most common.

Facsimile Machines

Facsimile (fax) machines used as specialized computer terminals have some of the same characteristics as OCR machines, which will be discussed in

The New York Stock Exchange is a large user of telecommunications. (Courtesy of New York Convention and Visitor's Bureau)

the next section. Instead of recording and coding individual characters, however, facsimile machines read and code patterns of light and dark areas on a sheet of paper, as was discussed in Chapter 4.

Facsimile machines are divided into four groups according to their technology and speed, as shown in Figure 6–4. Group I and II machines are internally analog in operation and the transmission of an 8 1/2 × 11 document takes several minutes. Group III and IV machines are internally digital and transmit documents in less than a minute. The speed with which Group IV machines can transmit documents is primarily limited by the speed of the communications line to which they are attached. On a digital line that transmits data at 56,000 or 64,000 bits per second, a Group IV facsimile machine can send an 8 1/2 × 11 inch document in under ten seconds.

Figure 6–4
Facsimile machines are grouped according to their technology and transmission speed.

Group I	Internally analog
	Transmits a page in about 6 minutes
	Rarely sold anymore
Group II	Usually internally analog
	Transmits a page in about 3 minutes
	Still sold because it is relatively inexpensive
Group III	Internally digital
	Fully automated operation
	Transmits an 8 1/2 × 11 page in under 30 seconds
	Transmits a 9,600 bps
	Much better copy quality than Group I or II
Group III Enhanced	Same as Group III but transmits at 14,400 bps with better error correction
	Most popular today
Super Group III	Same as Group III but transmits at 33,600 bsp with enhanced compression
Group IV	Internally digital
	Designed for digital phone lines
	Transmits a page in under 10 seconds
	Letter quality copy
	Expensive

The digital facsimile machines code a spot on the document (a pixel) as either a 1 or 0 depending on the amount of light that is reflected from it. The receiving machine produces a corresponding black or white pixel on the output document. The resolution or quality of the output depends on how many lines per inch are scanned by the transmitter—in other words, how many samples are taken. Standard Group III machines take 203 samples per inch horizontally across a page and 98 samples per inch vertically, though some models have "fine" or "superfine" modes that take 196 or 391 vertical samples for higher resolution. Most Group III machines are "smart" in that they only transmit what's on a page—they don't transmit white space as earlier machines did. Most facsimile machines apply algorithms to compress the 1s and 0s and reduce the number of bits that must be transmitted.

Group IV machines take 400 samples per an inch, both vertically and horizontally, so the resolution is much improved. Naturally the number of bits is higher too, but Group IV machines use a more sophisticated compression algorithm than Group III machines.

The digitized signal from a facsimile machine can be read into a computer and stored, since it is made up of bits. Conversely, a digitized image can be sent from a computer to a facsimile machine to produce hard copy output. Hardware boards, fax modems, and fax software for personal computers are widely available to provide this capability. The amount of storage required in the computer's memory or on its disk depends on the density of the image that is to be stored, but it ranges from

This Panasonic KX-FL501 facsimile machine with a handset has the ability to store frequently called numbers. It is designed to handle the needs of a small workgroup. (Courtesy of Panasonic Communications & Systems Co.)

25 to 120 kilobytes. You can imagine how quickly the hard disk on a personal computer can be filled if many facsimile images are stored!

If an appropriate all-points-addressable VDT is available, the facsimile image can be displayed. Facsimiles are to OCR terminals as all-points-addressable VDTs are to alphanumeric VDTs. In both cases, one is designed to handle only alphanumeric data, and the other is designed to handle images.

All analog Group III and Group IIIE machines pose a security risk. Anyone can attach a normal audio cassette recorder to a telephone line, record the incoming or outgoing fax tones, and play them back to another fax machine at a later time. A perfect reproduction of the fax will be obtained! There are encryption devices that will make the transmission unintelligible except to a fax machine that has an appropriate decryption unit.

Facsimile machines used as terminals find application where documents must be read into a computer and stored and/or transmitted to another location. They place a significant load on a communications line, however, but the key to reducing the load is sophisticated compression processing that reduces the number of bits to be transmitted without reducing the quality of the image as perceived by the recipient.

Specialized Terminals

Specialized terminals are not designed for a specific application or industry. Their use is more limited than that of a standard VDT or teletypewriter, but for certain applications they are extremely useful or even indispensable.

Telephone One important type of terminal in this category is the standard 12-key push-button telephone, which can be used to send digits and 2 "special" characters to a computer. The telephone is used in banking and other applications where the input is all numerical. A common use is for entering an account number, customer number, or product number used as the key to some information stored on a computer's disk. The computer looks up and reports particular information regarding the item identified by that key.

voice response unit

When the telephone provides the input, the companion output unit is often provided by an *audio response unit*, or *voice response unit*, on the computer that can form sounds from digital data stored on disk and "speak" the response back to the user. Audio response units produced in recent years can speak at varying rates of speed and with a pitch and inflection that is appropriate to the words being spoken. The use of an audio response is an example of the marriage of voice and data communications technology.

Several companies are using telephones and audio response units to allow their employees to select employee benefits in what is called a cafeteria-style benefits program. Employees can choose from several levels of medical insurance coverage, life insurance, contributions to savings plans, and so on. After dialing the special telephone number and providing a password through a Touchtone telephone, the employee hears the audio response unit giving the choices available for each of the benefit options. The employee makes his selection by pressing the appropriate digit on the telephone, and the selection is confirmed by a voice message. When all of the selections have been made, the audio response unit summarizes the choices for the caller before ending the telephone call. The computer application frequently confirms the choices in printed form a day or two later.

As was discussed in Chapter 5, the increasing popularity of cell phones is causing manufacturers to develop capabilities to use them as terminals, especially for accessing the Internet. In certain countries, especially Japan and in Scandinavia, cell phones are becoming the terminal of choice for sending e-mail, accessing stock reports, and viewing web sites that have been specially programmed to accommodate the cell phone's small screen.

Optical Recognition Another type of device is the *optical recognition* terminal, which can detect individual data items or characters and convert them into a code for transmission to a computer. This type of terminal uses a photo cell to sense areas of light or dark on paper or other media.

bar code readers

There are several types of optical recognition terminals. *Bar code readers* scan bars printed on merchandise or tags. The bars on grocery products are one example, but the use of bar coding is spreading to many industries. *Optical character recognition (OCR)* terminals detect and read individual characters of data. One type reads the characters on a typewritten page; another type can read handwritten data if it is clearly written. A simpler OCR device detects only marks on a page and is used to read survey forms, answer sheets from examinations, and medical questionnaires.

optical character recognition

Other Terminals Many other types of terminals are used for simple tasks, such as counting, weighing, measuring, and reporting results to the computer. On an assembly line, for example, a simple photoelectric cell may detect the passage of each item manufactured and report the event via communications lines to a computer. In a chemical plant, a device may measure the flow rate of a liquid through a pipe (or its temperature or pressure) and report this information to a computer. With advances in microprocessor technology, many devices that previously were unable to communicate are being given communications capability.

CLUSTER CONTROL UNITS

Cluster control units (CCU), sometimes called *terminal control units,* are used with some types of terminals as a way of sharing some of the expensive electronic components needed to support the advanced features of the terminals and the advanced communications software of today's multilayered networks. The IBM 3270 family of terminals is typical: Up to 64 VDTs may be attached to one CCU, and the control unit is attached to the communications line. This is illustrated in Figure 6–5.

CCUs are normally programmable devices, although frequently the programming is provided on diskettes by the manufacturer and cannot be changed by the user. CCUs may contain buffers for the terminals that can be shared. They may also perform code conversion and do error checking to ensure that data is received from the communications line correctly. Another use for the cluster control unit is to allow one or more printers to be shared among the terminals attached to it. A special key on the keyboard of each of the attached VDTs allows the image on the screen of the VDT to be printed immediately on a printer connected to the control unit without sending any data to or from the host computer.

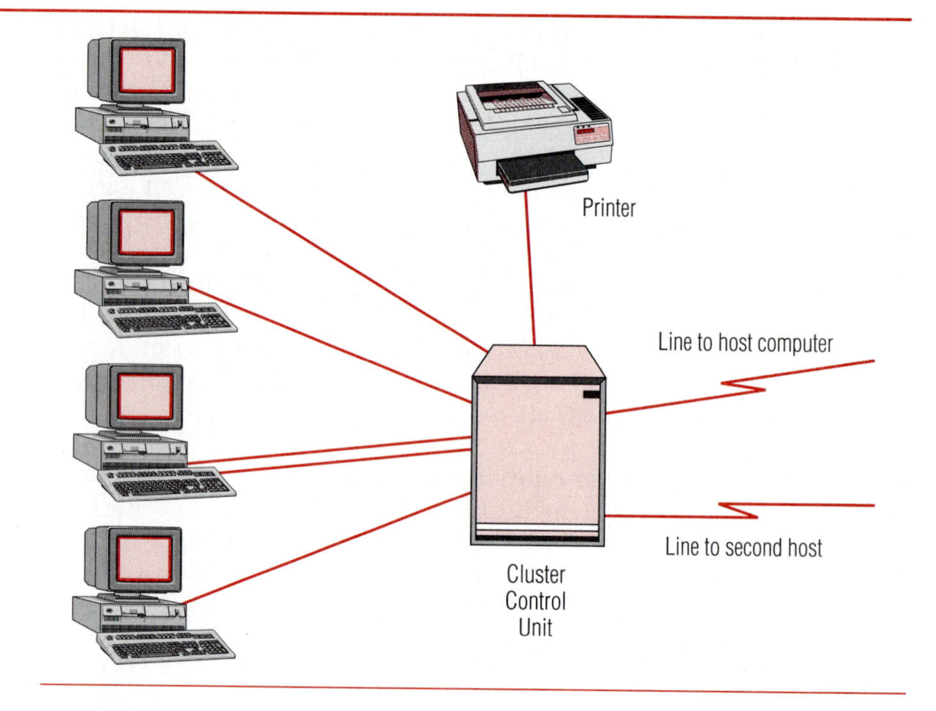

Figure 6–5
A cluster control unit with attached terminals.

Some CCUs can be attached to more than one communications line, allowing a terminal to access more than one computer. This is valuable to users who access multiple computers. Some CCUs allow the user to be logged on to more than one computer simultaneously. The VDT screen displays the information from one computer, but the user can switch to view information from the other computer using a key (or combination of keys) that is frequently called the *hot key*. Some CCUs allow a user to have multiple connections to one computer. Using this capability, the user can be logged on to more than one program on a single host computer and can switch back and forth between applications using the hot key. This capability is called *multiple sessions*.

multiple sessions

Cluster control units can have different degrees of intelligence ranging from simple buffering and translation facilities to being fully programmable controllers. One common use is for the CCU to provide the logic that allows the dumb terminals attached to it to participate in a network with a multilayer architecture of the type described in Chapter 3. Another example is the cluster controller that manages point-of-sale terminals in a retail store.

With the decline in the use of dumb terminals, cluster control units are not used as frequently as they were before PCs were so widely used. Hence, while there are many older units still installed and connected to networks, few new units are sold.

These two photos show bar code wands in use in typical manufacturing environments where they improve the accuracy of data input compared to inputting the data with a keyboard. (Top: Courtesy of McKesson Corporation. Bottom: Courtesy of PSC Inc.)

TERMINAL INTELLIGENCE LEVELS

With the reduction in the cost of integrated circuit chips, the desktop personal computer of today has as much power as a large mainframe computer had 20 years ago. If we were to classify all terminals on an intelligence scale, we would find that there is a range or continuum from fully

The IBM 3174 cluster control unit can attach up to 16 terminals or printers. Other models can attach up to 32 devices. The cluster control unit connects via a leased communication line to a host computer. (Courtesy of International Business Machines Corporation. Unauthorized use not permitted.)

intelligent, general-purpose computers used as terminals to completely unintelligent ("dumb") terminals. Some people classify terminals into three categories: *intelligent terminals, smart terminals,* and *dumb terminals.* However they are classified, today's terminals offer a wide range of capabilities.

Intelligent Terminals

The more intelligent a terminal is, the more able it is to participate in today's layered networks and in processing data. A fully intelligent terminal is in itself a general-purpose computer such as a personal computer. If a terminal can be programmed, some of the processing tasks can be performed locally on the terminal, in client-server mode, giving a degree of independence from the host. For many reasons, distributed processing is widely used in businesses. Computing tasks can be performed on the computer most suitable for the task. Screen formatting and data editing can be done by the intelligent terminal. However, where access to large databases or extensive computation is required, the power of the host processor can be brought into play. Using several computers also minimizes the risk of a massive computer failure, since users are not dependent on a single mainframe to do their processing.

When properly programmed, the intelligent terminal may be able to operate for long periods of time when there are failures in the data communications system or host computer. The terminal continues to operate without accessing the host and saves the results of its work. When the host computer becomes available, the terminal transmits the results of the processing it has performed.

Terminal intelligence is often present in less obvious ways. For example, a considerable amount of built-in intelligence is required for a supermarket terminal to read and interpret the bar code on a peanut butter jar passed through a laser beam at practically any angle or speed. Considerable intelligence also is required by bank or credit card terminals to read and interpret the magnetic stripe on a bank card, even though the stripe might have been damaged by abrasion from being carried in a person's wallet. The movement toward intelligent terminals is inevitable, depending primarily on their cost-effectiveness.

Smart Terminals

Smart terminals are not programmable, but they have memory that can be loaded with information. Their memory may be loaded by a transmission from a host computer, by keying data from an attached keyboard, or by reading it through some other attached device, such as a diskette reader.

One use for the smart terminal's memory is to store constant data, such as formats, that the operator uses repetitively. An order entry operator's terminal might be loaded with the format of the order form. This would save continually transmitting it from the host to the terminal each time the operator wanted to use the form. The operator would call up the blank format from the memory of the terminal to the display screen. The blanks in the format would be filled in with the data for the order. When the operator had checked the data, he or she would press the ENTER key. Usually only the data the operator entered, but not the order format, would be transmitted to the host, another saving of transmission time.

The smart terminal has a certain amount of independence from the host. The operator works with a format, completing information and making corrections until he or she is satisfied that the form is complete and correct. Only when the operator presses ENTER is there an interaction with the computer. Unlike the intelligent terminal, the smart terminal usually does not have the capability to save the data from an order if the communications line or host computer is down.

When a smart terminal is connected to a cluster control unit, some of its "intelligence" may be provided by the CCU. For example, the CCU might provide the storage for the formats either in its memory or on a disk.

Dumb Terminals

The advantages of unintelligent or "dumb" terminals are simplicity and low cost, but with today's sophisticated networks intelligence must be provided somewhere to allow dumb terminals to connect. Dumb terminals are typically totally dependent on a host computer for all processing capability and have either no storage or only limited, special-purpose storage for buffering and terminal control functions. Because of their low cost, however, dumb terminals are supported in most networks.

Because personal computers have become so ubiquitous and so inexpensive, there is little call for smart and dumb terminals anymore. Like cluster control units, there are still many in use, but few are manufactured or sold. To paraphrase an old song, anything a dumb or smart terminal can do, a PC can do better—and nearly as inexpensively.

WORKSTATION ERGONOMICS

In recent years, with the increased use of terminals in the workplace, a greater emphasis has been placed on workstation ergonomics, and much has been published about the total workstation environment for terminals. One meaning of the word *workstation* is the place where a person sits or stands to do work. It contains the working surface, terminal, chair, and any other equipment or supplies the person needs to do a job. It is the place where the computer meets the user. Having information about workstation design is useful for telecommunications people because they are often called on to advise others in the company about terminal use and environment. A drawing and some information about an ergonomically designed workstation are shown in Figure 6–6.

Figure 6–6
Ergonomically designed furniture allows for natural movement and allows workers to change position to prevent fatigue and stiffness.

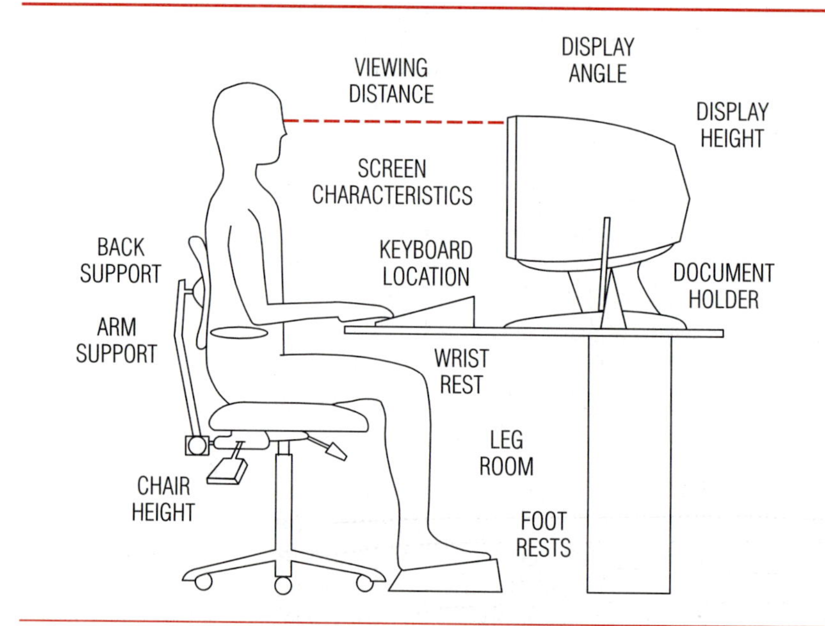

Considerations for Ergonomically Designed Workstations

1. The VDT screen should have a nonglare screen that can be tilted back 10 to 20 degrees.
2. The screen of the terminal should be about 18 inches from the operator's eyes.
3. The top of the screen should be no higher than eye level.
4. The chair should have a seat that adjusts up and down and a backrest that adjusts separately. When seated, the operator's feet should rest flat on the floor or on a footrest.
5. The keyboard should be attached to the rest of the terminal by a cable so it can be moved on the work surface. The keyboard should be positioned so that the operator's lower arms are parallel to the floor and the upper arms are perpendicular to it.
6. Nearby blinds should be closed and other sources of glare eliminated. Small adjustable task lights are usually preferable to overhead lights.

No matter how perfectly the workstation is set up, it is important to take a periodic break from concentrating on the VDT. There is a real incidence of headaches, fatigue, muscle aches, and temporary eyestrain resulting from prolonged use of VDTs under poor conditions. None of these problems are permanent, and they are minimized when VDT workstations, environments, and job structures are properly designed.

WORKSTATIONS FOR THE DISABLED

Telecommunications and computing are fields that disabled persons can actively participate in and contribute to with great success. Sometimes special adaptation must be made in order for a disabled person to access a workstation or other telecommunications device. Some people can't operate keyboards but can use a mouse. Displaying the letters and numbers of a keyboard on the screen and having the person select characters with the mouse is one solution that, while slow, may allow the disabled person to enter data. Voice entry, through voice recognition software, is another option that is applicable in some situations.

The key to finding a solution for a disabled person is ingenuity, creativity, and some assistance from others who have gone before. One source that provides a wealth of useful information is sponsored by the Assistive Technology Industry Association (ATIA) and is located on the Internet at www.atia.org. Searching the Internet and talking with occupational therapists and agencies that provide assistance to disabled people will locate many good ideas and other sources of information. People with small or large disabilities can clearly be productive users of telecommunications and computers.

THE VDT AND HEALTH CONCERNS

Much has been written about potential health hazards stemming from working with VDTs over long periods. The traditional questions are:

- Is there any radiation hazard in working with the VDT?
- Will it affect my eyesight?
- What is the most comfortable method of working with the VDT?
- How can I relieve the stress associated with constant use of a VDT?

Many health-related concerns about working with VDTs involve the radiation emitted from the display tube. Numerous studies by universities and government agencies worldwide have shown that the amount of radiation emitted from the terminal is well below established safety standards and, in most cases, is nearly undetectable. Other studies have shown no significant association between VDT use and vision defects or eye abnormalities.

This workstation is ergonomically designed to provide the operator with comfort, proper lighting, and functionality. (Courtesy of Steelcase)

TERMINAL SELECTION

When you select a terminal, it is important to identify the requirements of the applications for which the terminal will be used. The following questions can help you determine users' needs:

- Is a special application-oriented terminal required, or can a general-purpose terminal be used?
- Is hard copy required, or can a video display terminal be used?
- What types of operators will use the terminal? Will they be highly trained or novices who use the terminal only occasionally?
- Are the proposed terminals compatible with the existing computer?
- How much intelligence is required in the terminal?
- Is a personal computer required, or can a less intelligent, less expensive terminal be used?

Once these questions have been answered and the requirements determined, the next step is to compare alternative terminals from several vendors, looking at required and optional capabilities, vendor service and support, and cost. When all this information has been gathered, a selection can be made.

On a pragmatic basis, the type of terminal selected may be essentially predetermined by the type of computer to which it will be attached or by other terminals that the company has on hand. In many cases, a simple approach may be to assume that the terminal is going to be a standard PC, unless it is shown why some other type of terminal is required. PCs represent a good mix of capability and reliability. They are effective in most applications and are available today at a very low cost.

TERMINAL STANDARDIZATION AND TOTAL COST OF OWNERSHIP

The growth in the capability and complexity of personal computers has created a difficult situation for many corporations and other organizations. They find themselves with a variety of workstations that use different levels of technology, or even different technologies, and with users demanding to have the latest upgrades and features. Organizations can find themselves "chasing technology," trying to install every software release, and implementing every PC hardware advance in an effort to satisfy users.

total cost of ownership

Several research organizations, most notably the Gartner Group, have conducted research for many years on the *total cost of ownership* of personal computers in organizations. Their research tries to identify all of the costs associated with PC acquisition and usage during its lifetime. The most frequently published numbers from their studies show that a PC costs a company from $6,400 to over $13,000 during its lifetime. The actual hardware and software cost is only about 20 percent of the total. User operational costs, including learning to use the workstation, and training are about 46 percent, technical support costs are about 21 percent, and administrative costs about 13 percent. While these numbers are averages, and do not take into account the benefit that the organization receives because they have the workstations, the general conclusion is that these costs are too high.

The most significant opportunities to reduce the costs are in the areas of administration and support, including training. One way to reduce these costs is to have a workstation strategy, typically centralizing the management of the workstations and standardizing them. Companies that have developed aggressive workstation management programs have frequently standardized the hardware and software configurations of the workstations and established a policy to only upgrade or replace them periodically, typically every 3 or 4 years. For example, a company might move from a collection of workstations from a variety of vendors, which run various versions of software, to a standard configuration of two to four PC models, all from the same vendor and all running the same version of control software, such as Windows 98. Furthermore, these PCs would typically all have standard application software, such as the Microsoft Office suite of programs. Centralizing the decision making and standardizing the hardware and software results in less choice, but the organization benefits by being able to negotiate discounted bulk pricing for both the hardware and software. Users do not have to relearn software as often because software upgrades are installed infrequently. In these ways, the total cost of ownership of personal computers is significantly reduced.

Organizations that have adopted such programs are, in general, very satisfied with the results, and the users, after some initial complaining about losing the freedom to choose any hardware or software they want, usually agree that the standardization allows them to more easily exchange e-mail, data files, and other information with their coworkers, and is beneficial.

TYPICAL TERMINAL SCENARIO

This discussion is intended to give a practical example of how terminals are connected to networks, servers, and mainframe computers in typical businesses throughout the world. Although hundreds of options are available, certain configurations are more widespread than others.

The top part of Figure 6–7 shows personal computers connected to a local area network (LAN), which is discussed in Chapter 12. LANs operate at very high transmission speeds, but for limited distances. Also connected to the LAN is a server that has a printer attached and which may be shared by all of the PCs when they need to print. The server may have a large disk to provide mass storage capacity for large, shared files. A mainframe computer may also be attached to the LAN to provide additional processing or storage capacity.

The bottom part of Figure 6–7 shows dumb terminals and PCs emulating dumb terminals connected to a mainframe through a cluster control unit. The IBM 3270 series of computers is normally connected this way. The IBM 3270 series is still in use because approximately 80 percent of the mainframe computers in use are made by IBM or are IBM compatible. The IBM 3270 series contains alphanumeric and APA graphics terminals, which can display data in monochrome or color. The terminals are attached to the cluster control unit by cable and the CCU is typically connected to the mainframe computer with a communications line. The CCU also has a printer attached, which, like the printer connected to the server in the first example, may be shared by all of the terminal users.

SUMMARY

A wide variety of terminals are available today with a range of capabilities and prices. For the standard office environment, personal computers are by far the most dominant because they have the capability to do processing on their own, as well as to share processing with a server or mainframe computer. Special purpose terminals are optimized for a particular application, and their special features provide productivity benefits to their users, typically in the form of faster, more efficient data entry. Ultimately, the selection of terminals needs to fit the needs of the application and users.

Figure 6–7
Typical ways that personal computers and terminals are connected in a business setting.

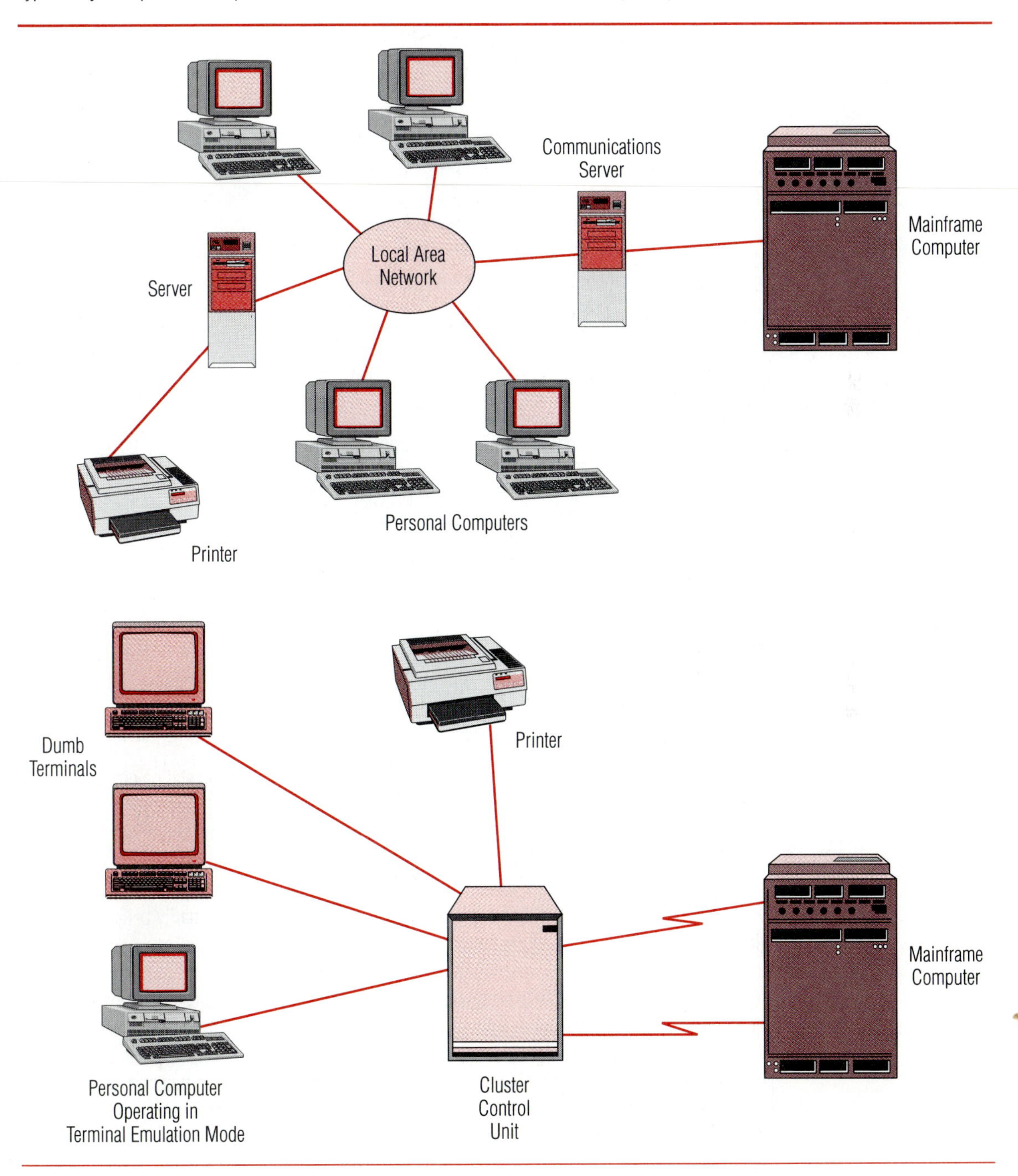

CASE STUDY

Evolution of Data Terminal Purchases at Dow Corning

The use of data terminals in Dow Corning started in 1953 when the first teletypewriters were installed for message communications between the corporate headquarters and several of the larger sales offices. These teletypewriters were not connected to a computer but simply connected with one another on point-to-point and multipoint lines. Early in 1969, the first video display terminals (VDTs) were installed. There was steady growth through the 1970s, but by 1980 there were just over 200 terminals installed in the company for about 5,000 employees. By then, they were located in many departments, but only a few people in each department used the computer applications. By 1989, there were over 3,000 VDTs, but by 1993 the number had decreased to around 2,000 as the number of personal computers increased. Today the only VDTs that are still in use are in the telecommunications control center, where they are used to monitor the status of certain communication circuits.

In 1983, the first personal computers were installed, and their growth has been rapid. The company began aggressively managing the acquisition of personal computers in 1984. A personal computer policy was established, and a list of PCs approved for use within the company was published. The growth in the number of PCs has been rapid.

1983—50
1987—800
1989—2,000
1993—4,600
1996—6,000
2000—11,000

Essentially all of them are connected to the telecommunications network.

In 1997, the last of the dumb terminals were replaced by a global workstation program, which placed standardized personal computers and software on the desks of all employees who require them for their jobs—virtually all employees. A request for price quotation was submitted to several PC vendors asking them to quote on a contract to supply nearly 8,000 standardized PC workstations with a standard set of software to all Dow Corning locations worldwide. After evaluation, Dow Corning determined that IBM could best meet the requirements, primarily because of IBM's global presence. Dow Corning negotiated a global 3-year lease contract for the workstations and embarked on an ambitious program to roll out the new workstations in a little more than 1 year. In 2000, when the original lease expired, up-to-date personal computers and software replaced all the original 1997 vintage workstations.

The company gained significant productivity benefits and cost savings by standardizing the hardware and software tools that employees use. Employee training was simplified, and the ability to share documents, graphs, and other information among employees was greatly enhanced. Also, the company found that it could leverage its negotiations with the potential hardware and

Dow Corning's torn tape message switching system was controlled from this location. Telex machines can be seen at the right. (Courtesy of Dow Corning Corporation)

A terminal user at Dow Corning is scheduling a meeting using one of the personal computers. (Courtesy of Dow Corning Corporation)

The personal computer is used in many Dow Corning departments for spreadsheets, word processing, and project management. (Courtesy of Dow Corning Corporation)

software suppliers because of the large quantity of hardware and software it acquired at one time, so there was a distinct financial advantage to the company as well.

A few other types of workstations are used in the manufacturing and research functions of the company, especially where they are connected to laboratory instruments or plant automation equipment. Dow Corning uses Compaq (formerly DEC), Sun, and IBM computers in its plants and laboratories. Interconnection of plant computers and communications equipment occurred as manufacturing and laboratory systems were integrated with traditional business applications in the early 1990s.

The use of facsimile (fax) machines is also quite extensive. Decisions about the acquisition of fax machines are made at the local departmental level; there is no central management. There are over 150 stand-alone facsimile machines used in the United States and more in the foreign locations. A few people also have fax modems in their personal computer, but the exact number is unknown. Users have found fax machines invaluable for rapidly exchanging printed documents that are not stored in a computer. Another large fax application has been the direct output of faxes from computer applications. One example is Material Safety Data Sheets, which contain safety information about Dow Corning's products. These sheets are distributed to anyone who requests them, and several thousand pages are automatically sent by fax each month. Another example is purchase orders to Midland-area vendors, which are sent directly by fax from the relevant business applications.

QUESTIONS

1. Dow Corning's global workstation program replaced all of the existing personal computers with new standard machines and software. How do you suppose employees, who in the past had been able to choose their own PC, reacted to this program?
2. Dow Corning has over 11,000 PCs for 9,000 employees. What are some situations where a Dow Corning employee might have more than one personal computer for use at work?
3. Identify some of the situations where non-standard PC workstations would still be appropriate.
4. Can you think of situations where Dow Corning might still need or want to retain some dumb terminals? What would be the benefits in doing so?

REVIEW QUESTIONS

1. Distinguish between data terminal equipment and data circuit-terminating equipment.
2. What are some desirable attributes of a terminal that will be used by the general public on an occasional basis?
3. How may the higher cost of an intelligent terminal be justified compared to a standard VDT?
4. What ergonomic factors must be considered when you select a terminal?
5. What are the advantages and disadvantages of a buffered terminal compared to one that is unbuffered?
6. Under what circumstances would an APA terminal be used?
7. Describe several methods of highlighting information displayed on a VDT.
8. What is the purpose of a function key on a terminal's keyboard?
9. Describe the factors to consider when you select a VDT. 238
10. For what applications is a mouse most useful?
11. How is remote job entry usually performed today?
12. Describe three levels of communication sophistication that may be employed when connecting a personal computer to a host computer.
13. Distinguish between the capabilities of facsimile machines classified as Group III and those classified as Group IV.
14. If a facsimile modem is to be installed in a personal computer, what other capabilities must the PC have to effectively use the facsimile capability?
15. If a typical facsimile requires 100 kilobytes on a hard disk, how many facsimile images can theoretically be stored on a 200 MB hard disk drive? Of course, in practice, the hard drive is not completely available for the user's data because much of the space is used to hold system and application software. On the personal computer you use, how many facsimile images could practically be stored?
16. In order to display an image that has been read into a computer from a facsimile machine, a(n) _____ VDT is needed.
17. List the primary purposes of the cluster control unit.
18. Distinguish among intelligent, smart, and dumb terminals.
19. What is client-server processing? What are its benefits?
20. What is a computer virus? How are they detected and corrected?
21. What are the advantages and disadvantages of VDTs that use LCD technology?
22. What is a network computer?
23. Explain the purpose of the file transfer protocol.
24. What is the difference between uploading and downloading?

25. Why are companies concerned about the total cost of ownership of their terminals, and especially their personal computers?

26. Identify several ways an organization can reduce the total cost of ownership of its PC workstations.

PROBLEMS AND PROJECTS

1. Visit a company that has a large number of terminals installed. Find out what portion of the terminals are "dumb," "smart," and "intelligent." Does the company make a strong distinction in usage between the different intelligence of its terminals? Do programmers design computer applications to take advantage of the capabilities of intelligent terminals? Does the company forecast that the number of intelligent terminals will grow in the future? If so, how fast? Will intelligent terminals ever totally replace dumb and smart terminals in the company?

2. Do you expect the use of terminals in most businesses to grow to the point where there is one terminal for every employee? What are some examples where this might not be the case? Can you think of a situation where there might be more than one terminal for every employee?

3. Vendors of color terminals claim that color improves the operator's productivity. How can a terminal improve productivity? What effect does color have? If color does, in fact, improve productivity, think of some applications where it would be especially useful.

4. Think of some applications that require a user to have a terminal that prints all of the interactions with a computer.

5. Visit a stockbroker and use one of the special "application-oriented" terminals to display current stock prices. How does the special design of the terminal make it easier to get the information? How much more difficult would it be to get stock prices if a general-purpose VDT were used instead? Talk to the broker and find out what other special capabilities the terminals have.

6. Do some research on the latest findings about the potential health hazards of radiation from VDTs.

Vocabulary

terminal
hardwired
data terminal equipment (DTE)
data circuit-terminating equipment (DCE)
dot matrix
unbuffered
buffered
buffer
video display terminal (VDT)
video display unit (VDU)
monitor
cathode ray tube terminal (CRT)
flat panel display
all-points-addressable (APA)
picture element (pixel, pel)
cursor
cursor control keys
intensifying
blink
reverse video
color
line-by-line mode
page mode
formatted mode
function keys
numeric keypad
mouse
trackball
joystick
touch sensitive
microcomputers
graphical user interface (GUI)
multitasking
terminal emulation program

- uploading
- downloading
- file transfer protocol (FTP)
- network computer
- client-server computing
- client
- server
- computer virus
- workstation
- computer-aided design (CAD)
- computer-assisted drafting (CAD)
- remote job entry (RJE)
- facsimile (fax) machine
- audio response unit
- voice response unit
- optical recognition
- bar code reader
- optical character recognition (OCR)
- cluster control unit (CCU)
- terminal control unit
- hot key
- multiple sessions
- intelligent terminal
- smart terminal
- dumb terminal
- total cost of ownership

References

Arar, Yardena. "Is an LCD Monitor in Your Future? Prices Are Shrinking, Screens Growing." *PC World* (March 1997): 72–74.

Brown, Bob. "Group IV Fax Grows Slowly but Steadily." *Network World* (August 7, 1989).

Chandler, Clay. "Japanese Bypass the PC On Their Way to the Web." *International Herald Tribune* (February 9, 2000).

Corson, Richard G. "VDTs—New Evidence Indicates Helpfulness over Harmfulness." *Data Management* (December 1986): 24.

Deixler, Lyle. "FAX Forges Ahead." *Teleconnect* (November 1996): 52–58.

Martin, James. *Design of Man-Computer Dialogues.* Englewood Cliffs, NJ: Prentice-Hall, 1973.

Sexton, Don. "Microsoft, Intel Boost Breed of 'Dumber' PCs." *The Japan Times* (March 14, 1997).

PART FOUR

How Data Is Communicated— The Technical Details

Part Four describes how communications systems work. Building on the background and vocabulary from earlier chapters, you'll be introduced to the technical details of communications systems.

Chapter 7 explores how data is coded for storage in a computer and for transmission on a communications line. Concepts and techniques of data compression and encryption also are discussed.

Chapter 8 deals with signal transmission on communications lines, and the interface of terminals to those lines. The transmission of analog and digital signals on analog and digital lines is discussed and the role of the modem is emphasized.

Communications lines—the transmission facility over which all of the communications takes place—are the subject of Chapter 9. The various media with which lines can be constructed are discussed, and several ways of classifying lines are explored.

Chapter 10 examines data link protocols, the rules by which data communications lines operate. Three categories of protocols are explained, and a specific protocol from each category is examined. You'll also learn about protocols for local area networks.

After studying Part Four, you will be conversant with all of the elements required to make a voice or data communications system operate. You will understand the pieces from which telecommunications networks are made in preparation for studying networks in the next several chapters.

CHAPTER

7

Data Coding

OBJECTIVES

After you complete your study of this chapter, you should be able to

- define what a code is;
- describe several different coding systems;
- describe three different types of characters that compose a code's character set;
- discuss typical functions of control characters;
- describe a method of error checking that is built into a code;
- explain the purpose of compression and how data is compressed;
- explain the purpose, pros, and cons of encryption;
- describe how symmetric and asymmetric key-based encryption systems work.

INTRODUCTION

In this chapter we will look at how data is coded when it is transmitted on telecommunications lines or stored in computers. The characteristics of several coding systems commonly used in data communications and computing will be examined in detail. You'll also learn about data compression and encryption, both of which are significant for efficient and secure communications.

TWO-STATE PHENOMENA

In introductory computer classes, you learned that many natural and physical phenomena are two-state systems. A baseball runner is out or safe; a basketball shot is made or missed; a light bulb is on or off; an electrical circuit is open or closed, resulting in a flow of electrical current or no flow. No ifs, ands, or maybes—it's one way or the other.

We refer to these two states as *binary states.* They can be represented using the binary digits 1 and 0. The term *binary digit* is abbreviated *bit.* When bits are used to represent the settings, we say that information has been coded. A 1 bit means the runner is safe, a 0 bit means the runner is out; a 1 could mean the basket is good or the current is flowing. Note that the 0 bit represents information just as the 1 bit does. It does not mean "nothing."

bit

CODING

A *code* is a predetermined set of symbols that have specific meanings. The key point is that the meanings are predefined; the sender and receiver must agree on the set of symbols and their meanings if the receiver is to make sense out of information that is sent. For data communications purposes, codes are assigned to individual characters. Characters are letters of the alphabet, digits, punctuation marks, and other special symbols, such as the dollar sign, asterisk, and equals sign.

A character code you may have heard of is the Morse code, shown in Figure 7–1. It is a binary code because it uses two code elements, a dot and a dash. Not all of the characters, however, have the same number of code elements. The code was structured so that the most frequently used characters require the fewest number of dots and dashes. In contrast, less frequently used characters are represented by more dots and dashes. The pitch (frequency) and volume (amplitude) of the coded signal are irrelevant to the meaning of the Morse code.

Morse code was designed for human use, and that is why the spaces between the letters and words have meaning. The translation of letters and numbers into Morse code is normally done by the human operator sending a message. Decoding from Morse code back into letters and numbers is done by the operator at the receiving end. With humans at both ends of the transmission, it is not important that all characters consist of the same number of dots and dashes or that the dots and dashes be perfectly formed and exactly consistent. Although officially the ratio of

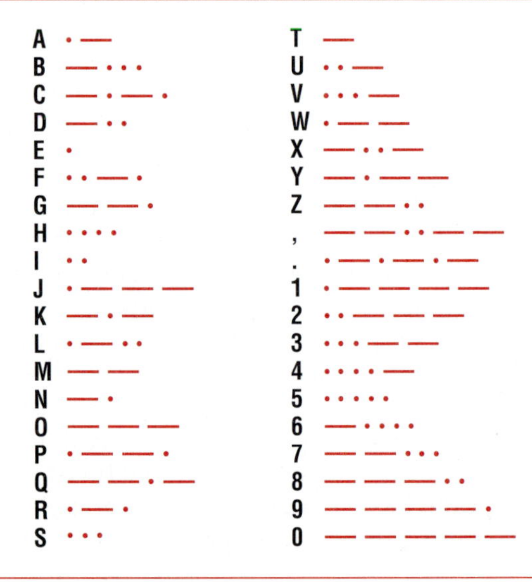

Figure 7–1
The Morse code.

length between a dot and a dash is 1:3, when a human is sending the Morse code using a telegraph key, this ratio varies widely. Expert telegraph operators can tell who is sending code just by listening to the way the characters sound.

MACHINE CODES

When two machines, such as a computer and a terminal, are used in communications, some of the attributes of the Morse code, or any other code systems designed for human use, are not desirable. It is much easier for a machine to process a code if the code has the following attributes:

- it is a true binary or two-state code;
- all of the characters have the same number of bits;
- all of the bits are perfectly formed;
- all of the bits are of the same duration.

A binary code works well for machines communicating by electrical means because 0 bits and 1 bits can be represented by a current flow that is on or off.

For transmission efficiency, it is ideal to have a coding system that uses a minimum number of bits to represent each character. How many bits are needed? With one bit we can have two states: 1 or 0. With two bits, there can be four combinations; 00, 01, 10, 11. With three bits, there can be eight combinations; 000, 001, 010, 011, 100, 101, 110, 111. Do you see a pattern? Each bit that is added doubles the number of combinations. If each combination of bits represents a character, then with three bits and eight combinations, eight characters can be represented. Mathematically, the number of unique combinations is expressed as 2^n where n is the number of bits.

Figure 7–2 shows a table of the powers of 2. From it you can see that using 5 bits gives 32 combinations, and with 6 bits there are 64 possibili-

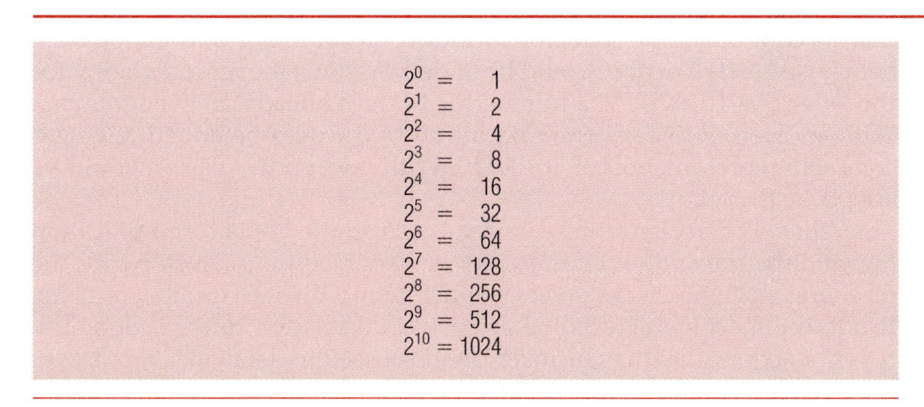

$2^0 = 1$
$2^1 = 2$
$2^2 = 4$
$2^3 = 8$
$2^4 = 16$
$2^5 = 32$
$2^6 = 64$
$2^7 = 128$
$2^8 = 256$
$2^9 = 512$
$2^{10} = 1024$

Figure 7–2
Powers of 2.

code points

ties. The number of possible combinations or characters in a coding system is called *code points.*

A code normally has unique groups of bits assigned to represent the various characters in a code. These unique sequences of bits are called *character assignments.* Different codes, even though they may have the same number of bits, use different character assignments. Character assignments must be made for three different types of characters: *alphanumeric characters, format effector characters,* and *control characters.* The alphanumeric characters are the letters, numerals, and symbols, such as punctuation marks and dollar signs. They are also referred to as *graphic characters* because they can be displayed on a terminal screen or printed on paper. Format effector characters control the positioning of information on a terminal screen or paper. Included in this group are tabs, backspaces, carriage returns, and line feeds. Control characters can be divided into two subgroups. Device control characters control hardware connected to a data processing or telecommunications system. A device control character might instruct a printer to skip to the next sheet of paper. Another device control character might change the color on a VDT screen. Transmission control characters control the telecommunications system and provide functions, such as identifying the beginning and end of the transmission or acknowledging that data has been correctly received.

PARITY CHECKING

parity bit

Many coding systems use an extra bit called a *parity bit.* The parity bit is added to each character representation for checking purposes. The parity bit is added so that the total number of 1 bits in the character will be an even (or in some cases, odd) number. From Figure 7–3, it can be seen that if the representation for the character *R* is 1010010, the number of 1 bits is 3, which is an odd number. If an even parity system is being used, which is the most common, a parity bit of 1 would be added to the character so that its complete representation would be 10100101. If odd parity were being used, a 0 parity bit would be added. Assume the representation for the letter *S* is 1010011. The number of 1 bits is already an even number. With an even parity system a parity bit of 0 would be added, giving a complete representation of 10100110. With odd parity, a 1 bit would be added.

When a character representation is transmitted on a communication line and the transmitter and receiver have agreed to use even parity, the receiving machine checks to see whether there is an even number of 1 bits. If not, the machine detects the error and takes appropriate action. The types of actions that may be taken are discussed in detail in Chapter 9.

The 7-bit ASCII code for the letter *R* (no parity bit)	**1010010**
• with even parity	10100101
• with odd parity	10100100
The 7-bit ASCII code for the letter *S* (no parity bit)	**1010011**
• with even parity	10100110
• with odd parity	10100111

Figure 7–3
If *even* parity is being used, the parity bit is set to 1 when necessary to make the total number of 1 bits in the character an *even* number. If *odd* parity is used, the parity bit is set to 1 when necessary to make the total number of 1 bits an *odd* number.

ESCAPE MECHANISMS

Most coding systems include a technique called an *escape mechanism.* One of the code points is assigned a special meaning, often called the escape or ESC character. When the ESC character is sent as part of the data, it means that the characters that follow are to be interpreted as having an alternate meaning. The concept is similar to the use of the SHIFT key on a computer terminal. When pressed, SHIFT changes the meanings of the other keys on the keyboard. Lowercase letters become uppercase letters, numbers become punctuation marks, and so forth.

In some systems, the escape character changes the meanings of all characters that follow it until a second escape character is sent. In other codes, the escape character changes the meaning of only the single character that immediately follows it. Different systems use escape characters in different ways. It is also possible to have several escape characters in a code. ESC1 might cause one meaning to be assigned to all of the following characters, whereas ESC2 might assign a different meaning.

The necessity to support escape codes complicates the design of equipment that is to code and decode the data. There is a trade-off between using escape codes to effectively obtain additional code points and adding another bit to the coding system to double the number of code points. If only a few more code points are needed, using escape codes may be an effective way to obtain them. Most coding systems in use today use 7 or 8 bits to represent each character but still include an escape mechanism to provide alternate meanings if required.

An ESC character in a coding system is normally not the same as the character that is generated when you press the ESC key on the keyboard of a personal computer or terminal. The ESC key generates a character that has a special meaning to the PC or its software, but rarely has a special meaning to the code system being used by the computer.

SPECIFIC CODES

Many different codes are used in telecommunications systems. One of the earliest was the original Baudot code invented by Emil Baudot, a Frenchman. His work produced a code structure that was eventually standardized as International Alphabet No. 1, but that code is not in use today. The name *Baudot code* has remained, however, and it is commonly but incorrectly applied to a code that was developed by Donald Murray. The Murray code was standardized as International Alphabet No. 2, and it is still in use. Since the name *Baudot code* is still so widely used, we will use it in this book to refer to the International Alphabet No. 2.

Baudot Code

The Baudot code uses 5 bits to represent a character and has no parity bit. With 5 bits, there are 32 unique code points, which are not enough to represent the alphabet, numerals, and punctuation marks. The escape mechanism used in the Baudot code is to assign two characters a unique function called the *letters shift* and *figures shift.* When a figures shift character is sent, all of the characters that follow it are treated as uppercase characters until a letters shift character is sent. Similarly, all of the characters following a letters shift are treated as lowercase characters until a figures shift is sent. The shift characters nearly double the possible characters that can be assigned. Since some characters must be recognized in either shift (examples are the letters shift and figures shift characters themselves and the carriage return and line feed control characters), the Baudot code actually has only 58 unique characters.

Figure 7–4 shows the Baudot code. The definition allows the uppercase characters to be interpreted differently depending on the country and to some extent the application. Figure 7–4 shows several different interpretations of the code points for different applications.

The Baudot code was originally developed for the French telegraph service and is still used today in older telegraph, teletypewriter, and telex communications. Most teletypewriters built before 1965 were designed around the use of the Baudot code. The mechanical design of the machine assumed that the 5-bit Baudot code would be used. Changing to another code would have meant completely redesigning the equipment.

American Standard Code for Information Interchange (ASCII)

The *American Standard Code for Information Interchange (ASCII)* grew out of work done by the American National Standards Institute (ANSI) and is the most widely used code in computers and telecommunications networks today. ASCII is a 7-bit code and therefore has 2^7, or 128, unique code points, as shown in Figure 7–5. The way to read the ASCII code chart

Figure 7–4
Baudot code.

• Denotes positive current

Start	1	2	3	4	5	Stop	Lowercase	Uppercase: Standard international telegraph alphabet No. 2	Uppercase: U.S.A. teletype commercial keyboard	Uppercase: AT&T fractions keyboard	Uppercase: Weather keyboard
	•	•				•	A	-	-	-	↑
	•			•	•	•	B	?	?	⅝	⊕
		•	•	•		•	C	:	:	⅛	○
	•			•		•	D	Who are you?	$	$	↗
	•					•	E	3	3	3	3
	•		•	•		•	F	Note 1	!	¼	→
		•		•	•	•	G	Note 1	&	&	↘
			•		•	•	H	Note 1	#		↓
		•	•			•	I	8	8	8	8
	•	•		•		•	J	Bell	Bell	'	↙
	•	•	•	•		•	K	(	(	½	←
		•			•	•	L	)	)	¾	↖
			•	•	•	•	M	.	.	.	.
			•	•		•	N	,	,	⅞	Ⓘ
				•	•	•	O	9	9	9	9
		•	•		•	•	P	0	0	0	ø
	•	•	•		•	•	Q	1	1	1	1
		•		•		•	R	4	4	4	4
	•		•			•	S	,	,	Bell	Bell
					•	•	T	5	5	5	5
	•	•	•			•	U	7	7	7	7
		•	•	•	•	•	V	=	;	⅜	⦶
	•	•			•	•	W	2	2	2	2
	•		•	•	•	•	X	/	/	/	/
	•		•		•	•	Y	6	6	6	6
	•				•	•	Z	+	"	"	+
						•	Blank				–
	•	•	•	•	•	•	Letters shift				↓
	•	•		•	•	•	Figures shift				↑
			•			•	Space				■
				•		•	Carriage return				<
		•				•	Line feed				≡

Note 1: Not allocated internationally; available to each country for internal use.

Figure 7–5
ASCII code.

First Three-Bit Positions (Bits 7, 6, 5)

Last Four-Bit Positions (Bits 4, 3, 2, 1)	000	001	010	011	100	101	110	111
0000	NUL	DLE	SP	0	@	P	`	p
0001	SOH	DC1	!	1	A	Q	a	q
0010	STX	DC2	″	2	B	R	b	r
0011	ETX	DC3	#	3	C	S	c	s
0100	EOT	DC4	$	4	D	T	d	t
0101	ENQ	NAK	%	5	E	U	e	u
0110	ACK	SYN	&	6	F	V	f	v
0111	BEL	ETB	′	7	G	W	g	w
1000	BS	CAN	(	8	H	X	h	x
1001	HT	EM	)	9	I	Y	i	y
1010	LF	SUB	*	:	J	Z	j	z
1011	VT	ESC	+	;	K	[	k	{
1100	FF	FS	,	<	L	\	l	|
1101	CR	GS	—	=	M	]	m	}
1110	SO	RS	•	>	N	∧	n	~
1111	SI	US	/	?	O	-	o	DEL

is to use the bits over the column head (high order bits) followed by the bits on the left side of the rows (low order bits). Thus, the ASCII code for the capital letter *P* is 1010000, and the code for the lowercase letter *s* is 1110011. Sometimes an eighth bit is added to the ASCII code. It can have two purposes: In some cases it is used as a parity bit to provide additional error checking capabilities. In other cases, the eighth bit is used as an additional data bit, which increases the number of code points to 2^8, or 256.

The ASCII code has uppercase and lowercase letters, digits, punctuation, and a large set of control characters that are discussed later. There are 96 graphic (printable) characters and 32 nonprintable control characters. Notice that the difference between the code for uppercase and lowercase letters is just 1 bit. Also, the last 4 bits of the code for the digits is their binary value. For example, the code for the digit 2 is 0110010. The last 4 bits, 0010, are the binary representation of the digit 2. Similarly, the last 4 bits of the digit 9, 1001, are the binary representation for the digit 9. These attributes are very useful when a computer manipulates data. The ASCII code is also designed for easy sorting by computer. Sorting by the binary value of the code yields a sequence that is meaningful to humans.

Figure 7–6
EBCDIC code.

Bits 0, 1, 2, 3	00				01				10				11			
Bits 4, 5, 6, 7	00	01	10	11	00	01	10	11	00	01	10	11	00	01	10	11
0000	NUL	DLE			SP	&	-									0
0001	SOH	SBA					/		a	j			A	J		1
0010	STX	EUA		SYN					b	k	s		B	K	S	2
0011	ETX	IC							c	l	t		C	L	T	3
0100									d	m	u		D	M	U	4
0101	PT	NL							e	n	v		E	N	V	5
0110			ETB						f	o	w		F	O	W	6
0111			ESC	EOT					g	p	x		G	P	X	7
1000									h	q	y		H	Q	Y	8
1001		EM							i	r	z		I	R	Z	9
1010					¢	!	|	:								
1011					.	$	,	#								
1100		DUP		RA	<	.	%	@								
1101		SF	ENQ	NAK	(	)	—	'								
1110		FM	ACK		+	;	>	=								
1111		ITB		SUB	|	—	?	"								

The international equivalent of the ASCII code is the International Alphabet No. 5, but ASCII also is widely used in many countries.

Extended Binary Coded Decimal Interchange Code (EBCDIC)

The *Extended Binary Coded Decimal Interchange Code (EBCDIC)* was developed by IBM. It is very widely used in IBM mainframe computers and older terminals. EBCDIC is an 8-bit code that has the 256 code points shown in Figure 7–6. This chart is read like the ASCII chart, described previously. The bits from the top of each column are combined with the bits at the left side of each row. Thus the EBCDIC code for the capital letter *P* is 11010111, and the code for lowercase letter *s* is 10100010.

Note the differences in the bit assignments between the ASCII and EBCDIC codes. The ASCII bits are numbered 7654321, whereas the EBCDIC bits are numbered 01234567. When data is translated from one coding system to the other, as is done frequently in data communications applications, it is important to be alert to these differing bit assignments to avoid incorrect translation.

Unicode

Whereas ASCII has 128 code points and EBCDIC has 256, neither of these coding systems provides enough code points to handle all of the world's languages. A very simple example is that ASCII does not have a unique code for the British pound sign. European languages have many unique characters such as the accented characters of French and the umlauted characters of German. But the languages of the Middle East and Asia don't even use the Roman character set and have thousands of unique characters. Think of all the special Greek symbols, and the Russian language, which uses Cyrillic characters. Or of Japanese, Chinese, and Korean, which use "picture" characters in addition to phonetic characters. Reading a Japanese newspaper requires a person to know about 2,000 Japanese kanji characters!

Unicode is a 16-bit character code designed to address the above problem and to provide enough code points so that all of the characters of all of the languages of the world have a unique 16-bit code point. Calculating 2^{16} yields 65,536 possibilities, which is enough to meet the requirement. No escape sequences or shift states are required in Unicode. Sixteen-bit characters are often called wide characters, although wide characters do not have to be Unicode characters. There are other 16-bit coding systems, such as IBM's *double byte character set (DBCS)*, which provide some of the functions that Unicode offers, but all of the other solutions are only partial and usually limited to a single vendor.

Unicode was developed by a group of companies called The Unicode Consortium. Members include IBM, Microsoft, Apple, Xerox, Sun, Digital Equipment Corp., Novell, Adobe NeXT, Lotus, and WordPerfect, an influential group indeed. With such backing, Unicode should have a good chance of becoming an international standard and of being designed into new hardware and software. Windows NT is the first major product to implement Unicode, but we can expect to see other products adopt it in the future. Unicode must replace ASCII, EBCDIC, and all other data codes if we are to achieve truly international systems and applications.

Other Coding Systems

Many other coding systems have been invented and used for data processing or communication purposes. However, most of them are not in wide use today. The *Binary Coded Decimal (BCD) code* is a 6-bit code that has 64 character combinations. It was used in the early days of data processing and grew out of the Hollerith code used to code data stored on punched cards. The BCD code was not standardized, and therefore one manufacturer's version was entirely different from another's.

BCD code

An *N-out-of-M code* is a system that helps detect the loss of the bit settings for a small number of bits. In this type of code, *M* bits are used to transmit each character, and *N* of these bits must be 1s. Therefore, the receiver has a way of detecting if certain characters are received incorrectly. IBM developed a 4-out-of-8 code in which 8 bits were used to represent each character. Four of the bits in each character were always 1s and 4 bits were always 0s. Although we won't go through the mathematics, the number of valid code points in the 4-out-of-8 code is 70. This limited the usefulness of the code. Although a 4-out-of-8 code detects errors more effectively than the use of a single parity bit, it is not as effective as some of the other techniques used for error detection discussed in Chapter 9.

N-out-of-M code

CONTROL CHARACTERS

In both the ASCII and EBCDIC code tables, you have seen that there are control characters—32 in ASCII and 27 in EBCDIC. These characters are used to control data transmission and the terminal devices connected to the lines. Some of the more common control characters and their meanings are as follows.

Transmission Control Characters

- SOH Start of Header is used at the beginning of the message's header. The header is a series of characters that indicates the address of the receiver and/or routing information for the message.
- STX Start of Text is a control character that precedes the main body or text of a message.
- ETX End of Text is used to terminate the text or main body of a message.
- EOT End of Transmission is used to indicate the end of a transmission that may have contained more than one set of headings and text.
- ACK Acknowledge is a control code sent by the receiver of a transmission as a positive acknowledgment to the sender.
- NAK Negative Acknowledgment is transmitted by the receiver as a negative response to the sender.
- NUL The null character in itself has no meaning. It is used to continue sending valid characters down the line for timing or other purposes. Note that the null character is not the same as the blank character. Blank has a different, unique binary code.

Device Control Characters

BEL Bell is used to sound an audible alarm, turn on a light, or otherwise indicate a need for operator attention or intervention at the receiving end.

DC1, DC2, DC3, DC4 These four device control characters are used in various ways by different communication systems. Most commonly, DC3 is also designated as the X-OFF character sent by a receiver to a transmitter when the receiver cannot accept any more data. When it is ready to receive again, the receiver sends the DC1 character, also known as X-ON.

Format Effector Control Characters

CR Carriage Return indicates to the receiver that the printing element should be moved to the first position of the current print line.

LF Line Feed indicates to the receiving device that it should move to the next print line on a printer or the next line on a display terminal.

HT Horizontal Tabulation moves the print element to the next predetermined (tab) position on a print line or display screen.

VT Vertical Tabulation moves the print element to the next predetermined vertical position on the print form or display screen.

CODE EFFICIENCY

Code efficiency is a measure of how few bits are used to convey the meaning of a character accurately. Efficiency of coding is important because an efficient code minimizes the number of bits that are transmitted on expensive communication facilities. If only numeric data is to be transmitted, a 4-bit code would be significantly more efficient than a 7- or 8-bit code. All of the digits can be represented with 4 bits, and the total data transmission would occur twice as fast as if an 8-bit code were used. Most data communications systems transmit only alphanumeric data and can therefore easily get by with a 7-bit code.

Another aspect of coding efficiency is the number of bits in a character that actually convey information. Bits that are used to determine the code points in a code are *information bits.* Any other bits are called *noninformation bits.* Parity bits, for example, convey no additional information and are not part of the original data. They are, therefore, noninformation bits.

Code efficiency is defined as the number of information bits divided by the total number of bits in a character. If an 8-bit code that includes one parity bit is being used, the code efficiency is calculated as $7/8 = .875$, or 87.5 percent. Codes that have more than one parity bit or in which only certain bit combinations are valid are less efficient than that. In addition, we will see that as data is moved to communications lines for transmission, other bits and even whole characters may be added for checking purposes. Because of checking that occurs during transmission (discussed in Chapter 9), the commonly used 7-bit version of the ASCII code and the EBCDIC code were designed with no inherent checking. They are dependent on other means to ensure that data is received accurately. Hence, both coding schemes have 100 percent code efficiency.

CODE CONVERSION

In most data communications systems, code conversion occurs from one coding system to another. Even though ASCII is the most widely used communications code, it is not universal. Although nearly all communication networks and personal computers use ASCII for both internal and communications purposes, most large IBM data processing equipment stores data in EBCDIC. Probably the most common conversion today is from ASCII to EBCDIC and back, which must be done whenever personal computers, which operate in ASCII, are connected to mainframe computers, which operate in EBCDIC. Data must be converted twice: from ASCII to EBCDIC when it arrives at the mainframe and from EBCDIC to ASCII when it is sent to the personal computer.

Code conversion is conceptually quite simple and is the type of task that a computer can perform readily. Usually, a table containing the target codes is stored in the memory of the computer. The binary value of the incoming character is used as an index into the table, and the target character is picked up as shown in Figure 7–7. The process gets more complicated when one code is converted to another code that uses a smaller number of bits. For example, when EBCDIC, with 256 unique characters, is converted to ASCII, with only 128 characters, many characters cannot be converted. Sometimes the ESC character must be brought into play so that one EBCDIC character translates into two ASCII characters—ESC and another character.

DATA COMPRESSION/COMPACTION

Compression or *compaction* is the process of reducing the number of bits used to represent a character, or shortening the number of characters before they are transmitted. The reasons for compressing data are to save storage space on the transmitting or receiving device or to save transmission time so that the message arrives faster and costs less to send. The results of successful compression are an apparent increase in transmission throughput and a reduction of storage or transmission cost.

In a typical application, a data compression device is employed at both ends of a communications line, as shown in Figure 7–8. At the transmitting end, the data is compressed using a set of mathematical rules called an *algorithm,* or a combination of algorithms. At the receiving end, a compatible device decompresses the data using the same algorithm(s) that were used to compress it. Most compression algorithms in use today

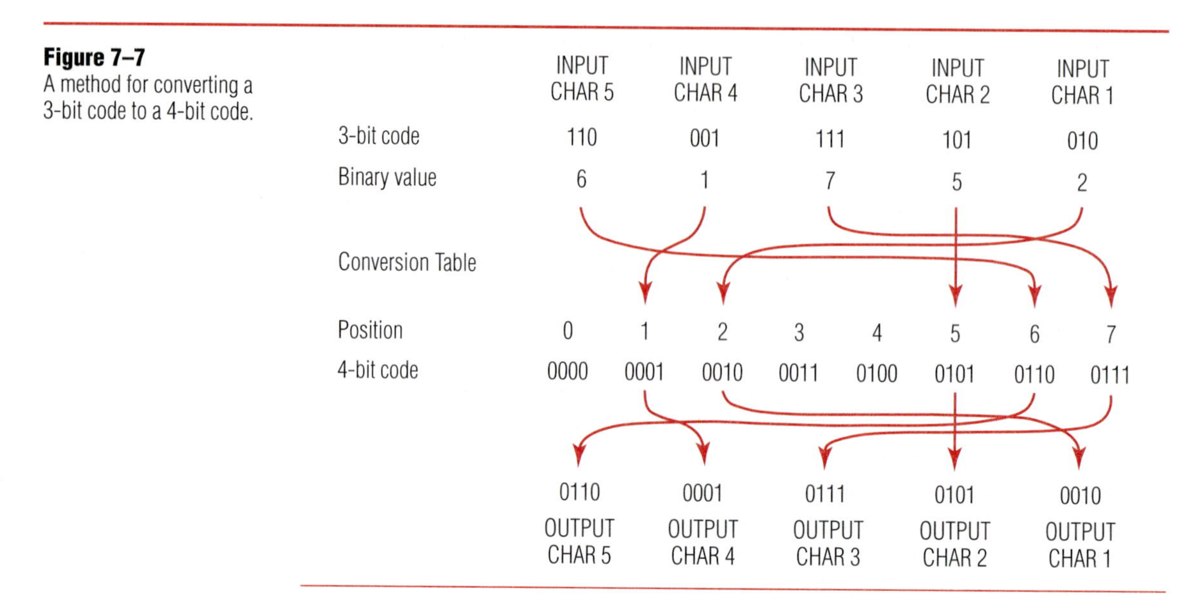

Figure 7–7
A method for converting a 3-bit code to a 4-bit code.

Figure 7–8
The locations of data compression devices on a telecommunications circuit.

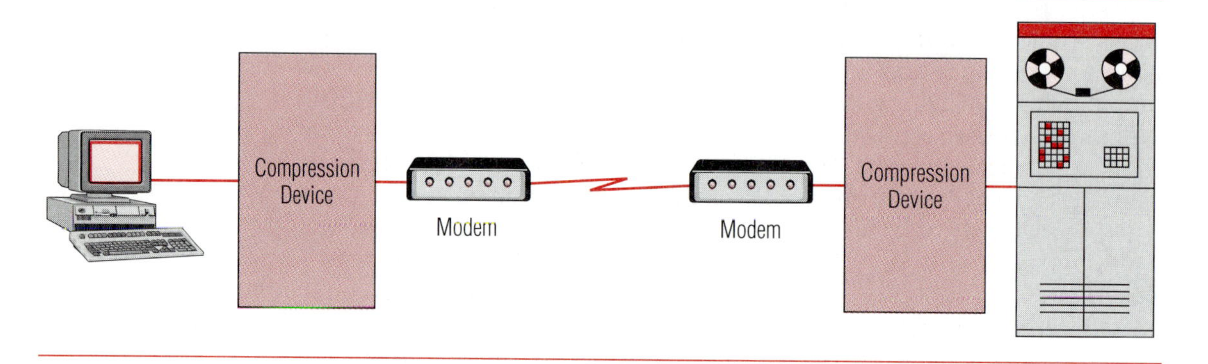

are proprietary, so it is necessary to use the same manufacturer's equipment for compression and decompression.

Three major types of data compression can be employed:

- character compression;
- run length encoding;
- character stripping.

Character Compression/Huffman Coding

character compression

Character compression, also called *Huffman coding,* consists of an algorithm that determines which characters are being transmitted most frequently and assigns a shortened bit configuration of perhaps 2 bits to those characters. Characters that are used less frequently are assigned longer combinations of bits. If performed successfully, character compression can reduce the total number of bits transmitted by a factor of about 2. This allows twice as much of a user's data to be sent down the communications line in a given amount of time, doubling the throughput rate. A more sophisticated method, called *adaptive Huffman coding*, updates the algorithm as the data is being transmitted to ensure that the fewest bits are always being assigned to the most frequently transmitted characters during a given time period.

Run Length Encoding

run length encoding

Run length encoding is a technique in which the data stream is scanned by the compression hardware looking for repetitive characters or repetitive groups of characters. The repetitive groups are replaced by a different, much shorter group of characters. For example, the sequence *XXXXXXXXXX* might become: *control character 10X*. This is shorthand notation that says that the letter *X* is to be repeated ten times. This shorthand reduces the original ten characters to four, including the control character, which is only 40 percent of the original number of characters. If a file of names and addresses was being transmitted and many of the addresses contained the same city name or zip code, many characters and much transmission time would be saved by substituting a one- or two-character sequence for the city name every time it appeared in the file. The shortened form would be transmitted, and at the receiving end, the special character sequence would be replaced by the original word.

Character Stripping

character stripping

Character stripping removes the leading and trailing control characters from a message and adds them back at the receiving end. Although in itself this technique may seem rudimentary, when combined with the other compression techniques it becomes very important. If the control characters were not removed from a message to be transmitted, they would be viewed by the character compression algorithm as being among the most frequently transmitted characters. This would reduce the effectiveness of the character compression on the main body of the message.

The most effective data compression devices use all three of these compression techniques, and others, in combination. First character stripping is performed, followed by character compression and run length encoding. To achieve maximum effectiveness, the compression algorithms must also be adaptive. That is, they must be constantly analyzing the data being transmitted and updating their internal tables of the most frequently sent characters or groups in order to assign the fewest bits to them.

One very popular compression algorithm, called MNP 5, is a combination of Huffman coding and run length encoding. MNP 5 is implemented in the hardware of many modems to effectively increase the transmission speed. It will be discussed again in Chapter 8.

Good data compression algorithms, such as the zip algorithm often used to compress data on personal computers, are highly mathematical in nature but can reduce the data transmitted in many business applications by a ratio of 4 to 1. Even more remarkable results are being obtained when digitized voice and video signals are compressed. The resulting savings in the cost of storage or transmission time must be weighed against the cost of the data compression equipment. The economics almost always works out in favor of doing as much compression as possible, especially for voice and video signals. For some companies, the most important benefit of data compression is the higher throughput rate, which may translate into improved response time for the users of an interactive application.

DATA ENCRYPTION

Another reason for manipulating the data stream before transmission is to encrypt the data to keep it private or secret. For many years, encryption was rarely used except by the government, but it is widely used in industry because of the concern for data security. Financial institutions are big users of encryption techniques to conceal account numbers and amounts in the data they transmit.

plaintext
ciphertext

Encryption is the transformation of the data from the meaningful code that is normally transmitted to a meaningless sequence of digits and letters that must be decrypted before it becomes meaningful again. Unencrypted information (often in readable form) is called *plaintext,* while encrypted data is called *ciphertext.* Simple encryption schemes use character substitutions. For example, a *B* may be substituted for an *A*, a *C* for a *B*, and so forth. Unfortunately, human analysts and computers can rapidly break this type of code. Therefore, more sophisticated techniques, which take advantage of the fact that machine codes are binary digits and can be manipulated mathematically, are used. The mathematical manipulation of bits is especially feasible when computers can be used to do the transformation.

Encryption is of particular interest in data communications because of the relative vulnerability of transmission over telephone lines or satellite. The environment is different from the computer room where data is transferred over cables that run a short distance and are often heavily shielded. Data transmitted using communications lines travels relatively long distances on various media and is essentially unprotected. Voice and data transmission are both subject to tapping or other unauthorized reception.

Modern encryption technology uses a mathematical algorithm and keys that are provided by the user. The encryption algorithm may be in the public domain, but the encrypted data can still be private because of the keys.

Simple encryption techniques are described as symmetric, which means that the unscrambling process is just the reverse of scrambling.

A mathematical formula scrambles or encrypts the data using a key that is provided by the user and which is usually a large number. To decrypt the data, the mathematical formula essentially processes the encrypted data in reverse, *using the same key* that was used to encrypt it.

A frequently used symmetric encryption technique is the *data encryption standard* (DES), which was developed by IBM and the U.S. government. The DES algorithm, in its original form, encrypts blocks of 64 bits using a 56-bit key which yields 2^{56}, or more than 72 quadrillion possibilities! The output of the algorithm is a string of bits that is transmitted. At the receiving end, the reverse process, *decryption,* occurs, using the same key.

decryption

Given the rapid increase in computing power in recent years, the DES encryption standard is vulnerable to a brute force attack. A powerful computer could be built that would have a good chance of determining a DES key in three or four hours. To solve the problem a new version called *triple DES* can be used. Triple DES doubles the key length to 112 bits. The block of data is encrypted with the first 56 bits of the key. Then it is encrypted again with the second 56 bits, and finally a third time with the first 56 bits. The resulting encrypted text is much harder and would require 2^{112} unique attempts instead of 2^{56} for standard DES.

Triple DES

Since the receiving end must know which key the transmitting end used, methods must be put in place to get the key from the transmitting end to the receiving end while protecting its confidentiality. This takes a combination of technical and management techniques. In many situations, managing the keys is a high cost overhead to the encryption. For example, couriers may have to be employed to deliver the keys, which are stored on magnetic tape. The couriers have to be trusted, and the tapes with the keys may need to be stored in secure boxes. The possibility that the keys could be disclosed is high.

Technologies are available that alleviate the situation by allowing keys to be transmitted electronically or by employing combinations of keys. However, key sharing still represents a weakness in the system. In 1976, an alternative encryption technique was developed at Stanford University. Messages are encrypted with one key that can be made public. The recipient uses a separate private key to decipher the text. This technique, called *public key encryption,* was implemented by a team at the Massachusetts Institute of Technology and is now marketed by RSA Security, Inc. The company's web site may be found at www.rsasecurity.com.

public key encryption

The RSA encryption system uses the concept of an *asymmetric key.* As the name suggests, the key used for encryption and the key used for decryption are not the same. A user, we'll call her Sue, picks her own set of keys, an encryption key and a decryption key. Sue publishes the encryption key so that everyone has access to it and it is referred to as the *public key.* She keeps the decryption key secret, and it is commonly referred to as the *private key.* When someone (call him Bob) wants to send her a message, he uses Sue's public key to encrypt the message. She can decrypt the message using

asymmetric key

her private key. Similarly if Al, Charlie, or Jim want to send Sue encrypted messages, they can use her public key, but only Sue can decrypt the message with her private key. The asymmetric key technique solves the problem of transporting the key between the sender of the message and the recipient.

The decryption keys used with the RSA algorithm are normally two very large prime numbers, and the encryption key is the product of the two. It turns out that if the prime numbers are sufficiently large, it is almost impossible to reverse the multiplication and deduce the original two numbers (the decryption key) if only the product (the encryption key) is known. In practical application, the prime numbers chosen are very large—in the range of 10^{130} to 10^{310}. Using the largest numbers, it has been estimated that the combined efforts of one hundred million personal computers would take one thousand years to crack such a key!

The great advantage of RSA encryption is that it solves the problem of key exchange. Sue doesn't care who knows her public key because it is only used for encrypting messages. She still needs to keep her private key, the one used for decryption, secret, but she can safeguard its identity. Because it is relatively invulnerable, public key encryption has been widely implemented in products such as Netscape Navigator and Lotus Notes, and it is in general use when data is encrypted on the Internet.

scramblers

In the voice world, we can identify a person we are talking to by the sound of his or her voice. In most situations, this provides adequate security. Voice encryption devices, commonly called *scramblers,* are available for voice transmissions. They make the voice transmission unintelligible to anyone without a descrambler, effectively rendering wiretapping useless. Scramblers are used to some extent in the government and Defense Department but are not widely used in industry.

The decision to encrypt data must be made carefully. The cost of the encryption hardware or software can be determined easily. The time that it takes a computer to encrypt or decrypt the data if software is used can be calculated and a monetary value placed on it. One must also calculate the throughput delays that occur during the encryption/decryption process and determine whether they are significant. Finally, the administrative or management costs of managing the keys, keeping them secure, and changing them regularly must be considered.

■ SUMMARY

This chapter looked at various alternatives for coding information. For machine communications, a binary code in which each character contains the same number of elements is desirable. The Baudot code was one of the earliest codes designed for machine-to-machine communication, but it was limited by a small number of unique combinations because it was only a 5-bit code. Most common today are the ASCII and EBCDIC codes, which contain 7 and 8 bits, respectively. Although ASCII is the most widely used code, EBCDIC is important because it is used by IBM in most of its products.

Compression and encryption are ways in which streams of bits are manipulated before being transmitted in order to eliminate redundancy or to provide privacy. The techniques of digitizing analog signals were examined. Of significance is the fact that all digitized signals look the same, whether they originated as voice, data, or images. These signals can, therefore, all be handled by the same equipment.

REVIEW QUESTIONS

1. Describe several attributes that are important in a code to be used for machine-to-machine transmission.
2. What is a code point ?
3. What are the advantages and disadvantages of an 8-bit code compared to a 5-bit code?
4. What is the minimum number of bits that would be needed to encode only alphabetic data? How many bits would be needed if numeric data were to be included?
5. Name the three different groups of character assignments that are made in a coding system.
6. Using the ASCII table in the text, convert your name to bits, 7 bits per character. Use upper- and lowercase characters as you would to write your name. Now, assuming even parity, add the parity bit to each coded character in your name.
7. Explain the meaning of the following control characters:

SOH	STX	EOT	ACK
NAK	CR	LF	

8. A certain coding scheme contains 5 information bits and 2 noninformation bits for each character. Calculate the code efficiency of this code.
9. What is the problem that the adoption of Unicode solves?
10. Describe three different techniques of data compression.
11. What is the most important data coding scheme? Why?
12. What is the difference between even parity and odd parity?
13. Explain the terms *asymmetric key, public key,* and *private key.*
14. Explain why triple DES is an improvement over DES.
15. Explain why data encrypted using the RSA algorithm and a sufficiently large key is virtually invulnerable to unauthorized decryption.

PROBLEMS AND PROJECTS

1. Amateur radio operators have modified computers to send Morse code automatically when a key on the keyboard of the computer is pressed. Describe some of the difficulties that would be encountered by a computer at the receiving end in interpreting the transmission for display on the screen.
2. Few words in any language have long strings of repetitive characters. When is data compression most useful?
3. Think of some applications where voice quality does not have to be as good as it is on the telephone system in order to be acceptable.

4. Develop a simple encryption technique to encrypt and decrypt alphanumeric data. Using your technique, encrypt this problem. If you have computer experience, write a program to encrypt data using the algorithm you developed. Compare the speed of your program to performing the encryption manually.

5. Do some research to find out about the current status of Unicode implementation.

Vocabulary

binary state
binary digit
bit
code
code points
character assignments
alphanumeric character
format effector character
control character
graphic character
parity bit
escape mechanism
Baudot code
letters shift
figures shift
American Standard Code for Information Interchange (ASCII)
Extended Binary Coded Decimal Interchange Code (EBCDIC)
Unicode
double byte character set (DBCS)
Binary Coded Decimal (BCD) code
N-out-of-M code
code efficiency
information bits
noninformation bits
compression
compaction
algorithm
character compression
Huffman coding
adaptive Huffman coding
run length coding
character stripping
encryption
plaintext
ciphertext
data encryption standard (DES)
decryption
triple DES
public key encryption
RSA public key encryption algorithm
asymmetric key
public key
private key
scrambler

References

Brostoff, George. "Utilizing Data Compression for Network Optimization." *Telecommunications* (June 1988):64.

Carey, John, and Peter Coy. "Duking It Out for the Decoder Ring." *Business Week* (November 22, 1993): 62.

Petzold, Charles. "Move Over, ASCII! Unicode Is Here." *PC Magazine* (October 26, 1993): 374–376.

Powell, Dave. "The Hidden Benefits of Data Compression." *Networking Management* (October 1989): 46–54.

Robinson, Sara. "Web Code-Cracking Sends Wake-Up Call to Business." *International Herald Tribune* (September 7, 1999).

Rothke, David. "Want to Encrypt Your Data? Try Triple-DES." *Datamation* (March 1997): 122–124.

Singh, Simon. *The Code Book.* New York: Doubleday, 1999.

Stipp, David. "Techno-Hero or Public Enemy?" *Fortune* (November 11, 1996): 173–182.

Vine, Andrea. "An Overview of The Unicode Standard 2.1." *MultiLingual Computing & Technology,* 10 (1): 50–52.

Zuckerman, M. J. "Decoding the Secrets of Digital Privacy." *USA Today* (October 6, 1999).

CHAPTER

8

Data Transmission and Modems

OBJECTIVES

After you complete your study of this chapter, you should be able to

- describe what is meant by the signaling rate of a circuit;
- describe the speed of a circuit measured in bits per second;
- describe three modes of data transmission;
- distinguish between asynchronous and synchronous transmission;
- explain how a modem works;
- describe several types of modulation used by modems;
- describe several types of interface between a terminal and the modem;
- describe several different types of modems and modem features.

INTRODUCTION

In this chapter we will study the way data is transmitted on a communications line. In Chapter 5 we examined the way analog telephone signals are transmitted on analog telephone lines. In this chapter we will look at three other cases:

- transmitting analog signals on a digital line;
- transmitting digital signals on a digital line;
- transmitting digital signals on an analog line.

For now, continue thinking of a communication "circuit" and "line" as being the same. Although this is not technically true (and it will be clarified in Chapter 9), the distinction is not important here.

The material in this chapter and in Chapter 9 describes the first layer of the ISO-OSI model that was introduced in Chapter 3. Layer 1, the physical layer, provides the physical path through which communications signals flow. In this chapter, you will study the way in which signals are converted and sent on the communications line. Although the material is somewhat technical, many basics of data transmission are introduced and explained. With the understanding of data transmission from this material, you will be well prepared for the material about data circuits, protocols, and networks presented in later chapters.

CIRCUIT SIGNALING RATE

Signaling rate

baud

Chapter 5 explained the concept of an analog circuit's bandwidth, which is the difference between the highest and lowest frequencies that the circuit can carry. You also saw that the standard telephone circuit has a bandwidth of 4,000 Hz. In 1928, Harry Nyquist of Bell Labs showed that the maximum signaling rate that can be achieved on a noiseless communication channel is 2B, where B is the bandwidth measured in Hz. The *signaling rate* is defined as the number of times per second that the signal on the circuit changes, whether in amplitude, frequency, or phase. The signaling rate is measured in *baud*. Thus, if the frequency, amplitude, or phase of a signal is changed 600 times per second, it is said to be signaling at 600 baud.

Nyquist's work suggests that on a circuit with a 4,000 Hz bandwidth, the maximum theoretical achievable signaling rate is 2 × 4,000 Hz, or 8,000 baud. In practice, the signaling rate that actually can be achieved often is significantly less than the theoretical maximum because in the real world, noise and other transmission impairments occur on every circuit. Furthermore, the time available to detect the signal changes at the receiving end becomes very small as the speed increases. At 2,400 baud, for example, the signal changes 2,400 times per second; therefore, the receiver must detect the signal change in 1/2,400 of a second, or 416.5 microseconds.

The more time the electronic circuitry at the receiving end has to detect a signal change, the more accurate the detection will be. Modem designers try to keep the baud rate low to provide as much time as possible for the signal change to be detected, but to achieve the highest possible throughput rates a signaling rate of 8,000 baud must be used.

CIRCUIT SPEED

bits per second

Circuit speed is defined as the number of bits that a circuit can carry in 1 second. It is measured in *bits per second (bps)*. The abbreviation bps is often incorrectly used interchangeably with the term *baud*. If only one bit is sent with each signal change on the circuit, the baud rate and the bps rate are the same, assuming all of the bits are of the same length. Today, however, sophisticated techniques allow more information to be encoded in each signal change. Therefore, the baud rate and bps rate of most data transmissions normally are quite different.

Suppose there were four unique signal changes on a circuit—that is, four unique changes of amplitude, frequency, or phase. With four possibilities, each of the four signal changes could represent 2 bits of information. For example, if a circuit could transmit four unique frequencies, one frequency could represent the bit combination 00, a second could represent 01, the third could represent 10, and the fourth frequency could rep-

Frequency	Dibit	Frequency	Tribit
f1	00	f1	000
f2	01	f2	001
f3	10	f3	010
f4	11	f4	011
		f5	100
		f6	101
		f7	110
		f8	111

Figure 8–1
If four different frequencies are used, each can represent 2 bits ($2^2 = 4$). If eight frequencies are used, each can represent 3 bits ($2^3 = 8$).

resent the bit combination 11. This is shown in Figure 8–1. When 2 bits of information are coded into one signal change, they are called *dibits.*

dibit

What if eight different frequencies were used? Then each frequency could represent 3 bits, called *tribits,* also shown in Figure 8–1. Using such techniques, dibits, tribits, and even *quadbits* can be encoded in each signaling change. Thus, circuit speeds can be achieved, as measured in bps, which are several times greater than the circuit's signaling rate as measured in bauds. When the signaling rate is 2,400 baud, it is possible to send 2,400, 4,800, 7,200, or even 9,600 bps, depending on how many bits are coded in one baud.

tribit

quadbit

On a data communications circuit, the data terminal equipment (DTE) at both ends of the circuit must send and receive bits at the same rate. Therefore, the bit rate of the circuit is the maximum bit rate that can be sent between the equipment. Of course, if the DTE is not generating data at the speed the circuit can handle, the actual bit rate sent will be less.

We have seen that bauds and bps are units of measure for different characteristics of a communications circuit. It is surprising that many people who work in the communications industry do not understand the difference between these two concepts and think that the terms are interchangeable. In fact, the only time that the baud rate and bps rate of a circuit are the same is when each signal change on the circuit indicates 1 bit. Although this is often the case in slow-speed transmissions of up to 1,200 bps, it is still incorrect to use the terms interchangeably.

MODES OF TRANSMISSION

Data transmissions can be classified in many ways. Three important ways are according to the data flow; type of physical connection; or timing. We'll look at each of these in detail.

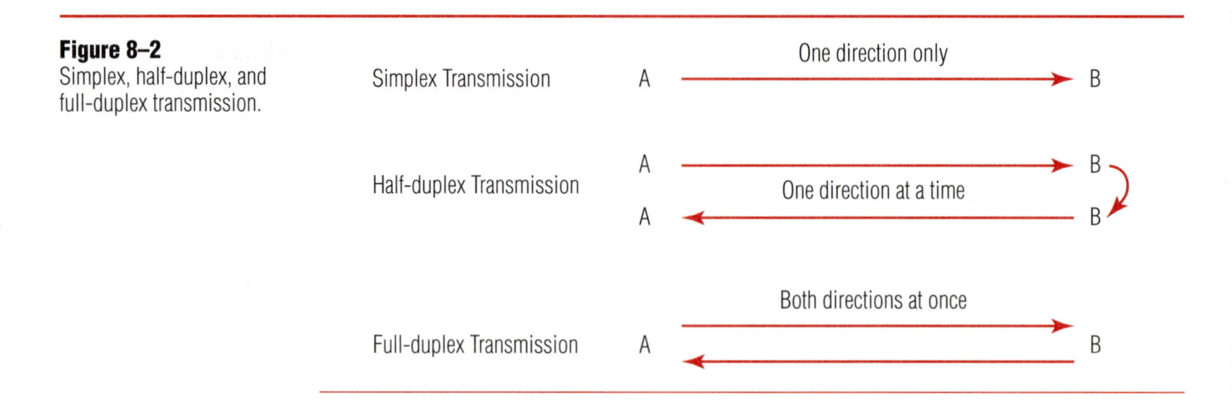

Figure 8–2
Simplex, half-duplex, and full-duplex transmission.

Data Flow

Simplex Transmission *Simplex transmission*, as shown in Figure 8–2, is data transmission in only one direction on a communications line. No transmission in the opposite direction is possible. Although simplex transmission is not what we usually first think of when we discuss data transmission, it is more common than you might imagine. In businesses, simplex transmission is used for monitors and alarms where a signal is sent from a sensor back to a central monitoring or control point. A similar application occurs in hospitals, where sensors on the patients send signals to the nurses' stations. Television and radio broadcasting are other examples of simplex transmission.

Half-Duplex Transmission *Half-duplex transmission (HDX)* is transmission in either direction on a circuit but only in one direction at a time. It is the most common form of data transmission, and it is commonly used in data processing applications where, for example, an inquiry is sent to the computer and then a response is sent back on the same circuit to the terminal. CB and amateur radio are other examples of half-duplex transmission.

Full-Duplex Transmission *Full-duplex transmission (FDX)* is data transmission in both directions simultaneously on the circuit. It requires more intelligence at both ends of the circuit to keep track of the two data streams. One place where FDX transmission is used is between computers. Computers have the intelligence and the speed to perform the necessary line control functions and to take advantage of the speed that FDX transmission offers. Whereas the computer power necessary to handle the complexities of full-duplex transmission once resided only in mainframe computers, now any personal computer can take advantage of FDX transmission. As a result, full-duplex transmissions are used more frequently than a few years ago, and the throughput benefits are realized more often.

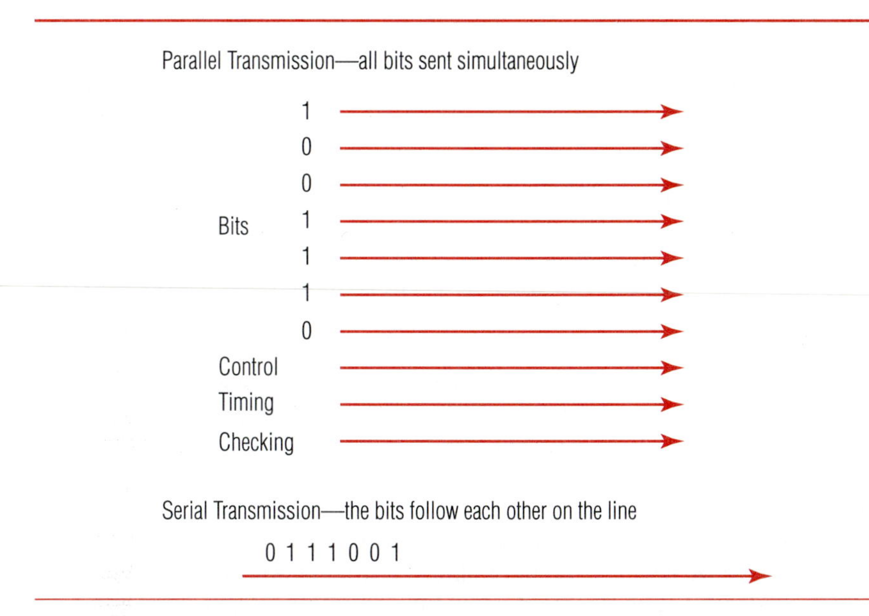

Figure 8–3
Parallel and serial transmission.

By the way, the definitions given above are those used in the United States. In Europe, the word *simplex* is applied to the above definition of *half duplex*, and *full duplex* is called simply *duplex*.

Type of Physical Connection

Parallel Mode When digital devices are in close proximity to one another, such as a personal computer and a printer, they are often connected by a cable that has one wire for each bit in a character of the data code being used. That is, if a 7-bit coding scheme, such as ASCII, is being used, the cable would have seven wires for the data bits as shown in Figure 8–3. There would also be several additional wires for control, timing, and checking. With multiple wires, all the bits of a character can be transferred between the devices at once. This is called *parallel mode* or *parallel transfer.* It is extremely fast, but because of the number of wires involved, it is also expensive and not practical over long distances.

Serial Mode As distance increases, *serial mode* or *serial transfer* is used. Each of the bits of each character are sent down a cable or communications line one after another, or serially. This type of transfer is illustrated in Figure 8–3 and is used for all telecommunications applications we will study. There is a trade-off, however. With serial transfer, the transmitter and receiver are more complicated because they have to decompose a character, serialize the bits for transmission, and reconstruct the bits into a character at the receiving end. With parallel transfer, the entire character is sent and received at one time.

Timing

start bits and stop bits

Asynchronous Transmission *Asynchronous transmission,* sometimes abbreviated *asynch,* is a transmission technique in which each character sent on a communications line is preceded by an extra bit called a *start bit* and followed by one or more extra bits called *stop bits.* It is sometimes called *start/stop transmission.*

Asynchronous transmission originated with the early mechanical teletypewriter terminals. For those teletypewriters, the start bit gave the mechanism in the receiving terminal time to start rotating and get up to speed and ready to receive the rest of the bits that made up the character. The stop bit gave the rotating mechanism time to get back to a known position and ready to receive the next character. Long ago, the convention was established that an idle line is one on which the signal for a 1 bit, also called the *mark* signal, is being sent. The opposite condition, the signal for a zero bit, is called the *space.*

mark and space signals

To start the mechanism of the receiver, the line is brought from the idle line, or mark state, to the 0, or space state, for one bit time, thus creating the start bit. For the next 7 bit times (when a 7-bit code is being used), the signal is changed between the 0 and 1 bits to represent the character being transmitted. Then the stop bit(s), a 1 bit, is sent. After the stop bit is sent, the line remains in idle mode (mark, or 1-bit, condition) until the next character is sent.

bit synchronization

The mark and space signals, coupled with the mechanical design of the early teletypewriters, allowed *bit synchronization* to be maintained. That is, the receiving device was able to determine just when to sample the communications line to detect a pulse or no-pulse condition. For maximum accuracy, it is important that the line be sampled at the center of the bit, not during the transition period between bits.

When asynchronous transmission is used, characters do not have to follow each other along the communications line in a precisely timed sequence. This is easy to visualize if you think about an amateur typist sitting at a terminal typing characters somewhat erratically. Ten or twelve characters may come rather quickly, followed by a long pause before the next character or group is sent. This is not a problem in asynchronous transmission because the start bit appended to the front of each character gives the receiving terminal some notice that a character is following.

transmission efficiency

Asynchronous transmission is relatively simple and inexpensive to implement. It is used widely by personal computers, inexpensive terminals, and commercial communications services. A penalty in terms of *transmission efficiency* is paid, however, because at least 2 extra bits are added to each character transmitted. The exact penalty depends on the number of bits that make up the character. The table below shows the transmission efficiency calculation when the Baudot, ASCII, and EBCDIC codes are used.

Code	Number of bits in code	Start/ stop bits	Total bits	Transmission efficiency
Baudot	5	2	7	5/7 or 71%
ASCII	7	2	9	7/9 or 77.7%
EBCDIC	8	2	10	8/10 or 80%

When the ASCII code is transmitted with an extra parity bit added for checking purposes, its transmission efficiency is 8/10, or 80 percent, like that of EBCDIC.

Asynchronous transmission does not require maintenance of precise synchronization between the transmitter and receiver for an extended period of time. When a start bit is sensed, the receiver knows that the next *n* bits (where *n* depends on the code being used) on the line make up a character. This is called *character synchronization.* After receiving a stop bit, the receiver simply waits for the next start bit. The only synchronization that has to occur is during the transmission of the 5 to 8 bits that make up a character so that the receiver looks at or samples the line at the right time. The sampling rate depends on the line speed, a rate that is predetermined and known by both the transmitter and receiving terminals. Without some form of character synchronization, a receiver might not know which was the first bit of a character, and the character would be misinterpreted.

character synchronization

Asynchronous transmission is used where equipment cost must be kept low. The transmission inefficiencies of asynchronous transmission were a big disadvantage when line speeds were primarily 300 bps or less. Now data can be transmitted asynchronously at over 28,800 bps, and the concern about the inefficiency is heard less frequently.

Synchronous Transmission The asynchronous transmission method is relatively inefficient because of the extra start and stop bits transmitted with each character; *synchronous transmission* is a more efficient transmission technique. When synchronous transmission is used, bit synchronization is maintained by clock circuitry in the transmitter and in the receiver. The timing generated by the transmitter's clock is sent along with the data so that the receiver can keep its clock synchronized with that of the transmitter throughout a long transmission.

With synchronous transmission, data characters usually are sent in large groups called *blocks.* The blocks contain special synchronization characters, with a unique bit pattern. They are inserted at the beginning and sometimes in the middle of each block. The *synchronization characters* perform a function similar to that of the start bit in asynchronous transmission. When the receiver sees the synchronization character, it knows that the next bit will be the first bit of a character, thus maintaining character synchronization.

synchronization characters

Synchronous communication has one to four synchronizing characters for each block of data. Asynchronous communication has two synchronizing bits for each character. If the ASCII code is used and 250 characters are to be sent, the number of bits sent by each transmission method is:

Asynchronous

250 characters × (7 data + 2 start/stop bits per character) = 2,250 bits

Synchronous

(250 data characters + 4 synchronizing characters)
× 7 bits per character = 1,778 bits

In this example, the synchronous transmission technique sends 21 percent fewer bits. It is, therefore, 21 percent more efficient than asynchronous communication in this example. The efficiency advantage of synchronous transmission improves as longer blocks of data are transmitted.

Because it has a higher transmission efficiency, especially when large blocks of data are being transmitted, synchronous communication is preferred over asynchronous data transmission when large amounts of data must be transmitted, especially when it can be done on dedicated communication lines. Sending data between servers in the same room that are connected by cable is one case where synchronous transmission may be useful in order to take advantage of its efficiencies. It is not as effective, however, when the public communication networks, such as the telephone system or the Internet are used, because these networks break the data into blocks according to their own needs, and the efficiencies of large block sizes are lost.

On the other hand, transmissions between two personal computers, or between a personal computer and a server in dial-up mode, are almost always asynchronous. The driving forces to use asynchronous instead of synchronous transmission are lower equipment cost and the fact that the amount of data to be transmitted is frequently relatively small.

ANALOG SIGNALS AND TRANSMISSION

As shown in Figure 8–4, there are two ways that data can be represented and two ways that transmission can occur on communications lines. The data, signals, and transmission are either analog or digital, and there are four combinations that you need to study and understand. While the technical details can get highly mathematical, we will keep the discussion at a high level. It is important that you understand the concepts and principles presented in this chapter because they are the foundation for further study of data communications in subsequent chapters.

In Chapter 5 you studied the first case shown in Figure 8–1, the analog transmission of analog signals. You learned that a telephone is a de-

Figure 8–4
The four combinations of analog and digital signals and transmission techniques.

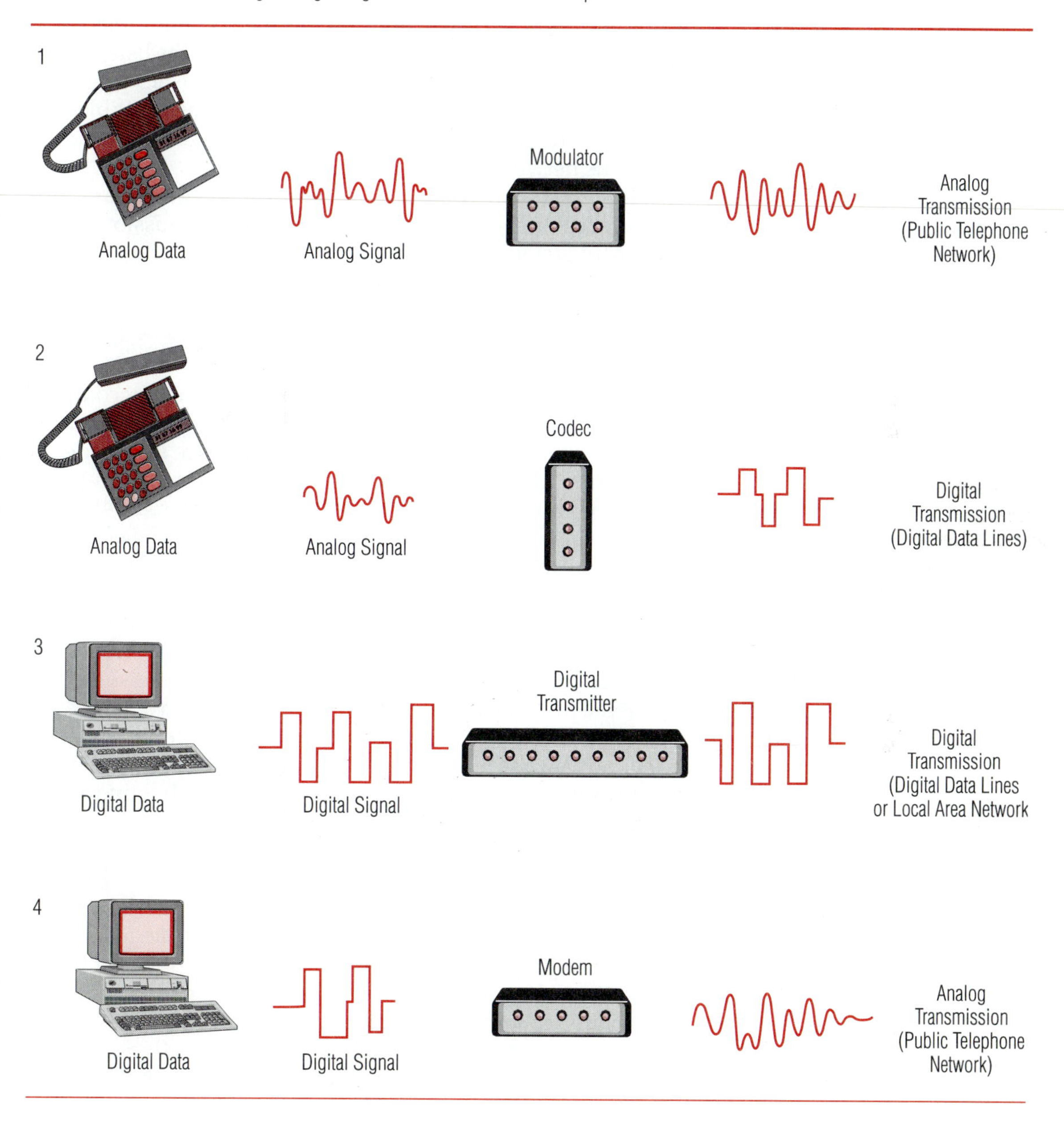

vice that converts the analog data (sound) to an analog electrical signal that can be transmitted. At the other end of the connection another telephone performs the reverse conversion, taking the incoming electrical analog signal and converting it to sound (also analog) that the listener can hear. You were introduced to the analog signal's characteristics, its bandwidth, and the meaning and purposes of frequency division multiplexing and modulation, and the way that analog signals can be transmitted over the public telephone network.

Now we'll look at the other three cases, and we'll begin by examining the characteristics of digital signals and digital transmission. Then we'll look at the conversions that are required.

DIGITAL SIGNALS

A typical analog signal generated by the human voice has a very complex wave form as was shown in Figure 5–2. In contrast, the signal generated by a terminal, computer, or other DTE is very simple, since it is made up of discrete, discontinuous voltage pulses. In simplest terms, each pulse represents one of the binary digits, 1 or 0, which in turn represent the coded data to be transmitted. This type of signal is called a *digital signal.*

There are several forms of digital signals, as shown in Figure 8–5. *Unipolar* signals are those in which a 1 bit is represented by a positive voltage pulse and a 0 bit by no voltage. *Bipolar, nonreturn-to-zero (NRZ)* signals have the 1 bits represented by a positive voltage and the 0 bits represented by a negative voltage. *Bipolar, return-to-zero* signals are similar to NRZ signals, but the pulses are shorter, and the voltage always returns to 0 between pulses. Unipolar signals rarely are used today. Bipolar signaling has the clear advantage of making the distinction between the 0 bit and a no-signal condition, a distinction useful in troubleshooting when problems occur.

NRZ coding techniques have certain inherent limitations which have to do with synchronization and become apparent at high speeds. To overcome these limitations other coding techniques have been developed, but two in particular, *Manchester coding* and *differential Manchester coding,* known as *biphase coding,* have become very widely used in digital transmission systems, including local area networks. Data coded with the Manchester code has a transition in the middle of each bit period. A low-to-high transition represents a 1 bit and a high-to-low transition a 0 bit. In differential Manchester coding, the transition in the middle of the bit period is still there, but a 0 is represented by a second transition at the beginning of the bit period, and a 1 is represented by no transition at the beginning of the bit period.

The benefits of biphase coding techniques are that since there is a predictable transition during each bit time, the receiver can synchronize on

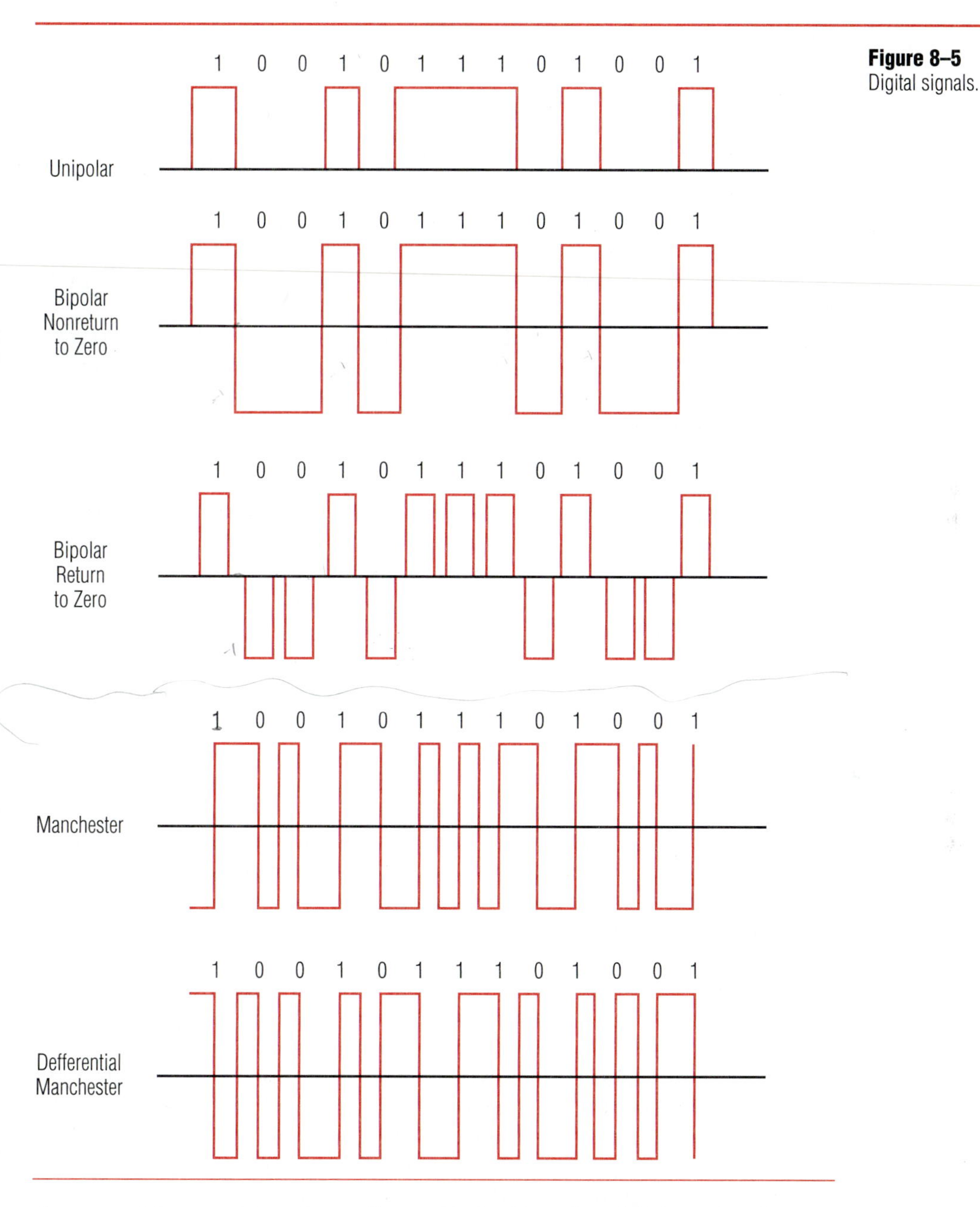

Figure 8–5
Digital signals.

that transition, making the coding self-clocking. In addition, the absence of the expected transition can be used as an error detection mechanism. Manchester coding has been specified for the IEEE 802.3 standard for data transmission and differential Manchester coding for the IEEE 802.5 for the 802.5 standard. Both of these standards will be studied in Chapter 10.

DIGITAL TRANSMISSION

You saw in Chapter 5 that the public telephone network is primarily an analog network, at least as far as the general public sees it. It was designed to do analog transmission because its original function was solely to transfer voice signals. Today, however, virtually all new telephone line and cable installations are being designed for digital transmission. One reason is because data transmission is growing more rapidly than voice, but a more significant reason is that digital transmission has many advantages compared to analog transmission. The benefits are so significant that it is normally worthwhile to convert analog voice signals to digital signals for transmission when digital transmission lines can be used. Most transmissions between telephone company central offices are already made in digital mode, and many private networks installed in companies and other organizations use digital transmission.

The reasons that digital transmission is superior to analog transmission are:

- **Better data integrity**—Digital transmission has fewer errors than analog transmission, and because the data is binary, the errors that do occur are easier to detect and correct.
- **Higher capacity**—It has become possible to build very high capacity digital circuits, including those that use fiber-optic cable. Multiplexing techniques used with digital transmission are more efficient than the frequency division multiplexing (FDM) that is typically used with analog circuits.
- **Easier integration**—It is much easier to integrate voice, data, video, and other signals using digital transmission techniques than it is with analog transmission.
- **Better security and privacy**—Encryption techniques can easily be applied to digital data.
- **Lower cost**—Large-scale integrated circuitry has reduced the cost of digital circuitry faster than the cost of analog equipment.

Extending digital transmission capability to the millions of homes, apartments, and businesses around the world presents a major challenge that will take decades to achieve. In the United States and some other countries, rapid progress is being made by using existing cable television systems, which can be upgraded to carry digital signals. Also, existing telephone lines can carry digital signals over limited distances using new transmission technologies, even though they weren't originally designed for digital transmission.

DIGITAL TRANSMISSION OF ANALOG SIGNALS

Now we'll look at how we would convert an analog signal, such as a telephone signal, for transmission on a digital circuit. You saw in Chapter 5 that when a user speaks into the microphone of a telephone handset, the resulting signal takes the shape of a continuously varying wave—an analog signal. If we look at the wave electronically and measure its height (voltage) at specific points in time, we obtain a series of voltages with numeric values. These values can be represented in binary form and transmitted as a series of bits. This is the way analog signals are digitized.

Figure 8–6 shows that at every unit of time, the height of the curve is measured by a special instrument called an *analog-to-digital (A/D) converter.* An A/D converter is essentially a digital voltmeter that can take many readings per second. Instead of reading actual voltages, however, the A/D converter uses a scale of integer values. In effect, this scale of integers is superimposed over the voltages so that the different heights of the curve (voltages) can be represented as integers. This is important

A/D converter

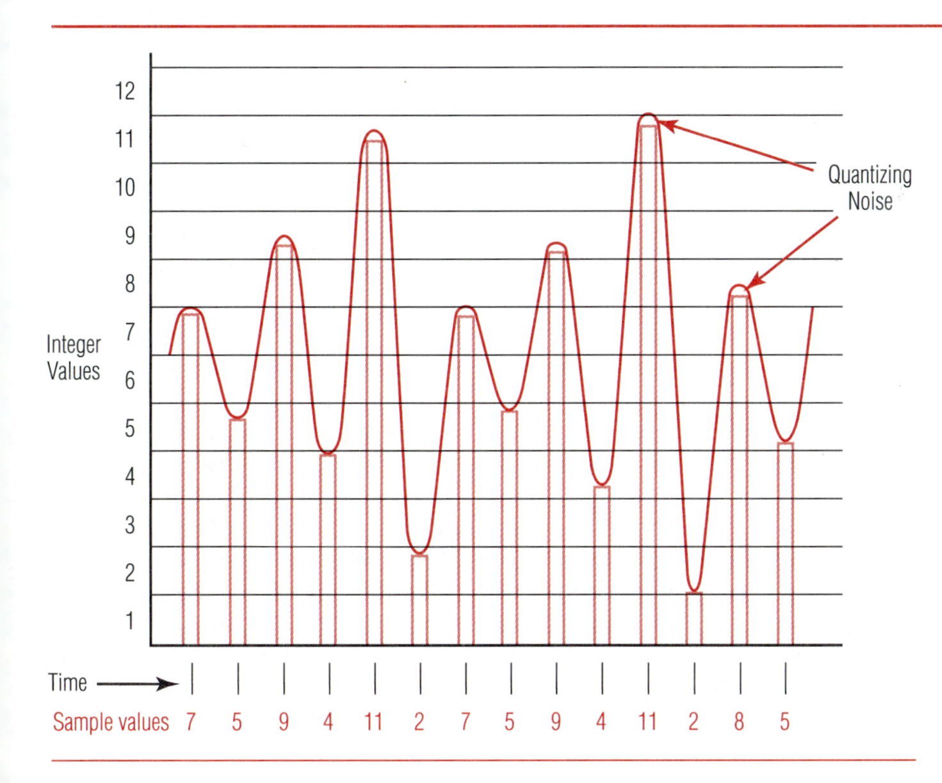

Figure 8–6
Quantization of an analog voice signal.

because integers can be converted accurately to binary numbers, whereas fractional numbers cannot. The process of approximating the actual analog signal value to the predetermined integer steps is called *quantization.* It is similar to rounding. The difference between the exact height of the curve when the sample is taken and the nearest integer value is called *quantizing noise* or *digitizing distortion.*

quantization

Although the diagram in Figure 8–6 only shows 12 values, there are typically 256 different integer values that can be obtained each time a sample is taken. If there are 256 different values, the integer is converted to an 8-bit binary number; there is no room for a parity bit. The 8 bits are then transmitted over the communications line. At the next unit of time, another sample is taken, and the process is repeated.

At the receiving end, the process is reversed. The bit stream is divided into 8-bit groups. The groups are interpreted as 8-bit binary numbers, are converted to voltages by a *digital-to-analog (D/A) converter,* and the original wave form is reproduced. The more frequently samples are taken, the more accurately the original wave form can be reproduced. The name that is commonly applied to A/D and D/A converters in the telecommunications world is *codec,* which stands for coder/decoder.

codec

One relatively old technique for digitizing voice signals is called *pulse code modulation (PCM).* PCM uses 256 integer values and samples the signal 8,000 times per second. Since 8 bits are used for each sample, the effective data rate is 8,000 times 8, or 64,000 *bits per second (bps).* This technique for digitizing voice was used by the telephone companies internally in the 1960s and 1970s, especially on long distance communications lines.

pulse code modulation

Later developments showed that good voice quality could be maintained at a lower sample rate of 4,000 samples per second. This means that the effective data rate is lowered to 4,000 times 8 bits, or 32,000 bps. As the sample rate is slowed, the effect is similar to compressing the bandwidth on an analog signal. If the sample rate is slowed to say 2,000 or 1,000 samples per second, voice quality is lost, and the reconstructed voice at the receiving end loses its distinguishing characteristics, though it is still understandable. At even lower sample rates, the reconstructed voice is unintelligible.

A more common technique for reducing the number of bits that must be transmitted is to keep the sample rate at 8,000 per second but reduce the number of bits transmitted for each sample. Since voice signals do not change amplitude very rapidly, adjacent samples usually are not very different in value from one another. Thus, if only the "difference" in the integer value of the samples is sent, it can be represented with 4 bits. This is illustrated in Figure 8–7 and is called *adaptive differential pulse code modulation (ADPCM).* ADPCM was adopted by the ITU-T in 1985 as the recommended method for digitizing voice at 32,000 bps.

adaptive differential pulse code modulation

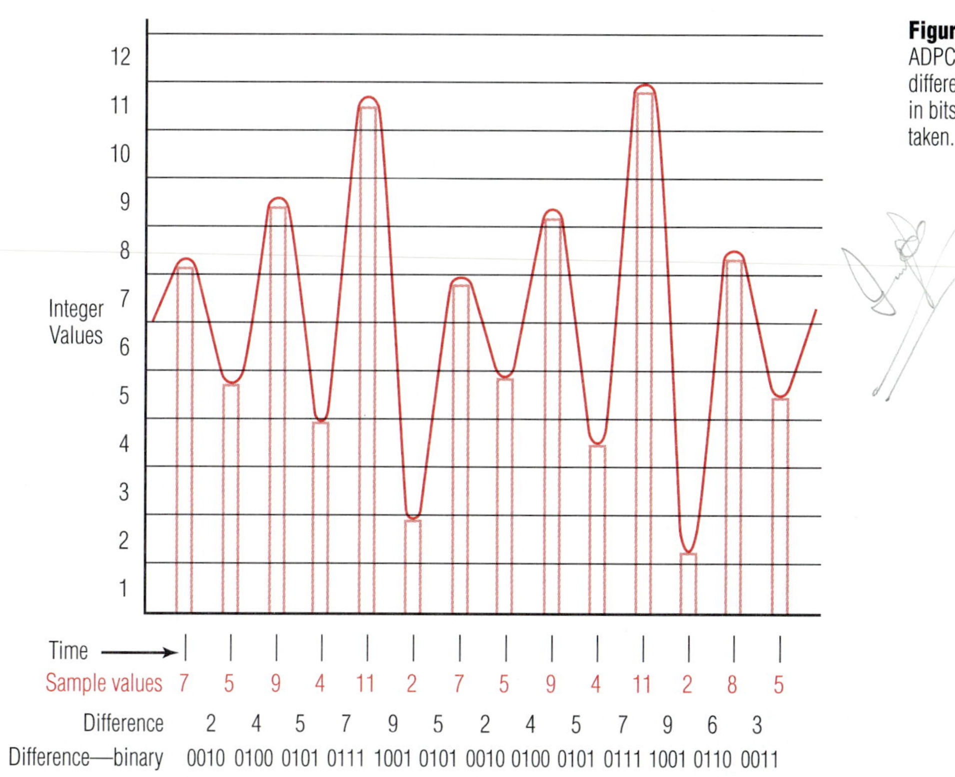

Figure 8–7
ADPCM codes the difference in signal strength in bits each time a sample is taken.

Another method of digitizing voice is called *delta modulation.* Delta modulation compares the analog signal level to the last sample taken. If the new value is greater than the previous sample, a 1 bit is sent; if the new value is less than the previous sample, a 0 bit is sent. To maintain good voice quality, delta modulation samples the analog signal 32,000 times per second. Since one bit is sent for each sample, the data rate is 32,000 bps, the same as for ADPCM.

delta modulation

Research is continuing into ways to reduce the data transmission rate required to send digitized voice signals while still maintaining acceptable quality.

Analog-to-digital converters that are small and cheap enough to fit in a telephone handset are available so that it is technically possible and affordable to transmit digitized voice directly from the telephone. Today, this capability is found frequently in telephones for private PBX systems.

When analog voice data is digitized, the resulting bit stream is the same as a bit stream of coded data. If someone were monitoring a digital communications line and knew nothing about what was being transmitted, it would be impossible to tell whether the bits on the line represented digitized voice, coded data, or an image from a digital facsimile machine. Another by-product is that the same mathematical techniques for compression and encryption can be applied to voice, data, or images, since, in digital form, they all look the same. Digital multiplexing equipment can also be employed to intermix the bits into a consolidated bit stream. Therefore, it is important that the recipient know how to interpret the bits of data received.

DIGITAL TRANSMISSION OF DIGITAL SIGNALS

baseband transmission

If we already have data in digital form, as shown in Figure 8–5, and want to transmit it digitally, the situation is relatively easy. Digital transmission, also called *baseband transmission,* is simply the transmission of the pulses of the digital signal in the form of electrical pulses. The pulses above the line in Figure 8–5 might be represented by a voltage of +3 volts and the pulses below the line by a voltage of −3 volts. This pulsing typically occurs at speeds of at least 1 million pulses per second, often much faster. The most common situation of this type that you will probably encounter is the connection of a personal computer to a local area network (LAN). We will discuss this subject in much greater detail in Chapter 12.

Digital Transmitter/Receivers

DSU/CSU

It may seem that the digital signal coming from a terminal, computer, or other DTE could be directly connected to a digital communications line, since both operate with electrical pulses. In practice, an interface unit, called a *digital transmitter/receiver,* sometimes called a *data service unit/channel service unit (DSU/CSU),* is required. The digital transmitter ensures that the digital data that enters the communication line is properly shaped into square pulses that are precisely timed. Sometimes it needs to convert the digital pulses coming from the DTE to a form that is suitable for the line. The digital transmitter also provides the physical and electrical interface between the DTE and the line. The primary purpose of the CSU portion of the unit is to provide circuitry to protect the carrier's network from excessive voltage coming from the customer's transmission equipment. The transmitter may also provide diagnostic and testing facilities. Since they do not perform digital-to-analog or analog-to-digital conversion, however, digital transmitter/receivers are much simpler than traditional modems.

ANALOG TRANSMISSION OF DIGITAL SIGNALS

The world's telephone system, and therefore most of its network, was originally designed to carry analog voice signals. Today, this network has largely been converted to digital operation internally, but in most cases the local loop connection between a business or home and the telephone company's central office, the so-called "last mile," still operates in analog mode. Hence, the digital signal from personal computers or other devices must usually be converted to analog form in order to get onto the public network, even though the signal is likely to be converted back to digital form for transmission within the network. Converting a signal from analog to digital, or from digital to analog is the function of a modem.

MODEMS

A *modem* (from MOdulation and DEModulation) is a specialized analog-to-digital and digital-to-analog signal converter that works by modulating a signal onto a carrier wave and demodulating it at the other end. A modem is connected in the communications network between the data terminal equipment and the transmission network, as shown in Figure 8–8. Modems are an example of data circuit-terminating equipment (DCE), a term that was introduced in Chapter 6. There are many modem vendors, and the modem market is very competitive.

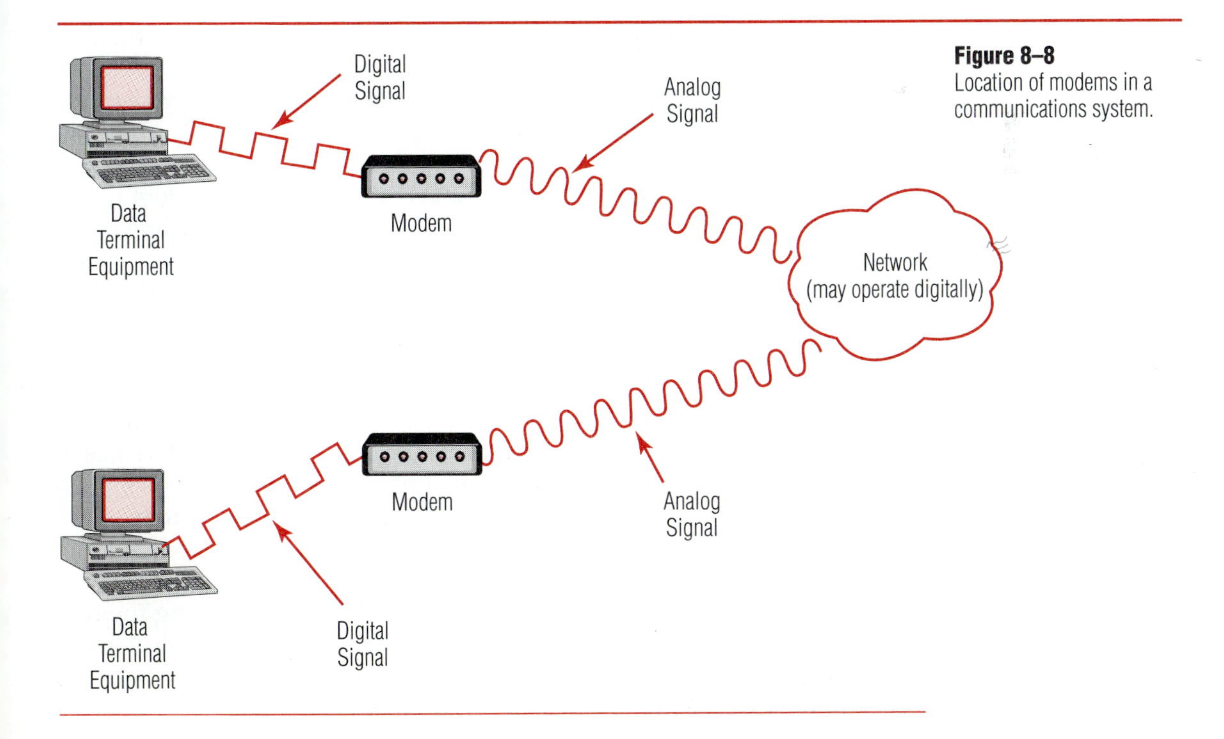

Figure 8–8
Location of modems in a communications system.

Figure 8–9
Block diagram representation of a modem.

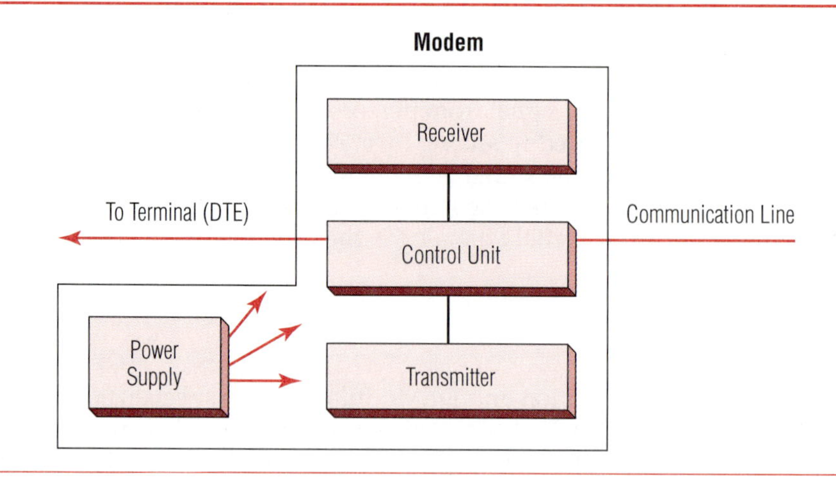

Figure 8–9 shows a general block diagram of a modem with its major components. The power supply takes standard 110 volt AC power (in the U.S.) and converts it to the lower voltages needed for internal operation.

In the transmitter, a modulator and digital-to-analog converter convert the digital bits (pulses) from the DTE to a modulated analog signal appropriate to the type of modulation being used. The output is a sine wave of the proper amplitude, frequency, and phase to represent the digital input signal.

equalizer

Equalizer circuitry compensates for variability in the actual transmission line used. Although telephone circuits have certain standard parameters and specifications, there is a range of acceptable characteristics. As a modem's speed increases, it becomes critical for the modem to detect and compensate for the exact parameters of the particular line being used. *Fixed equalizers* in the transmitter assume that a certain average set of parameters exists, and they shape the transmitted wave accordingly.

adaptive equalizers

More sophisticated *adaptive equalizers* are used in the receiving section of modems operating above 2,400 bps. A standard training signal of known characteristics is sent from one modem to the other. The received signal is examined, and based on its shape and any errors that have occurred, the receiver's circuitry adjusts itself to the exact characteristics of the incoming wave form. The training signal is then sent the other way, and the other modem adjusts its receiving circuitry appropriately. After the initial training time, which takes from 20 to several hundred milliseconds, the modems constantly monitor the quality of the incoming signal and make further equalization adjustments as necessary during the regular data transmission. The equalizing circuitry in high-speed modems is very complex and requires powerful signal-processing chips to perform the necessary calculations in a short amount of time.

The demodulator in the receiver converts the incoming analog signal back to digital binary bits by reading the wave form and detecting the phase, amplitude, and frequency. The demodulator also determines the rate at which the incoming signal is actually being received and passes this information to clock extractor circuitry.

The control unit of modems deals with the interfaces to the telephone line and the DTE. Since the control unit of most modems is a microprocessor, it can be programmed to provide a number of special features as well as the standard functions. The most common features provided in even inexpensive modems today are auto dial and auto answer capabilities, discussed later in this chapter.

How Modems Work

In its simplest form, a modem senses the signal from the data terminal equipment. When it senses a 0 bit, it sends an analog signal with certain attributes (amplitude, frequency, and phase). When it senses a 1 bit, it sends a different signal, with at least one of the attributes changed. For example, when frequency modulation is used, a 0 bit from the DTE causes the modem to turn on an oscillator that sends an analog wave of a specific frequency on the telephone line. When a 1 bit is sent, the modem turns on another oscillator, which generates a wave with a different frequency. At the receiving end, the process is reversed. The waves are converted to a digital signal (electrical pulses) representing the original 0 and 1 bits. This specific type of frequency modulation is called *frequency shift keying (FSK).*

frequency shift keying

We will look at the operational details of a particular type of modem, the Bell Type 103. This modem is very old and is not used any more, but it was widely used in its day. It was also a very simple modem and serves as an excellent learning tool for basic modem operation. The Bell Type 103 modem was used for asynchronous transmission at 300 bps and used FSK modulation. When two modems of this type communicate, one is designated as the "originate" modem and the other is designated as the "answer" modem. The originate modem transmits the 0 bits (spaces) at 1,070 Hz and the 1 bits (marks) at 1,270 Hz. The answer modem uses 2,025 Hz for spaces and 2,225 Hz for marks. These frequencies are well within the range of human hearing, and each signal sounds like a musical tone. In practice, the tones change so fast that if you listen to a telephone line when this type of modem is transmitting, the signals produce a warbling sound.

Looking at the signal diagrammatically in Figure 8–10, we can see where all of the frequencies fit within the bandwidth of a standard telephone line. This allows the transmissions from both modems to occur simultaneously, which is full-duplex transmission. Figure 8–11 shows how the wave form of the originate modem looks as it shifts back and forth from one frequency to another.

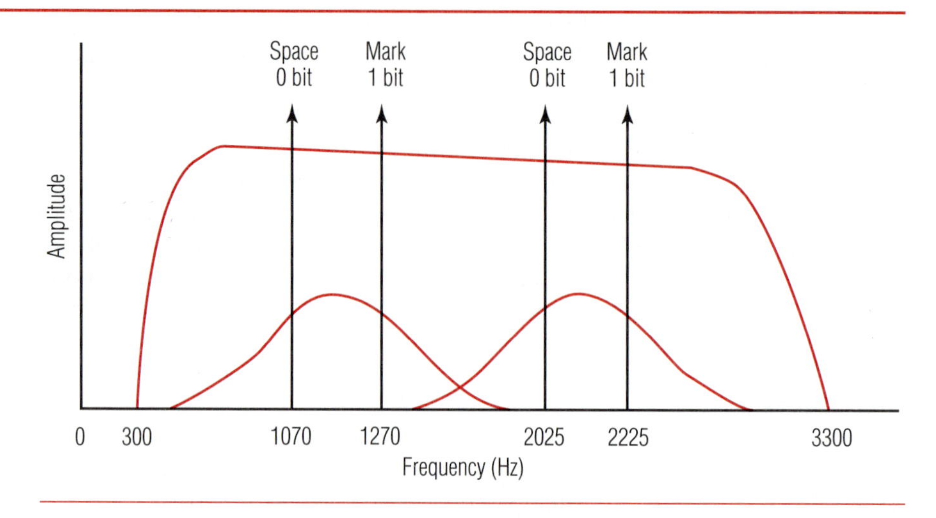

Figure 8–10
Frequencies used by a Bell Type 103 modem.

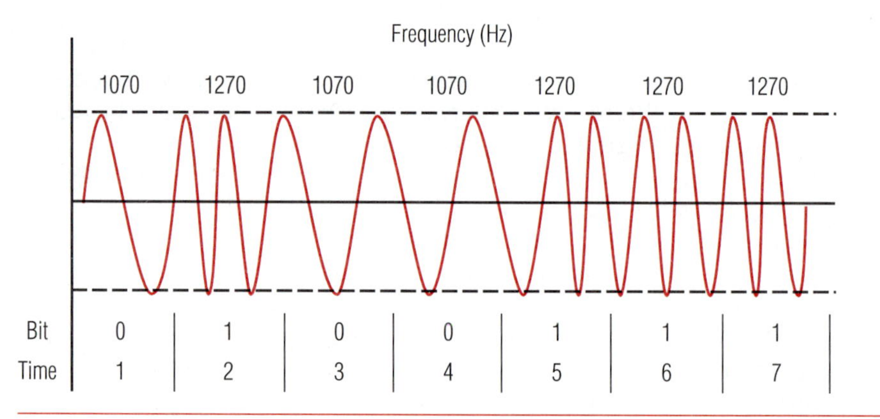

Figure 8–11
Frequency modulation in a Bell Type 103 modem.

Although the FSK modulation technique can be used for speeds higher than 300 bps, the full-duplex transmission is sacrificed. This is because FSK modulation requires a bandwidth of approximately 1.5 times the baud rate. At 1,200 baud, 1,800 Hz would be required for transmission in each direction, giving a total required bandwidth of 3,600 Hz. This is more than the 3,000 Hz available on a telephone circuit. Therefore, when FSK is used at 1,200 baud, the transmission is limited to half duplex.

Another practical difficulty is that as the speed increases, the frequency changes occur so fast that it becomes difficult for the receiver to detect them. At 1,200 baud, each tone is transmitted for only 1/1,200 of a second, or .0833 second. At higher speeds, the duration would be shorter. The problem is solved at 1,200 baud by widening the difference between

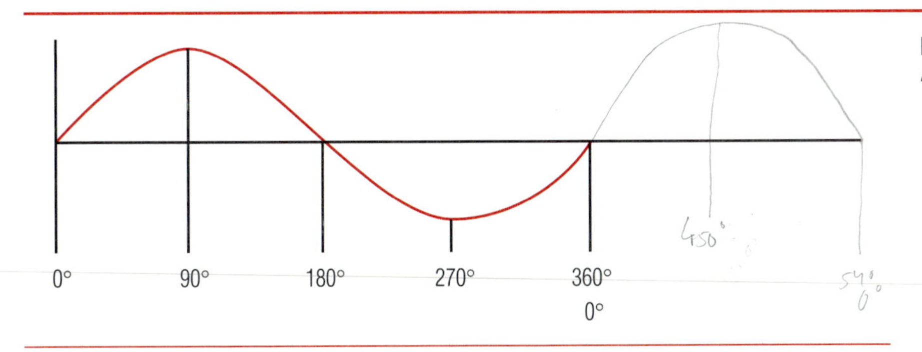

Figure 8–12
Angles of a sine wave.

the frequencies. With the greater difference, the modem can tell one frequency from the other more quickly. The Bell 202 type modem uses frequencies of 1,200 Hz for a 0 bit and 2,200 Hz for a 1 bit.

Amplitude modulation is an alternative, but in reality it is never used by itself in modems because noise on the transmission line makes it relatively unreliable. The third attribute of an analog signal that can be changed is the phase of the signal, which was briefly introduced in Chapter 5. It turns out that rapid phase changes of a signal can be detected more easily than rapid frequency changes, making phase modulation best suited for use in high-speed modems.

Phase Modulation (PM)

An analog signal's wave is called a *sine wave* because it is the shape generated by the geometric sine function. Figure 8–12 shows how it can be labeled with degree markings at any point on the X axis. An analog signal's phase can be thought of as a timing offset, as shown in Figure 8–13. If two waves are offset from one another, they are *phase shifted* a certain number of degrees. Figure 8–13 shows waves that are offset by 90, 180, and 270 degrees from one another. Of course, the phase shift could be any number of degrees; it does not need to be a multiple of 90 degrees.

phase shift keying
differential phase shift keying

In the simplest case, phase modulation is performed by shifting a sine wave 180 degrees whenever the digital bit stream changes from 0 to 1. It would shift 180 degrees again when the signal changed from 1 to 0. The wave would look like Figure 8–14. This type of modulation is called *phase shift keying (PSK).* The more usual type of phase shift modulation of this type is *differential phase shift keying (DPSK).* In DPSK, the phase is shifted each time a 1 bit is transmitted; otherwise the phase remains the same. This is illustrated in Figure 8–15. Notice that the wave form is the same in Figures 8–14 and 8–15, but the bits represented are different.

Suppose that instead of 180 degrees, the phase was shifted only 90 degrees. Now there are four possible shifts, 0, 90, 180, or 270 degrees. With a binary signal, each shift could represent a dibit. For example,

Figure 8–13
Phase shifts.

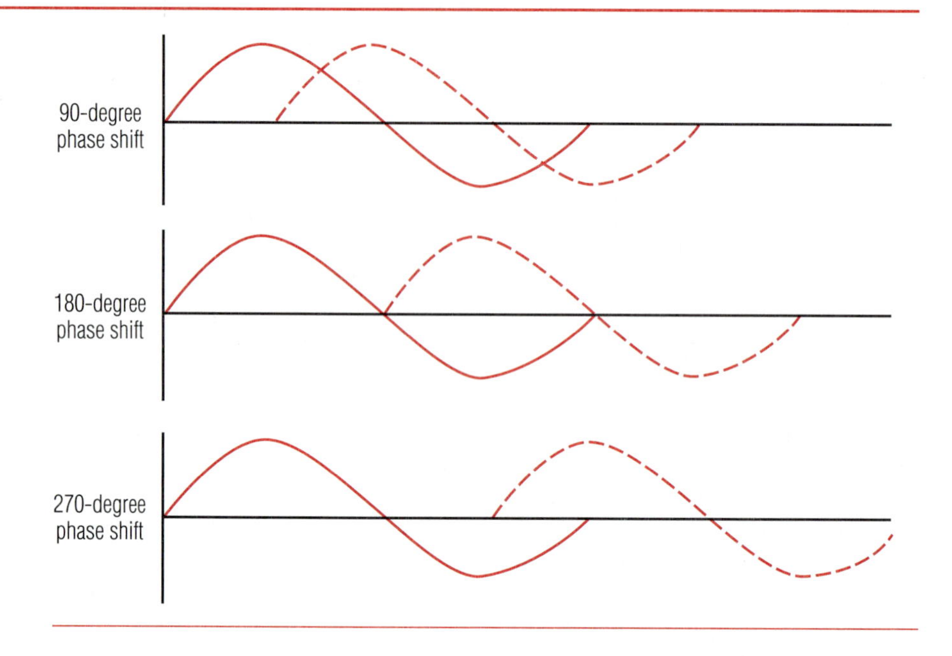

Figure 8–14
Phase shift keying (PSK).

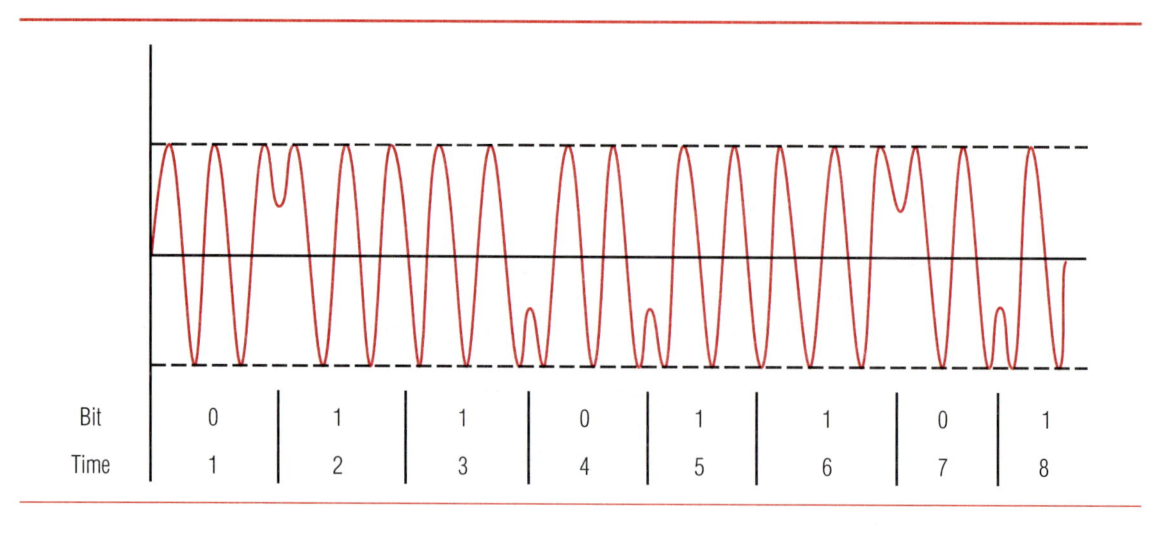

Figure 8–15
Differential phase shift keying (DPSK).

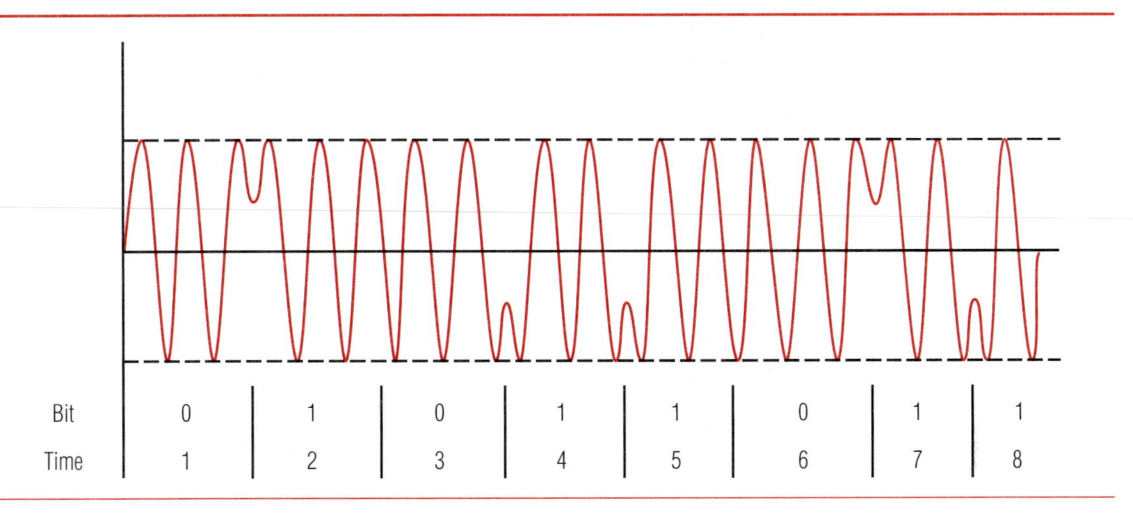

Phase Shift	Dibits
0 degrees	00
90 degrees	01
180 degrees	10
270 degrees	11

Now, whenever there is a signal change, 2 bits of information are transmitted. The data rate, measured in bits per second, will be 2 times the signaling rate, measured in bauds. If the phase is shifted in 45-degree increments, there are eight possible shifts, each of which can represent a tribit.

Phase Shift	Tribits
0 degrees	000
45 degrees	001
90 degrees	010
135 degrees	011
180 degrees	100
225 degrees	101
270 degrees	110
315 degrees	111

For each signal change, 3 information bits are transmitted. The bit rate of the circuit will be three times the baud rate.

One very popular modem that used a form of DPSK was the Bell 212A type modem. This modem had a combination of capabilities. It operated asynchronously at 300 bps using FSK to be compatible with the Bell 103 type modems. The 212A also operated asynchronously or synchronously at 1,200 bps, using a special form of DPSK. The ITU-T V.22

Figure 8–16
An example of the phase changes and amplitudes used in one type of modem that uses quadrature amplitude modulation.

Phase Change (degrees)	Relative Amplitude	Quadbit
0	3	0001
0	5	1001
45	$\sqrt{2}$	0000
45	$3\sqrt{2}$	1000
90	3	0010
90	5	1010
135	$\sqrt{2}$	0011
135	$3\sqrt{2}$	1011
180	3	0111
180	5	1111
225	$\sqrt{2}$	0110
225	$3\sqrt{2}$	1110
270	3	0100
270	5	1100
315	$\sqrt{2}$	0101
315	$3\sqrt{2}$	1101

specification is functionally identical to the Bell 212A characteristics, although variations of the V.22 standard have emerged in recent years.

Quadrature Amplitude Modulation (QAM)

If the phase is shifted less than 45 degrees, the receiver has a very difficult time detecting it. To get to higher speeds, modem designers have developed modulation techniques that use a combination of phase and amplitude modulation. The objective is to get 16 distinct combinations of phase and amplitude. This allows 4 bits, a quadbit, to be represented by each combination. There are several techniques by which this has been achieved, but all are called *quadrature amplitude modulation (QAM).* The most common technique is shown in Figure 8–16. Note that there are 8 phase changes and 4 amplitudes in use, but not all combinations are valid. Also note that the quadbits are not assigned in a neat, logical sequence. The reasons are very technical but have to do with ensuring that the receiver detects as many transmission errors as possible.

Trellis Code Modulation (TCM)

As the speed of a transmission is pushed above 9,600 bps, the techniques to achieve error-free transmission become very sophisticated. As the bit rate increases, noise and other impairments of the communications line have a more significant effect. The following table shows the number of bits and characters that are affected by a noise pulse on the communications line that lasts just .1 second.

Speed (bps)	Bits Affected	8-Bit Characters Affected
300	30	3+
1,200	120	15
2,400	240	30
4,800	480	60
9,600	960	120
19,200	1,920	240
28,800	2,880	360
33,000	3,360	420
56,000	5,600	700

Trellis code modulation (TCM) is a specialized form of QAM that codes the data so that many bit combinations are invalid. If the receiver detects an invalid bit combination, it can determine what the valid combination should have been. Trellis coding allows the transmission of 6, 7, or 8 bits per baud. A 2,400 baud signaling rate yields speeds of 14,400 bps, 16,800 bps, and 19,200 bps on leased lines.

As modulation techniques get more sophisticated, the amount of signal processing that the modem must do increases significantly. Indeed, many of the more sophisticated modems today depend on powerful circuit chips called *signal processors* to code the data for transmission and decode it at the receiving end.

signal processors

V.32 and V.32bis Modem Standards

A significant use of TCM has been to provide 9,600 bps transmission on dial-up telephone lines. The ITU-T published standard *V.32,* which specifies 9,600 bps, full-duplex operation using TCM with an echo cancellation technique. Echo cancellation permits high-speed signals traveling in opposite directions to exist on the same dial-up circuit at the same time and in the same frequency band. (Echoes on communications lines are discussed in Chapter 9.)

A further improvement was made in the way a modem encodes data for transmission, and an improved standard, *V.32bis,* was written. A modem following this standard can transmit data on a dial-up circuit at 14,400 bps.

V.34 and V.34bis Modem Standards

V.34 is the ITU-T standard for modems that operate at 28.8 kbps through a standard telephone line. In 1996, the ITU-T revised and upgraded the standard to allow a higher data rate of 33.6 kbps. The upgraded standard is *V.34bis* also known as *V.34+.*

V.34 and V.34bis assume that most of the data transmission occurs on digital lines and that analog lines are only used in the first and last part of the connection—the local loop. Because digital circuits are electrically cleaner and experience fewer errors, V.34 uses a baud rate up to 3,429 in combination with trellis code modulation. When V.34 modems first connect to each other, they send signals to each other to test the characteristics of the circuit and determine its quality. This exchange of signals is called *handshaking*. Based on the results of the handshaking, the modems decide the best combination of baud rate and modulation technique from among 59 pre-defined possibilities to produce the fastest throughput. If, during the transmission, the modems detect more errors than expected, they can change the combination dynamically, but this may result in a slower transmission speed.

V.34bis continues to use the same baud rate (3,429) as V.34 but uses an upgraded form of trellis code modulation that averages 9.8 bits per baud to achieve a throughput rate of 33.6 kbps. Experience has shown, however, that many telephone lines are not of sufficient quality to support the higher transmission speed, so many people using V.34bis will not actually get the 33.6 kbps data rate.

V.90 Modem Standard

In February 1998, the ITU-T adopted the *V.90* standard for 56 kbps modems, which effectively replaced two competing proprietary "defacto" standards, K56flex from Rockwell International, and X2 from 3-Com. V.90 technology assumes that at least one end of the communications line has a pure digital connection to the telephone network. In fact, the normal configuration that V.90 expects is shown in Figure 8–17, an example that shows a typical home connection to the Internet.

V.90 transmission is asymmetric in that the 56 kbps data rate is only achieved on the half of the transmission from the all-digital end of the connection, the Internet in this example. Transmissions from the analog end follow the V.34bis standard and occur at a maximum rate of 33.6 kbps. In most applications this difference in transmission speeds makes little difference because the amount of data flowing from the workstation to the Internet—keystrokes and mouse commands—is normally much less than the amount of data flowing from the Internet to the workstation as it downloads web pages or files. In this application you really need the speed for data coming from the Internet more than for the data flowing to it.

V.90 uses PCM for the digital half of the transmission. You might wonder why, since PCM is capable of 64 kbps, V.90 achieves only 56 kbps. There are several reasons, including noise on the analog portion of the circuit and the fact that some telephone companies use one of the bits in PCM for control signaling, which only leaves 7 bits available for data (7 bits × 8,000 samples per second = 56,000 bps). Another reason is that the FCC and other regulatory agencies have signal power level requirements. In order to meet agency requirements for maximum signal strength, the telephone companies must reduce the power of the data signal, and this reduction further reduces the maximum theoretical data rate to 54 kbps.

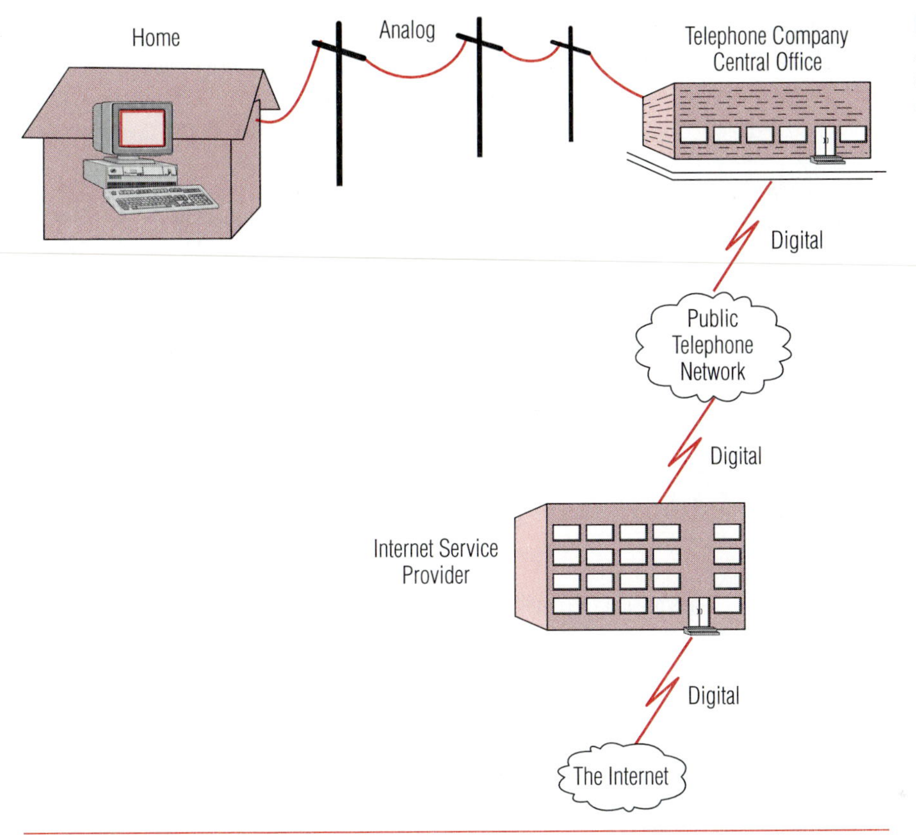

Figure 8–17
A V.90 connection is assumed to have this configuration.

In reality, if you have a reasonably good telephone line and are within three and one half miles from your telephone company central office, you should get transmission speeds between 40 kbps and 53 kbps. Some lines are electrically noisy or have other impairments that cause the actual transmission speed to be lower.

Higher Speeds

You can see the evolution of modems and transmission speeds from those that assume a purely analog circuit to those that assume a partially digital circuit, to the assumption that one end of the circuit is a direct digital connection. The next step is the case in which both ends are connected directly to a digital circuit, and when that occurs, transmission speeds into the millions of bits per second become possible. We'll discuss this case more in Chapter 9 when we look at telecommunications lines and circuits, but in the meantime, all four of the cases you've studied are illustrated in Figure 8–18. A summary of the modem standards is shown in Figure 8–19.

Figure 8–18
The ways that increasingly higher data transmission rates have been achieved on dial-up communication lines.

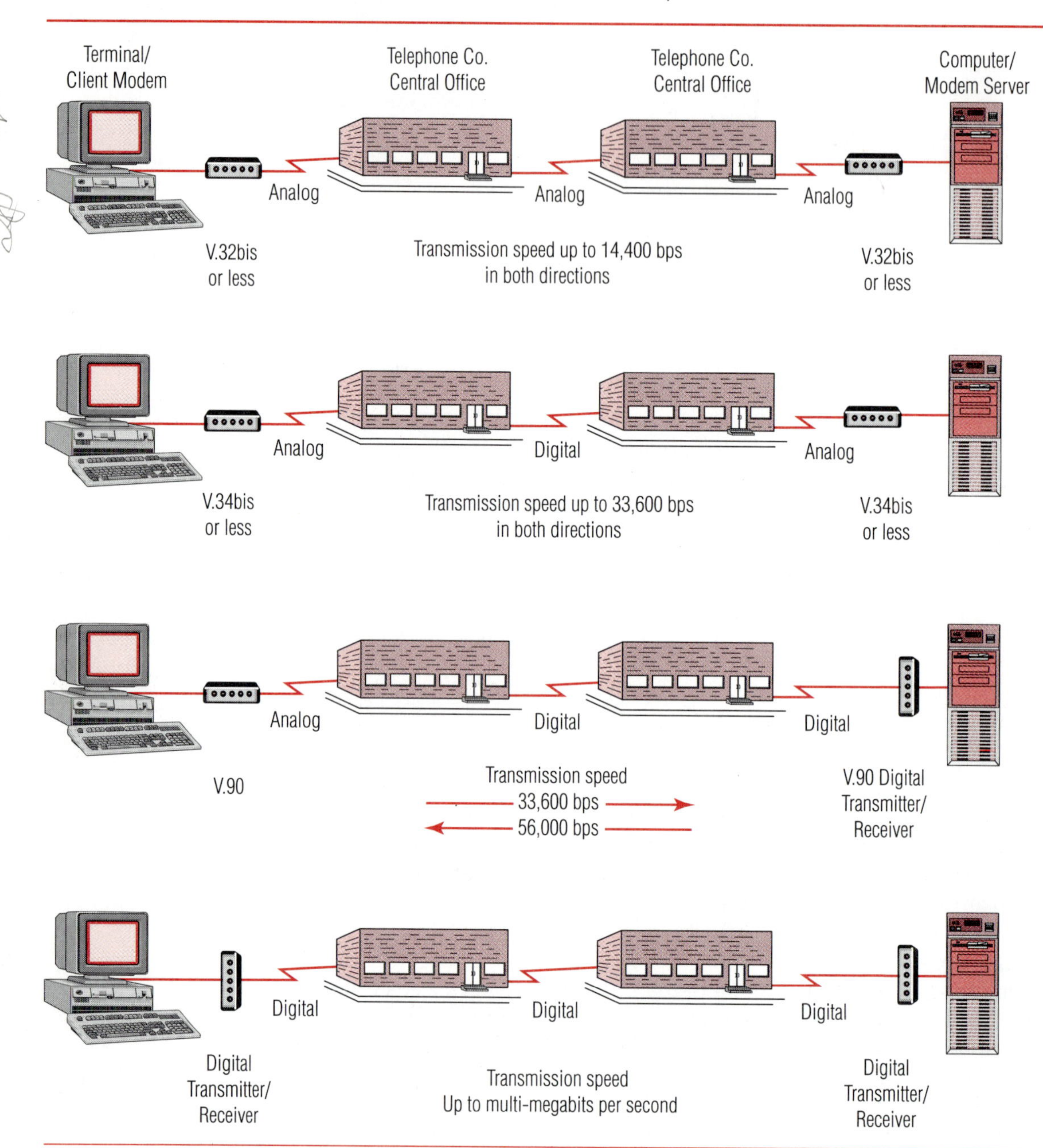

Name of Standard	Provisions of Standard	Comment
V.21	Transmission at 300 bps, full-duplex. Primarily used outside the U.S.	An old standard. Bell 103 is the comparable standard used in the U.S.
V.22	Transmission at up to 1200 bps. Full duplex. Primarily used outside the U.S.	An old standard. Bell 212A is the comparable standard used in the U.S.
V.22bis	Standard for transmission at up to 2400 bps, full-duplex. Used globally.	
V.23	Transmission at up to 1200 bps in one direction and 75 bps in the other direction (pseudo full-duplex).	Developed in the early 1980s as way to reduce modem costs. Used primarily in Europe.
V.32	Transmission at up to 9600 bps, full-duplex. Includes error correction and negotiation standards.	
V.32bis	Transmission at up to 14,400 bps, full-duplex.	
V.34	Transmission at up to 28,800 bps, full-duplex.	Assumes that most of the transmission is digital.
V.34bis	Transmission at up to 33,600 bps, full-duplex.	Assumes that most of the transmission is digital.
V.90	Transmission at up to 56,000 bps in one direction and 33,600 in the other.	One of the local loops must be digital.

Figure 8–19
Modem standards.

Modem Data Compression

The effective throughput of a data transmission can be improved by compressing the data before it is transmitted. ITU-T standard *V.42bis* is a data compression standard that works in combination with standard *V.42,* an error correction standard. V.42bis specifies how modems will compress data, typically by stripping off unnecessary bits, before it is transmitted. Used together the two can achieve a compression ratio of up to 2:1, thereby effectively doubling the throughput on a line.

Independently, a company called Microcom, Inc. developed its own error correction and compression algorithms, Microcom Networking Protocol, called MNP 5. MNP 5 uses a combination of run length and Huffman coding, compression algorithms that were discussed in Chapter 7. These protocols have become very popular despite the fact that they are not quite as efficient as the V.42-V.42bis combination. Microcom has enhanced their protocols with MNP 7 and MNP 10, which yield higher compression in particular circumstances.

Using the V.42bis compression technique the effective throughput on a line can normally be increased by a factor of between 3.5:1 and 4:1 assuming

This Courier modem is designed to operate at up to 56Kbps using the V.90 standard. Like most modems of this type, it can fall back to slower speeds when line conditions do not allow 56Kbps transmissions to proceed successfully. (Courtesy of 3Com.)

that the data has not already been compressed by a data compression program. Thus a V.34+ modem with a data rate of 33.6 kbps could provide an effective data rate of up to 134.4 kbps when V42bis is used, and a V.32 modem could provide an effective data rate of up to 38.4 kbps when upgraded to use V.42bis.

MODEM CLASSIFICATION

Modems usually are classified according to the type and speed of transmission they are designed to handle. Thus, we find modems designed for 300 bps, asynchronous, full-duplex transmission, such as the Bell Type 103 we looked at earlier. Others are designed for 2,400 bps, half-duplex, synchronous transmission; 9,600 bps, half-duplex, synchronous transmission; 28,800 bps, full-duplex, asynchronous transmission, and so on.

line turnaround

When a modem is designed for half-duplex transmission, additional circuitry is added to perform a process called *line turnaround.* Line turnaround occurs when one modem stops transmitting and becomes the receiver, and the receiving modem becomes the transmitter. The modems exchange synchronization signals to ensure that they are ready to operate in the new mode, and then transmission begins again in the opposite direction. Line turnaround can occur very frequently, and the speed of this process is an important factor in overall line throughput. Commonly, it takes 50 to 200 milliseconds for a pair of modems to turn the line around. This may not seem like a lot of time, but it can represent a significant portion of total line time.

This modem from US Robotics is very compact and still allows full 56Kbps transmission capability. (Courtesy of 3Com.)

Modems for Asynchronous Transmission

The earlier discussion about asynchronous transmission covered most of the specifics of asynchronous modem operation. As transmission speed increases, more sophisticated modulation techniques are used. Modems are available to communicate asynchronously at up to 33,600 bps on normal dial-up telephone lines and at higher effective speeds when data compression is used. Costs for asynchronous modems range from less than $100 for slow-speed units to more than $1,000 for high-speed modems with a high degree of sophistication, including compression and error checking.

One interesting modem type allows voice and data transmission, either alternately or together. Business people might want to discuss a file using the voice connection and then switch to data mode to transfer it. Alternately, they may want to simultaneously look at a shared spreadsheet and discuss it. Modems of this type can normally transfer data at 33.6 kbps in straight data mode or 19.2 kbps when voice is also in use. Some modems increase the data rate automatically when the voice channel is silent, for example, in pauses between sentences or during "think time."

Modems for Synchronous Transmission

Modems used for synchronous transmission are similar to modems used for asynchronous communications but they must have extra clock circuitry to time the release of bits onto the communication line. The clock is usually a crystal-controlled oscillator with a very high degree of accuracy. Synchronous transmission is primarily used at transmission speeds of 9,600 bps and up.

Acoustically Coupled Modems

An acoustically coupled modem, also known as an acoustic coupler, has two rubber cups into which the user places a telephone handset. This type

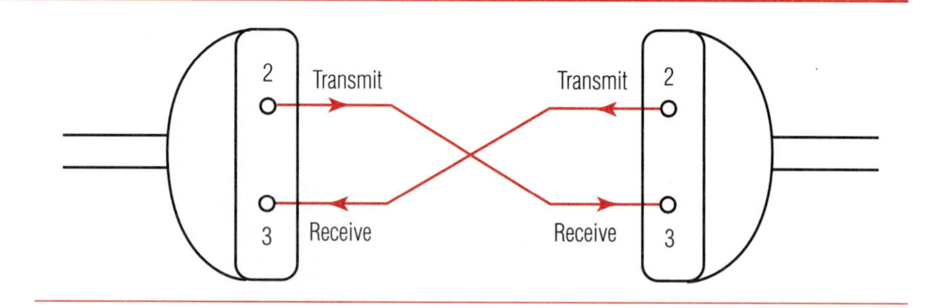

Figure 8–20
Modem eliminator (simplified).

of connection works adequately with FSK modulation at 300 bps and is appropriate for portable terminals where no direct wired connection is possible or desirable. For higher speeds, a direct electrical connection that is not subject to noise (sound) is necessary. Because other types of high-speed integrated modems are available for laptop computers, the acoustically coupled modem is rarely used anymore.

Limited Distance Modems/Short Haul Modems

Inexpensive modems are available for use when transmission distances are less than approximately 20 miles. There are trade-offs between speed and distance, so that a given modem might operate at 9,600 bps up to 5 miles, 4,800 bps up to 10 miles, and 2,400 bps between 10 and 20 miles. The primary advantage of this type of modem is cost. Savings of 50 percent to 80 percent compared to a conventional modem are possible.

Modem Eliminators/Null Modems

If distances are very short and the DTEs can be connected by a cable, modems may not be needed at all. Simply plugging a cable between the terminal and the computer won't work, however, because both devices deliver data to be transmitted on the same pin of the standard connector used for this type of connection. Conceptually, what is needed is a device with a cable that cross-connects the transmit and receive pins of the connectors so that the transmit pin on one device is connected to the receive pin on the other, and vice versa. A schematic diagram of this type of device called a modem eliminator is shown in Figure 8–20.

In practice, other pins must be cross-connected as well, and the exact implementation depends on whether the transmission will be synchronous or asynchronous. Modem eliminators are often used for terminals in close proximity to a computer or sometimes to connect two computers. Distances of up to several thousand feet can be achieved.

This AT&T direct acoustically coupled modem can be either directly attached to a telephone line or acoustically coupled via the rubber pads that cradle the telephone handset. (Courtesy of AT&T).

Facsimile Modems

The basic function of a *facsimile (fax) modem* is identical to that of a data modem; however, the ITU-T has written separate standards, called V.27, for 2,400 bps and 4,800 bps fax transmissions; V.29 for 7,200 bps and 9,600 bps fax transmissions; and V.17 for 12,000 bps and 14,400 bps fax transmissions. At the higher speeds, error correction and data compression algorithms are employed, but they are different from those used for data modems. Three different types of compression have been defined, but most of the installed fax modems only support the lowest level called *modified Huffman (MH) encoding*. The other two, *modified read (MR) encoding* and *modified, modified read (MMR) encoding*, are expected to be more widely used in higher speed fax modems of the future.

Fax modems commonly operate at 28,800 bps or slower speeds over dial-up lines and sell for under $200.

Modems for Fiber-Optic Circuits

When the medium carrying the communication circuit is a fiber-optic cable, a special optical modem is used to convert the electrical signals from the data terminal equipment to light pulses. Since light waves have the same characteristics as electromagnetic waves—albeit at a much higher frequency—the same concepts of bandwidth and modulation apply, but the transmission is entirely digital. The conversion that takes place is not from digital to analog, but from digital-electrical to digital-optical.

Cable Modems

A *cable modem* links a DTE to a cable television system cable. The use of cable television systems for data transmission is a rapidly evolving technology that has made great strides in just the last few years. Many cable television systems were not originally designed or engineered to carry data at the same time as television signals. Furthermore, cable television was primarily designed to carry signals one way—from the cable company's site, called the *head end,* to the home. Carrying data the other way is outside the design limits of many cable TV systems, so they need to be upgraded to handle two-way traffic.

Until recently, different cable systems required different cable modems. Now there is a standard cable modem for all systems. CableLabs, an industry technology organization, developed the standard, called DOCSIS, which stands for Data Over Cable Service Interface Specification. The standard was developed to ensure that cable modem equipment built by a variety of manufacturers is compatible.

Whereas traditional dial-up modems provide speeds of up to 56 kbps, a cable modem operates at speeds of more than 1 Mbps, about twenty times faster. When a cable modem is installed, a splitter is installed on the main television cable and a separate cable is run from the splitter to the cable modem, which is located near the customer's computer. The modem is typically connected to the computer through CAT-5 cable (discussed in Chapter 9) and an Ethernet card (discussed in Chapter 12) in the computer. Sometimes a connection is made through a computer's Universal Serial Bus (USB) port instead.

The benefits of the higher speed afforded by cable modems are especially apparent when large files, such as pictures, graphics, and audio or video clips are downloaded from the Internet. Another attribute of cable modems is that they are always on, so that whenever the customer's computer is running, it is connected to the Internet—there is no connection or logon procedure. You simply click on your browser and you're on the Internet. Many cable modem users tout the convenience of this capability as being as important as the increased speed.

Cable modem service is still relatively limited. As of this writing, in mid-2000, about 30 million homes in the U.S. and Canada are eligible to receive cable modem service. This is less than one third of all North American households. Market demand is motivating cable companies to upgrade their facilities and broaden the availability as quickly as possible. For more information and an up-to-date picture of cable modem technology and availability, consult the Cable Modem Information Network on the Internet at www.cable-modem.net or Cable Modem University at www.catv.org.

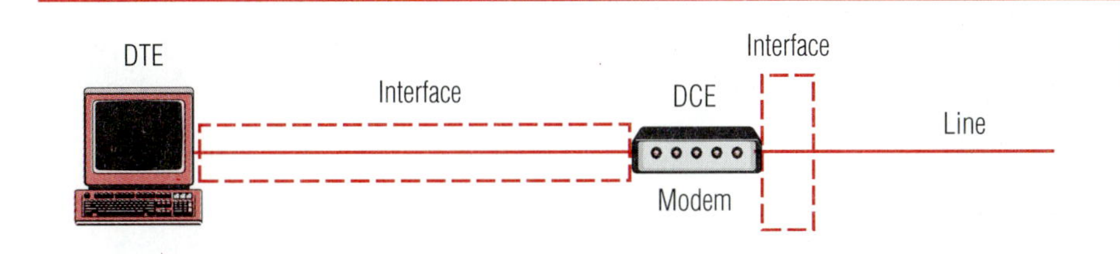

Figure 8–21
Modem interfaces.

These male and female 25-pin connectors meet the RS-232-C interface standards. (Courtesy of Black Box Corporation)

MODEM INTERFACES

A modem has two communication interfaces—one to the communications line and the other to the data terminal. The specifications of the interfaces include the mechanical characteristics (such as what type of connectors will be used and the number of pin connections), the signaling characteristics (such as how many signals there will be and what they will mean), and the electrical characteristics (such as what voltages and current level will be on each pin). Figure 8–21 shows where these interfaces occur in the communications system.

Interface Between the Modem and the Communications Line

The interface between the modem and the communications line is the simpler of the two. It consists of just two or four signaling wires with physical and electrical specifications that are prescribed by the telephone company. The connection is made with a short two- or four-wire cable that has standard RJ-11 plugs, like those used on telephones, on each end.

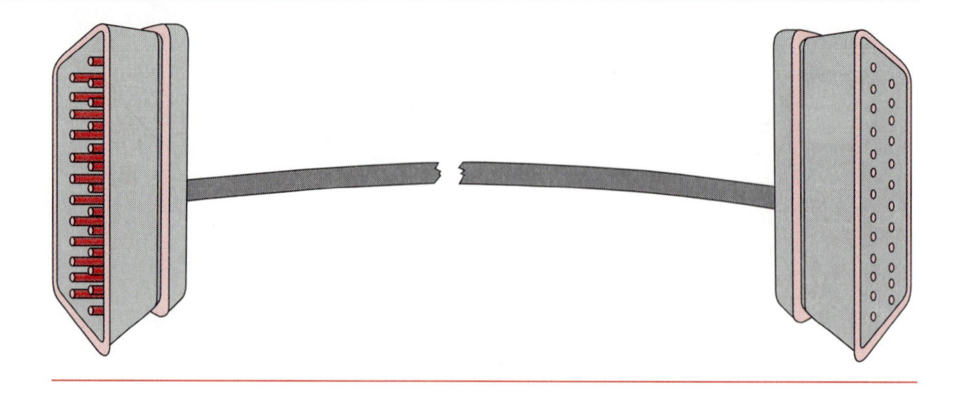

Figure 8–22
An RS-232-C cable with male and female 25-pin connectors.

Interface Between the Modem and the Data Terminal Equipment (DTE)

The interface between the modem and the DTE is more complicated. Several "standards" have been developed, but none is universally used. The most popular interfaces are discussed here.

RS-232-C Interface Virtually all computer and terminal equipment that uses telecommunications conforms to an interface that was standardized in the United States by the Electrical Industries Association (EIA). The EIA interface is called *RS-232-C.* It has been adapted for international use by the ITU-T under the name *V.24.* Although the RS-232-C and V.24 interfaces are not identical, they are functionally equivalent for most applications.

The actual hardware used to connect a DTE and DCE with the RS-232-C interface is a cable with 25 wires and a 25-pin connector at either end, as shown in Figure 8–22. The use of each of the wires or pins is specified in the RS-232-C standards document as well as the levels (voltages) of the signals. The signals sent across the connection are, in all cases, digital and serial. The standard also specifies the mechanical attributes of the interface, specifically that the DCE will have a female connector and the DTE will have a male connector. Another specification is a 50-foot maximum length for the cable connecting the DTE and DCE.

A list of the 25 signals and their functions is given in Figure 8–23. Note that not all of the pins have been assigned a use. This is to provide for future expansion. Note also that pins 2 and 3 are the ones on which the actual data is passed. All of the other pins are used for signaling between the modem and the DTE.

The RS-232-C standard has some limitations that are inconvenient at best and unacceptable in some applications.

Pin	Description
1	Protective ground
2	Transmitted data
3	Received data
4	Request to send
5	Clear to send
6	Data set (modem) ready
7	Signal ground
8	Received line signal detector
9	(reserved for modem testing)
10	(reserved for modem testing)
11	unassigned
12	Secondary receive line signal detector
13	Secondary clear to send
14	Secondary transmitted data
15	Transmission signal element timing
16	Secondary received data
17	Receiver signal element timing
18	unassigned
19	Secondary request to send
20	Data terminal ready
21	Signal quality detector
22	Ring indicator
23	Data signal rate selector
24	Transmit signal element timing
25	unassigned

Figure 8–23
RS-232-C connector pin assignments.

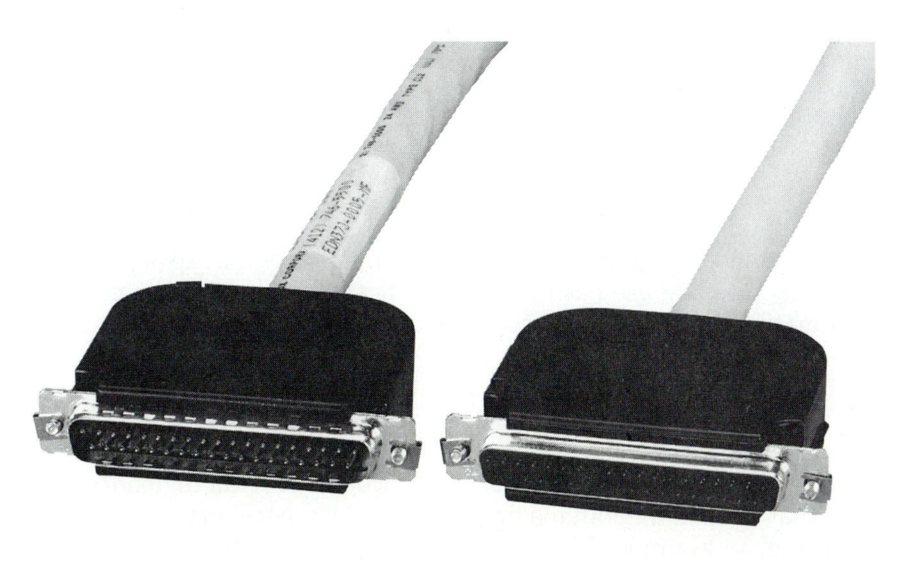

These male and female 37-pin connectors are used for the RS-449 interface. (Courtesy of Black Box Corporation)

- The cable length of 50 feet. At speeds less than 2,400 bps, a longer cable may work, but it is not guaranteed or supported by the equipment manufacturers. In actual practice, cable lengths of up to 100 feet at 9,600 bps are common.
- A technical problem with the way electrical grounding is handled that can lead to difficulties in detecting the difference between the 0 and 1 bit if the speed is high and the cable is long.

RS-232-D Interface In 1987, the RS-232-C interface standard was revised and renamed RS-232-D. *RS-232-D* made the following changes to the RS-232-C standard:

- specifications for the 25-pin connector;
- addition of a shield on the cable;
- redefinition of the protective ground;
- definition for pin 18 (local loopback), pin 25 (test mode interchange circuit), and redefinition of pin 21 from signal quality detector to remote loopback.

Although these changes solved some of the problems of the RS-232-C standard and made RS-232-D equivalent to the ITU-T V.24 standard, the RS-232-D interface has not seen widespread usage.

RS-449 Interface Another interface standard that overcomes some of the problems of RS-232-C is the *RS-449* standard. RS-449 uses 37 signal wires, as opposed to 25 for RS-232-C, with the extra ones added primarily for automatic modem testing. The RS-449 standard also overcomes the RS-232-C's length limitation; an RS-449 cable length can run up to 4,000 feet. One of the weaknesses of the RS-449 standard is that it makes no provision for automatic dialing between modems. Although the RS-449 standard was the intended successor to the RS-232-C interface, it has not been well accepted in the marketplace or widely implemented.

RS-336 Interface *RS-336* is an interface standard that does allow for automatic dialing of calls under modem control. However, it does not adequately provide for high-speed data transmission or the use of private circuits. Its primary use has been for applications in which the computer automatically calls numerous remote data terminals for data collection.

X.21 and X.21bis Interfaces The *X.21* interface standard was developed by the ITU-T and is widely accepted as an international standard. X.21 is based on a digital connection to a "digital" public telephone network. When the X.21 standard is used, data signals are encoded in serial digital form and kept that way throughout the transmission. All other signaling is in digital form, too; for example, digital dial tone is a series of + (plus) signs. The basic rate of a digital channel is 64,000 bps, which not only exceeds most dial-up data transmission speeds today but also can handle digitized voice.

Until digital networks are fully in place, the ITU-T has defined a temporary standard called *X.21bis,* which electrically is virtually identical to RS-232-C and V.24. Its application, however, is for connecting a terminal to a packet switch network via analog lines. Chapter 11 explores packet switching networks.

Current Loop Interface The *current loop* interface, although outdated, is mentioned because it is still in use today. Originally designed for use by teletypewriters, the current loop indicates 1 and 0 bits by the presence or absence of an electrical current. A 1 bit or mark is defined as the flow of either 20 milliamps or 60 milliamps of current, depending on the type of machine. Cables connecting current loop devices can be up to 1,500 feet long and can pass data at up to 9,600 bps. Although the current loop is very popular because it is simple and inexpensive to produce, it is totally nonstandard, which means that one manufacturer's equipment may not be able to connect to another manufacturer's equipment even though both are using a current loop interface.

OTHER MODEM FUNCTIONS AND CAPABILITIES

Reverse Channel

Some slow-speed modems, which do not use all of the available bandwidth of a communications line for primary signaling, use some of the remaining bandwidth to provide a slow-speed *reverse channel.* The modems can use a reverse channel for signaling one another. For example, the receiving modem can send a signal on the reverse channel indicating that it has received blocks of data correctly. The reverse channels typically operate at very slow speeds of 5 or 10 bps. The use of a reverse channel is not considered to be full-duplex operation because of the vast speed difference between the transmission on the primary channel and the reverse channel.

Auto Dial/Auto Answer

The rise in the use of modems with personal computers has led to the development of sophisticated auto dialing capability. *Auto dial* allows the operator to send commands to the modem through the terminal keyboard. With the auto dialer built into the modem, the user sends commands in the form of ASCII data characters from the terminal to the modem control section across the interface. These commands instruct the modem to perform the desired function. The command set has become relatively standardized around the commands developed in the early 1980s by Hayes Microcomputer Products, Inc. The commands tell the computer what number to dial, whether to use dial pulses or tones, how

This modem is designed to be mounted in one of the available slots inside the cabinet of a personal computer. Internal modems are usually a bit less expensive than external modems because no case is required. (Courtesy of 3Com.)

long to wait before answering, and so on. All Hayes commands begin with the letters AT, which stand for *attention.* When telecommunications software is run in a personal computer, the modem commands can be issued automatically. For example, the user could enter one high-level command such as "DS" for "dial stock," and the software would automatically issue all of the necessary detail commands to the modem to dial a stock quotation service, log the user on, request the particular stock prices of interest, log off, and hang up.

The *auto-answer* capability sets up the modem to automatically answer a telephone line when it detects the ringing signal. The incoming call can be handled by a computer or other device. The auto-answer capability is used by internet service providers and commercial communication systems, such as MCI Mail or CompuServe. It is also used by businesses that want to provide their employees with a way to dial in to the company's network.

The auto-dial/auto-answer feature also can be used to provide a dial-backup connection when a leased telephone line is normally used for transmission. If the modem detects that the leased line has failed, it automatically dials the receiving modem, establishes a connection through that modem's auto-answer capability, and resumes operation on the public telephone network. Often the backup connection operates at a slower speed than the original transmission. However, in many applications, the slower transmission is preferred to a total outage.

Internal Modems

An *internal modem* is not a feature but a modem contained on a single circuit card that is inserted in a slot inside a personal computer. The internal modem may have several indicators on a small control panel at the rear of the personal computer, but it usually does not have switches that allow the options and parameters of the modem to be easily changed. Internal modems cost approximately the same as stand-alone modems of the same capability. Some are slightly less expensive because they are not packaged in a cabinet.

Modem Diagnostics

Most modems today contain diagnostic routines that check the circuitry of the modem. Usually, the diagnostics run automatically when the modem is powered on and can be initiated by an operator anytime thereafter. In most modems, the diagnostics cannot run when data is being transmitted because they use the same circuitry. In some cases, diagnostics can be initiated remotely by a control operator or by a remote modem.

Loop Back One simple diagnostic is the loop back capability. It is a diagnostic capability that allows the transmitting modem to send a special signal to the receiving modem. The receiving modem sends the signal back to the originator, where it can be checked and analyzed for accuracy. If a terminal is having difficulty communicating on a circuit, the *loop back test* will help determine where the problem is occurring. If the loop back test runs successfully, the problem is not in the modem or circuit but likely to be in the terminal, the computer, or the cables that connect them to the modems.

MODEM SELECTION CRITERIA

Clearly, telecommunications professionals must select modems that match the requirements of the data communications application and the type of transmission that will be employed. The prospective purchaser must consider factors such as:

- whether the line to be used is analog or digital;
- whether the transmission will be asynchronous or synchronous;
- what transmission speed is required;
- whether the transmission will occur on dial-up or leased lines;
- what the distance between the modems will be;
- whether any special type of transmission line such as fiber-optic cables will be used;
- what is affordable for the application.

There may also be decisions about special features such as fax capability, simultaneous voice and data capability, or the ability to use the modem with a cellular phone. As always, there will be trade-offs to be made, and management may have to be involved in the decision making.

SPREAD SPECTRUM

A form of communication that does not easily fit into the other categories discussed in this chapter is known as *spread spectrum* transmission. The technique was originally developed by the military to make transmissions more difficult to intercept and therefore more secure, but now it is being used in various wireless technologies such as cellular telephones.

frequency hopping
direct sequence

Spread spectrum uses an analog signal to transmit either analog or digital data. The concept is that the signal is spread over a wider-than-usual bandwidth to make interception more difficult. Two techniques are used, *frequency hopping* and *direct sequence.* When frequency hopping is used, the signal is broadcast over a seemingly random series of radio frequencies, hopping from frequency to frequency at split second intervals. The transmission frequencies are determined by a pseudo random number generator. The receiver uses the same random number sequence so that it can determine what frequency to listen on to receive the signal. Would be eavesdroppers hear only short sections of the transmission because they don't know the frequency hopping sequence.

The direct sequence system combines bits from the original signal with bits generated by a pseudorandom bit stream generator using Boolean math. The receiver, using the same pseudorandom bit stream can reverse the Boolean math process and recover the original bits.

THE REALITIES OF TODAY'S NETWORKS

It is important for you to understand and remember that today's public networks almost always employ a combination of analog and digital technologies. Signals are frequently converted from one mode to another several times during their journey from the sender to the receiver. Using the telephone example, the analog signal generated by most telephone handsets may travel in analog form to the nearest telephone company office, but then is almost always converted to digital form for transmission within the telephone network. A home computer connected to the Internet generates digital signals that have to be converted to analog by the computer's modem so that they can be sent on the telephone network. The network probably converts the signal to digital for internal transmission, as described above, but may deliver it to the Internet service provider (ISP) in either analog or digital form, depending on what type of circuit (analog or digital) the ISP has installed. Similarly, the ISP's connection to the Internet may be analog or digital, although most ISPs would have a digital circuit because of the higher speeds that can be achieved. The point to remember is that analog to digital and digital to analog signal conversions occur frequently in the public networks.

SUMMARY

This chapter introduced many technical concepts of data transmission. Transmissions have been categorized in several ways. The important distinction between asynchronous and synchronous communications was explained, as well as the pros and cons of each type of transmission. Modems were defined as a specialized signal converter, and several modem standards and types of modems were examined in detail. The chapter also described contemporary interface standards between terminals and modems. Now you should have a good understanding of data transmission and should be ready to apply this knowledge to the study of various transmission lines, protocols, and networks.

CASE STUDY

Dow Corning's Use of Modems

Dow Corning acquires its modems from a variety of sources, although the number of modems in use has been steadily decreasing as the WAN has moved from being analog- to digital-based, and as LANs have been installed. Some transmissions occur over limited distances, and on those circuits, limited distance modems have been purchased to keep costs down. On the global digital data communications network, DSU/CSUs are used instead of modems.

Dow Corning's international network, which extends to all locations where the company has offices or plants, including Europe, Mexico, Canada, Brazil, Japan, Hong Kong, China, and Australia, is primarily digital so DSU/CSUs are used. Where there are modems, they must meet international standards. Dow Corning's circuits to Europe, Canada, and Asia are digital and use DSU/CSUs provided by WorldCom. Obviously, when installing modems or DSU/CSUs in international locations, there are many unique requirements that may dictate different solutions in each country. The corporate telecommunications staff works closely with its counterparts in Brussels, Australia, and Tokyo to determine the best configuration for each situation.

Dial-in data transmission in Midland operates at speeds up to 56,000 bps using V.34 and V.90 standard modems. The dial-in capability has been outsourced to UUNET, which provides a large pool of modems at its facilities. UUNET provides dial-up capability to Dow Corning on a global basis.

QUESTIONS

1. Is there any benefit in standardizing the modems a company acquires? If so, what would the benefits be?
2. What are some reasons a company would install modems that handle speeds slower than the maximum that technology will allow?
3. For security reasons, Dow Corning's dial-in security practices have not been discussed in the case. What would you imagine would be appropriate measures for the company to take to ensure that only authorized people access the network?
4. Because of the general movement in industry toward digital circuits, do you think that the use of modems will be obsolete in a few years?

REVIEW QUESTIONS

1. Compare and contrast the meanings of the terms *baud* and *bits per second.*
2. How many unique signal changes do dibits, tribits, and quadbits require in order to be transmitted?
3. Compare and contrast simplex, half-duplex, and full-duplex transmissions.
4. Distinguish between parallel and serial transmission.
5. Describe the purpose of the start and stop bits in asynchronous transmission.
6. Explain the terms *bit synchronization* and *character synchronization.*

7. Explain why synchronous transmission is more efficient than asynchronous transmission when long blocks of data are to be transmitted.
8. Using the ASCII code, a block of 1,000 characters of data is to be sent on a communications line. Calculate the relative efficiency of synchronous and asynchronous transmission methods. Calculate the efficiency again for a block containing 50 characters.
9. Why is synchronous transmission more expensive to implement than asynchronous transmission?
10. Explain the terms *mark* and *space*.
11. The purpose of the clock in synchronous transmission is ______.
12. Distinguish among unipolar, bipolar NRZ, bipolar return-to-zero and Manchester coded digital signals.
13. Explain why FSK modulation cannot be used for a 1,200 bps, FDX transmission.
14. Why is phase modulation better suited for high-speed transmission than frequency or amplitude modulation?
15. What are the differences between fixed equalization and adaptive equalization in a modem?
16. Describe how high-speed transmission has been achieved in modems that use the V.34 standard.
17. What are the upper limits to data transmission speed on analog lines?
18. How do high-speed modems achieve their high throughput rates?
19. What characteristics make a fax modem different from a data modem?
20. Explain the functions of DSU/CSU when lines designed for digital transmission are used.
21. Explain the concept of line turnaround in half-duplex transmission.
22. Identify the limitations of the RS-232-C interface.
23. When is the use of an acoustic coupler appropriate? What is its chief limitation?
24. When is the use of a limited distance modem appropriate? Why is it advantageous to use one whenever possible?
25. Describe the purpose of a loop back test.
26. What are the advantages of digital transmission compared to analog transmission?
27. Explain why, using PCM, it takes 64,000 bps to transmit a voice signal. How does ADPCM cut the data rate in half?
28. Describe the functions of an A/D converter.
29. Explain quantization.
30. Why is there interest in transmitting digital voice signals at slower data rates?
31. How is parallel transmission accomplished?
32. How does asynchronous transmission work?
33. Describe transmission efficiency.
34. Explain the differences among frequency division multiplexing, time division multiplexing, and statistical time division multiplexing.
35. What is a cable modem? What are the pros and cons of using one?
36. A signal made up of discrete, discontinuous voltage pulses is called a(n) ______ signal.
37. What is the purpose of using Manchester coding instead of NRZ coding?
38. List five advantages of digital transmission over analog transmission.
39. V.34 and V.34bis assume that most of a data transmission takes place on ______ lines.
40. Explain the unique characteristics of data transmission that follows the V.90 standard.
41. Why is the use of cable modems becoming more widespread?
42. Explain the purpose of V.42bis and MNP 4.
43. List the criteria to be considered when selecting a modem.
44. What is the primary advantage of using spread spectrum technology for data transmission?

PROBLEMS AND PROJECTS

1. A communications system is transmitting 100 character blocks of EBCDIC data at 4,800 bps, HDX. After each block of data is sent, the line must be turned around, and the receiving modem must either acknowledge correct receipt of the data (by sending an ACK character) or signal that the data was received incorrectly (by sending a NAK character). If the data was received correctly, the line is turned around, and the next block of 100 characters is sent. Line turnaround takes 50 milliseconds. Assuming that no transmission errors occur, calculate what percentage of the time the line is used for transmitting the data and what percentage of the time is spent performing line turnarounds.
2. Work problem 1 assuming the line speed is 9,600 bps instead of 4,800 bps. How does that change affect the results?
3. A personal computer sends blocks of ASCII data that are, on average, 30 characters long to a mainframe. The mainframe returns data blocks that average 500 characters in length. Calculate the transmission efficiency if asynchronous transmission is used, assuming 1 start bit and 1 stop bit per character. Calculate the efficiency for synchronous transmission assuming three synchronizing characters per block. Ignore the effects of line turnaround, and assume that no transmission errors occur.
4. How long will it take to transmit a file of 2 megabytes using a 7-bit code and a transmission speed of 28,800 bps?

Vocabulary

signaling rate
baud
circuit speed
bits per second (bps)
dibit
tribit
quadbit
simplex transmission
half-duplex transmission (HDX)
full-duplex transmission (FDX)
duplex
parallel mode or transfer
serial mode or transfer
asynchronous transmission (asynch)
start bit
stop bit
start/stop transmission
mark
space
bit synchronization
transmission efficiency
character synchronization
synchronous transmission
block
synchronization character
digital signal
unipolar
bipolar, nonreturn-to-zero (NRZ)
bipolar, return-to-zero
Manchester coding
differential Manchester coding
digitize
analog-to-digital (A/D) converter
quantization
quantizing noise
digitizing distortion
digital-to-analog (D/A) converter
codec
pulse code modulation (PCM)
bits per second (bps)
adaptive differential pulse code modulation (ADPCM)
delta modulation
baseband transmission
digital transmitter/receiver
data service unit/channel service unit (DSU/CSU)
modem
fixed equalizer
adaptive equalizer
frequency shift keying (FSK)
sine wave
phase shift
phase shift keying (PSK)
differential phase shift keying (DPSK)
quadrature amplitude modulation (QAM)
trellis code modulation (TCM)
signal processors
V.32
V.32bis
V.34
V.34 +

V.90
handshaking
V.42bis
V.42
line turnaround
facsimile (fax) modem
modified Huffman (MH) encoding
modified read (MR) encoding
modified, modified read (MMR) encoding
cable modem
head end
RS-232-C
V.24
RS-232-D
RS-449
RS-336
X.21
X.21bis
current loop
reverse channel
auto dial
attention
auto answer
internal modem
loop back test
frequency hopping
direct sequence

References

Alsop, Stewart. "The Cable Industry's Big Dream." *Fortune* (January 13, 1997): 147–148.

Camarro, Kenneth D. "Fax Modems Demystified." *Telecommunications* (July 1993): 49–54.

Davis, Beth. "Modem Speeds in Check." *Information Week* (February 24, 1997): 79.

Graves, Lucas. "Voice Modem Rising." *Teleconnect* (May 1996): 105–106.

Nee, Eric. "Is Excite@Home The AOL of Broadband?" *Fortune* (December 6, 1999): 156–163.

Overton, Rick. "Broadband or Bust." *PC World* (May 2000): 100–116.

Steers, Kirk. "Speed Up Dial-Up Modems." *PC World* (January 1997): 260.

Winton, Neil. "Cable Modems: Fastest Internet Access in the East and West." *PC World* (February 1997): 62.

______."Drinking Deep from the Net." *The Japan Times Weekly* (March 15, 1997): 13.

CHAPTER

9

Communications Circuits

OBJECTIVES

After you complete your study of this chapter, you should be able to

- distinguish among a communications line, circuit, and channel;
- discuss various types of communications circuits and their distinguishing attributes;
- describe the characteristics of the various types of media used to carry circuits;
- describe multiplexing and concentrating and tell when they are most productively used;
- describe the major types of errors that occur on communications circuits;
- describe the primary error prevention and detection techniques for communications circuits.

INTRODUCTION

Previous chapters discussed communications lines or circuits without defining them. In this chapter, circuits are defined more precisely, and the many alternatives for the physical media used in a circuit's construction are examined. Information about the ways in which circuits can be configured and used is also presented.

This chapter continues the discussion and description of the first layer of the ISO-OSI model that began in Chapter 8. Communications circuits provide the physical path on which the communications signals flow. The nature of the signals sent on the path is discussed in Chapter 10. The information in this chapter and the next is at the heart of telecommunications. You must understand the characteristics of circuits to learn how telecommunications can be applied to meet the requirements of a particular application. This chapter describes many types of telecommunications circuits, and it gives examples of how they are being used by companies.

DEFINITIONS

Authorities differ somewhat in the details of the definition of the word *circuit*. The commonly accepted definition of a telecommunications circuit, however, is the path over which two-way communications take place. A circuit may exist on many different types of media, such as wire, coaxial cable, fiber-optic cable, microwave radio, or satellite. The word *line* is often used interchangeably with *circuit,* although line gives a stronger implication of a physical wire connection. In fact, many circuits today do not run on wires at all. The longer the distance, the higher the probability that the circuit runs on at least one medium other than wire.

line
circuit

A *link* is a segment of a circuit between two points. For a telephone circuit, one link exists between the residence or business and the local telephone company central office. Other links exist between the central offices. The final link is from the remote central office to the remote residence or business. When the term *data link* is used, it almost always includes the data terminal equipment, modems, and all other equipment necessary to make the complete data connection—software as well as hardware.

link

Circuits are often subdivided into *channels,* which are one-way paths for communications. Channels may be derived from a circuit by multiplexing, or they may be an independent entity, such as a television channel. A data circuit is sometimes divided into two channels, one of which is high speed for data transmission and the other is low speed for control information. The data channel is called the *forward channel,* and the control channel is the *reverse channel.* As the name implies, the reverse channel carries information in the opposite direction from the data channel. The type of information carried on a reverse channel depends on the rules of communications or protocol used. (Protocols are defined and discussed in Chapter 10.)

channel

A *node* is a functional unit that connects to transmission lines. It also can be an end point on a circuit or a junction point of two or more circuits. Typical nodes are telephones, data terminals, front end processors, and computers.

node

As was discussed in Chapter 8, the data flow on a circuit can be in only one direction, called *simplex;* in both directions, but not simultaneously, called *half-duplex;* or in both directions simultaneously, called *full-duplex.* From the standpoint of the telephone company or other circuit provider, the way in which the circuit is actually configured may differ depending on the intended data flow.

■ TYPES OF CIRCUITS

Point-to-Point Circuits

A *point-to-point circuit,* illustrated in Figure 9–1, connects two—and only two—nodes. A typical circuit of this type connects two locations of a company or connects a computer to a terminal or a personal computer to a server. The standard telephone call between two locations is another example of the use of a point-to-point circuit.

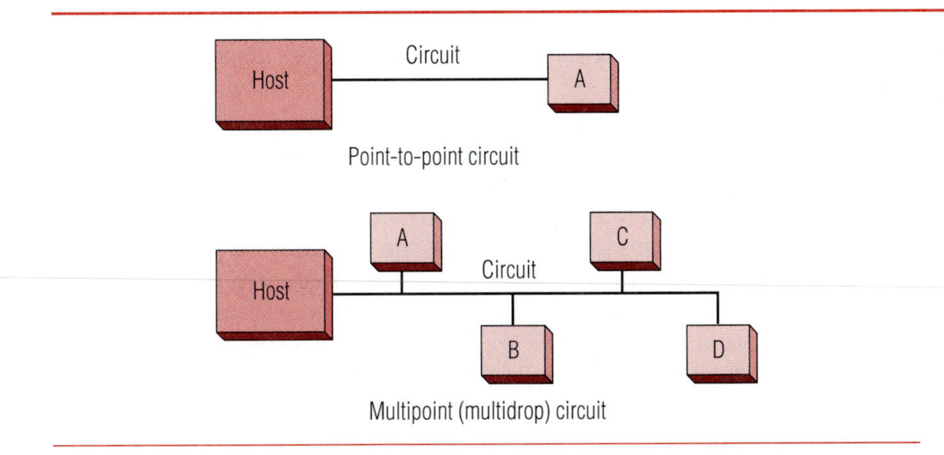

Figure 9–1
Types of circuits with nodes A, B, C, D, and the host computer.

Multipoint Circuits

If there are several nodes connected to the same circuit, as shown in Figure 9–1, it is called a *multipoint* or *multidrop circuit*. With multidrop circuits there is a clear distinction between the circuit and the links. The connections from the host to A, from A to B, and from B to C are links. The overall connection from the host to D is the circuit. In most cases, a multipoint circuit is less expensive than four point-to-point circuits, each connecting the host to one of the terminals (A, B, C, and D). Multipoint circuits are used to connect locations that have a relatively low volume of traffic and can share the line without interfering with one another and still get adequate performance.

Two-Wire and Four-Wire Circuits

Point-to-point and multipoint data circuits are normally implemented with either two wires or four wires connecting the points. Normally, two wires are required to carry a communication in one direction. Therefore, a *two-wire circuit* has traditionally been viewed as being a half-duplex circuit and a *four-wire circuit* as a full-duplex circuit. As mentioned in Chapter 8, however, some modems are capable of splitting a circuit into two channels through frequency division multiplexing or other techniques. In that case, it is possible to obtain full-duplex operation on a two-wire circuit. In most cases, however, four-wire circuits are preferable for data communications. With four-wire circuits, two wires provide the forward channel in one direction while the other two wires provide the forward channel in the other direction.

Standard dial-up telephone circuits are two-wire circuits. Four-wire circuits must be ordered from the telephone company and installed on a leased basis. The advantage is that the circuit is then available for full-time use. The disadvantage is that a leased circuit may cost more than a dial-up connection, and it may be uneconomical if relatively little data will be transmitted.

Analog Circuits

Because the public telephone system was originally designed to carry voice transmissions in analog form, most of the circuits that run to individual homes or small businesses today are analog. As we have seen in Chapter 8, using analog circuits to carry data requires using a modem to convert the digital signal from a terminal or other DTE to analog form before it is transmitted, and then using another modem to convert it back to digital form at the receiving end, before the data can be presented to a computer or other DTE. Analog circuits are inherently limited in the speed at which they can carry data, and are also more prone to noise and errors than digital circuits. Hence, communications carriers of all types are installing digital circuits as quickly as they can to take advantage of the benefits that digital transmission offers, which were also described in Chapter 8.

Whereas the modem usually determines the actual speed at which an analog circuit transmits data, the carriers offer analog circuits that are capable of certain speed ranges. Traditionally, these have been known as *low-speed circuits* or *subvoice-grade circuits, voice-grade circuits,* and *wideband circuits.* Subvoice grade circuits are designed for telegraph and teletypewriter usage but may also be used for low-speed signaling applications, such as fire alarms, burglar alarms, door opening indication, or for process monitoring systems where the data rate is low and/or infrequent. These circuits operate at speeds of 45 to 200 bps. They cannot handle voice transmission or data at higher speeds. In fact, subvoice-grade circuits are derived by dividing a voice-grade circuit into 12 or 24 low-speed circuits.

Voice-grade circuits are designed for voice transmission but they can also transmit data at up to 56,000 bps, when sophisticated modems are used. These are the type of circuits that have been most commonly described throughout this book so far. Wideband circuits are high-speed analog circuits designed to carry multiple voice or data signals. They are delivered to customers in a bandwidth of 48,000 Hz, which, with the use of appropriate multiplexing equipment, will carry 12 to 4,000 Hz voice channels. However, these circuits are rarely sold anymore because they have been largely superseded by digital circuits.

Digital Circuits

A *digital circuit* is one that has been designed and engineered expressly to carry digital signals. The direct digital transmission of data is simple and eliminates the signal conversions at each end. Digital transmission capability became available to end-users in the late 1970s, although the common carriers had been using digital transmission techniques and circuits for many years before that. Most new circuits being installed are designed especially for digital transmission.

One of the main advantages of digital transmission is that the distortion of pulses that inevitably occurs along the transmission path is easier

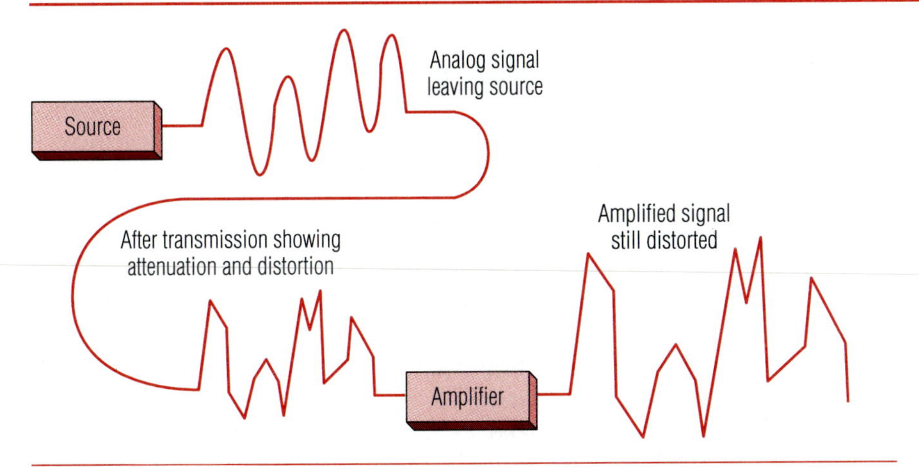

Figure 9–2
Amplification of an analog signal.

Figure 9–3
Regeneration of a digital signal.

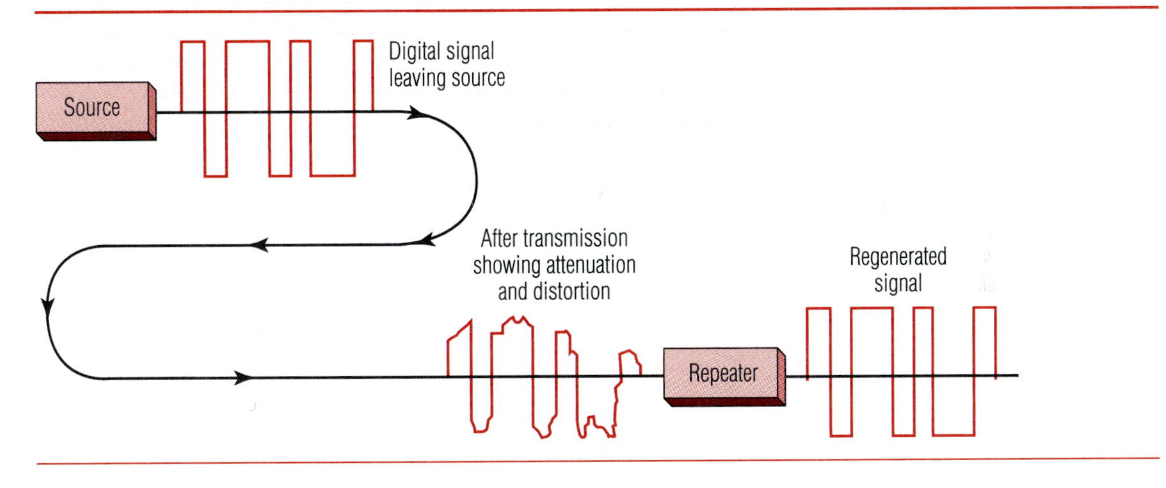

to correct than the distortion of an analog transmission. Analog signals are periodically amplified to increase their signal strength, and the distortions in the wave form are amplified as well as the original signal, as shown in Figure 9–2. Digital signals are made up of simple pulses and are not amplified but regenerated by a *repeater.* The regeneration eliminates any distortion that has occurred, as shown in Figure 9–3. Thus, the signal that arrives at the receiver is cleaner. The result is that digital transmissions have a lower error rate than do their analog counterparts.

repeater

Another advantage of using digital transmission is that no analog-to-digital conversion is required. Assuming digital links exist from one end of the circuit to the other, the digital output from the DTE can be simply shaped and timed to conform to the requirements of the digital network and transmitted directly. The device that performs the shaping and timing is called a *digital transmitter/receiver* or *data service unit/channel service unit (DSU/CSU),* which was discussed in Chapter 8. DSU/CSUs are much less complicated and considerably less expensive than modems.

DSU/CSU

Integrated Services Digital Network (ISDN)

The *Integrated Services Digital Network (ISDN)* is less of a network and more a set of standards than the name implies. The ISDN standards were developed by the ITU-T as a vision for the direction the world's public telecommunications systems should take. They believe that eventually ISDN will replace leased and switched circuits as we know them today.

ISDN can best be visualized as digital channels of two types, as shown in Figure 9–4. One type, the "B" (bearer) channel, carries 64 kbps of digital data. The other type, the "D" (delta) channel, carries 16 kbps of data and is used for signaling. These two types of channels are packaged, according to ISDN standards, into two types of access services. The *basic rate interface (BRI),* also known as 2B+D, provides two 64-kbps B channels and one 16-kbps D channel. Basic rate interface ultimately will be provided on the line side of a PBX or central office, thereby making it available in homes and offices on standard twisted pair wiring.

basic rate interface

primary rate interface

The other type of access is known as *primary rate interface (PRI),* or 23B+D. Primary rate access provides 23 64-kbps B channels for carrying data and 1 64-kbps D channel for signaling. The total capacity of 24–64-kbps channels happens to equal the carrying capacity of a T-1 circuit as it is

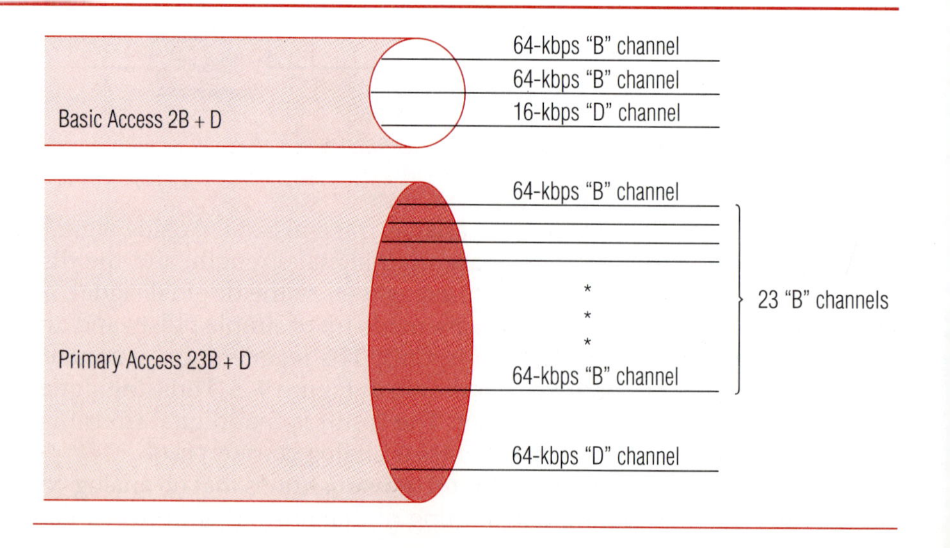

Figure 9–4
ISDN basic and primary access arrangements.

defined in the United States, Canada, and Japan. (T-1 circuits are described later in this chapter.) In Europe, primary access is defined as 30B+D, which matches the capacity of a European T-1 circuit. Figure 9–4 shows the two types of access schematically.

The large bandwidth provided by ISDN circuits can be used for digitized voice and data. With basic access in a home, a person could be having a telephone conversation and simultaneously transmitting data at 64 kbps from a personal computer. A business with a primary ISDN access group could subdivide the bandwidth in any way necessary to meet the needs of its application set. Half of the capacity, or 772 kbps, might be used for a television transmission while the other 772 kbps is further subdivided to support 24 voice conversations at 16 kbps each and 6 64-kbps data channels. At another time, the bandwidth could be configured differently—all under the control of the customer. In fact, these configuration changes could be programmed into a computer so that the ISDN capacity is automatically reconfigured at certain times of the day. There could be a daytime configuration, a nighttime configuration, or any other combination to meet the business needs of the company.

An upgrade to ISDN called *broadband ISDN (B-ISDN)* has been defined but has not yet been widely implemented. Broadband ISDN has three services:

- a full-duplex circuit operating at a speed of 155.52 Mbps;
- a full-duplex circuit operating at 622.08 Mbps;
- an asymmetrical circuit with two simplex channels, one of which operates at 155.52 Mbps and the other at 622.08 Mbps.

The first two options would be used as very high speed data circuits, whereas the third option would typically be used where the flow of data in one direction is much greater than in the other.

B-ISDN circuits are actually transported from one node to another using the asynchronous transfer mode (ATM) for transport. ATM will be discussed in Chapter 11, and at that point, B-ISDN will be mentioned again. Suffice it to say that ATM is transparent, and the user of a B-ISDN circuit is not aware that ATM is operating in the background. Furthermore, B-ISDN is backward compatible, meaning that a standard ISDN BRI or PRI transmission can be handled transparently on a B-ISDN circuit.

The benefits of ISDN are that

- it provides efficient multiplexed access to the public network;
- it has the capability to support integrated voice and data;
- it has a robust signaling channel, which is important for network management;
- it provides an open system interface that is internationally defined. This will go a long way toward making multivendor telecommunications systems a reality.

ISDN service has been implemented more slowly in the United States than in many other countries in the world, partly because the RBOCs and long distance carriers have had difficulty agreeing on precise standards, partly because they have not actively marketed the service, and partly because costs have been high. There have also been frequent problems in getting ISDN service to operate properly when it is first installed. In Japan and Australia, however, ISDN has been used by businesses for data circuits for many years, and now ISDN is being installed in many homes for little more than the cost of a regular analog telephone line.

T-Carrier Systems

A family of high-speed digital transmission systems, known as the *T-carrier systems,* has evolved within the carriers over the past 30 years. T-carrier systems are designated according to their transmission capacity, as shown in the table below.

Designation	Bit Capacity
T-1	1.544 Mbps
T-2	6.312 Mbps
T-3	44.736 Mbps
T-4	274.176 Mbps

The T-2 and T-4 circuits are used primarily by the carriers. T-1 and T-3 circuits are used by both the carriers and their customers.

A T-1 system uses a standard pair of wires for transmission. Repeaters are spaced about every mile to regenerate the signal and transmit it over the next link of the circuit. A T-1 system can carry 24 circuits of 64,000 bps (24 × 64,000 = 1.536 Mbps; the extra 8,000 bps are used for signaling). Multiplexing equipment is used to combine signals for transmission over the T-1 system and to separate them at the receiving end. A 64,000 bps channel can carry one or two digitized voice signals, depending on whether Pulse Code Modulation (PCM) or Adaptive Differential Pulse Code Modulation (ADPCM) is used to modulate the signal.

It should be noted that in Europe T-carrier systems are also in use. They are defined slightly differently, however. A European E-1 circuit is made up of 32–64 kbps channels for a total capacity of 2.048 Mbps. Other T circuits are multiples of the E-1 capacity.

Companies acquire T-1 or T-3 facilities when they have a high volume of voice, data, or video transmissions between two locations. One company found it economical to install a T-1 circuit to carry three 56 kbps data circuits and five voice tie lines. Even though the capacity of the T-1 circuit was not fully used, it was less expensive than if the individual voice and data circuits were leased separately. Either the carrier or the customer may provide multiplexing equipment to divide the capacity of the T-1 into usable circuits. Some carriers offer full T-1 packages that include the

This is one example from the family of Ascom Timeplex multiplexers, which comes in various sizes depending on the capacity required. (Courtesy of Ascom Timeplex, Inc.)

equipment at both ends of the circuit, configured to the customer's specifications, while other carriers offer the bare T-1 without the multiplexers. Several companies such as Cisco Systems, Inc., Timeplex, Inc., and General DataComm, Inc. develop and sell multiplexing equipment suitable for T-1 circuits.

The major reasons for using T-1 circuits are that they can save large amounts of money, can give flexibility in reconfiguring the T-1 capacity to meet different needs at different times of the day, and can improve the quality of voice and data transmission because the information being transmitted is digitized.

Fractional T-1 A newer communications offering is called *fractional T-1.* In the past, companies needing digital transmission service had no choice of speed between 56 kbps and T-1's 1.544 Mbps. Fractional T-1 provides companies with other transmission speed choices by subdividing a T-1 circuit into multiples of 64 kbps. Thus, the IXCs now offer leased circuits that operate at any multiple of 64 kbps, although speeds of 64, 128, 256, 512, and 768 kbps are most common. A company can select and pay for only the capacity it needs rather than having to lease a full T-1 circuit. This new capability is particularly interesting to companies with smaller networks that have not been able to justify the cost of a full T-1.

Fractional T-1 is provided by the IXCs, but potential customers must check to see if their local telephone company provides fractional T-1 service. If not, an alternative must be found to connect from the customer's premises to the IXC office.

Switched Multimegabit Data Service (SMDS)

connectionless service

Switched Multimegabit Data Service (SMDS) is a high-speed switched digital service offered by the carriers. Because the user does not have a dedicated line between locations, SMDS is called a *connectionless service.* Data to be transmitted via SMDS is broken down into 53 byte cells. Two SMDS speeds are available, either 1.544 Mbps (T-1 speed) or 44.736 Mbps (T-3 speed). The difference between SMDS and T-1 service is that a user must lease a T-1 circuit from point to point, all the way between two locations, whereas with SMDS the user leases a circuit to the nearest carrier's office at both ends, and the carrier handles the transmission in between using normal, shared communication facilities. For example, if a company wanted to connect each of its eight plants to headquarters at 1.544 Mbps, it could either lease eight T-1 lines, one from each plant to headquarters, or it could lease a 1.544 SMDS line from each plant to the nearest carrier's office, and then perhaps a 44.736 Mbps SMDS line from headquarters to the nearest carrier's office. Although a detailed cost analysis would have to be done for the particular locations involved, in most instances the SMDS configuration would be less expensive.

Digital Subscriber Line (DSL)

In 1987, in anticipation of competition with cable companies in the area of delivering video signals to homes, Bellcore developed a technology called *asymmetric digital subscriber line (ADSL),* which was originally designed to enable telephone companies to deliver digitized signals to subscribers at about 1.5 Mbps over existing twisted pair copper telephone wire. The general name for the service is *digital subscriber line (DSL),* and it has spawned many variants that collectively are also referred to as xDSL. While originally developed as a competitor for cable television, xDSL is turning out to be a boon for home and business users who want higher data communication speeds, especially for accessing the Internet.

xDSL services are dedicated, point-to-point, public network access over twisted pair copper wire on the local loop between a network service provider's central office (typically the telephone company) and the customer site. ADSL technology is asymmetric. It allows more bandwidth downstream from the central office to the customer's site than upstream from the customer to the central office. This asymmetry makes it ideal for Internet surfing, and indeed, ADSL is viewed as more of a consumer- than a business-oriented product offering.

DSL technology in general is very sensitive to distance and is not even available to customers who are located more than 3 or 4 miles from the central office. The speed of the service depends on the distance, but for ADSL is in the range of 16 to 640 Kbps upstream and 1.5 to 9 Mbps downstream, which, even at its slowest, is significantly faster than the typical customer has today. The actual speed that can be obtained de-

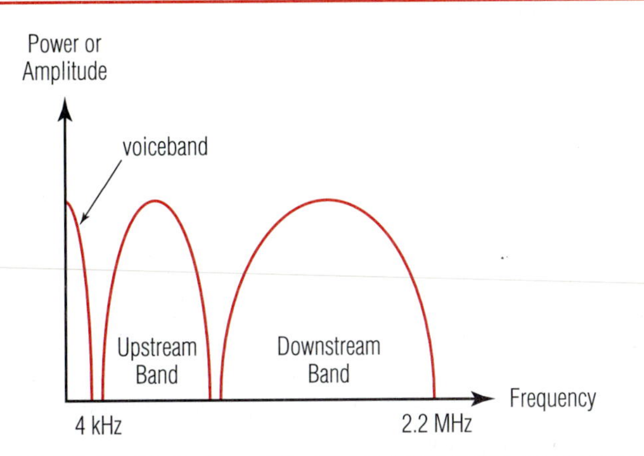

Figure 9–5
An ADSL circuit is divided into several channels and uses a much wider frequency range than a standard telephone circuit.

pends on a number of factors, including the distance from the central office, the wire size that is used, and the quality of the circuit.

ADSL circuits have a modem at each end, which divide the circuit into the high-speed downstream channel, the medium speed upstream channel, and a standard telephone channel as shown in Figure 9–5. Customers can, therefore, obtain their telephone service from the same wires that supply the ADSL service. The DSL provider almost always supplies the modems and includes a rental fee in the total price for the ADSL service. In the future, it may be possible for customers to purchase their own modems off the shelf at retail outlets.

There are three new offerings in DSL technology, G.Lite ADSL, SDSL, and VDSL. G.Lite ADSL is a newer standard ADSL service that offers a maximum speed of 1.5 Mbps downstream and 384 kbps upstream, and some carriers will allow the customer to order lower bandwidths for a lower cost. Because the speed is limited, the specifications for the wire are not as severe, and G.Lite is easier to install and maintain.

Another offering that is just becoming available is *SDSL, symmetric digital subscriber line.* SDSL provides equal speed channels in both directions and may not include a telephone channel. SDSL is capable of speeds up to 768 kbps in each direction and is targeted at business customers. Technically, it is easier for the network service provider to implement SDSL than ADSL because simpler, less expensive equipment is required at both the head end and the customer end of the circuit.

The newest development is *VDSL, very-high-rate digital subscriber line.* VDSL transmits data at top speeds of 51 to 55 Mbps over short twisted pair telephone lines of up to 1,000 feet and low speeds of 13 Mbps at 4,000 feet. Early versions of this technology are asymmetric, like ADSL, and have an upstream channel of 1.6 to 2.3 Mbps.

While DSL is a very interesting capability for many businesses and consumers at home, it is important to remember the distance limitation—the customer must be located within 3 to 4 miles of the central office. That means that DSL service will not be available to at least 30 percent of the potential users in the U.S. Hence, there will still be plenty of need for other high-speed access services, such as cable modems, ISDN, and direct satellites. For more information about the status of DSL development and implementation, check the web site www.2wire.com.

CIRCUIT MEDIA

conducted media

radiated media

This section discusses the various media used for communications circuits. In general, media can be divided into two categories: *Conducted* or *guided media* provide some type of physical path, such as wire, cable, or optical fiber, along which the signal moves from end to end. *Radiated* or *unguided media* employ an antenna and the transmission of waves, such as radio waves, microwaves, or infrared waves, through air, water, or a vacuum. Although there is a wide overlap in their characteristics and capabilities, each medium has found a particular niche in which it is most commonly used.

Twisted Pair Wire

twisted pair

crosstalk

The most common medium used for telecommunications circuits is ordinary wire, usually made of copper. Over the years, the common carriers have laid millions of miles of wire into virtually every home and business in the country, so wire is used for virtually all local loops. Originally, open wire pairs were used. However, they were affected by weather and were very susceptible to electrical noise and other interference. Today, the pairs of wires are almost always insulated with a plastic coating and twisted together. This type of wire is called *twisted pair* or *unshielded twisted pair (UTP);* it is illustrated in Figure 9–6. Wire emits an electromagnetic field when carrying communications signals, but twisting the pair together has the effect of electrically canceling the signals radiating from each wire. To a large extent, it prevents the signals on one pair of wires from interfering with the signals on an adjacent pair. This type of interference is called *crosstalk.* The wire used for inside applications and local loops normally is 26, 24, or 22 gauge. The smaller the gauge number, the larger the wire.

The Electrical Industries Association (EIA) has defined five categories of unshielded twisted pair wire for telephone and data transmission use. These categories, known commonly as *Cat 1* through *Cat 5,* are

Figure 9–6
Twisted pair wires are the most commonly used medium for communications transmission.

defined as follows. With the exception of Category 1, all wire is typically 24 gauge solid copper.

- **Category 1**—Basic twisted pair wire for telephone use. Not recommended for data.
- **Category 2**—Four unshielded, solid twisted pair. Certified for data transmission to 4 Mbps.
- **Category 3**—At least three twists per foot. About the same as normal telephone cable installed in most office buildings. Certified for data transmission to 10 Mbps but with proper design and over limited distances, may work to 16 Mbps. If there is more than one pair in the same jacket, it must not have the same number of twists per foot in order to minimize interpair crosstalk.
- **Category 4**—Similar to Category 3, but more twists per foot, and certified to 16 Mbps.
- **Category 5**—Three to four twists per inch. Data-grade cable certified for data transmission to 100 Mbps.
- **Category 5E (5 extended)**—An extension of the standards for Category 5 cable designed to handle data transmission up to 1 Gbps. Recommended for all new cable installations in office buildings.

Categories 3 and 5 have received the most attention, since Cat 3 wire is the type that is installed in most existing office buildings, and Cat 5 is the type that companies are choosing for new installations. Many companies are finding that if the distance isn't too great, they can successfully use it at speeds greater than the 100 Mbps that is specified.

In addition to the six existing standards for UTP cable, there are two other developing standards that you may hear about. *Category 6 (Cat 6)* is an emerging standard for cable consisting of four twisted pairs separately wrapped in foil insulators and twisted around one another, and intended to support high speed LANs operating at speeds up to 1 Gbps. *Category 7 (Cat 7)* is a developing standard for shielded twisted pair wiring also designed for data rates up to 1 Gbps.

Where wire enters a building, it is connected to a terminating block, sometimes called a *punchdown block,* with lugs or clips. This terminating block marks the demarcation point between the common carrier and the building owner, who is responsible for providing and maintaining all wiring within the building. Before divestiture in 1984, the Local Exchange Carriers (LECs) provided and owned all inside wiring. This wiring represented a substantial asset on the accounting books of the LECs. Over time, this large base of previously installed inside wiring is being sold or given to the owners of the buildings where it is installed.

As wires leave a building, they may be directly buried in the ground or suspended from overhead poles, as shown in Figure 9–7. As they approach the central office, they are grouped together in cables that get

Figure 9–7
Wires leaving a building may be buried or hung from poles above ground.

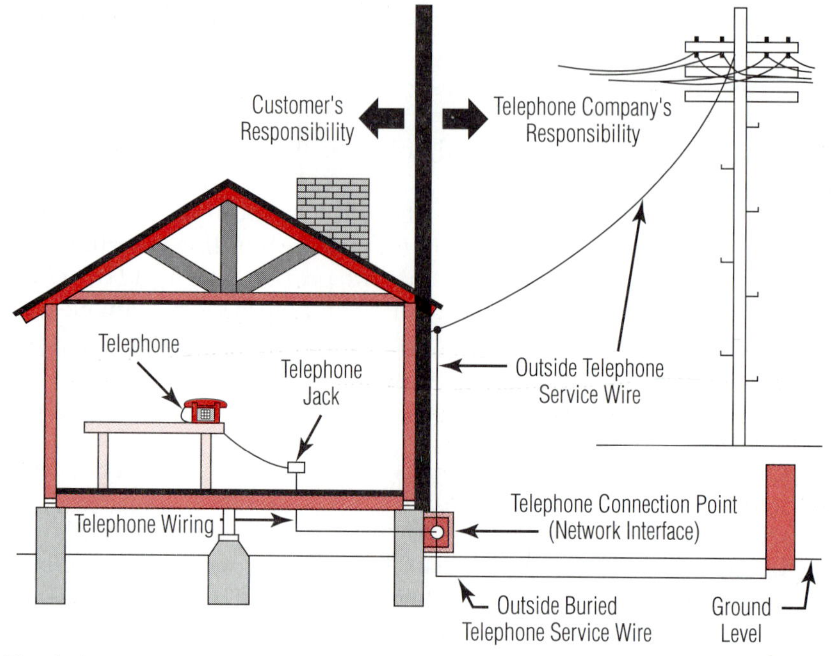

The telephone connection point (network interface) may be located inside or outside the building.

Twisted pair wires are connected together at this punchdown block in the telephone equipment room.

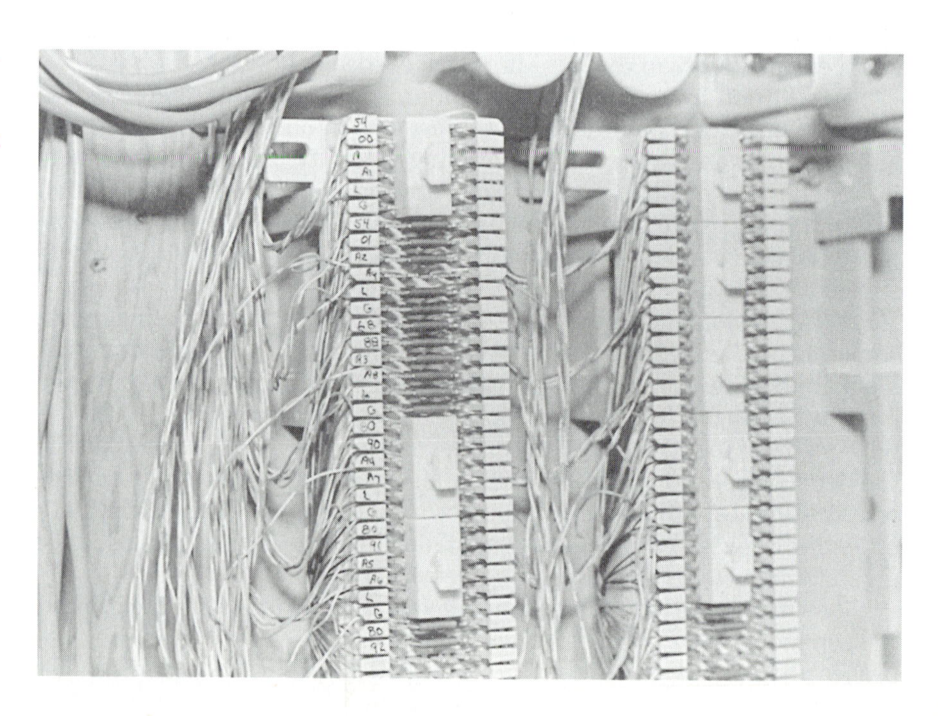

larger the closer they get to the office. Up to several thousand pairs are grouped together into large cables. They are sometimes surrounded by a wire shielding to provide protection from electrical interference or a heavy metal armor for physical protection. On local loops, one pair of wires usually is dedicated to one telephone or data circuit.

Shielded Twisted Pair Wire

A variation of twisted pair wiring is *shielded twisted pair.* Twisted pair wire is placed inside a thin metallic shielding, similar to aluminum foil, and is then enclosed in an outer plastic casing. The shielding provides further electrical isolation of the signal-carrying pair of wires. Shielded twisted pair wires are less susceptible to electrical interference caused by nearby equipment or wires and, in turn, are less likely to cause interference themselves. Because it is electrically "cleaner," shielded twisted pair wire can carry data at a faster speed than unshielded twisted pair wire can. The disadvantage of shielded twisted pair wire is that it is physically larger and more expensive than twisted pair wire, and it is more difficult to connect to a terminating block.

Coaxial Cable

Coaxial cable, as the name implies, is cable made of several layers of material around a central core, as illustrated in Figure 9–8. The central conductor is most often a copper wire, although occasionally aluminum is used. It is surrounded by insulation, most typically made of a type of

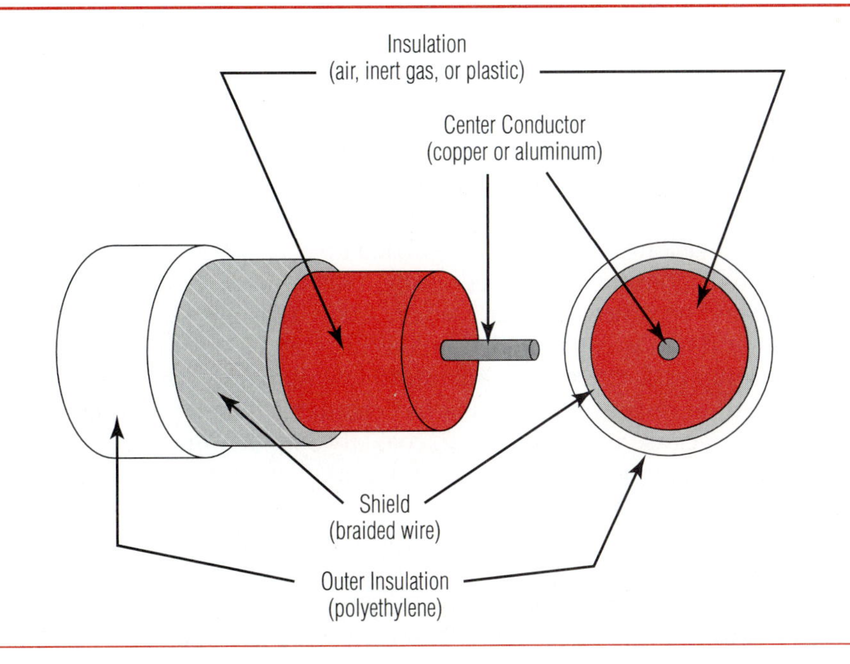

Figure 9–8
Parts of coaxial cable.

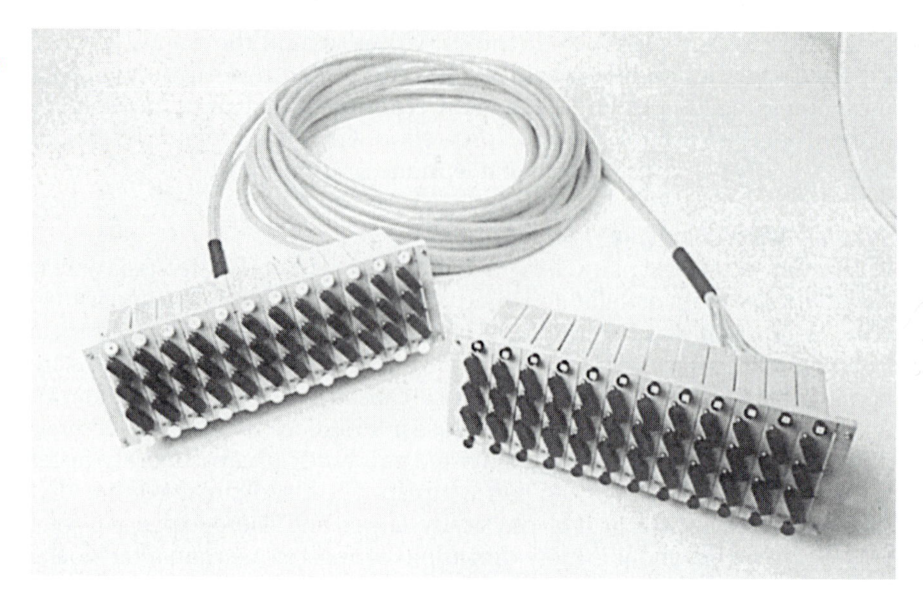

This shielded twisted pair cable is useful in electrically noisy environments or where high data rates must be guaranteed. (Courtesy of International Business Machines Corporation. Unauthorized use not permitted.)

plastic. Sometimes spacers are put in the cable to keep the center conductor separate from the shielding, and in that case, the insulation material is air or an inert gas. Outside of the insulation is the shielding, which is also a conductor, typically fine, braided copper wire. The shielding is surrounded by the outer insulation, which is almost always a form of plastic that also provides physical protection for the cable.

Coaxial cable has a very large bandwidth, commonly 400 to 600 MHz, and therefore a very high data-carrying capacity. The telephone industry uses pairs of coaxial cable in areas where the population density is high. One coaxial cable can carry up to 10,800 voice conversations when amplifiers, spaced about a mile apart, are used to boost the signal. The cable television industry uses coaxial cable extensively for carrying television signals from a central transmitter to individual homes or other subscribers. Over 50 television channels can be carried on a single coaxial cable.

Coaxial cable can be tapped easily. This is an advantage when it is used around an office or factory where many taps are needed but a disadvantage if one is concerned about security and illegal taps. The cable can be bulky and, therefore, difficult to install. Some cable has a rather large bending radius, which must be considered when planning the installation. Because of its shielding, coaxial cable is quite immune to external electrical interference, making it a good candidate for use in electrically noisy environments.

Optical Fiber

Optical fiber technology is one of the most rapidly advancing segments of telecommunications technology. The worldwide use of optical fibers has

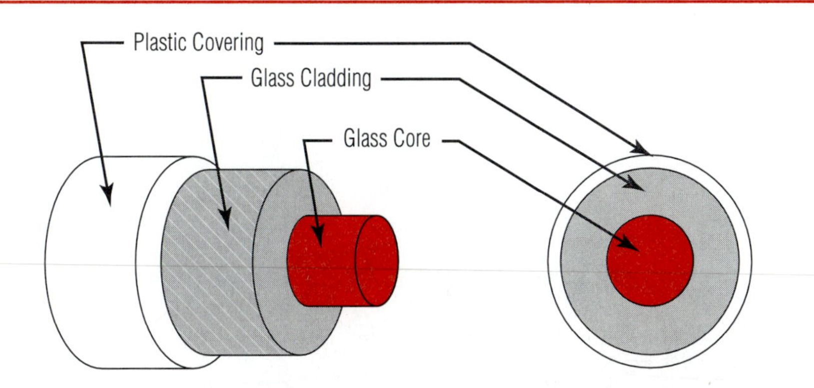

Figure 9–9
Parts of optical fiber cable.

been growing at a compounded growth rate of 20 percent per year since 1998, and the trend is expected to continue or accelerate until at least 2003. One of the byproducts of the rapid installation of optical fiber in the United States is that by 2001, the U.S. telephone infrastructure will have more than 80 times the capacity it had five years before. This fantastic buildup in capacity should be a boon for consumers and businesses alike, and at least one industry executive predicts that in the not-too-distant future, long distance calls may be offered free to consumers who buy a bundle of services including high-speed Internet access and wireless telephony. Every few months one reads of the latest developments, which frequently have to do with the bandwidth or speed that a single fiber can carry and the rapid conversion of networks to be optical fiber based.

The optical fiber itself, illustrated in Figure 9–9, is a very thin glass fiber of high purity. The glass *core* at the center provides the transmission-carrying capability. It looks like a very fine fishing line but is in fact a very pure, clear strand of silica glass. The core is surrounded by another type of glass called *cladding*, which is reflective and acts like a mirror to the core. The cladding is covered by a protective covering, usually of plastic. The total diameter of the fiber is less than that of a human hair. Individual fibers often are bundled together in groups around a central metallic wire that provides strength for pulling the fiber cable through conduit when it is laid.

light source

Data is placed on the cable with a light source, either a *light-emitting diode (LED)* or a *laser*. The laser is more powerful and is used when the distance between the transmitter and receiver is greater than 5 to 10 miles. For shorter runs, the less expensive LED can be used. The light stays in the core because the cladding has a low refractive index. When the light beam hits the edge of the core, it is reflected back toward the center by the cladding's mirrorlike surface. The light output of the LED or laser is modulated to provide the variations in the signal that can be interpreted at the receiving end. Without special techniques, a fiber carries a signal in one direction only, so fibers are normally used in pairs.

Several coaxial cables are often packed together as shown in this photograph. The 4½-inch cable on the left can carry as many as 40,300 telephone conversations. The optical fiber cable on the right is only ½ inch in diameter and can carry more than 43,600 telephone conversations. (Courtesy of AT&T Bell Laboratories)

single mode

Two primary types of fiber are used. One is called *single mode* and uses a fiber with a glass core approximately 5 microns (.005 millimeter) in diameter. With this very small core size, the light beam travels down the center of the core with little reflection from the cladding. Because the core is so small, however, it requires a very concentrated light source to get the signal into the fiber with adequate strength for long distances.

multimode

The other type of fiber is called *multimode.* The usual core size is 62.5 microns (.0625 millimeter) in diameter. With the larger core size, it is easier to get the light into the fiber, but there is more reflection from side to side off the cladding. Some of the light rays travel essentially straight down the center of the fiber, whereas others reflect at various angles. Those that travel straight through the fiber arrive at the destination faster than those that reflect, a phenomenon called *dispersion.* The effect of dispersion is that it causes the square pulses of a digital signal to become rounded and effectively limits the signaling rate that can be achieved. The trade-off, then, is that implementing a multimode fiber system costs less than a comparable system built with single-mode fiber. The cable and light source cost less, but the signal-carrying capacity is also less than in a single-mode fiber system.

One of optical fiber's most notable characteristics is its high bandwidth. A strand of fiber can carry 320 Gbps of data per second, enough to support 5.7 million PCs with 56 kbps modems simultaneously. Stated another way, the total bandwidth of radio is 25 GHz, whereas an optical fiber has a bandwidth of 25,000 GHz. Signals do not travel faster in fiber than copper, but the density or data capacity of fiber is much greater. Light has higher frequencies and hence shorter wavelengths, so more bits of information can be packed in the same space. Furthermore, photons, the base element of light, can occupy the same space. To visualize this concept, think of two flashlight beams that cross each other. The light from one passes through the other, and both are unaffected. Utilizing this principle, and a technique called *wavelength division multiplexing (WDM),* many light beams of different wavelengths can travel along a single fiber simultaneously without interfering with one another. Each light beam can carry many individually modulated data streams, hence the high data rates that optical fibers can achieve.

wavelength division multiplexing

Optical fiber cables are very difficult to splice, requiring specialized tools and skills. They are best suited for long point-to-point runs in which few or no splices are required. Splices can be detected using a reflectometer, an instrument that sends a light wave down the fiber and measures the reflection that comes back. From a security standpoint, this difficulty in splicing optical fiber cables and the ability to detect unwanted taps is an advantage. In addition, since the transmission is optical rather than electrical, the cables do not radiate signals, as all electrical devices do. Given these attributes, optical fiber cables are excellent in situations that require very high security.

Another advantage of optical fiber cables over wire or coaxial cables is that the fibers are so thin that a optical fiber cable has a very small diameter and is lightweight. This makes the cables easier to install, since they can be pulled through smaller conduits and bent around corners much more easily than coaxial cable. They can even be installed under carpeting or floor tiles, giving additional flexibility in laying out an office.

Costs of optical fiber cable and its related components continue to drop and are expected to continue doing so for the foreseeable future. These cost reductions are the result of improved manufacturing techniques and an ever-expanding market size. In addition, a lower cost optical cable made of plastic instead of glass is available. Although it doesn't have the capacity of glass cable and can't carry data as far without a repeater, it may be an inexpensive alternative for bringing fiber and its high bandwidth to the desktop in some offices.

The first undersea telephone cable between the United States and Europe was installed in 1956 and could handle 36 simultaneous telephone calls. In late 1988, the first optical fiber undersea cable, TAT-8, was completed on approximately the same route and could handle 40,000 simultaneous calls using two pairs of fibers. That capacity was subsequently upgraded with advances in the shore-based electronics. The

SIDEBAR 9–1

THE CABLE UNDER THE SEA

Underwater fiber-optic cables handle most international voice calls and Internet traffic calls (satellites handle most of the broadcast video). Each fiber-optic cable, as thin as a human hair, can carry at least 20,000 simultaneous calls. The C.S. Global Link, one of the ships that lays the cable, has accommodations for 138 crew, technicians, and guests.

How the cable is laid

1. At least a year before the ship goes out to sea, topographical surveys are conducted to plan the cable route, taking into account such factors as underwater earthquake faults, canyons and shipping routes.
2. Over the course of several weeks, thousands of miles of cable are manually coiled into the ship's storage tanks.
3. While the ship is still anchored, the cable is floated out to the shore and connected to the shore cable station. Cable running close to the shoreline or near a continental shelf is buried in a tunnel dug by a plow.
4. Once past the continental shelf, burying the cable isn't necessary. Guided by shipboard computers that communicate with global satellites, the ship begins dropping cable, which rests on the ocean floor, four or five miles deep. Two cable engines, one in the bow and one in the stern, lay out cable at the proper tension.
5. As the cable is lowered into the sea, buoys mark the location.
6. During installation, engineers continually test the cable system, which is powered and operating as it is laid.
7. If cable needs to be repaired, a remote-control robot submarine tethered to the ship dives to the bottom of the ocean and hauls the cable to the surface, where repairs are made.

Inside the cable

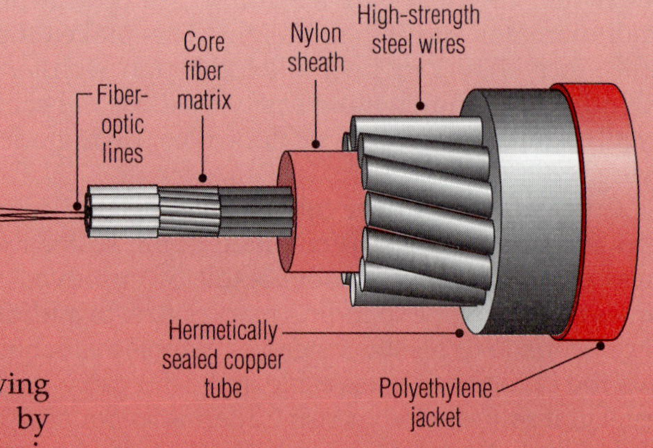

Sources: Tyco Submarine Systems, TeleGeography.

newest trans-Atlantic cable, called "Atlantic Crossing," can handle 2.4 million voice conversations at one time. In 1998, Lucent Technologies unveiled the capability to transmit as many as 10 million calls over a single fiber by dividing the strand into 80 separate wavelengths.

Literally hundreds of optical fiber cables connect the continents. Figure 9–10 shows most of the world's undersea cables, a few of which are the older coaxial cable type, but all of the new ones are optical fiber. As of the end of 1999, there were over 400,000 miles of fiber optic cable on the floors of the world's oceans, enough to circle the globe over sixteen times. These cables are sponsored and funded by the carriers of the countries served, working in cooperation with one another. The network continues to grow as the world's telecommunications users continue to demand new services and seem to have an insatiable demand for bandwidth.

In summary, the characteristics of optical fiber are

- high bandwidth—very high data-carrying capacity;
- little loss of signal strength—depends on the details of the cable construction, but the overall characteristics are excellent;
- immunity to electrical interference—since it operates in the optical part of the spectrum, electrical noise is not an issue;
- excellent isolation between parallel fibers—crosstalk between fibers does not exist;
- small physical size—lightweight;
- very secure—difficult to tap and splice and does not radiate electrical signals.

One large company in New York recently installed several fiber-optic cables to connect several locations in its community. In one case, the fiber cable was run along a railroad track, and in another situation, the fiber was run through some abandoned underground conduit formerly used for electrical wires. The fibers carry voice traffic today to allow a single PBX to serve the company's multiple locations around the city and high-speed data traffic between local area networks.

Synchronous Optical Network (SONET) *Synchronous Optical Network* is a standard for transmitting data on optical fibers that was originally created to allow easier connection between carriers that were using different vendor's products for their optical networks. SONET has become the de facto standard for carrying voice and data traffic over an optical network, and the American National Standards Institute (ANSI) has approved it. SONET provides for transmission at gigabits per a second (Gbps), as shown in Figure 9–11. Europe uses a similar standard, as approved by the ITU-T, but it has different designations for the various circuit speeds.

While SONET was originally developed by the carriers for their own use, it is now available to users who need the capacity it offers. The carriers

Figure 9–10
The world's undersea cable network.

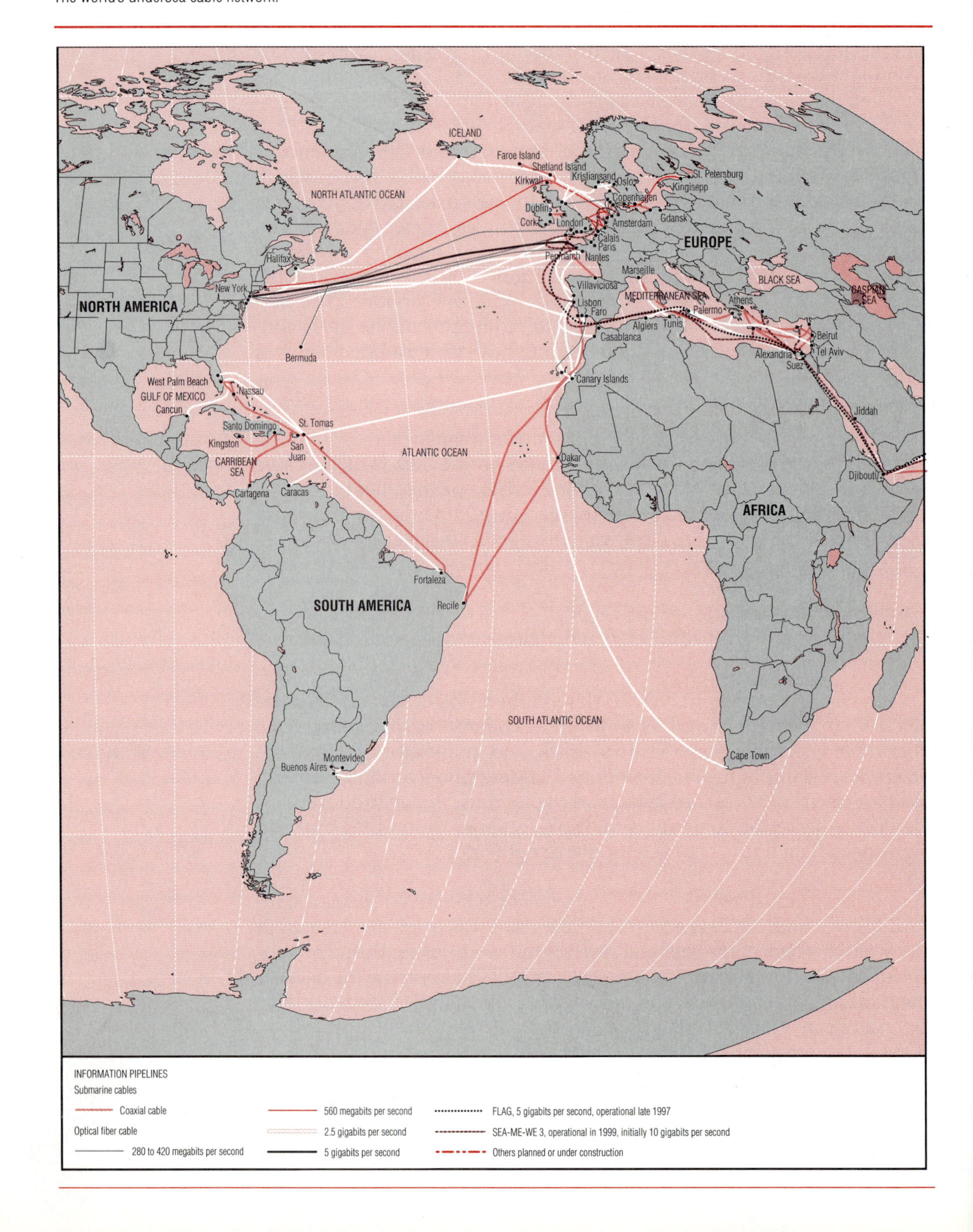

Figure 9–10
Continued

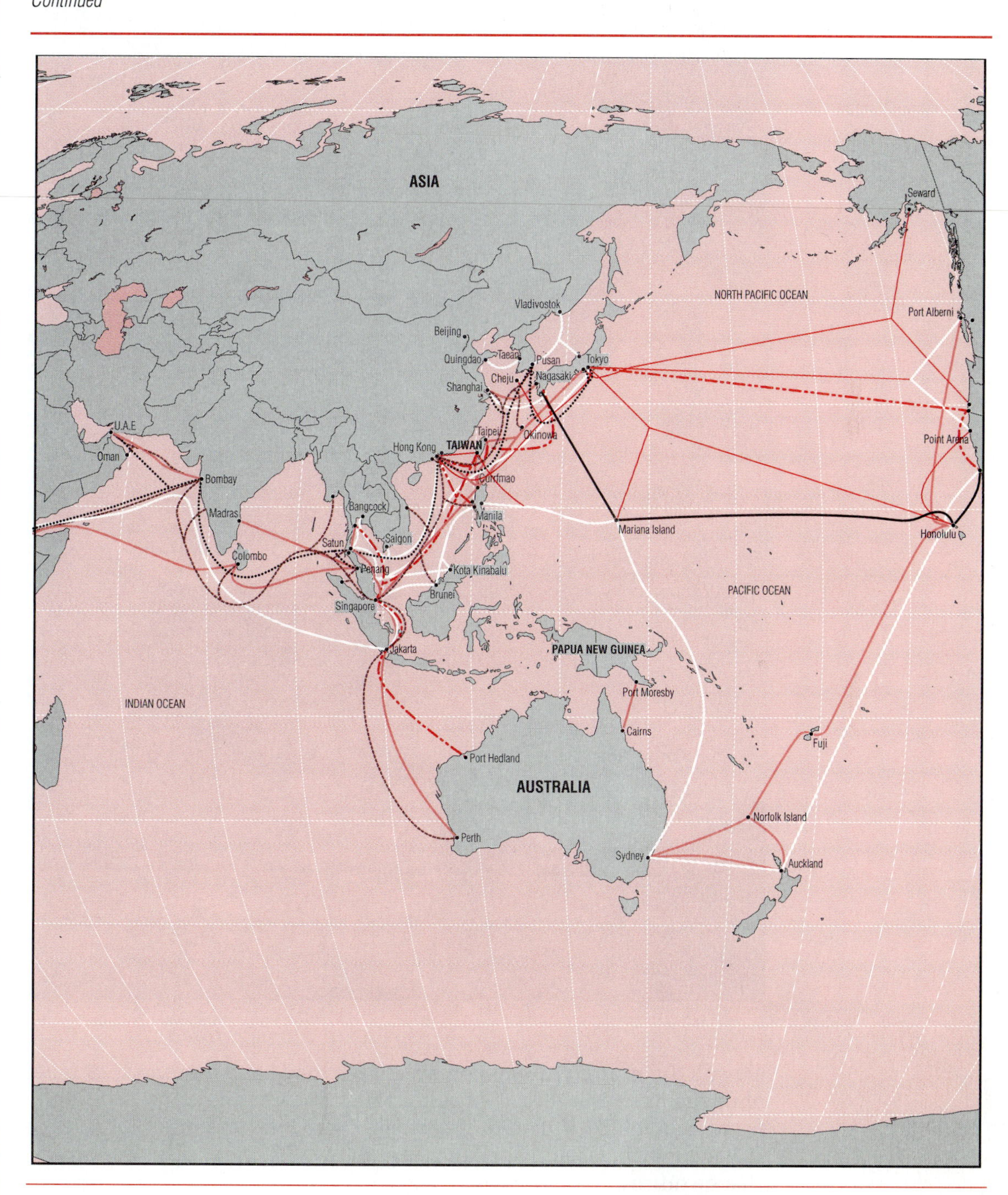

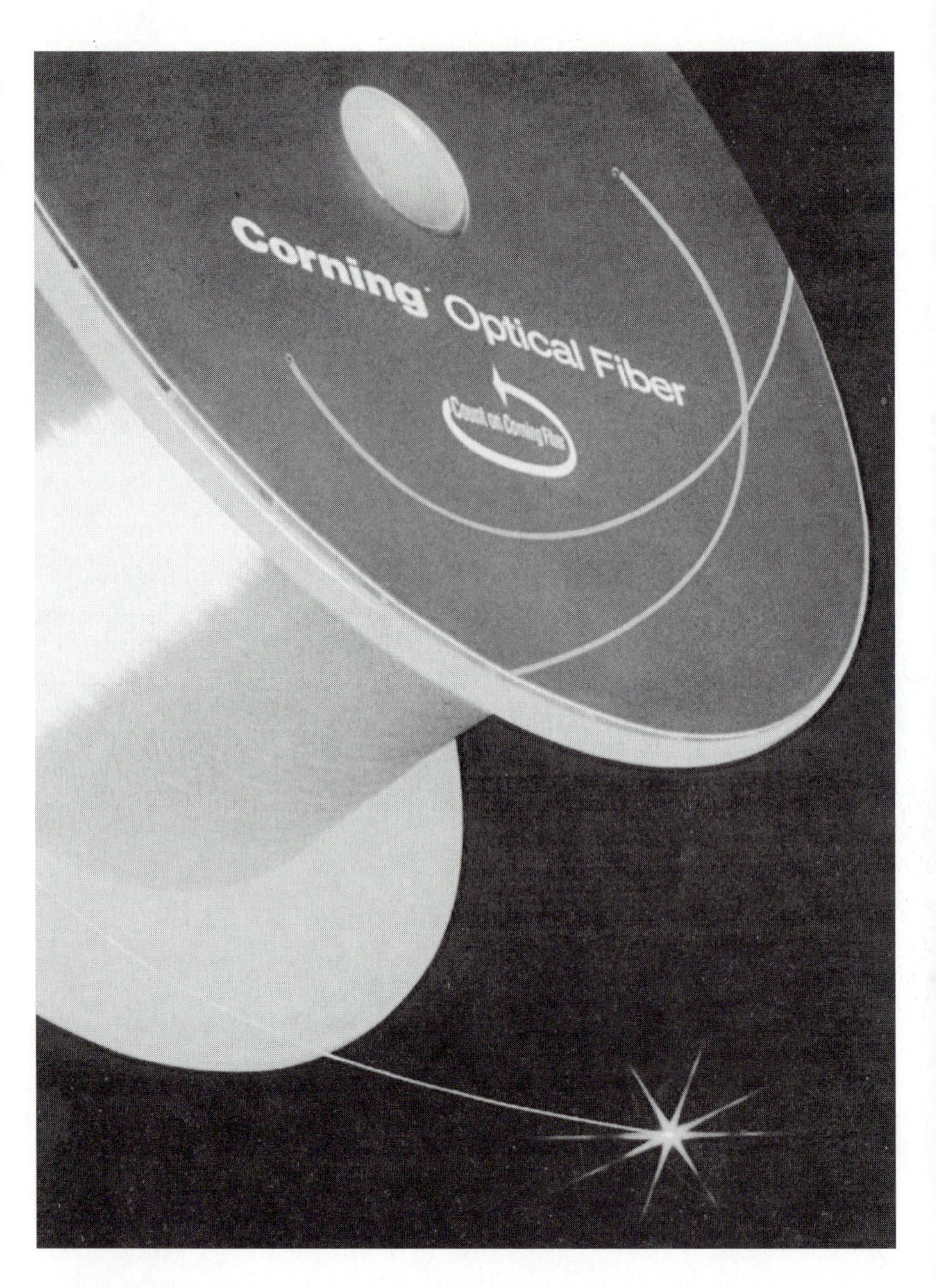

The flexible optical fiber shown in this photograph is approximately the diameter of a human hair but can carry signals for miles without repeaters. (Courtesy of Corning, Inc.)

are still the primary users, however, since they have thousands of miles of optical fiber installed. Notice that the basic OC-1 data rate is slightly faster than the T-3 data rate. Telephone companies and other carriers are replacing their T-3 and T-4 services with SONET, especially between exchange offices.

Synchronous Optical Network (SONET)	ITU-T Designation	Data Rate
OC-1		51.840 Mbps
OC-3	STM-1	155.250 Mbps
OC-9	STM-3	466.560 Mbps
OC-12	STM-4	622.080 Mbps
OC-18	STM-6	933.120 Mbps
OC-24	STM-8	1.244 Gbps
OC-36	STM-12	1.866 Gbps
OC-48	STM-16	2.488 Gbps
OC-192	STM-64	9.953 Gbps
OC-255		13.219 Gbps

Figure 9–11
Comparative data rates for the SONET and ITU-T optical fiber transmission standards.

Microwave Radio

Microwave radio was the medium most used by the common carriers for long distance communications transmission before optical fibers became so prevalent. Microwave radio transmissions occur in the 4 to 28 GHz frequency range. Specific frequency bands are set aside, and channels are allocated within the bands. Up to 6,000 voice circuits are carried in a 30 MHz-wide radio channel.

At the frequency range in which they operate, microwave radio signals travel in a straight line. Therefore, the transmitter and receiver must be in a direct line of sight with each other. Because of the curvature of the earth, microwave antennas are usually placed on high towers or building roofs to extend the line of sight to the greatest distance possible before another antenna is required. Practically speaking, a range of 20 to 30 miles between towers is common if the terrain is not too hilly. Where the distance to be covered is short, microwave antennas can be placed on the side of a building or even in an office window.

Microwave signals may carry data in either analog or digital form, but analog is more common. Voice, data, and television signals are carried, but each is given its own channel. Depending on the frequency used, some microwave signals are subject to interference by heavy rain. When this occurs, the channel is unusable until the rain subsides. Therefore, when a microwave system is designed, provisions must be made for this possibility. Temporarily stopping the transmission, sending it on an alternate path, and transmitting via a different medium are alternatives to be considered.

Microwave is sometimes installed privately by companies to connect locations that are near but not adjacent to one another. A radio license must be obtained from the Federal Communications Commission (FCC), but no right-of-way permits are necessary. Several companies in New York City have private microwave links connecting offices in different

This microwave tower is located in New Jersey. Both parabolic and horn antennas can be seen. (Property of AT&T Archives. Reprinted with permission of AT&T.)

parts of Manhattan. Similar private microwave installations can be found in most major cities.

Microwave systems should be considered when T-1 circuits are not available, when there is a financial advantage or other requirement to have a privately owned transmission system, or when alternative routing is required for certain critical communications links. Sales engineers from the microwave equipment vendors help to obtain the required transmission license, as well as configuring the equipment for the specific terrain, transmission speed, and reliability requirements. Major vendors of microwave equipment are Digital Microwave Corporation, Motorola, Inc., and Rockwell Communication Systems.

Satellite

Transmission using an earth satellite is a particular type of microwave radio transmission. In a typical satellite system, a microwave radio signal is transmitted from an antenna on the ground to a satellite in an orbit 22,300 miles above around the earth. At that distance, the circular speed of the satellite exactly matches the speed of rotation of the earth, and the satellite appears to be stationary overhead. This is called a *geosynchronous orbit.* An antenna on the earth can be aimed at the satellite, and because the satellite appears stationary, the aim doesn't have to be changed. Although the distance is great, the antennas are definitely in sight of each other. The microwave radio signal is beamed to the satellite on a specific frequency called the *uplink,* where it is received, amplified, and then rebroadcast on a different frequency, called the *downlink.* This is illustrated in Figure 9–12.

geosynchronous orbit

Because of its distance from the earth, a geosynchronous satellite can see and be seen from approximately one-third of the earth. The signals broadcast from the satellite can, at least theoretically, be picked up by any antenna and receiver in that area. This broadcast attribute is an advantage in some applications and a disadvantage in others. It is advantageous for organizations that want to use satellites to reach a mass market. Home Box Office (HBO) uses the broadcast capability to distribute movies to cable TV companies. Financial companies use satellite transmission to distribute stock market information to brokers all over the country.

For companies that want to use the satellite for point-to-point transmission, the broadcast capability presents a security concern. Anyone with the proper equipment can receive the broadcast as it is transmitted down from the satellite. When this concern is serious, encryption devices can be used to code the data before it is transmitted to the satellite. This makes it difficult to interpret the broadcast on the downlink.

Another characteristic of satellite transmission is that because of the great distance involved—22,300 miles up to the satellite and 22,300 miles back—there is a noticeable delay from the time a signal is sent until it is received. This delay, called *propagation delay,* exists for all communications circuits and radio broadcasts. It is a function of the fact that light signals or radio waves travel at a maximum speed of 186,000 miles per second in a vacuum. A signal traveling 4,000 miles across the United States on a terrestrial circuit will travel somewhat slower than 186,000 miles per second, but an approximation of the propagation delay can be calculated as

propagation delay

Figure 9–12
Satellite transmission.

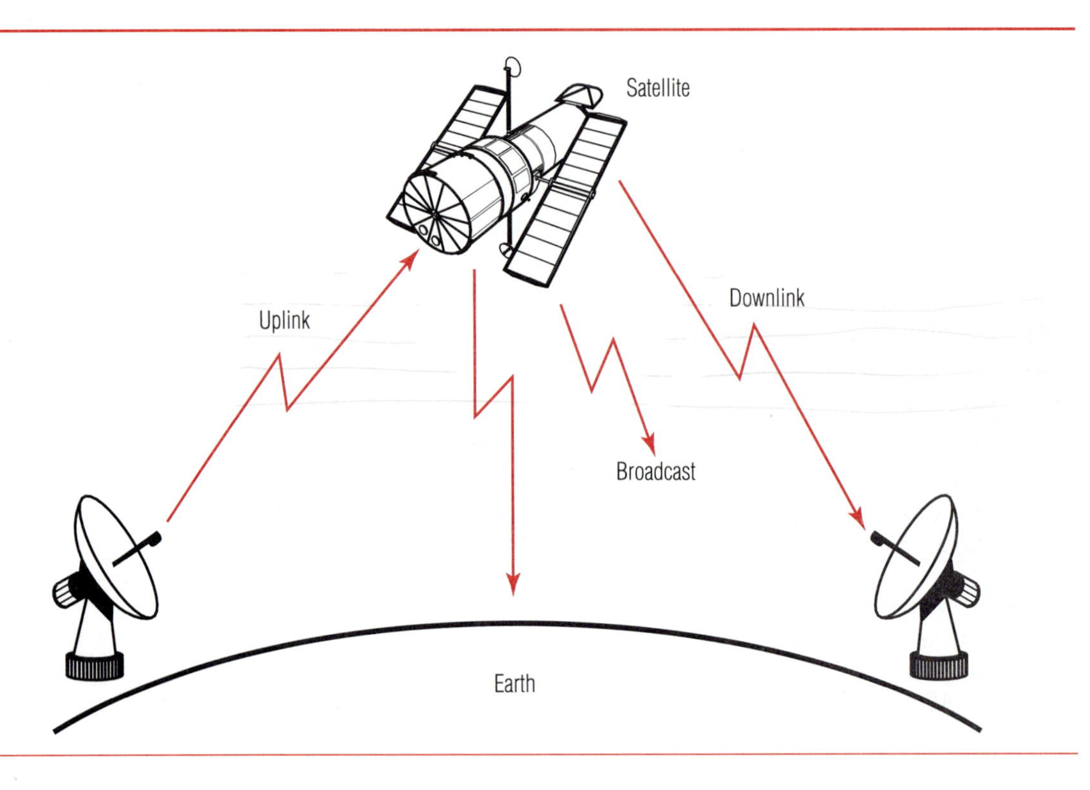

4,000 miles/186,000 miles per second = .0215 second (21.5 milliseconds)

When the signal goes via satellite, the propagation delay is

44,600 miles/186,000 miles per second = .2398 second (239.8 milliseconds)

The satellite signal takes 11 times longer to reach its destination. If a return signal is required, the same amount of delay would be encountered a second time. You may have noticed this satellite delay on some domestic long distance telephone calls a few years ago, and it still can be heard on some international telephone calls today. Most people find the delay annoying, and the common carriers in the United States have switched voice transmissions back to terrestrial media.

The propagation delay when using a satellite can be extremely significant when data is being sent. Since almost a quarter of a second is added to each transmission in either direction, the transmission time becomes a significant portion of the total response time in an interactive application. We will look at the effect of satellite delay on data transmission in more detail when transmission protocols are discussed in Chapter 10.

This satellite earth station is about 1 meter in diameter. Other dish antennas are as small as 18 inches. (Courtesy of Scientific-Atlanta, Inc.)

Satellites use various frequencies for transmitting and receiving. That's one reason why satellite dish antennas vary greatly in size. The other main reason has to do with the power output of the satellite transmitter; those with higher powered transmitters put out a stronger signal and the receiving antenna on the ground can be smaller. Satellites that transmit in the *Ku band* of microwave frequencies have such a short wavelength that the earth station antennas, called *very small aperture terminals (VSAT)*, can be as small as 18 inches. One use is by news services to communicate between reporters in the field and the main studio. A different application is the *direct broadcast satellite (DBS)* system commonly used to send television programming via satellite directly to homes. DBS antennas are available in the United States for as little as $100 when purchased with one year of programming.

VSAT

DBS

DBS, also sometimes known as *digital satellite service (DSS)*, is also a competitor to cable television and DSL for high speed Internet access. DSS is the most widely available of the three technologies and has been in use by millions of Americans since 1997. As long as you have a clear line of sight toward the sky above the equator, you are a potential DSS customer.

DSS service is receive-only, meaning that the customer must also use a telephone line and modem to send data to the Internet. Also, while the download speeds from the satellite are in the range of 350 kbps, the overall service is much slower. When you click a link to pull up a web page, your signal travels via the phone line to your ISP and then to the Internet. The information you requested is sent 22,300 miles up to the satellite, and back down 22,300 more miles to your DSS dish. There is a noticeable

This satellite dish antenna is used for receiving signals from a direct broadcast satellite. (Courtesy of Stock Boston.)

delay in the response, which is okay for file downloads, but probably not acceptable to most people if they are trying to chat interactively with another person. As usual, the suitability of this technology depends on the application for which it will be used.

In addition to geosynchronous satellites, there are other types of satellite systems in use and on the drawing boards. Medium earth orbit (MEO) satellites tend to circulate at 6,000 miles above the earth, while low earth orbit (LEO) satellites orbit at between 300 and 1,000 miles high. Because of their lower orbits, both of these types of satellites are in motion relative to the rotation of the earth. Nine to twelve MEO satellites or about seventy LEO satellites must be in orbit to ensure that at least one can be seen from anywhere on earth at all times. Users of MEO or LEO satellites need to have omni-directional antennas in order to receive the signal from the satellite as it passes overhead in its orbit.

MEO and LEO satellite systems have potential use for global telephone systems that would allow one handset to be used from any location on earth. However, the costs of such a system are very high, and one company has already failed because it could not get the costs of the telephone calls made through the satellites low enough to compete with terrestrial-based telephone systems.

	Twisted Pair (Cat 5)	Coaxial Cable	Microwave Radio	Satellite	Fiber
Transmission speed	100 Mbps	500 Mbps	275 Mbps	90 Mbps	2 Gbps
Ease of installation	Easy	Moderate	Difficult	Difficult	Difficult
Cost	Least	Moderate	Moderate	Moderate	High
Maintenance difficulty	Low	Moderate	Low	Low	Low
Skill required to install	Low	Moderate	High	High	High
Most common uses	Within buildings	Campus Multidrop	Point-to-Point Short distance	Point-to-Point Long distance	Point-to-point
Advantages	Inexpensive, familiar	Carries more information than twisted pair	Speed	Speed, availability	Speed, secure
Disadvantages	Subject to interference	Bulky	Can be intercepted	Delay, can be intercepted	Difficult to splice, cost
Security	Good	Good	Poor	Poor	Excellent
Notes	Shielded twisted pair allows higher speed	Broadband use is more maintenance intensive	Requires radio license from FCC	Private systems not common	Higher speeds coming

Figure 9–13
Comparison of the attributes of the media most commonly used for business telecommunications transmission.

Infrared

Infrared transmission uses light waves below the visible spectrum. Infrared waves can reflect off a light-colored surface such as the wall of a room, but generally speaking the transmitter and receiver must be in direct line of sight. The signals can also be blocked by fog, smoke, or even heavy rain. Companies sometimes use infrared transmission between nearby buildings. Other applications are to provide wireless communication between personal computers and printers, and between remote control units and television sets.

Summary of Media Characteristics

The chart in Figure 9–13 summarizes the characteristics of the main media used for telecommunications transmission.

CIRCUIT ACQUISITION AND OWNERSHIP

Another way of classifying telecommunications circuits is by ownership and method of acquisition.

Private Circuits

Private circuits are those installed and maintained by a company other than a common carrier. For example, a company may run coaxial cable between its buildings on a manufacturing site and wire within the buildings to form a data communications network. The typical situation where private circuits are installed is within a building or in a campus environment, where all the property is owned. It becomes more complicated if the circuit must cross property owned by others because permission to cross the property must be obtained. It is even more complicated when public roads or highways must be crossed, but it is possible to obtain per-

This photograph of the Advanced Communications Technology Satellite (ACTS) was taken from the space shuttle *Discovery* shortly after the satellite was released. (Courtesy of NASA)

mission to cross or bury cable under them. Permission is obtained from the agency in control of the road; it may be a city, county, state, or the federal government.

In some states, it is possible for private companies to get permission to run wire or cable on public utility poles. Usually, the utility company charges a small fee for the use of the pole. Private circuits also have been installed on the shoulders of public roadways, alongside railroad tracks, and on pipeline routes.

When a company installs a private circuit, it is totally responsible for its design, engineering, installation, operation, and maintenance. But the circuit is available for the company's full-time, exclusive use, and once the circuit is installed, it is usually very inexpensive to operate. One company, which had many locations in a city, received permission to run a private cable on the poles of the local electrical utility company. The rental charge was $6 per year per pole, an amount the company considered very reasonable. Using the cable, the company was able to connect all of its locations in the city into a common data network.

Leased Circuits

Leased circuits are circuits owned by a common carrier but leased from them by another organization for full-time, exclusive use. Leased facilities are attractive when some or all of these conditions are present:

- it is impossible or undesirable to install a private circuit;
- the cost of the leased circuit is less than the cost of a dial-up connection for the amount of time required;
- four-wire service is required (four-wire service cannot be obtained on dial-up connections);
- high-speed transmission is required.

The primary advantage of a leased circuit is that it is engineered by the carrier, installed, and left in place so that the same facilities are always used. This means that once the circuit is adjusted and operating correctly, it will continue to operate the same way for long periods of time. For most business data transmission, this consistency and reliability are significant benefits because they go a long way toward helping ensure that the communications service is reliable and trouble-free.

When a leased circuit does fail, the carrier that provides the circuit performs the diagnostic and maintenance work required to restore it to service. Carriers have special testing equipment located at their central offices with which they can examine all of the parameters of a circuit to determine the cause of failure. Furthermore, the technicians at all of the central offices through which the circuit passes can communicate with each other to determine which link in the circuit is experiencing the problem. In most instances, failures are isolated and corrected and the circuit is returned to normal operation within hours. Often, service is restored in minutes.

The price of leased circuits is based on speed and distance. For voice-grade circuits there is a charge for the local channel from the customer premises to the serving central office. The local channel is acquired from the Local Exchange Carrier (LEC). If the circuit crosses Local Access and Transport Area (LATA) boundaries, it must be carried by an interexchange carrier (IXC). In that case, the local channel and its associated charges extend through the serving Central Office (CO) to the IXC's Point of Presence (POP). Added to the local channel charge is the interoffice channel charge for the circuit connecting the POPs. For point-to-point circuits, the carrier computes the shortest airline mileage between the two POPs and bases the monthly charge on that distance. For multipoint circuits, the carrier computes the shortest airline mileage between all the points on the circuit.

A leased four-wire circuit costs about 10 percent more than a leased two-wire circuit. Since additional throughput can be obtained, most users pay the additional cost for four-wire circuits when they want to use half-duplex or full-duplex transmission.

One final note: Leased circuits often are called "private" circuits in common use. The distinction between the two types that has been made in this book frequently is not made by people in the communications industry. It is common to hear someone talk about "our private line" when they really mean "our leased line." For many purposes, the distinction makes no difference, but if in doubt, it is best to request a clarification.

Bypass In its simplest form, *bypass* involves installing private telecommunications circuits to avoid using (to bypass) those of a carrier. The usual reason for considering bypass is to reduce costs, although in some cases a company may consider bypass in order to obtain a capability the LEC cannot provide.

One application of bypass is to connect a company's facility directly with an IXC, bypassing the LEC. An example where this might be necessary is when the LEC cannot provide fractional T-1 service. A full T-1 circuit or microwave link could carry the voice and data traffic directly from the company's location to the IXC, providing access to the IXC's fractional T-1 service. In another case, bypass circuits might be installed to eliminate the local access charges for circuits that the LEC would normally impose. An economic analysis would have to be performed to determine over what time period the elimination in local access charges would offset the cost of installing the bypass circuit. Another practical consideration may be the company's relationship with the LEC, since the elimination of circuit revenue caused by the bypass will certainly not make the LEC happy.

Switched (Dial-Up) Circuits

Switched or dial-up circuits come from the standard public telephone network, and using them for data is similar to making a normal telephone call. A temporary connection is built between DTEs as though there were direct wires connecting the two. The circuit is set up on demand and discontinued when the transmission is complete. This technique is called *circuit switching.* An obvious advantage of circuit switching is flexibility in network configuration. Transmission speeds of up to 56 kbps can be obtained on normal switched circuits. Charges for switched circuits are based on the duration of the call and the distance, just like a standard telephone call.

circuit switching

One factor that must be dealt with when using switched circuits is that when a dial-up connection is made, the actual carrier facilities used in routing the call depend on the facilities that are available at the moment. For this reason, circuits are variable in quality and may be very good on one connection and marginal on another. We have all experienced this phenomenon on long distance telephone calls. Sometimes it seems as if the person we are talking to is just next door; other times the volume of the person's voice is low; during other conversations, crosstalk can be heard. On a dial-up connection you may find that one time the modems are able to transmit at 33.6 kbps, while the next time, 26.2 kbps is the fastest they will transmit.

When an organization uses switched circuits, it exposes itself to certain security risks. If the organization has a dial-in capability, allowing employees or others to call a computer, it faces the possibility that unauthorized people may dial in and access the computer, either accidentally or maliciously. Of course there are standard security precautions, such as user-IDs and passwords that can be implemented, but a clever hacker can often bypass these measures fairly quickly. One method that has been used is to dial in with a computer that is programmed to try all possible combinations of letters and numbers for the user-ID or password to find a combination that works.

One technique to combat unauthorized dial-ins is to use a *callback* or *dialback unit.* The authorized user dials in and identifies him- or herself to the callback unit. The connection is immediately broken, and the callback unit looks up the user's ID in an internally stored table and calls the user back at the prestored telephone number. Of course this means that the user always has to call in from the same place; the technique does not work well for travelers. Another problem is that the cost of the telephone call now falls on the central location that is being called.

dialback unit

CIRCUIT IDENTIFICATION

Whether a particular facility is a circuit, link, or channel depends to a certain extent on one's point of view. From the user's point of view, a multipoint circuit from Los Angeles to Dallas and Miami has links from Los Angeles to Dallas and from Dallas to Miami. From the carrier's point of view, the user's circuit is a part of a higher-speed facility onto which it is multiplexed. The carrier also sees many more links, in fact, probably thinks of the circuit as being made up of a series of links between each central office through which the circuit passes, plus the links at each end (local loops) connecting the serving central office to the customer premises. As the circuit passes through high-speed, long distance optical fiber or microwave facilities crossing the country, it is likely to be viewed by the carrier as a one-way, two-wire channel within the microwave link. The other channel, running in the other direction, may be on some other fiber or link and running totally independently.

The different viewpoints lead to occasional difficulties when communications customers and carriers talk to each other because the parties may use the same terms to mean different things or different terms to mean the same thing. To partially address this problem, the carrier attaches a circuit number to each circuit for identification purposes. The carrier has blueprints or other documentation that identifies every link making up the circuit and every piece of equipment through which it passes. Customers and carriers use the circuit number when they discuss problems or talk about changes to the circuit.

MULTIPLEXING AND CONCENTRATING

Chapter 5 discussed frequency division multiplexing (FDM). When FDM is used, signals are shifted to different parts of the frequency spectrum so that a single pair of wires can carry more than one transmission. Data signals in analog form are shifted using FDM techniques. With FDM, a channel has use of a limited range of frequencies all of the time. Other channels use the other frequencies.

When transmissions are in digital form, time division multiplexing is normally used. With time division multiplexing, a channel can use the entire frequency range the channel allows, but only for specified periods of time. Other channels use the other time slots. Figure 9–14 illustrates this concept.

Time Division Multiplexing (TDM)

Time division multiplexing (TDM) is a technique that divides a circuit's capacity into time slots. Each time slot is used by a different voice or data signal. If, for example, a circuit is capable of a speed of 9,600 bps, four terminals, each transmitting at 2,400 bps, could simultaneously use its capacity. A TDM takes one character from each terminal and groups the four characters together into a *frame* that is transmitted on the circuit. This process is shown in Figure 9–15. At the receiving end, another TDM breaks the frame apart and presents data to the computer on four separate circuits.

frame

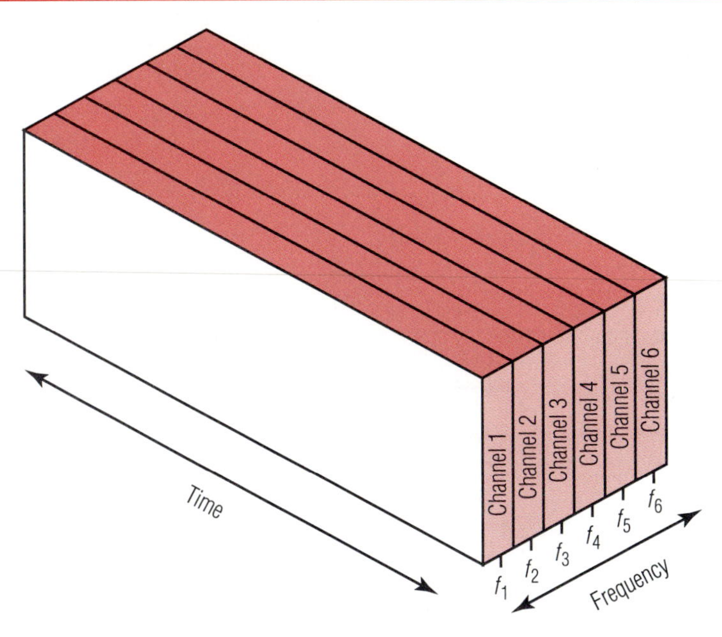

Frequency division multiplexing (FDM)

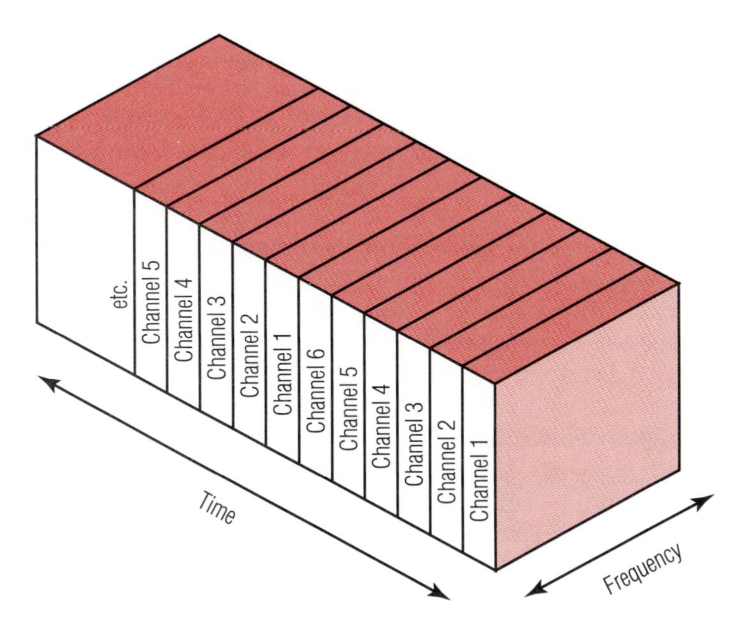

Time division multiplexing (TDM)

Figure 9–14
FDM channels have full time use of a limited range of frequencies. TDM channels can use the full range of frequencies but only during predetermined time slots.

Figure 9–15
Time division multiplexing.

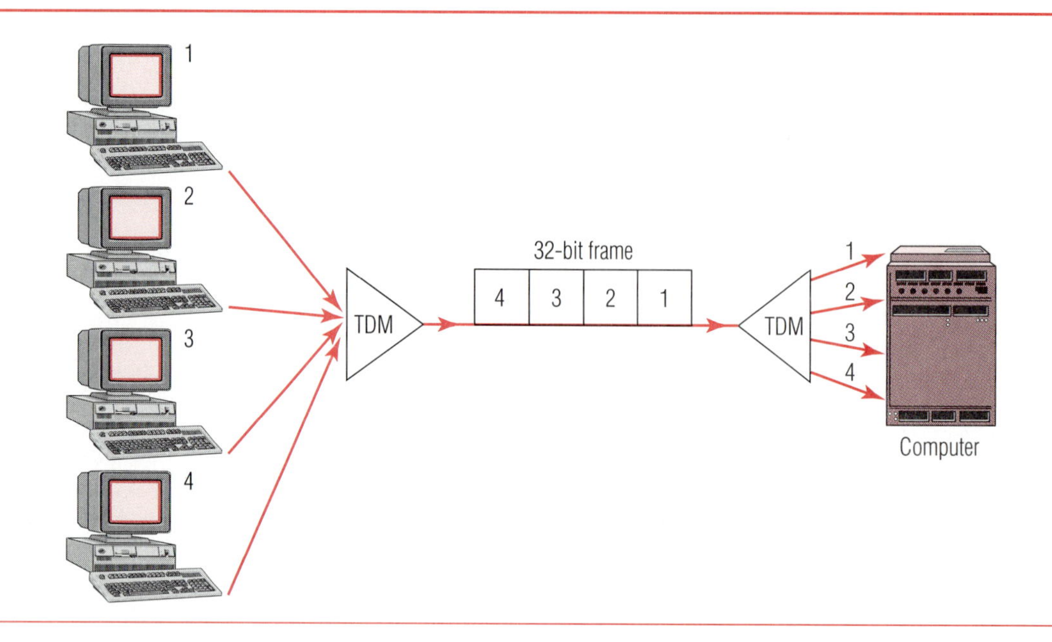

Time division multiplexing is totally transparent to the terminal, the computer, and the user. If a terminal has nothing to send at any point in time, its time slot in the frame is transmitted empty. A typical application for TDM is to have multiple slow-speed terminals at one location communicating to a computer at another location. Although four terminals are shown in Figure 9–15, more than four could be multiplexed (however, multiples of four are most common). Eight terminals transmitting at 1,200 bps would work just as well on a 9,600 bps line with the proper TDM equipment.

Another technique for TDM takes 1 bit from each terminal instead of one character and transmits a frame of bits. The bits are assembled into characters at the receiving end. A third technique, frequently used when the transmission is synchronous, is to multiplex entire messages. In this case, a *message* is defined as a group of characters not exceeding some predetermined length, say 128 or 256 characters. When message multiplexing is used, the frame of data that is transmitted is much longer than when character or bit multiplexing is used.

Statistical Time Division Multiplexing (STDM)

If you think about how terminals are really operated, it is obvious that no terminal with a human operator is transmitting data continuously. In fact, the wait, or "think," time between transmissions may be much longer

Figure 9–16
The STDM tries to avoid having empty slots in a frame, thereby improving the line use. If a terminal has no data to send in a particular time period, the STDM will see if the next terminal has data that can be included in the time slot. When the STDM at the receiving end breaks the frame apart, it uses the terminal address to route the data to the proper device.

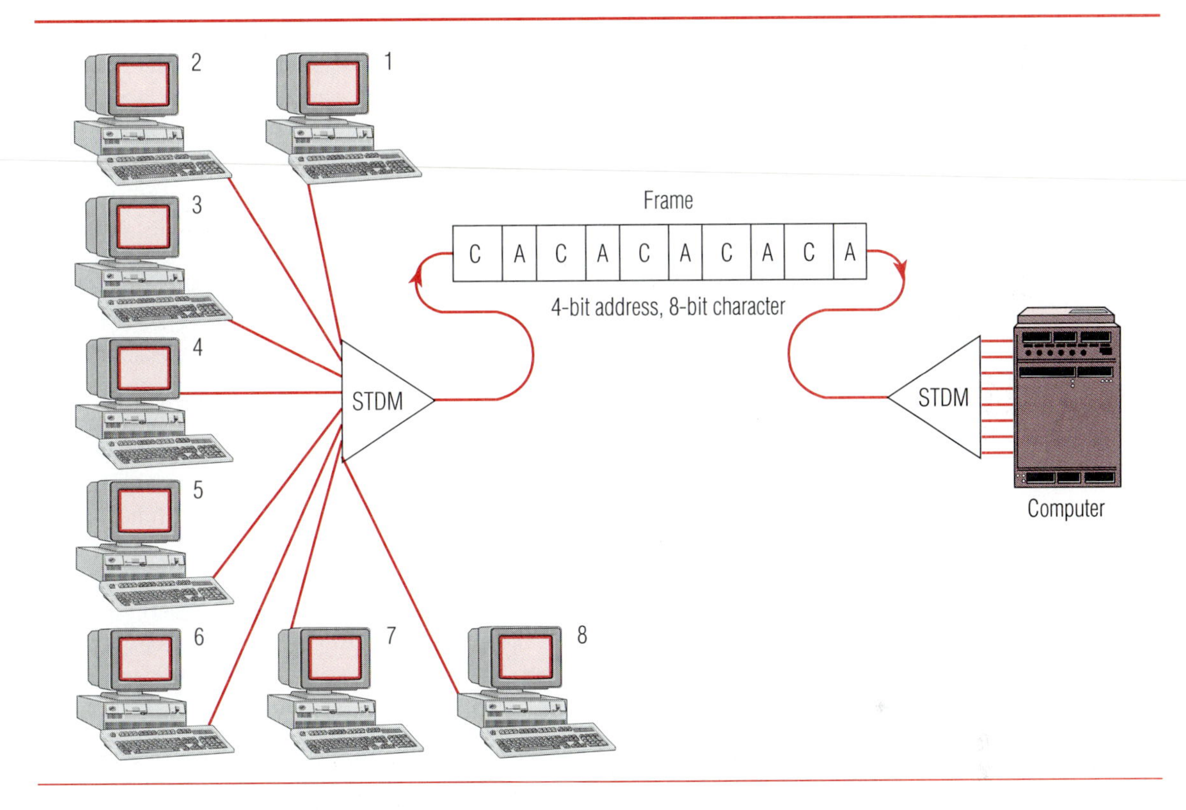

than the actual transmission time. With a time division multiplexer, this means that many of the time slots are transmitted empty and the capacity of the circuit is not fully used.

A *statistical time division multiplexer (STDM)* is a multiplexer that does not assign specific time slots to each terminal. Instead, the STDM transmits the terminal's address along with each character or message of data, as illustrated in Figure 9–16.

If the address field in an STDM frame is 4 bits long, there are 2^4 combinations and 16 terminals can be handled. With a 5-bit address, 32 terminals can be multiplexed. In any case, extra bits are required for the addresses when STDM is used. For most applications, this additional overhead is a good trade-off. Most of the time, the user will not notice any difference in performance or response time, and the line will be better used.

There are times when most or all of the terminals want to send data simultaneously. During these times, the aggregate data rate may be

buffer

higher than the circuit can handle. For these situations, the STDM contains a storage area or buffer in which data can be saved until the line can accept it. Buffer sizes of 32,000 characters and larger are common for this purpose. The user may experience a slight delay when buffering occurs.

STDM takes advantage of the fact that individual terminals frequently are idle and allows more terminals to share a line of given capacity. Students who are writing and debugging programs on a university timesharing system normally fit this model nicely. They spend some time typing the program into the computer and a great deal of time interpreting error messages and determining how to correct their program's problems. Using an STDM, 12 terminals running at 1,200 bps could be handled by a 9,600 bps line in most cases.

Leading statistical multiplexer manufacturers are Timeplex, Inc., Cisco Systems, Inc., and Tellabs, Inc. These vendors and others have a variety of products to handle varying numbers of terminals at diverse line speeds.

Concentration

Concentrators combine several low-speed circuits into one higher-speed circuit. A concentrator can be thought of as a circuit multiplexer. For example, six 9,600 bps circuits might be concentrated onto one circuit with a 56 kbps capacity, as shown in Figure 9–17. The intelligence and buffering in the concentrator take care of the fact that 6 × 9,600 bps = 57,600 bps, which is greater than the 56,000 bps capacity of the circuit. A primary reason for performing line concentration is economics. In most cases, it is less expensive to lease a 56 kbps circuit between two points than six 9,600 bps circuits.

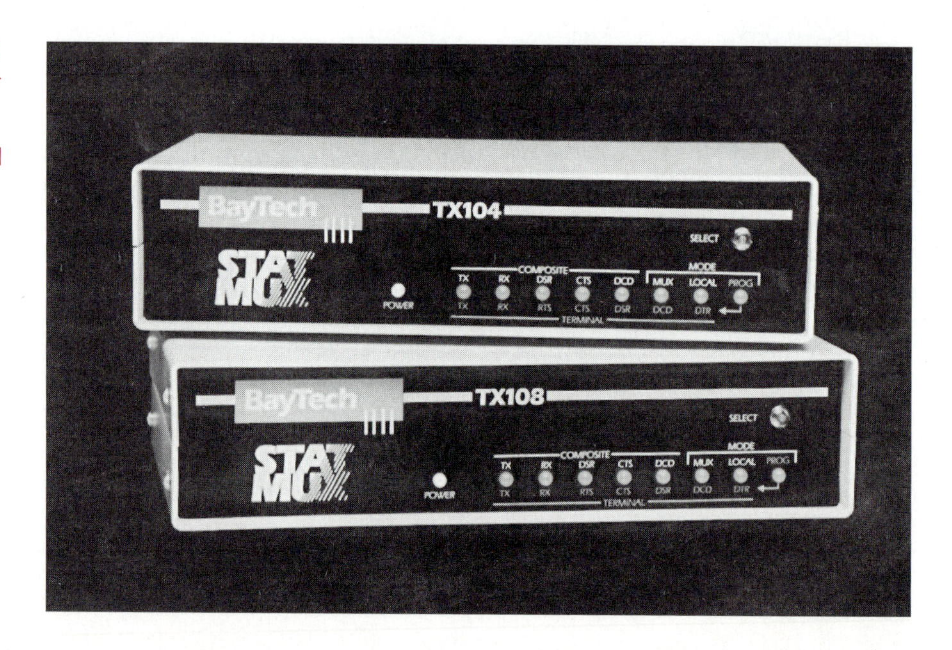

Two models of a statistical multiplexer that look virtually identical but provide different capacities. (Courtesy of Bay Technical Associates, Inc.)

Figure 9–17
Line concentration.

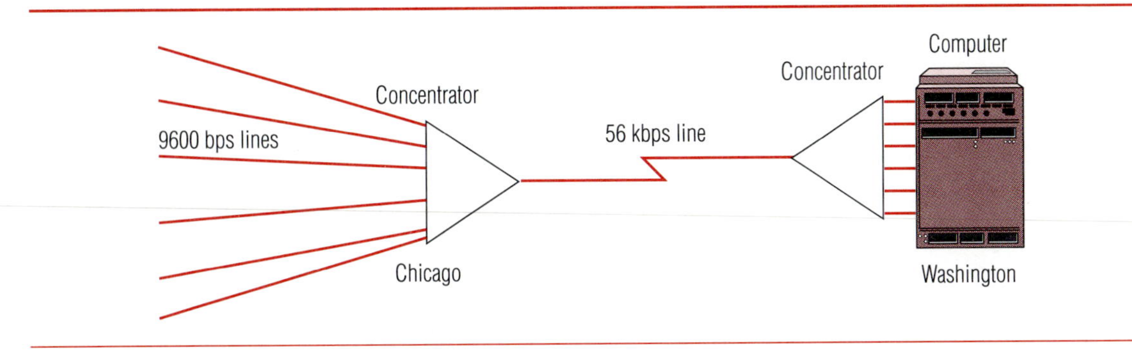

Figure 9–18
Inverse concentration.

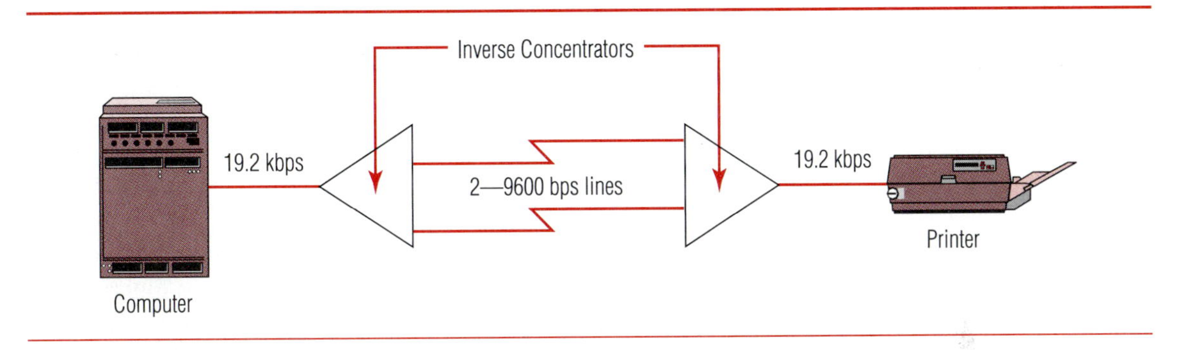

Inverse Concentration

In some cases, high-speed circuits are not available between two points, but it is desirable to provide high-speed service. An example is when there is a need to run a remote high-speed line printer. An *inverse concentrator* takes a high-speed data stream from a computer and breaks it apart for transmission over multiple slower-speed circuits, as shown in Figure 9–18. At the remote end, the slow-speed circuits are brought together again, providing a single high-speed data stream to the remote device.

CIRCUIT ERROR CONDITIONS

Communications circuits are subject to many conditions that cause degradation of the transmitted signal so that the receiving end cannot correctly determine what was sent. Some of these conditions arise because of normal characteristics of signal propagation, others are due to faulty circuit design, and still others are due to natural physical phenomena, such as electrical storms. The following section describes many conditions that can cause errors on a communications line. Of these, the most common are the various forms of noise, distortion, and attenuation.

Background Noise

Background noise, also known as *white noise* or *Gaussian noise,* is a normal phenomenon in all electrical circuitry. It results from the movement of electrons. It is present to some extent on every communications circuit. If the noise is at a high enough level, it can be heard as a hissing sound. It rarely represents a problem for either voice or data transmission because it is a known, predictable phenomenon, and the communications carriers have designed their circuitry and equipment to deal with it.

Impulse Noise

Impulse noise is a sudden spike on the communications circuit when the received amplitude exceeds a certain level. It is caused by transient electrical impulses, such as lightning, switching equipment, or a motor starting. You have probably heard it as annoying static crashes or clicking during a voice conversation. If the noise occurs during a data transmission, the impulse may cause one or more bits to be changed, invalidating the transmission.

Attenuation

Attenuation is the weakening of a signal over a distance, which occurs normally in all communications. Attenuation was illustrated in Figure 5–20. Just as your voice sounds weak at the back of an auditorium or a radio signal fades as you get farther from the transmitter, communications signals traveling through wires fade as the distance increases because of resistance in the medium. When the signal strength gets too low, it is impossible for the receiver to accurately pick out the individual signal changes. Amplifiers or repeaters are inserted in communications circuits often enough so that in normal operation, the signal's strength is boosted before attenuation causes a problem.

Attenuation Distortion

Attenuation distortion occurs because a signal does not attenuate evenly across its frequency range. Without special equipment, some frequencies on a circuit attenuate faster than others. Communications circuits are designed and special equipment is inserted on the circuit so that the signal attenuates evenly across the frequency spectrum. However, attenuation distortion can still occur if equipment is improperly adjusted or there is other maintenance activity.

Envelope Delay Distortion

Envelope delay distortion is an electrical phenomenon that occurs when not all frequencies propagate down a telecommunications circuit at exactly the same speed. The absolute propagation delay is not relevant—only the difference between the delay at different frequencies is relevant. Envelope delay distortion can be made worse when the signal passes through filters that are inserted in the circuit to filter out noise. Noise filters tend to delay certain frequencies more than others.

Phase Jitter

Phase jitter is a change in the phase of the signal induced by the carrier signal. It is especially problematic when phase modulation is used because the sudden shift in the phase of the received signal makes it difficult for the receiving modem to sense the legitimate phase changes that the transmitting modem sent.

Echo

Echo is the reversal of a signal, bouncing it back to the sender. Echo occurs on a communications circuit because of the electrical wave bouncing back from an intermediate point or the distant end of a circuit. Echoes are sometimes heard on a voice circuit when the speaker hears his or her voice coming back a fraction of a second after speaking. On a data circuit, echoes cause bit errors.

echo suppressors

Carriers install *echo suppressors* on switched circuits to eliminate the problems caused by echoes. Echo suppressors permit transmission in only one direction at a time. Since transmission in the reverse direction is prohibited, the echo cannot bounce back to the source. Although echo suppressors are a great help for voice transmission, they are a problem when data is sent because they take approximately 150 milliseconds to reverse and permit transmission in the opposite direction. Many modems turn signals around in 50 milliseconds or less, so if they send data, the first several hundred bits may be lost because the echo suppressors have not reversed.

The solution to this problem is to have the modem disable the echo suppressors at the beginning of a data transmission. When a connection is made, the modems send a tone that disables the echo suppressors. The suppressors stay deactivated as long as the modem sends a carrier signal on the circuit. When the communication ends and the carrier signal is no longer sent, the echo suppressors automatically reactivate. Leased data circuits do not have echo suppressors.

Crosstalk

As mentioned earlier, *crosstalk* is interference that occurs when the signals from one communications channel interfere with those on another channel. In a voice conversation, you may occasionally hear crosstalk as another conversation lower in amplitude than your own. Crosstalk can be caused by one signal overpowering another, by two pairs of wires that are in close proximity and are improperly shielded, or by frequencies of two or more multiplexed channels that are too close together.

Dropouts

Dropouts occur when a circuit suddenly goes dead for a period of time. Dropouts last from a fraction of a second to a few seconds, and then normal operation resumes. They can be caused by brief transmission problems, switching equipment, and other phenomena.

IMPACT OF ERRORS

Many of these errors, especially those caused by noise, occur in short bursts lasting from a few milliseconds to a few seconds. Often a short pause in the transmission or a retransmission of the block of data that is affected will circumvent the problem. An error of a given duration will affect more bits during higher-speed transmission than at lower speeds, as was shown in Chapter 8. One alternative for many types of transmission problems is to reduce the speed at which the data is being transmitted.

The effect of transmission errors is more significant in data transmission than in voice or television. In voice conversations, a crackle of static on the line does not usually cause a big problem. Human speech is very redundant, and the people at both ends of the circuit can interpolate and fill in a missing syllable or word. In television transmission, transmission errors often are seen as white flecks or snow on the screen, but a limited amount does not impair our ability to comprehend or enjoy the picture. In data transmission, on the other hand, an incorrect or missing bit can change the meaning of a message entirely. That is why it is particularly important to identify data transmission errors and take the necessary steps to correct them.

ERROR PREVENTION

Given that errors do occur in data transmissions, it is necessary to take steps to prevent, detect, and correct them. Economics must be considered because many of the error prevention techniques cost money to implement. The manager must judge whether the increased reliability is worth the cost. Certain standard techniques are in common use, however, and they are discussed here.

Line Conditioning

conditioned line

When a leased circuit is acquired from a common carrier, it is possible to request that it be conditioned. A *conditioned line* is one that meets tighter specifications for amplitude and distortion. Signals traveling on a conditioned circuit are less likely to encounter errors than when the circuit is unconditioned. The higher the level of conditioning, the fewer errors that will occur on the circuit. With fewer errors, faster signaling rates and, therefore, faster transmission speeds can be achieved.

Conditioning is accomplished by testing each link of a circuit to ensure that it meets the tighter parameters that conditioning implies. If necessary, special compensating or amplifying equipment is inserted in the circuit at the carrier's central offices to bring the circuit up to conditioned specifications.

Conditioning in the United States is specified as class C or class D. C conditioning adjusts a line's characteristics so that attenuation distortion and envelope delay distortion lie within certain limits. D conditioning deals with the ratio of signal strength to noise strength, called *signal-to-noise ratio,* and distortion.

With the improved sophistication of many newer modems, circuit conditioning is sometimes not required at all. Some modems have enough signal processing capability that they can tolerate the higher error rate of an unconditioned circuit. Modem manufacturers generally specify the type of conditioning that their modems require. In comparison to the total cost of the line, conditioning is relatively inexpensive. Therefore, it is almost always prudent to request it, especially if it is recommended by the modem manufacturer.

Conditioning cannot be obtained on switched circuits because the physical facilities used to make the connection vary from one call to the next. The variance is wide enough that conditioning facilities cannot be provided realistically or cost effectively.

Shielding

Shielding of a communications circuit is best understood by looking at Figure 9–8, which shows a coaxial cable. A metallic sheath surrounds the center conductor. Frequently, this shielding is electrically grounded at one end. Shielding prevents stray electrical signals from reaching the primary conductor. In certain situations, shielding on critical parts of a communications circuit may reduce noise or crosstalk. Electrically noisy environments, such as in factories or where circuits run near fluorescent light fixtures or elevator motors, are situations in which communications circuits can benefit from shielding.

Improving Connections

One rule of thumb for telecommunications wiring is "always check the cables and connections first." More problems seem to be caused by poor quality cables and loose, dirty, or otherwise poor connections, than anything else. Check to ensure that all cables and connectors are clean and properly seated, with retaining screws, where they exist, tightened. Avoid cable splices if at all possible. Splices in fiber cables can cause echoes, which are a source of interference and slow the throughput.

Electronic versus Mechanical Equipment

Though not entirely a user option, the replacement of mechanical equipment in the common carrier's central office or the company's equipment rooms with modern electronic equipment can lead to improved circuit quality. Not only is mechanical equipment more prone to failure, but it is also electrically noisier and more likely to induce impulse noise on circuits that pass through it. If a central office containing electromechanical switches is upgraded to an all-electronic configuration, it would be natural for the circuits running through that office to be of better quality and more trouble-free.

ERROR DETECTION

Even if circuits are well designed and good preventive measures, such as circuit conditioning, are implemented, errors will still occur. Therefore, it is necessary to have methods in place to detect these errors so that something can be done to correct them and data integrity can be maintained. *Error detection* normally involves some type of transmission redundancy. In the simplest case, all data could be transmitted two or more times and then compared at the receiving end. This would be fairly expensive in terms of the time consumed for the duplicate transmission. With the error rates experienced today, a full duplication of each transmission is not necessary in most situations.

A more sophisticated type of checking is to calculate some kind of a check digit or check character and transmit it with the data. At the receiving end, the calculation is made again and the result compared to the check character that was calculated before transmission. If the two check characters agree, the data has been received correctly.

Echo Checking

One of the simplest ways to check for transmission errors is called *echo checking,* in which each character is echoed from the receiver back to the transmitter. This is done in some timesharing systems in which a terminal operator can verify immediately that what he or she typed is what appears back on the screen as echoed from the computer. Of course, the original data transmission could be correct, and an error might occur on the transmission of the echo message back to the sender. In either case, the operator can rekey the character.

Vertical Redundancy Checking (VRC) or Parity Checking

The next most sophisticated error detection technique is called *vertical redundancy checking (VRC)* or *parity checking*. This technique was introduced in Chapter 7 and is illustrated in Figure 7–3.

Unfortunately, noise on the communications line frequently changes more than 1 bit. If two 0 bits are changed to 1s, the VRC will not detect the error because the number of 1 bits will still be an even number. Therefore, additional checking techniques are employed in most data transmission systems.

Longitudinal Redundancy Checking (LRC)

Horizontal parity checking is called *longitudinal redundancy checking (LRC)*. When LRC is employed, a parity character is added to the end of each block of data by the DTE before the block is transmitted. This character, also called a *block check character (BCC)*, is made up of parity bits. Bit 1 in the BCC is the parity bit for all the 1 bits in the block, bit 2 is the parity bit for all of the 2 bits, and so on. For example, assuming even parity and the use of the ASCII code, the BCC for the word "parity" is shown below.

block check character

	p	a	r	i	t	y	BCC
Bit 1	1	1	1	1	1	1	0
Bit 2	0	0	0	0	0	0	0
Bit 3	1	0	1	0	1	1	0
Bit 4	0	0	0	1	0	1	0
Bit 5	0	0	0	0	1	0	1
Bit 6	0	0	1	0	0	0	1
Bit 7	0	1	0	1	0	1	1
VRC	0	0	1	1	1	0	1

A VRC check will catch errors in which 1, 3, or 5 bits in a character have been changed, but if 2, 4, or 6 bits are changed, they will go undetected. Thus, VRC by itself will catch about half of the transmission errors that occur. When combined with LRC, the probability of detecting an error is increased. The exact probability depends on the length of the data block for which the block check character is calculated, but even when used together, VRC and LRC will not catch all errors.

Cyclic Redundancy Checking (CRC)

Cyclic redundancy checking (CRC) is a particular implementation of a more general class of error detection techniques called *polynomial error checking*. The polynomial techniques are more sophisticated ways for calculating a BCC than an LRC provides. All of the bits of a block of data are processed by a mathematical algorithm by the DTE at the transmitting end. One or more block check characters are generated and transmitted with the data.

polynomial error checking

At the receiving end, the DTE performs the calculation again, and the check characters are compared. If differences are found, an error has occurred. If the check characters are the same, the probability is very high that the data is error free. With the proper selection of polynomials used in the calculation of the BCC, the number of undetected errors may be as low as 1 in 10^9 characters. To put this in more familiar terms: On a 9,600 bps circuit transmitting 8-bit characters 24 hours per day, one would expect no more than 1 undetected error every 231+ hours, or 9.6 days.

Several standard CRC calculations exist; they are known as CRC-12, CRC-16, and CRC-CCITT. The standard specifies the degree of the generating polynomial and the generating polynomial itself. CRC-12 specifies a polynomial of degree 12, CRC-16 and CRC-CCITT specify a polynomial of degree 16. For example, the polynomial for CRC-CCITT is $x^{16} + x^{12} + x^5 + 1$, where x is the bit being processed. CRC-16 and CRC-CCITT generate a 16-bit block check character that can

- detect all single bit and double bit errors;
- detect all errors in cases where an odd number of bits is incorrect;
- detect two pairs of adjacent errors;
- detect all burst errors of 16 bits or fewer;
- detect over 99.998 percent of all burst errors greater than 16 bits.

Cyclic redundancy checking has become the standard method of error detection for block data transmission because of its high reliability in detecting transmission errors.

ERROR CORRECTION

In most applications, data validity and integrity are of prime importance, so once an error is detected, some technique must be employed to correct the data.

Retransmission

The most frequently used and usually the most economical *error correction system* is the retransmission of the data in error. Although there are many variations, the basic technique used is that when the receiving DTE detects an error, it signals the transmitting DTE to resend the data. This is called an *automatic repeat request (ARQ)* technique, and it is a part of the line protocol. In order for ARQ to work, the transmitting station must hold the data in a buffer until an acknowledgment comes from the receiver that the block of data was received correctly. Another requirement is that there must be a reverse channel for signaling from the receiver to the transmitter.

automatic repeat request (ARQ)

Stop and Wait ARQ In the *stop and wait ARQ* technique, a block of data is sent and the receiver sends either an acknowledgment (ACK) if the data was received correctly or a negative acknowledgment (NAK) if an error was detected. If an ACK is received, the transmitter sends the next block of data. If a NAK is received, the data block that was received in error and is still stored in the transmitter's buffer is retransmitted. No data is transmitted while the receiver decodes the incoming data and checks it for errors. If a reverse channel is not available, the line must be turned around for the transmission of the ACK or NAK and then turned around again for the transmission (or retransmission) of the data block. Stop and wait ARQ is most effective where the data blocks are long, error rates are low, and a reverse channel is available.

Continuous ARQ Using the *continuous ARQ* technique, data blocks are continuously sent over the forward channel while ACKs and NAKs are sent over the reverse channel. When a NAK arrives at the transmitter, the usual strategy is to retransmit beginning with the data block that the receiver indicated was in error. The transmitting station's buffer must be large enough to hold several data blocks. The receiver throws away all data received after the block in error for which the NAK was sent because it will receive that data again.

An alternate strategy is for the transmitter to retransmit only the block in error. In this case, the receiver must be more sophisticated because it must insert the retransmitted data block into the correct sequence among all of the data received. Of the two approaches to continuous ARQ, the first strategy, sometimes called "go back *N* blocks," is more commonly used.

Continuous ARQ is far more efficient than stop and wait ARQ when the propagation times are long, as they are in satellite transmission.

Forward Error Correction (FEC)

VRC, LRC, and CRC checking methods are effective in detecting errors in data transmission. However, they contain no method for automatically correcting the data at the receiving end. By using special transmission codes and adding additional redundant bits, it is possible to include enough redundancy in a transmission to allow the receiving station to automatically correct a large portion of any data received in error, thus avoiding retransmission. This technique is called *forward error correction (FEC).*

Research into FEC techniques has been conducted by organizations such as Bell Laboratories and the military. The military's interest lies in being able to make one-way transmissions to submarines or aircraft, knowing that messages will arrive with a predetermined but very low probability of error. Three well-known error correcting codes are the Bose-Chaudhuri code, the Hagelbarger code, and the Hamming code. The Bose-Chaudhuri code, in its original version, uses 10 check bits for every 21 data bits and is capable of correcting all double bit errors and detecting up to 4 consecutive bit errors. The Hagelbarger code will correct up to 6 consecutive bit errors if the group of bits in error is followed by at least 19 good data bits. The Hamming code, in its 7-bit form, allows single bit errors in each character to be corrected. However, only 16 unique characters are allowed in the character set. Other modifications to the Hamming code allow a larger character set, with a corresponding increase in the number of checking bits.

FEC codes have a high cost in terms of the number of redundant bits required to allow error correcting at the receiving end. In certain applications—particularly where only one-way simplex transmission is allowed or possible—the cost is well justified. Since the FEC techniques are sophisticated, a specially programmed microcomputer in the DTE or DCE normally is used to calculate the FEC codes and perform the error correction.

WIRING AND CABLING

Though not strictly a circuit issue, the subject of wiring and cabling within a building or facility is closely related to circuit installation and operation. Most buildings today contain several pairs of twisted wires, originally installed by the telephone company, running to each office. One pair is for the telephone, and the others are spares. If the office contains a computer terminal, it is, in all likelihood, connected to the server or network via a totally separate wire or cable that was installed by the data processing department. Where private television exists, a third cabling system—perhaps "owned" by the audiovisual department—also may be found.

A commonly stated requirement, which is being implemented in some organizations, is that all of these telecommunications wiring systems need to be merged and consolidated so that a single communications outlet is installed in each office. This outlet would provide jacks for connecting all of the communications equipment in the office. The wire behind the outlet would run to a local wiring distribution center or equipment room on the floor of the office building, as shown in Figure 9–19. The distribution centers would be connected via high-capacity cable, perhaps one of optical fiber, to the wiring center for the building or site. The communications wiring of the future should be similar to the electrical wiring of today, where, for most offices, a single type and size of wire runs to one or more conveniently located standard outlets.

To achieve this ideal standardized communications wiring plan, a long history of nonstandard communications wiring must be overcome. Each vendor of communications equipment traditionally has set individual standards for data communications wiring and for the media that connect terminals to its computers. Of course, these vendors' standards bear little or no relationship to the standards for telephone wiring that the telephone companies have used for years. Even within a single vendor's product line, many standards exist.

Figure 9–19
Simplified wiring diagram of an office building.

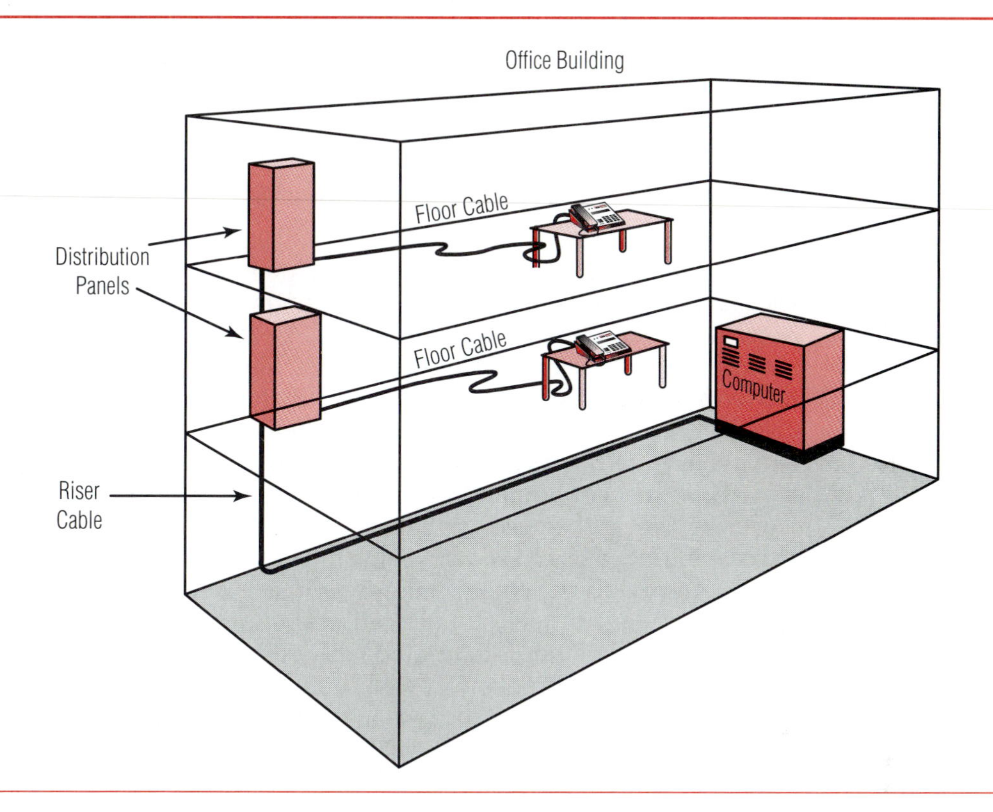

Within the past few years, there has been movement toward using standard Cat 5 twisted pair wire for connecting data terminals to computers and as the media for local area networks. Many companies are choosing Cat 5 for voice too because it gives them the flexibility to use the wiring for data later if circumstances change. Companies are also learning that it makes sense to install spare pairs of wires everywhere. The major cost of wiring is the labor cost to install the cable. Whether the cable contains two pair, four pair, or eight pair of Cat 5 wire makes little difference in the overall cost.

Some general guidelines for managers to follow when involved in a wiring project are:

- Estimate the capacity required and add at least 25 percent.
- Install at least two Cat 5 pair to each desk. Three or four pair is better.
- Consider fiber-optic cable for the backbone between floors and between equipment rooms.
- Choose a wiring contractor carefully, and get test results after the wiring is installed.

The major problem that most companies face is the cost and effort to replace all existing wiring with wire that meets the requirements of the new "standard." In most cases, the only practical, affordable alternative is to convert existing wiring when opportunities such as building renovations arise and to install the new type of wiring in all newly constructed buildings.

THE GROWING DEMAND FOR BANDWIDTH

Any way you view it, companies and people are demanding faster telecommunications circuits—higher bandwidth—at an accelerating pace. The good news is that the technology to provide the bandwidth, such as digital circuits, fiber-optic cables, higher speed modems, cable modems, and the like is progressing rapidly too. The bad news is that the ability of carriers and other companies to deliver the bandwidth to the potential customers is fraught with problems that will take time to solve. If they could, the carriers and their competitors would love to rewire the world and run high-speed digital service to every business and home. But the cost of doing so, at least in a short time frame, is prohibitive.

The rapid rise of Internet usage is unprecedented and has caught most observers and telecommunications suppliers by surprise, but it is providing a new impetus to upgrade the world's telecommunications infrastructure. The base of people who, directly or indirectly, realize the benefits of increased bandwidth is much larger than it was just a year or two ago. That means more potential customers who are willing to pay at least something in order to access the Internet faster.

Overall, research and experience suggest that 5 years from now, a significant percentage of the population of the U.S. will be receiving some or all of its communication, information, entertainment, and educational material by means of broadband Internet. Large numbers of businesses will be using it as well, and significant amounts of commerce will be conducted over the Internet. Users will come to view bandwidth as a *commodity* that they pay for according to how much they use, just like other utilities, but they'll want bandwidth to be available in essentially unlimited quantities, much as electricity is today in most parts of the world. This concept is called *bandwidth-on-demand.* It's not new, but the need for it is recognized more today than ever before. Cable TV operators, DBS broadcasters, telephone companies, and others will all be competing bandwidth suppliers, and applications will be thought of as appliances plugged into the bandwidth outlet. The path to commoditized bandwidth-on-demand will take some restructuring in the telecommunications industry and some years to achieve, but seems to have a very high probability of occurring within 10 if not 5 years.

bandwidth-on-demand

SIDEBAR 9–2

DOWNLOADING A "TITANIC" FILE

How much of the 3-hour, 14-minute movie 'Titanic' could be downloaded by these modems and data lines in 7 minutes, 23 seconds:

Descriptions

Cable modem—10 Mbps, 100%

This modem uses the cable TV wire, instead of a phone line, for data transmission. The coaxial cable can carry much more data than a copper phone line. It's often faster, as well as less expensive, than most alternatives. But customers in a neighborhood share the trunk lines, which means speed can slow when usage is high.

DSL—4 Mbps, 40%

Digital Subscriber Line technology uses existing phone connections with a DSL modem to provide service at relatively low cost. Speeds vary considerably, and it's often faster to download than to transmit. The variations depend on:

- Quality of the Internet service provider's equipment
- Level of service the customer orders
- Distance from the phone switching facility

T-1 line—1.5 Mbps, 15%

A fast phone technology using a dedicated line to serve several phones. But it's expensive to install, with prices rising the farther the user is from the nearest phone switching facility. Also, the monthly charge is high.

ISDN—128 kbps, 0%

An Integrated Services Digital Network connection can connect several phones. But many say it's not fast enough to be worth the special hardware and, usually, the dedicated phone line it needs. Service is limited to within one mile of the phone-company switching facility.

Phone-line modem—28.8 kbps, 0%

This inexpensive modem—the most common now—connects to an ordinary phone line. It transmits and receives data at one relatively slow rate.

In the meantime, as you have seen, there are many ways that a business or individual consumer can get high bandwidth network access today. Figure 9–20 shows many of the alternatives and the tradeoffs with each one.

Figure 9–20
Types of high speed network access and the tradeoffs between them.

Service	Availability	Typical Speed (downstream/upstream)	Tradeoffs	Typical Uses
Dial-up	Everywhere	56 kbps/33.6 kbps	Inexpensive. Easy to install. Relatively slow.	People who can't obtain or afford other alternatives. Travelers.
ISDN	Widespread	128 kbps/128 kbps	Twice as fast as regular dial-up. Costly. Prone to setup difficulties.	Small office and home users who can't get a faster service.
Satellite (DSS)	Widespread	384 kbps/33.6 kbps	Good downstream speed. Complex installation. Must have clear view of southern sky.	Businesses and home users in rural areas or who can't get other high-speed services.
Cable	Limited	384 kbps–5 Mbps/128 kbps–2.5 Mbps	Least expensive high-speed connection. Always on. Limited availability. Most businesses aren't wired for cable. Can't choose your ISP.	Home office. Telecommuters. Businesses that have cable installed.
DSL	Very limited	144 kbps–8 Mbps/64 kbps–8 Mbps	Uses ordinary telephone line. Always on. Requires technician to install. Expensive.	Businesses that can't justify T-1 service. Home users who can't get cable.
T-1–T-3	Widespread	56 kbps–45 Mbps/56 kbps–45 Mbps	Business oriented. Speed guarantees. Quick repair. Expensive.	Medium size to large businesses that can afford the cost and can't afford service problems.

SUMMARY

This chapter looked at the characteristics of telecommunications circuits, various ways of classifying them, and the media with which they can be implemented. The telecommunications network designer is faced with a wide variety of choices for circuits. Only by understanding the characteristics and trade-offs between various circuit types and media can the designer reach a reasonable solution for a given network or application.

More often than not, a company's diverse communications needs indicate that several different circuit types are required. There may be a combination of private, leased, and dial-up circuits. Usually, there is a mixture of transmission speeds to accommodate differing applications and transmission volumes. Different media may be used to satisfy still other needs. All in all, the design of a comprehensive communications network is a complex task requiring both telecommunications and application knowledge.

CASE STUDY

Dow Corning's Data Communications Circuits

One of Dow Corning's philosophies about its data communications system is that it must deliver excellent response time to its users. For that reason, the circuits that make up the system operate at the highest speeds that the company believes it can justify economically.

In 1997, Dow Corning entered into an agreement with WorldCom to provide and manage its international data communications circuits. At the same time, WorldCom proposed converting the network to one based on frame relay technology (which will be discussed in Chapter 11). This means that between locations, WorldCom provides all of the circuits and backup capability, which are mainly high-speed digital circuits. Dow Corning specifies the capacity it needs and the desired response times, and WorldCom configures the network to meet the requirements. There is, of course, ongoing discussion between Dow Corning and WorldCom about the costs as the service requirements change.

WorldCom digital circuits that operate at 128 kbps or higher connect Midland with the larger U.S. plant locations. Internationally, Dow Corning's data communications circuits tie headquarters in Midland to the following locations:

Brussels, Belgium	512 kbps;
Barry, Wales	512 kbps;
Tokyo, Japan	512 kbps;
Sydney, Australia	512 kbps;
Hong Kong	128 kbps;
Toronto, Canada	128 kbps;
Mexico City, Mexico	128 kbps;
Sao Paulo, Brazil	128 kbps.

To avoid the delays associated with satellite circuits, all of these circuits operate on terrestrial cables. While circuits can be envisioned as a star network fanning out from Midland, in reality it is a mesh subnetwork within the overall WorldCom global network.

Dow Corning uses 30 T-1 circuits in the Midland area and about 60 globally. Three T-1s connect the corporate center with the Midland plant 5 miles away, and others run to nearby plants and other office buildings. A T-1 circuit also runs directly to Ameritech's central office and is used for voice circuits. All of the T-1 circuits are leased from Ameritech, which also did the installation. The circuits are less expensive than the multiple voice and data circuits that would otherwise be required. Timeplex multiplexing equipment is also used by Dow Corning on a limited basis, to divide the T-1 capacity into video, voice, and data circuits, but in most cases Ameritech provides the multiplexing equipment as a part of the T-1 service.

The company has a broadband coaxial cable transmission system in the Midland plant. Midland plant personnel installed it many years ago in response to unique requirements for television and data transmission, and it is operated and maintained entirely by plant people. The broadband cable originally carried data from instruments to process control computers throughout the laboratories and plant, and it was also used for data transmission between the IBM terminals on the site and a concentrator. The concentrator forwarded the data on high-speed circuits to the mainframe computers at the corporate headquarters. A third use of the cable was for a television channel providing news and other information to plant employees. In the early 1990s, the use of the cable for data transmission was phased out, and now it is only used for intraplant television transmissions. The data traffic is carried on LAN circuits, either twisted pair wire or optical fiber cable, either of which are easier to maintain than the broadband cable.

In 1986, Dow Corning installed its first private fiber-optic circuit. The fiber cable connects the corporate headquarters building with a building located 1.1. miles away. The fiber was justified because of the voice and data requirements at the site and because it could provide a stronger security capability than would otherwise have been possible.

GTE won the bid to install the cable and buried it in the ground on Dow Corning property. Installation was fairly simple because all of the property between the two buildings is owned by Dow Corning, and only one road had to be crossed. Permission was obtained from the county to bury cable under the road. The optical fiber cable carries data communications between buildings as well as several television channels. Based on the successful experience, other fiber cables were installed, mainly to connect local area networks with each other.

QUESTIONS

1. Dow Corning's philosophy of providing excellent response time means that some of its data circuits are lightly utilized. Explain why this is so. Do you think the company is wasting money by having low circuit utilization?
2. Is the company missing out on service or cost-saving opportunities by shunning satellite circuits?
3. What difficulties do you imagine arise because the entire global network is configured and specified centrally by the telecommunications staff in Midland? Would there be advantages or disadvantages to having the staffs in the outlying locations configure the network for their areas of responsibility?
4. The line speed, or bandwidth, of Dow Corning's circuits has grown faster than the number of employees. Why would this be so? Do you suppose the growth will continue in the future?
5. What are the tradeoffs of entering into a global network contract such as Dow Corning did with WorldCom?

REVIEW QUESTIONS

1. Explain how a line, a circuit, a link, and a channel differ.
2. Describe a multipoint circuit.
3. Why is a four-wire circuit preferable to a two-wire circuit for data transmission?
4. Compare and contrast the functions of a modem and a DSU/CSU.
5. What is the data-carrying capacity of a T-1 circuit? Can a T-1 circuit be used to carry analog voice signals?
6. What is the normal maximum data transmission speed on an analog voice-grade circuit?
7. Identify five different media used for carrying communications signals and discuss under what circumstances each is most appropriately used.

8. What are the advantages of shielded twisted pair wire compared to ordinary twisted pair?
9. Under what circumstances would it be most appropriate to use a broadband coaxial cable instead of a fiber-optic cable?
10. List the characteristics of optical fiber.
11. What are some potential disadvantages of using microwave radio for data transmission?
12. Explain the term *propagation delay* and why it is important in satellite transmission. 361
13. Under what circumstances would a company consider installing a private fiber-optic link connecting two of its locations?
14. Under what circumstances would a company consider bypassing its local telephone company?
15. What are the advantages of STDM over traditional TDM?
16. How does dialback improve security on a switched communications line?
17. Discuss the circuit attributes of attenuation, envelope delay distortion, phase jitter, and crosstalk.
18. Why are communications line errors more significant when data is being transmitted than when voice is being transmitted? If voice is transmitted digitally, do line errors become more significant?
19. What is the purpose of communications line conditioning?
20. What is the difference between LRC and CRC checking? 381
21. Compare and contrast the stop and wait ARQ and continuous ARQ techniques.
22. Why is it desirable for a company to manage its communications wiring?
23. What are the advantages of ISDN circuits? 340
24. Explain the difference between a private circuit and a leased circuit.
25. What is SMDS?
26. Distinguish among T-1, T-2, T-3, and T-4 circuits.
27. What is Cat 3 wiring? Why is it important?
28. What is Cat 5 wiring? Why is it important?
29. Distinguish between T-1 and E-1 circuits.
30. What is noise on a transmission line? What causes noise?
31. What is a cyclical redundancy check? How does it work? 381
32. What is the difference between a repeater and an amplifier?
33. Describe several ways to get high-speed access to the Internet. Are all of those methods available today?
34. Distinguish between the capabilities of SDSL and ADSL.
35. What is the major use for SONET in the United States?
36. What is the difference between unguided and conducted media? Give two examples of each.
37. Identify several reasons why a business might and might not select satellite service for high-speed access to the Internet.

PROBLEMS AND PROJECTS

1. Using the VRC and LRC parity checking techniques and the ASCII code, calculate the parity bit and block check character for your last name.
2. A company has a leased satellite circuit between its New York and San Francisco locations. The circuit is routed on a satellite. When 500-character blocks of data are sent from New York to the San Francisco office using a stop and wait ARQ technique, what percentage of the line time is used for actual data transmission and what percentage is spent waiting for acknowledgments and line turnarounds? Assume that
   - line turnaround takes 50 milliseconds;
   - no errors occur during the transmission;
   - line speed is 9,600 bps;
   - an acknowledgment message is 5 characters long.

How does the percentage change if 5,000 character blocks are transmitted?

3. Identify the trade-offs a company would have to consider when deciding whether to implement dial-up data or leased communications circuits. The company has locations throughout the United States. The applications are primarily basic business transactions, such as customer order entry, shipping, purchasing, accounts receivable, and accounts payable.

4. Visit a company that has installed fiber-optic links. Why did they install optical fiber? What difficulties did they have when installing the fiber? What error rate are they experiencing for data transmission on the fiber? How often has it failed? Overall, how has the fiber operated compared to the previous communications technology that was installed?

5. What type of wiring would you suggest for a small service business that just moved into a building recently vacated by a grocery store? The business does no manufacturing, so employees will be sitting in an office environment. The company expects to install a LAN with a data rate of either 10 or 16 Mbps, and will use the telephone company's Centrex system for telephone service.

6. Investigate the status of ISDN service in your community. Is the service available? Try to find someone who has installed ISDN at home. Talk with them about the ease or difficulty of installing it, and how it has been operating since it was installed.

7. Investigate the alternatives for getting high-speed access to the Internet in your home. What services are available today where you live? If you have several alternatives to choose from, which would you select? Why?

Vocabulary

circuit
line
link
data link
channel
forward channel
reverse channel
node
simplex
half-duplex
full-duplex
point-to-point circuit
multipoint circuit
multidrop circuit
two-wire circuit
four-wire circuit
low-speed circuit
subvoice-grade circuit
voice-grade circuit
wideband circuit
digital circuit
repeater
Integrated Services Digital Network (ISDN)
basic access
primary access
broadband ISDN (B-ISDN)
T-carrier system
fractional T-1
connectionless service
asymmetric digital subscriber line (ADSL)
conducted media
guided media
radiated media
unguided media
twisted pair
unshielded twisted pair (UTP)
crosstalk
punchdown block
shielded twisted pair
coaxial cable
optical fiber
core
cladding
light-emitting diode (LED)
laser
single mode
multimode
dispersion
wavelength division multiplexing (WDM)
Synchronous Optical Network (SONET)
microwave radio
geosynchronous orbit
uplink
downlink
propagation delay
Ku band
very small aperture terminals (VSAT)
direct broadcast satellite (DBS)
digital satellite service (DSS)
infrared
private circuit
leased circuit
bypass
circuit switching
callback unit
dialback unit

time division multiplexing (TDM)
frame
message
statistical time division multiplexing (STDM)
concentrator
inverse concentrator
background noise
white noise
Gaussian noise
impulse noise
attenuation distortion
envelope delay distortion
phase jitter
echo
echo suppressor
dropout
conditioned line
signal-to-noise ratio
shielding
error detection
echo checking
vertical redundancy checking (VRC)
parity checking
longitudinal redundancy checking (LRC)
block check character (BCC)
cyclic redundancy checking (CRC)
polynomial error checking
error correction system
automatic repeat request (ARQ)
stop and wait ARQ
continuous ARQ
forward error correction (FEC)
commodity
bandwidth-on-demand

References

Chatterjee, Samir and Suzanne Pawlowski. "All Optical Networks." *Communications of the ACM* (June 1999): 75–83.

Conover, Joel. "Sorting Out Cabling Standards." *Network Computing* (February 21, 2000): 91–93.

Crotty, Cameron. "New Flavor of DSL Brings Faster, Cheaper Web Access to Your Door." *PC World* (September 1999): 62–64.

Fowler, Thomas B. "Internet Access and Pricing: Sorting Out the Options." *Telecommunications* (February 1997): 41–69.

Kaplan, Mark. "Voice over SDSL: Effectively Combining Voice and Data." *Telecommunications* (February 2000): 79–80.

McCracken, Harry. "Bandwidth on Demand." *PC World* (March 1999): 109–118.

Mills, Mike. "A Wealth of Data on Ocean Floors." *International Herald Tribune* (March 10, 1998).

Peterson, Kerstin. "So You Want to Wire a Building." *Teleconnect* (March 1997): 168–173.

Ramo, Joshua Cooper. "Welcome to the Wired World." *Time* (February 3, 1997): 37–48.

Rosenbush, Steve. "Charge of the Light Brigade." *Business Week* (January 31, 2000): 62–66.

Sekar, Richard. "G.lite: Pragmatic, Mass Market High-Density DSL." *Telecommunications* (April 2000): 35–36.

Willis, David, "Staying Sober at the xDSL Party." *Network Computing* (October 18, 1999): 49.

Woods, Darrin. "Shedding Light on SONET." *Network Computing* (March 20, 2000): 47–54.

———. "SONET from Scratch." *Network Computing* (May 15, 2000): 129–146.

———. "Too Much Long Distance." *Fortune* (March 15, 1999): 105–110.

CHAPTER

10

Data Link Control Protocols

OBJECTIVES

After studying the material in this chapter, you should be able to

- explain what a data link protocol is and the function it performs;
- describe desirable attributes of protocols;
- distinguish between contention and polling protocols;
- explain the distinguishing characteristics of character-oriented, byte-count oriented, and bit-oriented protocols;
- explain in general terms how several protocols work;
- describe how LAN protocols work;
- explain protocol conversion and situations in which it is likely to be useful.

INTRODUCTION

Chapter 9 looked at telecommunications circuits in detail. The circuits are like the roads in a highway system; they provide a mechanism over which communications can occur. In themselves, however, circuits have no "rules of the road." Just as there must be rules for the use of the highways to ensure efficient and safe transportation, there must be rules for the use of the circuits to ensure that data is transported efficiently and accurately.

Many protocols have been defined for the various layers of the ISO or TCP/IP hierarchies. In this chapter we will look at the protocols of the data link control layer, layer 2 of the ISO-OSI reference model. They are known as *data link protocols* and they define how data is structured into blocks or frames and surrounded by control characters before being transmitted on a communications circuit.

DEFINITION AND THE NEED FOR PROTOCOLS

In general terms, a *protocol* is a set of rules or guidelines that govern the interaction between people, between people and machines, or between machines. Protocols exist for all types of social situations. When two people meet on the street and one says, "Hello, Jim," to which Jim responds, "Hello, Bill, how are you?" they are following a simple protocol. This simple exchange gets the conversation going, and after the initial words of greeting, the two men launch into any of a variety of topics. Similarly, when a telephone rings, the convention or protocol is that the answering party says "Hello" and the calling party identifies herself by saying something like, "Hello, Mary? This is Jane Smith." How the conversation proceeds from there depends on the relationship between the individuals and the reason for the communications.

In data communications, protocols are the rules for communicating between data terminal equipment, which for our purposes can be between terminals and computers, between computers, or between terminals. In order for communications between equipment to occur, several types of protocols need to be in place. This chapter only concerns the data link protocol—the rules for operating the circuit and sending messages over it.

DATA LINK PROTOCOL FUNCTIONS

Data link protocols must have rules to address the following situations that occur during data communications.

- Communications startup—There must be rules to specify how the communications will be initiated: whether there is automatic startup, whether any station can initiate the communications, or whether only a station designated as the "master" station may initiate communications.

- Character identification and framing—There must be a way for the data terminal equipment to separate the string of bits coming down the communications line into characters. Furthermore, there must be ways of distinguishing between control characters, which are a part of the protocol, and data characters, which convey the message being communicated.
- Message identification—The data terminal equipment must separate the characters on the communications line into messages.
- Line control—Rules must exist to specify how the receiving terminal signals the sending terminal whether it has received data correctly, how and under what circumstances the line will be turned around, and whether the receiving terminal can accept more data.
- Error control—The protocol must contain rules specifying what happens when an error is detected, what to do if communications suddenly and unexplainably cease, and how communications are reestablished after they are broken.
- Termination—Rules must exist for ending the communications under normal and abnormal circumstances.

In normal conversation between two people, there are rules or protocols that specify how all of these situations are handled. Because of the human intelligence at both ends of the communications, the rules need not be precisely specified, and there is room for considerable variation. Whether we end a telephone conversation by saying "Goodbye, Mr. Smith," "See 'ya later," or "Bye" matters little. The intention is clear and understood by both parties. Data communications between two pieces of equipment, however, must be somewhat more precise and rigid in the application of the rules because even the growing intelligence of workstations and computers is still no match for the human intellect.

DESIRABLE ATTRIBUTES OF DATA LINK PROTOCOLS

In addition to the functional requirements specified earlier, there are several other desirable attributes for a data link protocol to have.

- Transparency—It is very desirable for data terminal equipment (DTE) to be able to transmit and receive any bit pattern as data. The complication occurs because certain bit patterns are assigned to represent control characters that have a specific meaning within the protocol.
- Code independence—It is desirable for the protocol to allow the transmission of data from any data coding system, such as ASCII, EBCDIC, Baudot, or any other. (Some protocols require a certain code to be used.)
- Efficiency—The protocol should use as few characters as possible to control the data transmission so that most of the line capacity can be used for actual data transmission.

PROTOCOL IMPLEMENTATION

A data link protocol is implemented by transmitting certain characters, or sequences of characters, on the communications line. The specific characters are determined by the code and protocol being used. If you look at the table of ASCII characters shown in Figure 7–5, you will see that there is a whole set of characters on the left side of the table that begin with the bits 000 and 001. These are control characters in the ASCII code. Some of them are used to implement the data link protocol.

control characters

By way of contrast to the other characters in the table, which are sometimes called the graphic characters, control characters usually are not printable or displayable by a terminal or printer. Not all of these characters are used in the data link protocol. Some of them are used for other control purposes, such as skipping to a new line on a terminal, tabbing, and printer control.

PREDETERMINED COMMUNICATIONS PARAMETERS

Many communications parameters used to establish compatibility between terminals or between a terminal and a computer are set manually by switches or are specified as parameters via software. The data code (ASCII, EBCDIC, or some other) usually is predetermined. The code determines how many bits make up a character, such as seven in the case of ASCII. Whether there will be a parity bit and whether odd or even parity will be used is often determined by a switch setting in the DTE. The modem usually automatically determines the transmission speed by examining the first one or two characters that are transmitted.

Preestablishing these parameters is necessary so that the control characters of the protocol may be correctly received and interpreted, allowing the communications to be established and maintained.

PROTOCOL CONCEPTS—A GENERAL MODEL

This section explores protocols in a generic sense. The elements that make up a protocol will be introduced, and they will serve as a general model that can be used for comparison when specific protocols are examined later in the chapter.

Line Access

Before a DTE can begin communicating, it must gain control of (access to) the circuit. There are three primary ways in which access can be obtained: contention, polling, and token passing.

Contention Contention systems are quite simple. A DTE with a message to send listens to the circuit. If the circuit is not busy, the DTE begins sending its message. If the circuit is busy, the DTE waits and keeps checking periodically until the line is free. Once a DTE gains access to the circuit, it can use the circuit for as long as necessary. In simple contention systems, there are no rules to limit how long a DTE may tie up the circuit.

The contention approach works best

- on point-to-point circuits on which either DTE is capable of sending data;
- on multipoint circuits when message traffic is not too heavy;
- when the speed of the circuit is relatively fast so that traffic can be sent and the circuit can be cleared quickly.

The third point suggests that contention systems work best on local area networks, because LANs operate at much higher speeds than other networks, and indeed, that is where they are used the most. This will be discussed more fully later in this chapter.

Polling In Chapter 4 we first looked at polling without calling it a line access technique. Now we'll look at the concept in more detail.

Roll Call Polling *Roll call polling* is the most common implementation of a polling system. One station on a line is designated as the master, and the others are slaves. The master station sends out special polling characters that ask each slave station in turn whether it has a message to send. If the slave has no messages, it sends a control character indicating so. If the slave does have a message, it sends the message. Each station on the line responds only to its unique polling characters, which usually correspond to its address. In the simplest case, the master station polls each terminal in turn. In most systems there is a sequenced list, called the *polling list*, which specifies the sequence in which terminals are polled. Terminals that normally have more data to send may be polled—and given the opportunity to use the circuit—more frequently than other terminals with less traffic.

polling list

Fast Select Polling *Fast select polling* speeds up the polling process because a slave station with no traffic isn't required to return a character to the master station. The master station polls several terminals on the circuit and then waits long enough for any of the slave terminals to respond. If a message is not received, the master station continues polling the other stations on the circuit. Using the fast select technique can significantly reduce the time it takes to poll all stations.

Hub Polling Another type of polling is called *hub polling*. It is most easily visualized if one thinks of a multipoint circuit. One station begins by polling the next station on the line. That station responds with a message if it has one. If it does not have a message, it passes the polling message on to the next station on the line, and so on. Hub polling reduces the amount of polling character traffic on the line. It requires more complicated circuitry in each terminal because each is performing a part of the overall line control (a function that is concentrated in the master terminal when the roll call polling technique is used).

Polling is a suitable technique if there is a master station on a network; however, many of today's networks don't have a master station—rather, all of the stations (terminals, clients, servers) are peers. In that case another way to gain control of the line must be used.

Token Passing Token passing is similar in concept to hub polling, but the access technique is more complicated. A single *token*, which is a particular sequence of bits, is passed from node to node. The node that has the token at any point in time is allowed to transmit, and a node with a message to send waits until a free token passes. The node changes a bit in the token, thereby changing the token's status from "free" to "busy," and attaches its message to the "busy" token.

All of the stations on a token passing system are considered equal, but at least one of the stations must have logic to ensure that there is always one, but only one, token circulating on the circuit. The advantage of a token passing technique is that the response time of the network is deterministic or predictable. Token passing is discussed again later in this chapter and is referred to again in Chapter 12.

Message Format

Once the line has been seized, whether by contention, polling, or token passing, and the DTEs have both indicated they are ready, the communications begin.

header
text
trailer

A data message normally consists of three parts: the *header*, the *text* of the message, and the *trailer*, as shown in Figure 10–1. The header contains information about the message, such as the destination node's address, a sequence number, and perhaps a date and time. The text is the main part of the message—it's what the communication is all about. In most protocols, the trailer is quite short and contains only checking characters.

In some protocols, each part of the message begins and ends with special control characters, such as the *start of header (SOH)* and *start of text (STX)* (which also marks the end of the header) characters. In other protocols, the header may be of a fixed length, in which case a special end of header character is unnecessary. The end of the text is either marked by a special character, such as the *end of text (ETX)* character, or its location can be calculated from a "text length" field in the header. In many protocols, the trailer consists of only a *block check character (BCC)*, which is generated by the error checking circuitry or software.

During the transmission of a message or a group of messages, error checking and synchronization occur at points in time determined by the

Figure 10–1
Message format.

Header | Text | Trailer

Start of Header | Start of Text | End of Text | Block Check Character

SOH Header STX Text ETX Trailer BCC

Direction of transmission

specific protocol in use. Error checking may occur at the end of each block or message, at which time the line is usually turned around and the receiving terminal sends an ACK or NAK to the sending terminal, indicating whether the data block that was just received was correct. If a continuous ARQ technique is being used, the line turnaround may only occur periodically, and blocks of data can be sent essentially continuously until an error occurs.

Synchronization characters (SYN) may be inserted in the data stream from time to time by the transmitting station in order to ensure that the receiver is maintaining character synchronization and properly grouping the bits into characters. The synchronization characters are removed by the receiver and do not end up in the received message.

ASYNCHRONOUS DATA LINK CONTROL PROTOCOLS

Asynchronous data transmission is the oldest form of data communications. It evolved from the early days of teletypewriter operation, where operators sat at the teletypewriters at both ends of a point-to-point circuit. With operators at both ends and with most transmissions occurring manually, little protocol was needed. Usually, it was as simple as the operator with a message to send looking to see if his or her teletypewriter was printing a message. If it wasn't, the operator could assume that the circuit was free. He or she would then type the message, and the message was printed simultaneously on the teletypewriter at the receiving end of the connection. Error checking consisted of the receiving operator reading the message, and if it was understandable, acknowledging it, perhaps by sending the sequence number of the message back to the sender. If some of the characters were garbled, the receiving operator would ask for a repeat of all or portions of the message until the characters were correct. Frequently, sending operators would automatically repeat the words or numbers, such as dollar amounts or addresses, when the message was originally sent as a way of ensuring that they would be received correctly.

Although communications have become more automated, many asynchronous protocols have remained quite simple. They deal with individual characters rather than blocks of data. Typically, the only protocol used is the start and stop bits surrounding each character and the parity checking performed if certain codes are used. Although the ASCII code defines special control characters, they are used only when more elaborate protocols are employed. The typical asynchronous transmission today is still character oriented.

As terminals came to have more intelligence and communications between terminals and computers rose in popularity in the 1960s and early 1970s, synchronous communications actually grew at a faster rate than asynchronous transmission because it is a more efficient way to transmit data. With the development of the personal computer and the desire to add communications capability to it, asynchronous communications began regaining favor. Asynchronous communication enjoys increasing popularity primarily because it provides adequate capability for most personal computer communications, and requires less sophisticated—and therefore less expensive—equipment.

As the desire for more automatic verification of messages and file transfers between personal computers occurred, it became necessary to define and implement more elaborate protocols. The more elaborate asynchronous protocols allow the transmission of data in blocks. They perform additional checks on the blocks of data to ensure that transmission is correct.

The XMODEM Protocol

One widely used asynchronous protocol is the *XMODEM protocol* developed for use between microcomputers, especially for transfers of data files between them. Using this protocol, one microcomputer is designated as the sender, and the other is designated as the receiver. The receiver indicates that it is ready to receive by sending an ASCII NAK character every 10 seconds. When the transmitting system receives a NAK, it begins sending blocks of 128 data characters surrounded by a header and trailer. The header consists of a start of header character, followed by a block number character, followed by the same block number with each bit inverted. The trailer is a checksum character, which is the sum of the ASCII values of all of the 128 data characters divided by 255.

checksum character

At the receiving end, the message is checked, usually by software, to ensure that the first character was an SOH, that the block number was exactly 1 more than the last block received, that exactly 128 characters of data were received, and that the checksum computed at the receiving end is identical to the last character received in the block. If all of these conditions are true, the receiver sends an ACK back to the transmitter, and the transmitter sends the next block. If the data was not received correctly, a NAK is sent, and the transmitter resends the block of data that was in error.

The entire message or data file is sent in this way, block by block. At the end, the transmitter sends an end of text character that is acknowledged by the receiver with an ACK, and the transmission is complete.

Since the XMODEM protocol is quite simple, its error checking is not very sophisticated. Therefore, the reliability of the received data is not as good as with other protocols. XMODEM is considered to be a half-duplex protocol because the sender waits for an acknowledgment (stop and wait ARQ) of each block of data before sending the next block. Thus, if a full-duplex transmission facility is used, XMODEM can only use it inefficiently.

Other Asynchronous Protocols

There are many other asynchronous protocols in use today. They have been developed by a variety of sources and are available from the developers on bulletin boards, the Internet, and sometimes they are included in other software packages. Some of the more popular asynchronous protocols and their distinguishing features are listed below:

- *XMODEM-CRC*—This was the first evolution of XMODEM and is assumed by most programs when XMODEM is specified. It improves the error handling of the XMODEM protocol by replacing the checksum with a one-byte *cyclical redundancy check (CRC)* character (CRC-8). As a result it is much more robust in detecting transmission errors.
- *XMODEM-1K*—The next step up in sophistication, XMODEM-1K increases the efficiency of XMODEM-CRC by transmitting 1,024-character blocks instead of 128-character blocks.
- *YMODEM*—YMODEM is essentially XMODEM-1K with a 2-byte CRC (CRC-16). It allows multiple files to be transferred with a single command.
- *YMODEM-G*—A variant of YMODEM that expects the modem to provide software error correction and recovery. Therefore, this protocol should only be used with modems that support hardware error correction such as V.42 or MNP. Because it does not have error correction and recovery, YMODEM-G is very fast.
- *ZMODEM*—A different protocol, not an upgrade from XMODEM. ZMODEM uses a four-byte CRC (CRC-32) and adjusts its block size depending on line conditions. ZMODEM is also very fast and provides good failure recovery. If a transmission is canceled or interrupted, it can be restarted later and the previously transferred information does not have to be resent.
- *X.PC*—Developed by Tymnet especially for connecting asynchronous devices to packet switching networks. Allows several logical connections (sessions) to take place simultaneously on one circuit by interleaving the packets from different sessions.
- *KERMIT*—Developed and copyrighted by Columbia University, KERMIT is a very popular protocol, especially in a university setting. Normally it uses 1,000-byte blocks and a three-byte CRC (CRC-24), but the block size can be adjusted dynamically during a transmission, based on line conditions. Whereas most other asynchronous protocols only handle 8-bit data codes, KERMIT can also handle 7-bit codes. The protocol is error checked and very fast, and it works equally well in PC-to-PC and PC-to-mainframe connections.
- *Serial Line Internet Protocol (SLIP)*—SLIP is a very simple full-duplex protocol for carrying IP over an asynchronous dial-up or

leased line. It is designed for communication between two machines that have been previously configured, so it does not do any initial negotiation with the receiver when the transmission is started. SLIP has no error detection or correction. SLIP was designed as a temporary protocol for use between a workstation and an Internet Service Provider (ISP), but because of its limitations it has mostly been replaced by the point-to-point (PPP) protocol.

- *Point-to-Point Protocol (PPP)*—PPP is primarily used by PC workstations to dial in to a TCP/IP-based network (e.g. when dialing in to an ISP for connection to the Internet), although leased lines may also be used. PPP supports full-duplex transmission and performs authentication, data compression, error detection and correction, and packet sequencing. PPP used CRC-16 for error detection. The protocol also supports synchronous as well as asynchronous transmission. Where a choice is possible, PPP is almost always preferred over SLIP.

There are of course other protocols available, and as personal computers grow in power and capability, it is reasonable to assume that the protocols will grow more sophisticated to meet more stringent requirements.

SYNCHRONOUS DATA LINK PROTOCOLS

Classification

Synchronous data link protocols typically deal with blocks of data, not individual characters. Synchronous protocols may be divided into three types according to the way the start and end of the message are determined. The three types are character-oriented protocols, byte-count-oriented protocols, and bit-oriented protocols.

Character-Oriented Protocols A *character-oriented protocol* uses special characters to indicate the beginning and end of messages. For example, the SOH character is used to indicate the beginning of a message and the ETX character to indicate the end. The best known character-oriented protocol is the Binary Synchronous Communications protocol, also known as BSC or BISYNC.

Byte-Count-Oriented Protocols *Byte-count-oriented protocols* have a special character to mark the beginning of the header, followed by a count field that indicates how many characters are in the data portion of the message. The header may contain other information as well. It is followed by the data portion of the message. The data portion of the message is followed by a block check character or characters. The best known byte-count-oriented protocol is Digital Equipment Corporation's Digital Data Communications Message Protocol (DDCMP).

Bit-Oriented Protocols A *bit-oriented protocol* uses only one special character, called the *flag character,* which marks the beginning and end of a message. Within the message, the header and the fields within it are of a predefined length, and the header is followed by the data field with no intervening control character. No special control character is used to mark the beginning of the trailer segment of the message, if one exists. The flag character also marks the end of the message. The receiving terminal knows that the bits preceding the flag are the check characters for the message. The best known bit-oriented protocol is IBM's Synchronous Data Link Control (SDLC), which is a proper subset of the International Standards Organization's High-Level Data Link Control (HDLC). Other bit-oriented protocols include the CSMA/CD (discussed later in the chapter) and token protocols that are used on local area networks.

We will now look at three of the most common synchronous data link protocols—BSC, DDCMP, and HDLC—in more detail.

Binary Synchronous Communications (BSC, BISYNC)

The *Binary Synchronous Communications (BISYNC or BSC)* protocol was introduced by IBM in 1967. While it is a very old protocol, it was implemented by many companies for a wide variety of equipment and applications. Despite its age, BISYNC is still in use, and serves as an excellent example of a character-oriented protocol. Because it is relatively straightforward and easy to understand, it is a useful illustration of many concepts that apply to all protocols.

BISYNC supports only three data codes: the 6-bit transcode (SBT), which is rarely used anymore, ASCII, and EBCDIC. Certain bit patterns in each code have been set aside for the required control characters: SOH, STX, ETB, ITB, ETX, EOT, NAK, DLE, and ENQ. Some additional control characters are really two-character sequences: ACK0, ACK1, WACK, RVI, and TTD. All of these control characters and sequences are defined in Figure 10–2.

BISYNC operates in either nontransparent or transparent mode. *Transparent mode* means that any bit sequence is permissible in the data field, even if it is the same as a character used for control. Transparency is needed when transmitting binary data, such as computer programs and some data files. In transparent mode, the data link escape (DLE) character is inserted before any control characters. DLE STX initiates transparent mode, and DLE ETX or DLE ETB terminates the block. Since the combination DLE ETX could also occur in the middle of the data portion of the message and inadvertently terminate the transmission, the transmitter scans the text portion of the message. Whenever it finds a DLE character, it inserts an additional DLE. On the receiving side, the data stream also is scanned. Whenever two DLEs are found together, one is discarded.

Figure 10–2
Binary Synchronous Communications control characters.

Symbol	Description	Purpose
SOH	Start of header	Marks the beginning of the header of a transmission.
STX	Start of text	Marks the end of the header and the beginning of the data portion of the message.
ITB	End of intermediate	Marks the end of a data block, but does not reverse the line or require the receiver to acknowledge receipt. A block check character follows the ITB for checking purposes.
ETB	End of text block	Marks the end of a data block. Requires the receiver to acknowledge receipt.
ETX	End of text	End of data block and no more data blocks to be sent.
EOT	End of transmission	Marks the end of transmission that may have contained several blocks or messages.
ACK0 ACK1	Positive acknowledgment	Previous block was received correctly. ACK0 is used for even-numbered blocks; ACK1 for odd-numbered ones.
NAK	Negative acknowledgment	Previous block was received in error. Usually requires a retransmission.
WACK	Wait before transmit	Same as ACK, but receiver is not ready to receive another block.
SYN	Synchronization	Sent at beginning of transmission to ensure characters will be received correctly.
ENQ	Enquiry	Requests use of the line in point-to-point communication.
DLE	Data link escape	In transparent communication, creates two-character versions of ACK, WACK, and RVI.
RVI	Reverse interrupt	Positive acknowledgment and asks transmitter to stop as soon as possible because receiver has a high-priority message to send.
TTD	Temporary text delay	Transmitter uses this to retain control of the line when it is not ready to send data.

The receiver also knows that when it sees the sequence DLE DLE ETX in the data portion of the message, the sequence is not to be interpreted as a set of control characters indicating message termination, but only as data.

Synchronization of Characters and Messages Character synchronization is accomplished in BISYNC by sending SYN characters at the beginning and periodically in the middle of each transmission. The exact number of SYN characters is somewhat dependent on the hardware. Generally, at least two are transmitted. The hardware scans the line searching for the SYN character (e.g., 00010110 in ASCII). Once the SYN character is found, character synchronization is established, and the following characters can be interpreted correctly.

Message synchronization in BISYNC is accomplished with the SOH, STX, ETB, and EOT characters. These control characters mark the beginning and end of each message being sent.

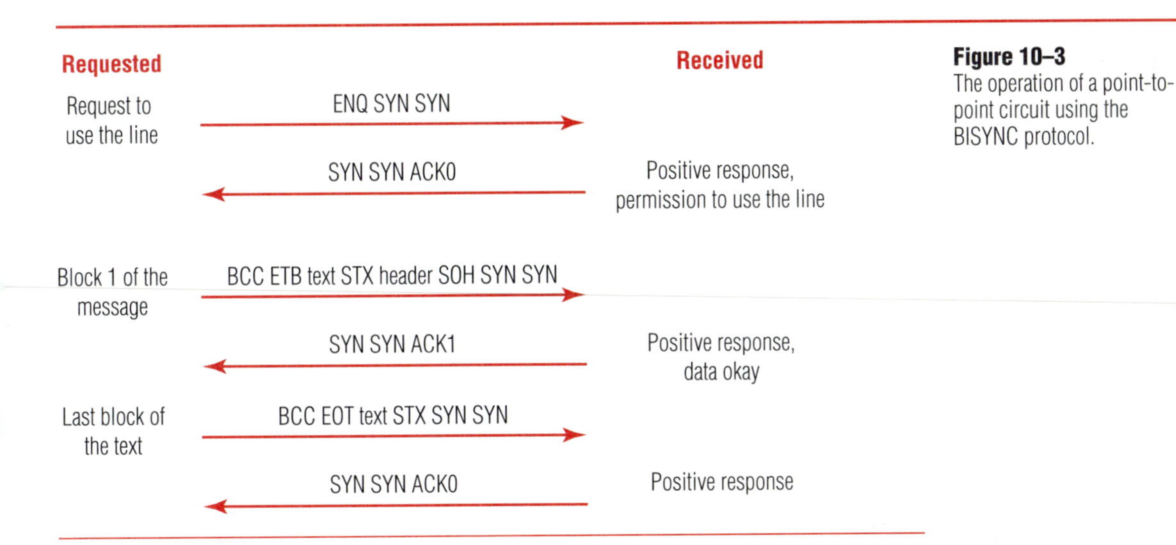

Figure 10–3
The operation of a point-to-point circuit using the BISYNC protocol.

Block Checking BISYNC uses various techniques for error detection and an ARQ approach for error correction. When ASCII code is being used, a parity check (VRC) is performed on each character and an LRC (horizontal parity) is performed on the whole message. When EBCDIC or 6-bit transcode is being used, no parity check is made, but a CRC is calculated for the entire message.

In all cases, if the block check character or characters transmitted with the data do not agree with the characters calculated by the receiver, a NAK is sent back to the transmitter, which is then responsible for resending the block in error. When the block check characters agree, a positive acknowledgment is sent, ACK0 for an even-numbered block and ACK1 for an odd-numbered block. The alternation between ACK0 and ACK1 provides an additional check against totally missing or duplicated blocks of data. If two ACK0 acknowledgments are received in succession, the transmitting station knows that a block of data has been lost.

Transmission Sequences BISYNC has well-defined rules that govern the sequence of transmission, acknowledgment, and placement of control characters in various situations. Three common sequences are described here.

Point-to-Point Operation In point-to-point operation, a station with a message to send first requests permission to use the line from the other station. Once permission is granted, the transmission proceeds as shown in Figure 10–3. After the positive acknowledgment of the last block of data, the line is again available, and either station may request its use with the SYN SYN ENQ sequence.

Multipoint Operation In multipoint operation, polling solicits input from the stations on the circuit. This is illustrated in Figure 10–4. When

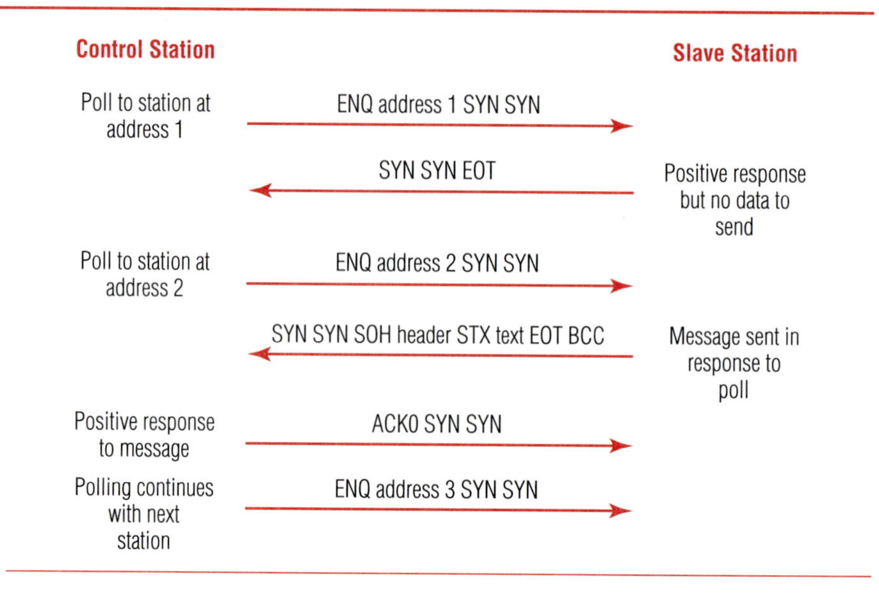

Figure 10–4 The polling operation on a multipoint circuit using the BISYNC protocol.

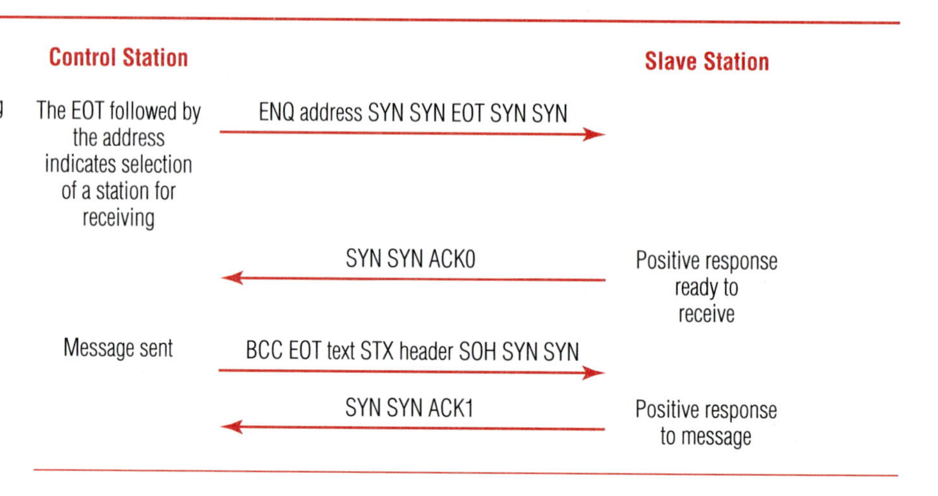

Figure 10–5 The addressing operation on a multipoint circuit using the BISYNC protocol.

the control station has a message to send, it uses the addressing sequence shown in Figure 10–5. After the last acknowledgment, the line is again free.

Overall, BISYNC is a relatively efficient protocol that is easy to understand and implement. This has led to its wide popularity among computer and terminal vendors. Its primary drawbacks are

- it is not code independent;
- it is a half-duplex protocol and cannot take advantage of full-duplex circuits;
- its implementation of transparency is cumbersome.

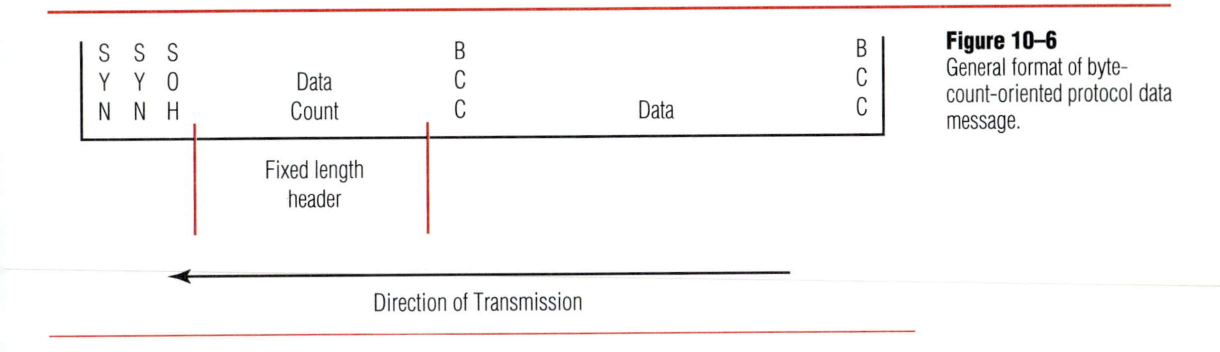

Figure 10–6 General format of byte-count-oriented protocol data message.

Despite these drawbacks, BISYNC is likely to be in widespread use for years to come.

Digital Data Communications Message Protocol (DDCMP)

Digital Data Communications Message Protocol (DDCMP) was developed by Digital Equipment Corporation (now Compaq Computer Corporation) as the data link protocol for its Digital Network Architecture (DNA). Over the years the protocol was enhanced several times to make better use of high-bandwidth and high-latency (satellite) links, but the basic principles of the protocol remained the same.

DDCMP is a byte-count-oriented protocol. The general format for byte-count-oriented protocol messages is shown in Figure 10–6. The header of the message is a fixed length, preceded by at least two SYN characters and a special SOH character. The SOH character is the only unique character required in the protocol. Therefore, the implementation of transparency is relatively easy. One of the fields in the header indicates the number of characters that the message contains, and the data can be a variable length. Both the header and the data portions of the message are checked with a block check character.

DDCMP has two message types. The data message is similar in format to the general format for byte-count-oriented protocols shown in Figure 10–6. Additional fields in the DDCMP header, shown in Figure 10–7, allow the data message sent in one direction to also acknowledge the receipt of a data message sent in the other direction. (More on this later.) The other DDCMP message type is the control message. Control messages are of fixed length, the same length as the header of the data message. Control messages are used to initiate communications between two stations, to send an ACK or NAK about a previously received message, and to request an acknowledgment of previously sent messages.

DDCMP operates in point-to-point or multipoint configurations on either half-duplex or full-duplex lines. In full-duplex operation, both stations can transmit simultaneously. As previously mentioned, the data

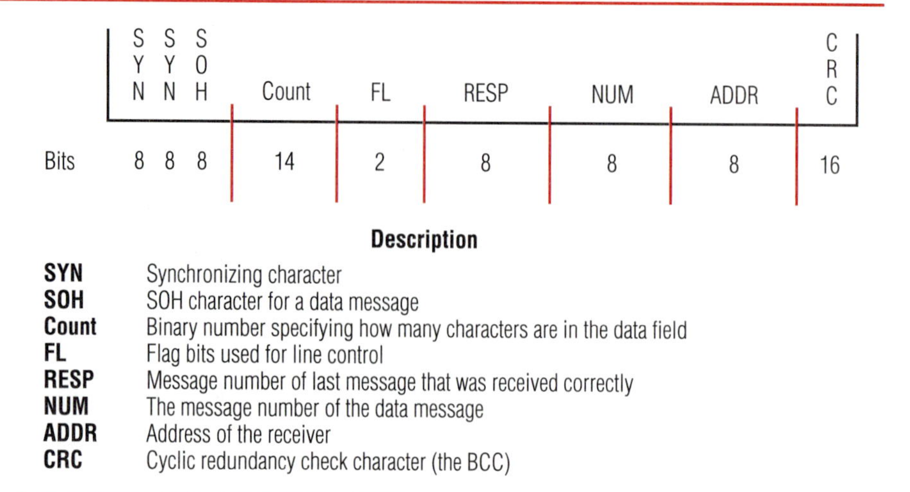

Figure 10–7
The fields in the header of a DDCMP data message.

message from one station can contain an ACK or NAK for a message received from the other station. A station can send several messages in sequence without receiving an acknowledgment, but if the receiver cannot handle them, they may have to be retransmitted. However, there can never be more than 255 outstanding messages before an acknowledgment is made.

Message Sequence Numbers One significant difference between DDCMP and BISYNC is the implementation of a message sequence number. With DDCMP, each transmitted message is assigned a unique and increasing sequence number. When the receiver provides an acknowledgment, it only needs to indicate the sequence number of the last message received correctly. This tells the sender that all messages up through that message have been received and checked. The implication is that not all messages need to be specifically acknowledged. This adds to the efficiency of the DDCMP protocol.

Compared to BISYNC, DDCMP and other byte-count protocols implement transparency in a much more efficient manner. Furthermore, with message sequencing, full-duplex operation is easily implemented. An example of the DDCMP protocol in a half-duplex, point-to-point environment is shown in Figure 10–8.

High-Level Data Link Control (HDLC)

High-Level Data Link Control (HDLC) is the most important single data link control protocol. Not only is it widely used, but it is also the basis for several other important protocols that use the same or similar formats. HDLC is based on IBM's *synchronous data link control protocol (SDLC),*

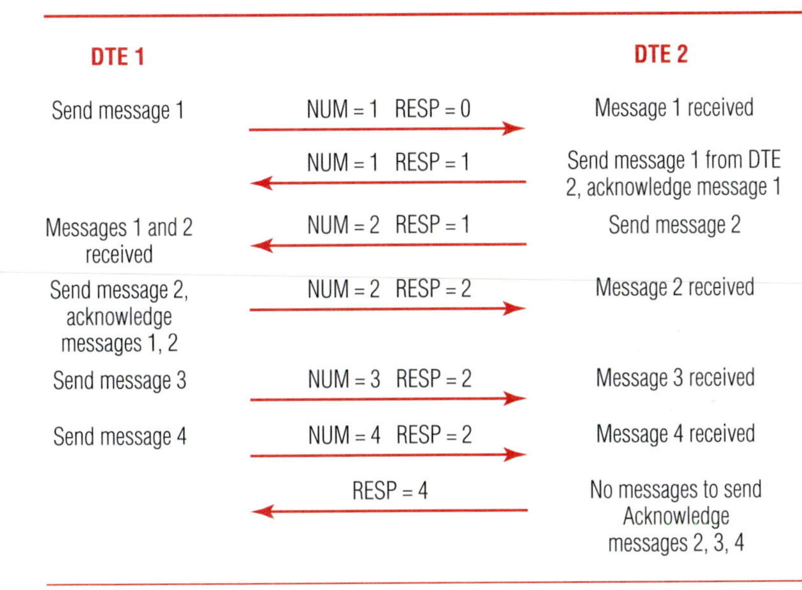

Figure 10–8 DDCMP protocol: half-duplex, point-to-point.

which is widely used in IBM mainframe environments. After developing SDLC, IBM submitted it to various standards organizations and ISO modified it to create HDLC. The ITU-T modified HDLC to create the LAP and LAPB protocols, and the IEEE modified HDLC to create the IEEE 802.2 protocol. All of these protocols will be discussed later in the text.

Operating Modes HDLC operates in one of three modes:

Normal response mode (NRM)—NRM is used when there is a primary node and one or more secondary nodes on a circuit. The primary node polls the secondary nodes. Secondary nodes may only send data in response to a poll. Normally used on multipoint but may also be used on point-to-point circuits. SDLC only operates in normal response mode.

Asynchronous balanced mode (ABM)—ABM is used with nodes that are peers: that is, either node may initiate a transmission on the circuit without receiving permission from the other. ABM is the most widely used of the three modes because it makes the most efficient use of full-duplex lines because there is no polling overhead.

Asynchronous response mode (ARM)—ARM is used when there are primary and secondary nodes, however the secondary nodes may initiate transmission without permission from the primary. The primary node still retains responsibility for the circuit, including initialization, error recovery, and disconnection. ARM mode is rarely used.

Figure 10–9
HDLC frame format.

Fields	Beginning Flag 01111110	Address	Control	Data	Frame Check Character	Ending Flag 01111110
Bits	8	8 or more	8 or 16	8 × n Any multiple of 8 bits	16 or 32	8

Frames The basic operational unit for HDLC is a *frame,* as shown in Figure 10–9. Frames are sent across a network to a destination node that verifies their successful arrival. HDLC defines three types of frames:

the supervisory or S format;

the information or I format;

the unnumbered or U format.

Supervisory frames provide the mechanism for error handling. Information frames contain the information that is being transmitted across the network. Unnumbered frames contain supplemental link control information and are used for setting operating modes and for initializing transmissions.

Flag HDLC and all bit-oriented protocols use a special character to mark the beginning and ending of frames. This special character is called a *flag,* and its unique pattern is 01111110. Because of the number of consecutive 1 bits in the flag character, it also serves as a synchronization character, so no SYNs are required. To ensure its uniqueness, no other sequence of six consecutive 1 bits can be allowed in the data stream. To accomplish this, the bit stream of all transmissions is scanned by the hardware. Using a technique called *bit stuffing,* a 0 bit is inserted after all strings of five consecutive 1 bits in the header and data portion of the message. At the receiving end, the extra 0 bit is removed by the hardware. Bit stuffing is illustrated in Figure 10–10.

Other Fields The other fields in an HDLC frame are:

- Address Field: Identifies the destination node for a frame. Normally 8 bits long, but may be longer.
- Control Field: Identifies the type of frame as either supervisory, information, or unnumbered. The control field in information frames contains send and receive sequence numbers that may be 3 or 7 bits long. Transmitting nodes increment the send sequence number, and receiving stations increment the receive sequence. As in the

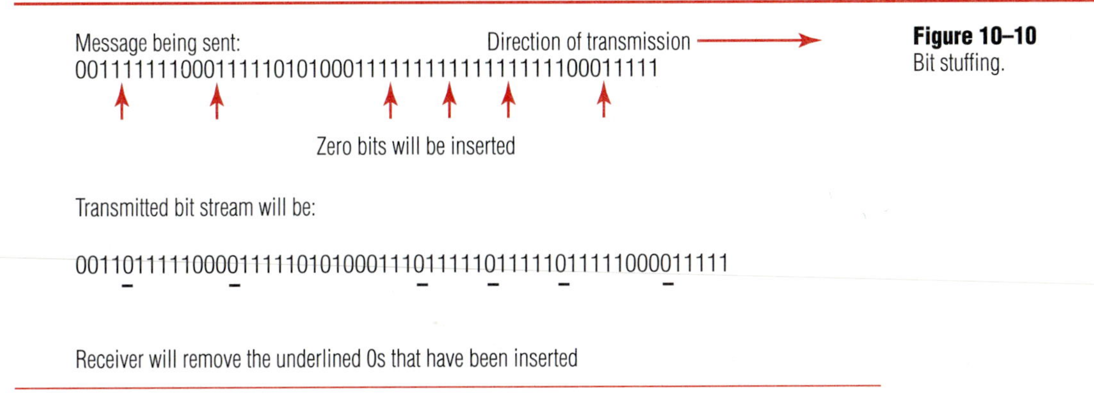

Figure 10–10
Bit stuffing.

byte-count-oriented protocols, these sequence numbers provide an additional check to ensure that no frames are missing or duplicated.

- Information Field: The information field is only present in I frames and some U frames. It is always a multiple of 8 bits in length up to a network-defined maximum. This does not imply, however, that an 8-bit coding scheme must be used. If seven 7-bit characters are sent, the 49 data bits are passed with an additional 7 bits to bring the total number of bits to 56, a multiple of 8.
- Frame Check Sequence (FCS) Field: The FCS filed is the error-detecting field that normally contains a CRC-16 or CRC-32 character.

Frame Flow The flow of frames in HDLC depends on which mode (NRM, ABM, or ARM) is used, but in all cases there is an initialization phase, a data transfer phase, and a disconnect phase. The flows can become quite complicated, especially on full-duplex circuits when both nodes may be sending and receiving data simultaneously.

OTHER DATA LINK CONTROL PROTOCOLS

There are several other important data link control protocols that you need to be familiar with. The first four are based on HDLC, while the fifth, ATM, is completely different.

Link Access Procedure, Balanced (LAPB)

The *link access procedure, balanced (LAPB)* protocol is a subset of HDLC that provides only the asynchronous balanced mode (ABM) of operation. LAPB operates in full-duplex, point-to-point mode. It is most commonly used between an X.25 DTE and a packet switching network. (Packet switching will be discussed in Chapter 11.)

Figure 10–11
A comparison of HDLC-based synchronous data link control protocols.

Protocol	HDLC Subset	HDLC Mode	International Standard	Uses
HDLC		NRM, ABM, ARM	Yes	Many types of networks when subsets are included.
SDLC	Yes	NRM	No	IBM mainframe-based networks
LAPB	Yes	ABM	Yes	Workstations connecting to a packet network
LAPD	Yes	ABM	Yes	ISDN circuits
LLC	Partial		Yes	Local Area Networks
LAPF	Partial	ABM	No	Frame relay high-speed networks operating on essentially error free digital lines
ATM—cell based	No		No	ATM high-speed networks operating on essentially error free digital lines

Link Access Procedure, D-Channel (LAPD)

The *link access procedure, D-channel (LAPD)* protocol standard was issued by the ITU-T as a part of its recommendations for ISDN networks. LAPD is a subset of HDLC and provides data link control on an ISDN D channel in AMB mode. LAPD always uses a 16-bit address, 7-bit sequence numbers, and a 16-bit CRC.

Link Access Procedure for Frame-Mode Bearer Services (LAPF)

The *link access procedure for frame-mode bearer services (LAPF)* is a data link protocol for frame relay networks. Frame relay will be discussed in Chapter 11, but basically, the frame relay technology is designed to provide streamlined capability for high-speed packet switching networks operating on digital circuits, which are assumed to have very low error rates. LAPF is actually made up of a control protocol, which is similar to HDLC, and a core protocol, which is a subset of the control protocol. The control protocol uses 16- to 32-bit addresses, 7-bit sequence numbers, and a 16-bit CRC. The core protocol has no control field, which means that there is no mechanism for error control, hence streamlining the operation of the network.

Asynchronous Transfer Mode (ATM)

Asynchronous transfer mode (ATM) is also designed to provide streamlined data transfer across a high-speed, digital, error free network. Unlike frame relay, ATM uses a completely new protocol based on a cell rather than a frame. ATM will be discussed in Chapter 11.

Figure 10–11 shows a comparison between the various HDLC-based synchronous data link protocols that have been discussed.

PROTOCOL CONVERSION

With the proliferation of data link protocols, it is often desirable to convert a data transmission from one protocol to another. This is the function of the *protocol converter,* which can be implemented using hardware, software, or a combination of both. This equipment or software takes a transmission in one protocol and changes it as required to a different protocol. The most common protocol conversion being performed today is from the protocol of one local area network to the protocol of another as a transmission passes through several networks. That conversion is typically done by one of the servers on the LAN. Another popular conversion is from an asynchronous protocol to a synchronous protocol, such as one of the variations of HDLC. This particular conversion occurs frequently because it allows personal computer workstations using an asynchronous protocol to log onto networks that use HDLC.

protocol converter

Protocol conversion involves, at a minimum, changing the control bits or characters that surround the data. Because there is not always a one-for-one correspondence of control characters in different protocols, the process is more complex than a simple character translation. Sometimes blocks of data must be reformatted or changed to a different length. The protocol converter may have to receive an entire message, made up of several blocks, before it can do the reformatting. Protocol converters also frequently perform error checking on the incoming data stream and request retransmission if an error is detected.

Another common function of protocol converters is code translation from one transmission code to another. When personal computers that use ASCII communicate with IBM mainframes that use EBCDIC, the code must be converted from ASCII to EBCDIC or EBCDIC to ASCII on each transmission.

Dedicated Hardware Protocol Converters

Protocol converters can be implemented in several ways and located at the host, the terminal, or an intermediate location in a network. This is illustrated in Figure 10–12. A *dedicated hardware protocol converter,* such as the IBM 3708 shown in the accompanying photograph, has as its sole purpose the conversion of protocols. Such devices can often handle multiple streams of data simultaneously; that is, several terminals can be connected and in operation at the same time.

Add-In Circuit Board Protocol Converters

An *add-in circuit board* may be installed in a microcomputer to perform protocol conversion. This approach is frequently used when microcomputers are used as terminals on a mainframe-based network that operates with a protocol such as SDLC. A very popular version of this type of circuit board

Figure 10–12
Different methods and locations of protocol conversion.

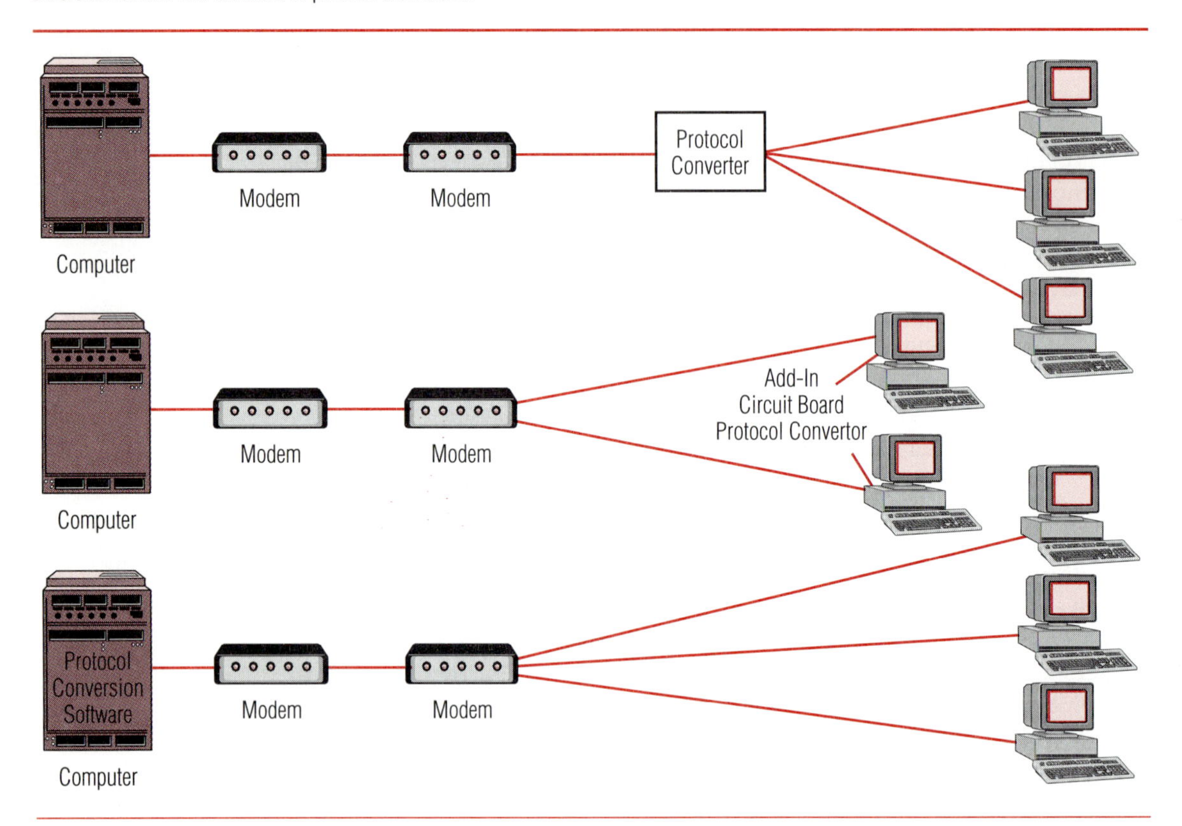

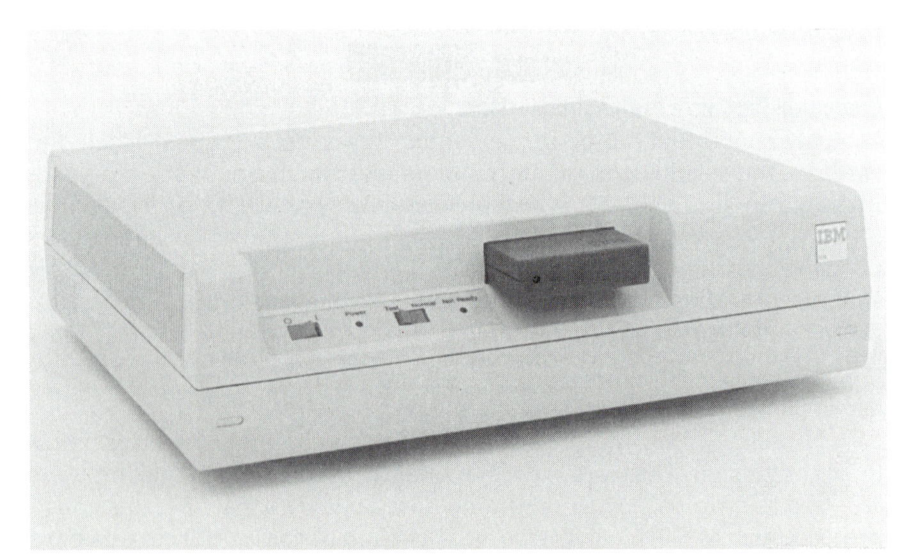

The IBM 3708 protocol converter handles up to eight dial-in lines simultaneously. Many conversion options are available, but most conversions are now done by software. (Courtesy of International Business Machines Corporation. Unauthorized use not permitted.)

makes the personal computer emulate the IBM 3270 terminal. Such boards are available from a number of vendors.

Software Protocol Converters

Software protocol converter packages are also available. These commonly run in the mainframe computer and handle the protocol conversion for all of the terminals on the network that require it.

Gateways as Protocol Converters

LAN gateways, which will be discussed in Chapter 12, often perform protocol conversion as data is passed from one local area network to another or from a local area network to a wide area network inside or outside the organization.

It should be pointed out that some protocols cannot be completely converted to other protocols. The implication is that some of the capabilities of a particular terminal may not be translatable and, therefore, may be unavailable after the translation process. This usually is not a big problem, however, it is something to be aware of and taken into consideration.

SUMMARY

This chapter looked at the rules or protocols under which data communications circuits operate. The need for circuit control was discussed, and circuit access control techniques were examined. The details of several specific protocols, all of which are in widespread use throughout industry, were explored. Protocols unique to LANs also were described. Data link protocol standardization is paving the way toward greater interconnection of terminals and computers from all manufacturers.

CASE STUDY

Protocols Used in Dow Corning's Data Network

The protocols used in Dow Corning's data communications network have undergone a significant change over time. The company's decision in the mid-1970s to adopt IBM's SNA network architecture meant that for many years SDLC was the primary protocol on the business portion of the network. At the same time, the plants and laboratories were primarily using Digital Equipment computers that communicated with the DDCMP protocol.

In the early 1990s, Dow Corning's Network Architecture Board (NAB) examined the architectural direction that Dow Corning's network should take and determined that the ability to support multiple protocols on the network would be necessary. After considerable discussion and study, the final architecture specified that for the future Dow Corning's networks would support SNA-SDLC, DECNET-DDCMP, TCP/IP, and the Ethernet and Token ring LAN protocols. The architecture specified that there must be interoperability between parts of the network using different protocols, so that, to users, the network looks and operates like a single entity. As individuals considered network upgrades, extensions, and enhancements, this architecture provided a strong guideline on which to base decisions.

In 1995–1996, the rise of the Internet and its TCP/IP protocol lead to another overhaul in Dow Corning's network architecture. For interoperability, most vendors began making TCP/IP available on their equipment, and over a 2-year period, Dow Corning changed most of its equipment to use the TCP/IP, so that today it is the main protocol in use. The FDDI protocol is used on some of the optical fiber-based backbone LANs that interconnect LANs in nearby locations, especially in the Midland area.

At the same time, the company embarked on a change of its local area networks from being primarily token ring-based to being primarily Ethernet-based. That conversion was completed in mid-2000 with the installation of new standardized personal computer workstations, which were not configured to support token ring.

Manufacturing data is processed and exchanged by a variety of computers, almost all of which are connected to Ethernet LANs. Numerous gateways and bridges are installed between LANs and they provide protocol conversion when required. When one network is connected to another, inside or outside the company, the TCP/IP protocol is used.

The details of the protocol are not something that Dow Corning's data communications people concern themselves with unless there is a problem. In that case, Dow Corning's technical communications specialists get the appropriate vendors involved and may use an instrument called a *protocol analyzer* to get a detailed listing of all of the activity on the line in order to identify the problem.

QUESTIONS

1. What factors caused the company to maintain a single network protocol for many years? What technological and business factors caused the company to modify its strategy to a multi-protocol network and then back to a single protocol network?
2. Why is the company now using a single protocol (TCP/IP) almost exclusively?
3. What costs are incurred when changing a LAN from the token ring to the Ethernet protocol?

REVIEW QUESTIONS

1. Define the term *protocol.*
2. Why are data link protocols required?
3. Identify the six functions of a data link protocol.
4. Explain the term *transparency* as it relates to data link protocols.
5. What is meant by "predetermined communications parameters"?
6. Compare and contrast contention techniques for accessing a circuit with polling techniques.
7. In a list of terminals to be polled, why might a terminal's address be listed more than once?
8. What is the function of the header of a message?
9. Why is the XMODEM protocol not a particularly reliable means of transmitting large blocks of data?
10. What advantages does the ZMODEM protocol have over the XMODEM protocol?
11. Can the BISYNC protocol transmit Baudot code? Why or why not?
12. How is character synchronization established in BISYNC?
13. Describe BISYNC's approach to error correction.
14. Give some reasons why DDCMP is more efficient than BISYNC.
15. What is the purpose of the flag in SDLC?
16. What is bit stuffing?
17. Describe the characteristics of the PPP protocol.
18. What advantages does the PPP protocol offer compared to the SLIP protocol?
19. Is the SDLC protocol an offshoot of HDLC? Explain your answer.
20. What is the purpose of a protocol converter? Why do protocols need to be converted?
21. Why are codes sometimes translated during protocol conversion? Give an example of when code translation would be needed.
22. Explain four methods of protocol conversion.

PROBLEMS AND PROJECTS

1. Make a chart of the way the XMODEM protocol operates similar to the charts shown in the text for BSC and DDCMP.
2. In what situations or applications is a predictable, known rate of access more important than having the fastest access?

Vocabulary

data link protocol
protocol
roll call polling
polling list
fast select polling
hub polling
token
header
text
trailer
start of header (SOH)
start of text (STX)
end of text (ETX)
block check character (BCC)
synchronization character (SYN)
XMODEM protocol
XMODEM-CRC

cyclical redundancy check (CRC)
XMODEM-1K
YMODEM
YMODEM-G
ZMODEM
X.PC
KERMIT
Serial Line Internet Protocol (SLIP)
Point-to-Point protocol (PPP)
character-oriented protocol
byte-count-oriented protocol
bit-oriented protocol
flag character
Binary Synchronous Communications (BSC, BISYNC)
transparent mode
Digital Data Communications Message Protocol (DDCMP)
High-Level Data Link Control (HDLC)
synchronous data link control (SDLC)
frame
flag
bit stuffing
link access procedure, Balanced (LAPB)
link access procedure, D-channel (LAPD)
link access procedure for frame-mode bearer services (LAPF)
asynchronous transfer mode (ATM)
protocol converter
dedicated hardware protocol converter
add-in circuit board protocol converter
software protocol converter
protocol analyzer

References

Bartee, Thomas C. *Data Communications, Networks, and Systems.* Indianapolis, IN.: Howard W. Sams & Company, 1985.

Lane, Malcolm G. *Data Communications Software Design.* San Francisco: Boyd & Fraser, 1985.

Stallings, William. *Data and Computer Communications,* 6th ed. Upper Saddle River, NJ: Prentice Hall, 2000.

PART FIVE

Communications Networks

Part Five brings the basic telecommunications technical concepts from Part Four together into an integrated whole, the telecommunications network.

Chapter 11 covers the fundamentals of telecommunications networks. You'll learn how communications lines are connected together to form networks. Networks will be classified several ways, and you'll see that the classifications frequently overlap. You'll see the ways in which the network can be connected to a computer. The functions of front-end processors are discussed. Two real-world network architectures are introduced.

Chapter 12 explains local area networks in detail. The reasons why local area networks have become popular are discussed, as are the hardware and software that are required to make a local area network work. The need for and roles of the network administrator are also discussed.

Chapter 13 delves into network management and operations. The network operations and technical support organizations, the groups responsible for running the communications network on a day-to-day basis and solving problems as they arise, are introduced. The necessity of taking a broad approach to the responsibilities of network management is explained, and six specific responsibilities are discussed.

Chapter 14 describes the process of communications network analysis and design. A phased approach to the activity is used with emphasis on understanding and defining the requirements for the network before any design work is done.

After studying Part Five, you will have an excellent knowledge of the different types of telecommunications networks that an enterprise might use and the choices to be considered during their design, construction, and operation. A good knowledge of the material will place you on a level with many communications professionals who are working in industry today.

CHAPTER

11

Network Fundamentals

OBJECTIVES

After you complete your study of this chapter, you should be able to

- explain how communications circuits are connected to form networks;
- explain the major types of communications networks as characterized by their topology, ownership, purpose, type of transmission, and geography;
- distinguish among WANs, MANs, and LANs;
- discuss several ways that networks can be connected to one another;
- describe the functions of TCP/IP for connecting networks;
- describe several ways that messages are routed within a network;
- discuss alternative methods of attaching a circuit to a computer;
- explain the functions of the various pieces of software that help a communications network operate;
- explain the importance of checkpoint/restart processing.

INTRODUCTION

In Chapter 9, you studied the characteristics and attributes of single telecommunications circuits of various types. Now you will expand your horizon and look at how several circuits and other communications equipment can be connected and arranged into a network. The material in this chapter describes and explains layers 3 and 4 of the ISO-OSI model. These layers, the network control and transport control layers, explain how a transmission is addressed and routed through the network from its source to its destination.

It should be noted that throughout this chapter there will be references to the term *protocol*. You should recognize that throughout a network there are many protocols. Data link protocols, which you studied in Chapter 10, are the protocols of layer 2 of the OSI model. Other protocols exist at the higher layers of the model, and you will be introduced to protocols of layers 3 and 4 in this chapter.

NETWORK DEFINITION AND CLASSIFICATION

A *network* is one or more communications circuits and associated equipment that establishes connections between nodes. Usually when we think about a communications network in a business sense, we think about the way in which users or locations in the company are connected. Many companies have several networks. If the organization is large, there may be a data network connecting several locations. Within a location, for example, an office or laboratory, there may be a different network connecting personal computers or other workstations. The company may have a voice network that is separate from its data network, or the two may share certain facilities, such as T-1 or higher speed lines. There may also be a facsimile network that consists of facsimile machines connected when needed by dial-up telephone calls. With the rapid growth in the use of communications terminals in the last few years, most businesses are finding that they have multiple communications circuits and nodes to arrange into one or more networks.

Networks may be classified in several different ways, as shown in Figure 11–1. In most companies, these network types overlap. That is, one company's network may have a combination of network types. Wide area networks may be connected to local area networks. The wide area portion may be arranged into a basic star—but with some mesh connections—while the local area networks are arranged in a bus or ring configuration. You'll see how this is possible in the following pages.

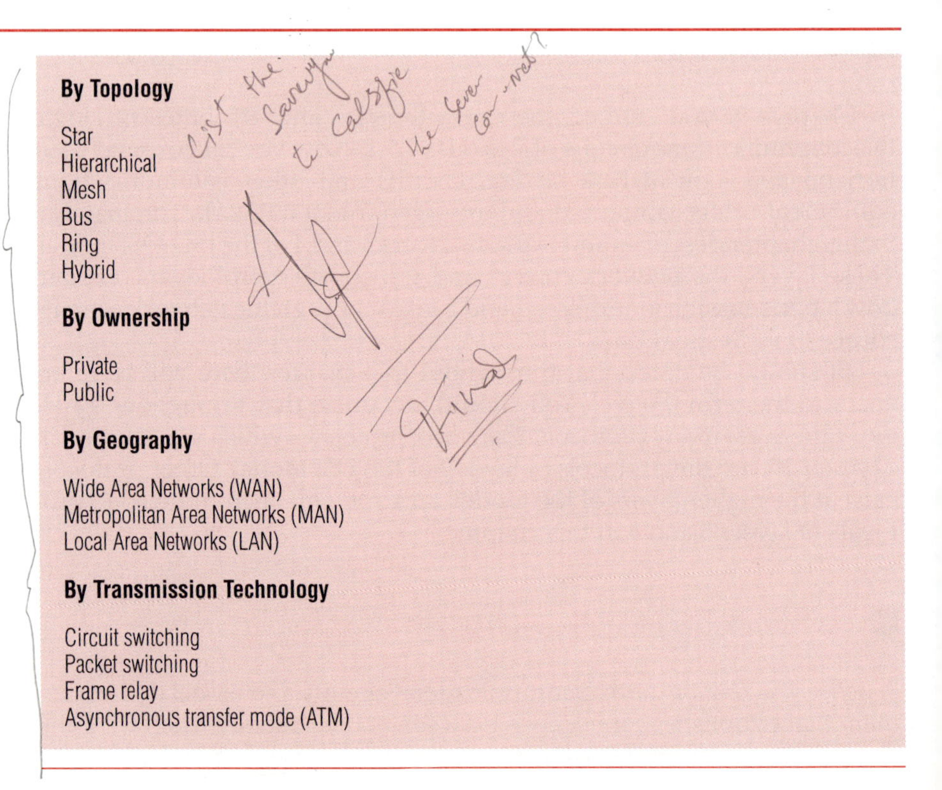

Figure 11–1
Several ways to classify communications networks.

One other concept for you to be familiar with at this point is that of a *backbone network.* A backbone network is the main network in an organization or a location. It may be the network that carries the most traffic or is the most critical. It is likely to connect other, slower speed networks to one another.

backbone network

Backbone network is a relative term. Within a manufacturing building, it may be a local area network that connects all of the machines together. That network may, however, be connected to another plantwide network that connects all of the buildings in the plant site. In turn, the plantwide network may be connected to a companywide network that connects all of the company's locations in the world. Depending on your perspective, any one of the networks might be a backbone. To the building manager, the local area network in his building would "break his back" if it were down for long. To the vice president of manufacturing who is responsible for many plants, however, the failure of a network in one plant might not be a "back breaker." In general, however, people think of the backbone network as the largest, most important network on which they regularly depend.

NETWORKS CLASSIFIED BY TOPOLOGY

The way circuits are connected together is called the network *topology.* If a map is drawn showing how all of the circuits and nodes are connected to one another but without regard for the geography of where they are located, the topology of the network can be seen. Network topologies fall into six major categories.

Star Network

A *star network* is illustrated in Figure 11–2. All circuits radiate from a central node, often a host computer or a cluster of servers. The circuits can be point-to-point, multipoint, or a combination of both. The star network puts the central point in contact with every other location, which makes it easier to manage and control the network than in some other configurations. There is no limit to the number of arms that can be added to the star or the length of each of the arms. Thus, it is relatively easy to expand a star network by adding more nodes. On the other hand, since all transmissions are controlled by the central node and they typically all flow through it, the central node is a single point of failure. If it is down, the entire network may be out of service. Another potential problem with star networks is that in times of peak traffic, the central node may become overloaded and unable to keep up with all of the messages that the outlying stations want to transmit.

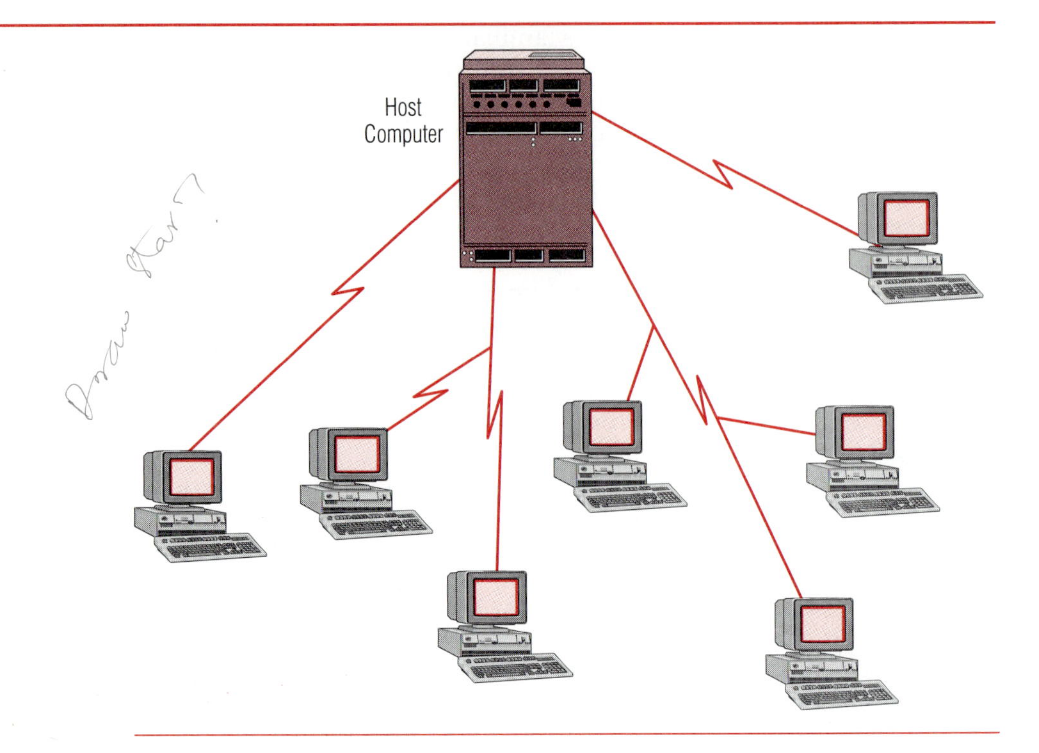

Figure 11–2
Star network.

Despite these shortcomings, many business data communications networks have a star topology. Where there is a central computer that supports many terminals, the star configuration is easy to implement. Since the central computer is the master node and controls the network, the rules for network operation are relatively simple. The star topology also applies to PBXs, to which all of the telephones connect.

Hierarchical Network

A *hierarchical network* is illustrated in Figure 11–3. It is sometimes said that this type of network has a tree structure, and the top node in the structure is called the root node. You may notice that the hierarchical structure mirrors a typical corporate organization chart, and in fact it is in this setting that a hierarchical configuration is most likely to be found. Compare, if you will, the root node to the corporate headquarters on an imaginary organization chart and the nodes immediately below that to a divisional level. Under each division, there are nodes corresponding to plants, in the case of manufacturing divisions, and to district offices, in the case of sales and marketing divisions. Under the district offices, you might find nodes in local sales offices.

This type of network would most likely be implemented where the lower level nodes at the second or third level are in themselves computers. You can envision a district office computer being connected to the lo-

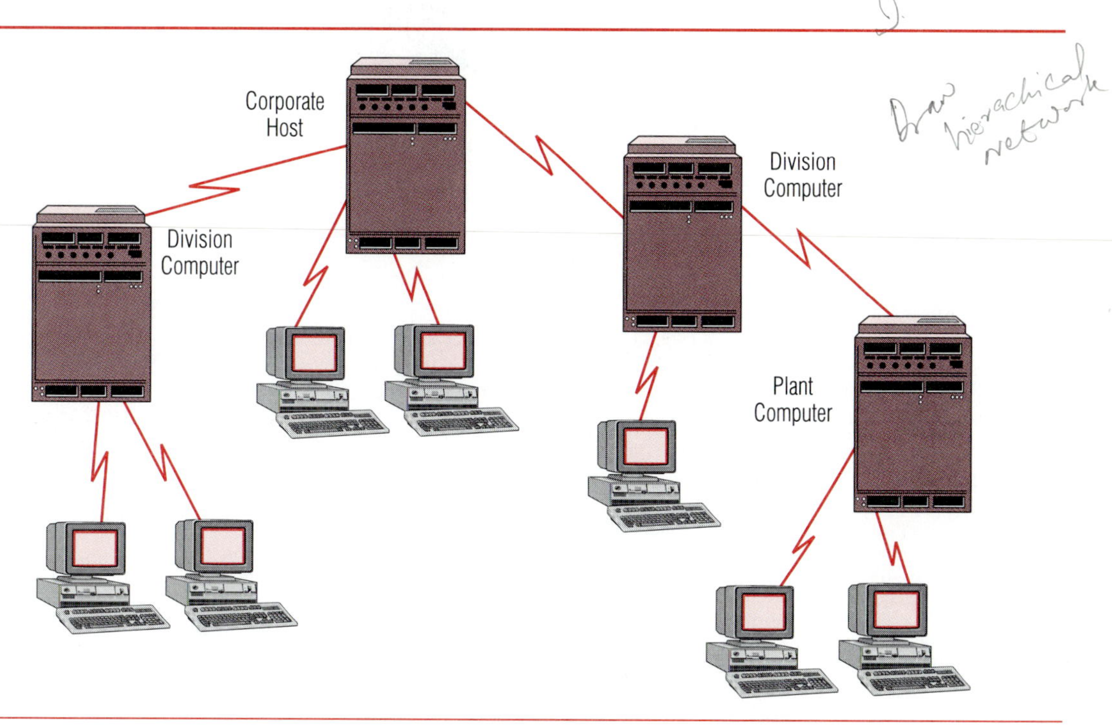

Figure 11–3
Hierarchical network.

cal sales office computers and collecting data from them. The district data would be consolidated with similar data from other districts at the divisional level. Finally, all of the divisional data would be consolidated at corporate headquarters. In this configuration, there is no single point of failure in the network. If one division's computer or network fails, the other divisions are not affected. Even if the root node at corporate headquarters fails, the divisions can go on doing their daily processing and send and receive data from lower levels in the hierarchy. The results of divisional communications and processing can be transmitted to the higher level in the network when it is restored to service.

Mesh Network

A *mesh network,* illustrated in Figure 11–4, is similar to a hierarchical network except that there are more interconnections between nodes at different levels or even at the same level. In fact, levels may not exist at all. In a fully interconnected mesh, each location is connected to every other location; however, for cost reasons, this is seldom implemented. The major nodes are usually connected, whereas minor nodes are connected to one or more locations depending on their need and criticality. The public telephone network is an example of a mesh network that, because of good design and a high level of re-

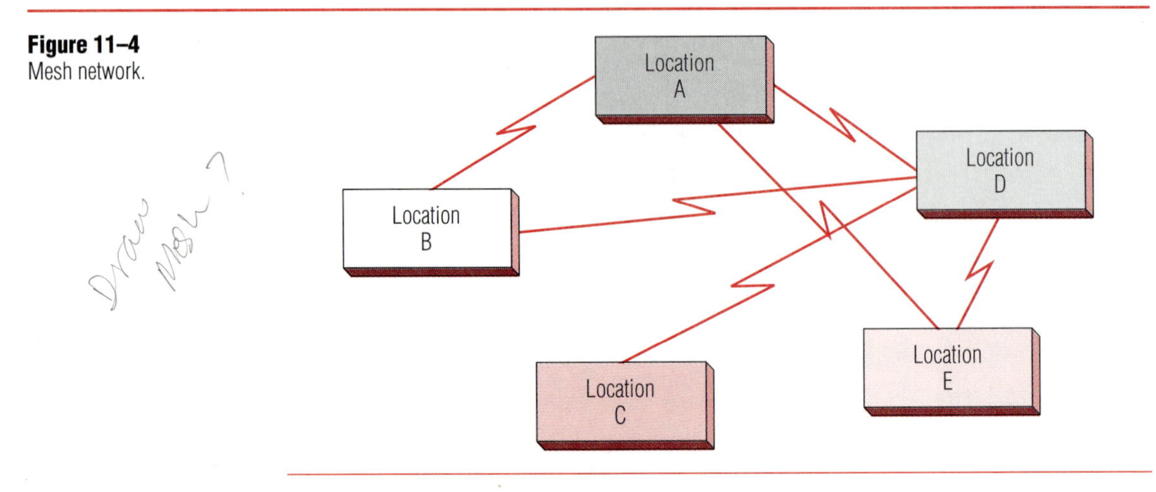

Figure 11–4
Mesh network.

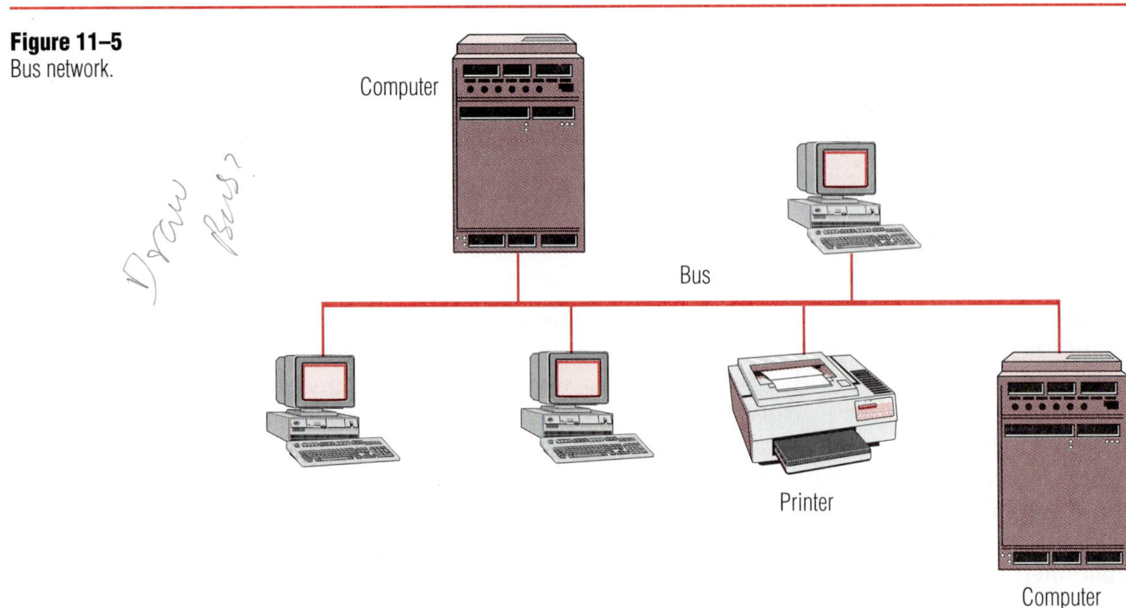

Figure 11–5
Bus network.

dundant connection, provides many alternate paths between nodes. The heavy interconnection makes the telephone network virtually fail-safe.

Bus Network

A *bus network* is shown in Figure 11–5. Conceptually, a bus is a telecommunications medium to which multiple nodes are attached. The term *bus* implies very high-speed transmission, and bus networks are usually implemented in situations where the distance between all of the nodes is limited, such as a local area network (LAN) within a department or building.

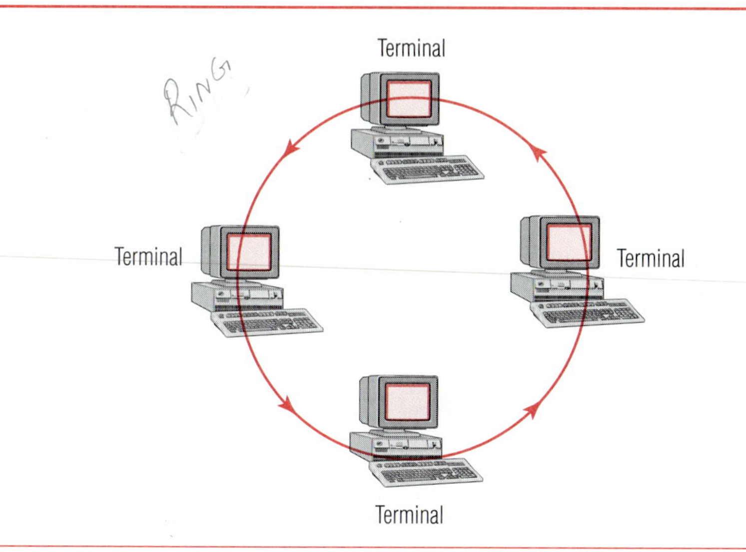

Figure 11–6
Ring network.

Devices are attached to the bus by a tap connection that breaks into the bus cable. Each tap causes a certain amount of signal loss on the cable, which is one reason bus networks have a limit as to the number of devices that can be attached to them. Another problem with bus networks is that when problems do occur, faults can be difficult to locate. Because all of the devices are connected serially on the bus, each device may have to be checked in sequence to locate the problem.

Although the bus may look similar to a star or hierarchical network, the major difference is that on a bus all stations are independent of one another. There is no single point of control or failure as in a star network. The loss of a single node on a bus has no impact on the other nodes, and, in fact, unless the bus itself fails, the reliability of a bus network is excellent. However, because of their high-speed operation, bus networks usually are limited in the distance they can traverse, which is why the topology is normally found only in local area networks.

Ring Network

A *ring network* is illustrated in Figure 11–6. Ring networks are also usually associated with networks where the nodes are relatively close together such as LANs. Each device is connected to the ring with a tap similar to the type found on a bus network. As communication signals pass around the ring, a receiver/driver unit in each device checks the address of the incoming signal and either routes it to the device or regenerates the signal and passes it on to the next device on the ring. This regeneration is an advantage compared to bus networks; the signal is less subject to attenuation. Furthermore, each device can check the signal for errors, which allows for more sophisticated error control and network management.

Figure 11–7
Ring network with two channels.

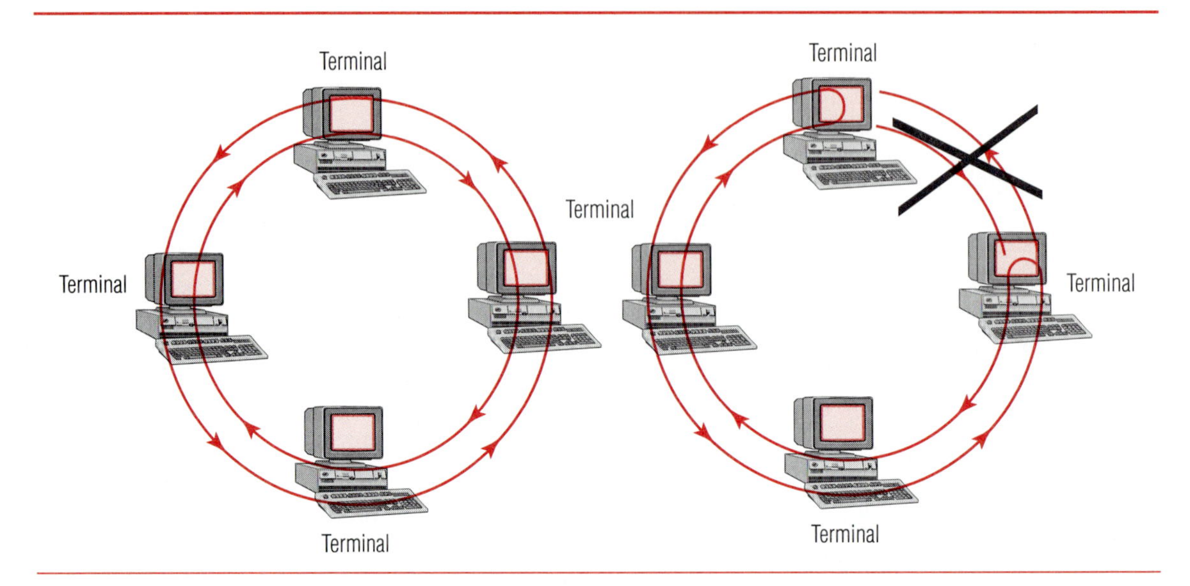

All of the stations on the ring are essentially equal and must be active participants to make the ring work. If one of the nodes fails, the potential exists for the entire ring to be out of service because messages cannot be passed through the failing node. Figure 11–7 illustrates a ring with two channels that transmit the data in opposite directions, which is the way rings are normally implemented. This allows transmission to continue even if one of the links or nodes fails. When that happens, the nodes on either side of the failure reconfigure themselves to begin using the second ring to transmit data in the opposite direction, as shown in the figure. All stations can be reached from either direction by transmitting on one ring or the other.

Hybrid Networks

The network configurations discussed so far can be combined into *hybrid networks.* For example, a star network might have a ring on it, as shown in Figure 11–8, or a hierarchical network could have a star radiating from one of the nodes. As long as all of the networks use the same protocol and basic operating technology, connecting them is easy. If they use different protocols, conversion devices are needed (they will be discussed in Chapter 12).

In practice, most networks are hybrid networks of one type or another. Only very small networks, such as might be found in a small office, use a single topology. The different topologies, combined with other network attributes that will be discussed in the next sections, meet diverse telecommunications needs, and most organizations find that they require a combination of capabilities.

Figure 11–8
Hybrid network.

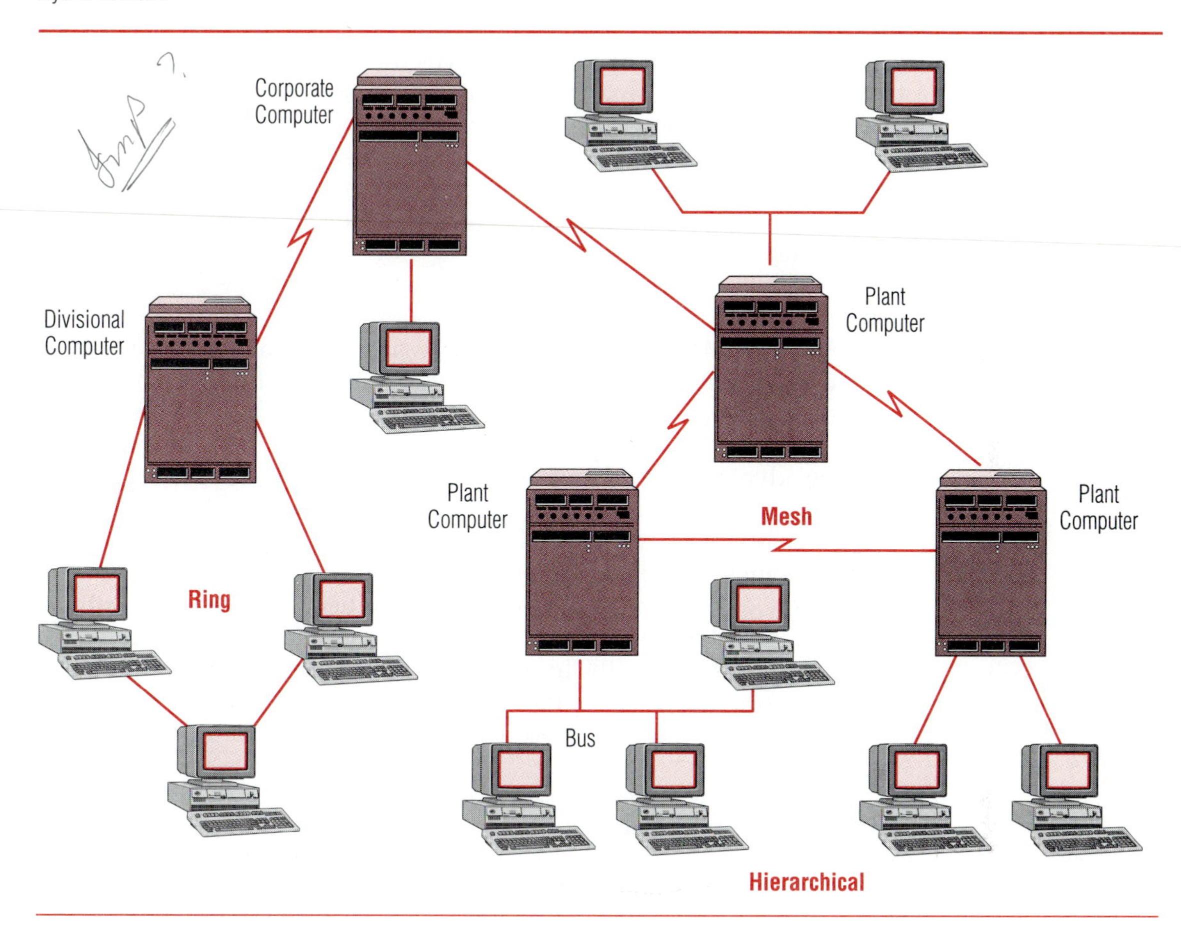

NETWORKS CLASSIFIED BY OWNERSHIP

Another broad way of classifying networks is by their ownership. Two broad categories exist—private networks and public networks. We will look at each category and its variations.

Private Networks

A *private network* is built by an organization, normally for its exclusive use. The network is built from circuits available from a variety of sources and may include a combination of privately installed and operated circuits and leased or switched facilities. There also may be connections to one or more of the public networks. One of the advantages of a private network is that it can be designed to specifically address the data and/or voice communications requirements of the organization. Since it is built around particular traffic patterns or communications flows, it can potentially make better use of circuits than a public network could. Another advantage is that it gives the company full control of the network's operation and potentially better security. Communications using a private network may, however, be more expensive than comparable communications on public networks. Because of the flexibility to tailor the private network to a company's exact requirements, however, most major companies have one or more private networks within or connecting their locations.

Public Networks

A *public network* is a network built and owned by a communications company or a common carrier for use by its customers. The two most familiar examples of public networks are the one we use to make telephone calls, the Public Switched Telephone Network (PSTN) and the Internet.

The advantage of a public network is that it provides services or access to locations that a company might not otherwise be able to afford. Companies frequently use public data networks to exchange messages with their small locations, such as a single-person sales office, where the installation of a private line cannot be justified. The communications company or common carrier may have many similar customers in the area and can spread out the implementation and operational costs of providing service among all of its customers. The carrier can achieve good utilization of its network while at the same time providing high-quality service to infrequent users.

Value Added Networks (VANs)

A *value added network (VAN)* is a particular type of public data network that, in addition to offering transmission facilities, contains intelligence that makes the basic facilities better suited for satisfying the communications needs of a particular type of user. The intelligence might provide code or speed translation, or it could store messages and deliver them at a later time (store-and-forward). This intelligence provides the "added value" from which the generic name for this type of network is derived.

The intelligence in a VAN is provided by computers located at network nodes that may assist in the routing of messages or perform other communications-related processing, such as code translation or speed conversion. The computer may also perform more sophisticated processing related to the business of its subscribers.

Because computers are employed for code conversion and storing messages in the public telex network, it has become a VAN. Other examples of industry-oriented VANs include the SWIFT network that connects international banks, the SITA network that connects the airlines' networks, the IVANS network connecting many U.S. insurance companies, and, in the United Kingdom, the Tradanet service that connects major retailers and their suppliers. Each of these networks contains the intelligence to perform certain types of processing for its users.

NETWORKS CLASSIFIED BY GEOGRAPHY

Yet another way of classifying networks is by the geographic expanse they cover. The most common designations are the wide area network (WAN) and the local area network (LAN). A third category, which is less well defined, is the metropolitan area network (MAN).

Wide Area Networks (WANs)

Wide area networks (WANs) are generally understood to be those that cover a large geographic area, require the crossing of public rights-of-way, and often use circuits provided by a common carrier. Wide area networks may be made up of a combination of switched and leased, terrestrial and satellite, and private microwave circuits. Since carrier facilities are used, almost any circuit speed can be used. The usual topology of WANs is star, hierarchical, or mesh.

The WAN for a large multinational company may be global, whereas the WAN for a small company may cover only several cities or counties. One large bank has built a private WAN to link its offices in Massachusetts, Hong Kong, London, Munich, and Sydney. It reports that circuits operating at speeds of 56 or 64 kbps provide a 3-second response time anywhere in the world. Most international financial organizations have similar networks and must have them if they are to compete. Manufacturing companies that operate on a national or international scale have similar networks to link their offices and plants. Information about sales orders, forecasts, inventory, expenses, and so on is routinely exchanged on a daily or more frequent basis. Every industry has numerous examples of WANs used to make organizations more productive and competitive.

Metropolitan Area Networks (MANs)

A *metropolitan area network (MAN)* spans an area ranging from a few buildings to 30 miles or so, though the distances are not precise. MAN technology has been standardized by a committee of the IEEE, which developed a standard for MANs known as IEEE 802.6. The work grew out of the committees that defined LANs and their unresolved concerns about how LANs could be connected across distances of a few kilometers or greater. Figure 11–9 shows a typical MAN connecting several LANs in locations scattered around a city. Note that voice traffic is also carried on the MAN in the figure, a capability that is more common with MANs than with either WANs or LANs. High bandwidth between the locations on

Figure 11–9
A metropolitan area network (MAN).

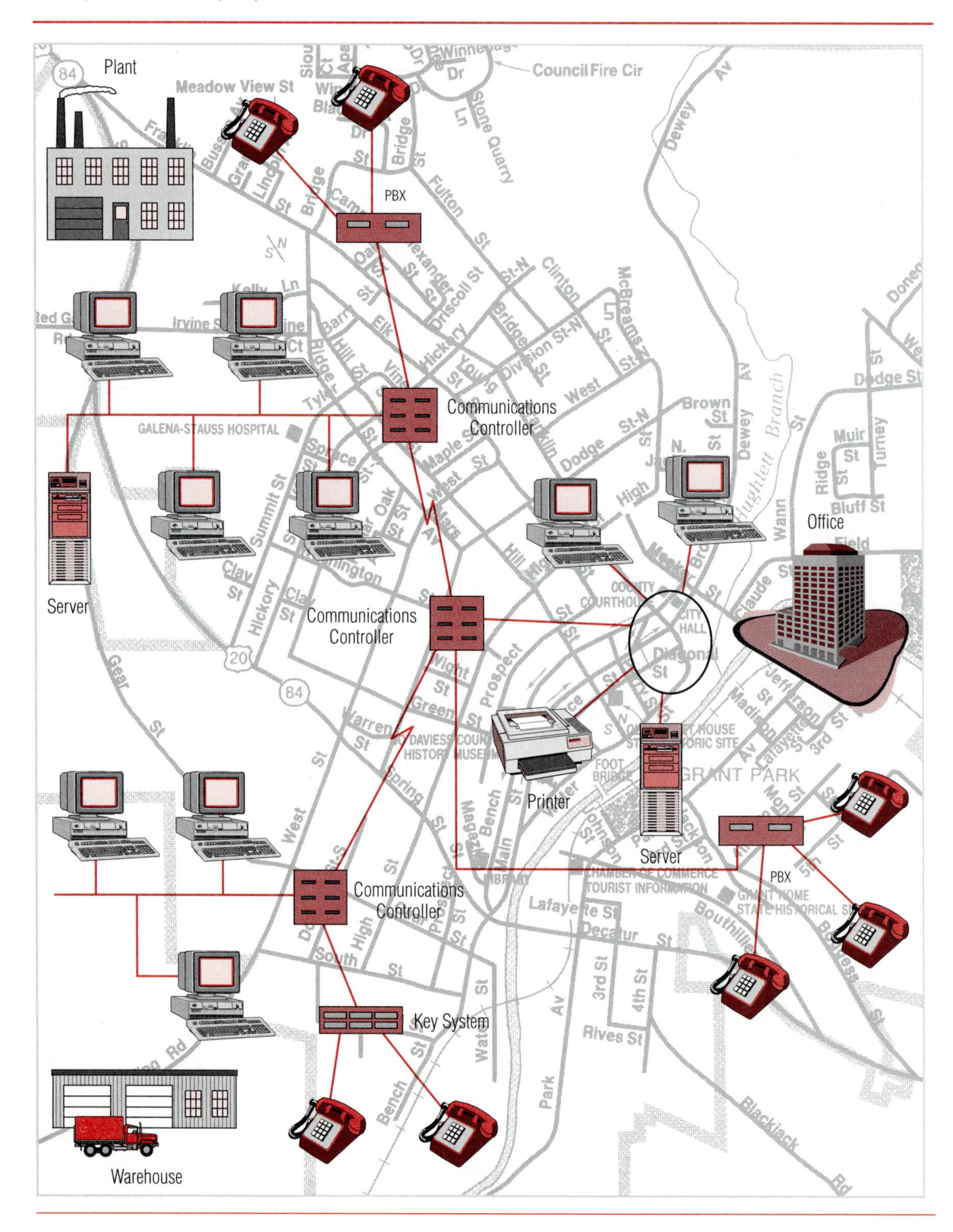

the MAN is almost always a requirement, especially if the MAN is connecting LANs, which always operate at high speed. Voice traffic would also demand high bandwidth, depending, of course, on how much is carried. T-1, T-3, or fiber circuits are common, but cost is a consideration. A MAN may be owned and operated by a single organization or it may be shared by several organizations.

In the past few years, a number of small companies have begun business by installing optical fiber cable in metropolitan business areas. Rights-of-way are obtained in many ways—on telephone poles, in subway tunnels, under city streets, and so on. These companies primarily sell high-speed digital bandwidths and typically offer no value added or other services. If the media used is optical fiber cable, the term frequently applied is *dark fiber* because only the fiber cable itself is supplied. The customer is responsible for providing all equipment to drive and operate the circuit, including multiplexers, CSU/DSUs, and so forth.

dark fiber

The primary market is the customer that needs a lot of high-speed digital service, such as multiple T-1 or T-3 circuits within a relatively small, often metropolitan, geographic area. These customers are often large companies with multiple offices or other locations within a city. The MAN providers typically offer lower prices than the telephone companies and offer diverse routing, providing backup in emergency situations. They also claim to offer quicker installation and better service than the telephone companies.

Local Area Networks (LANs)

Local area networks (LANs) have become extremely popular and are almost certainly the most common type of network. Because of the capabilities they offer, they are widely installed in organizations of all sizes. The discussion of LANs is extensive and deals with many topics beyond those usually associated with WANs. For this reason, Chapter 12 is dedicated to complete coverage of LANs.

NETWORKS CLASSIFIED BY TRANSMISSION TECHNOLOGY

Networks can also be classified by the type of transmission technology they use. Since you will be studying LANs in Chapter 12, this discussion will focus on the technologies that are used in wide area and metropolitan area networks. To help visualize some of the concepts, it is useful to refer to the network shown in Figure 11–10.

Circuit Switching

Transmission of data (including voice, image, and video) in wide area networks normally occurs on a series of communication links that connect

Figure 11–10
Circuit switching networks build temporary connections between devices that wish to communicate.

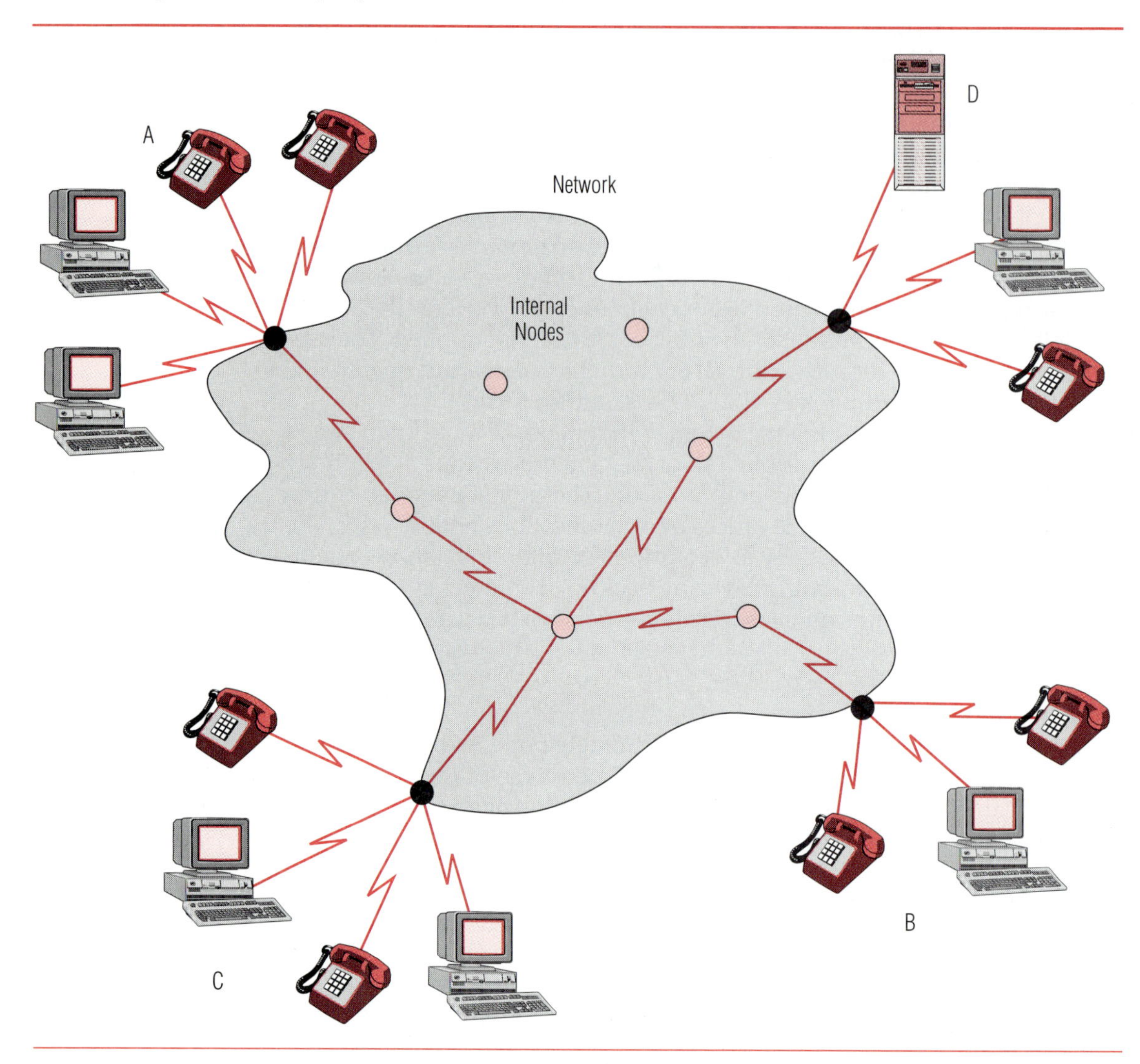

nodes in the network. Some of the nodes have user stations connected to them, represented by the small dark circles in the cities in Figure 11–10. Other nodes, represented by the open circles, are internal to the network and do not serve user stations directly. Communication by *circuit switching* means that there is a dedicated communications line between the two stations, but perhaps only for a short time. This is in fact how the telephone network works. When you make a call, there is a period of time after you dial the number, called call setup time, when the switches in the network determine how to route your call and then make the connections to establish a temporary dedicated path for your call to use. This is represented by the connection between telephone A and telephone B in the drawing. Similarly, switched circuits can be established for data transmission, as shown by the connection between terminal C and computer D. At the end of the call, the connection path is "torn down" and the links become available for other calls.

Circuit switching is reasonably efficient for voice calls, because conversation occurs almost constantly during most calls—although a voice conversation does not use the entire capacity of a circuit by any means. For data calls, the circuit is used less efficiently, which you can imagine if you think about a terminal-to-computer connection. Even if the typist is reasonably fast and the computer's response time is good, the circuit is still idle a lot of the time.

Circuit switching was developed for the public voice network, and of course it is used for data transmission as well, but there are more efficient ways to transmit data.

Packet Switching

Packet data networks (PDN) are based on *packet switching* technology in which messages are broken down into fixed-length pieces called *packets* and sent through a network individually. Packet switching networks are designed so that there are at least two, but usually several, alternative high-speed paths from one node to another, as shown in Figure 11–11. A message from Detroit to San Francisco might normally be routed through Chicago. However, if the Chicago node or the link from Detroit to Chicago is down or extremely busy, the computer node in Detroit would know, using internal tables, to route the message via St. Louis instead. This provides a redundant, fail-safe capability and implies that the route a specific message takes is in part a function of the condition and traffic on the network when the message is sent.

packets

Packets and Packetizing When packet switching is used, messages are segmented into packets of a predetermined size before they are transmitted. This process is called *packetizing*, and packets are most commonly less than 1,000 bytes long. The packetizing is performed by hardware or software called the *PAD*, which stands for *packet assembly/disassembly*. The PAD may exist in the users' DTE equipment or in the PDN's node. At the receiving end, another PAD assembles the message from the packets, which may, by the way, arrive out of sequence.

packet assembly/disassembly

In addition to the original data, each packet contains an address field, which is added by the PAD to give the packet's destination or a data stream identification field that tells which data stream the packet belongs to. Other checking or control fields may also be added to the packet to ensure data integrity.

Figure 11–11
A hypothetical packet switching network (heavier lines indicate multiple or higher speed circuits).

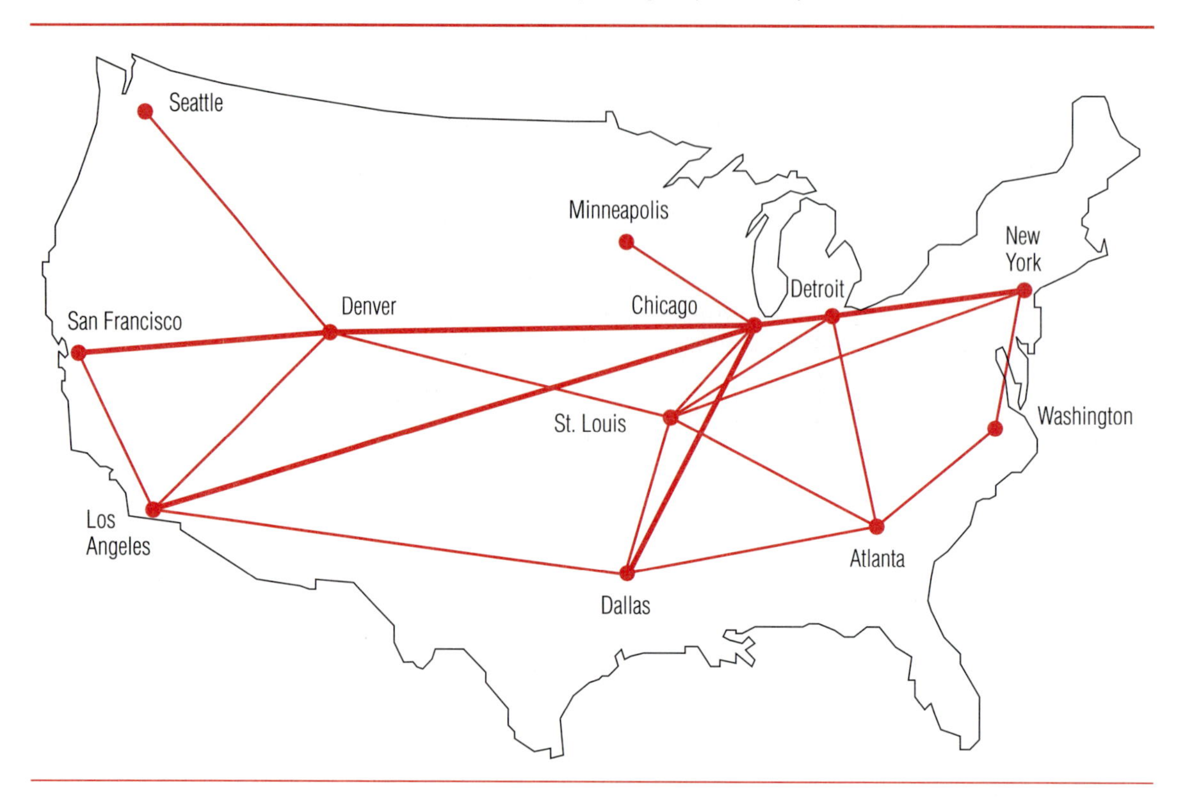

virtual circuit

Connection Types Messages can be sent through the network in one of three ways. Most commonly, the user or terminal equipment establishes a dial-in (switched) connection with the nearest node of the PDN. The destination node is identified, and the PDN establishes a *switched virtual circuit* between the sending and destination nodes. This virtual circuit exists only for the duration of the session, which may be for one message or for a long exchange between a user and a host computer. When the session is finished, the virtual circuit is dissolved.

If two nodes communicate frequently, a *permanent virtual circuit* may be established. Permanent virtual circuits save the time it takes to establish a connection between two nodes, a useful attribute if communication traffic is heavy and frequent. Many companies have permanent virtual circuits between their private host computers and one or more PDNs. The PDNs provide access to electronic mail systems or to other companies for electronic data interchange, electronic mail, or other transactions.

datagram

A third type of connection called *datagram* service is available in PDNs. Datagrams are short messages of one packet or less, which can potentially

This packet node uses interchangeable cartridges to provide specific operating characteristics for packet networks. (Courtesy of Black Box Corporation)

traverse the network at high speeds. However, according to international standards, neither the delivery nor the arrival order of the packets is guaranteed. The user is responsible for detecting missing packets and initiating retransmission if necessary. Because of these characteristics, datagram service is not widely used outside of a few specialized applications.

Packet Transmission Through the PDN Packets are usually sent through the network somewhat autonomously. Packets from the same message may be sent on different routes and may arrive at the receiving node out of sequence. Computers at each network node route the packets through the network toward the destination node. The PAD at the receiving node is responsible for reassembling the packets into the correct sequence before passing them on to the user.

Packets from different messages—indeed, from different users—are interspersed on the network at any point in time. A packet switching network does not establish an actual physical link between two nodes that wish to communicate. The only direct connection is between the user's terminal and the nearest node of the packet network at each end of the connection.

Since a direct connection is not established between the sender and receiver on a packet switching network, there can be delays in communications. Therefore, the use of packet switching networks for interactive access to computers must be studied carefully to be sure that the response time that the packet network can deliver will be adequate. If the primary application is electronic mail, messages that do not require instantaneous delivery, the use of a packet switching network may be ideal.

Packet switching technology has been standardized by the ITU-T's recommendation X.25, which defines the interfacing between terminals and the PDN operating in packet mode on a public network, or layers 1 to 3 of the OSI model. The X.25 standard also applies when an organization establishes a private PDN. When two PDNs interface with one another, they follow the ITU-T X.75 standard, and there are several other ITU-T standards relating to PDN networks.

Packet switching networks have been operational since 1969, having been first developed by the Advanced Research Projects Agency (ARPA) of the U.S. Department of Defense. The first packet switching network was known as ARPANET, and it was used to connect ARPA research centers. The ARPANET later evolved into the Internet. The first commercial packet switching network, known as Telenet, became available in the United States in 1975.

Many public packet switching networks have been established in the world. In the United States, Tymnet and General Electric Information Services (GEIS) are two of the principal networks. In Canada the packet switching network is called Datapac; in France it is called Transpac; and in Japan the network is called DDX-2. All of these networks provide access in major cities in their country, and virtually all of the networks are interconnected, following the X.75 standard, so that it is possible to send messages to subscribers in other countries. Users of these networks generally pay according to the amount of traffic (number of packets) they send. There may be a fixed subscription fee to enable access to the network, but the basic charge is variable.

PDNs provide another alternative for an organization's message or data communication. If, for example, a company wanted to provide its customers with access to its computer or to information, it could install a leased circuit (permanent virtual circuit) from its computer to the nearest PDN node. Customers all over the world could contact their local PDN node, usually via a dial-up telephone call; and, with proper authorization, they would be able to sign on and retrieve information. Similarly, salespeople or other employees with terminals could access the company computer while traveling simply by making a telephone call to the nearest PDN node and then logging on.

Yet another alternative for a company is to install a private PDN network. Several vendors sell node computers and software allowing a company to build its own network that uses packet switching. Leased or dial-up lines may be employed, depending on the traffic volumes between specific locations. Many organizations have chosen this method for handling their internal messages or data traffic, particularly where rapid response time is not required.

Public packet networks provide many capabilities that make them attractive for a variety of applications:

- reliable service—because of multiple connections and alternate routing between nodes;
- nationwide service—public packet switching networks may be accessed from most locations in the world with a local telephone call. For high-volume users, leased lines may be connected directly to the packet switching node;
- low error rates—within the packet switching networks, the data is fully checked and retransmitted when it is incorrect. The actual error rate is primarily dependent on the local loops at the ends of the connection;
- variety of transmission speeds—although originally designed for 300 bps or 1,200 bps asynchronous communications, packet switching networks today are capable of transmitting data at 64,000 bps and higher using both asynchronous and synchronous techniques;
- cost-effectiveness—the user of a packet switching network pays a connection charge for each fraction of an hour of connection to the network and a packet charge for each packet transmitted. Other charges exist for leased-line connection to the network and for dedicated dial-up ports. In many cases, however, the total cost of using a packet switching network is less than the cost of other alternatives.

Frame Relay

Packet switching and PDNs were developed in the 1960s and early 1970s when data transmission circuits were not as reliable as they are today. Hence, packet switching technology has a large amount of error checking and control built in, which often proves to be unnecessary today, especially with the use of digital circuits with low error rates.

Frame relay, which was briefly introduced in Chapter 10, was developed to reduce the overhead of packet switching and provide more efficient data transmission, and it was designed to operate on multi-megabit speed lines. It is a data link (layer 2) protocol that divides messages into variable-length *frames* for transmission through the network. A frame has a 1-byte header, 2 addressing and control bytes, a variable-length data field (from 1 to 64,000 bytes), a 16-bit cyclic redundancy check, and a terminating byte. Frame relay hardware at each end of a connection builds and disassembles frames out of other kinds of data streams, such as asynchronous terminal input, local area network packets, and digitized voice signals, and writes them out serially on the frame relay network. The variable-length frames give the equipment great flexibility in dynamically allocating the bandwidth to multiple data streams and changing priorities as required by the applications.

Frame relay has approximately one-quarter of the overhead of basic packet switching. Unlike packet switching, which requires an acknowledgment for every packet from every node through which it passes, frame relays keep track of the addresses of all of the nodes through which they travel, and the final destination sends an acknowledgment back through the same path on the network by looking at the node addresses stored in the received frame.

circuit congestion

committed information rate

Because the frame relay protocol has been streamlined to make it fast and efficient, it has less ability than some other protocols to deal with *circuit congestion*—too much traffic on the circuit. Queuing theory shows that if traffic on a circuit gets too heavy the throughput drops quickly and dramatically, so it is in the best interest of all users to keep circuit congestion under control. Frame relay systems accomplish this control by applying the concepts of a *committed information rate (CIR).* A company that subscribes to frame relay service leases a circuit of a certain speed, sometimes called the *port speed,* to the carrier's office. The port speed is the maximum rate at which data can be transmitted. The customer contracts with the carrier for a certain CIR measured in bits per second, which is usually up to one-half of the port speed, and the price paid is determined by the CIR. The basic mechanism for relieving congestion in a crowded frame relay network is to discard frames. In normal operation, the user can send data up to the port speed but if the network gets congested, the vendor may discard frames that exceed the CIR. For example, a frame relay user might acquire a circuit with a port speed of 256 kbps and a CIR of 64 kbps. When necessary, data can be sent at up to 256 kbps, but anything over 64 kbps is subject to being discarded if the network gets congested while that data is being sent. Any data sent at less than 64 kbps is within the CIR and normally gets through, although it should be pointed out that there is no guarantee, and if the network becomes extremely heavily congested, even frames within the 64 kbps CIR may be discarded. If the network is properly designed, this situation should be rare, however.

The majority of frame relay networks deployed today are provided by telecommunications carriers that offer data transmission services. This is referred to as *public frame relay service,* and the switching equipment is located at the central offices of the carrier. Subscribers are charged based on their use of the network but are relieved from administering and maintaining the frame relay network equipment and service. Some organizations are deploying *private frame relay networks.* In private frame relay networks, all of the equipment is owned by the enterprise, and the network's operation, maintenance, and administration are its responsibility too.

Asynchronous Transfer Mode (ATM)

Frame relay dramatically increases the throughput of packetized data transmissions; however, there can still be delays in the network that frame relay systems cannot prevent. Some data, such as realtime voice or video, is not very tolerant of delays. You can imagine how you would feel about the telephone system if it inserted delays into your speech and the person you were talking to thought that you were stuttering or talking with very unnatural short pauses!

Asynchronous transfer mode (ATM), sometimes called *cell relay,* was developed to effectively eliminate the problem of the delays. (Note: Don't confuse this ATM acronym with the one discussed earlier that stands for automatic teller machine, a type of terminal used in the banking industry!) ATM is an evolution of frame relay. It reverts to fixed-length packets of 53 bytes called cells. Fixed-length *cells* can be assembled and disassembled more quickly than variable-length frames, thus improving on the efficiency of frame relay. ATM is designed to operate with circuits that operate at 45 Mbps (T-3) or higher speeds.

The high-speed circuits, fast-switching speeds, and fast processing of the fixed-length cells combine to effectively eliminate delays in transmission and make ATM suitable for voice processing. A constant rate data channel can be guaranteed even though packet switching technology is being used. If a user needs to deliver synchronized video and sound, ATM is a technology that will meet the requirement. Digital voice, data, and video information can simultaneously travel over a single ATM network at high speeds using varying amounts of bandwidth as needed.

ATM networks are typically designed to be able to handle many types of traffic simultaneously, such as voice, video, and realtime data. Although each data stream is handled as a stream of 53 byte cells, the way in which the flow of cells is prioritized and handled depends on the characteristics of the flow and the needs of the application. For example, voice or realtime video must be handled with almost no delay.

To meet the transmission requirements of diverse applications, ATM has five service categories that are used by an end system to identify the type of service required from the ATM network. The categories are:

Realtime service

- Constant bit rate (CBR)
- Realtime variable bit rate (rt-VBR)

Non-realtime service

- Non-realtime variable bit rate (nrt-VBR)
- Available bit rate (ABR)
- Unspecified bit rate (UBR)

The *constant bit rate (CBR)* service is used by applications that require a fixed and continuously available data rate such as would be available on a leased or private circuit. Examples of CBR applications include telephone connections, videoconferencing, and others that involve people at both ends of the connection communicating in realtime. Delays on these kinds of communications are annoying and often make the communication unproductive and frustrating to the participants.

The *realtime variable bit rate (rt-VBR)* service is also used by applications that require minimal delays in the transmissions, however the distinction is that the transmission rates tend to be somewhat bursty, such as might be found with compressed video transmissions.

Applications that are candidates for non-realtime variable bit rate (nrt-VBR) service are those such as automobile or airline reservation systems and financial transaction systems. These applications have a requirement for fast response time, but, compared to voice applications, can tolerate some delays, especially if they are infrequent. With this service, the user specifies the peak and average cell rate that is required and a measure of how bursty the traffic is. With this information, the network can allocate resources to provide relatively low delay and minimal cell loss.

After the available circuit capacity is allocated to the above three services, the remainder is available for the *available bit rate (ABR)* service. An application that uses ABR specifies the peak cell rate it will use and the minimum cell rate it requires. The network allocates resources to ensure that the ABR application receives at least its minimum cell rate. An example of an application that is a good candidate for ABR is a LAN-to-LAN connection. If the ATM circuit cannot immediately handle the traffic, it can be buffered in the LAN until circuit capacity is available.

After ATM circuit capacity is allocated to all of the above services, there may still be some capacity available. This unused capacity is available for the *unspecified bit rate (UBR) service,* which uses the ATM circuit on an "as available" basis. UBR traffic is forwarded on a first-in-first-out basis using capacity not consumed by other services, so both delays and cell losses are possible. This service is suitable for applications that can tolerate those delays and possible cell losses, which is typical of TCP-based traffic.

ATM combines the strengths of traditional packet switching—bandwidth efficiency—with those of circuit switching—high throughput, low delay, and transparency. The technology and capabilities of ATM networks are still evolving, and they merit close attention by organizations that have a need for high capacity circuits with a wide variety of data flow types (e.g., voice, video, computer data). Because of their high capacity and flexibility, ATM circuits tend to be expensive, so the cost compared to the needs of the applications must be understood before the use of an ATM circuit can be justified.

ROUTING MESSAGES IN THE NETWORK

Many wide area networks today have several paths that a message could take from its source to its destination. Routing messages through the network is one of the important responsibilities of the network layer (layer 3)

	Static	Dynamic
Centralized	Relatively simple Least flexible Single point of failure Potential performance bottleneck	More flexible Reacts to changing traffic conditions
Distributed	No single point of failure Routing table updates may be a burden on the network	Most flexible Most complicated to implement
Broadcast	Simplest Adequate with small network	Ineffective with moderate to heavy traffic loads

Figure 11–12
Advantages and disadvantages of various network routing techniques.

of the OSI model. There are many ways to route messages, ranging from fairly simple techniques suitable for small networks to elaborate, adaptive techniques that only large networks need and can afford.

Connection-Oriented and Connectionless Routing

Some messages sent through a network can be sent as a single unit because they will fit in a single packet or frame, but many messages must be broken into pieces to fit the parameters of the transmission method. The application doesn't care how the transmission layer handles the messages, but the application at the receiving end does expect to receive an entire message, regardless of how it might have been divided for transmission. In some networks, as we have seen, a *virtual circuit* is built between the sending and receiving nodes and then the entire message, or all of its packets if it has been divided, can be sent at one time on the virtual circuit. This is called *connection-oriented routing.* In other networks, each packet is sent independently, and all may travel to the destination on different routes. This is called *connectionless routing,* and when it is used, the network layer at the receiving end must reassemble the message as the packets arrive, including handling those that arrive out of sequence.

In either case, the network hardware and software must decide how to build the virtual circuit or how to route the individual packets—the process called *routing.* Figure 11–12 shows the attributes of the major routing techniques.

Broadcast Routing

One very simple technique, *broadcast routing,* used by the CSMA/CD protocol, is to broadcast all messages to all stations on the network. The station for which the message is intended copies it. All other stations ignore it.

Centralized Routing

As networks grow beyond a few stations, however, broadcast routing becomes impractical. The next level of sophistication is *centralized routing*. A central or control computer keeps a table of all the terminals on the network and the paths or routes to each. All traffic flows to the central computer, which then uses the table to determine the best route for the message's destination.

Centralized routing is used in star or hierarchical networks, where a central computer controls all communications flows. All but some of the newest IBM-SNA networks use this approach, since for years it was the only routing technique available in SNA. Centralized routing is relatively simple, but it is normally static. Routes are established when the software table is created, and they do not normally change based on network traffic loads or communications line failures. Furthermore, the central computer can become a performance bottleneck or, worse, a point of failure. Despite these disadvantages, centralized routing is used in many networks and can be very successful. Planning is required, however, to ensure that the proper routes are built into the software table.

Distributed Routing

Distributed routing places the responsibility for building and maintaining routing tables on at least some of the nodes in the network. Each node that performs routing is responsible for knowing which paths or links are attached to it and their status. The node also must advise other nodes in the network when the status of paths changes so that they can update their tables. This transmission of path status information can put a communications burden on the network.

Distributed routing avoids the single point-of-failure problem associated with centralized routing. However, it is more complicated to implement, especially when the practical problems of updating routing software in each of the nodes are considered.

Static and Dynamic Routing

Static routing means that the route between two nodes on the network is fixed and can only be changed when the network is taken down and software tables are modified. Messages must always be sent on the predefined route. If the network is down, messages must be held. Some static routing schemes allow alternate paths to be statically defined so that a limited backup routing capability exists.

Dynamic routing is considerably more flexible in that each node chooses the best path for routing a message to its destination. Consecutive messages to the same destination can be routed by different paths, if necessary, and the node can adapt to changing network traffic volumes, error rates, or other conditions. As would be expected, dynamic routing is more difficult to implement, particularly when it is combined with distributed routing, but the benefits may make the implementation cost worthwhile.

In reality, many of today's networks use a combination of dynamic and distributed routing, especially when several LANs and WANs are involved. Very frequently there is more than one path that a packet or message can take between the source and destination nodes. Routing techniques have been the subject of considerable research for many years, and there are many tradeoffs that must be considered when trying to design an appropriate routing strategy. One of the major contingencies that routing algorithms must deal with is congestion on one or more of the circuits due to heavy traffic. Another situation that must be handled is circuit failure, and alternate routing when primary circuits are out of operation. Additionally, any routing technique requires some processing at each node in the network. More elaborate routing algorithms require more processing, and the designer must be sure that the sophistication of the algorithm and resultant processing does not impair the efficiency of the transmission by delaying it unnecessarily. The penalty of such overhead needs to be less that the benefit accrued.

INTERCONNECTING NETWORKS

Companies frequently have several networks that they would like to connect into a single logical network. Additionally, organizations frequently want to connect their networks to the networks of other organizations. From a user's point of view, she wants to be able to communicate with other people or processes on a wide variety of networks, both inside and outside of the organization, and not worry about how the networks are connected. In other words, users want transparency so that they can access all of the resources they need for their job or other purposes in essentially the same way. WANs, LANs, and MANs all need to be interconnected, and as the ability has become available, the term *internetworking* or *internet* has been applied.

internet

Transmission Control Protocol/Internet Protocol (TCP/IP)

Connecting a wide variety of networks, which have been developed to meet different standards, and by different vendors, was a challenge the U.S. government undertook in the late 1960s and early 1970s under the Advanced Research Projects Agency (ARPA). The transmission control protocol/internet protocol (TCP/IP), protocols that operate at OSI layer 3, the network layer, and layer 4, the transport layer, were developed by ARPA to connect incompatible computers and networks used by government agencies, the military, government suppliers, and researcher institutions. As you saw in Chapter 3, TCP/IP has a layered structure that is similar in concept to the OSI model, but different in implementation.

internet

subnetworks

An *internet* is an interconnected set of networks that may simply appear as a larger network from a user's point of view. Since the interconnected networks may maintain their own identity, they are usually referred to as *subnetworks.* In internet terminology, each subnetwork supports the devices connected to it, known as *end systems.* In addition, subnetworks are connected by devices known as *intermediate systems* which provide a communication path and perform necessary communication, relaying, and routing of messages. A simple internet is illustrated in Figure 11–13. Two types of intermediate systems are *bridges* and *routers,* which will be described in Chapter 12.

The *transmission control protocol (TCP)* functions at OSI layer 4, the transport layer and is implemented at each end of a connection. TCP ensures that entire messages are delivered error-free with no loss or duplication. For example, if an application wants to send a file over the Internet, the TCP divides the file into one or more packets, numbers the packets, and then forwards them individually to the internet protocol (IP) program layer. At the other end, TCP reassembles the individual packets and waits until they have all arrived correctly before sending the file to the application program or user.

The format of a TCP packet is shown in Figure 11–14. Whereas IP only handles packets of data, TCP has the responsibility, on the application's or user's behalf to achieve reliable transfer of the entire message. In addition, TCP allows users or applications to indicate the urgency or priority of their transmission and to assign a security classification to it.

The *internet protocol (IP)* is also implemented in the end systems and each intermediate system of an internet. IP performs the functions of OSI layer 3, the network layer. Each computer or node on an internet has at least one address that uniquely identifies it from all other computers on that internet. On the sending side, IP treats a packet that is passed to it from TCP as data and surrounds it with its own header and trailer information. At the receiving end, IP strips off the header and trailer information and passes the rest to TCP so that at both ends of the transmission, TCP is working with the same packet. Two formats of an IP packet are shown in Figure 11–15. IP version 4 (IPv4) is still in wide use today, but IP version 6 (IPv6) is replacing it. The primary reason for the change is to increase the size of the address from 32 to 128 bits in order to accommodate the already widely used and still rapidly growing use of internetworking. The IP packet format has also been simplified in version 6 to allow it to be processed more quickly and efficiently by the nodes on the network.

Because a message is divided into a number of packets, each packet can, if necessary, be sent by a different route across the internetwork. Packets can arrive in a different order than they were sent. IP just delivers them. It's up to TCP to put the packets back in the right order.

As was seen earlier, networks may communicate using either connection-oriented or connectionless techniques. IP is inherently connectionless,

Figure 11–13
A simple internet.

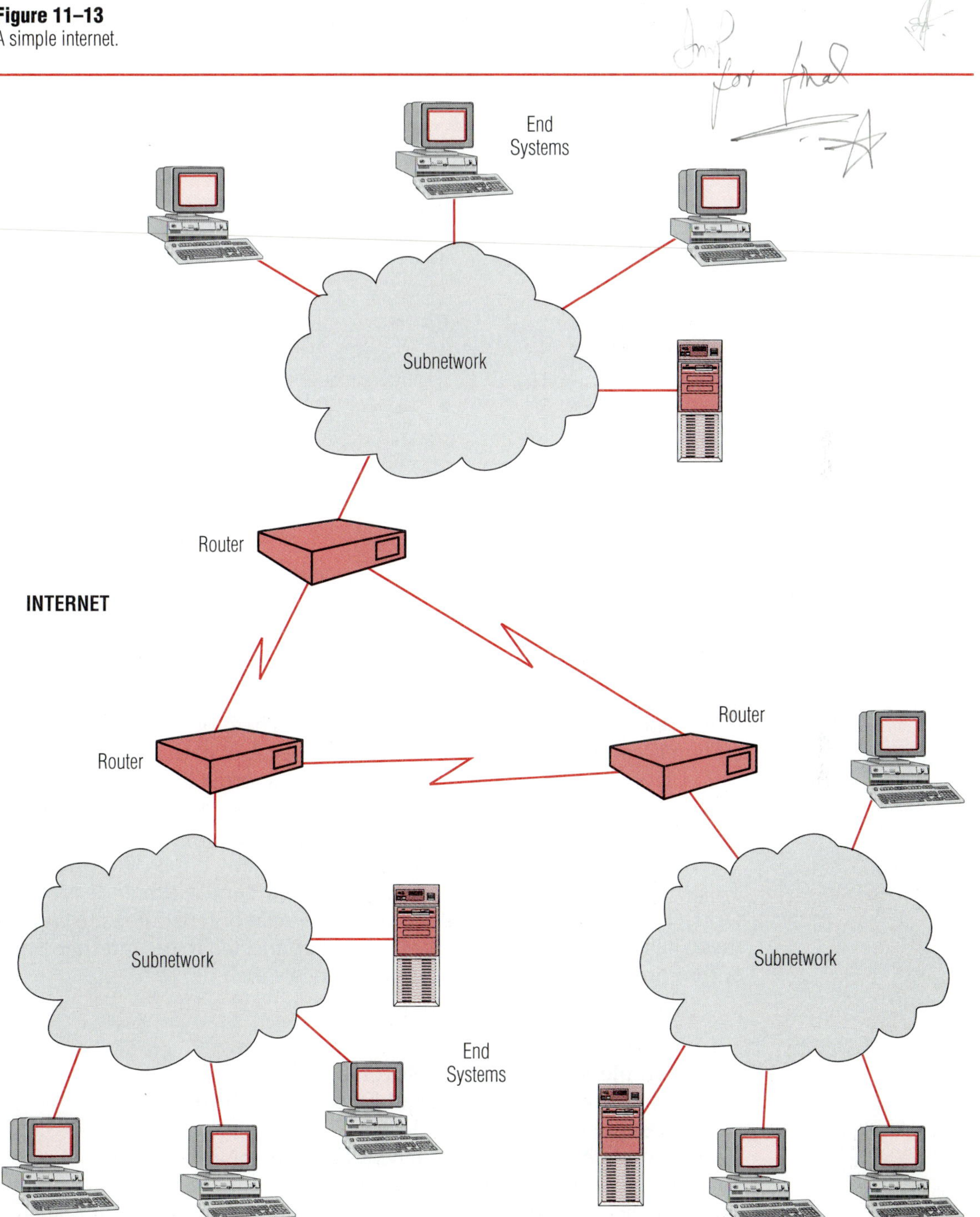

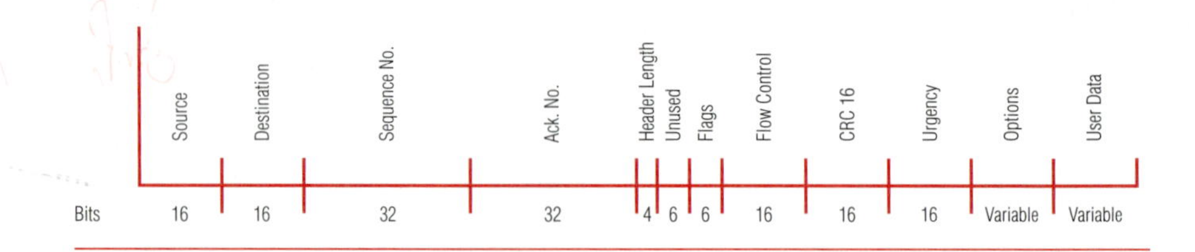

Figure 11–14
TCP packet formats.

Version 4

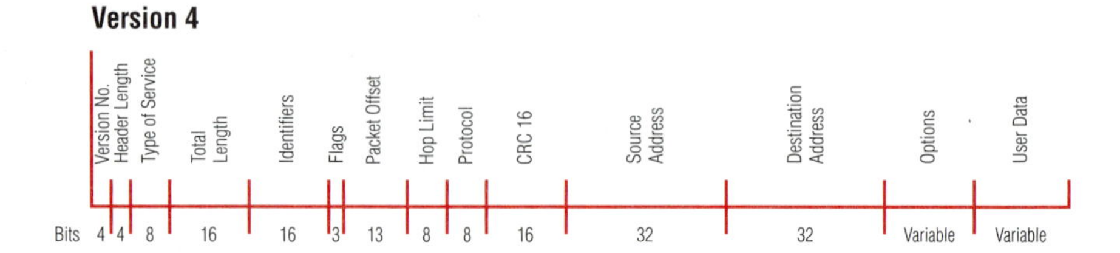

Version 6

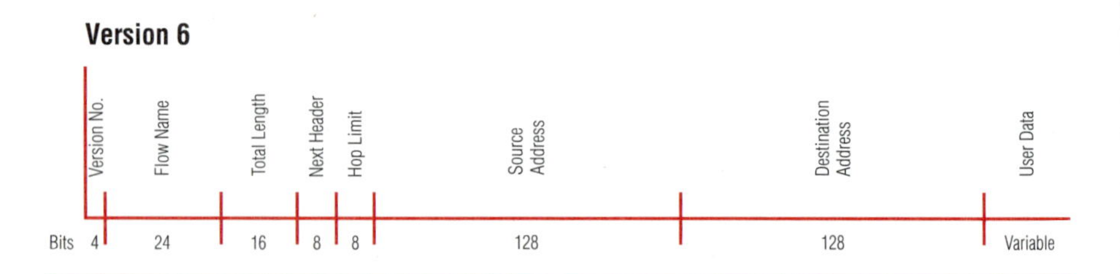

Figure 11–15
IP packet formats.

while TCP is inherently connection-oriented. Therefore, if the network operation is connectionless, only IP need be used, but in that case, the application layer is responsible for performing the error handling and other functions that the TCP layer normally performs. When connection-oriented routing is used, both TCP and IP are required.

IP is an efficient layer 3 protocol used for many types of applications. This efficiency allowed IP to function over low bandwidth wide area network (WAN) links, which led many commercial organizations to connect their sites using TCP/IP, even before they connected to the Internet. Some of the common TCP/IP application protocols include:

- File Transfer Protocol (FTP);
- Telnet;
- Simple Mail Transfer Protocol (SMPT);
- Multipurpose Internet Mail Extensions (MIME);
- Remote Procedure Call (RPC);
- Simple Network Management Protocol (SNMP).

These application protocols enable common network applications such as file sharing, electronic mail, and network management.

SIDEBAR 11–1

THE HOME DEPOT: NETWORKING TO BUILD BETTER CUSTOMER SERVICE

"In our business, customer service is everything," said Dave Ellis, vice president of Information Services for The Home Depot. "We need to provide the sales associate with all of the tools necessary to provide superior service . . . and meet the customer's need."

Founded in 1978, The Home Depot operates more than 950 stores in the United States, Canada, Puerto Rico and Chile, and plans to double that figure by 2003. The stores handle products from 11,000 vendors, with 50,000 SKUs per store and another half-million specialty-item SKUs sold regionally. The numbers are staggering and communications needs continue to grow at a phenomenal pace.

The challenge facing the company was how to supply more bandwidth to the stores, provide employees with access to existing information, and facilitate the introduction of new tools and resources while keeping costs under control. The solution was found in AT&T's Integrated Network Connection Service.

"We are now positioned to use 100% of our bandwidth 24 hours a day in the most leveraged way possible," explains Ellis. "The AT&T Integrated Network Connection Service infrastructure provides the means, distributing information from our store support center to the associate. And once you get the information in a format that's readily available, you see an exponential increase in the use of it."

AT&T Integrated Network Connection Service has become the core fabric for communications to Home Depot's U.S. locations, with nearly all applications converged over the network, allowing Home Depot to enhance network performance, simplify operations and reduce communications costs.

"With AT&T providing the equipment on our premises, we've been able to forget about the technology piece of it and really focus on the business benefit," said Dan Haumann, senior manager, Information Services, The Home Depot. "The type of applications that we're going to be able to put in the store environment over the next couple of years is just mind-boggling."

Home Depot's new sales training process is just one example of how the company has reacted to the service. Image-rich applications include product pictures and specifications, multiple vendor information, and electronic catalogs. Sales associates can now quickly and easily place special orders or include special instructions that require communications with suppliers, such as contacting an outside source for carpet installation.

Another key example is Home Depot's Special Services System, which handles special orders for products outside a particular store. According to Haumann, 99.5% of special-order information used to be in-store. Now, nearly the opposite is true.

"At the store, you enter customer information once and it follows through every step of the transaction," said Haumann. "Therefore, the transaction goes much faster. The customer is more satisfied, and you can take care of more orders. The lines aren't as long and it's a better use of people, which affects product price. With Integrated Network Connection Service, technology takes time out of the process."

The Internet

The *Internet* was conceived in the era when timesharing was widely used on large mainframe computers, but has survived into the era of the personal computer and workstation, client-server and peer-to-peer computing. It was designed before local area networks existed but has evolved to accommodate LANs, as well as more recent frame relay and ATM services. It started as the creation of a small group of researchers and has grown to be a commercial service with billions of dollars invested annually.

If it is possible to categorize the many diverse uses that people make of the Internet, the following four may provide a summary:

- electronic mail;
- remote log in to various computers;
- discussion groups;
- information search and retrieval.

Electronic mail (e-mail) has been discussed previously, but basically, anyone who is connected to the Internet may send e-mail to anyone else if the e-mail address is known. E-mail addresses have the general form of userid@computer.domain. User IDs may generally be chosen by the person requesting the ID, so last names and initials are common. Computer names are assigned when the computer is registered to be connected to the Internet. Common domain names in the United States are:

- edu for educational institutions;
- com for commercial organizations;
- org for nonprofit organizations;
- gov for government organizations;
- mil for military units;
- net for organizations that have an administrative responsibility for the Internet.

Other countries sometimes assign similar domain names, but also often add a country code, so a typical Australian internet address might be hmobson@dccomp.co.au, where *au* indicates Australia.

telnet

The *telnet* command is used to log on to another computer on the Internet. Using telnet is similar in concept to dialing in to the remote computer using a modem. After the connection is made, the user must know a user ID and password that is valid for the remote computer. Some organizations do not allow telnet capability because of the security threat it implies, so it is not automatically available on every computer connected to the Internet.

newsgroups

Discussion groups, as the name implies, are groups of people who get together to talk about particular topics—in fact, virtually any topic is discussed on the Internet! *Usenet newsgroups* are the most common form of discussion group. After joining one, each time you log on you are told about new messages since the last time you connected. You can read any or all of the messages and respond to them if you want. The discussions are made available to every computer on the Internet, but network administrators for each computer have the option of making them available to users or not.

Information search and retrieval has been occurring on the Internet for many years, but it was not until the advent of the World Wide Web (WWW) in 1995 that it became popular with the general public. The reason is that the WWW made searching and retrieval much easier and intuitive by adding a graphical interface to what previously had been pretty much a text-only capability. Search programs such as Gopher and Archie, which were described in Chapter 4, were supplemented by more powerful searching tools, called *search engines,* which are far easier to use. Commonly used search engines include Yahoo, Alta Vista, and InfoSeek.

search engine

The main components of the WWW are *web servers* and *web browsers.* Web servers are computers that store data in *pages* that are formatted with a tool called *Hypertext Markup Language (HTML).* HTML is a very flexible formatting tool that allows the author of a web page to specify colors, fonts, the inclusion of objects such as photographs, images, or movies, and to define links to other pages, which may be stored on the same server or anywhere on the WWW. Web pages are addressed with a *Uniform Resource Locator (URL),* an address that specifies the location and format of the page. (URLs were introduced and described in more detail in Chapter 4.)

HTML

URL

Web browsers are programs that can read and display HTML-formatted pages. A user enters a web site at the top page in the hierarchy, called the *home page.* The home page indicates what information is available, and then, by using the HTML links, the user can jump to other pages at the site or to other sites that have information of interest. The WWW is easier to use than it is to describe! Common web browsers available on the market are Netscape and Microsoft Internet Explorer.

home page

The multitude of web servers on the Internet contain information on every imaginable topic. What has surprised many people is how quickly the business world has adopted the Web as a new way to communicate with its customers or potential customers. Thousands of companies, large and small, have home pages on the Web, and you find them advertising their URL alongside their telephone and fax numbers.

Access to the Internet and the WWW can be made in a variety of ways. Most universities and many companies have dedicated connections to the Internet, and many of them make the capability freely available to their students or employees. The advantage of these dedicated connections is that they usually operate at fairly high speeds, which makes the transfer of information relatively quick. This is an important factor when using the WWW because the information on web servers is highly graphical and requires a large bandwidth for good response time. Dial-up connections to the Internet are also available through online service providers such as America Online, CompuServe, and hundreds of other companies that have

ISP
firewall

direct connections to the Internet. Collectively, these organizations are known as *Internet service providers* or *ISPs*. They typically charge a fee for each minute or hour of Internet access, and many of them have a flat rate for unlimited use. The limiting factor of these services is the speed of the dial-up connection, which is commonly 28.8 or 33.6 kbps, unless of course you can access an ISP with Cable TV or satellite access, in which case the connection speeds will be much faster. The slower speeds give adequate but not extremely fast response time from the WWW, and it seems that the more one uses the Internet, the less satisfied you are with the slower speeds.

Intranets

The use of web servers and browsers has been extended within organizations for private use. Many organizations have built *intranets* using the same technology used on the Internet. The difference is that the access to intranets is limited to members of the organization or others who are authorized. Intranets typically use the same network that an organization uses for other purposes, including accessing the Internet. Usually, a computer with special software, called a *firewall,* is installed between the Internet and the rest of the organization's network, as shown in Figure 11–16, to prevent unauthorized people from coming through the Internet

Figure 11–16
The location of a firewall, which provides protection against unauthorized access to an intranet from the Internet.

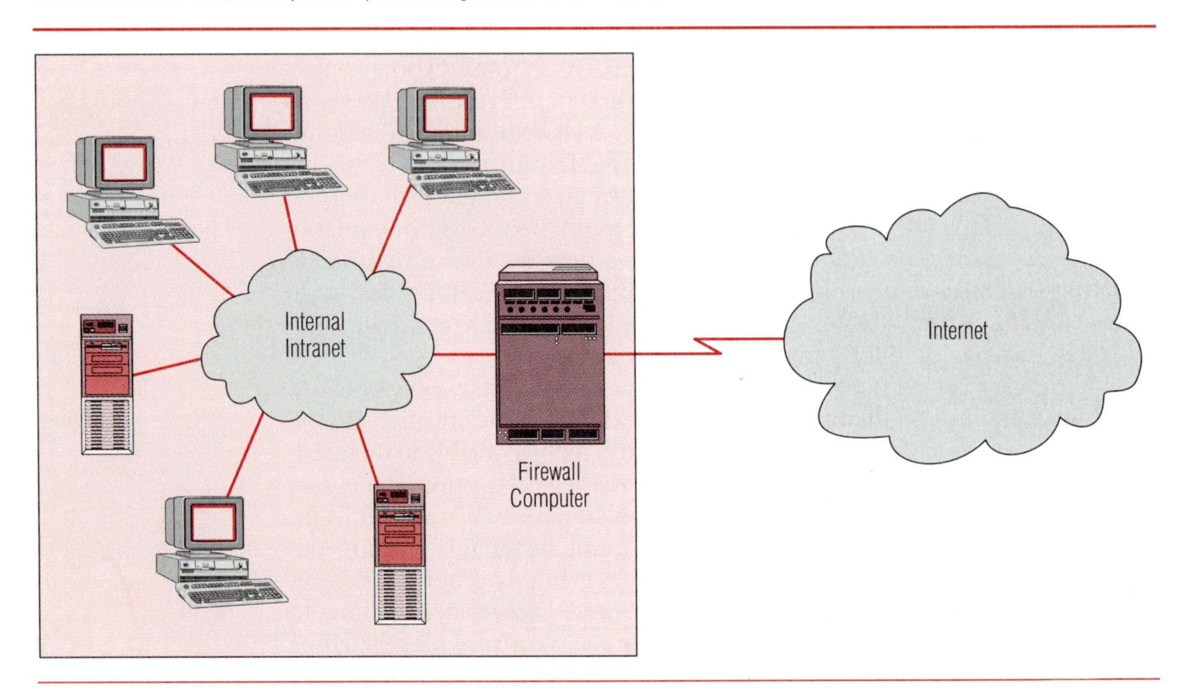

into the organization's private network and accessing the intranet or other information. The term *extranet* is sometimes used when a set of people such as customers or suppliers is allowed to pass through the firewall and access authorized sections of the intranet.

Intranets have proven to be very useful in many organizations as a communications tool for disseminating news and other information to employees for knowledge sharing, conducting internal discussions, bulletin boards—in general, anything the Internet is used for, but on a private, internal basis.

CONNECTING THE NETWORK TO THE COMPUTER

Now we will look at how a network is connected to a computer and the functions that the computer hardware and/or software must perform to make the network operate. Depending on the size, complexity, and type of network, these functions may be performed in a general-purpose computer or one of several special-purpose computers, called front-end processors, routers, or switches.

Circuit Termination Alternatives

A telecommunications circuit may be connected to a host computer in three primary ways. The first is a direct connection between the circuit and the computer so that as bits arrive they are stored directly in main memory. This is the approach used in many microcomputers. It works well when only one or a small number of circuits are connected.

The advantage of this approach is low cost. The problem is that, although there is usually specific circuitry dedicated to assembling groups of bits into characters, the CPU must be interrupted to store each character into main memory. These interruptions, called *cycle stealing*, can put a significant drain on the resources of the computer and slow down other work it is doing.

cycle stealing

The second type of connection is through a *network interface card (NIC)*, a printed circuit board that is inserted in a slot in a personal computer or other device. The NIC takes data from the transmitting workstation, forms it into the specific packet or other format required by the data link protocol used on the telecommunications line, and presents it to the circuit. On the receiving end, the process is reversed.

network inferface card

The third type of connection happens when several circuits terminate at one computer, frequently a *front-end processor (FEP)*, which is also sometimes known as a *telecommunications control unit* or *transmission control unit*. The FEP is a specialized computer that, at minimum, assembles bits into characters and characters into blocks before feeding them to the main computer. Therefore, the computer is interrupted a fewer number of times. For outbound transmission, the transmission control unit accepts

front-end processor

blocks of characters from the computer and splits them into bits for transmission. FEPs are normally used with large mainframe computers or minicomputers that operate large networks.

The FEP may go one step further and assemble blocks of incoming data into complete messages. This requires the FEP to be a programmable computer itself. If the FEP is programmable, it also can be programmed to do additional editing of incoming messages before passing them on to the computer. This approach further minimizes the number of interruptions to the main computer, but, of course, requires more elaborate hardware and software in the FEP.

Although these three alternatives have been described as if they were discrete there is, in fact, a range of possibilities between them. The line connection approach that is chosen depends to a large extent on the size of the host computer, the techniques the computer uses, and even the philosophy of the computer manufacturer. Microcomputers most often attach circuits directly to the processor through an interface card. Servers often have special circuitry in the CPU dedicated to the FEP function. Large mainframe computers usually have a separate piece of hardware that provides the FEP function.

Another set of alternatives that must be considered by the computer designer is whether the front-end function will be provided in hardware, software, or a combination of the two. On the one hand, the advantage of implementing it in the software is flexibility and adaptability. Software can be modified or corrected much more easily than hardware. New protocols or terminal types can be supported by adding program codes. On the other hand, the advantage of implementing the function in the hardware is speed of operation. Once a function has been designed and thoroughly checked out, it can be put into a circuit chip for very little cost and will operate much faster than a software implementation.

FRONT-END PROCESSORS (FEPS)

Regardless of which implementation alternatives are chosen, certain capabilities must be provided to interface communications lines with the computer. This discussion assumes that the interface has been implemented as a separate piece of FEP hardware because it is relatively easy to visualize. It is important to remember, however, that each vendor decides how it will actually implement the communications line interface.

A full-capability, stand-alone FEP is a specialized computer specifically designed to control a data communications network. It links the host computer with the network as shown in Figure 11–17. FEPs are most often provided by computer manufacturers, although there are other third-party vendors in the market. Clearly, the computer manufacturer is in the

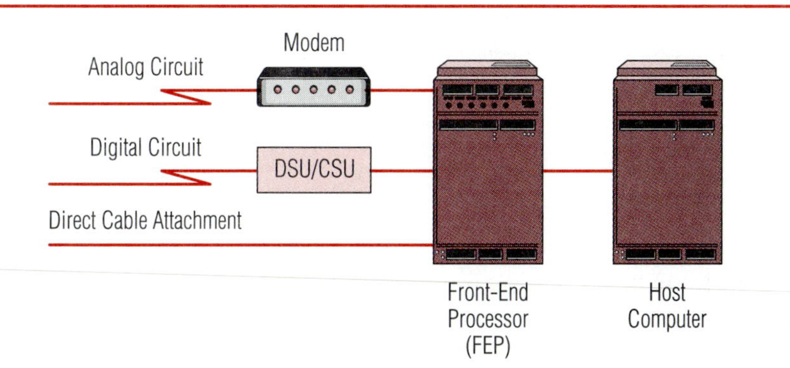

Figure 11–17
Location of the front-end processor in the network.

best position to understand what is required to make the computer and FEP operate compatibly.

Figure 11–18 shows the basic internal structure of a typical front-end communications processor. The heart of the device is a computer, the communications processing CPU. The CPU requires memory in order to store the software programs and data used during communications processing. The console may take several forms. It may simply be some lights and switches on the control panel of the FEP, or it may be a separate personal computer through which an operator can check status and enter commands.

On the network side, the FEP provides interfaces with the telecommunications circuits. Each interface in the FEP is called a *port*. Stand-alone FEPs range in capacity from 10 to 500 or more ports. The port capacity usually depends on the speed of the circuits that will be connected because the FEP has an inherent limit to the number of bits per second it can process. In other words, the slower the speed of the connected circuits, the more that can be handled. With higher speed circuits, fewer can be connected.

port

Normally, the FEP manufacturer also provides the software to drive it. This programming, commonly called the *network control program (NCP)*, is very complicated. Just as users seldom write computer operating systems, they seldom write NCPs, although they may modify the software provided.

Front-End Processor Functions

The functions that an FEP performs are:

1. Circuit control
   a. polling and addressing terminals
   b. adding protocol to outgoing messages; removing protocol from incoming messages
   c. answering dial-in calls; automatically dialing outgoing calls
   d. converting protocol—such as an asynchronous protocol to SDLC

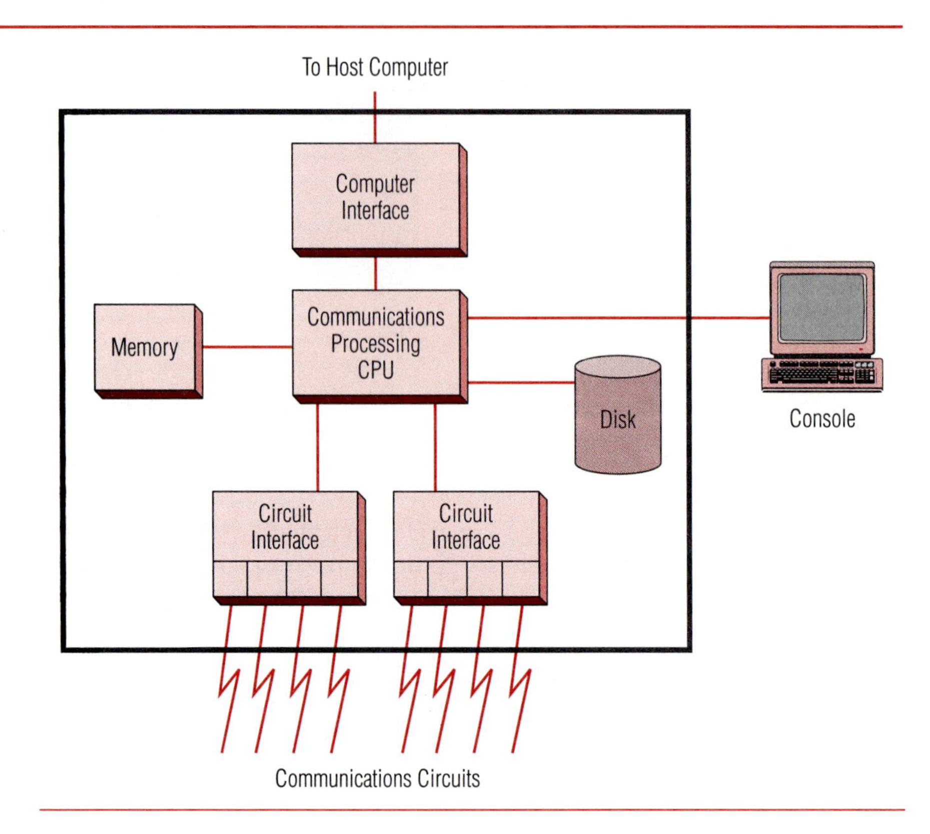

Figure 11–18
The internal structure of a front-end processor.

   e. converting codes—for example, ASCII to EBCDIC
   f. recognizing characters, including the distinction between control characters and data characters
   g. accommodating circuit speed differences
   h. multiplexing
2. Assembly of characters into blocks or messages
   a. assembling incoming bits into characters
   b. assembling characters into blocks of data or complete messages, depending on the design of the interface with the host computer
3. Message queuing or storing
   a. holding outgoing messages when they arrive from the computer faster than the communications lines can accept them
   b. holding incoming messages from the communications lines when the computer is down or otherwise unable to accept them
   c. adjusting the sequence in which messages are passed to the computer or the line so that higher priority messages are handled first

4. Error control
   a. checking VRCs, LRCs, and CRCs
   b. requesting retransmission of blocks that contain errors
5. Administrative functions
   a. noting when terminals or circuits are not operating correctly and sending a message to a network control operator
   b. logging all messages onto a disk
   c. keeping statistical records of line use, terminal use, and response time
   d. keeping records of the various types of transmission errors
   e. performing online diagnostics of circuits or remote terminals
   f. switching all circuits to a backup computer if the primary computer fails

It is worth noting the wide difference in speed between communications lines and computers that FEPs must manage. A communications line operating at 9.6 kbps is capable of passing

9,600 bits / 7 bits per character = 1,371 ASCII characters per second.

Disks typically pass data to a computer at 1 to 3 million characters per second. Even if the communications line speeds up to 56 kbps, it can only pass ASCII data at 8,000 characters per second. Clearly, to a computer, a communications line operates at a very slow speed. The FEP (being a computer) can easily manage many communications lines and have time left over for other network management functions.

THE ROLE OF THE COMPUTER

The functions performed by the computer are complementary to those performed by the FEP. To some extent, what the FEP does not do, the computer must do, and vice versa. However, many functions are performed exclusively in the computer, and virtually all of them are controlled by software.

Functions of the Software

Telecommunications software can be packaged in many different ways within the computer. It may be integrated into the *operating system,* such as Microsoft Windows. It may be in separate programs called *drivers* that are added to the operating system only when and if needed, or it may be a part of separate software packages that run under the control of the operating system. Regardless of how it is packaged, the software has many functions to perform that are described below.

Security The software provides security to ensure that only authorized users or terminals are able to use application programs and to access data. Techniques used to ensure security include

- validating user identification codes and passwords entered by operators at terminals;
- checking the user's authority against tables or files stored in the computer to determine whether the user can use the specific program she is requesting;
- checking the same tables to determine whether the terminal operator has the authority to read, write, update, or delete data files or can do only some or one of these functions;
- accepting the user ID and password from an operator on a dial-up circuit; then terminating the connection and calling the operator back at a predetermined telephone number.

Communications security is discussed in more detail in Chapter 15.

Buffer Management Incoming messages are stored in portions of the computer memory called *buffers*. Buffers are almost invariably of a fixed length, whereas messages are almost always of a variable length. Consequently, managing the buffers is a significant job for the software. Incoming messages must be assigned to buffers as they arrive and before they are passed to the application program for processing. Response messages from the application are stored in buffers until they are passed to the FEP for transmission. Once the messages are sent, the buffers are freed and added to a queue of available buffers. Efficient buffer management can help ensure good message-handling performance and can reduce the storage requirements.

Routing a Message to a Program Messages that come into the software may be destined for one of several application programs. The software can examine the contents of the incoming message and determine for which application program it is destined.

Queueing Messages When messages arrive faster than they can be processed by the application program, they are held in a queue or sequential waiting line (which may be in the computer's main memory or on disk) until the application program can handle them. In some systems, incoming messages are held in a queue until a specified number have arrived. The application program then processes all of the messages in the queue at one time. Queuing may also occur if an application program only operates during certain hours of the day, whereas messages for it arrive anytime. Messages are held in a queue until the application program is ready to accept them.

Scheduling Application Programs The operating system must decide when to bring an application program into memory for execution. Its scheduling may be based on elapsed time since the program was last scheduled, or it may be dependent on the arrival of incoming messages. In other words, when the queue of waiting messages reaches a certain level, it may trigger the operating system to schedule the application program to process those messages.

Providing Continuity between Parts of a Transaction Some transactions arrive in several parts. For example, a customer order-processing transaction may require the terminal operator to first enter the customer number and order number. This data is transmitted to the computer, where the operating system schedules the application program. The application program examines the customer number and looks up the customer's name and address in a file. The name and address are then transmitted back to the terminal operator for verification. The operator then enters the identifying number of the product the customer wants to order. When the number arrives at the computer, the software must determine that it is associated with the first part of the order that was processed previously. The way that this is achieved is to save the parts of the transactions in memory or on disk and effectively build an order from the data in the transactions as they come in.

Another type of continuity may be required if an operator is in the middle of processing a complicated transaction but stops for an extended period of time to answer the telephone, go to lunch, or take a break. Undoubtedly, the operator will want to resume processing the transaction after the break without having to reenter previously submitted data. The software can provide continuity in this situation.

Message Formatting Another function of the software is to convert the format of messages from that required or generated by the application program to or from the format required by the specific terminal. This message formatting is also called *presentation services,* which corresponds to layer 6 of the OSI model.

The objective of the message-formatting function is to provide a *virtual terminal* interface with the application and its programmer. The idea is to allow the application program to pass messages to the software in a standard format and without regard to the specific type of terminal for which it is ultimately destined. Thus, if new terminals become available and are added to the network, only the software must be changed to format messages for the new terminal; the application programs do not have to be modified.

The virtual terminal concept works in reverse, too. Incoming messages from different terminal types may arrive in different formats. The responsibility of the software is to put them into a standard format before presenting them to the application program for processing.

Checkpoint/Restart The software has a responsibility for providing the ability to restart processing after a failure of the FEP, the host computer, or the software has occurred. Whereas with traditional batch computer processing the usual approach is to simply rerun the jobs that were processing when the failure occurred, this technique is not adequate or even possible in communication-based message-processing systems. You cannot tell terminal operators at 4 P.M. on a Friday afternoon, "The computer has just failed and we have restarted it, but you are going to have to reprocess the entire day's work. Please resubmit all of the transactions you have sent since 8 A.M."

The concept behind *checkpoint/restart* processing is to periodically record on disk the status of the computer processing, including the contents of memory, the programs that were active, the messages that were being processed, and the database or file records that had been accessed. This status information is called the *checkpoint record.* After a failure occurs and the computer is restarted, the checkpoint record is loaded into memory and used to reset the computer to the conditions that existed when the checkpoint record was taken. The only processing that has to be redone is any that occurred since the last checkpoint record was written.

Determining the frequency with which to take checkpoint records requires knowledge of the telecommunications network and the applications. The determination normally is made by a group of telecommunications and data processing people working together. On the one hand, if checkpoints are taken too frequently, checkpoint processing can place a significant load on the host computer. Furthermore, since in many systems all other processing must stop while the checkpoint record is being written, terminal operators may see unusually long response times or other unexplainable delays. On the other hand, if checkpoints are not taken frequently enough, many transactions may have to be reprocessed after a failure occurs, causing frustration and redundant work on the part of many people.

Preventing Messages from Being Lost or Duplicated The two biggest problems that can occur in communications-driven business systems are missing messages and repeated messages. On the one hand, if a message containing a customer order for two tank cars of an expensive chemical is lost or otherwise not processed, not only will the customer be angry, but also the supplier will have lost the opportunity to sell some product and enjoy the resulting profit. On the other hand, processing the message twice in an automated order-processing system and sending four tank cars of the chemical to the customer is equally undesirable. Duplicate shipping costs: the cost to reship the material to another customer, and the potential lost opportunity to sell it during the erroneous shipment eat into the product's profitability. In addition, the supplier looks somewhat foolish in the eyes of the customer.

Telecommunications software written to handle business applications must contain adequate controls using techniques, such as checkpoint/restart and sequence numbers, to ensure that all incoming messages are processed once, but only once!

Application Programs

Application programs perform basic business functions, such as customer order processing, inventory control, and accounts receivable. If the telecommunications software, operating system, and network-control program are properly designed and well written, writing an application program that performs transaction processing from terminals connected via communications lines should be no different from writing any other program. To the application programmer, messages to be processed should be obtained with simple GET or READ commands, much like obtaining data from a disk or tape file. After processing, the response message should be sent with a PUT or WRITE command. With proper implementation of the virtual terminal concept, the application program should be shielded from the unique attributes, characteristics, and idiosyncrasies of the telecommunications terminals. An application programmer should not need to be any more aware of the operation of a telecommunications circuit than of the operation of a channel on a computer that connects a disk drive to the CPU.

Software for Network Management

Other types of telecommunications software that exist, at least partially, in the host computer are programs for network management. Like most other communications software, there are alternatives for where these programs are located. They may reside entirely in the host computer, or they could be in the FEP. Alternatively, they may be shared between the two. If the network-management software is in the host, it may be a part of the operating system. It could also be a separate program that is able to work closely enough with the other software to direct the network's operation.

Network-management software performs several functions. One is to monitor the status of the network and display data pertinent to the network's operation. Some networks have large status maps with lights controlled by the network-monitoring software. When a line or control unit is operating normally, its corresponding light is green. If a problem exists, the light is changed by the software to yellow or red, depending on the severity of the problem. Other systems display the network status on a VDT screen that may be dynamically updated as the status changes. Circuits or nodes that are fully operational may be shown in green. Those with some problems might be displayed in yellow. Units that are out of service might be shown in red.

Another function of the software is to log pertinent network operational data to a data file stored on magnetic disk. The log file may contain statistical records about the network's operation. These records can be analyzed to calculate performance measures, such as network availability and response time. The log file also may contain error records. These records help to pinpoint problems, especially those that occur infrequently but over a long time span. For example, an intermittent line problem might cause transmission errors two or three times a day and might not be noticed by users of the network, especially if the transmission errors are automatically corrected by the hardware. An analysis of the error records would show the pattern of errors and provide data for service personnel.

A third function of network management software is to provide the network control operator with commands to control and monitor the status of the communications network. Using these commands, the network operator can manually control certain aspects of the network's operation.

Network Control Commands Some of the types of commands required by the network control operator are

- start line—this command tells the software to start polling, addressing, or otherwise using a communications line. It could be used when new lines are installed but is used more frequently to resume operation after a circuit has been out of service;
- stop line—this command provides the inverse capability of start line and instructs the software to stop using one of the circuits. Another use for this type of command might be to shut down operations on a circuit at the close of a business day. If a host computer located in Seattle has a network that stretches to the East Coast, the network operator may want to stop the East Coast circuits at 5 P.M. or 6 P.M. EST because the terminal operators will have gone home. Since it is only 2 P.M. or 3 P.M. in Seattle, the local terminals would continue to operate for several more hours;
- start terminal/stop terminal—these commands are similar to the start and stop line commands, but they only affect a specific terminal on a circuit. They might be used to stop a terminal that is failing and to restart it after the problem has been corrected;
- check status—the check status command may have many variations that allow the network control operator to check any part of the communications network's operation. The operator could use one variation to check the status of a particular terminal. Another variation checks the status of a line, and still other variations could allow the links or contents of message queues to be examined;
- start network/stop network—these commands are used to start normal network operations at the beginning of a business day and to stop operations at night. All of the stop commands usually examine the status of terminals and circuits and allow any traffic in

progress to be completed normally. Since this process can take some time with long messages, there are some variations of the stop commands, such as stop normal—which was just described—and stop immediate, which instantly interrupts all transmissions and brings the network to a crashing halt.

Other commands are available in most networks. Some software allows *user written commands* to be defined by the network operators. Using this capability, a network operator can write commands unique to the particular requirements of each company's network.

The subject of network management and operation is discussed in more detail in Chapter 13.

MANUFACTURERS' ARCHITECTURES

Chapter 3 identified the fact that several computer manufacturers developed their own models for network architecture. Now that you have studied all of the components that make up a network, we will look at two of the manufacturers' architectures in detail.

IBM Corporation's Systems Network Architecture (SNA)

Historical Basis IBM's *Systems Network Architecture (SNA)* was announced in 1974. Product development followed, and by 1978 when the OSI model was announced, IBM had already implemented a number of SNA hardware and software products. IBM's largest customers were beginning to jump on the SNA bandwagon, and IBM saw little reason at that time for changing SNA to meet the OSI model and the international standards.

Certain countries in Europe developed public packet switching networks based on the X.25 standard in the late 1970s. Some of IBM's large European customers wanted to use these networks to connect their IBM equipment, and they put pressure on IBM to support X.25 as a part of SNA. In 1980, IBM announced support for X.25, available to European customers only. A few years later, the X.25 support for IBM hardware and software also became available from IBM in the United States.

SNA Concepts SNA is conceptually similar to the OSI model but not directly compatible with it. Like OSI, it can be viewed as a seven-layer architecture, but some of the layers are defined differently, as shown in Figure 11–19. Rather than making an exact layer-by-layer comparison between the OSI model and the SNA layers, we will examine some concepts and terminology that are basic to SNA and important to its understanding.

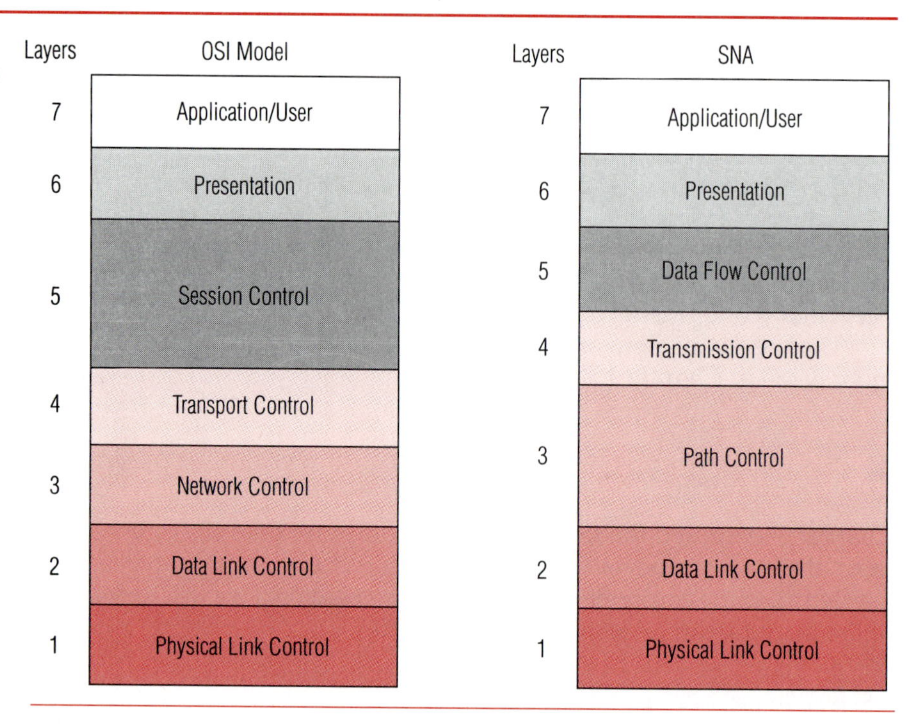

Figure 11–19
ISO-OSI layers compared to IBM's SNA layers.

Physical Units SNA views hardware as being of four specific types called *physical units (PUs)*. The four types of physical units are numbered 1, 2, 4, and 5 with no type 3 currently defined. The four types are

Physical Unit Type	Type of Hardware
1	terminals
2	cluster controllers
4	front-end processors
5	host computers

Figure 11–20 shows how this hardware might be connected in an SNA network and identifies some of the IBM model numbers associated with the different types of equipment.

Logical Units SNA users can be either people at terminals or application programs. Users are represented in the system by entities called *logical units (LUs)*, which are implemented in software. The communication between two system users is really a communication between LUs and is called a *session*.

Sessions LUs can request several different types of sessions, specifically terminal-to-program, terminal-to-terminal, or program-to-program. Terminal-to-program sessions are requested when a user wants to use an application program on a host computer. Terminal-to-terminal sessions are established when one user wants to communicate directly

Figure 11–20
Typical IBM SNA network.

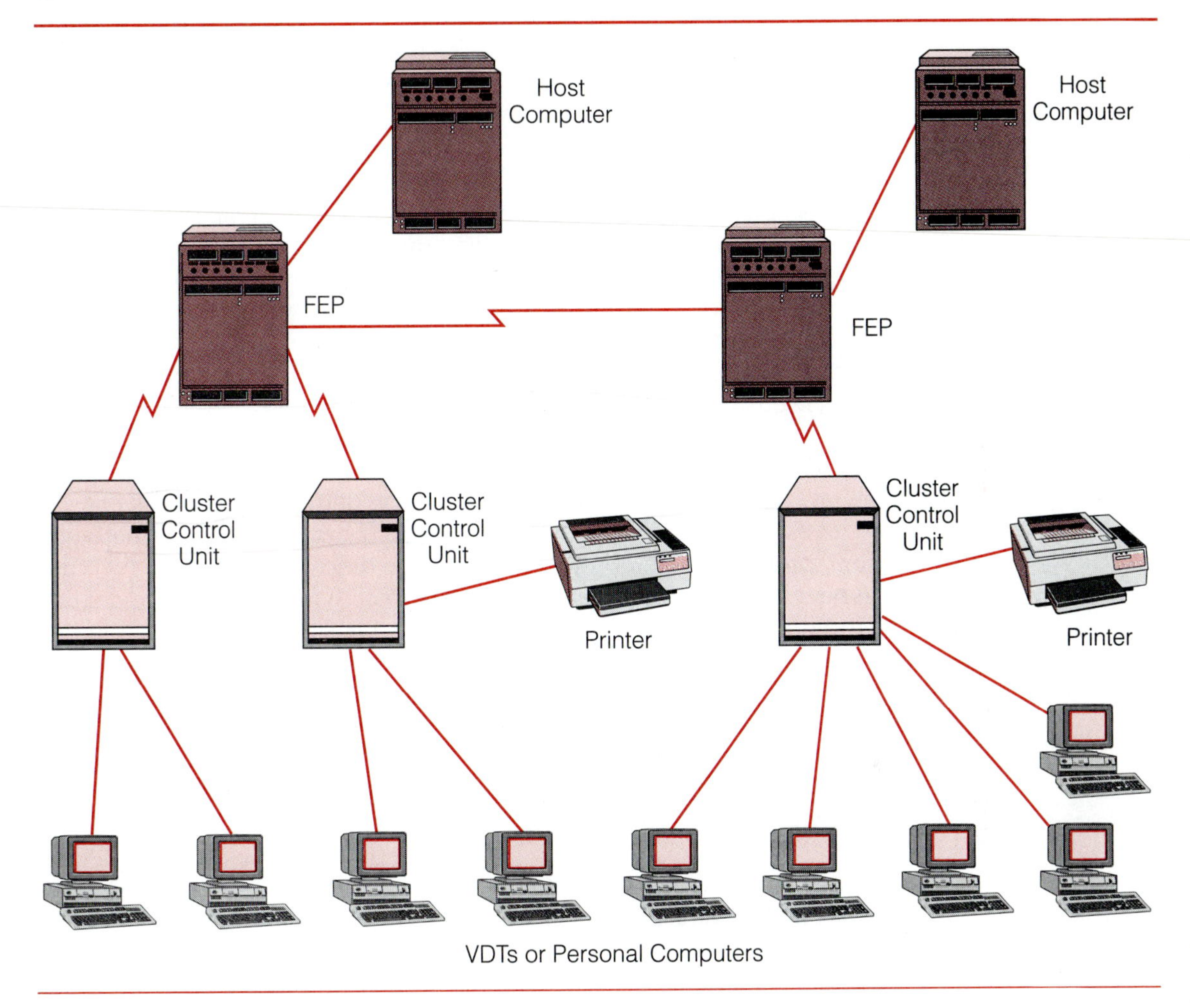

with another. Program-to-program sessions are established when one program needs to pass data to another.

Sessions also are classified as interactive, batch, or printer sessions. In addition, a user can have multiple simultaneous sessions in progress, each with its own logical unit. This gives the user at a terminal the ability to communicate with two or more computers or with two or more programs on the same computer simultaneously.

Network Addressing In SNA, all LUs and PUs are called *network addressable units (NAUs)*, and each has its own unique network address. SNA addresses are 24 bits in length, which allows for a very large number of nodes to be connected.

Data Link Protocols The primary data link protocol used within SNA is Synchronous Data Link Control (SDLC), developed at the same time as the SNA architecture. In fact, the terms SNA and SDLC often are confused and incorrectly used interchangeably. However, SNA can also operate with the BISYNC and X.25 protocols. SDLC is a proper subset of OSI's HDLC.

SNA Software Although the physical link control layer and part of the data link control layer are implemented in hardware in the IBM front-end processors, most of the SNA layers are implemented in software. The network control program software in the front-end processor is called the Network Control Program, or NCP. The other major software component is the telecommunications access method, which resides in the host computer. It is called VTAM, which stands for Virtual Telecommunications Access Method.

Several supporting pieces of software are available to help SNA customers manage the network. The most significant is a product called Netview, which provides an interface to the network operator's console and gives that operator with statistics about the operation of the network, errors, problems with equipment, response times, and so forth. Netview was never extended to help manage the internetworking environment that many companies are now facing and hence it has largely been supplanted by products from other vendors that provide a broader capability.

SNA received wide acceptance in the marketplace. At one time, in the late 1980s and early 1990s, there were reportedly more than 36,000 SNA networks installed worldwide. SNA users included the world's largest corporations, and as a result there were many thousands of workstations and users communicating using SNA.

Digital Equipment Corporation's Digital Network Architecture (DNA)

Digital Equipment Corporation (DEC) defined an architecture very similar to the OSI model. *Digital Network Architecture (DNA)* is the framework for all of Digital's communications products and was implemented in a family of hardware and software collectively called *DECNET.* DECNET establishes a peer relationship between all nodes in the network; no central controlling node is required. Three data link protocols are supported by DECNET: Digital Data Communications Message Protocol (DDCMP), X.25, and Ethernet for use on local area networks. DEC also supplies an interface to IBM's SNA architecture in the form of a combination of hardware and software called the SNA Gateway.

DECNET contains five layers as shown in Figure 11–21. The physical link control layer, data link control layer, transport layer, and network services layer correspond almost exactly with the lowest four layers of the OSI model. DECNET's application layer is a combination of the OSI pre-

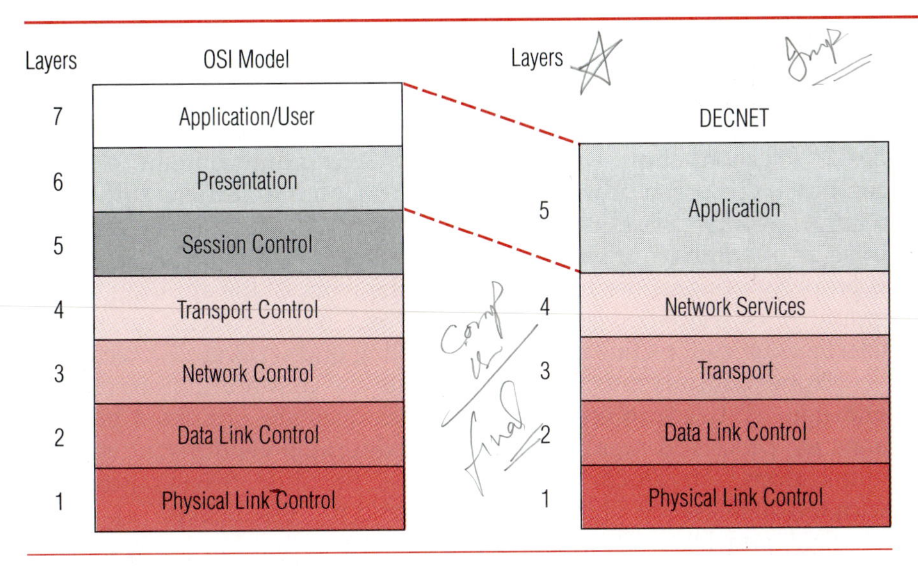

Figure 11–21 ISO-OSI layers compared to Digital Equipment Corporation's DECNET layers.

sentation and application layers. DECNET does not have a separate session layer.

The data link control layer can use any of the previously mentioned protocols interchangeably. As a result, DEC networks have a very well-developed capability to interconnect DEC hardware, other vendors' hardware that subscribes to the X.25 standard, and equipment connected to local area networks that communicate using Ethernet. DEC also introduced DECNET/OSI, which conforms to the OSI model more closely.

SUMMARY

Communications networks come in many shapes and sizes. Networks can be classified several different ways: by their topology, their ownership, their geography, and their transmission technology. The OSI layers that deal with networks must perform many functions, and there often are differences as to how the functions are performed in different types of networks.

The telecommunications network designer is faced with a wide variety of choices for building a network. Only by understanding the characteristics and trade-offs between various types of networks can the designer reach a reasonable solution for a particular application. More often than not, a company's diverse communications requirements indicate that a hybrid network is required. There may be a combination of WANs and LANs, each with a different topology.

Internetworking has become very common and the TCP/IP protocol is widely used to connect networks. The Internet provides an unprecedented example of internetworking, with thousands of networks connected and millions of people using it worldwide.

We have also looked at the ways in which a communications network is interfaced with a host computer. The computer manufacturers make design decisions about which functions will be implemented in the host computer, the FEP, hardware, and software. Communications software is relatively complex and is most often written and provided by the computer or operating system manufacturer. If the communications software is properly designed and implemented, the work of the application programmer is simplified so that reading and writing messages from a telecommunications line is similar to reading and writing disk records.

Proper controls must be built into both the communications and applications software to ensure that all messages are processed but that none are processed more than once. Although the vendor-written communications software can simplify this task, the ultimate responsibility is with the business for which the network and processing exist.

SNA and DECNET are major architectures that have shaped how thousands of data communications networks are organized and operated. Although neither of these architectures exactly matches the OSI model, the basic principles and concepts are the same.

CASE STUDY

Dow Corning's Wide Area Data Communications Network

Dow Corning's wide area data communications network, as shown in Figure 11–22, is conceptually a classic star configuration with the center in Midland, Michigan. In actuality, short circuits run from Dow Corning's locations to WorldCom's global frame relay network, which provides the data transport. Dow Corning entered into a contract with WorldCom in 1997 for a global network based on frame relay technology. The contract allows the speed or level of service to be specified at each Dow Corning location, so that high capacity can be provided to large locations and less capacity provided at smaller sites. As additional sites are opened, a new connection with the WorldCom network is installed and added to the contract. While this arrangement is fairly new for Dow Corning, it has been working quite satisfactorily by most measures. In the past, the company has found that a major overhaul or redesign of the network has been necessary about every 8 to 10 years, and it will be interesting to see whether that experience will repeat itself in the middle of this decade.

The company does not think in terms of MANs, but the data communications manager points out that in the Midland area there are multiple locations in a several square mile radius that are connected with high-speed circuits that effectively form a MAN. So, although MANs are not consciously planned or discussed as a separate entity, one has been implemented.

Dow Corning provides global dial-in capability to its traveling employees through its contract with CompuServe, the company that actually provides the service. By dialing in, travelers can access the Dow Corning network, including data they may have stored on LAN servers, computer applications, and e-mail. Of course, operations on the dial-in network occur at slower speeds than when the employee is in his or her office, but most of Dow Corning's traveling employees highly value the ability connect to the company's network from their home, their hotel, from airport lounges, and a multitude of other locations.

A primary goal of Dow Corning's network is to allow users at any Dow Corning location, anywhere in the world, to log on to any server for which he or she is authorized. This goal has been achieved, and employees routinely make intercontinental connections to computers in other countries as a part of their jobs.

QUESTIONS

1. Dow Corning's wide area network mainly uses circuits leased from a single vendor. Discuss the relative merits of having a single vendor, with which the customer has influence and the "clout" to demand good service, versus having multiple vendors and "playing them off" against one another.
2. Dow Corning's network has changed from being mainframe-based and largely self-engineered to one that is essentially outsourced and run by a network provider (WorldCom). What are some of the costs the company incurred as it moved network architecture to another?
3. What are the advantages and disadvantages of outsourcing the wide area network?
4. With the change in the technology and structure of the network, the skills of the telecommunications staff need to be constantly upgraded. What are some of the methods the company can use to keep the staff's knowledge fresh and relevant?

Figure 11–22
Dow Corning's U.S. data communications network has a classic star topology with the center at the company's headquarters in Midland, Michigan.

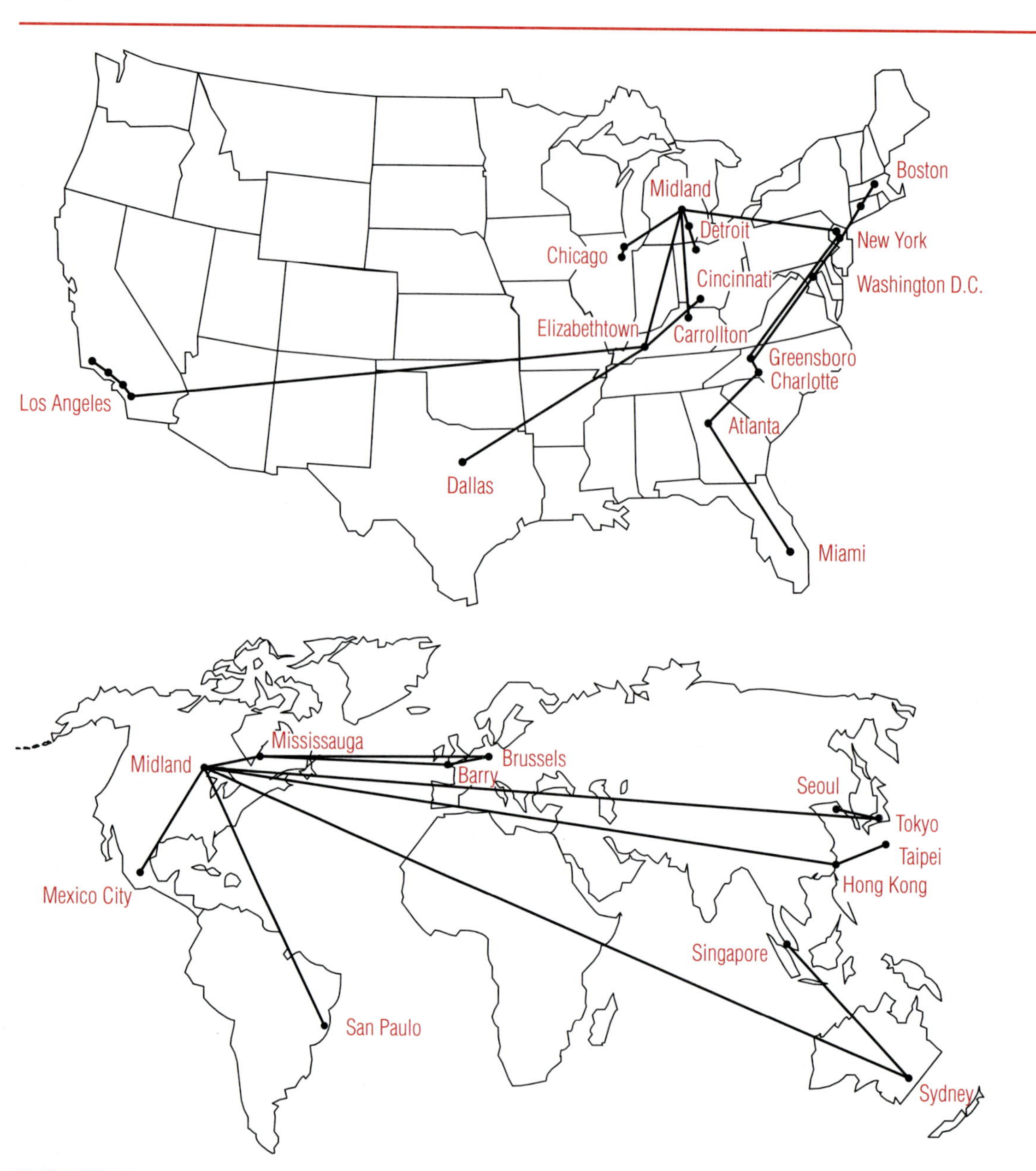

REVIEW QUESTIONS

1. Define the word *network.*
2. Explain what a backbone network is.
3. Describe four ways that networks can be classified.
4. Explain the characteristics of a star network and a ring network.
5. Describe some situations in which public networks may be more appropriate than private networks.
6. Describe how a packet switching network operates.
7. What advantages does frame relay have over standard packet switching?
8. Why is ATM so efficient and fast?
9. What is the function of the TCP/IP protocol?
10. What are the advantages and disadvantages of a centralized, static routing system?
11. What are the advantages and disadvantages of a distributed, dynamic routing system?
12. Why is it unlikely that cycle stealing would be used on a mainframe computer that is controlling several communications circuits?
13. What are some of the factors a computer/FEP manufacturer would consider in deciding whether to implement the line interface function in a separate piece of hardware or in the CPU?
14. Explain the term *dark fiber.*
15. Explain the difference between ATM's CBR and UBR services.
16. Describe the functions of checkpoint/restart processing.
17. Describe some situations in which queueing would occur in a telecommunications system.
18. Explain several security techniques that might be employed to ensure that users of a telecommunications network are properly authorized.
19. Explain the concept of a virtual terminal.
20. Why is there so much concern about ensuring that messages are processed only once?
21. Are SNA and SDLC different names for the same thing? Explain your answer.
22. Explain the function of an SNA logical unit.
23. Compare and contrast star and mesh network topologies.
24. Describe the functions of a front-end processor.
25. What is X.25?
26. Describe frame relay.
27. Describe both types of ATM.
28. Explain the term *circuit switching.*
29. What is a URL?
30. Explain the purpose and function of the internet protocol (IP).
31. How is HTML used?
32. Explain why it is important for any routing technique to have alternate routes available to send messages.
33. Explain the functions of the TCP.
34. What is the purpose of an ISP?

PROBLEMS AND PROJECTS

1. What type of routing would be used in a public packet switching data network? Why?
2. Think of an application where a PDN's datagram service could be used successfully, considering that the PDN does not guarantee the arrival of datagrams or their arrival sequence.
3. Research the Internet and the ways in which it is being used by companies. Report on it to your instructor or your class.
4. Use your imagination to "invent" several commands other than those listed in the text that would be useful for the network control operator to have when operating the network or diagnosing problems with it.
5. Can you think of some situations in which it would not make any difference if a message were processed two or three times by a communications application? What if it were processed 500 times?
6. Visit a computer installation that has a telecommunications network. Find out what security techniques are being used to ensure that users of the system are properly authorized.
7. SNA and DECNET assume that a fundamentally different network structure will be used. Explain what characteristics each architecture assumes the network will have.
8. Describe the functions that a gateway between DECNET and SNA must perform.

Vocabulary

network
backbone network
topology
star network
hierarchical network
mesh network
bus network
bus
ring network
hybrid network
private network
public network
value added network (VAN)
wide area network (WAN)
metropolitan area network (MAN)
dark fiber
local area network (LAN)
circuit switching
packet data networks (PDN)
packet switching
packet
packetizing
packet assembly/disassembly (PAD)
switched virtual circuit
permanent virtual circuit
datagram
frame relay
frame
circuit congestion
committed information rate (CIR)
port speed
public frame relay service
private frame relay networks
asynchronous transfer mode (ATM)
cell relay
cell
virtual circuit
connection-oriented routing
connectionless routing
routing
broadcast routing
centralized routing
distributed routing
static routing
dynamic routing
internetworking
internet
subnetwork
end systems
intermediate systems
bridge
router
transmission control protocol (TCP)
internet protocol (IP)
the Internet
electronic mail (e-mail)
telnet
discussion group
Usenet newsgroups
search engine
web server
web browser
page

Hypertext Markup Language (HTML)
Uniform Resource Locator (URL)
home page
Internet service provider (ISP)
intranet
firewall
extranet
cycle stealing
network interface card (NIC)
front-end processor (FEP)
telecommunications control unit
transmission control unit
port
network control program (NCP)
operating system driver
buffer
presentation services
virtual terminal
checkpoint/restart
checkpoint record
user written commands
Systems Network Architecture (SNA)
physical unit (PU)
logical unit (LU)
session
network addressable unit (NAU)
Digital Network Architecture (DNA)
DECNET

References

Cypser, R. J. *Communications Architectures for Distributed Systems.* Reading, MA: Addison-Wesley, 1978.

Czubek, Donald. "What Are the Differences Between SNA and DECNET?" *Communications Week* (February 9, 1987): 54.

Digital's Networks: An Architecture with a Future. Maynard, MA: Digital Equipment Corporation.

Gurley, J. William, and Michael H. Martin. "The Price Isn't Right on the Internet." *Fortune* (January 13, 1997): 152–154.

Jainschigg, John. "Voice over Frame Relay Saves Big." *Teleconnect* (December 1996): 77–79.

Leiner, Barry M., et al. "The Past and Future History of the Internet." *Communications of the ACM* (February 1997): 102–108.

CHAPTER

12

Local Area Networks

OBJECTIVES

After studying the material in this chapter, you should be able to

- describe the general characteristics of LANs;
- discuss the reasons why organizations install LANs;
- describe LAN topologies, media, and transmission techniques;
- describe the IEEE standards that apply to LANs;
- describe the Ethernet and token ring LANs and their methods of operation;
- describe the role and function of LAN servers;
- describe the role and function of LAN software;
- describe the factors affecting LAN performance;
- describe the characteristics of some common LANs;
- explain the necessity of and some of the problems of managing LANs;
- discuss the necessity for having LAN security measures.

INTRODUCTION

Local area networks (LANs) are one of the fastest-growing and most exciting parts of the telecommunications scene. The technology that makes LANs possible and practical is relatively new compared to many WAN technologies, and sophisticated network control and application software are making LANs productive for millions of workers in small offices and large organizations alike. Because of the "local" nature of the network, and because it can help them do their job better, most people feel a strong ownership in the capability a LAN provides.

As a student, you are more likely to have used a LAN than any other type of data network, except perhaps the Internet, because they have been installed in thousands of schools for classroom use and as an aid for faculty and administration. Whereas you may also have used a wide area network (WAN)—for example, to dial in to a service such as America Online, or CompuServe—you probably don't feel as strong a sense of the network that was involved to make that connection. But if you've used a LAN at school or at work, you probably have a much stronger sense of the network itself, how it works, and its advantages and problems. Let's now look at LANs in more detail.

DEFINITION AND CHARACTERISTICS OF A LAN

A LAN is a high-speed data network that covers a relatively small geographic area. It typically connects personal computers, other types of workstations, printers, and servers. More specialized LANs may also connect a group of automation devices in a manufacturing plant, telephones, facsimile machines, or combinations of the above types of terminals. A key characteristic is the *limited distance* that LANs cover. We usually think of LANs serving a single department, a building, a plant, or in some cases several buildings on a campus if they are in close proximity to one another. In some cases LANs may extend a few miles, but special equipment, called *repeaters,* are usually required to regenerate the signal.

limited distance

Another characteristic of LANs is that they usually operate at high data rates—from 2 Mbps to 1 Gbps. This is possible because of the relatively short distances and the resulting low error rates. These characteristics mean that users of LANs usually have very good response times compared to users of WANs. Response time on a LAN is less likely to be limited by the speed of the transmission facility than it is on a WAN, but there may be other factors that limit the response time, such as the speed of the serving computer or the number of users sharing the LAN.

high speed

LANs are almost always privately owned and installed, and, therefore, a regulatory agency is not involved. As long as the LAN does not cross a public right-of-way, such as a highway, no governmental agency is involved with the LAN's establishment or operation. Of course, some LANs connect to wide area networks, which are subject to regulation, but the LAN itself is generally exempt.

private

The typical university campus, hospital, corporate headquarters, manufacturing site, and research center are good candidates for a LAN installation. Some LANs are installed within departments so that users of personal computers can share expensive hardware, such as laser printers or large disk storage units. Other LANs are installed simply to provide a fast data path and good response time from terminals or personal computers to a large central computer.

REASONS WHY LANs ARE POPULAR

There are several reasons why LANs are so popular and have become widely used.

Information Sharing

Sharing information within workgroups or departments is the primary reason why many LANs are installed. Studies have shown that a very high percentage of information generated within an organization is distributed and used close to home. Whether it is electronic mail or analytical reports, approximately 50 percent of all information is used only within the originating department. Another 25 percent is distributed and used within nearby departments, whereas only 15 percent is sent to other people in the organization. Only about 10 percent is sent outside the company. Financial people process and use financial information; marketing and sales people use customer, market, and order information, and so on. Normally only summarized information is sent to other departments. This distribution pattern suggests a need for a high-speed, efficient local information distribution system, a role a LAN can effectively play.

Hardware Sharing

LANs provide an opportunity to share relatively expensive pieces of equipment, such as laser printers, document scanners, or large hard disks. This sharing of resources holds down the cost of equipping a department or other group with a full range of computer equipment. The sharing usually works well, since most personal computer users do not use a laser printer often enough to keep it busy and/or to justify its cost. If shared by a group of people, however, the cost may be easily justified.

Of course there are other ways to share printers, such as with switches, but they usually impose more stringent distance limitations. That is, the users must be located within 50 or 100 feet of each other. LANs do not have such stringent limitations.

Another type of hardware that can be shared are large hard disks. It is almost always less expensive to buy a single large disk than multiple medium-sized disks, so again, economies of scale play a significant role. Furthermore, with a large shared disk, it may be possible to buy only one copy of certain software and allow it to be shared; or to store all of a department's data files on a common disk that is accessible to all department members and that can be regularly backed up.

Other devices may be shared as well, and the savings may help pay for the cost of installing and operating the LAN.

Software Sharing

There are two aspects to software sharing. The first is that certain vendors will allow an organization to buy a single copy of a piece of software designated to be shared by the users of a LAN. For example, the department may be able to buy a single spreadsheet package that all members who are connected to the LAN may use. The other type of sharing is related to a new type of software called *groupware,* which is designed to be used by a group of people. Using such software, the group may find that they communicate more efficiently and easily and that because of a software-imposed filing discipline they are able to find and recall information more quickly than before.

groupware

Service

LANs may provide better service in the form of response time or availability to their users than do mainframe-based networks. Although careful design and control are required, it is inherently easier to provide consistent service to a small group of users than to a large group. Since the transmission rates are high, most users should be able to get low-second or subsecond response time unless they request excessive processing, extensive image handling, or voluminous output.

Local Control

LANs provide the ability for the local department to operate and make the decisions about the LAN. This may be more effective than working through a centralized network management group, which may be located miles away and which may not be familiar with local needs and problems. However, as will be discussed later, with this decision-making authority comes a certain responsibility for proper LAN management, which the local department may or may not want to assume.

LAN APPLICATIONS

Organizations install LANs for many purposes, but frequently it is for the purpose of using a new application or using an existing application in a new way. For example, data on a computer that is attached to a LAN may be more easily shared between users of personal computers without copying the data to a diskette and carrying the diskette from one office to another. In one company, the tax department—with the help of the computer department—set up a LAN so that several tax accountants could use an application package designed to help in the preparation of the company's tax returns. Several tax accountants were able to work on different tax schedules at the same time, all sharing a common data file of basic corporate financial data on which the return was based.

As mentioned previously, groupware can connect many people at the same time for sharing information, holding electronic group discussions or debates, and exchanging electronic mail. Groupware software can collect these conversations and produce a written transcript at any time. Some groupware programs can handle graphics, images, or even video in addition to words. Because information can be shared with hundreds of people simultaneously, groupware networks can give office workers intelligence that was previously available only to their bosses.

Application software packages, such as word processing, spreadsheet, and graphics programs, must be written so that they can take advantage of the features that a LAN offers. If the application has been programmed in such a way that it cannot use the LAN's disk or printers, the LAN is of little use. The biggest problem that most applications have is the data integrity of shared files. Most applications were not originally written with the idea that multiple people might be simultaneously accessing or updating the data. They must be modified to provide a new level of access control, which allows for the possibility of several people accessing the same files at the same time. For these reasons, many application software packages have separate versions that are designed for LAN use.

When software is available for a LAN, it is tempting to purchase one copy, install it on the LAN, and let everyone use it simultaneously because the software cost savings would be significant. However, this is illegal, and most companies have policies that prohibit the practice. Most software contracts require the customer to purchase one license for each person who will be using the software simultaneously. Because this is difficult to keep track of, some programs have a *software metering* routine that keeps track of the number of simultaneous users and doesn't allow more users than the number of licenses that have been purchased. Other software companies sell *site licenses* that allow an unlimited number of people to use the software for a single flat fee.

software metering

site license

LAN PROTOCOLS AND ACCESS CONTROL

In Chapter 3 we studied the ISO-OSI model for data communications networks. LANs follow the OSI model but require a modified view of its lower two layers because they have different configuration requirements. LAN standards have, for the most part, been developed by a committee of the Institute of Electrical and Electronics Engineers (IEEE), known as the IEEE 802 committee, and subsequently adopted by the International Organization for Standardization (ISO). The IEEE 802 committee developed a slightly modified version of the OSI reference model, which is shown in Figure 12–1. Notice the underlying reference to the medium in the IEEE 802 model. It's

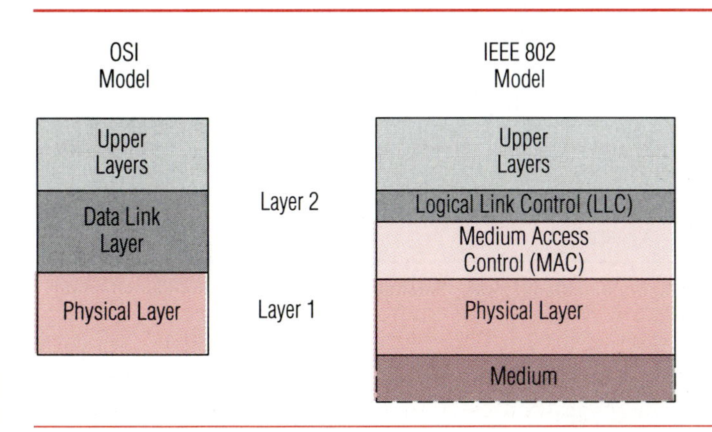

Figure 12–1
A comparison of the OSI and IEEE 802 models.

there because the transmission medium and its topology have a major impact on LAN design and so they are specified in the IEEE standards. Layer 1 specifies the physical connections to the medium that are specific to the LAN environment. Layer 2 is split into the LLC and MAC sublayers. The MAC layer contains the logic for access to a shared medium, typically CSMA/CD or token access, which is not contained in OSI layer 2. Therefore, for the same LLC, two MAC standards, CSMA/CD or token access, can exist.

Logical Link Control (LLC)

The *logical link control (LLC)* sublayer performs the functions of assembling and disassembling frames, providing flow and error control, and interfacing to higher level layers functions that are normally associated with layer 2 in the OSI model. LLC, or IEEE 802.2 as it is better known, is a bit-oriented data link protocol that is similar but not identical to HDLC. LLC's frame is called a *protocol data unit (PDU),* and its format is shown in Figure 12–2. The destination address identifies the node to which the information field is to be delivered, and the source address identifies the node that sent the message. The control field contains the commands, responses, and sequence numbers necessary to control the data link. The information field can contain any multiple of 8 bits, and any combination of bits (transparency) is acceptable. The LLC sublayer performs the functions of flow control, message sequencing, message acknowledgement, and error checking.

Carrier Sense Multiple Access with Collision Detection (CSMA/CD)

CSMA/CD stands for *carrier sense multiple access with collision detection* and it is also known as IEEE 802.3. CSMA/CD is the primary form of contention access to a circuit that is used in local area networks. It was originally developed by Xerox Corporation and has been refined by Digital Equipment Corporation and Intel. It is the contention scheme used on the Ethernet LAN, which is described later in this chapter.

CSMA/CD is a broadcast protocol. There is no master station on a network that uses CSMA/CD; all stations are equal. When a terminal has a message to send, it examines the carrier signal on the network to deter-

Figure 12–2
LLC protocol data unit (PDU) format.

	Header	Destination Address	Source Address	Control Field	Data	Trailer
Bits	*	8	8	8 or 16	$8 \times n$	*

*The format and length of the header and trailer depend on the media being used.

Figure 12–3
The frame format for the CSMA/CD (802.3) protocol.

	Preamble	Start Frame Delimiter	Destination Address	Source Address	Length	Data	Pad	CRC
Bits	56	8	16 or 48	16 or 48	16	≥ 0	≥ 0	32

Preamble—56 bits of alternating 0s and 1s used to establish bit synchronization
Start Frame Delimiter—The bit sequence 10101011
Destination Address—Specifies the station for which the frame is intended; 16 or 48 bits is an implementation decision
Source Address—The address of the station that sent the frame
Length—The length of the data field
Data—The data unit being transmitted
Pad—Bits added if necessary to ensure the frame is long enough for proper collision detection
CRC—A 32-bit cycle redundancy check

mine whether or not a message is already being transferred. If the network is free, the station begins transmitting, indicating with an address the destination terminal that is to receive the message. All connected terminals monitor the network at all times but act on messages only if they see their address characters in the message. Figure 12–3 shows the frame format for the CSMA/CD protocol.

At times, two stations on the network may decide to transmit simultaneously, causing a data collision that garbles the transmission. When the collision is detected, the stations that caused the collision wait a random period of time, determined by circuitry in their communications interface, and then try transmitting again. The length of the random delay is critical because, on the one hand, if it is too short, repeated collisions usually will occur. On the other hand, if the delay interval is too long, the circuit remains idle.

CSMA/CD works quite well when the traffic is light. Its biggest weakness is the nondeterministic nature of its performance when the network is heavily loaded. As the number of terminals or the amount of traffic grows, the number of collisions increases, and the network delays become unpredictable. Thus, CSMA/CD is not a good protocol to use when response time must be consistent, such as in a manufacturing plant where control signals need to be sent to a machine. CSMA/CD can work well, however, in an office application where inconsistencies in response time (as long as they aren't too great) can be tolerated.

Token Passing Protocol

As was described earlier, a small frame called a token circulates on the circuit until a station with a message to send acquires it, changes the token's

Figure 12–4
The frame format for the token passing (802.5) protocol.

General Frame Format

	Starting Delimiter	Access Control	Frame Control	Destination Address	Source Address	Data	CRC	End Delimiter	Frame Status
Bits	8	8	8	16 or 48	16 or 48	≥ 0	32	8	8

	Starting Delimiter	Access Control	Frame Control
Bits	8	8	8

Starting Delimiter—Indicates start of frame. The actual format depends on the type of signal encoding on the medium
Access Control—Variable format depending on type of frame
Frame Control—Indicates whether this is an LLC data frame. If not, bits in this field control operation of the token ring MAC protocol
Destination Address—Specifies the station for which the frame is intended: 16 or 48 bits is an implementation decision
Source Address—The address of the station that sent the frame
Data—The data being transmitted
CRC—A 32-bit cyclic redundancy check
End Delimiter—Contains an error-detection bit, which is set on if any repeater detects an error
Frame Status—Contains certain redundant error-checking bits

status from "free" to "busy," and attaches a message. The token and message move from station to station, and each station examines the address of the message to determine whether it is the intended receiver. When the message arrives at the intended destination, the station copies it. If the CRC check is okay, the receiving station sends an acknowledgment to the sender. The token and acknowledgment return to the originating station, which removes the acknowledgment from the circuit and changes the token's status from "busy" to "free." The token then continues circulating, giving other stations an opportunity to use the circuit. Figure 12–4 shows the frame format for the token passing protocol.

The performance of a circuit using the token passing protocol is predictable. However, if the number of stations on the circuit is large, it may take a long time for the token to get around to a station that has a message to send. Therefore, although response time is predictable, it may be longer than on a CSMA/CD circuit.

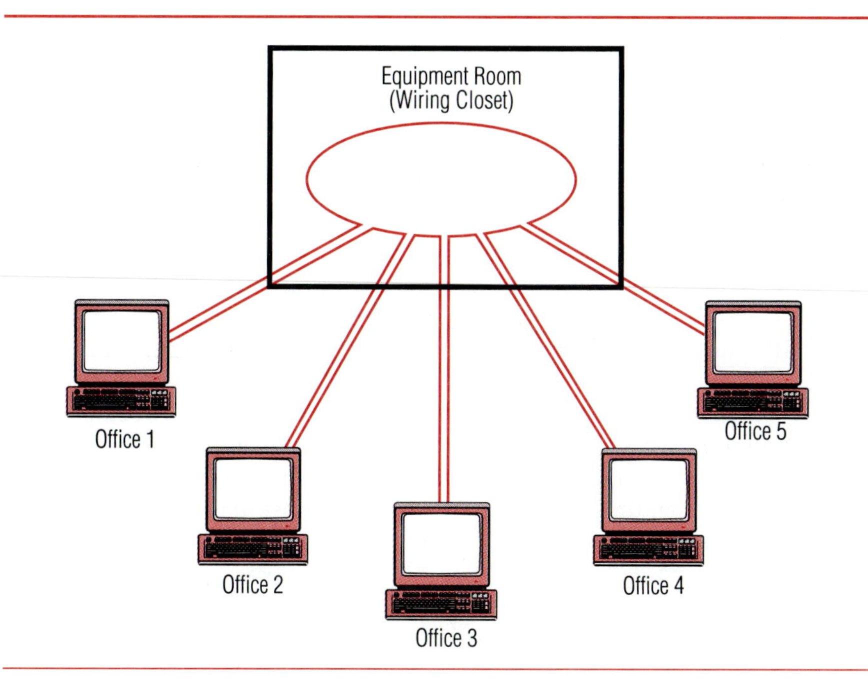

Figure 12–5
A ring LAN installed as a physical star.

LAN TOPOLOGIES

LANs are almost always implemented using either a bus or a ring topology as was illustrated in Figures 11–5, 11–6, and 11–7. Because of the speeds at which LANs operate, there are distance limitations that must be observed. Equipment manufacturers specify the distances over which their equipment will operate, but the limitations have a physical basis related to the propagation delay of the signal on the circuit and the distance the signal can travel without being amplified or regenerated.

LANs with a ring topology are most frequently installed as a physical star connected into a logical ring as shown in Figure 12–5. In most office buildings, there is an equipment room on each floor where the telephone and other network wiring and equipment are located. Wire is run from the equipment room to each office in a star fashion. In the equipment room, the wire is connected to form the ring. The advantage of this type of installation becomes evident when problems occur. Rather than having to trace a ring all over the floor of the building, the technician can work with the ring in the equipment room. Once the technician determines which part of the ring is causing the problem, that leg, running to an office, can be followed or traced. This type of installation has similar advantages in a factory, hospital, or almost any other setting.

LAN MEDIA AND CABLING

Manufacturers design their LAN equipment and software to use specific media according to standards that will be discussed later in the chapter. When a LAN technology is selected, the choice of media is immediately dictated by the requirements of the particular LAN type.

The characteristics of unshielded twisted pair, shielded twisted pair, and coaxial cable were discussed in Chapter 9. For LAN installations, unshielded twisted pair (UTP) wire is very popular and probably the most frequently used. UTP is inexpensive, and in many cases wiring that was originally installed for telephone use can be used for a LAN.

Coaxial cable is also used with many LANs, although it is more expensive than UTP and difficult to work with. It does, however, provide faster transmission speeds than twisted pair wiring. Optical fiber cable is used only in the largest, most geographically dispersed LANs because of its cost. However, the cost of the fiber and its installation is continually dropping, and it is almost assured that optical fiber will be extensively used for smaller LANs in the future.

balun

Normally the same type of wire or cable is used throughout a LAN; however, it is possible to connect different types together. One device that accomplishes the interconnection is a *balun,* which stands for balanced-unbalanced. Baluns are small transformers that allow twisted pair wire to be connected to coaxial cable. They are relatively inexpensive, and are also frequently used where there is a need to connect one type of wire to a device such as a terminal that was designed to connect to a different type of cable.

Hubs

hub

At the place where all of the wires and cables come together, typically in the wiring closet, a device called a *hub* is normally installed. Hubs provide an easy way to connect all of the wires and cables, and most of them allow different types to be interconnected. Hubs also serve as a repeater to boost the signal strength, thereby allowing longer cable runs out to individual workstations or servers. Many hubs have intelligence to detect errors and provide assistance to a technician in locating a failing component such as a cable with a high error rate, a cut cable, or a failing workstation. Hubs are sometimes called *multistation access units (MAU), multiplexers,* or *concentrators.*

Wiring Cost

The cost of installing the wire and cable for a LAN can be a significant portion of the overall cost of the LAN. The least expensive alternative is to use previously installed UTP telephone wire if it is of adequate quality and if the LAN type permits it. If new wiring must be installed, it is important to remember that the major portion of the cost of installing new wiring is generally not the cost of the wire or cable, but the cost of the labor to install it. Therefore, remembering the steadily increasing demand

This is an example of a local area network hub and is manufactured by Cisco Systems. (Courtesy of Cisco Systems, Inc.)

for bandwidth, it is far better to install higher quality cable or more pairs of wires than seems to be necessary at the time the wiring is designed. Many companies today install four pairs of Category 5 wiring to each desktop when they are doing new wiring installations. One pair is designated for the telephone, one pair for data (the LAN), one pair as a spare, and one pair as—another spare! Maybe it will be used for a full-duplex LAN in the future, perhaps for desktop video conferencing, or some other as-yet-unimagined application.

Wiring Documentation

The LAN wiring should be documented with a diagram that shows where each pair of wires or cable runs and to which piece of equipment it connects. Usually a *cabling plan* that specifies all of this information is prepared before the wire is installed, but inevitably, during the course of installing the wire, changes are made. It is very important to go back and update the cable plan to show what was actually installed. Engineers sometimes call this the "as built" drawing.

cabling plan

The wire and cable should be physically labeled with tags at each end that relate back to the wiring diagram. Because LANs are prone to frequent change, reconfiguration, and expansion, it is very easy to let this documentation get out of date, and only strong commitment on the part of the LAN owner/administrator, with management insistence and support, will ensure that the wiring documentation is kept up to date.

WIRELESS LANs

If you are a mobile person, you can't normally just roam around your campus or office with your laptop computer and still use your LAN. The adapters and wires keep you tethered fairly closely to your desk or

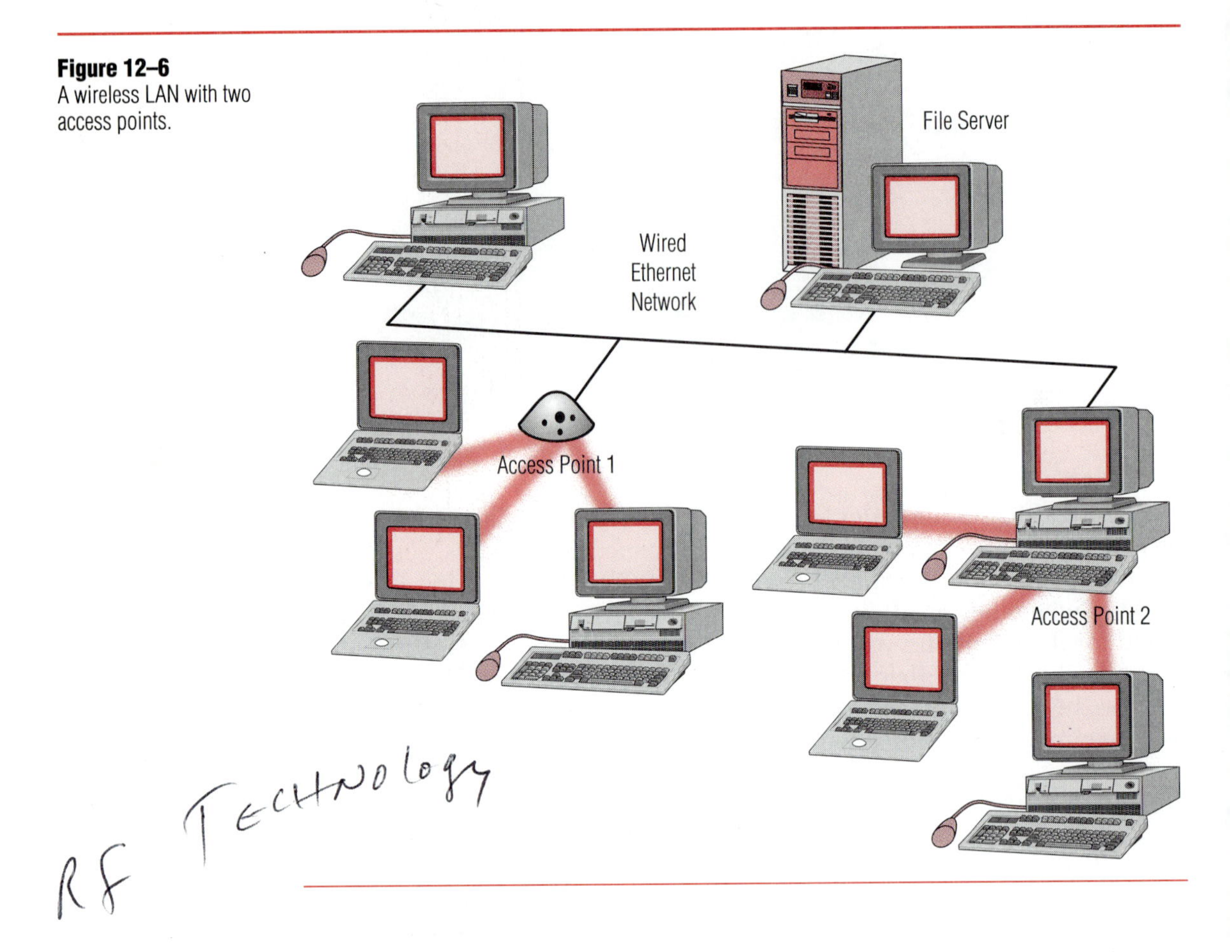

Figure 12–6
A wireless LAN with two access points.

workspace. Wireless technology can be a solution for this problem in certain situations.

A *wireless LAN* uses radio frequency technology as an extension of, or as an alternative to, a wired LAN. In a typical wireless LAN configuration, a transmitter/receiver (transceiver), called an *access point,* connects to a wired LAN using standard cabling as shown in Figure 12–6. An antenna is usually mounted high enough to give radio coverage in the desired area. The access point acts like a hub, receiving, buffering, and transmitting data between the wireless workstations and the wired network. A single access point can support a small group of users within a range up to several hundred feet. Many real-world applications exist where a single access point services 15 to 50 user devices at a range on the order of 500 feet indoors and 1,000 feet outdoors.

Users access the wireless LAN through wireless LAN adapters, which are implemented as PC cards in their computers. The LAN adapter provides the interface between the computer and the radio signal via an

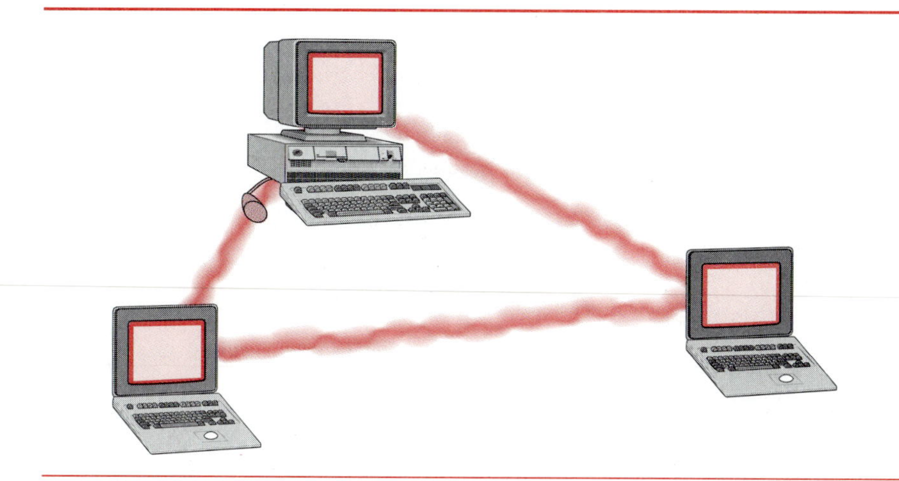

Figure 12–7
A peer-to-peer wireless LAN network.

antenna. The computer and its software is unaware of the fact that the connection to the LAN is wireless rather than through a standard cable.

In another configuration, a number of computers, each equipped with a wireless LAN adapter card, may communicate with each other without going through an access point as shown in Figure 12–7. This is called a *peer-to-peer wireless network,* and the users can share files and printers but are not able to access wired LAN resources.

In a large facility or in a campus setting, multiple access points can be installed providing wide coverage. Access point positioning is determined by means of a site survey. The goal is to blanket the coverage area with overlapping coverage cells, as illustrated in Figure 12–8, so that users can move freely without disruption to their LAN communication. This capability, called *roaming,* operates transparently to the users, much like a cellular telephone system. The wireless networking hardware automatically shifts to the access point with the best signal.

roaming

Wireless LANs usually use either frequency hopping or direct sequence spread spectrum technology, which were discussed in Chapters 5 and 9. Most manufacturers have adopted the IEEE 802.11 standard for transmission, allowing different brands of equipment to work together. It is always a good idea, however, to check with the vendors to verify that their equipment will interoperate.

Wireless LAN technology has been successfully implemented in a variety of situations. Doctors and nurses in hospitals can update patients' data from their terminals while visiting with the patients in their rooms. Students have established study groups to share data as they work together. Warehouse workers use wireless LANs to exchange information with central databases, updating inventory counts as they pull or store merchandise. Wireless LANs are becoming more widely recognized as a general-purpose connectivity alternative for a broad range of situations.

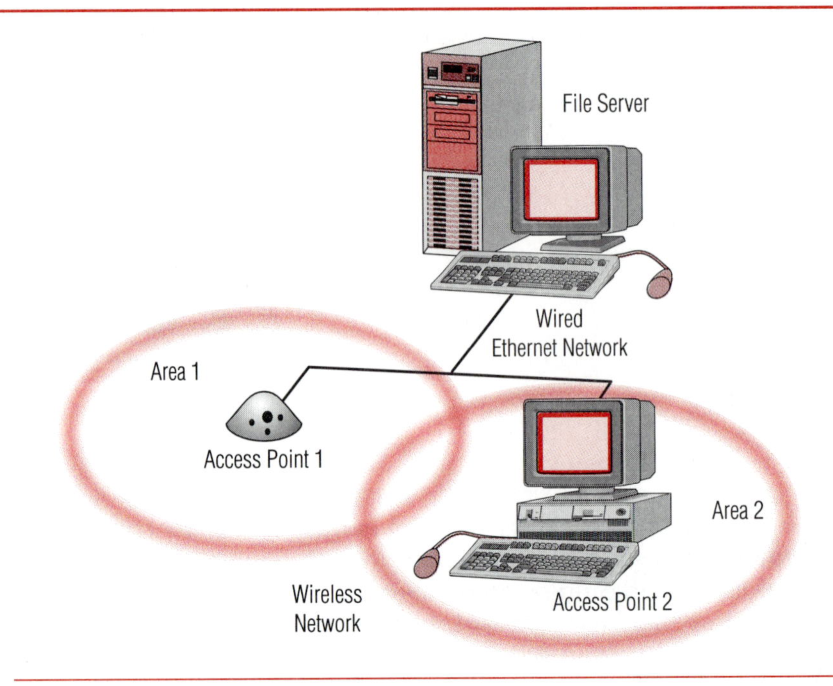

Figure 12–8
Multiple access points with overlapping coverage.

LAN TRANSMISSION TECHNIQUES

baseband

There are two transmission techniques used for LANs. One technique, called *baseband transmission,* uses a digital signal. When baseband transmission is used, the medium is directly pulsed, and the entire bandwidth is used for a single signal. Baseband transmission typically occurs at speeds of 1 Mbps and higher. Baseband transmission is used when data is being transmitted, but it is only suitable for voice or television if the signals are digitized.

broadband

The other transmission technique is called *broadband transmission,* and the signal is transmitted in analog form. The capacity of the cable is subdivided, using frequency division multiplexing, into whatever circuits or channels are required for the particular applications. The difference between broadband and baseband transmission is in how the bandwidth of the circuit is used, not necessarily in the capacity or the medium. A chief advantage of broadband transmission is that many different kinds of communications can be going on simultaneously. The broadband system can be used for multiple purposes, such as data, voice, and television, without having to install separate lines for each type of traffic.

The majority of today's LANs use baseband transmission. LANs often are used to connect personal computers so that they can communicate with one another or so they can share a large disk or a printer. Since only one transmission can be carried at a time, baseband techniques rely on handling each transmission very quickly using the high speed at which baseband operates. A transmitting node gets control of the medium and transmits its message. Since the transmission speed is high, the transmission is completed quickly, and the medium becomes free for another node to use.

Broadband transmission is typically found in an environment where there are diverse requirements for many types of transmissions. The broadband system can be divided as if it had several baseband channels inside it. In addition, broadband systems typically carry transmissions that do not meet the technical qualifications of a LAN. In a manufacturing plant, for example, a broadband system might be used to connect

- robots to a minicomputer serving a production line;
- VDT terminals to a host computer;
- laboratory instruments to a central computer used for quality testing;
- a series of personal computers that are connected to each other in LAN fashion (on one channel of the broadband system);
- a plant television studio to television sets located around the plant for broadcasting plant news, notice of job availabilities, and other information of interest to the employees.

All of these transmissions could occur simultaneously on a broadband system, and each would operate at a speed appropriate for the type of transmission. Thus, a broadband system has a greater ability to handle a wider range of signals than a baseband system.

Broadband and baseband systems should be seen as complementary, not competitive. Both have distinct advantages. Baseband systems are simpler; new terminals can be attached by simply tapping into the cable. Broadband systems require modems to modulate the signal to the proper frequency range. Thus, they are more expensive to implement. On the one hand, broadband cable is normally physically much larger than baseband cable; hence, it is more expensive and difficult to install. On the other hand, broadband systems can handle more diverse communications requirements.

LAN STANDARDS

The Institute of Electrical and Electronics Engineers (IEEE) has done a great deal of work for many years to standardize the definitions of LANs and how they operate. They recognized that different LAN applications might have different technical requirements, so they established several standards for LANs with differing characteristics. The IEEE 802 committee has published a number of LAN standards, and committee work continues as the technology evolves. A complete list of the IEEE 802 subcommittees is shown in Figure 12–9.

Figure 12–9
The LAN-related IEEE standardization subcommittees.

Standard Number	Committee's Purpose
802.1	Higher layer LAN protocols
802.2	Logical Link Control
802.3	Ethernet
802.4	Token Bus
802.5	Token Ring
802.6	Metropolitan Area Network
802.7	Broadband
802.8	Fiber Optics
802.9	Isochronous LAN
802.10	Security
802.11	Wireless LAN
802.12	Demand Priority
802.13	Not used
802.14	Cable Modem
802.15	Wireless Personal Area Networks
802.16	Broadband Wireless

Four standards in particular, numbered 802.2, 802.3, 802.4, and 802.5, form the heart of LAN network standardization. These define the logical link control (LLC) protocol and the contention and token methods of accessing a LAN. Manufacturers have designed their hardware and software to conform to these standards, enabling equipment from a variety of manufacturers to communicate. The interrelationship of these four standards is shown in Figure 12–10, which also shows the four major attributes that describe a LAN. Figure 12–11 identifies the alternatives for each of these attributes and shows combinations used by some real-world LANs.

LAN SYSTEMS

Now we will look at specific information about the LANs that are implemented most widely. Ethernet LANs, token ring LANs, and the others described in this section comprise the vast majority of LANs installed in organizations throughout the world.

Ethernet

The original work on LANs was done at Xerox Corporation in the 1970s. In 1980, Xerox teamed with Digital Equipment Corporation and Intel, and jointly announced a LAN product called *Ethernet,* the oldest of the LAN technologies. Ethernet is an 802.3 LAN and, in its basic form, operates at 10 Mbps using baseband transmission with Manchester data coding on a

Figure 12–10
LAN standards. From William Stallings, Data and Computer Communications, 5th edition, © 1997, p. 367. Adapted by permission of Prentice Hall Publishing Company.

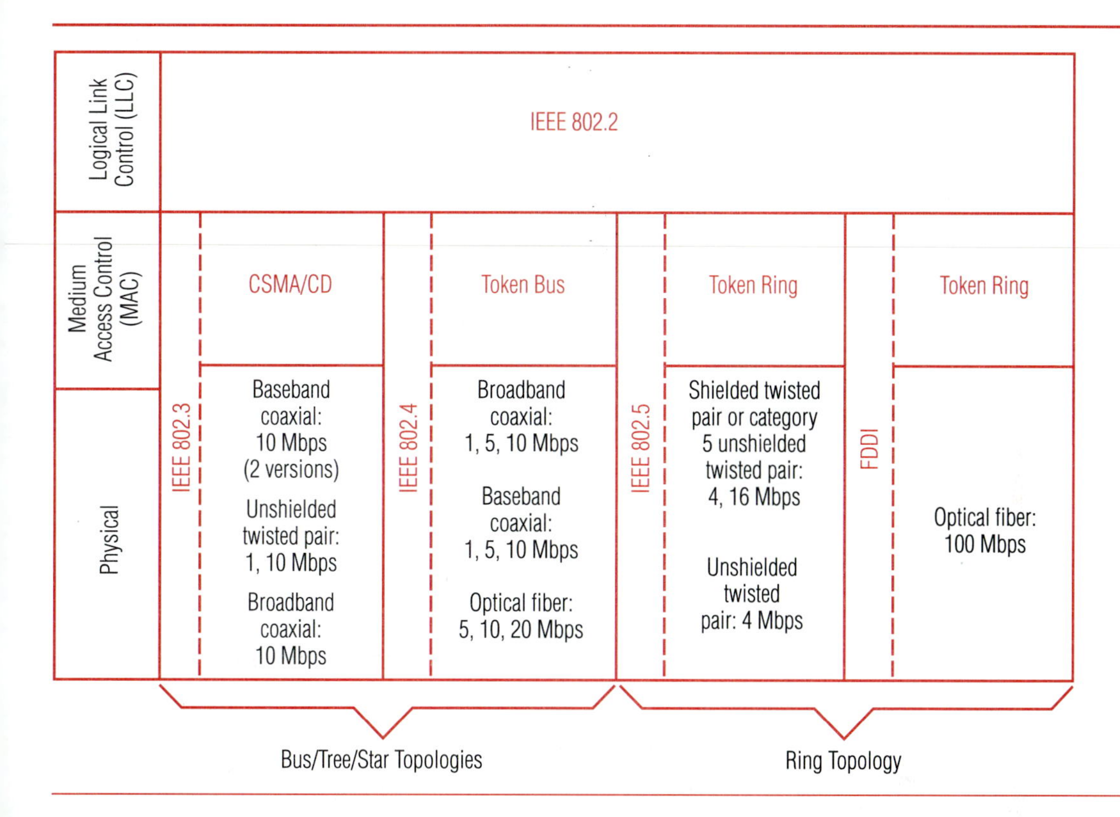

Figure 12–11
LAN attributes and options, and the combinations used by certain commercially available LANs.

LAN Attributes	Options	Ethernet	Token Ring	Appletalk	Arcnet
Transmission Technique	Baseband	X	X	X	X
	Broadband				
Access Control	CSMA/CD	X		Non-std.	Non-std.
	Token		X		
Topology	Bus	X	X	X	
	Star		X	X	
	Tree				
	Ring		X		
Medium	Twisted Pair	X	X	X	X
	Coaxial Cable	X	X	X	X
Speed		10 Mbps 1 Gbps	4 Mbps 16 Mbps	230 kbps	25 Mbps 20 Mbps

Figure 12–12
An Ethernet LAN—conceptual view.

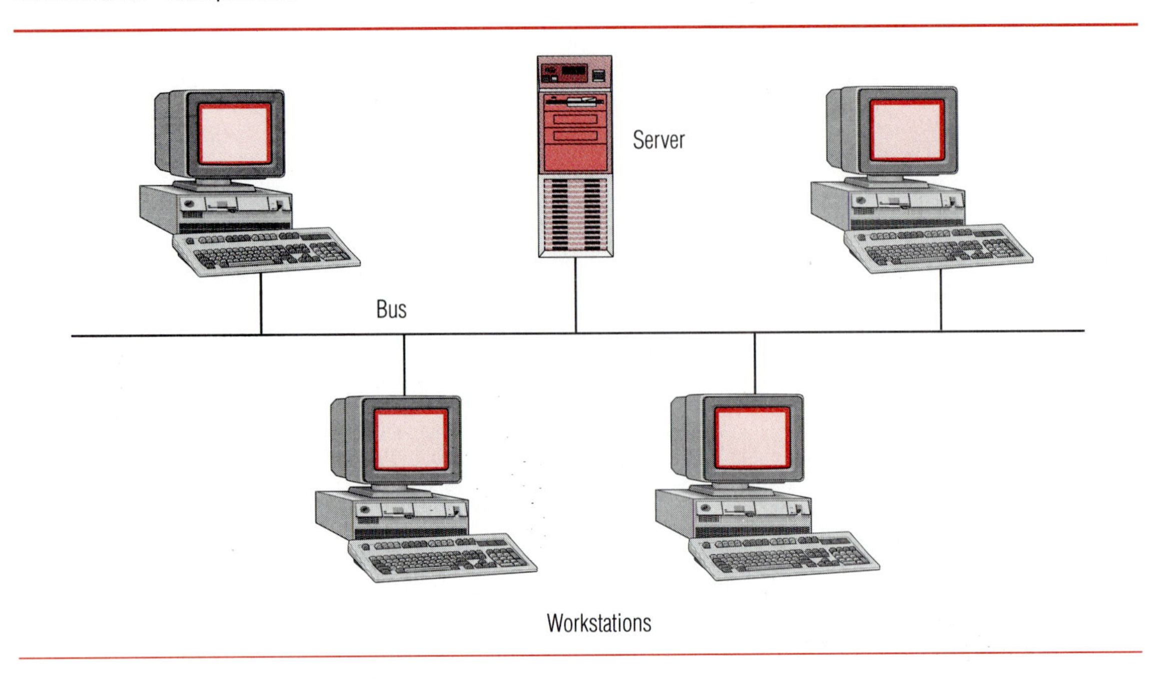

bus

bus. The term *bus* implies a high-speed circuit of limited distance, and a bus is normally implemented within a single building. Figure 12–12 shows how a picture of an Ethernet LAN is usually drawn, but in actuality, networks are more generally implemented as shown in Figure 12–13 because it is desirable to confine as much critical cabling and hardware as possible to an equipment room and to connect the wire or cable to a hub to simplify testing and debugging, and to improve security.

segment

Ethernets were originally implemented using standard coaxial cable, which is 0.4 inch in diameter. They were given the name 10Base5 Ethernet, which means a speed of 10 Mbps, using baseband transmission, for a maximum distance of 500 meters. The 500-meter distance limitation constitutes a *segment* of a 10Base5 Ethernet LAN, and if the distance needs to be exceeded, a second segment must be installed and the two segments connected with a bridge or switch. Today Ethernets implemented with 0.4-inch diameter cable are frequently called "Thick Ethernets." Now, more flexible cable that is only .25 inch in diameter is commonly used because it is cheaper and easier to install. The specification is 10Base2, 10 Mbps over 200 meters, and the common names are "Thin Ethernet" or "Cheapernet." The most common type of Ethernet, however, is 10BaseT, which yields 10 Mbps over twisted pair wire, but normally for a distance of only 100 meters per segment.

Figure 12–13
Ethernet LAN as typically installed.

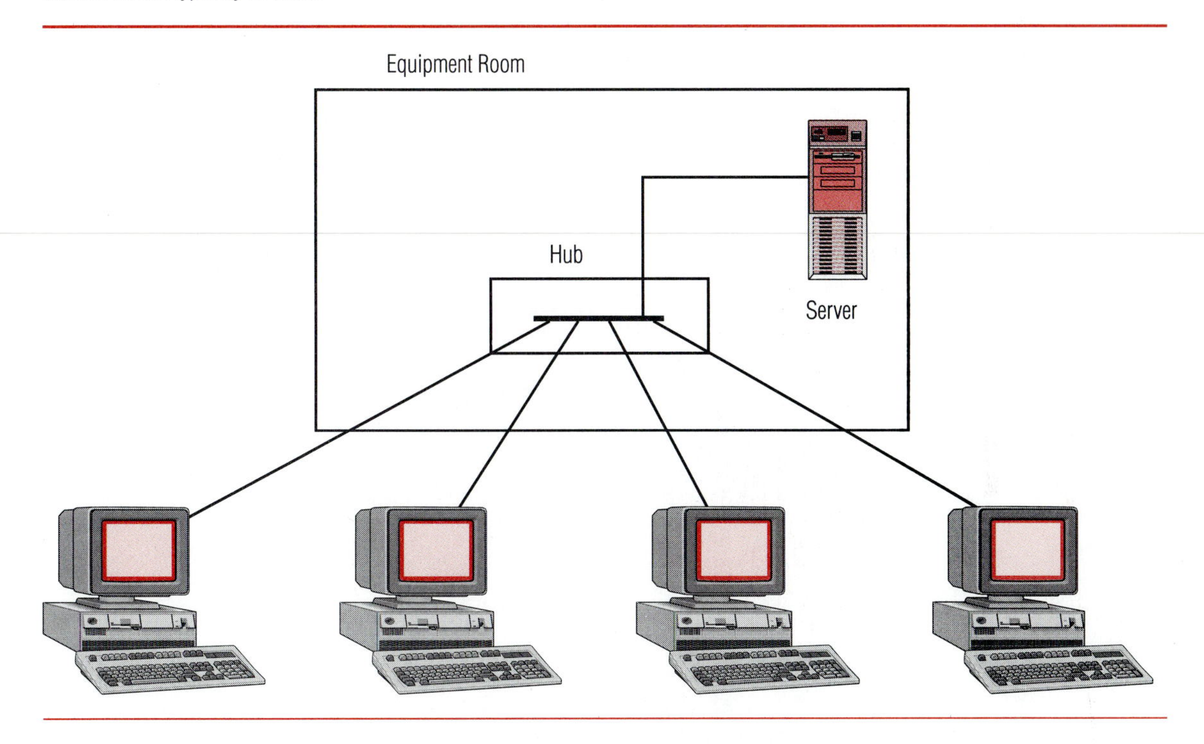

The never-ending demand for more bandwidth and higher transmission speed has pushed vendors to develop higher speed Ethernets. Most common among these today is 100BaseT, a 100 Mbps Ethernet that operates over two pairs of unshielded twisted pair cable, known as 100BaseTX, two optical fibers, known as 100BaseFX, or four pairs of Category 3 UTP, known as 100BaseT4. 100BaseT retains the IEEE 802.3 frame format, size, and error detection mechanism. It supports all applications and networking software running on 802.3 Ethernet networks, so it is relatively easy for network managers to upgrade to the higher speed, especially if they have Category 5 cabling installed.

Gigabit Ethernet, also known as 1000BaseT, is an extension of the IEEE 802.3 standard called IEEE 802.3ab. This standard was approved by the IEEE in mid-1999. Running over distances of up to 100 meters, 1000BaseT uses four pairs of Category 5 cable, and is used for applications that have very high bandwidth requirements. And if that weren't enough bandwidth, IEEE committee 802.3ae is expected to a have a standard for 10 gigabit Ethernet running on optical fiber by the end of 2001! The characteristics of all of the forms of Ethernet are summarized in Figure 12–14.

Figure 12–14
Characteristics of various Ethernet LAN technologies.

Characteristic	10Base5	10Base2	10BaseT	10Broad36	100BaseTX	100BaseFX	100BaseT4	1000BaseT
Medium	Coaxial cable	Coaxial cable	Two pairs of Category 5 UTP	Coaxial cable	Two pairs of Category 5 UTP	Two optical fibers	Four pairs of Category 3 UTP	Four pairs of Category 5 UTP
Topology	Bus	Bus	Star	Bus	Bus	Bus	Bus	Bus
Signaling	Baseband	Baseband	Baseband	Broadband	4B/5B NRZI	4B/5B NRZI	8B/6T NRZI	NA
Transmission speed (Mbps)	10	10	10	10	100	100	100	1000
Maximum segment length (meters)	500	185	100	1800	100	400	100	100

Token Ring

Token ring technology was developed by IBM and has been standardized as 802.5. It is second only to Ethernet in general popularity. As the name implies, token ring LANs use a token passing access technique and protocol, described in Chapter 10, on a ring topology. Data is encoded using the differential Manchester technique. Figure 12–15 shows the conceptual view of the way token rings are usually pictured, and Figure 12–16 shows the way they are normally installed. Token ring LANs operate at 4 Mbps on most unshielded twisted pair wire or 16 Mbps on shielded twisted pair or Category 5 unshielded twisted pair. Generally, UTP wiring allows the

Figure 12–15
Token ring LAN—conceptual view.

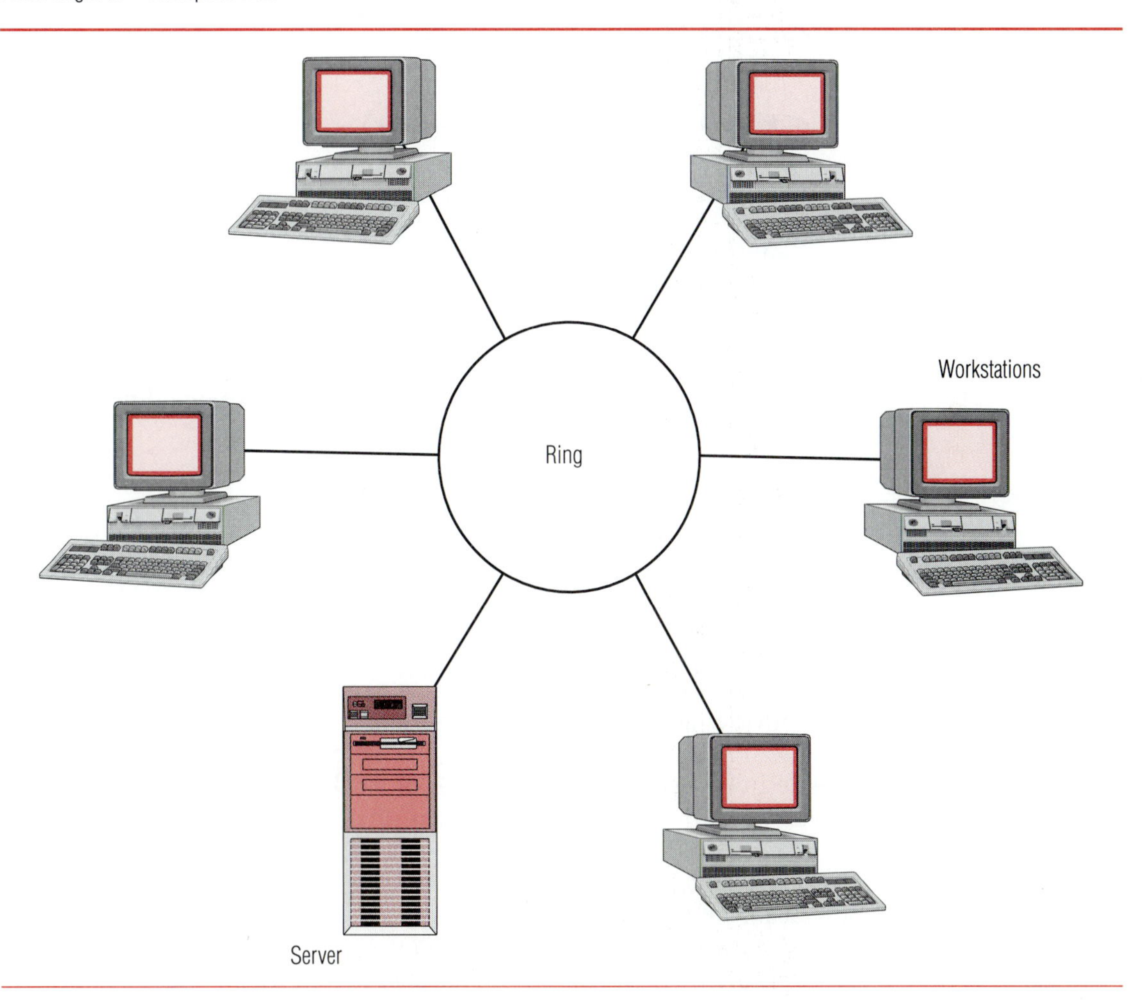

Figure 12–16
Token ring LAN as typically installed.

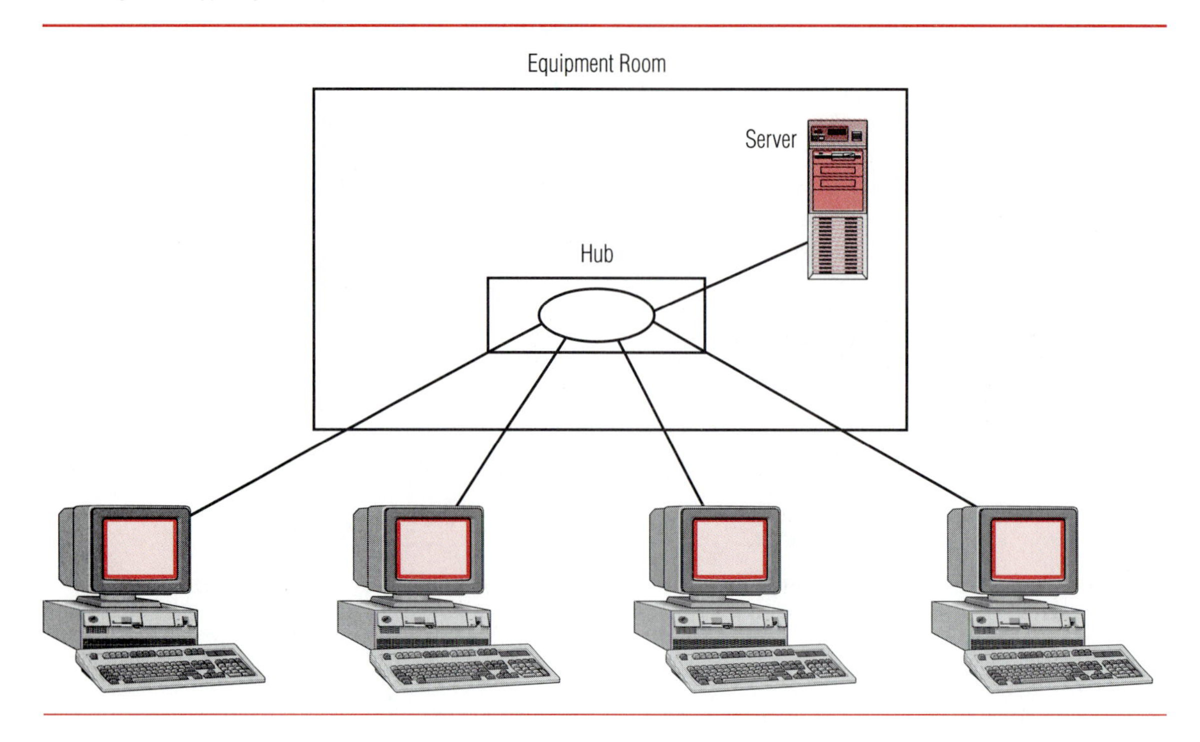

connection of up to 72 workstations per ring, while shielded twisted pair allows up to 260 connections.

100VG-AnyLAN

100VG-AnyLAN was developed as another way to improve the throughput on LANs. It is an extension of Ethernet, but allows both Ethernet and token ring packets. 100VG-AnyLAN does not use CSMA/CD for media access but rather a new technique called *demand priority* or *demand priority access method (DPAM)*, which is like roll call polling (discussed in Chapter 10), except that there is also a provision for computers to issue high-priority access requests and get control of the LAN more quickly. 100VG-AnyLAN transmits data at 100 Mbps and uses four pairs of unshielded twisted pair, Category 3, 4, or 5 cable, and fiber-optic cable can also be used.

The Fiber Distributed Data Interface (FDDI) Standard

The *Fiber Distributed Data Interface (FDDI)* standard was developed by a subcommittee of the American National Standards Institute (ANSI) and

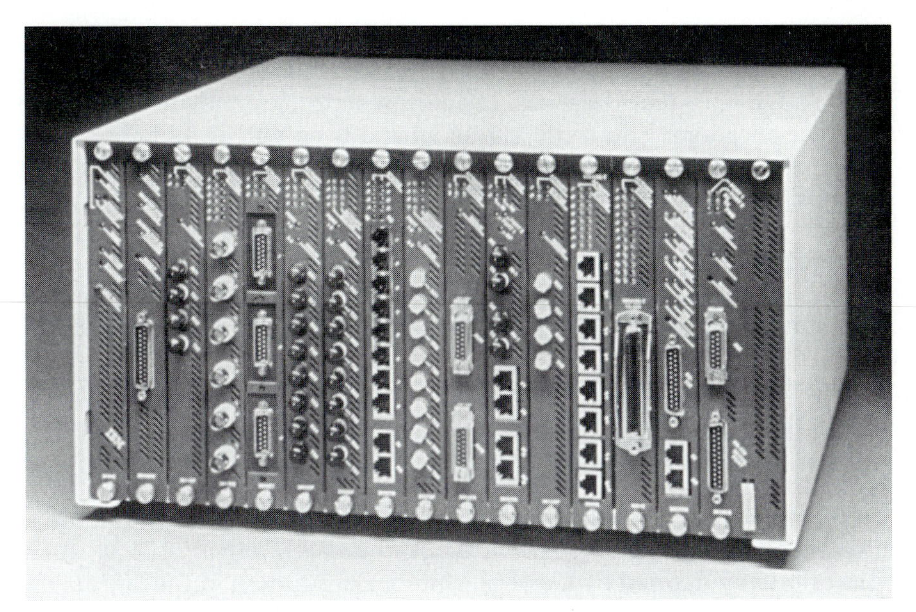

This IBM multi-station access unit (MAU) is an intelligent hub for token ring LANs. (Courtesy of International Business Machines Corporation. Unauthorized use not permitted.)

was completed in 1990. As LANs based on the IEEE 802 standards reach capacity, optical fiber LANs based on the FDDI standard become an alternative growth path. In their first implementations, FDDI LANs were used to provide high-speed backbone connections between other LANs. Now, however, FDDI is being used for normal LANs that need to operate at very high speed, or which have a large number of users. It is interesting to note that FDDI has also been implemented on wire using two pairs of Category 5 cable. Other than the media, the implementation is the same.

The FDDI standard defines a LAN based on two counterrotating 100 Mbps token rings. One of the rings is designated as the primary ring, and the other is the secondary ring. All stations on the LAN are connected to both rings. In normal operation, data is passed on the primary ring while the secondary ring is idle. Should the primary ring fail, the stations on each side of the failure point reconfigure themselves, as shown in Figure 11–7, so that data flow continues using a combination of the primary and secondary rings. The token passing technique defined by FDDI is similar to the one defined by IEEE 802.5 but with significant enhancements in the areas of fault tolerance and topology.

The FDDI standard allows up to 500 stations to be connected to the ring in a priority system. High-priority stations can access the ring for longer periods of time. The maximum length of the ring can be up to 200 kilometers, and stations must be located no more than 2 kilometers apart.

FDDI-II is a newer standard that provides additional capability beyond that offered by FDDI. FDDI-II provides the ability to handle circuit-switched traffic in addition to the packet-switched traffic handled by the original FDDI standard. FDDI-II also provides the ability to provide a constant data rate connection between two stations, which is not possible with FDDI. Constant data rate connections are required for voice and video applications. You can imagine what a telephone call might sound like if the voice data were not able to get through the network at a constant rate!

It is easy to see that FDDI and FDDI-II greatly expand the speed and operating distances of LANs. Now it is possible to connect a large campus or corporate complex into a single LAN or to easily connect smaller LANs in buildings or departments where that is desirable. Eventually, there will be a need to have FDDI speed on many desktops to support applications, such as full motion, color video, high resolution graphics and photographs, traditional data, and voice—all concurrently!

Map

The other standardized token ring architecture is 802.4, token ring on a broadband bus, called *Manufacturing Automation Protocol (MAP).* MAP was originally defined by General Motors Corporation in the late 1970s, and the work was later standardized in IEEE 802.4. MAP LANs use broadband coaxial cable at 1, 5, 10, and 20 Mbps. However, MAP goes beyond the IEEE 802.4 standard and defines all except the presentation layer of the OSI model. In that context, MAP's use of the word *protocol* is really the definition of a complete set of rules for the entire communications process, rather than just a data link protocol. MAP standards are especially attuned to the needs of manufacturing plants. MAP networks provide for the connection of many dissimilar devices, guaranteed response times, ease of maintenance, and high reliability. The large number of companies that support the MAP standards provides a large market for vendors of MAP hardware.

AppleTalk

AppleTalk LANs gained wide popularity among the users of Apple Macintosh computers because they are inexpensive, easy to install, and the required hardware can be easily added to Macintosh computers. AppleTalk LANs are best suited to small groups of workstations that are in close proximity to one another.

The AppleTalk protocol, which is built into all Macintosh computers, is a nonstandard CSMA/CA medium access protocol. An AppleTalk workstation sends a small 3-byte packet on the network signaling its intent to do data transmission. When other terminals on the network see this packet, they stand by until the data from the first terminal has been sent.

AppleTalk networks may be configured in a bus or star topology using Apple's cabling system called *LocalTalk.* The medium can be twisted pair, coaxial cable, or fiber-optic cables, but twisted pair wiring is most commonly used because of its low cost and availability. Transmission speed is 230 Kbps, which is quite slow compared to other LANs. Up to 32 workstations can be connected to an AppleTalk LAN with a maximum total distance of about 1,000 feet. Multiple AppleTalk networks can be connected using routers or other intermediate devices.

The major problems with AppleTalk are its slow transmission speed, by today's standards, and its use of nonstandard protocols. Most networked Apple computers, especially the newer iMacs, are now linked with Ethernet.

Arcnet

Attached Resource Computer NETwork (Arcnet) is a LAN architecture that was developed by Datapoint Corporation in 1977. Arcnet became popular because of its low cost, ease of installation, and early availability compared to other LANs. Because of its popularity, it became a de facto standard of its own. However, since it does not follow any of the IEEE 802 LAN standards, it has lost popularity in recent years to Ethernet and Token Ring.

Arcnet is a baseband, token passing architecture. Networks may be established with either a bus or star topology using unshielded twisted pair, coaxial cable, or optical fiber media. Transmission speeds are 2.5 Mbps, or, using a newer Arcnet Plus architecture, 20 Mbps on coaxial cable.

Repeaters are called active hubs, and they allow the connection of multiple workstations in addition to performing signal amplification. Passive hubs allow the connection of multiple workstations without the signal amplification function. Using combinations of active and passive hubs, Arcnets may be arranged in many different configurations with varying distance limitations. A typical Arcnet configuration is shown in Figure 12–17.

EXPANDING THE LAN

Only the smallest organizations find that one LAN meets all of their requirements. An organization may have offices in different towns or several miles apart in the same town. Or a token ring LAN is installed in the office, but an Ethernet LAN is required in the factory. Or a large organization may have a LAN in each department. In many cases, the organization finds that there would be benefits from connecting the LANs to allow people who use one of the LANs to communicate with people on another LAN—or on all of the other LANs.

If the networks are a long distance apart, the network designer must carefully consider the speed of the communications circuit connecting the LANs. Since LANs operate at speeds measured in millions of bits per second, a metropolitan area network made up of one or more high-speed communications lines, such as T-1 or T-3 circuits, is desirable in order to maintain high throughput. However, the cost of high-speed circuits may be

Figure 12–17
Typical Arcnet configuration.

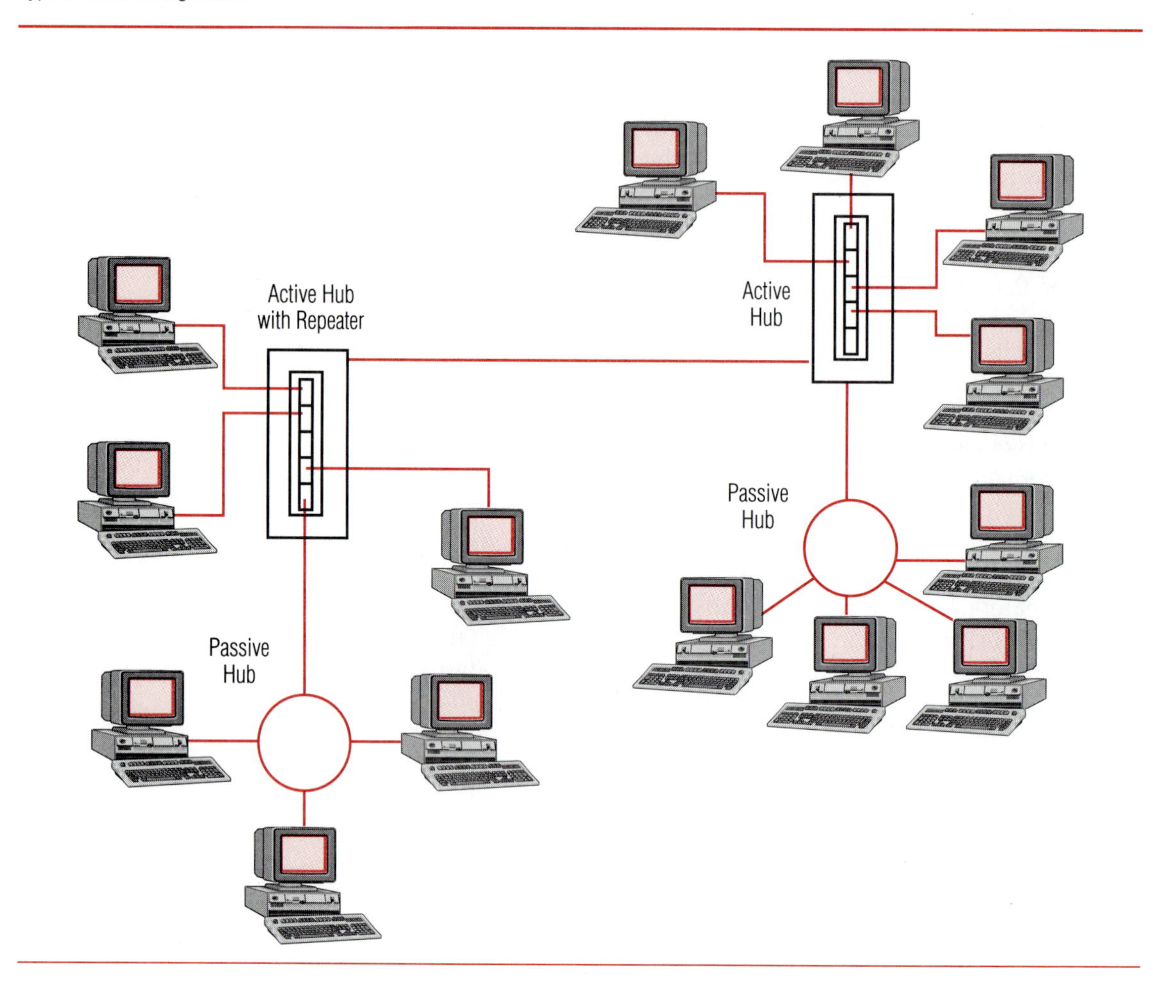

prohibitive for all but the largest organizations, in which case a lower speed circuit with the resulting performance compromise will have to be tolerated.

LAN interconnection can be done in several ways, depending on considerations such as the LAN technologies, the distance between the LANs, and the volume of communication that will flow between the LANs. As you might guess, connecting two LANs that use the same data and network protocols, for example, two Ethernet LANs, is easier than connecting LANs that have dissimilar technology. Depending on the factors, LANs may be directly connected to each other using one of several devices which are described on the next page, or LANs may be connected to a high-speed backbone LAN, which might, for example, use optical

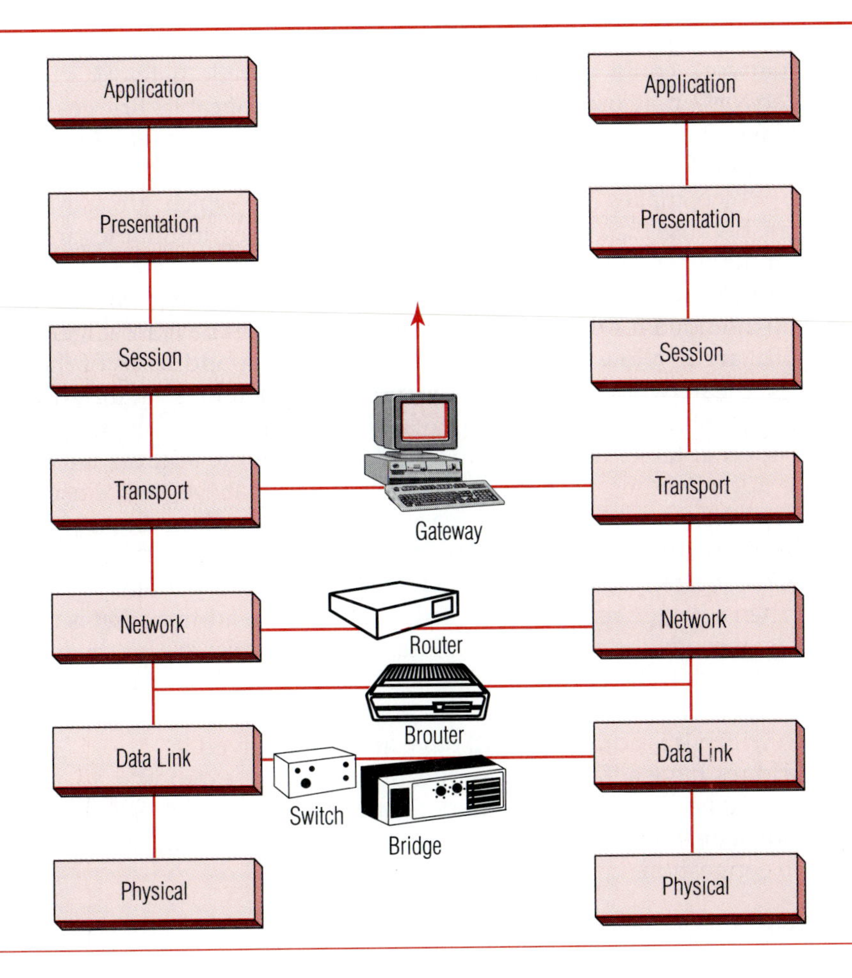

Figure 12–18
Network interconnection can occur at various layers of the OSI model.

fibers and the FDDI protocol to transport data between the lower speed LANs. Let's look at the options in more detail.

There are five types of network interconnection hardware:

- bridges;
- switches;
- routers;
- brouters;
- gateways.

Each operates at a different layer of the OSI model, as shown in Figure 12–18.

Bridges

Bridges connect two or more network segments that use the same protocol. For example, two Ethernets could be connected to each other with a bridge allowing data to be sent from one to the other. Workstations on each network could communicate as though a single network existed. The networks may use the same or different types of cable. Bridges operate at layer 2, the data link layer, of the OSI model and are usually implemented in software running on specialized hardware or a personal computer, which makes them slower than switches, which are hardware devices. Because they perform no translation, bridges are simpler and operate at higher speeds than routers or gateways.

Bridges "listen" to all of the traffic on one network segment and pass on data intended for the other segment. Most bridges have intelligence in the software to allow them to "learn" the data link addresses of the devices on the network. When a bridge receives data, it looks at the address of the device that sent it and compares the address to those already stored in an internal table. If the device is not already in the address table, the bridge adds it, along with the network segment that the device is on. In this way, the table gradually increases in size and reflects the addresses and locations of all of the devices on the network. Using this table, the bridge can more quickly determine where to send future data.

When a bridge receives data with a destination address that is not in its table, it sends the data out on all network segments except the one on which it came. If the destination address is on the same network segment that the data came in on, the bridge can discard the data, because the device to which it is addressed will have already received it.

Bridges have a variety of applications, such as extending a LAN to longer distances and greater numbers of ports. Another important use is to divide a busy LAN into two smaller LANs to improve throughput and reduce congestion.

Switches

A *switch* is a device that also operates at layer 2, the data link layer, and connects two or more network segments. Unlike the bridge, switches are implemented in hardware and allow all of the connections to operate simultaneously, making them very fast. Another difference is that switches do not learn addresses; they need to have the addresses of devices defined for them.

If, for example, a switch connects four LAN segments, A, B, C, and D, as shown in Figure 12–19, segments A and B can be communicating, while simultaneously segments C and D are communicating. The switch interconnects the segments as needed on a packet-by-packet basis. The fast switching and simultaneous connection ability gives the overall throughput a significant boost compared to other configuration options.

cut-through switch

There are two types of switches to be familiar with. A *cut-through switch* only looks at an incoming packet's address and immediately sends it out to the destination LAN segment. However, if that segment is in use, a collision will occur and error recovery must be invoked. A *store-and-forward switch*

store-and-forward switch

Figure 12–19
A switch allows simultaneous connection of LAN segments.

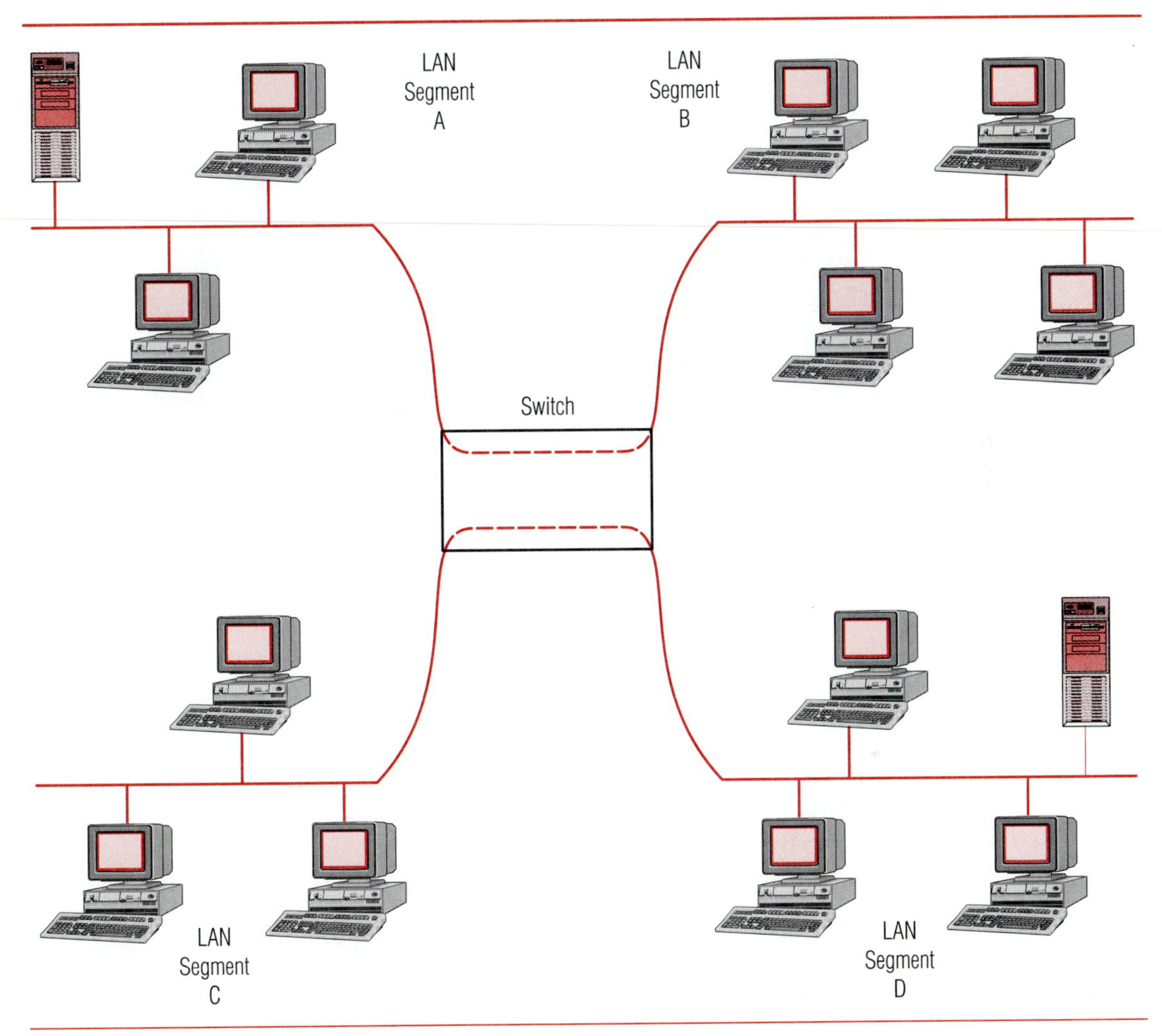

brings each incoming packet into memory. The switch examines the destination segment, and if it is busy, the switch holds the packet until the segment is free and then sends it out. Store-and-forward switches are, in general, slower and more expensive, because of their memory, but the buffering results in fewer errors on the LAN.

Routers

Routers perform the function of passing messages from one network to another and translating the destination address to the format required by

This family of routers from Cisco systems provides varying capacities and capabilities depending on the requirements of the customer. (Courtesy of Cisco Systems, Inc.)

the network receiving the message. Routers operate at layer 3, the network layer, of the OSI model, and the networks they connect may or may not be similar at that layer.

A router performs two basic activities: determining the optimal routing paths and transporting data through the network. The optimal path may be measured several different ways, such as the fewest number of links to the destination, the least cost, or the speed of the circuits along the way. The router uses a packet's destination address and a routing table stored in its memory to determine how to forward the packet. Routers can keep track of several possible routes to a destination and forward the packet along an alternate path if the primary route is busy or out of service.

Routers communicate with one another and maintain their routing tables with the latest information about the status of the network. By analyzing routing updates from other routers, a router can maintain an up-to-date picture of the network topology.

The logic that routers use to determine how to forward data is called a *routing algorithm.* Routing algorithms have been proposed and described in computer science literature for many years. They are designed to be simple, yet robust, stable, and flexible. They need to be capable of selecting the best route for a packet, which depends on the metrics, such as cost, number of links, bandwidth, delay, and traffic load. As you can gather, routing messages in a large network is a very complex topic, and routers perform a very critical function.

Brouters

A *brouter* performs the functions of a bridge and a router on a selective basis. For example, it may bridge certain messages and route others, depending on the needs of the source and destination networks.

Router and brouter technology is developing rapidly. As more companies find a need to connect their diverse networks or to connect to networks of other organizations, router/brouter technology will play an increasingly important role.

Gateways

If two networks operate according to different network protocols, a *gateway* is used to connect them. Gateways operate at OSI layers 4 and up and basically translate the protocols to allow terminals on the two dissimilar networks to communicate. Some gateways also translate data codes, for example, from ASCII to EBCDIC. This would be useful on a LAN when a communications server routes traffic from a PC-based network using ASCII to an IBM mainframe that uses the EBCDIC code.

Like bridges, gateways are combinations of hardware and software. They may be implemented on a specially designed circuit card or by using specialized software in a standard personal computer. Gateways can suffer from slow performance because of the protocol translation, so their performance must be considered and tested when a gateway installation is contemplated. Gateways perform an important role in allowing an organization to interconnect different types of LANs so that, to the user, the network appears as a single entity. Figure 12–20 illustrates the difference between gateways and bridges.

As the number of LANs or LAN segments that needs to be interconnected increases, the design approach that is often taken is to connect the LANs to a high-speed backbone LAN as shown in Figure 12–21. The backbone may use any one of several media and protocols, but using optical fibers and the FDDI protocol is common. The backbone-LAN design works well, but is subject to the limitation that the backbone is still a LAN and its capacity must be shared by all of the LANs connected to it.

A different approach is to connect all of the LANs to a switch instead of a backbone LAN, as shown in Figure 12–22. The switch has intelligence and multiple paths through it, so if, for example, LAN A wants to send a message to LAN B, the switch can make the connection. If, at the same time, LAN C wants to send a message to the server on LAN D, the switch can also make that connection—simultaneously. Now, instead of all of the LANs sharing the 100 Mbps speed of the backbone LAN (if it is Ethernet), they all essentially have their own, dedicated link. Throughput goes up dramatically, in practice by as much as 200 percent. Token ring LANs can be switched in a similar way.

Taking the concept one step further, if additional intelligence is added to the switch, it can forward messages to a metropolitan area

Figure 12–20
Bridges connect networks that use the same protocols. Gateways connect networks that use dissimilar protocols.

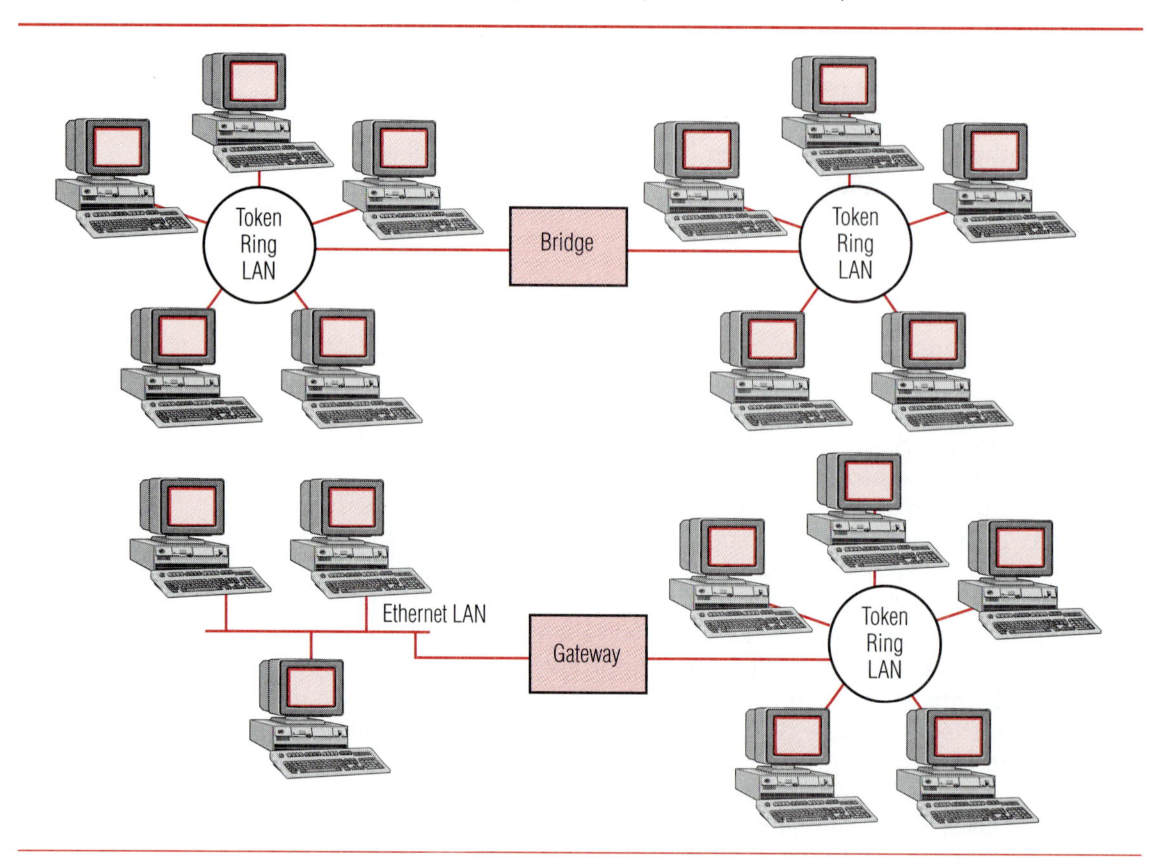

network or onto the wide area network. This will give the benefits of fast switching to users who need to make internet connections outside of their immediate LAN network.

Sounds good, but what's the downside? There are two: First, if the switch fails, the entire ability to interconnect the LANs goes down. Switches are pretty reliable pieces of hardware, since they are electronic, but still, backup plans must be made for the eventuality that the switch may fail. Second, in most real-world implementations, it requires more wire or cable to connect all of the LANs to the switch. Considering that the installation cost for cable is high, the economics must be compared to the cost and benefits of the switch. Suffice it to say that many companies have done the economic evaluation and concluded that the benefits of the faster throughput far outweigh the cost. Several vendors have great faith in the future of switching technology and are working on ways to increase

Figure 12–21
LANs interconnected with a backbone LAN.

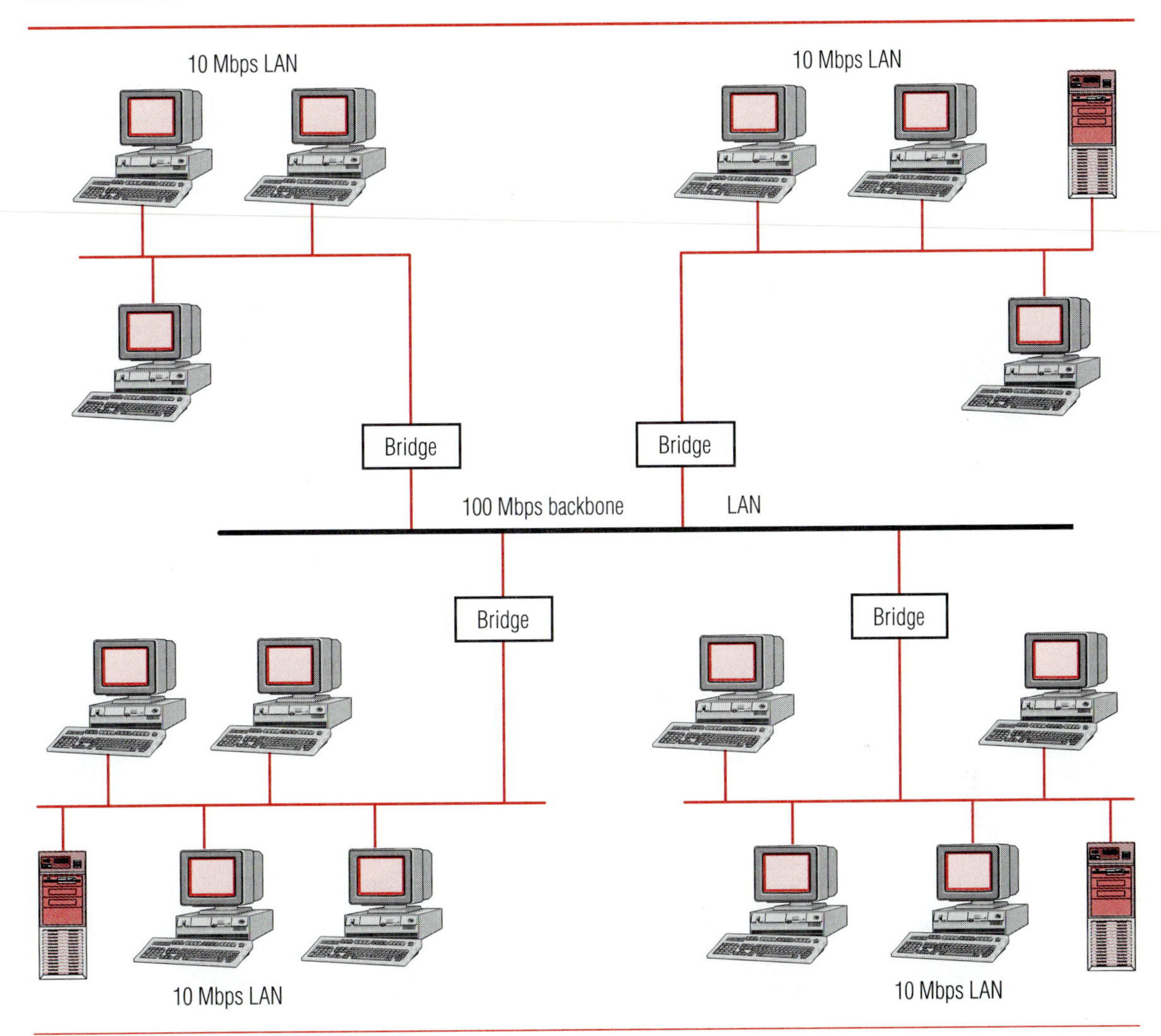

the capability of switches, including the incorporation of switch, bridge, router and gateway capability into one piece of hardware.

One caution for you to keep in mind as you begin to work in the telecommunications field is that the terminology and technology used by telecommunications vendors is not always consistent and as distinct as the examples in this book. The distinction between hubs, bridges, switches, routers, brouters, and gateways is blurred, and vendors are working hard to combine the capabilities into single pieces of equipment.

Figure 12–22
This diagram shows how a LAN switch can be used to connect several LANs together and to connect LANs to a WAN or MAN.

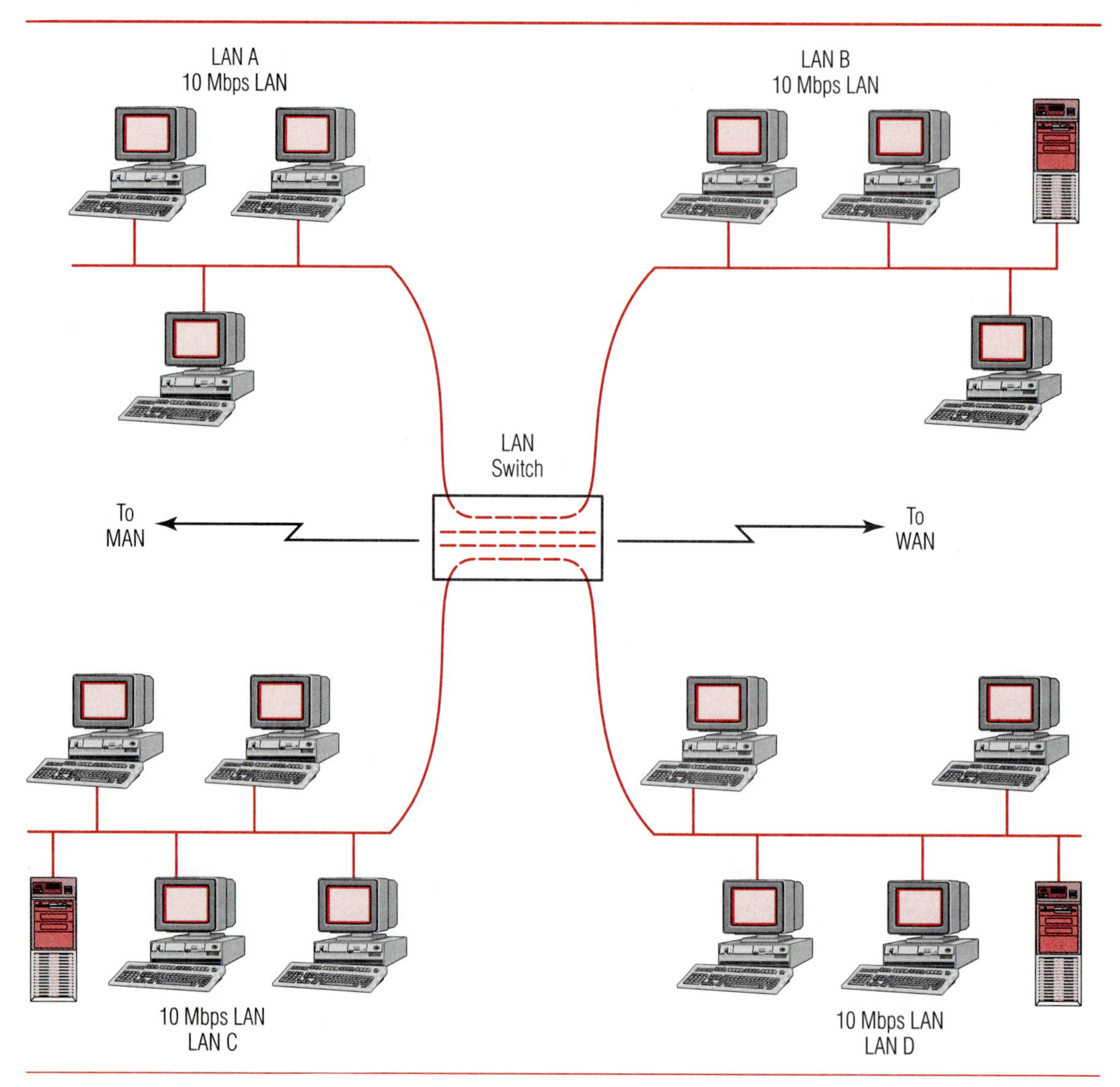

For example, you might encounter terms such as "intelligent bridge" to describe a device that combines the features of a bridge and router, or *"intelligent router"* for a device that combines the functions of a router and a gateway. Your defense against these vagaries of the marketplace is to clearly define and understand what networking problem you are trying to solve and then ask a lot of questions to determine if a vendor's proposed solution will meet your requirements.

WORKSTATIONS AND SERVERS

Many types of terminals or workstations can be connected to a LAN, including those normally found in an office, laboratory, or factory. The most common type in an office setting are personal computers because they have the intelligence to take advantage of a LAN's capabilities, which can be extended to a client-server environment if desired. More information about various types of workstations was given in Chapter 6, so they will not be discussed again here.

On most LANs, one or more special computers called *servers* are attached. Servers provide unique capabilities that can be shared by all other devices on the LAN. The most typical types of servers are *file servers* that allow users to share one or more large capacity disk drives, *print servers* that provide access to one or more printers, *communications servers* that provide access to other LANs, host computers, or dial-up lines, and *applications servers* that provide processing capacity for applications that are shared by many people. The server function can be performed by a microcomputer that is also used as a normal workstation, or by a dedicated computer whose only purpose is to provide server capabilities to other users on the LAN. Dedicated servers are able to provide better performance because there is no interference from a user trying to use the same computer for other purposes. Dedicated servers are the norm in most LAN configurations. Server computers have special software that will be discussed later in the section on LAN software.

Client-Server Computing

When a network has computers attached whose sole or primary function is to act as a server, the other workstations on the network are often called *clients.* The outgrowth is the concept called *client-server computing,* which has gained broad recognition and enormous popularity in recent years largely because of the availability of powerful, low-cost computers that can act as servers and sophisticated software that can manage the client-server environment. The basic concept is that a client's processing or data may come from a number of different machines located in widely scattered locations. The objective is to have software manage the environment in such a way that the user does not know or care where the data being used is located or where the processing is performed. True client-server applications may in fact have some of the processing performed on the client computer and some on the server. If little or no application processing is done on the client computer, it is called a *thin client* approach. If all or almost all of the processing is done on the client, it is termed a *fat client* approach. The terms *thin* and *fat* refer to the size of the client processor that is required to do the work. In either case, the implementation of client-server computing requires a fast communications network and applications that are designed to work in the client-server environment.

client

File Servers

File servers allow a large-capacity disk to be shared by the users of a LAN. Users may have multiple files, limited only by the overall capacity of the disk. File server software normally provides the capability for users at workstations to use the disk of the server as though it were their own. That is, a personal computer user may have a hard disk of 10 gigabytes on his or her PC, but when it is attached to a LAN with a file server, the user may have access to a 500-gigabyte server disk. The user is able to store programs or files on the server disk and access them as easily as if they were on the PC's disk.

Because file servers are shared among many users, they need to be fast, powerful machines with fast, reliable hard disk drives. Personal computers based on fast or multiple Pentium processors are good candidates to be servers because they can provide good service to the LAN users.

With easy access to the server's disk, new possibilities emerge. The user can access a program on the server's disk and run it. Files stored on the server disk can be designated as shareable, thereby allowing other users to look at the data or, if authorized, to change it. Files can be transferred easily to other individuals, either by giving them access to a copy on the server disk or by allowing them to transfer a copy of the file from the server's disk to their own personal computer's disk.

File sharing leads to some new data integrity problems, however. Since two or more users could be given access to update a file on the server's disk, some form of protection must be instituted to ensure that two users don't update the same data at the same time. If both user A and user B access the same record in a file simultaneously and make changes to it, the last one who updates the record will wipe out the changes made by the first person. Good file server software provides a mechanism that prevents two or more users with update authority from accessing the same record in a file at the same time. If one user has the record and a second tries to access it, the second user receives a message that the record is unavailable.

A variation on the file server notion is a *disk server.* Disk servers divide the capacity of their large disk into smaller disk volumes. Workstation users have their own private disk volumes on the disk server. The administration of disk servers can be complex, however, and there are security and data integrity considerations that are more difficult to manage than on file servers.

Print Servers

Print servers are microcomputers that allow one or more printers to be shared by users on the LAN. Printers managed by the print server may have unique capabilities, such as high speed for printing large reports, high quality, or color. These types of printers are often too expensive to attach to each workstation, and the print server allows the capability to be available and the cost to be shared.

Since more than one user may try to use a printer at the same time, some type of queuing capability must be provided by the print server software. The queuing capability is called *spooling* or *background printing*. It allows data that will be printed to be stored temporarily on the print server's disk if the printer is busy printing another user's output. When the printer finishes printing the first user's output, the spooling software takes the second user's output from the disk and prints it.

spooling

Spooling software is transparent to the user or application program that requests the print operation. Data is accepted by the print server regardless of whether the printer is busy or not. Good spooling software provides data integrity by taking care of operational problems, such as the printer running out of paper. The software ensures that all data is printed properly.

Communications Servers

Communications servers provide the capability to communicate with other networks or with other computers that are not connected to the LAN. Communications software on the server provides the interface with the other networks or computers and surrounds the data with the appropriate protocol for transmission. Communications server software most often works in conjunction with network interface hardware, such as routers, bridges, and gateways.

Application Servers

Application servers are computers that are dedicated to application processing. They may be large mainframe-class machines or personal computers. Organizations typically require a mix of application servers because some applications may be used enterprise-wide, while other applications are used by only a few people or departments. Also some applications require different operating systems, such as Unix of Sun's Solaris, and each of these environments would require a different computer. Application servers may have *database servers* attached to them, if the database is large or complex. The need for dedicated database servers is dictated by the application design and the amount of data to be stored and processed.

A diagram of a typical LAN with servers is shown in Figure 12–23.

Network Attachment

Terminals with intelligence, such as personal computers, normally attach directly to the LAN, whereas dumb terminals may be connected through a terminal control unit that provides the intelligence necessary to handle the LAN protocols and that is shared by several terminals. These connections are illustrated in Figure 12–24. Either the terminal or the control unit contains an electronic interface circuit that is frequently called the *network*

Figure 12–23
A ring LAN with servers attached.

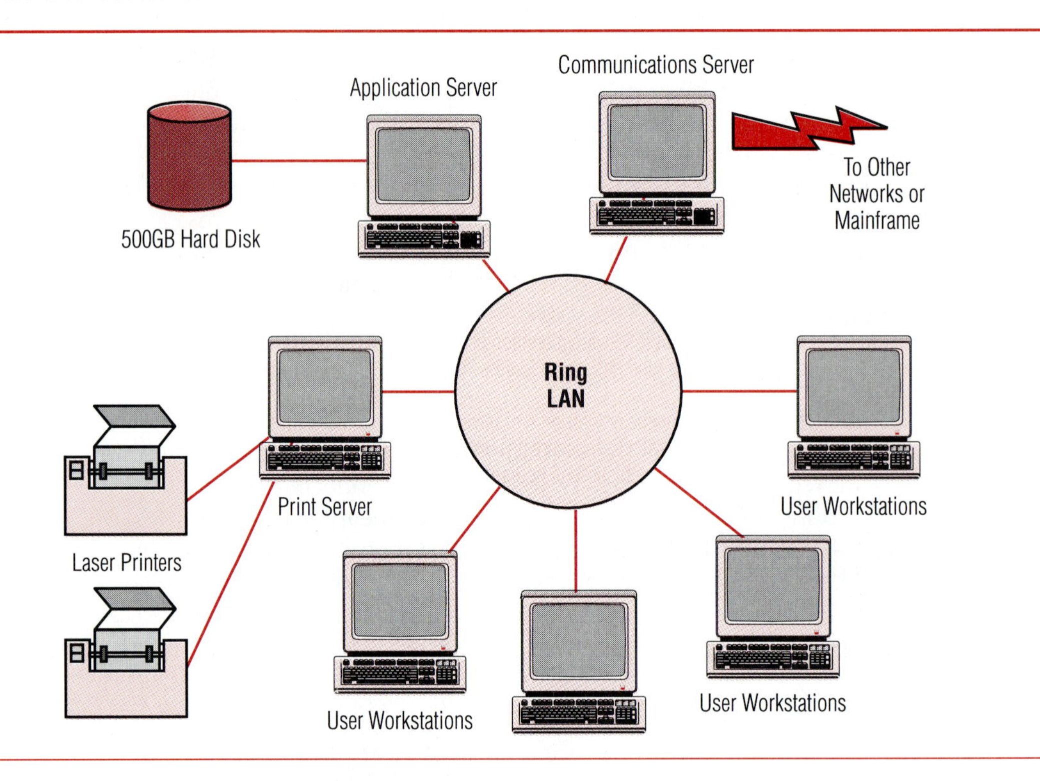

interface card, or *NIC.* This circuit card provides the electrical interface between the workstation and the network and is attached to the network with a cable. When an organization installs a LAN, it must purchase NIC cards for all terminals, workstations, and terminal control units that will be connected to the LAN. The cost of the cards can be substantial and must not be forgotten when the LAN is planned and the costs of implementing it are estimated.

LAN SOFTWARE

Simply connecting several microcomputers according to the rules of one of the previously described architectures is not sufficient to have a functioning LAN. Like the wide area networks you studied in the previous chapter, LANs must have software to control their operation. In fact, it is software that provides most of the capabilities that we associate with a LAN, such as servers and shared applications.

Figure 12–24
A LAN showing the network interface cards (NICs).

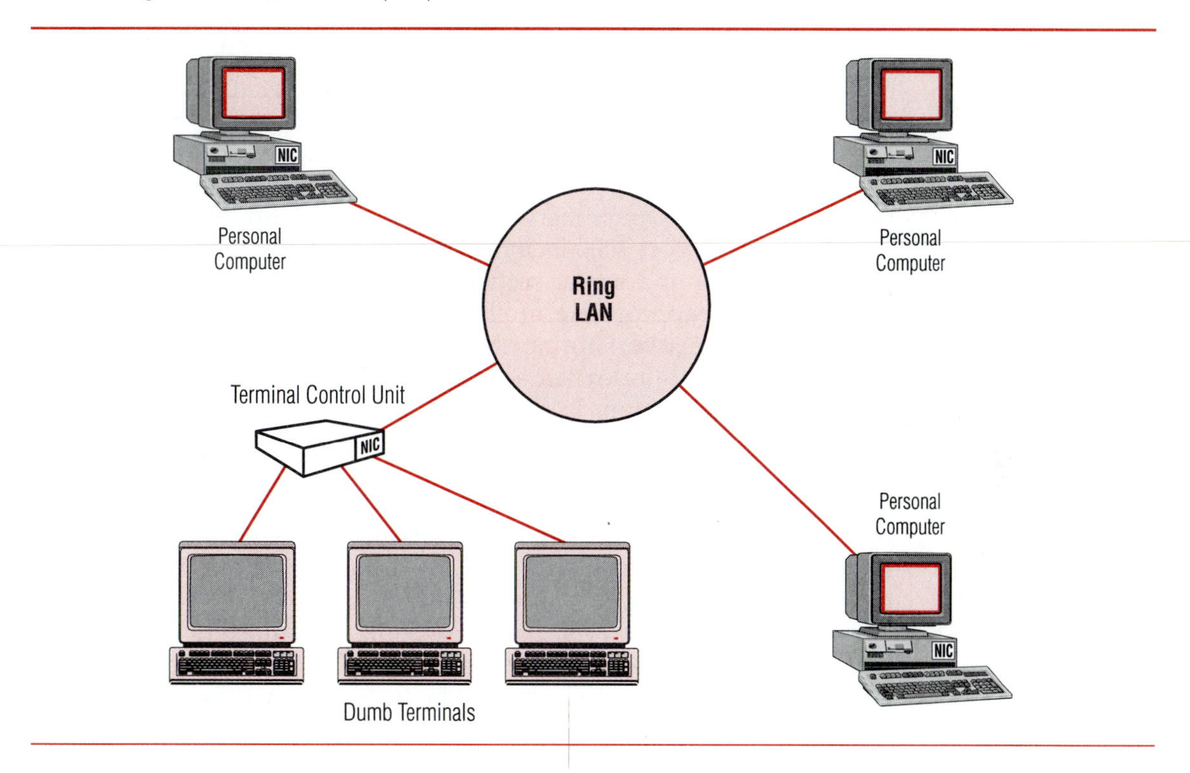

At the heart of any LAN is a variety of software that provides much of the intelligence that tells the hardware how to operate. Each user workstation connected to the LAN must have software that tells it how to communicate with the servers. Servers require software to perform their function. Applications software for a LAN may be different from the software for a single personal computer because the probability exists that the software, and perhaps data files, will be accessed by more than one user at a time. Finally, there is a need for management software to provide commands and status information to the person who manages the LAN. Each of these types of software is discussed in more detail in the following sections.

Software for the Workstation

For personal computers, the basic communications software for a LAN can be viewed as an extension of the operating system that provides the data link and network layer capability. Applications programs communicate with the LAN through software called the *application program interface (API).* In the IBM DOS world, the operating extension or API is called

NetBIOS, which stands for *Network Basic Input-Output System.* NetBIOS operates at OSI layers 2 and 5, the data link control layer and the session control layer. At the data link layer, it provides for applications programs to send and receive short messages, but without acknowledgment. At the session layer, it supports two-way, reliable communications between applications programs, with acknowledgments that messages have been received. At a higher level, a LAN workstation program uses NetBIOS and provides the capability for application programs in the personal computer to use servers and otherwise share network resources.

Windows 98 and 2000 include software modules to interface with a variety of LANs and server software. Normally a user selects the one module that is appropriate for the network he or she will be using. Including basic networking capability with a general-purpose operating system began with an earlier version of Windows called Windows for Workgroups, also known as Windows 3.11. Versions before Windows 3.11 did not include networking and required the user to purchase additional software, such as Microsoft's LAN Manager product to add to the operating system. Today the inclusion of networking software is standard.

One of the most notable developments that affects networking software for the workstation is the rise of TCP/IP to widespread use. If the TCP/IP protocol is included in the workstation software suite, it provides connectivity to almost any network that is also using TCP/IP software. In the recent past, it was almost mandatory to acquire all of the LAN software from one vendor in order to ensure compatibility. Novell's workstation software could only communicate with Novell's server software; Microsoft's workstation software with Microsoft's server software, and so forth. With TCP/IP being so widely used, it is now quite easy to build networks that use heterogeneous software rather than the homogeneous networks of the past.

Dumb terminals need only basic communications software that lets them interface with the LAN. Since these terminals cannot run programs, they are not interested in sharing files on file servers or using other functions that personal computers need. Thus, their software is much simpler and usually is provided by a cluster control unit to which the dumb terminal is attached.

Software for the Server

Software for the server, often called the *network operating system (NOS),* can be viewed as providing the overall control for a specific LAN. There is no standard configuration for a NOS, and different vendors implement NOS capability in different ways. Adding the LAN capability on top of an existing operating system was the approach used by several vendors a few years ago, including IBM and Microsoft. Developing a separate operating system specifically designed to support LANs, networking, and client-server configurations is the approach that Novell and, more recently, Microsoft have used.

The implication is that you do not simply select a LAN architecture; the software must be selected at the same time. In fact, you can argue that the software should be selected first; the hardware architecture is not really selected but depends on what the selected software will support. Frankly speaking, most users don't care whether their data is being transmitted using a token ring or an Ethernet. They are much more interested in parameters, such as ease of use, reliability, and speed.

Major suppliers of NOS software for LANs include Novell, Microsoft, IBM, and Banyan. There are others, of course, but these companies constitute a very high percentage of the NOS market. We'll look at their NOS products in the following sections.

NetWare NetWare is a network operating system developed by Novell, Inc. that provides a full range of distributed network services, including printer sharing, remote file access, and application support. It provides standard operating system functions as well as NOS capabilities. That is, the NOS provides full control of the server—no other operating system is required. NetWare specifies the upper five layers of the OSI reference model and as such runs on virtually any data link protocol. Additionally, NetWare runs on virtually any size computer system from PCs to mainframes. NetWare has its own layer 3 and layer 4 protocols, called *internetwork packet exchange (IPX)* and *sequenced packet exchange (SPX),* and provides full-function interfaces to networks using other protocols such as TCP/IP, IBM's SNA, and other LANs. NetWare provides a complete set of LAN capabilities, including software for the management and operational monitoring of the LAN.

Windows NT Server (NTS) Microsoft's premier NOS product for LANs is the operating system called *Windows NT Server.* Like NetWare, it is a complete operating system that was designed to support LAN's networking. NTS is a 32-bit operating system that provides a full set of server functions, including disk mirroring, whereby two copies of data are written to two different disk drives. If one disk fails, the other copy of the data is still available. NTS also provides comprehensive security features. One of Windows NT Server's strengths is its support of applications, and it has become one of the leading pieces of software for applications servers. NTS supports TCP/IP, NetBIOS, AppleTalk and IPX/SPX protocols, so it can work with most networks and workstations, regardless of what protocols and software they use.

VINES Banyan's NOS is called *Virtual Integrated Network Services (VINES)* and is based on a proprietary protocol family derived from Xerox Corporation's Xerox Network Systems (XNS) protocols. VINES uses the *VINES internetwork protocol (VIP)* to support layer 3 work, including internetwork routing. VINES is well known for its directory service called

StreetTalk. StreetTalk provides a directory of users, hardware, and files across interconnected local area and wide area networks, simplifying the life of users and network administrators. This directory is stored in a database on each server in the network, making it immediately available to users wherever they are. VINES runs under the UNIX operating system and is known for its support of large networks.

LAN PERFORMANCE

The performance of a LAN is based on several factors, including the protocol that is used, the speed of the transmissions, the amount of traffic, the error rate, the efficiency of the LAN software, and the speed of server computers and disks.

Protocol

The CSMA/CD protocol provides quite good throughput as long as the overall traffic load on the network doesn't exceed a certain threshold. When the load gets too high, more collisions occur, which means that a higher percentage of the traffic must wait before it can be sent. When traffic is below the critical point, however, messages can be sent almost immediately, so response time can be very fast.

Token passing protocols have the advantage of giving a guaranteed, calculable response time. This is possible because the time it takes a token to circulate on the ring or bus and the amount of time a workstation is allowed to keep the token or send data is known and controlled. However, because there is an inherent delay while the token circulates; the best response time on a token passing network may not be as fast as on a CSMA/CD network. Token passing protocols thus have an inherent delay as a trade-off for their predictability.

Speed of the Transmission

Obviously a LAN that transmits data at 100 Mbps can deliver a given message faster than a LAN that sends the data at 10 Mbps. In practice, however, the throughput of the former LAN is not ten times faster than the latter LAN because many other factors are involved in the overall LAN performance. One company reported that in its office environment the installation of a 16-Mbps LAN to replace a 4-Mbps LAN was scarcely noticeable to most users who were doing transaction processing. The higher-speed LAN did, however, make a difference during file transfers and in situations in which it was used as a backbone, connecting several other LANs.

The wide use of high-speed backbone LANs and LAN switches is helping to improve the performance of the network component of

LANs—or maybe it is better viewed as allowing the performance to keep up with the increasing demands of a larger number of users!

Amount of Traffic

If a LAN is extremely busy, delays will invariably be introduced into some or all of the transmissions. Users' workstations may have to wait until collisions are resolved or the token is passed to get control of the LAN so that their data can be sent. Some of the traffic may tie up the LAN for relatively long periods of time, causing still further delays, and so on. Generally speaking, a lightly loaded LAN will deliver better performance than one that is heavily loaded. Splitting a LAN into segments, each with fewer users, and connecting the segments with a switch is one way that may help performance and throughput.

The ever-increasing demand for bandwidth, fueled by intranets and the Internet, continues to push network designers and vendors to find innovative and creative ways to obtain more capacity from existing network facilities and to install new, higher-speed facilities at accelerating rates.

Error Rates

When errors occur, data must normally be retransmitted. For the user whose data was in error, there will be a delay. If the error rate and the number of retransmissions is high, all users may be affected—even those users whose data is sent correctly—since the apparent traffic load on the LAN will be higher. In most LANs, however, because distances are relatively short, high error rates is not a big problem like it can be with analog lines in a WAN.

Efficiency of the LAN Software

As with any software, some LAN software operates more efficiently than others and can deliver better throughput with a given set of hardware and LAN protocols to work with. Reviews in personal computer or network magazines often measure the throughput and efficiency of various LAN software. However, vendors are continually improving their products with new versions of the software, so the inefficiencies of a given piece of software may be corrected within a few months when the next version of the software is released.

Speed of the Server Computers and Disks

As you would guess, a server using a 1GHz Intel Pentium microprocessor will operate faster and generally deliver better performance than a server using a 500 MHz Pentium. Similarly, disks with faster access times will provide faster delivery of data in response to a user's query than will slower disks. The speed and efficiency of the server hardware/software combination will have a significant impact on the overall speed and response time of a LAN.

LAN SELECTION CRITERIA

Choosing a LAN from the many alternatives available is not an easy decision for several reasons. First of all, LAN technology is continuing to evolve at a rapid pace, and a system that is up-to-date and fully featured today may only be an average system in 1 or 2 years. More important than the technology change, however, are the needs of the users or potential users of the LAN. Their needs also continue to evolve. LANs are often first installed because a single department or office has a need or a desire for the capabilities that a LAN can offer. Once the LAN is installed and operating, however, other uses for it are recognized, and other departments or offices may become interested in having their own LAN, too. What started as a simple LAN providing service to a few people begins to grow in ways that were previously unanticipated. It is important, therefore, to do some "long-range" thinking, planning, and forecasting for how the LAN might evolve before any decisions are made about the initial implementation.

wide participation

Whereas there may be an initial sponsor or champion for the installation of a LAN, it is desirable to get participation and buy-in from a variety of people. Forming a committee or "LAN selection team" to study and recommend the specific LAN may be the best way to ensure participation and understanding of the implications of the LAN. Participants should include prospective users of the LAN, technical and managerial representatives from the computer department, and one or more representatives from general management. A broad perspective of the organization's business requirements along with the specific needs of one or more user departments is a desired balance in the group.

checklist

Next, it is very useful—and highly recommended—for people to make a managerial-oriented checklist of the criteria to be considered when choosing a LAN. A sample checklist is shown in Figure 12–25. The items in this sample list are not necessarily complete nor in priority sequence, since each organization has different considerations. For large organizations, it may be very important to ensure that the new LAN can be connected to other LANs or WANs that already exist. This obviously would not be a factor for a smaller company installing its first network. Each organization or even each department has to make its own list customized to its specific circumstances.

technical alternatives

The most important considerations, however, relate to the organization. Its requirements need to be defined, understood, and agreed upon first. Once the requirements are understood, the technological alternatives for meeting those requirements (for example, what type of LAN, what type of cabling, and which NOS) can be determined. Frankly, some individuals want to install a LAN just because they've heard about them from friends or read about them in magazines. Their motives need to be tested against real business requirements because installing a LAN is po-

- ☐ Objectives of the LAN: Why is it being proposed and what problems is it expected to solve?
- ☐ Number of users and their geographic spread
- ☐ Applications to be used
- ☐ Performance required
- ☐ Cost constraints
- ☐ Security requirements
- ☐ Availability of wiring that can be used, or must new wiring be installed?
- ☐ Availability and sophistication of technically trained people to install and maintain the LAN
- ☐ Vendor support and training provided
- ☐ Expected expansion of the LAN in the future
    - in this department
    - in other departments
- ☐ Workstations to be used: Readily available or need to be purchased
- ☐ Other LANs in the organization
- ☐ Required interfaces to other networks

Figure 12–25
Managerial-oriented LAN selection criteria.

tentially expensive, and it certainly can take a lot of time and effort. The decision should not be taken lightly.

Another point that should be carefully considered is who will be responsible for the LAN on an ongoing basis after it is installed. LANs have an amazing way of growing and changing, and a LAN administrator is needed to act as a "clearinghouse" for changes and day-to-day problems that may arise. The identification of a LAN administrator and his or her responsibilities should be done before the LAN is installed, ideally while the evaluation and selection is being done.

LAN administration

One other point that deserves special mention is cost. Costs of new workstations, servers, and software are always on people's minds when LAN costs are considered. Two costs that are often underestimated or not even considered, however, are the cost of installing the wiring or cabling and the cost of upgrading existing workstations to operate on a LAN. An organization is fortunate if it can use existing telephone wiring for the LAN it plans to install. In most cases, however, new wiring must be installed, and the cost of doing that in existing offices can be high.

LAN cost

Some LAN planners forget to include the cost of the network interface cards that have to be added to the existing workstations to connect them to the LAN. These cards have varying costs, but they can be as much as one-fourth of the total cost of the workstation. In total, the costs of the cabling and the NICs can be a significant portion of the total cost of the LAN.

One option for reducing or eliminating the cost of NICs is the so-called *zero slot LAN*, which does not require the installation of a NIC in a personal computer. Zero slot LANs use the PC's standard serial or parallel ports for communications. This limits the speed of transmission on the LAN to rates measured in thousands rather than millions of bits per second, but that trade-off is acceptable in many situations. It is also possible

zero slot LAN

Figure 12–26
LAN cost considerations.

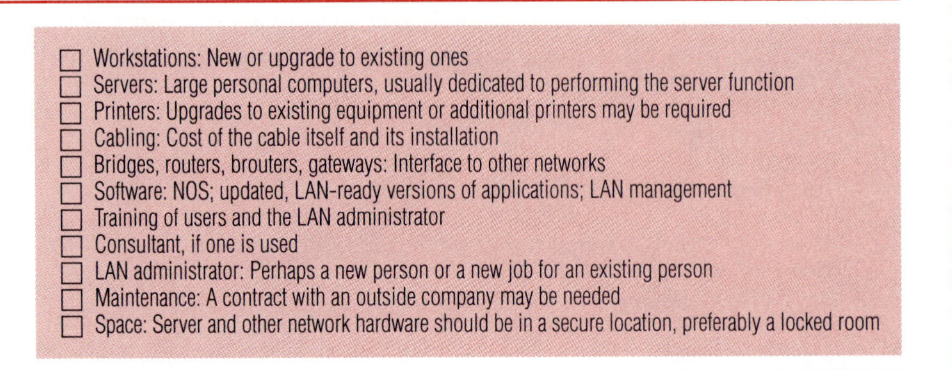
- ☐ Workstations: New or upgrade to existing ones
- ☐ Servers: Large personal computers, usually dedicated to performing the server function
- ☐ Printers: Upgrades to existing equipment or additional printers may be required
- ☐ Cabling: Cost of the cable itself and its installation
- ☐ Bridges, routers, brouters, gateways: Interface to other networks
- ☐ Software: NOS; updated, LAN-ready versions of applications; LAN management
- ☐ Training of users and the LAN administrator
- ☐ Consultant, if one is used
- ☐ LAN administrator: Perhaps a new person or a new job for an existing person
- ☐ Maintenance: A contract with an outside company may be needed
- ☐ Space: Server and other network hardware should be in a secure location, preferably a locked room

to save money by buying "slow-speed" NICs that operate at less than 1 Mbps. Again, depending on the way in which the LAN will be used, this may be a perfectly acceptable trade-off.

The low-cost alternatives need to be studied carefully for other limitations, however. Some of them limit the number of workstations that can be attached to the LAN. Others have distance limitations. Still others may inhibit the amount of file sharing that can occur. As you might expect, to a certain extent, you get what you pay for!

It is very important to think about future growth requirements when planning the LAN. LANs do grow! Saving money in the short term but inhibiting the ability to expand the LAN may be short-sighted and cost-ineffective. One of the worst scenarios is the case of installing a LAN, and then a few months later having to tear it out and start over with new cabling, NICs, and software because the original LAN could not be expanded.

A standard industry guideline is that a LAN costs approximately $1,000 per node to install, assuming an organization already has the PC workstations. This may seem surprisingly high to you, but in fact the estimate is probably a good rule of thumb for a fully featured, high-capacity, high-capability LAN. It may be possible to install simple LANs for a small number of users for less.

Some of the costs of LANs that need to be considered are shown in Figure 12–26.

After all of the LAN selection criteria are identified and prioritized, they should be weighed. Some of the criteria will be more important than others, and it is important to gain an understanding of which criteria the organization feels should carry the most weight in making the final decision about the LAN. Using the weights, points can be assigned to the LAN alternatives that are being considered as a way of quantifying what can otherwise be a fairly emotional decision.

After considering all of the factors, the LAN selection team should make its recommendation, including cost and benefit estimates, to management. The final decision will be made by a manager with sufficient authority who, hopefully, is aware of the overall impact the LAN will have on the organization.

INSTALLING A LAN

Once the LAN selection has been made, the LAN must be installed. This work usually falls upon someone within the organization, although consultants can be hired to do all or most of the work and to oversee the portions that they may not be able to do themselves, such as installing wire or cable.

In the course of evaluating and selecting the LAN, contact has probably been made with several suppliers or dealers of LAN hardware and software. These companies may offer installation and maintenance services, and negotiating the installation may have been a part of the LAN's total cost calculation.

Even if the basic work is to be done by an outside company, it is very desirable to have the organization's LAN administrator involved in the

These rack mounted servers handle the processing needs for a large local area network. (Courtesy of Sun Microsystems)

installation process. By doing so, the administrator will gain detailed knowledge of the LAN and its components, and will be in a better position to handle problems and make changes after the LAN is installed and operational.

project plan

A list of typical LAN installation tasks is shown in Figure 12–27. As with any project, it is desirable to have a project plan that identifies the tasks in detail, the person or group responsible for performing the task, the date for starting the task, and the target date for completing the task. The sophistication of the project management process should of course be geared to the size of the project. Sufficient planning and control should be in place without burdening the project team with unnecessary overhead.

software configuration

After the LAN software is installed, it normally needs to be configured. Configuration entails telling the software, through parameters or tables, what resources are available and what capabilities each user should have. The hard disks on personal computers other than the server are frequently considered "LAN resources" and are available for access and sharing. The LAN software needs to be told exactly what hardware is available. An important part of the LAN administrator's job is establishing and maintaining these resource tables.

Finally, at some point after the installation is complete and the users have been trained, the LAN is ready to be used. Despite the careful planning and testing that occurred during the installation, some problems will be apparent only after the LAN begins to be used. Some users may not be able to access data they need. The connection to another network may not work properly. There could be problems sharing certain hardware or software. Performance for a certain user or group of users may be poor or worse than expected. The LAN administrator, installation company, and consultant need to be available to solve the problems that arise as soon as possible. Unless the LAN is very small, having a central place or telephone number for users to call to report problems is desirable. It may also

Figure 12–27
The tasks of installing a LAN.

- ☐ Install: New workstations
  NICs on existing workstations
  Wiring or cabling
  Server hardware
  Bridges, routers, brouters, or gateways
  LAN software
- ☐ Determine the access and capability required by each user
- ☐ Document the LAN's hardware and software configuration
- ☐ Train the users
- ☐ Begin using the LAN and its new capabilities
- ☐ Troubleshoot any startup problems

be desirable to keep a list of all outstanding problems and to update it daily (or as often as required) as problems are solved and new problems are identified. Gradually—and hopefully quickly—the problems will be solved, the LAN operation will smooth out, and the benefits of the LAN will begin to be realized.

MANAGING THE LAN

When companies or departments consider installing and operating a LAN, one of the most underestimated functions is the ongoing LAN management. LANs are often proposed and installed by people who have very good technical skills but who lack an understanding of the necessity to manage the LAN on an ongoing basis and to provide appropriate policies and procedures. Many times, the need to physically protect the LAN hardware, to provide backup for the data stored on the file servers, and to establish procedures for authorizing access to the data are ignored. The need may only become apparent after a problem has occurred. LANs have a way of growing and changing that mandates that some controls must be in place.

Any time a LAN is established, the following management items should be considered.

- Organization and management: A LAN administrator or manager should be formally designated. Policies and procedures should be established and communicated to all LAN users.
- Physical safeguards: LAN hardware, such as the cable and servers, should be physically protected from unauthorized tampering. Keeping the servers in a locked room is a preferred technique.
- Documentation: The LAN must be documented. Documentation should include the network topology, types and locations of attached workstations, names and version numbers of installed software, authorized users and their capabilities, and so on.
- Hardware and software control features: Hardware and software control features should be acquired or developed. These features should be used and documented to help ensure that the LAN's use and operation is well controlled.
- Change control: Procedures should be established for making changes to the network or its software. These procedures should include notification to users, testing, and fallback plans.
- Hardware and software backup: Effective backup provisions and contingency plans for both the hardware and software must be made. Users will depend on the LAN to do their job.
- Access to network facilities: Adequate security mechanisms should be provided to restrict access to network hardware, software, and data to authorized people.

- Network application standards: There should be adequate controls and training regarding applications to ensure compatibility, integrity, and effective application usage.
- Network performance monitoring: Network performance monitoring mechanisms should be established to ensure effective network throughput, load leveling, and overall performance reporting.

From the preceding list, it is evident that establishing and operating a LAN is a task that needs good planning and control. In many ways, operating a LAN is like operating a small computer center. The LAN manager must consider and develop plans for most of the same items that the company's data processing organization considers for the mainframe computer. For that reason, in many organizations, the installation and control of LANs are the responsibility of a central group. Although this would seem to take away some of the departmental or workgroup autonomy often associated with LANs, it does help to ensure that the LANs are managed properly and are protected from unforseen problems.

LAN Management Software

It is very desirable for LAN software to provide commands and information that assist in the daily operation of the LAN. This capability is usually provided by the NOS. Users may not be in close proximity to the LAN's servers, or the servers may be located in a locked room for security reasons. Therefore, it is helpful if the users can inquire as to the status of the servers. Knowing how many jobs are in the print queue may give some indication of when one's output will be printed. Knowing how many users are logged on to the LAN may give an indication of the response time and other performance that can be expected. Having the ability to send a message to all users on the LAN may be useful so the LAN manager can inform people that the LAN will be taken down for maintenance.

In addition to these daily operational types of commands, the LAN software also should collect statistics that show usage over periods of time, such as a day, week, or month. This helps the LAN manager monitor growth and anticipate the need for additional disks, faster printers, or even a faster LAN.

LAN SECURITY

Providing appropriate security for the LAN, its data, and its users is a combined responsibility of management, the LAN administrator, and the LAN software. Management must set the environment by providing security policies and by communicating the message to all employees that security is important. They may also insist that regular audits of the LAN be performed by internal or external auditors to ensure that internal control requirements are being met.

LAN hardware, such as servers and communications controllers, should be kept in a locked room where it cannot be accessed by unauthorized people. The LAN administrator is normally responsible for this hardware, and it is his or her responsibility to ensure that the equipment is physically protected and secure.

proxy server

If the LAN is connected to the Internet, the installation of a *proxy server* should be strongly considered. A proxy server acts as an intermediary between the Internet and Net-equipped computers. Whenever a net-equipped workstation wants to access a page on the Internet, it contacts the proxy server. The proxy server stands in for the PC and communicates with the web directly so others on the Internet do not know the actual terminal address of the workstation. Conversely, outsiders cannot get into the company's network without going through the proxy server. Another benefit of a proxy server is that it can give network administrators a level of control over the Internet habits of users on the network by filtering certain sites or restricting employees' access to non-business-oriented web sites.

The LAN administrator is also the person who is responsible for maintaining the security and access control tables that are part of the LAN software. However, the administrator should establish and change those tables only with appropriate authority from management. A form should be used that shows the level of access for each user, and it should be approved by the user's manager. Subsequent changes should also be approved, and the LAN administrator should not make changes without proper approval.

Other policies and procedures may be required to ensure the right level of security and LAN management. Much depends on the needs of the organization. Items to be considered include

- password policies regarding the use and frequent changing of passwords to be used when logging on to the LAN;
- a sign-off policy, requiring users to sign off the LAN when they leave their workstation;
- the encryption of data sent on the LAN or stored on the servers;
- regular backups of the server disks, usually a responsibility of the LAN administrator;
- policies for downloading data from other computers or networks to the LAN disks;
- policies regarding the scanning of disks for viruses;
- policies regarding dial-up access to the LAN;
- policies stating that only legal, licensed software may be used on the LAN.

By now it should be clear that operating a LAN in a safe, secure manner is a responsibility that requires careful attention at both the managerial and technical levels of the organization.

SUMMARY

Local area networks are an important, fast-growing subset of the total network environment. The technology is developing rapidly, and as quickly as new standards are finalized others are under development. The LAN designer is faced with a wide variety of choices for building a network. Only by understanding the characteristics and trade-offs between various types of LANs can the designer reach a reasonable solution for a particular organization or application.

After the LAN is installed, it must be managed. Security implications must be considered, the LAN performance must be monitored and tuned, and a steady stream of changes are invariably required by the users. A recognized LAN administrator who has the backing of management is a necessity for ensuring that the LAN is well controlled and operates effectively to meet the needs of its users.

CASE STUDY

LAN Selection at Nippon Chemics

Tsuyoshi Akiyama, research manager at Nippon Chemics, a medium-sized Japanese chemical company, was very frustrated. For weeks he had been trying to convince Hiroshi Watanabe, the MIS manager, that an AppleTalk LAN should be installed in the laboratory to allow his research staff to share certain software and data that was related to their work. Watanabe-san (Mr. Watanabe) was not opposed to a LAN, but he had been arguing that a token ring LAN should be installed to comply with the company's general standards for networks. "We've had token rings installed in our offices for over two years now," Watanabe-san said. "And they have run very well. Furthermore, my staff in the MIS section knows how to install token rings and manage them."

"But we won't need your help," Akiyama-san replied. "My research staff has been using their Apple Macintosh computers for six months, and the dealer who sold them to us tells us that connecting the wires for an AppleTalk LAN is very easy. I'm expecting my people to do it themselves and to work with the dealer if they have problems."

"Well, I think you'll be surprised at how muzukashii (difficult) it can be to get a LAN operating," Watanabe said. "The dealers always make it sound simple, but the experience of my staff is different. There are always some wiring problems, and often the software has some bugs. If you'll let my staff help your people, I'm sure we can get a token ring LAN installed quickly."

"But if we use a token ring, I'll have to buy some expensive adapter cards for all of my Macintosh computers," Akiyama-san replied. "And we'll have to install expensive cabling too, won't we?"

"Well, yes," Watanabe said, "you would have to get the token ring interface cards. Of course, the other option that I've suggested to you is that you replace your Macintosh computers with NECs. Many of their machines have the adapter cards already built in."

"No, that is totally impractical," Akiyama-san said, somewhat irritated. "The reason we got the Macintosh computers in the first place was because of the chemical design software that is only available for Macs. That software has saved my staff hundreds of hours in the last few months—why, I'm not sure if we could operate without it anymore! How important are these company standards for LANs anyway? It seems to me that they were written considering the needs in the office but without thinking about our different situation here in the laboratory. Does anyone besides you really care if we just go ahead and install the AppleTalk? We're not planning to connect our LAN to anyone outside of our building so I can't understand why it is so important if our LAN is different . . . especially if it is saving my people a lot of time."

The discussion between Akiyama-san and Watanabe-san continued, but we will take leave to consider a few of the points they made.

QUESTIONS

1. Do you think Akiyama-san participated in the development of the company's LAN standards? Should he have?
2. Considering that the lab people are using some software that is only available for Macintosh computers and which is apparently making them very productive, do you think the LAN standards should be enforced in the lab?
3. From Watanabe-san's perspective, people from almost every department he talks to feel that they are different and that exceptions to the LAN standards should be made for them. Of course he wants his colleagues to be productive, but his job is to try to ensure that someday all of the LANs can be connected into a companywide network. What are some steps he can take to address this dilemma?
4. Akiyama-san says that he thinks his research people will be able to hook up the LAN. What do you think about his idea? Do you think he has considered the work involved in managing and operating the LAN after it is up and running?
5. What would be your recommendation to resolve this situation?

CASE STUDY

Dow Corning's Use of Local Area Networks

Dow Corning's use of LANs has grown rapidly. Originating in the manufacturing plants and research laboratories, where Ethernet LANs were installed with the DEC VAX computers, LANs are now installed in virtually every company location. Dow Corning's wide area network can essentially be viewed as one that interconnects all of the company's LANs.

LANs originally came to be used at Dow Corning for a variety of different purposes:

- At the corporate center (headquarters), the original purpose of the token ring LANs was to provide better response time from the IBM mainframe computers by taking advantage of the 16 Mbps transmission rate;
- The research department used some Apple Macintosh computers and AppleTalk LANs because of certain chemical-related software that was uniquely available on the Macintosh computers;
- Departmental LANs were installed to meet unique needs. The treasury, tax, and legal departments all installed their own LANs so that their people could share certain data that was being worked on by more than one individual.

Many people originally thought that they could install a LAN and forget it. After some experience they realized that wasn't the case, and LAN administrator jobs began to appear. For a time in the early 1990s, it was a company requirement that every LAN had an administrator identified and responsible for the LAN's day-to-day operation, for assisting users when necessary, and to ensure that proper LAN management procedures are followed.

In late 1995, the company embarked on an ambitious program to architect, design, and install a complete, global LAN-based network for the entire company. The program had different implications for each geographic region. The Americas and Europe already had a large base of installed LANs, so conforming to the architecture of the new LAN-based network involved significant conversion of older hardware and software to meet the new standards. In Asia there were almost no LANs installed, so a significant investment to install LAN hardware and software was required. In addition, the telecommunications staffs and employees in the region had to be trained to use the new capabilities. Despite these costs, corporate agreement and support was reached because the expected benefits were far in excess of the costs. The master plan called for the upgrade program to be completed worldwide by the end of 1997, and that plan was achieved.

The upgraded network provides Ethernet LANs in all locations and the WAN circuits between locations were upgraded in speed when the WorldCom frame relay network was implemented to provide good response time for LANs in widely separated locations. Backbone LANs using optical fiber and the FDDI transmission technique interconnect some of the LANs that are located near each other. Some token ring LANs remained for a few years, but no new ones were installed; and by mid-2000 all of the token ring LANs had been replaced.

The director of Global Information Technology Services at Dow Corning stated that the network and personal computer upgrade program reduced the total cost of personal computer ownership by providing standardization throughout the company and by providing a uniform base for building employees' skills in the use of computer-based software tools.

Tests of wireless LANs are also being conducted. One application of the technology that looks very promising is for people who take their laptop computers to meetings. They can connect into the LAN using wireless technology, eliminating the need to have LAN wiring with multiple outlets at the conference tables. Since the company has many meetings, and more people are finding benefit in taking notes or making presentations using their laptops, wireless technology will make it much easier by allowing people to walk into the conference room just before the meeting begins and log on to the LAN without having to hook up cables.

QUESTIONS

1. If you were a manager in Dow Corning and thought you had the need for a LAN in your department, would you maintain your independence and install a LAN for your needs, or would you work with the corporate telecommunications group to determine the optimum solution for your department and the company as a whole? Why?
2. How do you explain to a long-time user of Apple's AppleTalk software that the company has a new directive to use Ethernet LANs and standard (non-Apple) PCs, and you want him or her to change the LAN and PC workstations to meet the new corporate standards?
3. What did the Director of Global Information Technology Services mean when he stated that the "total cost of personal computer ownership" would be reduced in the company by installing the upgraded network and new personal computers?
4. What benefits do you think Dow Corning's employees received by being able to access LAN servers from anywhere, inside or outside the company?

REVIEW QUESTIONS

1. What are the main characteristics of a LAN?
2. What are some of the reasons that organizations want to install LANs?
3. Describe the characteristics of the main LAN topologies.
4. Describe the advantages and disadvantages of broadband and baseband transmission.
5. Why do LANs have distance limitations?
6. Distinguish among the IEEE 802.3, 802.4, and 802.5 standards.
7. What are the main characteristics of the FDDI and FDDI-II standards?
8. Compare and contrast Ethernet with a token ring LAN.
9. What are the functions of a print server on a LAN? A file server?
10. Compare and contrast the functions and capabilities of bridges, switches, routers, and gateways.
11. Why must personal computer applications software frequently be changed in order to operate on a LAN?

12. Describe the functions of an NOS.
13. What are the elements of a LAN that can affect its performance?
14. What are the key elements in determining the cost of a LAN? How can LAN costs be reduced?
15. Why is it important for a company to emphasize LAN management?
16. What are management's responsibilities for LAN management?
17. What are the responsibilities of a LAN administrator?
18. What are some of the security considerations that LAN administrators must deal with? Why is management's support required?
19. Explain the word topology.
20. Compare and contrast a LAN, MAN, and WAN.
21. What are the advantages and disadvantages of mainframe-based networks and client-server networks?
22. How does FDDI differ from IEEE 802.5?
23. What is a switched Ethernet?
24. Explain the difference between baseband and broadband.
25. What is a hub?
26. What is a server?
27. What is an NIC?
28. What are the advantages and disadvantages of using a switch in a LAN configuration?
29. What do the terms 10Base2, 10BaseT, and 10Broad36 mean?
30. What are the pros and cons of Ethernets compared to token rings?
31. What are the key items to consider when selecting a LAN?
32. Explain how the token passing protocol works.
33. A token passing access technique is said to be deterministic. Explain this concept and why a CSMA technique is non-deterministic.
34. What is the function of a proxy server?
35. What is the function of *logical link control?*
36. What does the access point on a wireless LAN do?
37. Explain the concept of *roaming.*
38. Distinguish between the functions of a database server and a file server.
39. Describe how a server's disk speed can affect the performance of a LAN.

PROBLEMS AND PROJECTS

1. Calculate the theoretical amount of time it would take to transmit a 1-megabyte file from one location to another on networks that operate at the following speeds: 9,600 bps, 64 kbps, 4 Mbps, 100 Mbps. Why will the actual transmission times be longer than the ones you calculated?
2. Draw diagrams and describe the operation of LANs implemented using the 802.3, 802.4, and 802.5 standards.
3. Think of applications where the token passing bus might not provide fast enough response time. In what applications might a CSMA/CD LAN not provide consistent enough response time?
4. Visit a company with a LAN to study the LAN management techniques that are in place. Report your findings to your instructor or the class.
5. Using the sample LAN selection criteria checklist shown in Figure 12–25, develop a complete list for a LAN at your organization or your school. Put the items on your list in priority sequence and explain why you ranked them the way you did.

6. Create a list of all of the security controls that are in place for the LAN that you use. Comment on the effectiveness of each of the controls. Can some or all of the security be bypassed by the average user, by a knowledgeable user, or by a hacker?
7. Find out about the LAN (or LANs) at your school, answering questions such as: What is its architecture? How many workstations are connected? How reliable is it? Who is the LAN administrator and what are his or her responsibilities? If there is more than one LAN are they interconnected? How? Does the school have a disaster recovery plan for its network?
8. Do some research to find out how the disks on LAN servers are most commonly backed up.
9. For fun, see how many relevant words you can identify that have the letters LAN in them. For example, if you live in the capital city of Michigan, LANsing would have meaning to you. Or, if your network is very simple with no features or frills, you might describe it as bLANd . . . and so on!

Vocabulary

local area network (LAN)
repeater
groupware
software metering
site license
logical link control (LLC)
protocol data unit (PDU)
carrier sense multiple access with collision detection (CSMA/CD)
token passing
balun
hub
multistation access unit (MAU)
multiplexer
concentrator
cabling plan
wireless LAN
access point
peer-to-peer wireless network
roaming
baseband transmission
broadband transmission
Ethernet
bus
segment
demand priority
demand priority access method (DPAM)
Fiber Distributed Data Interface (FDDI)
Manufacturing Automation Protocol (MAP)
LocalTalk
bridge
switch
cut-through switch
store-and-forward switch
router
routing algorithm
internetwork address
brouter
gateway
server
file server
print server
communications server
application server
client
client-server computing
thin client
fat client
disk server
database server
spooling
network interface card (NIC)
application program interface (APC)
Network Basic Input-Output System (NetBIOS)
network operating system (NOS)
NetWare
internetwork packet exchange (IPX)
sequenced packet exchange (SPX)
Windows NT Server (NTS)
Virtual Network Integrated Services (VINES)
VINES internetwork protocol (VIP)
StreetTalk
zero slot LAN
proxy server

References

Abramowitz, Jeff. "Wireless LANs: A Status Report." *Telecommunications* (March 1997): 70.

Anderson, Ron. "Messaging in the Next Millennium." *Network Computing* (December 13, 1999): 80–88.

Axner, David. "Gigabit Ethernet: A Technical Assessment." *Telecommunications* (March 1997): 31–34.

Bairstow, Jeffrey. "GM's Automation Protocol: Helping Machines Communicate." *High Technology* (October 1986): 38.

Davis, Beth. "Wireless LANs Ready to Hit the Fast Lane." *Information Week* (January 27, 1997): 122.

Derfler, Jr. Frank J. "Long Line the Switch." *PC Magazine* (October 5, 1999): 187–204.

Glass, Brett. "Understanding NetBIOS." *Byte* (January 1989): 301–306.

Schonfeld, Erick. "The Technology That May Save the Net." *Fortune* (February 17, 1997): 32–36.

Utell, Michael J. and Asad Irshad. "Wireless Bridges Span the Divide." *Network Computing* (May 1, 2000): 77–96.

CHAPTER

13

Network Management and Operations

OBJECTIVES

After you complete your study of this chapter, you should be able to

- explain why it is necessary to manage the network;
- describe the functions of the network operations group;
- describe how network problems are diagnosed and repaired;
- explain why problems are escalated;
- explain three levels of problem resolution;
- describe how network performance is monitored and measured;
- explain how the network configuration is documented and controlled;
- discuss the importance of having change management procedures;
- describe the functions of the communications technical support group.

INTRODUCTION

This chapter deals with the tasks required to manage and operate a communications network. The operation of a communications network is a complex, interrelated set of activities, some oriented toward normal or routine operations and others that are performed when problems occur. The various facets of network operations must work together cohesively to provide the user with consistent, reliable service.

The communications technical support group is often a part of network operations. This group has the responsibility of supporting the communications software and solving the most complex network problems. Technical support specialists must understand the interrelationships between the network hardware and software.

DEFINITION OF NETWORK MANAGEMENT

Network management, sometimes called *network operations,* is the set of activities required to keep the communications network operating. The proper view of the scope of network operations is the same view that the user has of the network. It is an all-encompassing view in which all of the elements that are required to deliver communications or computing service to the user are equally important. If any of the elements are not functioning properly, from the user's perspective, "the network is down." If the central computer or one of the application programs running in it is not operating, the user cannot receive the service he or she expects. The network operations group should be concerned about the problem and involved in getting that service restored. From the user's perspective, anytime he or she wants to use a terminal on the network, it should be available.

Communications users have become accustomed to this level of service because of the extremely good reliability of the telephone system. For most of us, waiting for a dial tone when we lift the handset rarely occurs. We expect that when we want to make a telephone call, the network will be ready to serve our needs. Unfortunately, some business data communications networks do not operate at that same high level of reliability. In many companies, several network failures per day or erratic response times are considered normal or "good enough."

This implies that all business communications networks should operate at the same level of availability and reliability as the public telephone network. From a practical standpoint, the service requirements of the business must be defined and the network designed to meet those requirements. If 24-hour-per-day availability of the network is not required, it is a waste of money to design the network for 24-hour operation. If an outage or two each week is not critical (albeit inconvenient for the users), perhaps it is not worth spending money to provide more reliable service. The key concept is that each company must define the service level it expects from its communications network.

Network management people have the challenging responsibility of ensuring that the defined requirements for availability and reliability are met. In most companies, the service requirements also include requirements about consistently good response time and fast problem resolution. The network management group can only meet these service objectives if the scope of its responsibility is defined broadly.

In some companies, however, the network management group is viewed more narrowly as only having the responsibility for lines and modems. By limiting the scope, major pieces of the network may be left with no management. The proper scope of network management responsibilities includes

- user workstations;
- cluster controllers;
- modems;
- communications lines;
- line concentrators;
- multiplexers;
- front-end processors;
- communications software.

In addition, the network management personnel need to be familiar with the computers and applications software that are attached to and use the network. Admittedly, there is some potential overlap with the functions of the computer operations staff, and in each company, the lines of responsibility must be defined. However, the network management people are in the best position to view the entire process of delivering information from the computer to the user and are, therefore, in the best position to communicate with the user and to see that problems are resolved.

The same network management staff may also be responsible for the proper operation of the voice network and the services it provides. Although historically the voice and data operations in a company have been handled by two different groups, there is no organizational or technical reason why they shouldn't be brought together. There may be strong political reasons for not combining the two groups, but it is in the best interest of the company to overcome these objections and move toward a single network management organization.

WHY IT IS IMPORTANT TO MANAGE THE NETWORK

One might ask why it is worthwhile to spend any time or effort managing the communications network. There are three primary reasons:

1. The network is a corporate asset.
2. The network is a corporate resource.
3. The network is growing rapidly.

The communications network as well as computers and applications are assets of the business that in most cases are vital to the company's operation. Perhaps this can be most easily understood by thinking about the importance of the telephone to businesses and what would happen if it were not available. Similarly, companies with computers and terminals are finding that they are increasingly dependent on the reliable operation of their data communications system in order to do business. This dependence is similar to the company's need for good people, sufficient financial resources, suppliers, and customers. Each of these assets must be managed, and the communications network is no exception. It must be managed, too.

Viewing the communications network as an asset implies that its creative use can give the company advantages in its business operations. The communications facilities might be employed to make it easy for customers to contact the company and get product information. With the increase in doing business on the Internet, some companies are totally dependent on it and their own internal communications networks to do business. Networks might be used to give production people instantaneous access to material requirements or to give service people access to a database containing solutions to customer problems. Communications facilities also may be used to improve the productivity or effectiveness of employees by making it easier for them to exchange information, schedule meetings, or gain approvals. Imagination is the major factor limiting the creative use of communications facilities.

network growth versus budgets

Computer and communications networks in many companies are growing at a rate of 15 percent to 50 percent per year, compounded. The average communications budget for companies in the United States is growing 10 percent to 20 percent each year. The breadth of service being provided and the amount of money being spent on communications makes network management virtually mandatory. Without management, there would be chaos. In particular, network operations management must ensure that the service provided to existing users of the network does not degrade as new users are added. Furthermore, they must ensure that the change (growth) is implemented in a controlled manner.

Network Management in Small Organizations

Even in small companies, communications networks must be managed. Although small organizations cannot afford to have very many people working on network management tasks, at least one person must be given the responsibility for the network. Successful small companies may be experiencing a lot of change and/or rapid growth, and in that environment, network management is especially important. Furthermore, someone must be responsible for the day-to-day operation of the network and workstations, because, just as in large companies, the small organization may be dependent on the network for its day-to-day business operation.

Many of the tasks in the rest of this chapter may be handled quite informally in a small organization, but the basic framework and techniques that are described are just as relevant as in larger organizations. With a high percentage of employees now having experience with personal computers, either at home or on the job, the situation can easily arise where "everyone is an expert" and has his or her own thoughts about how the network should be run, problems should be solved, and so forth. That's another reason why it is especially important to be clear in the small organization about who has the responsibility and authority for network-related decisions.

THE FUNCTIONS OF NETWORK MANAGEMENT

The business objective of network management is to satisfy users' service expectations by providing reliable service, consistently good response time, and fast problem resolution. Network management is made up of the following six activities:

1. network operations;
2. problem management;
3. performance measurement and tuning;
4. configuration control;
5. change management;
6. management reporting.

Each of these activities interrelates with the others. In large companies, each of the activities might be handled by a separate department, whereas in small organizations one person might perform all of the functions. We will study each activity separately as though it were done by a separate department, and we'll examine where the interrelationships occur.

Network Operations Group

The network operations group is the heart of network management. It is most easily visualized as a group of people who reside in a place called the network operations center. The network operations group has many similarities to a traditional computer operations group. In some companies, the two groups are combined or at least report to the same supervisor. The network operations group is responsible for the management of the physical network resources, activation of components, such as lines or controllers, rerouting of traffic when circuits fail, and execution of normal and problem-related procedures.

starting and stopping the network

If the network does not operate 24 hours per day, 7 days per week, there is a daily routine of starting and stopping the network. To start the network, communications software is loaded in all of the computers and communications controllers that are software-driven. Then commands must be issued to instruct the software to activate each circuit. A command such as Start Line might, for example, instruct the software to begin polling the terminals on a line. In some cases, individual terminals may need to be activated with a Start Terminal command. This might occur when terminals are in different time zones, and the terminals in each time zone are activated just before the users come to work.

At the end of the day, there is a similar process for shutting down or deactivating the network. Individual terminals may be stopped as people leave work, circuits may be individually deactivated, and finally, after the entire network is shut down, the communications software may be removed from the computers, freeing the memory for use by other programs. In most communications systems, the individual commands required to start up and shut down the network are aggregated into a macrocommand that might be called Startnet or Stopnet. When these macrocommands are entered, all of the individual line and terminal commands are executed automatically. In some systems, even the issuance of the Startnet and Stopnet commands can occur automatically when triggered by a clock at a certain time each day.

monitoring the network

When the network is operating, the network operations staff is responsible for monitoring its behavior. Because communications circuits and terminals, especially on a WAN, exist in relatively uncontrolled environments (as compared to the controlled environment of a computer room), unusual conditions and problems are a certainty. Good telecommunications software provides regular status information for the network operations group. It identifies which components of the network are operating normally and which are having problems. Often the status is in the form of a visual display on a workstation and is updated every few seconds, giving a current, realtime picture of the network's operation. Well-designed network monitoring hardware and software should quickly alert the network operations group when problems occur, so that appropriate actions can be taken promptly. Problem identification and resolution are discussed extensively in the next section.

collect statistics

Another responsibility of the network operations group is to collect statistics about the network's performance and to watch developing trends that may affect performance. The statistics-gathering process should be a routine part of the operation, and software is available that automates and simplifies the task. When performance or other problems occur, additional information should be gathered. Although the first responsibility is to get the problem solved and the network operation back to normal, the secondary responsibility is to gather other data that can be analyzed later to prevent the problem from recurring. In networks where the performance data is gathered automatically by hardware and software, the network operations group must ensure that the data-gathering mechanisms are working properly.

The network operations staff normally has other responsibilities, such as problem management, configuration control, or change management, each of which will now be discussed in detail.

Problem Management

Problem management is the process of expeditiously handling a problem from its initial recognition to its satisfactory resolution. One of the important subgroups of network operations is the *help desk.* The help desk is the single point of contact with users when problems occur. Ideally, a single telephone number is established, and users are instructed to call the help desk whenever they have problems with any of their communications equipment.

help desk

The first responsibility of the help desk personnel is to log each problem that is reported. The information may be kept manually, but software is available that allows entry of the problem into a database where it can be tracked until the problem is resolved. A sample of a manual log sheet

The help desk serves as a single point of contact for users with network-related problems. Help desk operators have access to a variety of network status displays and other tools, which allows them to monitor the network's performance and assist in problem diagnosis. (Courtesy of Dow Corning Corporation)

is shown in Figure 13–1. The types of information that are recorded include the date and time, the name of the user reporting the problem, the type of terminal being used and its identification, and the symptoms of the problem being reported. The advantage of an automated problem-logging system is that the information recorded in it is available to those who have a need for it. Another advantage is the system's ability to easily sort and report the problems in various sequences for later analysis.

In many cases, the help desk operator will be able to offer immediate assistance while the user is on the telephone. If, for example, a communications circuit or LAN with 10 terminals on it has failed, it is likely that the help desk will receive several calls from users at terminals on the circuit. After the first call, the help desk operator will be familiar with the problem and will be able to assure subsequent callers that action is being taken. Other types of problems, such as certain types of terminal errors, tend to be repetitive in nature. The help desk operator, on hearing the symptom, may be able to tell the user a particular sequence of key strokes that will clear the problem or correct the error. Even in such simple cases, the problem should be logged. Later analysis may show that many users are having the same difficulty and that additional training or documentation is required.

Help desk operators should be provided with a script or at least a standard list of questions to ask all callers. Some of these questions are

Figure 13–1
Help desk log of each reported problem.

HELP DESK PROBLEM LOG						
Date	Time	Name of Caller	Phone No.	Terminal Type	Symptoms	Resolution

designed to ensure that the proper data about the problem is gathered, such as "What is your name, employee number, and telephone number?" or "What type of terminal are you using?" Other questions are diagnostic in nature, such as, "Is the green light on your terminal lit?" or "Have you checked to be sure your terminal is plugged in?" or "Are any other people sitting near you having a similar problem?" Some companies have developed flowcharts or decision trees of questions that assist the help desk operator in diagnosing even relatively complex problems. The help desk person asks the user a series of questions, and the answer to one question narrows the range of possible causes and determines which question will be asked next. Diagnostic tools such as these are a simple form of artificial intelligence, and they themselves are subject to automation.

trouble ticket

In large organizations, a *trouble ticket* is opened for each problem, again often on a computer-based system that allows the trouble to be recorded in a database. In fact, the system that generates the trouble ticket may be the same as the problem log that was referred to earlier. If the trouble ticket is online, other technicians can access the ticket and update it as they take corrective action. Everyone involved with resolving the problem works with the same "ticket" that is online and updated in realtime. Each person can be given the authority to view or update certain parts of the information. When the problem is resolved, the person who corrects it is usually the one who closes the trouble ticket and completes the form. The software may be programmed to automatically notify the help desk and the user. The ticket is then stored in the database where it can be referred to later, if necessary. An analysis of trouble tickets can identify trends. For example, a growing number of problems on a LAN may indicate a cabling or software problem that is the root cause of all of the troubles. It is always desirable to find and fix the root cause of a series of related problems.

Problem Resolution Levels One approach to problem tracking and resolution uses the notion of *levels of support.* The exact definition of the levels varies from organization to organization, but the idea is that the person who initially takes the telephone call about the problem has enough knowledge to be able to quickly resolve a high percentage—say 80 percent to 85 percent—of all problems reported. Problems that cannot be resolved within a few minutes at this level, called level 1, are passed to level 2, which is made up of technicians who have a higher degree of skill, experience, or training. Level 2 handles all of the problems remaining from level 1 and those that must involve a vendor. It is expected that level 2 would solve 10 percent to 15 percent of the total problems that are reported.

levels of support

All problems not solved at level 2 are passed to level 3, which is made up of communications technical support specialists or vendor specialists. Level 3 usually receives only about 5 percent of the problems—those that are extremely complex, difficult to identify, or difficult to solve. Problems that reach this level may require software modifications, hardware engineering changes, or other corrective actions that take a long time to put in place.

The advantage of the leveling approach to problem solving is that the problem is solved by the person with the lowest skill level who is able to do so, and highly skilled technical people are reserved to work on the most difficult problems.

Escalation Procedures Procedures need to be in place to escalate the status of a problem if it has not been resolved within a predetermined period of time. Certain problems are more critical than others, and these are the ones that need more formal and rapid escalation. A single inoperative terminal might be escalated if it has not been repaired within 24 hours, whereas a front-end processor outage that brings down the entire network may be escalated immediately.

Problem escalation takes two forms. One type of escalation is to bring additional technical resources to bear in order to help solve the problem. The other type of escalation is to make users and management aware of the actions being taken to resolve the problem. Often these escalations proceed at different rates. Figure 13–2 shows a sample of a generic escalation procedure. It is generic because it does not distinguish between types of problems that might be escalated at different rates. In many companies, problems are first ranked according to their severity. Severe problems are escalated very quickly, whereas problems with limited impact or those that only affect one or a small group of users are escalated more

problem escalation

Technical:

1. Level 1 (help desk) works on problem for a maximum of 15 minutes. If problem is not resolved, pass it to level 2.
2. Level 2 (technician) works on problem for up to 1 hour. If the problem is not resolved, notify the help desk supervisor and continue working on the problem.
3. If the problem is not resolved in 4 hours, get the appropriate level 3 (network specialist) involved. Level 2 retains "ownership" of the problem. It is level 2's responsibility to monitor the progress and to keep the user and the help desk supervisor informed about the status of the problem every 2 hours after level 3 gets involved.

Managerial:

1. The help desk supervisor is notified by level 2 if problem has not been resolved in 1 hour.
2. If the problem is not resolved in 2 hours, the help desk supervisor notifies the supervisor of network operations.
3. If the problem is not resolved in 4 hours, the supervisor of network operations notifies the manager of telecommunications. The manager of telecommunications calls the manager in the user department to discuss the situation and decide on any extraordinary action to be taken.
4. If the problem is not resolved in 8 hours, the manager of telecommunications notifies the chief information officer (or equivalent) and discusses the actions taken to date, and the future plans to get the problem resolved. The discussion should also include the possibility of contacting vendor management if appropriate.

NOTE: These sample procedures do not account for actions to be taken if the problem continues after normal working hours. The actions to be taken depend on the nature of the problem and the criticality of telecommunications to the company.

slowly. The example in Figure 13–2 shows the differences between technical and management/user escalation.

From time to time, all open problems or trouble tickets need to be reviewed by the network operations supervision. Typically, this is done each day with an eye toward spotting unusual problems or those for which the normal problem escalation procedures may not be applicable. On an exception basis, supervisors can make decisions about the relative priorities of outstanding problems and whether to take extraordinary steps, such as accelerating the normal escalation procedures, to resolve the problem.

problem tracking meeting

Another technique used by many companies is a periodic *problem tracking meeting*. In this meeting, people from network operations, software support, vendor organizations, and perhaps computer operations and user groups review all outstanding problems. This type of meeting is an excellent communications vehicle, and if the computer operations people are involved, it gives them an opportunity to prioritize all of the unresolved problems, whether they are computer- or network-related.

Bypassing the Problem If a problem is caused by a piece of equipment or circuit failure, the ideal solution from the user's point of view is to bypass the problem, enable the user to keep working, and then diagnose and fix the problem later. A failing terminal or modem might be "fixed," from the user's perspective, by replacing it with a spare that is available for just such a purpose. The failed equipment can then be repaired or returned to the vendor for service. When it is repaired, it can either become a spare itself or be returned to the original user.

Reconfiguration In the network control center where all of the circuits come together, spare equipment can be substituted by using hubs, switches, and *patch panels.* Depending on what equipment needs to be substituted, analog or digital switches can be used to switch a workstation to a different port on a hub, a circuit from a failing modem to a good one, or to bring a new server online. Usually, software switching must also occur to allow workstations to be addressed on the alternate ports, modems, or circuits. A patch panel is a piece of equipment on which each circuit has one or more jacks. Using cords with plugs, the network technician can connect spare equipment into the circuits temporarily by inserting a plug into the appropriate jack at the patch panel.

substituting equipment

Another technique for bypassing a failure is the use of a dial-up line as the backup for leased circuits. This technique is called *dial backup.* Certain types of modems can handle both leased and switched connections and are designed to automatically make a dial-up connection if the leased circuit fails. The dial-up connection often operates at a slower speed, but for many applications, slower speed operation is preferable to being totally out of service.

dial backup

Problem Diagnosis and Repair Because it is not so easy and usually not economically attractive to have spare circuits, front-end processors, or other expensive equipment available, the diagnosis and repair of many types of equipment problems are essential. On the one hand, routers and switches do not normally fail often, and because the diagnosis and repair require specialized equipment and training, they are almost always left to the vendor's maintenance people. On the other hand, since communications circuits are more prone to failure, many companies find it desirable to have some testing equipment available to assist in diagnosing circuit problems.

Depending on whether an analog or digital circuit is to be tested, different types of equipment are required. Analog signals are analyzed with simple speakers, tone generators, butt sets, and conductive probes. The techniques and equipment are similar for all analog lines, whether they are used primarily for voice or for data. Digital circuits or the digital sides of analog circuits require the use of breakout boxes and oscilloscopes. Some modems are capable of running tests on the communications circuits to which they are attached. The types of problems that this equipment can identify include frequency response, dB loss, and various types of distortion (problems described in Chapter 9).

This technician is testing parameters of a data circuit to ensure that they fall within the specifications required for high-speed data transmission. (Courtesy of Agilent.)

The carrier that provides the circuit obviously has extensive testing equipment as well and can be relied on to diagnose and resolve circuit problems. In many cases, however, additional information provided from tests at the user's end of the circuit greatly assists the vendor in problem resolution. Test data can pinpoint the exact location of the problem and help resolve it.

With a LAN, the wiring and cabling that comprise the circuits are usually privately owned, so either organization's network technicians or an organization that has been contracted to do the diagnosis and maintenance must do the analysis and problem determination work. LAN hubs, bridges, switches, and routers are often relatively "smart" and have built-in diagnostic capability to help in locating the source of trouble, noisy or cut cables, and other problems. The extra cost for this smart network equipment is usually well worth the cost for the assistance it can provide when trouble occurs, and there is pressure to make the network operational again.

Most modern communications equipment has diagnostic capabilities built into its basic design. Modems usually have self-test routines that can be activated manually through the front panel or, under certain circumstances, that will be activated by the modem itself. Terminals frequently have diagnostic routines that the user can activate through a key on the keyboard. One of the first questions the help desk should ask when a problem is reported is whether the appropriate terminal diagnostics have been run.

When all hardware appears to be working correctly but communication is still impossible, it may be necessary to employ a device called a *protocol analyzer.* A protocol analyzer is an electronic instrument that can look at the actual bits or characters being transmitted on the circuit to determine whether the rules of the protocol are being followed correctly. Some protocol analyzers are designed to handle a specific protocol, such as IP, BISYNC, or SDLC, whereas other equipment can examine and diagnose multiple protocols. When protocol analysis is performed, it is usually necessary to have software specialists involved because few hardware technicians are trained to understand the detailed characteristics and sequences of the protocols and data being transmitted.

protocol analyzer

A basic decision that companies must make is how much problem diagnosis and testing they will perform themselves and how much they will leave to vendors or other maintenance organizations. Often the answer to the question is to use a combination of people, some from inside the company and others from the vendor. In that case, coordination of the work is important.

One reason why it is important to have trained technicians and appropriate diagnostic equipment is that in today's multivendor networks, it is not always clear which vendor should be called when a problem occurs. For example, a WAN line from Colorado to California has two Regional Bell Operating Companies, a long distance carrier, modem manufacturers, and workstation manufacturers involved. A LAN might have router, switches, and workstations from different manufacturers.

Sometimes technicians must perform tests on the network to determine exactly where the problem lies. Customers may be paying for service of a certain quality. They may want to test the circuit periodically to ensure that they are receiving the quality of service they are paying for. Even if a vendor's equipment is clearly at fault, it may be possible to perform tests that further pinpoint the problem while the vendor's technician is on the way. This may greatly speed the process of repairing the failing component and may keep the cost of the technician's time to a minimum.

Performance Management

Performance management is the set of activities that measure the network performance and adjust it as necessary to meet users' requirements. The people responsible for network performance management use the statistical data gathered by the network operations staff. The data is analyzed and summarized, and the network performance is reported to management. This work also triggers performance adjustment activities or performance problem resolution, which is normally done by the communications technical support group. It also provides utilization data to network capacity planners. The location of the performance management responsibility depends on the size of the telecommunications department, the skill levels of the staff, and the preference of the

These technicians can view the status of the entire network or zero in on specific circuits to isolate problems when trouble occurs. (Courtesy of Agilent.)

telecommunications manager. Performance tuning will be discussed in more detail later in the chapter because it is most frequently performed by the technical support people.

service level agreement

In order to determine whether the performance of a network is adequate, as measured by its reliability and response time, preestablished performance objectives are required. Performance objectives may be established for the entire network or for a specified set of equipment or users. When the objectives are established for a group of users, such as a department, the objectives are commonly called a *service level agreement.* Service level agreements often involve stating that of the *X* hours of the day or week that a user's terminal and applications will be available, the reliability of those applications will be measured as a certain percentage of *X*. It normally also contains a statement about the response time the user can expect to receive at the terminal. A typical service level agreement might read like the following:

- The order entry application and customer service terminals will be available from 7 A.M. to 8 P.M. EST each business day.
- The reliability will be 98 percent as measured at the user terminal. Downtime will not exceed 2 percent.
- Response time, measured from the time the user presses the ENTER key until the response is received at the terminal, will be less than 1.5 seconds 90 percent of the time and less than 10 seconds 100 percent of the time.

This type of service level statement must be negotiated between the user and the network and computer operations people. The starting point should be to identify the user's requirements. The network and computer people may have an option to provide a better grade of service at an increased cost if the user is willing to pay for it.

The other part of performance measurement is the actual tracking of the network's performance. Performance data about a data communications network can be gathered by hardware, software, or a combination of the two. Recording the time that the network actually started in the morning and closed at night is normally a software function. Measuring the availability and response time from the user's point of view must be done at or close to the terminal.

performance tracking

Most network management architectures use a similar basic structure. End stations, such as workstations, computers, and other network devices, run software that enables them to monitor and compile information about the state of the device in which they reside. This information is stored in a *management information database (MIB)*. The software can also send *alert messages* to a central network management server or computer when it recognizes problems. The central network management server receives the alerts and is programmed to react by executing appropriate actions, including operator notification, event logging, and automatic attempts at repair.

The central network management server can also proactively poll end stations to check the values of certain variables. Polling can be automatic or operator-initiated, and the software in the end station devices responds to the polls by accessing data stored in the MIB. These proactive or reactive interchanges between the software in the end station and the central network management computer occur according to the rules of a *network management protocol*. Well known network management protocols include the Simple Network Management Protocol (SNMP) and Common Management Information Protocol (CMIP), which will be discussed later in the chapter.

For voice networks, statistics are gathered by the carrier that provides the circuits, as well as by PBXs and many key systems. The carrier is usually willing to provide the statistics collected to the customer. Working together, the carrier and the customer can ensure that the network is properly configured for the actual traffic loads. In some PBX systems, the statistics collection can be tailored to meet the particular requirements of the company. Key systems usually have less flexibility and provide a predefined set of standard statistics about a call: its duration, the extension number of the caller, and perhaps the outgoing circuit or trunk identification if one was used.

Data from the entire network is normally stored in a central database where operators can examine it on a realtime or periodic basis. Summaries and statistical reports can be developed that show the entire network's performance as well as highlight potential problem areas. The data can be compared to service level agreements to see if performance objectives are being met.

performance history

In any type of performance measurement system, it is important to keep a history of performance statistics so that trends can be examined. One useful technique is to plot trend charts that show how certain network parameters, such as availability, response time, or circuit utilization, are changing. Some examples are shown in Figure 13–3. The trend charts should also show the standard or desired level of performance. Trend charts are particularly useful for watching parameters that change gradually over time. While a gradual increase in response time might not be noticed for a long time if the daily performance reports were the only information available, a trend chart would show the increase in response time more quickly.

Some of the parameters of the network that need to be tracked are

- response time—by hour of the day, by application, by circuit, all measured from the user's point of view;
- circuit utilization—by circuit and by time of day;
- circuit errors—by circuit;
- transaction mix—by time of day, over time;
- routing—utilization of each circuit when multiple routes to the same destination are possible;
- buffer utilization—in the FEP, in cluster controllers, and in the computer;
- queue lengths—in routers, in multiplexers, in the FEP, and in the computer;
- processing time—the time to process a transaction or message in the computer.

Figure 13–3
These trend charts show several parameters of a network's operations. The standard level of performance is shown on many of the charts.

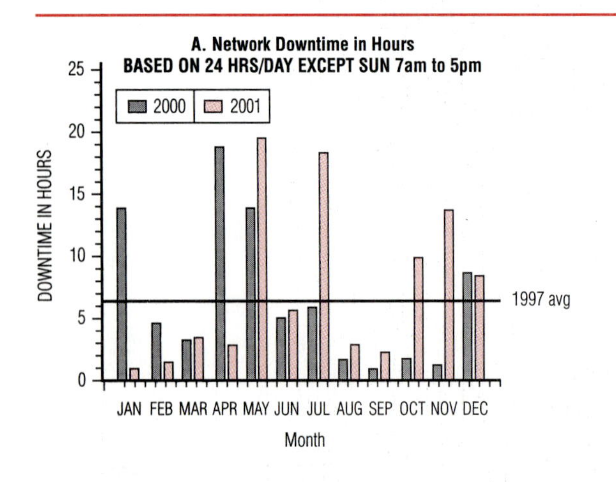

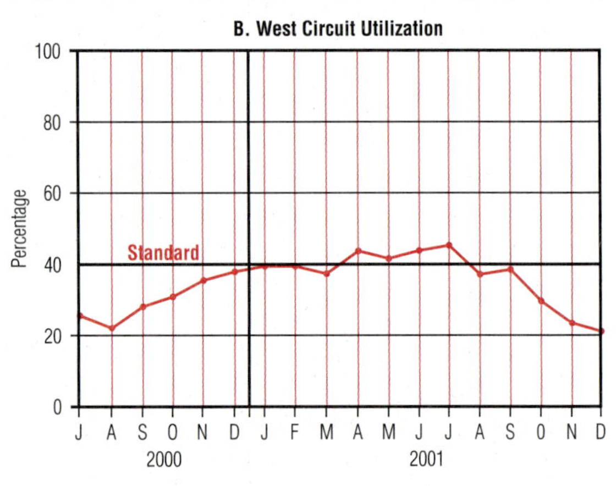

Figure 13–3
continued

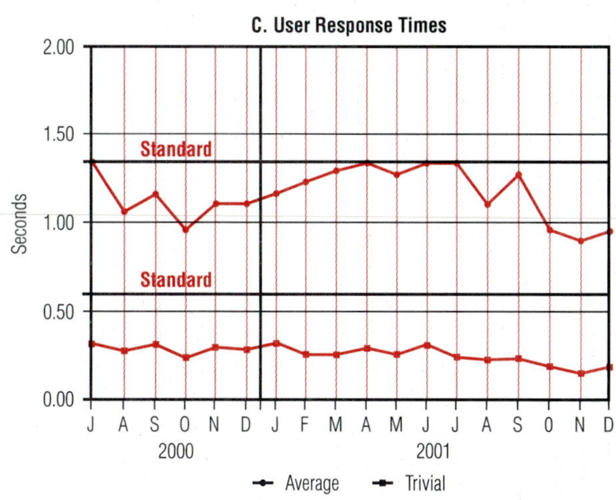

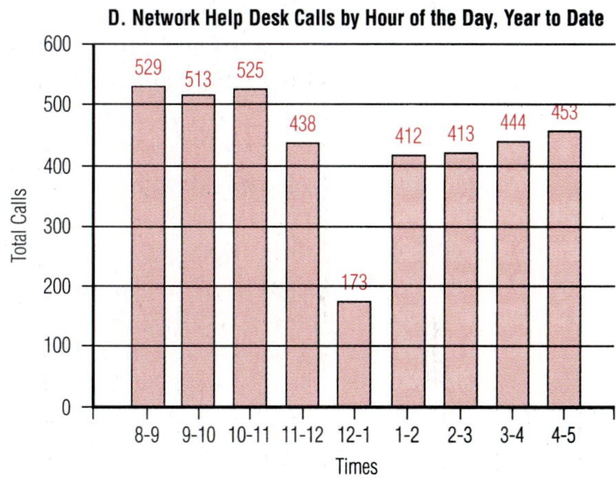

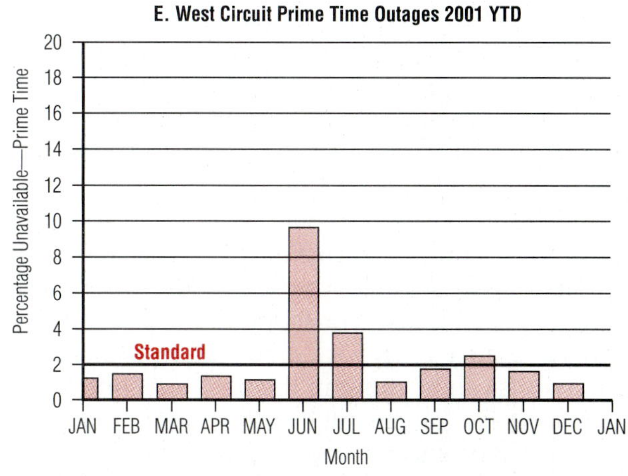

Each of these statistics gives a unique view of network operation. Two or more will often correlate, as, for example, when response time gets longer because there is a shortage of buffers or because the message queues are long.

normal values

Over time and with experience, the network operations personnel will begin to know what the normal values are for their network. They develop a feel for when the network is performing normally and when it isn't. More formal statistical techniques can establish control limits on the values so that it is easy to determine whether day-to-day variations are normal or whether they indicate that a statistically significant change has occurred.

Configuration Control

Configuration control is the maintenance of records that track all the equipment in the network and how it is connected. One required record is an up-to-date inventory of all network hardware and software. Diagrams showing how the pieces of equipment and circuits connect are another useful form of documentation. Configuration control may be performed by the network operations group or the administrative support group. As with all of the organizational alternatives, the definition and assignment of responsibilities are a management function.

Equipment Inventory The network equipment inventory is like any other inventory system in that it is updated to keep track of the installed equipment and circuits as well as all additions and deletions. If the inventory is kept in a computerized system, the data can easily be manipulated, sorted, or reported in different ways. Network operations personnel might want to see a report of all of the workstations of a certain model or to find the location of a server with a certain serial number. Management might be interested in data about the total value of all of the terminals or the total costs of all circuits. Such information is obtained easily from a properly designed, computerized inventory of network equipment. An example of a report from a network equipment inventory system is shown in Figure 13–4.

inventory software

Software can be purchased to keep these types of inventory records, or a simple system may be developed in-house. As an alternative, the company may decide that its network equipment inventory will be kept in the company's property ledger, the accounting system that tracks all of the company's physical assets.

maps

Network Diagrams One of the most useful forms of displaying the topology of the network is a map that shows all of the network locations, controllers, and computers, and the circuits connecting them. Examples of this type of map are shown in Figure 13–5. Such maps can be produced in many sizes. Wall-sized versions can be hung in the network control center. Versions of an 8½-by 11 inch size can be put in notebooks or made into transparencies for use in presentations or at meetings.

Account 7045-336 Detail by Device Code
Run Date 06/25/2001

Device Type		Model	Serial	Site Code	Account Billed	Vendor Charge
COMMCTL	8911	A54	00951A	CMD	7045-336-4X3	5228
ROUTER	2070	31A	12506	CMD	7045-336-4X3	1182
	2070	31B	D0055	CMD	7045-336-4X3	1182
	2590	41C	D0056	MID	7045-336-4X3	1182
	2590	41C	D4528	MID	7045-336-4X3	1182
	2590	41C	D5477	CMD	7045-336-4X3	1182
SWITCH	611A		G1843	CMD	7045-336-4X3	36
	690D		W3090	CMD	7045-336-4X3	36
PC	XL-500	P-100	P0319	MID	7045-336-4X3	38
	XL-500	P-100	P7802	CMD	7045-336-403	38
	XL-500	P-100	Q1611	CMD	7045-336-403	38
	XL-650	P-200	T0173	CMD	7045-336-4X3	38
	XL-650	P-200	T1157	MMS	7045-336-4X3	38
	XL-950	P-200	AE958	CMD	7045-336-403	38
	XL-950	P-200	VI904	MID	7045-336-4X3	38
	XL-950	P-200	T4790	MID	7045-336-4X3	38
	XL-950	P-200	B1140	MMS	7045-336-4X3	
PRINTER	2700	A	32255	CMD	7045-336-4X3	822
	2700	A	32305	CMD	7045-336-4X3	822
	2700	B	32431	CMD	7045-336-4X3	822
	9412	4112	40435	CMD	7045-336-4X3	723
	9412	4112	40560	MMS	7045-336-4X3	723
	8000	5	40938	CMD	7045-336-4X3	416
	8000	5	41039	CMD	7045-336-4X3	416

Figure 13–4
Inventory list of telecommunications equipment.

circuit charts

Another level of detail, as shown in Figure 13–6, is a listing of each circuit and the devices attached to it. Charts of this type usually indicate the circuit number assigned by the carrier as well as the names or network addresses of the controllers and terminals. Equipment model numbers and serial numbers are also generally included.

wiring diagram

A third level of detail is the wiring diagram. The wiring diagram is drawn for a single building or floor of a building. One version is a type of map that shows the actual locations of the devices on the floor or in the building as well as the connections between them. Another type of wiring diagram has no bearing on the actual location of the equipment but shows in detail the exact cable runs and types of wire, twisted pair numbers, and pin numbers of each component in the network.

Figure 13–5
Maps of typical wide-area communications networks.

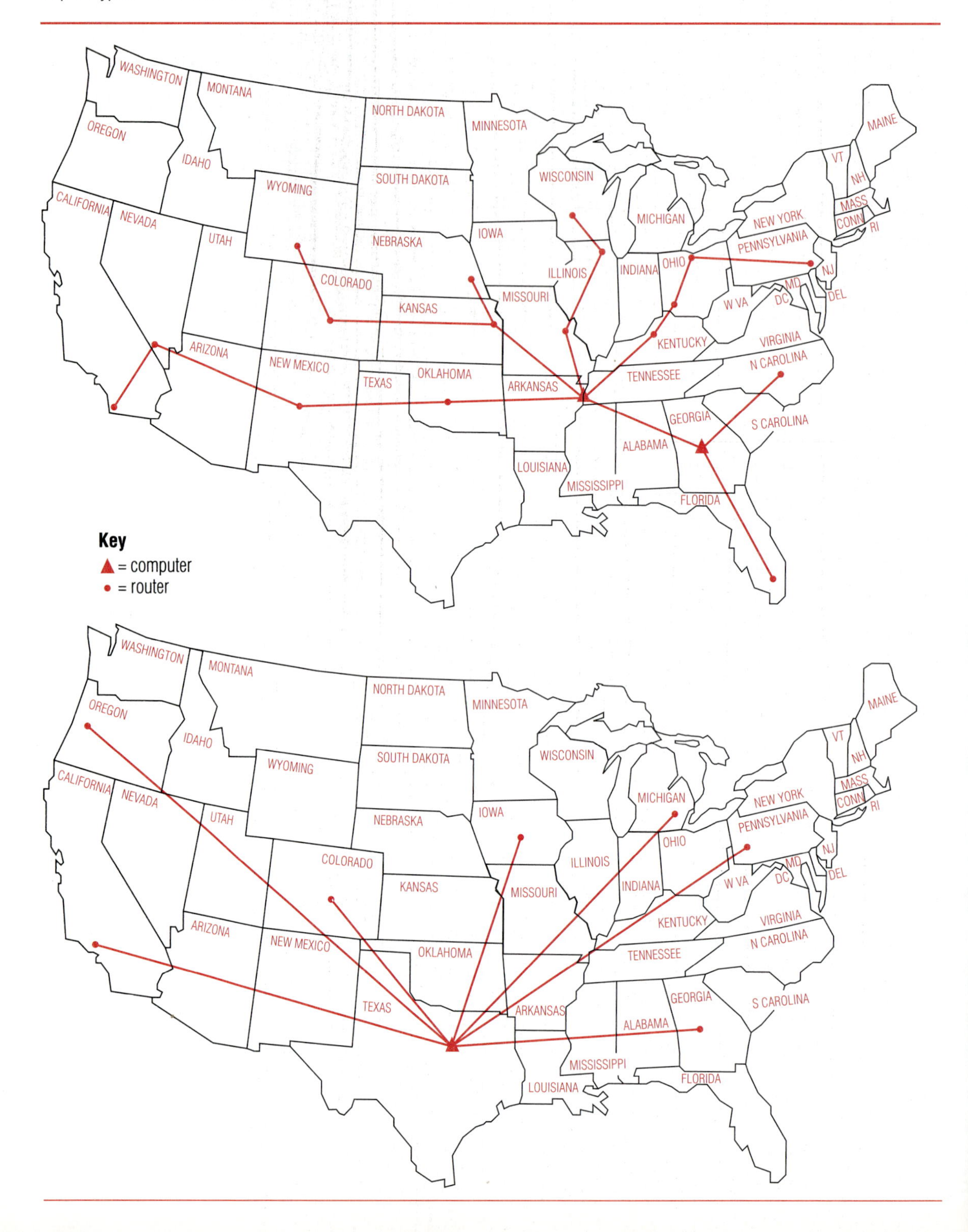

Circuit Name	Circuit Number	Router Type	Model No.	Serial No.	Network Address	Location
West	FDA3856B	2520	4A	135269	3E1C	Detroit
		2503	3	426578	3E9D	Chicago
		2503	2	890124	3E1A	Denver
		2501	2	182930	3E4F	San Fran
South	FDA3472	2520	4A	135269	3E1C	Detroit
		2503	3	42457	1A46	Toledo
		2503	3	675645	1A3D	Memphis
		2502	2	276578	1A88	Mobile
		2502	2	764190	1A6A	Miami

Figure 13–6
Listing of circuits and attached controllers.

Drawings and diagrams of the network are most easily developed and maintained on a computer-aided drafting (CAD) system that captures the drawing in a computer database and simplifies the maintenance. CAD software with enough capacity to handle most networks is now available for personal computers, so the tool is easily afforded by most telecommunications departments.

computerized drawings

Other Documentation Other network documentation includes

- software listings for all telecommunications software;
- the name and telephone number of a contact person at each location on the network;
- vendor manuals for all hardware and software;
- disaster recovery plans;
- vendor contact names and telephone numbers for all equipment and software;
- problem escalation procedures;
- routine operating procedures.

The network documentation should be backed up. Whether it is kept online in electronic form or printed, duplicate copies of all lists and procedures should be kept, preferably off-site. Furthermore, a procedure to keep the backup copies up-to-date must be in place.

Change Management

Change management is an activity for monitoring all changes to the network and coordinating the activities of various groups in such a way that there is minimal impact on network operations and user service when changes are made. The change management activity is often called *change control*. However, the word *control* tends to imply that many networ'

Figure 13–7
A change request form that would be completed by the person requesting the change.

Change Request Form

Request Date: ________ Request No. ________

Implementation Date: ________ Requestor: ________

Type of change: Addition ____ Modification ____ Replacement ____
Removal ____ Correction ____

Description of the change: ________

Assigned to: ________

Approvals:

Supvr. Network Operations: ________
Supvr. Technical Support: ________

Implementation Data:
Date Change Made: ________
Comments: ________

changes are unnecessary and could be eliminated. Although many operations people would like to eliminate change to the network to preserve stability, it is hardly realistic to do so in a telecommunications environment. However, change must be implemented in a planned, coordinated way.

Most change management systems require that a "request for change" form, such as the one shown in Figure 13–7, be filled out by the person who is sponsoring the change. A change may be the installation of a new terminal, the reconfiguration of a circuit to support a new location, or the installation of an updated version of the telecommunications software. In most companies, all of the change request forms go to a single person, the change coordinator, who screens the requests to find any obvious conflicts between changes or other inconsistencies. The change coordinator can also keep a master list of all planned changes, such as the one shown in Figure 13–8.

change coordination meeting

From time to time, the change coordinator convenes a *change coordination meeting* that is attended by representatives from network operations, software support, computer operations, vendors, users, and others who may be interested in or affected by the proposed changes. At the coordination meeting, the changes are reviewed and discussed. The plan, which shows when each change is to occur, the nature of the change, and

Change Coordination Plan as of 10/25

Change No.	Description	Planned Date	Responsibility	Date Completed
1047	Add Dallas to network	10/27	Vandewoe	
1048	Maintenance to NCP software	10/29	Morris	
1049	Add 5 VDTs in marketing	11/15	Vandewoe	
1050	Upgrade west circuit to 64 kbps	11/28	Aymer	
1051	Add router to administration's LAN.	12/10	Aymer	
1052	Update network management software.	12/12	Morris	

Figure 13–8
A change coordination plan for a small network. Large networks would have a much more extensive list of changes.

who is responsible, is updated. Assuming no major obstacles are discovered, the plan is approved and the changes are implemented according to the plan.

In many organizations, the change coordination meeting is held weekly. Many network changes are implemented on the weekend to avoid disrupting the network during normal business hours. Software changes in particular often require deactivating the entire network so that the new software can be installed and tested, and usually the only time this can be done is at night or on the weekend.

When major changes to the network are anticipated, such as a complete reconfiguration of the circuits or the provision of a totally new service, the network operations people should participate with the network analysts and designers in the planning and design process. By participating in the development project, the operations people can stay in touch with coming events that will affect the work they do or the procedures that are in place.

documentation updates

It is extremely important to keep the network documentation up-to-date as changes are made. Of course, if the documentation is kept online in a word processing system, the maintenance process is simplified. The proper way to keep network documentation up-to-date is to make it a part of the change management process so that it is done routinely at the time that the changes are implemented. Although this statement may seem obvious, it is surprising how many organizations ignore the documentation in the rush to implement changes.

Management Reporting

Management reporting of network operations activities and statistics is discussed here as a separate topic, although a piece of it is likely to be done by each group within the network operations department. Several different types of reporting need to be done for different levels of management. Generally, the higher the level of management, the more the data should be summarized and the more the reports should focus on the exceptions rather than routine detail. User management is primarily interested in data about the network performance, particularly when it is compared to agreed-upon service levels. Network operations management is equally concerned about the performance compared to the service levels because their objectives are based on providing the agreed-to service. They are also interested in basic operational statistics, such as the number of problems that were logged, the length of time it took to resolve problems, and the overall use of the network facilities. Graphs normally are the most useful way to convey this type of information.

management summaries

Different types of information are reported on different time schedules. In most networks other than the smallest, a daily status review meeting is held to discuss the results of the previous day and any outstanding problems. Plans for the coming day also may be discussed. This type of meeting is held most frequently early in the morning and is kept quite brief—normally 15 minutes or less. Some companies connect remote locations into these meetings via audio teleconferencing so that status information can be distributed widely and quickly.

daily status review

Network performance and utilization statistics are sometimes calculated and distributed monthly or they may be continuously available online. A useful variation of this approach, however, is to calculate performance numbers daily. If service levels are not being achieved, it is considered a problem and is handled through the problem tracking system. Since open problems are discussed at the daily status meeting, the performance problems get immediate attention.

Statistics in a voice network are similar to those in a data network. Centrex and PBX systems can generate statistics and measurements, such as trunk utilization and average call length. Much of the focus of telephone statistics is on long distance calling, which is frequently the most expensive part of the voice network. The number and duration of long distance calls, evidence of call abuse, utilization of FX lines or tie lines, and call queuing times, where appropriate, are all relevant to the efficient and effective operation of a voice communications network.

NETWORK MANAGEMENT FOR LOCAL AREA NETWORKS (LANS)

Managing the operation of a LAN would intuitively seem to be easier than managing a wide area network (WAN) because the LAN covers a smaller geographic area and because the cabling is generally in a protected environment and therefore less subject to transmission errors. It is true that when a LAN is first installed it is often quite simple, in some cases being entirely the product of a single vendor who uses equipment and software designed to operate well together. However, as LANs grow, it is typical to add equipment and software from other vendors, such as inexpensive workstations or special-purpose software. Furthermore, it often becomes desirable to link the LAN to other LANs, to metropolitan area networks (MANs), or to WANs, in which case bridges, switches, routers, or gateways must be added. Although each of these additions may be small in itself, the cumulative effect can be that the original, small LAN grows into a complex piece of an extensive network. Add to this the fact that if the LAN is successful (and if it is growing, it probably is successful), the organization grows increasingly dependent on the LAN, so failures and outages are likely to disrupt the company's business operation.

This growth in complexity and increasing dependence places a premium on good LAN management practices and efficient problem resolution techniques. The LAN administrator normally plays a critical part in both of these activities, and hence the importance of his or her role is magnified.

The basic techniques of LAN management, which you studied in Chapter 12, are the same as for managing MANs or WANs, but many specialized tools are designed especially for LANs. One example is the NetWare Management System, designed for Novell's LANs. This software allows a single workstation to manage an entire LAN no matter how extensive or geographically dispersed it may be. Other LAN vendors have similar products.

It is fair to say that LAN management, as a part of the broader subject of network management, is receiving a great deal of attention from vendors and standards organizations because of the rapid growth of LANs and the growing reliance of organizations on all types of LANs.

NETWORK MANAGEMENT SOFTWARE

Managing a network of any size—other than the very smallest, single-location LAN—requires assistance in the form of automation tools. Medium and large networks simply have too many hardware and software components for an individual or group of people to keep track of manually. When coupled with rapidly changing status conditions, the monitoring and alerting activities are good candidates for automation. There are many analogies between a network's operation and the process monitoring and control in a manufacturing plant.

Network monitoring cannot be done using software alone. There is a need for certain hardware capabilities in workstations, terminal control units, bridges, routers, gateways, FEPs, and computers to monitor and collect network performance information and to pass it to software for analysis and action. What users want is the ability to select the best vendors for each of the network components and to be assured that they all

hardware monitoring

will be managed by a network management package. Furthermore, if the devices are on different, but interconnected, networks that use different technologies and protocols, there is an even stronger need for a common network management protocol that can be used across all networks.

Simple Network Management Protocol (SNMP)

In the diverse world of telecommunications, it may not surprise you to learn that two network management protocol standards have emerged, somewhat in competition with each other. The first is called the *Simple Network Management Protocol (SNMP),* and it is a de facto standard that originated in conjunction with the Internet to manage networks and devices using the TCP/IP protocol. The second is called *Common Management Information Protocol (CMIP)* and was developed by the International Organization for Standardization (ISO).

An SNMP managed network consists of three key components: *managed devices, agents,* and *network management systems (NMS).* A managed device is a network node that contains an SNMP agent and resides on a managed network. Managed devices collect and store management information and make this information available to NMSs using SNMP. Managed devices can be routers, access servers, switches, bridges, hubs, computers, or printers. Information about the device is collected using another SNMP standard called *remote monitoring (RMON).* RMON is a standard monitoring specification that defines a set of statistics and functions that can be exchanged between RMON-compliant devices or software. Nine types of statistics are specified, and a vendor can select which of the nine types to implement, normally those most relevant to the device being monitored. An agent is a network management software module that resides in a managed device. An *agent* collects the information provided by RMON and translates it into a form compatible with SNMP. An NMS executes applications that monitor and control managed devices. NMSs provide the bulk of the processing and memory resources required for network management. One or more NMSs must exist on any managed network. The relationship between these components is shown in Figure 13–9.

Hardware that uses SNMP, for example, routers, gateways, and switches, collects information about itself and the circuits or other equipment connected to it and stores the information in a database called the *management information base (MIB).* Some of the devices have their own MIB, but many of them store their information in a central MIB on a server. The network manager has a workstation that runs the network management application software that can access the MIBs, collect the information, process it, and produce management information and reports. The network manager can also send commands or instructions to the devices and effectively manage and control the network. Although SNMP is a standard, many vendors have added their own extensions, so the MIBs from one vendor's equipment may not be exactly compatible with the

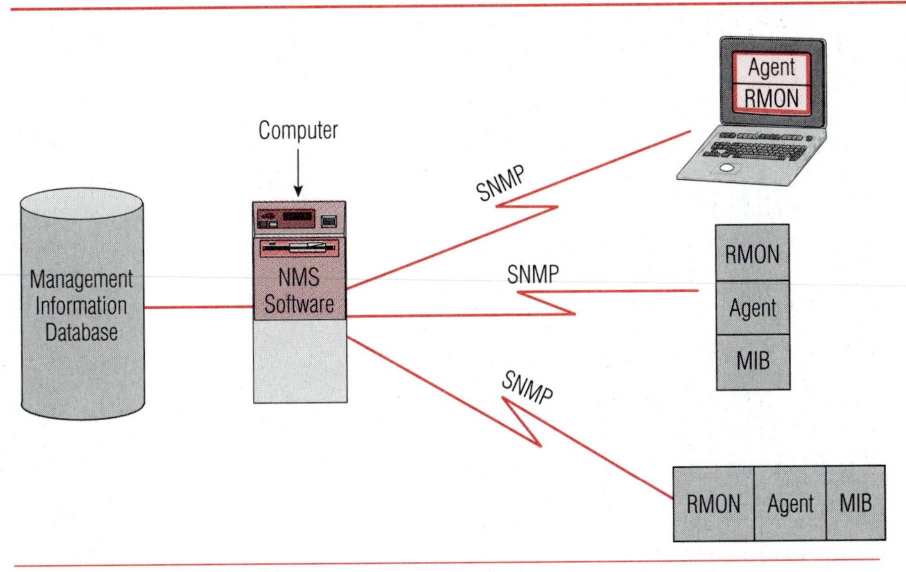

Figure 13–9
The structure of an SNMP managed network.

MIBs from another. Hence, the user needs to be careful when selecting network management hardware or software from different vendors, both of which are supposed to be SNMP compliant.

Two versions of SNMP exist: SNMP Version 1 (SNMPv1) and SNMP Version 2 (SNMPv2). Both versions have a number of features in common, but SNMPv2 offers enhancements such as additional protocol operations, making the two versions incompatible. SNMPv2 messages use different header and protocol data unit (PDU) formats than SNMPv1. SNMP is an older standard than CMIP and has some command and security limitations that CMIP addresses. But CMIP has not been implemented in very many products and is not widely used, so SNMP has become the de facto standard for the management of multivendor networks.

Proprietary software packages that are designed to help the telecommunications staff manage a data network are also widely available. They collect data from various sources, such as the hardware described earlier, and store the data in databases for analysis and action. These packages provide an array of status displays, commands, error analysis programs, performance measurement, capacity management, and reporting tools.

IBM's *Netview* and Netview/6000 network management products were originally programmed to assist in the management of IBM's SNA networks, but now there are several versions that can monitor and manage most kinds of wide and local area networks. Netview automates many network management tasks, including restarting devices that have been taken offline, perhaps because of temporary errors, sending alert messages when certain error threshold values are exceeded, and reporting network performance statistics. Netview is a part of a more comprehensive network management architecture called *Open Network Architecture (ONA)*, which include the capability to track service levels, do capacity planning, handle network financial administration, and provide other management support for heterogeneous networks.

Novell's NetWare management system (NMS) was written to help network managers manage Novell LANs. Each server on the network must have NMS software to gather statistics and monitor the status of the network and the devices attached to it. The network operator has a wide variety of commands at his or her disposal with which to gather the statistics, monitor the network in realtime, issue commands, and generate reports. NMS modules are available to interface to devices that use the SNMP protocol and to IBM's Netview.

There are also many software packages on the market to assist the telecommunications staff with such activities as keeping equipment inventory, drawing network diagrams and maps, keeping track of changes, and charging costs to users. Software is available that runs on either mainframes or personal computers, depending on the customer's requirement. Some vendors have integrated their software in modular packages so that the customer can buy only the pieces that are needed. Most organizations should be able to find commercially available network management software that meets their needs.

NETWORK SECURITY

In addition to the traditional operational responsibilities, telecommunications management has a very important security responsibility. Organizations have become dependent on information stored in computers and the data transmission facilities to access this information. No longer can they operate the business without having the information available. At the same time, terminals and data communications systems have made access to computers much easier. With the rapid growth of interest in the Internet, the private networks of corporations and other organizations are frequently connected to an ISP, which provides a much more open environment but also makes the organization's network more vulnerable to unauthorized access from the outside. The fact that information and tools needed to penetrate corporate networks are widely available has provided further impetus for a heightened concern about network security by companies throughout the world. Since information is viewed as a business asset, it must be protected like other assets and surrounded with proper controls and the appropriate security to protect it. For a company to adequately manage information security either on the computer or on the network, it must have

- a network security policy that clearly defines the reasons why curity is important to the company;
- a security implementation plan that describes the steps to imp ment the policy;
- clearly defined roles and responsibilities to ensure that all aspe of security are performed;
- an effective management review process to periodically ensı that the security policies and standards are adequate, effective, and being enforced.

Each of these elements must be present. The policy is management's statement of importance and commitment. The plan describes exactly what practices will be in effect. A network security officer is often appointed and is responsible for carrying out the security plan. The security officer also investigates violations of the security policy and makes recommendations about additions to the security plan or changes to be made. The management review process is a periodic check that the security program is operating properly. The initial step in the review process may be a security audit performed by the company's inside or outside auditors. The audit report serves as the basis for the management review.

Types of Network Security

There are three primary ways that someone can gain unauthorized access to a network:

1. using a workstation located on the organization's premises;
2. dialing into the network;
3. accessing from another network.

Specific measures can be put in place to make it more difficult for someone to gain unauthorized access to the network, as described below, but you should always bear in mind that stopping a determined hacker from breaking into a network is extremely difficult. Most security measures are designed to raise the threshold of difficulty to the point where the hacker will lose patience and go elsewhere.

Communications security takes four primary forms:

1. physical security of the telecommunications facilities, including the network operations center, equipment rooms, and wiring closets;
2. access control to prevent unauthorized use of the telecommunications terminal circuits, telephone systems, or computers to which they are attached;
3. personnel security, such as security checks on prospective employees, training, and error prevention techniques;
4. disaster recovery planning, to ensure that the organization can continue operations in the event of a disaster.

Physical Security

The primary emphasis of physical security is to prevent unauthorized access to the communications room, network operations center, or communications equipment that could result in malicious vandalism or more subtle tapping of communications circuits. It may also be necessary or desirable to inspect the facilities of the common carrier through which the communications circuits pass. Physical security can be thought of as the lock and key part of security. The equipment rooms that house telecommunications equipment should be kept locked. Terminals may be equipped with locks that deactivate the screen, keyboard, or on/off switch. In lieu of actual locks and keys, magnetically encoded cards that resemble credit cards can be used to activate terminals or to unlock doors. These cards could also serve as a personnel identification badge.

One interesting facet of communications network physical security is the crime rate against ATM users. There have been a number of robberies shortly after people have withdrawn money from an ATM. One significant legal question is how far banks must go to protect ATM users. Banks maintain that as long as they locate the machines in well-lighted areas or on major streets, they aren't liable when crimes occur. They also say that consumers should bear part of the risk when they use the machines after dark. Consumer advocates suggest that banks should have guards posted by the machines. This situation exemplifies the type of societal problems that will continue as communications systems become more widely used by the general public.

Network Access Control

Telecommunications access adds another dimension to the security concern. With telecommunications, data can be accessed from terminals and networks outside the computer room and a new set of questions emerges:

- How do we know who is at the terminal?
- Once we know who is at the terminal, is the person authorized to access the computer?
- What operations is the terminal user authorized to perform?
- Is it possible that the telecommunications lines could be tapped?

Several types of people try to gain unauthorized access to corporate networks. The most infamous and widely publicized are dedicated professional hackers who break into networks and computers with malicious intent, perhaps to destroy data, commit fraud, or in the case of government networks, commit espionage. A second group, almost as well publicized, is people who break in mainly to prove they can do it. If they are successful, they may do harm, such as destroying data, or they may simply leave messages that in effect say, "I was here." People in this category may also be skilled, professional hackers who are experts on network security, or they may be casual users who simply surf the Internet looking for networks or web sites that are easy to break into, perhaps because the network or web site managers have "left the door open," so to speak. An additional type of intruder is employees who have legitimate access to an organization's network but who try to access unauthorized information.

Network access control techniques begin with ensuring that all legitimate users of a network have a unique identification code and secret password. Users should be required to log in each time they use a workstation on the network by entering their identification code and password.

This information, along with the date, time, and terminal identification, should be automatically recorded by the software in a central file on a computer so that a complete record is kept of all users of the system. When someone tries to log on and enters an incorrect password a predetermined number of times, the user's ability to log on should be disabled, recorded in a security log, and a security officer notified. Reactivation of the user identification code should be done only when it has been determined that the user really is authorized. Supervision usually is required to reauthorize the user, and a new password is issued. Similar control techniques should be put in place to limit the access to long distance voice facilities, such as 800 service lines.

Dial-up data lines are especially vulnerable to unauthorized access. The most common security techniques used with dial-up data circuits are *call back* and *handshake.* With the call back technique, the user dials the computer and identifies himself or herself. The computer breaks the connection and then dials the user back at a predetermined number that it obtains from a table stored in memory. The disadvantage of this technique is that it does not work very well for salespeople or other people who are traveling and calling from various locations. The handshake technique requires a terminal with special hardware circuitry. The computer sends a special control sequence to the terminal, and the terminal identifies itself to the computer. This technique ensures that only authorized terminals access the computer—it does not regulate the users of those terminals.

In addition to these techniques, it is important that network operations management monitor the use of dial-up computer ports for suspicious or unusual activity. This usually is done by reading a printed log of all dial-in accesses or attempts. In particularly sensitive applications, such as the transmission of financial transactions or data about new product developments, encryption, which was discussed in Chapter 7, can be employed to scramble the data so that if unauthorized access to the computer or data is obtained, the data is still unreadable without further work to decrypt it.

A relatively effective technique for limiting unauthorized access to the corporate network from outside networks to which it is connected is to install a *firewall.* A firewall is a combination of hardware and software that enforces a boundary between two or more networks, for example,

firewall

between the corporate network and the Internet to which it is connected. Without a firewall, anyone on one network could theoretically get into the other network and pick up or dump information. The primary purpose of the firewall is to provide a single point of entry and exit from the corporate network allowing controlled access from one network to another. A very frequent application of firewalls is between a corporate network and the Internet. The firewall allows access to resources on the Internet from inside the organization and controlled access from the Internet to the internal networks, computers, and data. One company whose employees all use PC workstations allows all of its people to pass through the firewall and access an Internet web site that is managed by another organization that contains information about the employees' 401k retirement savings program.

While there are several ways to implement a firewall, traditionally the firewall is implemented using a router, a server, or both in combination. A router running *packet-level firewall* software can examine all network traffic at the packet level allowing or denying packet passage from one network to the other based on the source and destination addresses. This technique is called *packet filtering*. A server can act as an *application-level firewall* if it examines and controls data at the application level. The server looks at entire messages and does a more detailed analysis of the appropriateness before making a decision to let the traffic pass.

A *proxy server* is a form of application server that works differently. Proxy servers change the network addresses from one form to another so that users or computers on one network do not know the actual addresses of nodes on the other network. Using the Internet example, computers or users on the Internet would think there is only one node on the corporate network, the proxy server. Internet users would not know that other nodes exist behind the proxy server or the addresses of those nodes. Only the proxy server would have the table that translates between the real internal node addresses of the workstations and computers and the fake addresses that are used on the Internet.

Firewalls normally log all of their activity so that information about network access and denial is available for later analysis, either because of a problem or for routine security audits. The firewall must also provide a method for a security administrator to configure and update its access control lists in order to establish and modify the rules for access according to the security policy.

Personnel Security

Having employees who are motivated, security conscious, and well trained in the use of proper security tools is a very effective security technique that organizations should not overlook. Trained employees may not prevent a hacker from breaking into the corporate network, but the likelihood that an intrusion is discovered and reported quickly is greatly increased. Furthermore, having employees who routinely do not leave their PC workstations logged on and unattended and who protect their laptop computers when they are traveling reduces the possibility that a number of the simpler security breaches will occur.

Personnel security involves using one or more of the following techniques:

- security checking or screening of prospective new employees before they are hired;
- the identification of employees and vendor personnel through badges or identification cards;
- having active security awareness programs that constantly remind employees about their security responsibilities and the company's concern for security;
- ensuring that employees are properly trained in their job responsibilities;
- having error prevention techniques in place to detect accidental mistakes, such as keying an erroneous amount.

Network operations personnel should be charged with the responsibility to see that these measures are in place, and employees' security knowledge and awareness should be checked from time to time, perhaps as a part of a security audit.

Disaster Recovery Planning

As the network becomes a vital part of a company's operation, plans must be made for recovery steps to be taken if a natural or other disaster destroys the network operations center or part of the network. Organizations that have centralized computing capability must also be concerned about disaster planning for the computer center. The factors to be considered when planning for the recovery of a company's communications facilities were discussed in Chapter 4.

It has become evident in recent years that another type of disaster must be planned for—a disaster at a telephone company office that serves the company. In May 1988, a large fire struck a telephone company central office in Hinsdale, Illinois, leaving more than 35,000 customers, including many businesses, without telephone or data communications service for more than a week. Some companies that were dependent on telephone service actually went out of business. Others were very inconvenienced. In August 1989, workers of NYNEX were on strike for several weeks. Customers were not able to get new services installed, and service calls took far longer than normal. Clearly, these types of situations will occur again in the future. Companies must protect themselves by having adequate plans for emergency communications service if a disaster strikes their communications company. Management's responsibility is to ensure that the proper disaster recovery planning is done and that the telecommunications capabilities can be restored.

All of these techniques are important because they provide an environment in which employees and others know that security is important to management.

The Physical Facility

network operations center

The network technicians are often housed in a *network operations center (NOC)*. The network operations center contains the consoles or terminals used to operate and monitor the status of the network and other telecommunications equipment such as modems, multiplexers, and front-end processors. Because it contains this equipment, much of which has the same environmental requirements as servers and mainframe computers, the network control center is often located in or adjacent to the company's computer room. Depending on the needs of the particular equipment being used, special power and air conditioning may be required.

flooring

A raised floor, at least 12 inches high, is extremely desirable because of the large number of cables and other wiring that connects all of the telecommunications equipment. If the wiring cannot be run under the floor, it can be run above the ceiling, but it is less accessible there. If that option is not available, special cable trays or racks can be obtained. Without proper planning, it is very easy to end up with wires and cables that are difficult to trace and troubleshoot.

smoke and heat detectors

Another similarity between the network operations center and the computer room is that smoke and fire detection and extinguishing equipment must be installed. Smoke, water, and heat detectors are usually mounted under the raised floor and on the ceiling of the room. They are designed so that if any two detect smoke or heat, the fire extinguishers are set off. The fire extinguishers can use water or carbon dioxide as an extinguishing agent. Water sprinklers are the least expensive, but a discharge of water is almost certain to damage some of the electronic equipment. Carbon dioxide is effective but messy.

configuration

A key attribute of the network operations center is that it must be flexible so that it can be expanded and rearranged easily. Given the dynamic, growing nature of most telecommunications networks, the configuration of equipment in the network operations center seldom stays the same for long. The layout that seems ideal today will seem hardly workable in six months. Network operations people should be involved in the design of the network control center because they will have the best understanding of the requirements and the implications of seemingly small changes.

Most communications equipment is designed to be mounted in standard 19-inch-wide racks and cabinets. Some small equipment, such as modems, if not designed to be rack mounted, can sit on shelves in the cabinets. Many communications equipment cabinets have glass doors in the front so that the operators can see the lights and meters on the equipment.

These modem cabinets contain many modems. Lights on each modem display its status. The cabinet also has a door in the back for easy access to the power cables and communications circuits. (Courtesy of Dow Corning Corporation)

labels

There are also doors in the back for access to the wiring that connects the various pieces of equipment. There is some debate about whether individual pieces of equipment should be labeled, particularly if they might be swapped for a spare piece of equipment when a problem occurs. If labels have to be changed, it delays the problem resolution. An alternative approach is to label the shelves on which the equipment sits or into which the rack-mounted equipment slides. Many cabinets have a space for such labels on the front of each shelf.

Cables should be labeled at both ends, and the cable identification should be typed or written twice on the label so that it can be read without twisting the cable to an unnatural position. This approach also ensures that the label is readable if one of the identification codes becomes smeared or otherwise unreadable.

Other pieces of equipment that should be located in the network operations center are the PBX, patch panels and switching equipment to bypass failing components of the network, and the network test equipment.

The help desk does not need to have access to all of the equipment. The help desk can be located in a traditional office environment outside the network operations center in less expensive floor space. Help desk operators need telephones, perhaps with headsets, access to one or more PC workstations for monitoring network status and problem conditions, and access to the tool (perhaps the same workstation or another one) on which all incoming problems are recorded. If trouble tickets are filled out, the help desk operators must have a way of transmitting or transporting them to the technicians who will work on the problems.

Depending on the hours of operation, the network operations center and help desk may need to be staffed for a long shift, say from 7 A.M. to 8 P.M., or for multiple shifts. If a long shift is required, the problem may be solved by having some people start early, for example at 7 A.M., and work until 4 P.M., and others start later, say at 11 A.M. and work until 8 P.M. If second or third shift staffing is required, it may be possible to share people with the computer operations group. This can be done if the network operation center, help desk, and computer room are located in close proximity to one another.

STAFFING THE NETWORK OPERATIONS GROUP

The staff of the network operations group is made up of people with a variety of skills. There are two philosophies of staffing the help desk. One suggests that it be staffed by the most senior, most experienced people available, such as experienced technical support people. These people can resolve virtually all of the reported problems on the spot. Proponents of the other staffing philosophy believe that the help desk can be staffed by relatively new, inexperienced people who, with a minimum of training, can solve a high percentage of the problems and, by asking a set of predetermined questions, often called *scripts,* can properly route the other problems to the right person. The advocates of this approach say that recent high school graduates with good verbal communications and telephone skills can be trained to perform well on the help desk in a few weeks. Proponents suggest that this is a good entry-level job into the telecommunications or data processing organization.

Routine network operations work, such as starting, stopping, and reseting the network, circuits, or terminals, also can be handled by help desk personnel in all but the largest networks. Since they are handling trouble calls, the help desk operators are usually the most closely in touch with the state of the network at any time.

Hardware technicians, the second level in the problem-solving hierarchy, are people with electrical and electronics training and aptitude. They are often recruited from the local telephone company or the military, but they might also come from a local electronics school. They typically do not have a four-year degree. The hardware technicians can also be used to install equipment, such as terminals or personal computers, where some initial setup as well as connection and checkout are required.

Network operations supervisory personnel should have a production orientation. They should be focused on meeting the numerical service objectives and constantly on the lookout for techniques to improve the efficiency and effectiveness of the network operations group. In addition, they should have good verbal communications skills because they may have to deal with users under stressful situations. Ideally, they should have education and experience in data processing and data communications as well as some general supervisory experience. The network operations supervisor and computer room supervisor are in many ways similar jobs, and each may provide a backup for the other.

It is important that network operations personnel at all levels spend 5 to 15 percent of their time each year in formal or on-the-job training. Because of the rapid changes in the telecommunications field and the changes in most companies' communications network, this type of educational investment is necessary to keep the staff's knowledge and skill level current.

NETWORK MANAGEMENT OUTSOURCING

One option that management needs to consider is the feasibility and desirability of *outsourcing* the organization's network management work. There are many companies whose business is to provide network management services. A wide variety of possibilities exist, including network design services, acting as a general contractor for network installation, and network operation. For a fee, a service company will contract to perform any of those services and will provide liaison with vendors and other service organizations. Some businesses probably would not feel comfortable turning over the design and operation of their network to a service organization, but for companies that have difficulty hiring skilled telecommunications people or that simply don't want to manage their own network, outsourcing is an option.

Some items to consider when studying the outsourcing alternative are:

- location where the outsourcing company's people will sit;
- service guarantees;
- what to do with existing network staff—reassign, transfer to outsourcing company, outplace;
- cost/benefits of the service.

Several years ago the state of Wisconsin entered into a contract whereby AT&T designed, built, and managed the state's communications network. Patterned after the way AT&T manages its nationwide long distance network, the Wisconsin Network Management Center was designed to be the nerve center of the network that tied together more than 1,800 state, county, and municipal government locations and served 50,000 employees. The NMC was staffed around the clock by AT&T personnel and was located in an AT&T building near the state capitol in Madison.

COMMUNICATIONS TECHNICAL SUPPORT

Communications technical support specialists manage the communications software and handle other nonroutine situations or problems that arise during the design or operations of a communications network. To be most effective, they require a broad understanding of all facets of the communications system, both hardware and software.

In some companies, the communications technical support group is a separate department reporting to the manager of telecommunications, as will be illustrated in Figure 15–8. In other companies, communications technical support is a part of the network management group. The actual organization depends on the size of the company, the skills of the particular individuals, and the preference of the telecommunications manager. Either organization can work effectively.

The major functions of communications technical support are

- supporting communications software;
- providing third-level problem solving;
- assisting in network analysis and design;
- network performance analysis and tuning;
- hardware evaluation;
- programming;
- consulting and general problem solving.

Each of these activities is discussed in the following sections.

Supporting Communications Software

The communications technical support group is responsible for installing and maintaining the communications software that runs the network.

This includes the network control program that operates in the front-end processor and the communications access method and communications monitor software that reside in the mainframe computer. Communications software is generally delivered by the vendors in various levels or releases. Each succeeding release contains additional capability and may correct software problems (program bugs) of prior releases. The technical support staff and its management must decide whether to install each new release of the software as it becomes available. This decision largely depends on whether the new release cures problems the network has experienced and whether the new capability is required.

Testing communications software on a personal computer or small LAN is relatively simple, but testing the network control program, which runs a large multiline network with several thousand terminals, is extremely complex and difficult. Because it is usually impossible to test all of the terminals and combinations of communications possibilities, the communications technical support specialist's experience is critical in determining which parts of the software have changed and should be tested.

Another aspect of software maintenance is to reconfigure the software as necessary to support new circuits or terminals as they are installed. In some systems, this process must be done each time a circuit or terminal changes. In other systems, the software itself senses new equipment and adapts to the changes.

A third aspect of software support is diagnosing problems that occur with the software, obtaining fixes for the problems from the vendor, and applying and testing those fixes to solve the problems. Communications software problems are frequently very difficult to diagnose because it is impossible to specify or repeat the exact conditions under which a problem occurred.

Vendors have different approaches to software maintenance. Some vendors take the approach of sending out "fixes" for all known software problems and asking the users to apply them as *preventive software maintenance.* Other vendors adopt the strategy "if it's not broken, don't fix it"—fixes are only applied for the specific bugs that occur at each site where the software is installed.

Providing Third-Level Problem Solving

The help desk and network technicians usually solve approximately 95 percent of all network problems. Some problems are so unusual or difficult that they require a higher degree of skill to resolve. The communications technical support group serves as the third level for problem resolution. In this role, the group stands behind the help desk and network technicians ready to jump in and work on the most difficult problems.

Whereas the network technicians are very familiar with communications hardware, they are not usually trained in communications software. The communications technical support specialists' knowledge of both hardware and software becomes very important because, in many of the most difficult problems, it is hard to tell whether hardware or software is at fault. The technical support person may also call on the vendor's technical specialists to work on the problem because in many cases a problem experienced at one company has been seen and solved elsewhere.

Assisting in Network Analysis and Design

Communications technical support specialists often are called upon to assist when network designs are being reviewed and updated. Depending on the size of the communications department, the technical support specialist may even perform the network analysis and design. In any case, the detailed knowledge of hardware and software capabilities makes the technical support specialist a valuable member of the network design project team.

Network Performance Analysis and Tuning

The regular gathering of performance data can be performed by the network operations staff according to procedures established by the technical support specialist, but the analysis and interpretation of the performance data require considerable skill, judgment, and experience. The technical support specialists are usually charged with the responsibility of making sure that the network is delivering the performance for which it was designed and for spotting trends, such as increases in message or transaction volume or error rates, that could lead to capacity problems and performance degradation. The technical support specialist can make some adjustments in the software, such as varying the number of buffers in the network control program or changing the priority of certain transactions to improve performance. In other cases, he or she may recommend that existing circuits be reconfigured or additional circuits be added.

tuning

The process of *tuning* a network consists of monitoring its performance, making adjustments to improve the performance, and then monitoring the results of those adjustments. This is illustrated in Figure 13–10. The results of the initial adjustments may suggest that further adjustments are necessary or may show that performance has returned to the desired operating range. Network performance monitoring and tuning should be an ongoing process that is a routine part of the technical support job.

Hardware Evaluation

Another function of the communications technical support group is testing and checkout of new equipment being considered for installation on the communications network. Most manufacturers claim that their equipment conforms to certain specifications or standards, such as RS-232 or SNA. However, there is enough variation in the way the standards are implemented that testing in the actual environment where the equipment is to be used is desirable. Experienced telecommunications people insist

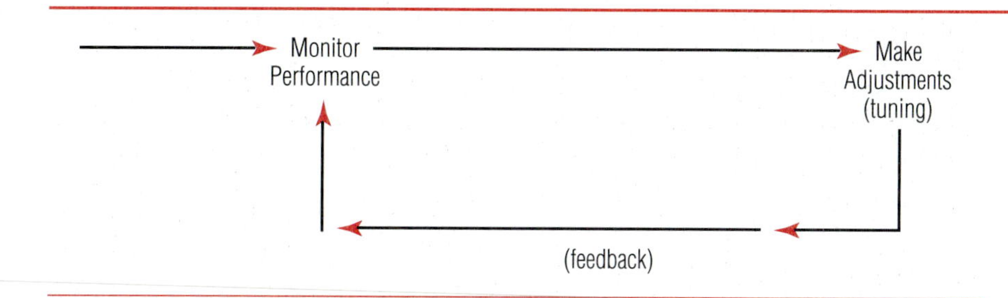

Figure 13–10
Diagram of network tuning process.

that the vendor provide a piece of equipment for testing purposes. The job of connecting the equipment to the network and testing it falls to the technical support specialist.

When the specialist tests new equipment, it is important to test the equipment's error handling capability as well as its ability to operate under normal conditions. The device may work fine when no errors occur, but it is important to know how it will perform in all situations.

Programming

In some companies, communications technical support specialists are communications programmers who design and write communications software. This type of programming is complicated and difficult, and most organizations believe it is best to avoid writing such programs if possible. Occasionally, however, a unique capability or function not provided by any vendor's software is needed, and the only way to get it is to do the programming in-house. Software of this type needs to be especially well documented because very few people in the organization will understand how it works well enough to be able to maintain or debug it.

Consulting and General Problem Solving

Good communications technical support specialists are called on to help solve a surprising variety of communications and data processing problems in an organization. This is because they are highly skilled, analytical, and by nature good problem solvers. They may find that a high percentage of their time is spent on ad hoc or spur-of-the-moment activities that are outside their control. When managers plan the specialists' time, it is important to recognize that some time must be left unscheduled and available for such activities.

STAFFING THE COMMUNICATIONS TECHNICAL SUPPORT GROUP

Communications technical support specialists have many of the same characteristics and much of the same training and experience as the system programmers in the data processing organization. In fact, in many companies there is a single technical group for both communications and data processing.

Communications technical support people should be intelligent, analytical, technical, curious, self-motivating, and independent. A study, performed by IBM, has shown that communications technical support specialists who work for IBM's largest customers have the following education and experience:

- 100 percent have a high school education;
- 52 percent hold a bachelor's degree or higher;
- 84 percent have formal software training;
- an average of 8.1 years experience in data processing or communications and 3.1 years experience on their current job;
- 50 percent are familiar with four or five IBM software products.

Good technical specialists are excellent problem solvers who often prefer to work independently. Their independence and self-motivation serve them well when they are called on to solve difficult problems. They have a "can-do" attitude and enjoy the challenge and self-satisfaction that comes from diagnosing and solving a problem when others have failed.

SUMMARY

This chapter studied the network operations and communications technical support organizations that must be in place to properly operate the communications network of a company. We have looked at why it is important to manage the network, and we have seen the six major activities that the network operations function comprises. The chapter discussed the different skills required of the network operations group and described the physical facility, called the network operations center, which is the focal point for network operations and trouble shooting.

Technical support people install and maintain the communications software that drives the network. The communications technical support group provides detailed technical assistance and consultation, heavily focused on software but with a large amount of hardware expertise required as well. Because of the education and experience requirements, good technical support people usually are in great demand.

CASE STUDY

Dow Corning's Network Management and Operations

The overall management and operation of Dow Corning's network resides in the corporate telecommunications group, which is a part of the Global Information Technology Services (ITS) Department, as shown in Figure 1–17. There are sub-groups responsible for the wide area network, the local area networks, and the voice network, and coordination is generally very good. The help desk reports to the Manager of Learning and Support, who is also responsible for user training. This manager reports to the Director of Global Electronic Workplace Services, but cooperation between the departments is excellent.

Dow Corning has had a network operations center and help desk since 1968. The original help desk evolved from the message center in the days when the company had a teletype-based torn tape message switching system. When the message switching was taken over by the computer in 1968, the message center became the company's first help desk, and the staff also monitored the operation of the communications circuits. The help desk today is different from help desks at many other companies in that it provides all-inclusive support for data and voice network users. Integrated service covering problems related to networks, computer applications, personal computer problems, security issues, voice service, and voice mail are provided by the same help desk specialists.

The help desk had always been located in the same area as the mainframe computers until early 1993, when it was realized that it wasn't important for the specialists to be located next to the computer. As a result, the help desk is located in a building away from the computer servers, in a place where space is more available. Dow Corning's network is essentially operational 24 hours per day, seven days a week. Parts of the network are sometimes taken down for a few hours for maintenance, often on weekends, but there is always an awareness of the business hours in Dow Corning's locations around the world. For example, when people in Australia come to work on Monday morning, it is 1 P.M. EST on Sunday afternoon in Midland.

The global help desk has operations in Midland, Sydney, Tokyo, and Brussels. In Midland, the help desk is staffed 24 hours a day, 7 days a week, while in the other locations, staff is on duty during the business hours of the regions for which they are responsible. For example, the Sydney help desk, which is responsible for all of Asia except Japan, is staffed from 9 A.M. until 9 P.M. Sydney time to cover locations in different time zones in Asia.

An operator takes incoming calls to the help desk, unless all operators are busy, in which case a recorded "all lines are busy" message is played. It reminds the users to have their employee number available to give to the help desk operator when he or she answers the call. The operator logs the call, the problem symptoms, the workstation identification, and the user's name and telephone number in software called "Service Center." Approximately 60 percent of the calls are relatively routine and can be handled by the person taking the call within minutes. Callers who have more difficult problems are passed to a level 2 technician who has additional training in the problem area. In the most difficult cases, communications technical support or vendor personnel are contacted and asked to help resolve the problem. The global help desks are currently handling 11,000 to 13,000 calls per month.

The help desk is staffed almost entirely by contractors rather than Dow Corning employees. In effect, the help desk service is outsourced, although Dow Corning employees very closely supervise it. Dow Corning employees also develop procedures for help desk operations and work with the contractors to develop scripts that provide standard problem solving approaches to the most common types of problems that users experience.

Dow Corning has a problem tracking and change control system in place for both the network and the Midland computer center. A meeting is held each Wednesday morning at 8:30 to discuss all outstanding problems and upcoming changes. This meeting is intended to ensure that difficult problems are getting the proper attention and that all changes to be made to the computer, software, or network are communicated to network and computer personnel and to vendors.

Central communications equipment at Dow Corning is primarily located in the computer centers. A conscious effort has been made in the past few years to get the communications equipment relocated into a few secure areas rather than being scattered throughout the headquarters buildings. Moving communications equipment takes a great deal of planning to minimize the disruption to users and unscheduled outages. In plants and other outlying locations, communications equipment normally is kept in a locked equipment room.

Dow Corning's communications technical support group consists of seven people. These people handle the communications software maintenance, get involved in difficult communications problems, and act as consultants on communications matters. There is also a technician, a former telephone company employee, who handles the physical connection of equipment and who is also a specialist in the T-1 and fiber optics multiplexing equipment.

Since many of the network changes are made during a short time window on Sunday morning, one or more of the technical people normally take part of Friday off and work part of the Sunday. The computer technical support people work a similar schedule, so having a plan for the Sunday activities is mandatory to ensure that the staff can get their work done without interfering with one another. The technical support people are also on call when severe network problems develop at night or on weekends. Dow Corning has found that having talented people who are flexible about their schedules is a necessity for this type of work.

QUESTIONS

1. Since the communication and computer industries have become so tightly interwoven, sharing similar products and problems, would it make sense for Dow Corning to merge its telecommunications and computer operations and technical support staffs into a single group? What advantages or disadvantages would accrue?
2. Would there be benefits to managing the LAN operations in a more distributed manner than the centralized way they are currently managed? What technical problems would have to be solved to make this happen? What control issues would have to be overcome?
3. Given the rapid growth in the telecommunications industry and the demand for skilled telecommunications people, what should a company like Dow Corning do to ensure that its communications technical support people stay with the company and don't leave soon after they get trained?
4. What attributes would you look for in a potential employer if you wanted to ensure that you had good opportunities for advancement, reasonable job security, and the possibility of having a long-term career?

REVIEW QUESTIONS

1. What is the function of the network help desk?
2. Distinguish between the help desk and the network operations center.
3. In a star network, why should the network operations group be concerned about the reliability of the computer processing?
4. What are some of the start-up and shut-down activities the network operations group performs if the network is started and stopped each day?
5. What is the purpose of having a daily status meeting?
6. Is it possible to design a network that is totally redundant and, therefore, does not need to be managed? Why?
7. What are some ways to bypass problems in a network?
8. Who should attend a problem tracking meeting?
9. Describe the process of problem escalation.
10. Explain the level 1, level 2, and level 3 method for problem resolution.
11. What is a service level agreement?
12. What are some of the statistics about network operations that the network operations supervisor would be interested in seeing each day?
13. Some companies institute a moratorium on all network and computer changes at the end of the year. Why might such a moratorium be desirable?
14. What information should be kept in an inventory file of telecommunications equipment?
15. List several techniques to ensure that when telecommunications change is introduced in a company, it is implemented in a controlled, well-managed way.
16. What is the purpose of telecommunications management software packages?
17. Why is the operational management of a LAN more or less complicated than managing a WAN?
18. What is the function of agent software in an SNMP-managed network?
19. Describe the purpose of RMON capability in an end device on a network.
20. Explain the functions of a firewall.
21. Distinguish between the capabilities of a packet-level firewall and an application-level firewall.
22. Frankly, preparing and updating network documentation is tedious. No one likes to do it, and few companies do it well. If you were the manager of telecommunications,

what incentives can you think of that would motivate your people to keep the documentation up-to-date? What guidelines would you develop to ensure that the documentation is kept up-to-date?

23. Prepare a list of management reports that would be useful for the help desk supervisor, network operations supervisor, and telecommunications manager to keep them informed about network operations. Describe the information that would be included in each report and how frequently it would be created. Also consider the possibility of providing some of the information via online inquiries.

24. What are the factors to consider when a manager decides whether or not to apply maintenance to the network software?

25. What are the advantages and disadvantages of using an outside service organization to manage a company's telecommunications network?

26. How can the communications technical support people assist with network analysis and design?

27. Discuss the factors to be considered when deciding whether to include the communications technical support group in the network operations group or create a separate group in the telecommunications department.

28. Joe Brown, a senior network technician, has expressed a strong interest in getting into your company's communications technical support group. He has nine years of communications experience with the telephone company and four years with your organization. He graduated from high school and completed two years of college, where he studied data processing and programming. He has been a top-rated performer and has shown a great deal of initiative in solving hardware problems. What attributes and skills would you look for in assessing whether Joe has the capability to work successfully with the other technical support specialists and in deciding whether to give him an opportunity in technical support?

PROBLEMS AND PROJECTS

1. Describe some advantages of having all telecommunications personnel spend some time working in the network operations group.

2. Write an appropriate service level agreement for
   a. an airline reservation system;
   b. an ATM-based banking application;
   c. an online inventory system in a chemical company.

3. To some extent, making decisions about future network growth and configuration by looking at past network statistics is like steering a boat by looking at the wake. Discuss how the statistics about past performance can be properly used and misused for forward planning.

4. Draw a floor plan of a network operations center containing a central console with three VDTs, four modem cabinets, a front-end processor, two cabinets of test equipment, and any other equipment that would be appropriate to have in the room.

5. When a three-level approach (as described in the text) is used for problem resolution, it is possible that none of the people that have worked on a problem feel totally responsible for it. Level 1 passes the problem off to level 2, and level 2 may pass it to level 3. As the manager of network operations, what procedures would you institute to ensure that the responsibility for a problem is always well defined and that the help desk always knows its status in case it gets questions from the users?

6. What types of information would a good communications line trace program capture?

7. Most vendors keep a database of known problems with the hardware and software it sells. Do you think a vendor would ever let its customers have direct access to such a database? What would be the advantages and disadvantages of giving customers such access?

8. The text stressed the importance of doing ongoing network performance monitoring and tuning. Why do you think many companies don't do this important activity regularly?

9. Draw a flow diagram showing the flow of information from hardware devices to a network management program. Identify the data that needs to be captured by each piece of hardware, and show how the data is used by the network management software to alert operators of problems in the network and to monitor network performance.

10. Good technical support specialists find that communications programming is challenging work. What are the trade-offs for a company in letting them develop communications software? As the manager of technical support, how would you convince the specialists that writing communications software is generally not in the best interests of the company?

Vocabulary

network management
network operations
problem management
help desk
trouble ticket
levels of support
problem escalation
problem tracking meeting
patch panel
dial backup
protocol analyzer
performance management
service level agreement
management information base (MIB)
alert messages
network management protocol
configuration control
change management
change control
change coordination meeting
Simple Network Management Protocol (SNMP)
Common Management Information Protocol (CMIP)
managed devices
agents
network management systems (NMS)
remote monitoring (RMON)
Netview
Open Network Architecture (ONA)
firewall
packet-level firewall
packet filtering
application-level firewall
proxy server
network operations center (NOC)
outsourcing
preventive software maintenance
tuning

References

Derfler Jr., Frank J. "Network Management: Behind the Scenes." *PC Magazine* (December 7, 1993): 335.

Head, Joe. "How SNMP Is Breaking Down Barriers as an Enterprise Integration Standard." *Telecommunications* (November 1993): 65–68.

Leonhart, James. "Wisconsin Is Wired." *Communications News* (August 1989): 42.

Quillan, Robert. "The Integration of Network and System Management." *Telecommunications* (August 1993): 27.

CHAPTER 14

Network Design and Implementation

OBJECTIVES

After you complete your study of this chapter, you should be able to

- describe the process of designing a telecommunications network;
- discuss each of the phases of the network design and implementation process in detail;
- explain the importance of understanding the requirements for a network before beginning its design;
- describe the differences between the design of a voice network and the design of a data network;
- describe several project management techniques that are relevant to the network design and implementation process.

INTRODUCTION

This chapter describes the process by which communications networks are designed and implemented. Proper design requires a detailed understanding of how the network will be used. Once the requirements are understood, the network designer can investigate alternative ways to design the network. The designer must understand different telecommunications technologies so that he or she can match them with the requirements to provide the optimal design at minimal cost. After the network is designed and the design is approved, the circuits and other equipment are ordered, and the network is implemented.

Network design and implementation are a structured process composed of several phases, each of which will be examined in detail in this chapter. In most phases, several techniques are used to accomplish the work. After reading this chapter, you will have a good understanding of the network design and implementation process, as well as the specific tasks required.

THE NETWORK DESIGN AND IMPLEMENTATION PROCESS

Communications network design and implementation exemplify the more general systems analysis and design process used to design and implement any computer-based system. In fact, computer systems analysis and design have many similarities to the architectural design and engineering processes followed when any new structure is built. The architectural and engineering processes for designing and constructing buildings have been in existence for many years and are, therefore, widely understood and almost universally applied.

Network analysis and design together form a much newer discipline. Although the process is becoming more scientific with each passing year, there are still many aspects that are somewhat "artistic" in nature. Furthermore, compared to architectural design, or computer application analysis and design, where the resultant building or computer application is expected to have a lifetime of many years, the telecommunications environment is much more dynamic. Rapid growth in traffic volume in networks means that sometimes networks that were designed to last several years run out of capacity in a few months. The traditional network analysis and design that worked well in the past, when networks had a longer lifetime, is heavy and cumbersome by today's standards.

Network analysis and design are processes for understanding the requirements for a communications network, investigating alternative ways for implementing the network, and selecting the most appropriate alternative to provide the required capability. Understanding the requirements with great precision is a time consuming exercise, so it must be decided whether a detailed, precise analysis will be done, or less precise data will be gathered or estimated. Establishing less precise estimates can often be done relatively quickly, thus speeding up the network design process. Also, as the cost of bandwidth keeps decreasing, it is often economically feasible to install extra capacity on a LAN or WAN. The extra capacity can provide a margin of safety if less precise traffic estimates are made.

Network implementation is the process of installing a network and making it operational. Network analysis, design, and implementation are normally performed by a project team that has the overall responsibility to analyze the requirements, design the new or upgraded network, and implement the solution.

Most aspects of network analysis and design are similar for voice and data networks; the basic process is the same. This chapter primarily describes data network design but includes a section that examines the differences between voice and data network design.

In every network design, certain work must be performed, much of it in a particular sequence. That is why traditional network design normally is broken into discrete phases that are completed in sequence. The phases are called by various names in different companies, and the dividing lines between phases vary a little from company to company. Overall, however, every network design requires the same general work. The goal of the network analysis and design process is to ensure that the network satisfies user requirements at an appropriate cost.

Since network design is not an activity that most companies do every day, management may want to consider outsourcing the analysis and design work or bringing in consultants to assist. The rapidly changing technology of telecommunications means that new capabilities become available frequently, and a specialist who follows communications technology and regulatory matters may be aware of solutions to communications situations or problems that company insiders don't know. It is, of course, possible to turn over the entire network design and implementation project to another company, and many small businesses that don't have people with telecommunications expertise take advantage of that possibility, sometimes using, for example, a small, local company that specializes in personal computers and LANs for small businesses.

outsourcing

Project Phases

The phases of communications network analysis, design, and implementation are

1. the request;
2. preliminary investigation and feasibility study;
3. understanding and definition of the requirements;
4. investigation of alternatives;
5. network design;
6. selection of vendors and equipment;
7. calculation of costs;
8. documentation of the network design and implementation plan;
9. management understanding of the design and approval of implementation;
10. equipment order;
11. preparation for network implementation;
12. installation of equipment;
13. training;
14. system testing;
15. cutover;
16. implementation cleanup and audit.

Although this may seem like a large number of steps, on most network projects, many of them can be completed very quickly, especially in smaller networks. Essentially all of the steps must be completed, but for many of them the work and outcome is almost intuitive. For example, in a small company, the manager who can approve the design and give the go-ahead to implement the project may be part of the project team. Therefore, gaining approval to implement is a relatively minor phase. Nonetheless, at some point (whether implicitly or explicitly), management approval to implement is obtained.

Another example of a trivial activity may be the response when a business competitor comes out with a new, innovative communications-based service. In an effort to quickly provide a similar service, the preliminary investigation stages to respond to the competitor's edge may be very abbreviated. In addition, because the technical viability of the new system has already been demonstrated by the competitor, the feasibility study may be minimal.

NETWORK ANALYSIS AND DESIGN

This discussion explores the various phases of network analysis and design in detail.

The Request Phase

For the network analysis and design process to begin, a person or department must ask for work to be done. The request phase begins when the communications department is formally asked to perform some work. There are generally four sources of requests for projects. The most common is a request initiated by a user group or department that needs a communications service or facility to meet a business situation.

sources of requests

A second type of request can come from senior managers who, through their industry contacts or other sources, get an idea for a new communications-based project. This type of request is relatively rare.

The third source of requests is from outside organizations. Customers may insist on being able to check inventory and prices and enter orders themselves. The competitive environment in some industries requires an e-business solution so that customers can browse catalogs and enter orders on the Internet. Vendors may request that purchase orders be sent electronically. A change in the law may, in effect, be a government request (requirement) for the transmission of certain types of data. A bank may suggest that payroll data be transmitted electronically so that it can be quickly deposited in employees' accounts.

The fourth type of request comes from the communications department itself, which may proactively initiate a project to improve or enhance the existing communications services. A typical request of this type is an increase in an existing network's capacity to handle increasing business transactions. A similar type of project might be aimed at improved response time or higher reliability. Requests generated by the communications department should be evaluated like any other project, and the benefits of the work compared to the cost.

form of request

Ideally, the request for work should be somewhat formal and in written form. Usually, the idea for a new project begins rather informally. Perhaps it comes from a conversation between two managers, a discussion over lunch, or an idea spawned by a comment made in a seminar. Often there are several meetings between the initiator of the idea and the communications department that are, in effect, preliminary feasibility discussions. The purpose of these meetings is to make an initial determination of whether the potential project has any merit. During these discussions, it may be determined that some formal work is required. If so, a formal request memo that documents the idea and the discussions should be written. The potential costs of the project and the potential benefits of the project to the company should also be outlined in the request memo.

project prioritization

In many companies, there are far more requests for work than can be satisfied by the staff of the communications department in a reasonable period of time. A prioritization process determines which projects the communications staff will work on. The most effective way to prioritize projects is to solicit the involvement of management from various departments, especially those who want to have work done. The management group should listen to a presentation that describes the potential projects and their potential costs and benefits in order to have information on which to base its decision. If properly conducted, this prioritization process should highlight projects that are potentially of the most benefit to the company. Some of the projects will drop from consideration, and the remaining projects will compete for the required dollar and staffing resources. If a prioritized list of projects cannot be established at the middle management level, the list of projects may have to be presented to senior management to obtain a decision.

outcome

Ultimately, the outcome of the prioritization process must be a list of projects for the communications department to work on. The communications department can estimate how long it will take to complete the prioritized list of projects with the existing staff. Frequently, the estimated time is quite long, and the departments that are sponsoring the projects are unhappy. An alternative may be to add additional people or to allocate more money so that a greater number of projects can be worked on simultaneously and the elapsed time to complete projects may be shortened. The decision to apply these additional resources ultimately rests with senior management.

Without taking away from the desirability of having formal documented requests and a priority setting process, it is important to recognize that in the real world, the request phase is often very short and undocumented. Often a few key managers will, in a brief conversation or meeting, determine what work needs to be done and assign the people to do it. There is a risk that one or more of the people involved will have a misunderstanding of the project scope and desired outcomes, but the time pressures of the workplace often dictate quick decisions and fast action based on verbal agreements.

Preliminary Investigation and Feasibility Study

The two primary objectives of the preliminary investigation and feasibility study phase are to gain an understanding of the current communications system and its problems and to determine whether some of the problems can be solved with a new or improved communications system. This phase usually is performed by a small number of people, perhaps an analyst or two from the communications department and one or two people from the user department. These people constitute the *project team.*

project team

The project team looks at how the existing communications system works and the problems that are prompting requests for changes or improvements. If, for example, the feasibility of a new telephone system were being examined, the team would look at the existing telephone system and its operation. The team members would talk to the people responsible for administering and operating the current systems, and they would meet with telephone users in various parts of the company to get their input about the existing telephone system. Similarly, if the feasibility of improving the response time on an interactive computer system were being examined, the team would talk to the users to determine what their perception of the response time was. They would also measure the actual response time, talk to the network and computer operations people about the response time problems, and try to determine whether the response time delays were really caused by the communications network or by other factors, such as slow processing in the computer.

study's objectives

The feasibility study attempts to determine whether implementing a solution to the identified problems makes sense from both a technical and a nontechnical standpoint. Clearly, if the cause of the response time delay in an online system is slow computer processing, increasing the speed of the communications lines will have little or no effect on the response time. Similarly, the team investigating a new telephone system might find that the perceived problems with the existing system are not widespread. They may be primarily focused on one or two departments that have unique telephone requirements. In that case, the unique telephone requirements of those departments might be met with special equipment or circuits rather than by replacing the entire telephone system. Chances are that the customized solution for the two departments will be more effective and less expensive, too.

The project team might also determine that solving a particular communications problem is not feasible. There may be technical or cost reasons why a solution cannot be implemented. For example, providing dial-up 56 kbps data communications service to a disaster backup site may not be feasible if the local telephone companies do not have the service available. Although a solution using leased lines could be implemented, the costs are likely to be prohibitive. It is important to make an assessment of the feasibility of any potential communications project before recommendations are made.

A secondary objective in the preliminary feasibility stage is to use a minimum amount of people resources to study the situation. Properly performed, the phases of a telecommunications project require a *creeping commitment of resources* so that if, for one reason or another, the project is stopped, a minimum amount of personnel hours or dollars will have been spent. That is why the early phases of the project are performed by small numbers of people, and others are added to the team as the project progresses. This is also known as the *project life cycle*. Figure 14–1 illustrates this creeping commitment.

creeping commitment of resources

project life cycle

A preliminary estimate of the costs and benefits of completing the project should also be done in this phase. Since relatively little work has been done and the scope of the new or revised communications system has not been defined, the cost estimate may not be extremely accurate. It will be refined in subsequent phases, as a more thorough understanding of the project requirements is obtained. Identifying the benefits of the project work also is extremely important. Since communications systems frequently are installed for specific groups of users or departments in the company, their assistance in estimating the benefits should be obtained. Furthermore, benefit estimates made by the users are likely to have more credibility with senior management than those made by the communications staff.

comparing costs and benefits

Finally, some performance or success criteria should be established that will help people determine if the new network or communications service that is being designed is successful. The criteria should be quantifiable, if at all possible, because it is easier to measure whether quantifiable criteria are being achieved. Sample criteria might include:

- There will be a workstation on every person's desk in the Seattle office.
- The workstations will be operational 99.9 percent of the time between 7:00 A.M. and 9:00 P.M. local time, Monday through Friday.

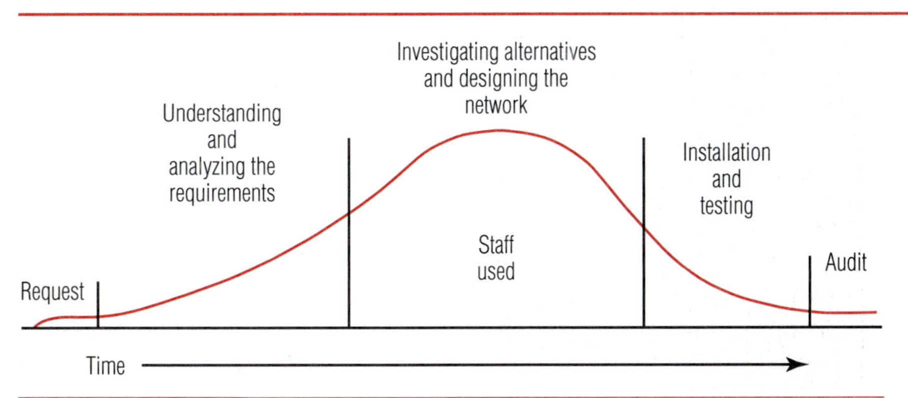

Figure 14–1
Creeping commitment of staffing as a project progresses. This concept is also known as the project life cycle.

- When making inquiries on the patient database on the local server, response time will be less than 1 second 95 percent of the time and never more than 5 seconds.
- When accessing the mainframe computer at company headquarters in Phoenix, response time will be no more than 2 seconds 90 percent of the time and never more than 5 seconds.
- Data on the local server will be backed up every night.
- Employees will be trained to use the new LAN within two days before it becomes operational.

Sometimes the solution to a network problem is almost self-evident at the time the project is initiated. For example, if the throughput on a LAN has slowed to a crawl because a lot of new users were added to it, it may be pretty obvious that either the speed of the LAN needs to be increased or the LAN needs to be split so that there are fewer users on each segment. It may also be known that speeding up the LAN will require new faster NIC cards in all of the attached PC workstations, but because some of the workstations are older, they cannot operate the higher LAN speed. Hence, the only practical, short term solution may be to split the LAN. This work may only require a few hours to complete.

When the preliminary investigation and feasibility study have been completed, however simple or complex it may be, they should be documented with a written report that summarizes the findings. In many companies, the project team is also expected to make an oral presentation to one or more groups of management or users. Ultimately, the team needs to obtain a decision about whether the project will proceed to the next phase. The creeping commitment of resources concept suggests that it is far better to terminate infeasible projects that have marginal benefits early in the cycle before a lot of work has been done or money spent.

Understanding and Definition of the Requirements

Assuming that the project team received a "go" decision at the close of the preliminary investigation and feasibility phase, it proceeds to the next phase. The team needs to gain a detailed understanding of the users' requirements. Often the project team is expanded in size at this point. As the phase name implies, the work is significantly more detailed and time-consuming than in previous phases. Additional people may be needed to help get the work done in a reasonable amount of time.

To help organize the requirements, it is common to break them into several categories. Organizationally, one member of the project team might be assigned to work on each category of requirements. Alternately, subgroups of the project team could be formed, each with responsibility for identifying the requirements in several of the categories.

The following sections explain the work that must be done in a maximum case where a new, large wide area network or upgrade to an existing network is being studied. Simpler networks or network upgrades, such as a departmental or single-building LAN would only require that some of this work be done, and hence the definition of requirements can be completed more quickly.

Geographic Requirements The geographic scope of the network must be determined. It may be an international network serving company locations in several countries or a national network serving just one country. Regional networks serve a part of a country, a state, or perhaps a city. A local network serves a building or campus and is usually a candidate for private, company-owned circuits or a LAN. Frequently, the network requirements don't fall cleanly into one category. An example of overlapping requirements would be a network connecting warehouses in many locations throughout the country to a central manufacturing plant. Within the plant there are many terminals also using the network. This system requires a combination of national and local networks.

It is quite probable that each geographic piece of the network has unique requirements. For example, there may be a requirement that terminals within the plant have a very fast response time. On the one hand, some of the in-plant terminals may be production controllers that need consistently fast response from a computer in order to control machinery. These terminals might best be served by a local area network operating at a very high speed (such as 100 Mbps) and perhaps using the MAP protocol. On the other hand, the terminals in the warehouses may be used by people to enter or access information about shipments. Standard leased telephone lines operating at 64,000 bps may be perfectly adequate to meet their requirements.

Traffic Loads One of the most important parts of this phase is to understand the amount of traffic that will flow on the network and develop definitive statements of requirements. A typical statement of requirements might be:

> A circuit between Seattle and the computer center in St. Louis must be capable of carrying 15 million characters per day during normal business hours (8 A.M. to 5 P.M. Seattle time) with 3 million characters being transmitted during the peak hour of 10 A.M. to 11 A.M. Seattle time. The average response time for all transactions should be less than 5 seconds.

To develop this type of requirements statement, the network designer must know the anticipated number of messages and their lengths, the users' response time requirements, and the business hours at the two locations.

When designers plan data networks, average message rates are most frequently used to calculate the traffic load. However, the designers must

peak loads

pay particular attention to those situations in which a high percentage of the message traffic occurs in a relatively short period. These peak periods can play havoc with the responsiveness of the network if they extend for very long. One of the most significant problems that can occur in communications systems is the inability to handle the transmission load during peak periods. The only way the problem can be avoided is by knowing what traffic the network must carry and designing a network with enough capacity to handle the peaks.

busy hour

It is especially important to determine how much traffic will be carried during the busiest hour of the day, called the *busy hour.* Busy hour traffic analysis is at the heart of both voice and data network design. It is not uncommon in business systems for busy hour peaks to occur around 10 A.M. and 2 P.M., as shown in Figure 14–2, although peaks may also occur at other times.

If, for example, a network carries 5,000 transactions a day and 80 percent of them occur between 3 P.M. and 5 P.M. each afternoon, the network must be designed to accommodate the volume during that peak period. The requirements are quite different if the same number of transactions is spread out over an 8-hour business day. An organization such as the New York Stock Exchange would have yet a different pattern.

Usually a table is prepared, like the one shown in Figure 14–3, to assist the analyst in gathering data about the expected traffic loads. If a totally new computer system is being designed, it may be difficult to get information about the number and length of transactions or messages. If exact information is not available, the analyst must make an estimate.

Note that calculating the number of messages and their lengths is much easier in a transaction processing system, where the quantity and length of each type of transaction can be ascertained relatively easily. In an Internet-like environment, where many of the response transactions

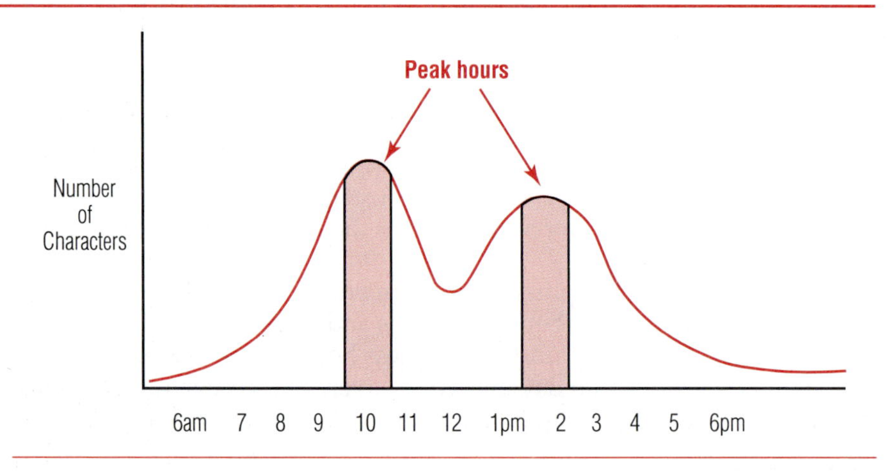

Figure 14–2
Traffic load by hour of the day on a typical data communications network.

Message Type	Average Length in Characters	Total Number of Messages per Day	Number of Messages in Peak Hour	Number of Characters in Peak Hour
Customer				
—Inquiry	30	170	45	1,350
—Response	425	170	45	19,125
Product				
—Inquiry	19	525	220	4,180
—Response	1,545	525	220	339,900
Inventory Update	220	1,750	175	38,500
(list all of the transactions)				
Total		**3,140**	**705**	**403,055**

Figure 14–3
Analysis of expected traffic loads in a network.

may contain images, sound, or even video, calculating the traffic volume is much more difficult. Simulating the workload may be helpful, but overdesigning the network by putting in circuits and LANs with extra capacity may be the prudent thing to do. Cost must always be considered, however.

Traffic Flow Patterns Another aspect of requirements analysis is to determine message flow patterns. In a star network with a computer at the hub, it is obvious that all message traffic flows into and out of the central computer. In a mesh network, however, the traffic flows are not so simple. The network analyst must determine which message traffic goes to which location. If a map showing the locations on the proposed network has not been drawn, this is an appropriate time to do so. Depending on the geographic scope of the network, it may be necessary to have world, national, city, or building maps or a combination of all four. Typical maps of this type are shown in Figure 14–4. It is useful to connect the locations with lines and identify the number of characters each must handle. At this point, the lines on the map do not correspond to circuits in the network but only serve to indicate the traffic patterns and volumes.

Depending on the geographic characteristics of the network, it may be necessary to make traffic estimates on a location-by-location basis. For example, a network with a hub in Kansas City might be connected to sales offices in El Paso, Phoenix, Los Angeles, San Francisco, and Salt Lake City. It would be important to know if the traffic volume from the Los Angeles office were significantly greater than the volume from the other offices. If so, it might be necessary to run a separate circuit just to carry the Los Angeles traffic load.

Figure 14–4
Typical network maps that are useful during system design and for network documentation.

Availability Requirements Closely related to understanding the peak traffic period is knowing the range of hours during which the network will be operational and available to the users. The ability to carry 10 million characters in an 8-hour period requires quite a different network than one that must carry the same number of characters spread evenly over 24 hours. The problems created by having terminals in different time zones is even more severe in networks that handle traffic from Europe or Asia. Japan's business day is 13 or 14 hours ahead of eastern time in the United States, and Europe is 6 or 7 hours ahead, depending on the time of year. If a single network is serving locations in the United States, Europe, and Asia, it must essentially be operational 24 hours a day, 6 days per week, in order to cover the normal business hours on the three continents.

Response Time and Reliability Requirements The project team must understand the true requirements for response time and reliability. Two studies, referenced in Chapter 1, have shown that for certain types of interactive computer-based systems, subsecond response time is desirable and leads to increased productivity by the users. In other applications, however, such as the delivery of electronic mail on a store-and-forward basis, it may be acceptable for the message to be delivered within minutes or even hours.

economics of reliability

Designing a network for high reliability can lead to some serious economic issues. It may be possible to achieve extremely high reliability by installing redundant hardware and circuits. It is important to assess from an objective point of view, however, whether the application really requires such high reliability. Although users like to think that their application is critical to ongoing company operations, it may turn out that the occasional loss of the network for several minutes or even several hours, though inconvenient and irritating to the users, is not critical to company operations. For most business applications, the greater than 99 percent reliability of standard leased circuits is perfectly adequate.

Type of Terminal Operators It is important for the network design team to understand the characteristics of operators who will be using the network's terminals. Full-time, highly trained operators who sit at terminals all day will put a more constant and overall higher demand on a network than will occasional users or the general public. The nature of the traffic load when the users are full-time operators is fairly easy to estimate.

Whereas the overall amount of traffic from consumer-operated terminals, such as ATMs, is lower, it tends to be "bursty" and to have high peaks at certain times of the day or perhaps on certain days of the month. A network that can handle the peak load from a collection of ATMs at lunch hour and right after work may be nearly idle the rest of the day.

Future Growth Projections—Capacity Planning Understanding the expected traffic growth over a 6-month to 5-year period is an extremely important part of network design. The single most common mistake made by network analysts and designers is underestimating the growth in traffic volume and failing to provide adequate extra capacity in the communications circuits and equipment. This causes great problems later when the network must be expanded "unexpectedly" and rapidly. One situation in which this occurs frequently is failing to adequately project the growth in the number of new employees who will require telephones when planning a telephone system. Inadequate cables may be run or a PBX that is too small might be purchased.

new employees

superhighway effect

Data communications systems often fall victim to the *superhighway effect* when a new terminal-based computer system provides an especially useful function. When superhighways are built, it is often observed that traffic increases well above the level that existed on the old highway because travel on the new highway is so easy and convenient. More people travel to new destinations than ever before. As a result, a highway that was planned to last 15 or 20 years runs out of capacity in 5 years.

planning for growth

A similar phenomenon occurs in data communications systems when more users than were anticipated want to use the new system and they enter more transactions than were planned for. As a result, data circuits and other network hardware run out of capacity sooner than expected. Although extra circuits can always be ordered from the common carrier, all voice and data hardware, such as key systems, PBXs, and routers are designed to handle a certain maximum number of circuits, telephone calls per minute, or aggregate data rate. Once that capacity is exceeded, there is usually no alternative but to replace the hardware with a larger model. Although it may seem absurd, companies have spent thousands of dollars for telephone systems or network switches only to find that they are out of capacity within 6 months. Walking in to tell senior management that the company's "new" telephone or data communication system is now obsolete is not a pleasant experience!

With the explosion in the use of online terminals, many companies have found that their data communications traffic is growing at a rate of 40 percent or 50 percent per year, compounded! This means that the rate of traffic is doubling every two years. A communications controller that is only 50 percent used this year will be out of capacity next year. Unless the network has been designed to handle such rapid growth and management has been primed to know what to expect, the company can be in for some nasty surprises. Of course, a 50 percent growth in the communications traffic usually has wider implications, such as a similar growth in computer use and hopefully increased revenue! If managers expect a computer to last 4 years, they will be shocked to find out they need to make major upgrades every year or two.

Rapid growth rates emphasize the desirability of designing the network in a modular fashion so that parts of it can be upgraded incrementally as the need arises. Modular upgrades can be implemented more quickly and cost less than major network overhauls. Fortunately, the decreasing cost of bandwidth and network components makes it relatively practical to upgrade portions of the network. Of course one must remember the old adage that fixing one bottleneck in the system only uncovers the next one. Splitting a LAN to have fewer users on each segment in hopes of improving their throughput and response time may uncover the fact that the hubs or switches that connect the segments are at or near capacity and will have to be upgraded.

Capacity planning is not an exercise to be done only when a new communications system is designed. Traffic loads should be reviewed at least annually and more frequently in systems with a high growth rate. Many companies produce graphs of their communications traffic on a monthly basis. Firms that are highly dependent on communications may look at statistics hourly. Capacity projections should be made and presented to management regularly, even if network upgrades do not need to be made immediately. In some companies, capital expenditures must be planned ahead for more than 1 year. The capacity plan may be an important input to the annual budgeting process. If major new facilities will be needed in the coming year, it is important to be sure that the financial plans (budgets) are made appropriately so that money is available when needed.

Date the New Service Must Be Available The network design team must understand when the new network or communications service must be installed and ready for use. In many situations, especially for international networks, it takes at least 3 months from the time communications equipment is ordered until the service is operational. Yet equipment cannot be ordered until the network is designed and approved. Although there is a wide variability depending on exactly what work must be done, it is not unusual for communications projects to take a year or more from inception to implementation. If major new network capabilities are required in less time, it is sometimes possible to speed the ordering and installation process, but trying to do so can be risky.

Constraints The typical constraints implicitly placed on a network design are that an existing network must be expanded or that the new network must be compatible with the old one. Preexisting computer hardware or software presents a similar type of constraint. For example, if a SUN server is already in place and is going to process the transactions, the data network must be designed to interface to the SUN hardware and to work compatibly with it.

compatibility

Another type of constraint that may be placed on the network design is a cost constraint. Placing a direct limitation on the cost of the new network or an expansion occurs less frequently because the initial focus is usually on meeting new business requirements. Only after some design

cost

work has been done and the costs start to become evident are the costs usually given serious attention.

As the analyst gathers the data and analyzes the network requirements, inevitably he or she begins thinking about solutions to the communications problems. These "mental designs" are a natural part of the iterative analysis and design process, and these early, tentative thoughts or sketches may prove useful later. It is important, however, for the analyst to avoid the natural tendency to jump to conclusions and to design the network solution before thoroughly understanding the problem. Spending adequate time to ensure that problems and requirements are completely understood pays big dividends later in terms of a faster, more accurate network design and implementation.

Investigation of Alternatives

Another facet of the work to be done before designing the network is to investigate alternatives for addressing the communications requirements. Everyone tends to do this work based on personal experience. Although experience is extremely useful, in a field that is changing as rapidly as telecommunications, it is important to continually assess options that may not have been available when the existing network was designed. This is particularly true as the use of workstations connected to LANs, the Internet, and other communication networks continues to expand. With a steadily growing market, vendors are producing a constant stream of new products and enhancements that need to be considered when designing a new network or upgrading an old one.

Depending on the geographic locations to be served, certain vendors or communications service options may not be available. Obviously, if all of the locations are in major cities, more alternatives are likely to exist than if some of the facilities are in small towns or other less populated areas. Some companies like to have as few vendors as possible involved in their communications networks. A rule of thumb says that the number of operational problems in a communications network increases as the square of the number of vendors. This is because at each point where two vendors' equipment interconnects there is a potential incompatibility. The fewer the number of vendors, the fewer the number of these interfaces.

Many vendors choose to market their products only in large metropolitan areas where there is a substantial potential customer base. They do not make their products available outside the areas where their marketing and support exist. The network designers must understand what products and services are available at the locations to be served by the network.

There are numerous other alternatives that must be investigated. Can the Internet be used? Are public switched facilities available, or should private lines be used? Will the network be centralized in a star configuration or highly distributed? Is the use of multiplexers or concentrators appropriate? Do packet switching networks serve the network locations, and should they be used? Are satellite services appropriate? Can the network be outsourced to a carrier?

When alternatives are being investigated, costs must be considered. Often there are trade-offs between costs and service. In some cases, an alternative that is technically viable may be eliminated because its cost is prohibitive. The basic cost information gathered while the analyst evaluates alternatives is useful for estimating the total cost of the network being planned.

It is helpful in this stage to eliminate some alternatives so that attention can be focused on the ones that are most relevant.

The objective of this phase of activity is to be sure that all relevant alternatives are considered and that inappropriate options are eliminated. After considering the alternatives, the design team must decide which alternatives to select. The products and services selected must meet the users' requirements.

Network Design

The entire process of network design from information gathering through ordering the equipment is iterative, and the detailed design or layout of the network is also an iterative process. Often new insights are gleaned or new information is obtained while analysts and designers go through the design process. When this occurs, it frequently is desirable to redo some of the work performed in previous phases to take the new insights or information into account. The result of the rework is usually a better network design.

If the locations the network must serve have not yet been laid out on a map, that should be done now. As the old saying goes, a picture is worth a thousand words. Many times the relative locations of the network nodes suggest certain network configuration possibilities when they are viewed on a map. Assuming for a moment that the network is not going to be a local area network with a ring or bus topology, the first cut at the circuit configuration should be made by connecting the nodes to the central computer (if one exists) in a star topology. If the network is configured around multiple computers instead of a single central computer, the initial configuration would normally have the computers connected together and the terminals connected to the nearest computer.

using a star network

From the traffic load figures developed in the data gathering and analysis phases, the traffic load should be applied to each route showing the number of characters to be transmitted in a given period of time—typically a business day. The number of characters during the peak hour should also be identified. A map of this type is shown in Figure 14–5.

When the amount of data to be transmitted and the available hours are both known, designers can calculate the speed of the line required to carry the traffic. This gives the capacity requirement but does not necessarily indicate the responsiveness. Although there are many factors that

Figure 14–5
Typical star network map showing the number of characters to be transmitted during the business day and during the peak hour (in parentheses).

make up the overall response time of a communications/computer system, the communications time to transmit the characters from the terminal to the computer and the response characters from the computer back to the terminal is, in most cases, a significant part of the overall response time. If the line is heavily loaded, the probability is high that queuing will occur somewhere in the network. Queuing occurs when the line is busy handling the data from one terminal and another terminal is waiting to send or receive data. The figures vary widely, but if the line use on a standard leased, multipoint, polled circuit is greater than 40 percent, it is very likely that some traffic is being delayed because of queuing. Queuing in itself is not bad, but it is important to understand that it introduces variability into the transmission time between a terminal and a computer. That is why most response time numbers are stated in probabilistic terms, such as "90 percent of the transactions will be processed in less than 5 seconds, 95 percent in less than 10 seconds, and 100 percent in less than a minute."

queuing

Detailed analysis of the effects of queuing in the overall performance of a network can be done with simulation tools. Several are commercially

network simulation

available. Creating a simulation model of a network, however, is a significant job, requiring a detailed understanding of both the network and the simulation tool. Once the simulation model is created, it must be validated for the set of assumptions used in its creation. Only then can it be used to project the performance of the proposed network.

A simulation model can give the answers to many "what-if" questions, such as "What if I change the line speed?" "What if I add more terminals to the line?" "What if I change the parameters of the network control program and increase the number of buffers?" Furthermore, after the network is operational, a simulation model can help predict when the network will reach capacity by increasing the simulated number of transactions until response time degrades to unacceptable levels. A simulation model not only helps to make the most of the available resources but also confirms that changes being planned for the network will have the desired effect.

If simulation is not used, it is still possible to estimate the responsiveness characteristics of the network. Response time to the user can be calculated by adding together the time for the transmission from the terminal to the computer, the processing time in the computer, and the transmission time of the response back to the terminal. This simple calculation assumes that no queuing exists and gives a "best possible" response time. If this simple model shows that the "best possible" response time is inadequate even though the circuit may have enough capacity to handle all the characters that will be transmitted in a day, two options exist. One is to increase the speed of the processing time on the computer, and the other is to increase the speed of the transmission of the circuit. The option selected depends on many variables, but often the existing computer is one of the constraining variables put on the system design. That is, a computer of a given size is already installed and will be used. In that case, the only viable option is to increase the speed of the communications circuit to improve performance even though, for capacity purposes, the slower speed circuit is sufficient.

Let's look at an example. Suppose that a communications line must transmit 100 million characters in an 8-hour business day while providing terminals with no longer than a 1-second response time. Since this is a simple example, we will assume that there is only one terminal on a line and, therefore, that no queuing exists. As shown in Figure 14–6, a 64,000 bps circuit can transport 100 million characters in 3.5 hours, clearly meeting the capacity requirement. Suppose a typical transaction consists of 1000 characters transmitted from the terminal to the computer, followed by .2 second of computer processing time and then 10,000 characters transmitted from the computer back to the terminal. The figure shows that the 64,000 bps circuit yields a response time of 1.6 seconds and does not meet the response time requirement. A 256 kbps circuit cuts the response time to .53 second and clearly meets the response time requirement. At the same time,

Figure 14–6
Response time at various line speeds.

Line speed	64,000 bps	256,000 bps
Line capacity	= 8,000 characters per second	= 32,000 characters per second
	(at 8 bits per character)	
If 100 million characters per day are to be sent, the total transmission time would be	3.5 hours	.875 hour
1,000 characters in to computer	.125 sec	.03 sec
Computer processing time	.20 sec	.20 sec
10,000 characters out to terminal	1.25 sec	.30 sec
Total response time to user	**1.575 seconds**	**.53 second**

it greatly exceeds the capacity requirement of the circuit because it can carry all of the day's traffic in only .9 hour. The network designer faced with these circumstances would probably want to consult with the user to determine whether the additional cost of the 256 kbps circuit was warranted for this application.

The map of the network also needs to be viewed with an eye toward the possible use of concentrators or multiplexers. For example, as shown in Figure 14–7, if the computer doing processing for a network is located in Saint Louis, it may be desirable to concentrate multiple lines serving the west in Denver and then run a single high-speed line from Denver to Saint Louis. Similarly, circuits from the southeast might be concentrated in Atlanta and circuits from Texas in Dallas.

In the course of analyzing the data about network traffic and the configuration alternatives, it may be that more than one network configuration emerges as a possible candidate for the final design. In fact, it is a good idea to try to develop several alternative configurations. Each may have slightly different assumptions, use different equipment or vendors, and deliver different performance. In all likelihood there will also be cost variations between the alternative designs. Software is available to assist in evaluating alternative network designs. Communications carriers often make this software available to their customers free of charge. Similar software can be purchased from independent vendors.

Selection of Vendors and Equipment

Once the network has been designed or when several viable alternative designs exist, the specific equipment, software, and vendors must be selected. This may be done on a relatively informal basis in which known or preferred vendors are contacted and asked for price quotes and specifications for specific pieces of equipment. Alternately, a more formal approach

Figure 14–7
Network map showing the use of concentrators to consolidate traffic and reduce cost.

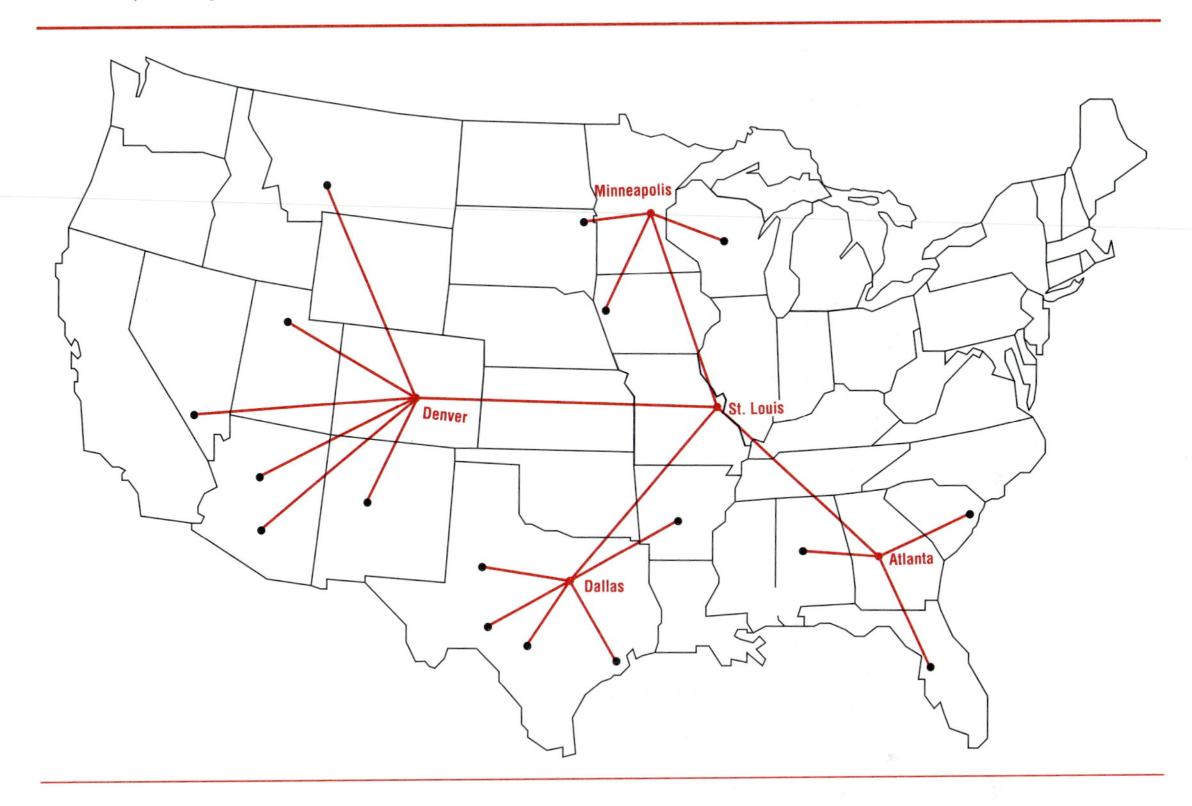

RFP or RFQ

can be used in which the company prepares a *request for proposal (RFP)*, also called a *request for quotation (RFQ)*, and sends it to several vendors.

Before contacting vendors or sending out RFPs, it is best to get an overview of the types of equipment available on the market. There are many sources of information about data communications equipment. Advertisements in data communications magazines and to some extent data processing magazines are one possibility. Buyer's guides, contain complete listings of data communications equipment by equipment type and vendor. Often these guidelines contain equipment ratings based on information provided by users of the equipment. Salespeople normally are happy to provide information about the equipment they sell over the phone or by sending product literature, but they usually prefer to make a direct sales call. Some vendors, such as Cisco, Inc., provide extensive information about their products on the Internet. Cisco has configuration tools, online support, educational information, and allows—even prefers—its hardware to be ordered by the customer through their Internet ordering screens. Other sources of information

Figure 14–8
The items to be included in an RFP document.

Introduction	A management overview of the company submitting the RFP and the problem or situation it wishes to have the vendor address
Table of contents	A detailed list of the contents of the RFP document
Description of the problem	A detailed description of the problem or situation that the vendor needs to address
Scope and requirements of the desired solution	A statement that frames the kind of response or solution the company expects to receive
Format, content, and due date of response	The information the vendor's response must contain, the format it needs to be in (e.g., electronic), and the date by which it must be submitted
Evaluation criteria	How the responses to the RFP will be evaluated
Estimated decision date	The estimated date the company will make a selection of which response, if any, it will accept
References	Several reference accounts for whom the vendor has performed similar work

about data communications equipment include data processing vendors, professional associates, and even friends. Most communications people are happy to share information and make recommendations about products they have used, especially if they are pleased with the results.

The RFP is a document that asks each vendor to prepare a price quotation for the configuration (or configurations) described in the RFP document. Some RFPs give the vendor considerable latitude in preparing responses or even ask for suggested solutions to specific design problems described in the RFP. Other RFPs are more structured and present a detailed outline for the vendors' response. Figure 14–8 lists the type of information that should be included in an RFP. Similarly, Figure 14–9 outlines the type of information that would be included in a typical response from a vendor. Usually the RFP is sent to several vendors so that the responses can be compared and alternative solutions considered.

When the RFP responses are returned, the network design team must evaluate them against the requirements presented to the vendors. Silly as it may seem, it is necessary to check to be sure that each vendor is proposing the type of network requested. Some vendors don't read RFPs very carefully! Vendor responses can be compared on the basis of the proposed solution, the price, the sales support provided, technical support offered, and product maintenance and repair service. A vendor's financial viability should be considered, because there are always some vendors who are here today and gone tomorrow. Sometimes after analysis, pieces of one

Introduction	A brief introduction to the company
Executive summary	A brief summary of the proposed solution
Table of contents	A detailed list of the contents of the response to the RFP
System design	A full description of the proposed solution
System features	The capabilities of the proposed solution that uniquely solve the identified problems and/or distinguish this vendor and solution from the competition
Growth capacity	An explanation of the spare capacity built into the proposed solution and the additional capacity (add-on modules) that can be purchased later
Installation and testing methods	How the solution is installed and tested
Maintenance arrangements	How maintenance support is provided, and its cost
Ongoing support	What support the vendor will provide after the solution is installed and operational
Installation schedule	When the proposed solution can be installed or implemented
Pricing and timing of payments	The cost and when payments are to be made
Warranty coverage	What warranty protection the vendor offers
Training and education	How customer training will be conducted, and the additional cost for training, if any
Other recommendations	Additional recommendations by the vendor that don't strictly fall within the scope of the RFP

Figure 14–9
The typical kind of information to be included by a vendor in a response to an RFP.

vendor's proposed solution are combined with parts of the proposal from another vendor to make a hybrid network that provides a unique combination of capabilities.

Calculation of Costs

With one or more network configurations and actual cost information provided by the vendors in hand, the costs of the network designs or alternatives can be recalculated and compared. Just as there are technical trade-offs between different types of lines, modems, and other equipment, there is a cost/performance trade-off between vendor proposals. The objective of the cost analysis is to finalize the configuration of the network that will meet the capacity, performance, and other requirements at the lowest possible cost. The cost can then be compared to the benefits that the new communications system will yield, and a decision can be made whether to move ahead to the implementation phase.

Network costs can be grouped into the following categories:

- circuit costs;
- modem costs;
- other hardware costs;
- software costs;
- personnel costs.

Calculating the cost of leased circuits is a complicated process that usually is left up to the communications carrier. The details of the process are described in Chapter 9. Personal computer programs are available to assist with the calculation if a company wants to calculate the cost internally.

The costs of modems, routers, switches, and other communications hardware and software can be obtained directly from the manufacturer or distributor of the equipment. Purchase prices usually are quoted by the vendor, since most companies buy (rather than lease) this type of equipment, but some vendors also provide rent or lease options for their equipment.

It is important to separate the costs of the personnel who develop the network from the costs of those who run it. The latter costs are ongoing operational costs, whereas the former are one-time development costs. The costs of the network analysts and designers are a part of the project costs but are not a part of the ongoing operational costs. Once the project is completed, the analysts and designers will work on other projects. Costs for the personnel to operate the network may be based on actual salaries and benefits or on a standard amount the company uses for all personnel cost estimation.

When all of the costs are calculated, they need to be assembled on a worksheet, such as the one shown in Figure 14–10. It may be helpful to build up a total cost for each circuit (or segment on a LAN) including the cost of the circuit itself as well as any hardware associated with it. Ongoing monthly rental or lease charges must be separated from purchase costs. The purchase costs are a one-time charge, but if the purchase is large enough, there may be an intracompany depreciation charge to be factored in. The cost for each of the circuits can be added to give the total cost for all circuits and equipment. To this is added the cost of network software and operational personnel. The sum of all of these costs is a total that can be compared to the proposed benefits of the new network and analyzed according to the company's financial criteria.

Although it is relatively easy to identify the costs of a new or upgraded network, it is often more difficult to quantify the benefits. However, benefit estimates must be made in order to determine whether or not the new network is cost justified. Users and financial people in the

	Circuit Cost ($ per mo.)	Modem Cost (purchase)	Other Hardware (purchase)
Circuit 1	$ 565	$640	$22,560
Circuit 2	1,884	885	18,290
Circuit 3	140	270	0
Circuit 4	998	490	12,010
Circuit Cost	$3,587	$2,285	$52,860
Network software	885		
Operating personnel	5100		
Total Operating Cost	$9,572 plus depreciation on purchased equipment		
	Monthly Cost		

Figure 14–10
Network cost worksheet.

company can be very helpful in identifying the benefits that will accrue, and their help and advice should be solicited. Some of the types of benefits that may accrue include

- improved customer service;
- increased sales;
- ability to sell a new product;
- reduction of inventory;
- reduction of time needed to process routine work;
- elimination of clerical personnel;
- more accurate information because of the improved ability to share data;
- improved employee morale.

Obviously some of these benefits are tangible and will increase revenue or save out-of-pocket expenses, whereas others are intangible, sometimes called soft benefits. As the list of benefits is identified, it should be quantified whenever possible.

Documentation of the Network Design and Implementation Plan

Once the network has been designed, it is important to document it in a form that can be understood by management for review and approval purposes. It is also necessary to document the design in technical terms for those who will implement it. The actual form the documentation will take depends to a great extent on the size of the network change and the formality or informality of the project. What is described in this section is fairly extensive documentation; many companies would do a lot less and be satisfied.

diagrams and maps

Network documentation takes two forms: words and pictures. The pictures may be maps that show the locations of nodes in the network, as previously discussed, or diagrams that show all of the nodes and the types of connections between them, although not in a geographical layout. The maps or pictures need to be accompanied by a narrative that describes the network. A circuit and equipment list, which shows a detailed breakdown of all of the network components, also should be included.

components list

specifications and model numbers

Detailed technical network documentation must show exact circuit specifications, including the types of conditioning and any special features. Hardware model numbers must be included, as well as any required accessories or cables. The equipment specifications must be detailed enough to be placed on a purchase order and understood by the vendor. In addition, wiring diagrams that illustrate equipment connections and where the wire and cable will run are included in the design package.

wiring diagrams

Detailed specifications must also be given for the software that will be required. Specific version or release numbers should be identified when possible. The number of copies of software required for workstations on a LAN should be identified as well as other software that must be upgraded to support the new or reconfigured network.

A final piece of required documentation is the implementation plan, preferably in a Gantt chart similar to the one shown in Figure 15–9. The Gantt chart should show the activities leading up to cutover in a time sequence so that it is clear which tasks or activities need to be done earlier than others. Activities shown in the plan may include ordering equipment, writing software, training users, testing the network, and actual cutover. The plan should show the elapsed time required and the estimated completion date of each of the required activities.

Management Understanding of the Design and Approval of Implementation

Once the network is designed and documented, the next step is to distribute the documentation to the users of the network, the network operations staff, and the senior management. As with most complex designs, the documentation itself rarely provides enough information to give the reader a complete understanding. It is common to have a series of presentations and meetings in which the designers elaborate on the design and answer questions. The ultimate objective of this review and approval process is to gain the support of the users and operations staff and then the approval of management to implement the system.

review process

The users should understand the design well enough to be able to state that if the network is implemented as described in the documentation, it will provide the required communications capability and/or solve the communications problems they have been experiencing. The network operations staff needs to understand the design of the proposed network well enough to ensure that they can operate it after it is implemented. Management must satisfy themselves that the implementation of the network is in the company's best interests and that the investment of staff time and money will yield the claimed benefits.

NETWORK IMPLEMENTATION

Once approval to implement the network has been obtained, the project moves into the implementation stage. Although the same project team that designed the network is likely to continue, the implementation of the network is a distinct set of activities in itself.

Equipment Order

Ordering equipment is normally one of the first actions taken after implementation approval is obtained. Some equipment and services have lead times of up to six months, and therefore it is important to place specific orders promptly after the project is approved. The two most important provisions in the equipment purchase order are the equipment specifications and the acceptance test. The specifications define the products and services being purchased, and the acceptance test provides an objective basis for determining whether the equipment installation, after cutover, meets the buyer's expectations. The order should specify the exact configuration of each piece of equipment, the date required, the location to which it is to be delivered, and any other terms and conditions the company's purchasing department requires. The more precise the specifications on the purchase order, the less likelihood of mistakes or misunderstandings when the equipment is built, shipped, or installed. Other important provisions of the purchase agreement should cover the vendor's warranty, maintenance arrangements, and the terms of the installation and cutover.

Ultimately the order for services or equipment is likely to result in a formal contract with the telecommunications supplier. A company's legal counsel should review any contract before it is signed, but it is also worthwhile for the telecommunications manager or project leader to be familiar with the contract so that he or she can work with and answer the questions of the legal counsel. Some points to consider when reviewing a contract are:

- Get the most overall flexibility possible. Usually the shortest possible deal is best, considering the fast-changing technology.
- Be certain that any special terms or equipment features are clearly specified.
- Be clear on warranty coverage and service commitments and cost.

- Insist on free installation.
- Be aware of penalties for terminating a service early.
- Know what others are paying for similar equipment or services.
- Bargain—ask for the best possible price.

Preparation for Network Implementation

Many tasks must be done after the equipment is ordered and before implementation. Depending on the nature of the work, there may be network operating policies and practices to establish. Standards may have to be written. Procedures for the network operations group may need to be developed. Purchased software may need to be modified to meet certain network requirements, or new software may have to be written.

tracking progress

Each of these activities should have been shown in the network implementation plan that was prepared earlier. Progress must be tracked against the plan to ensure that the target date for network cutover will be met.

Installation of Equipment

Before equipment is delivered, the project team must prepare for the installation. Physical planning is the process for determining where the equipment is to be located, ensuring that adequate power and air conditioning are available, and seeing that large pieces of equipment can be conveniently transported from the shipping dock where they are delivered to where they will be installed.

layout drawings

Drawings may be required that show where all of the equipment will be placed, as well as how power and communications cables will be routed to it. Most vendors have specialists who can assist in the physical planning, especially for large pieces of equipment, such as PBXs.

Once the equipment arrives, it must be installed. Depending on the nature of the equipment and the terms of the purchase agreement, it may be installed by the vendor or the company's own communications staff. Most frequently, the initial installation is performed by the vendor, who also handles any required training at the same time. Circuits are installed by the common carrier. When pieces of equipment from different vendors are connected, it is important to ensure that the vendors work together and test the equipment while all are still on the premises. Then if there is a problem, they can work together to resolve it.

Training

Both the users of the communications system and the operational and maintenance personnel must be trained in the characteristics of the new system. User training should be conducted by other users who have been trained by the network design team. Usually a combination of classroom

type of instruction

instruction and hands-on experience is provided. The classroom instruction gives background and overview material, but the heart of the training is performed at the terminal, with users actually entering transactions and receiving responses. This also gives users a feel for how the network will perform, so that they gain a sense of its characteristics and responsiveness.

The operational and maintenance personnel will be trained by both the members of the project team and the vendors who install the equipment. Ideally, the operations staff will have been involved in the project since its inception and will already understand the design of the network. The final training, then, is a matter of getting hands-on experience with the commands and the various real-life situations.

Operations people can get good experience with the network while the users are using it in a training mode. If the users do a good job of simulating real operation, the operations staff will get a good feeling for how the network will operate and perform when it is finally cut into production. Of course, not every unusual situation that will occur during real operation will occur during the training period.

System Testing

After all of the equipment and all of the circuits have been installed, the entire network must be tested. Ideally, each individual component of the network is tested under a variety of circumstances and conditions. However, with the size of some of the networks being installed today, this is virtually impossible. It is important that every terminal on the network exchange at least one message with another terminal or the host computer to verify that it is properly connected and working correctly. This type of test also exercises the modems, circuits, and multiplexers and gives some assurance that they are working properly.

The system's software must be tested separately. Initial tests may be conducted by network technical support specialists or computer system programmers. When communications software is included in the users' workstations, as in the case of personal computers, the users may perform some of the testing. Indeed it can be a good learning and training exercise for them to understand how to use the software. When the workstations are connected to a LAN, the proper operation of the network operating system (NOS) must be verified along with the proper operation and accessibility of servers by all of the workstations. Sometimes this testing can be accomplished during normal working hours, but in other cases it may be necessary for people to do the testing after business hours, in the evenings, or on weekends.

stress testing

Another type of testing required is *stress testing*. In stress testing, a heavy load is put on the system, usually by having a number of terminal operators simultaneously use the system. Stress testing indicates how the network will perform in real life and is also important as a final software test. Certain program bugs only show up when a high volume of transactions passes through the software in a short period of time.

workload generator

One way to stress test a communications system is to use a tool called a *workload generator,* also called a *network simulator* or *driver.* The workload generator is a computer program that generates transactions. They are fed into the network to see how it behaves. To the network, the transactions appear to come from terminals. Workload generator programs often run on a computer that is attached to the front-end processor. They may also run in one or more personal computers attached at various points on the communications lines. Good workload generator programs, such as IBM's Teleprocessing Network Simulator (TPNS), allow the network designers to specify the type and length of transactions generated so that networks can simulate various transaction mixes. The advantage of the workload generator program is that if it is properly set up, it can simulate full network operation without requiring users to be involved. Furthermore, the transaction mix and rate at which transactions are introduced into the system can be controlled more precisely than they could be with human operators.

testing error handling capabilities

The most important oversight of network testing plans is the failure to test the error handling capabilities of the individual components and the system as a whole. The acid test of the error recovery capabilities built into each of the components is whether their error recovery procedures will work together. When a failed circuit is restored to service, do the modems automatically resume communications, or must they be restarted manually? What about the telecommunications monitor and other computer software? Do they recover from circuit or modem errors automatically?

Sometimes, the error recovery procedures of one device can create errors in other devices. It is important to identify these situations early so that proper procedures can be written for the network operations staff describing how to handle different types of errors. Some error recovery procedures will invariably be exercised during normal network testing. To thoroughly test the error recovery, however, it may also be necessary to introduce errors in the network by disconnecting circuits, powering off modems, or injecting noise on a circuit with an appropriate test instrument.

Cutover

Cutting the new network into production use is a major milestone and an event that usually is very visible throughout the organization. If the planning and preparation have been done properly, all of the hardware and software components will have been installed and thoroughly tested. The users will have been trained and will be eager for the cutover to occur so that they can use the new network and accrue its benefits. Network cutovers frequently are scheduled to occur on a weekend. There may be a substantial amount of last-minute work to be done, and the extra time afforded by the weekend allows the work to be completed and a final round of testing to be conducted before all users come back to work on

Monday morning. Sometimes, the cutover is scheduled for a Thursday night if it can be physically accomplished in a short period of time. This gives one day of live network operation before the weekend, and then the weekend is available for correcting problems or other unforeseen circumstances that became apparent on the first day of real operation.

Despite the thoroughness of the testing, there may be some unanticipated startup problems. It is good practice to insist that technicians employed by the equipment vendors be on hand when the network or upgrade is cut over into production so that equipment problems can be resolved as quickly as possible. It is important that the help desk be in operation, and even overstaffed, in the first days of live operation so that incoming telephone calls can be handled promptly. The help desk should meticulously record all reported problems so that follow-up action and analysis can occur promptly.

vendor technicians

Implementation Cleanup

When planning a network project, it is prudent to allow some time after the cutover for cleaning up problems that are identified but not quickly resolved. If, for example, it becomes apparent that despite careful analysis, a certain circuit is not performing up to expectations, a higher-speed circuit or an additional circuit may have to be installed. The network control operators may find that some minor changes to their status displays would help them identify problems before the user sees them. Making the required changes may take several days of analysis and programming, but many times, a few changes such as these can make the difference between a good network and a superior network.

changing status displays

Another activity that should occur after the implementation phase is updating the documentation of the network. In most cases, the network will have been implemented slightly differently from the way it was designed. These changes should be reflected in the final documentation so that it is up-to-date and provides an accurate description of the actual installation.

revising documentation

Network Audit

Approximately 6 months after cutover, the new network or network upgrade should be audited by the company's internal auditors or a team of user and telecommunications management. The purposes of this audit are to determine whether the network is delivering the benefits that were promised, whether operational controls are in place and effective, and whether modifications need to be made to the network configuration, software, or application programs. The audit is conducted by interviewing network users and network operations personnel and by observing network parameters, such as response time at user terminals and the procedures in the network operations center. The audit is a good time to check the performance or success criteria that were established in the preliminary investigation and feasibility phase of the project. Six months after the network was implemented its operation should be stable. The bugs and procedural problems should have been worked out, so a fair measurement against the criteria can be obtained.

The audit team should write a formal report that describes its findings and makes recommendations for changes that need to be implemented. This report should be directed to senior management and appropriate telecommunications line management. Line management is responsible for writing a response to the audit report stating what actions will be taken based on the recommendations that were made.

LAN DESIGN

Although the process of designing a LAN is basically the same as for any other network, several items deserve special mention.

Traffic Analysis

Local area networks (LANs) operate at much higher transmission speeds than conventional wide area networks (WANs), so the length of specific messages or raw throughput for a single message on a LAN is rarely a major consideration. However, LAN users routinely transfer files of information and programs from servers to their workstation, and these files can be quite long compared to the messages of a transaction processing or electronic mail application. With well-designed LAN software that is properly configured, the transmission of long files should not tie up the LAN excessively. The files should be broken into blocks by the LAN software, and other users should have a chance to access the LAN between blocks of the file. However, if the software parameters are not set correctly, it is possible that a single user can dominate a LAN and cause poor service to all of the other users. It is important that the LAN designer understand enough of the technical details about the way a LAN operates to ensure that the LAN will have adequate capacity and responsiveness to meet the users' requirements.

Internetworking

LANs tend to be connected to other networks, if not initially when they are set up, then soon after. The implications of these connections need to be considered during the design process, even if the need is not immediately envisioned. The LAN designer should do a "what if" analysis, asking questions such as, "What if the proposed LAN needed to be connected to another LAN in the company?" and "What if the LAN needed to be connected to the Internet?" Connecting to other networks is a major way in which LANs grow, and planning for growth at the design stage will forestall problems that may arise if the growth is not anticipated.

LAN Administration

The need for a LAN administrator was thoroughly discussed in Chapter 12. Identifying the need for and cost of a LAN administrator during the design of a new LAN is an excellent way to help ensure that ongoing LAN administration and the need for an administrator get the recognition within the organization that they deserve. The LAN designer should include ongoing operational considerations and costs in the design report and make sure that management recognizes the ongoing cost.

Although this point is really no different from the need to recognize the ongoing operational cost of managing a WAN, LAN management and administration seem to be more easily overlooked, especially when the LAN is small and the company or department is inexperienced in operating a communications network.

VOICE NETWORK DESIGN

Most of the concepts and procedures that have been discussed thus far are applicable to both data and voice network design. After all, a circuit is a circuit, and both data and voice telecommunications needs are defined in terms of the amount of information transmitted. Voice conversations usually are quantified in terms of the minutes of conversation taken up by a call. Data transmissions are thought of as bit streams of a certain length that can be translated into transmission times.

When a designer creates voice systems, however, some unique techniques and terms are traditionally used. Whereas a data traffic study measures the number of data messages from the terminal to the computer and back and the length of each message in characters, a voice traffic study measures the number of telephone calls and their average duration. The objectives of the studies are the same—to gather data about traffic volume and peak loads, so that enough circuits with sufficient capacity can be installed to meet peak demands. Many companies routinely gather voice traffic information and have graphs showing the call statistics extending back for several months or even years. As with the data traffic study, the purpose is to obtain data to use in predicting future requirements.

busy hours and busy days

It is important to understand when the busy hours of each day occur, whether there are any days of the week that are busier than others, and whether there are any seasonal factors that cause unusual peaks during certain times of the year. There are some companies, for example, that do a heavy mail order business just before Christmas and employ extra telephone operators to handle the additional calls during the 2- or 3-month peak period.

holding time

The duration of a telephone call is called the *holding time.* Holding time is the time it takes the central office equipment to complete a telephone call plus the duration of the conversation. The number of telephone calls multiplied by the holding time measured in seconds is the number of seconds that the equipment is in use. For example, if there were 200 calls per hour each lasting 4 minutes, or 240 seconds, the equipment would be in use 200 × 240, or 48,000 seconds. Dividing this number by 100 gives the number of hundreds of call seconds, which is abbreviated *CCS*. CCS stands for *centa call seconds*, where *centa* is the common designation for hundreds. Another measure of equipment or circuit usage is the *Erlang*. One Erlang is 36 CCS, and because 36 CCS equal 3,600 seconds, both 1 Erlang and 36 CCS equal 1 hour of equipment usage. Both CCS and Erlangs are used frequently by telephone engineers when discussing telephone traffic.

centa call seconds

Erlang

Telephone systems are designed to provide sufficient equipment to handle traffic during the busiest hours with a certain grade of service, as was discussed in Chapter 5. The grade of service is the probability that a call cannot be completed because all of the equipment or circuits are busy. This probability is expressed as a percentage, such as P.02, which means that 2 percent of the calls during the busy hour are likely to be blocked. Telephone systems can be designed to provide any grade of service.

grade of service

Determining the number of circuits required to carry the traffic in a voice communications network and provide a given grade of service is done using capacity tables. The *Poisson capacity table* is based on the assumption that there is an infinite source of telephone traffic and that all blocked calls will be retried within a short period of time. A sample Poisson capacity table is shown in Figure 14–11. The table is used by finding the CCS usage in the appropriate grade of service column and reading the number that appears in the trunks column.

capacity table

Figure 14–11
Sample Poisson capacity table.

Trunks Required	Grade of Service P.01	P.02	P.05	P.10
2	5.4	7.9	12.9	19.1
4	29.6	36.7	49.1	63.0
6	64.4	76.0	94.1	113.0
8	105.0	119.0	143.0	168.0
10	148.0	166.0	195.0	224.0
15	269.0	293.0	333.0	370.0
20	399.0	429.0	477.0	523.0
25	535.0	571.0	626.0	670.0
30	675.0	715.0	773.0	836.0
40	964.0	1012.0	1038.0	1157.0
50	1261.0	1317.0	1403.0	1482.0

Usage in CCS

Another commonly used capacity table is the *Erlang B capacity table.* The Erlang B capacity tables are based on the assumption that the sources of telephone traffic are infinite, but that all unsuccessful call attempts are abandoned and not retried. A sample Erlang B capacity table is shown in Figure 14–12. This table is read by finding the traffic load measured in Erlangs in the appropriate grade of service column and reading the number in the trunks column. If the usage figures are converted to common units (either CCS or Erlangs) and the tables are compared, it will be seen that the Poisson table is slightly more conservative than the Erlang B table. That is, for a given traffic load and grade of service, the Poisson table will suggest that a slightly higher number of circuits is required to carry the load. The reason for the difference is the slightly different assumption the two tables make about blocked calls.

Considering the total costs of the people and machines that use a communications network compared to the costs of the lines themselves, there is some argument for using the most conservative capacity tables (Poisson) and slightly *overdesigning* the network. Overdesigning means that additional lines or trunks are added so that users will almost never experience a delay or a network-busy condition. U.S. telephone companies have overtrunked the public switched telephone network for years, and that is at least part of the reason why the network is so reliable and always available.

planned delay

In some cases, telephone systems are purposely designed so that some calls will experience a delay. The most common examples are reservation systems and the telephone systems of organizations that dispense information. For example, when people call the Internal Revenue Service

	Grade of Service			
Trunks Required	**P.01**	**P.02**	**P.05**	**P.10**
2	.153	.224	.382	.6
4	.870	1.093	1.525	2.0
6	1.909	2.276	2.961	3.8
8	3.128	3.627	4.543	5.6
10	4.462	5.084	6.216	7.5
15	8.108	9.010	10.63	12.5
20	12.03	13.18	15.25	17.60
25	16.13	17.51	19.99	22.80
30	20.34	21.93	24.80	28.10
40	29.01	31.00	34.60	38.80
50	37.90	40.25	44.53	49.60
	Usage in Erlangs			

Figure 14–12
Sample Erlang B capacity table.

to check on their tax returns, they frequently receive a recorded message saying that all tax agents are busy and that the call will be handled as soon as possible. Since the IRS is the only organization that has the information the caller wants, he or she will wait, even though the delay may be annoying. In a system such as this, the cost of providing a better grade of service is traded off against the probability that the call will be lost. In the case of the IRS, the probability that the caller will hang up and never call again is low. For an airline reservation system, however, the trade-off is not so clear because the caller may call another airline to make the reservation. Therefore, the balance between the cost of the network and the grade of service may be viewed from a competitive perspective.

FIFO queuing

Calls waiting for service are queued. Most telephone queuing systems queue the calls in the order that they arrive, and the customer gets service when his or her call reaches the head of the waiting line. This is known as *first-in-first-out (FIFO) queuing*. Some PBXs have the ability to queue calls, but in most cases, a separate piece of hardware—the automatic call distribution (ACD) unit—does the queuing and passes the call at the head of the queue to the next available agent.

Telephone systems with the queuing capability are designed using a different traffic capacity table, the Erlang C table. This table takes into account the number of calls to be held in the queue as well as the grade of service and predicted traffic load.

Other traffic capacity tables have been developed by communications engineers and consulting organizations. They are all based on the same concepts and probabilities, and they vary only in the assumptions they make and in their degree of conservatism in suggesting the number of lines to be used to carry a given traffic load at a given grade of service.

PROJECT MANAGEMENT

Implementing a new communications system is a project made up of many phases or steps. As we have seen, the steps are interrelated and are sometimes repeated as more information is gathered. It is extremely important that the project team use project management tools that are appropriate to the size of the work to be done.

A detailed project plan is mandatory for any size project. Gantt charts, which will be discussed in Chapter 15, are appropriate for any size project, but large projects may require the use of a computerized project management system such as Microsoft Project. This project management software runs on personal computers and provides a variety of views of the project, such as Gantt charts, task lists by person, task lists by week, and lists of critical items that must be completed if the project is to remain on schedule. Someone must be assigned to keep the data in the project management system up-to-date with the latest progress information and other changes.

Regardless of which project management tools are used, it is important that an initial plan for the project be prepared, key milestones of project activity be identified, and project progress be reviewed on a regular basis. Frequently, the project team will have regular review meetings, typically weekly, in which team members review their activities for the past week, problems that they have encountered, and plans for the coming week. The frequency of these meetings depends on the size of the project, the number of people involved, and the technical complexity of the network that is being implemented.

Most project teams find it appropriate to review their progress with management at least monthly. The purpose of such a review is to keep management apprised of the progress the team is making as well as any difficulties encountered. Properly done, the management review meetings ensure that no one is surprised by changes to the schedule, scope, or resource requirements.

SUMMARY

Designing and implementing a communications network is a multistep, iterative process. The degree to which an organization follows rigorous network analysis and design procedures will depend on the complexity of the network or network upgrade. Today, many networks are upgraded frequently and in small modular pieces, so the analysis and design can be less rigorous because the scope of the work and even the solution may be fairly obvious. Ultimately, management must determine the extent to which the work is formalized and documented.

Regardless of the size and complexity, after a request for work, a (sometimes small) team studies the feasibility of the new communications system or upgrade, and gathers more detail, so that members grasp the problems with the existing system and the requirements for the new system. Then the team members examine the various alternatives for the new network and prepare one or more network designs. Each alternative is subjected to a cost analysis, and one or more of the alternatives is presented to management for review and approval. After approval is given, the equipment is ordered and installed. After the network is tested, the users are trained and the system is cut over into production use. Time should be allotted for implementation cleanup activities, and an audit of the system should be conducted approximately six months after the network is installed.

Designing voice systems is a similar process, but special terms and traffic tables are used. Based on the traffic load and the grade of service to be provided, these tables help determine the number of circuits required.

In any network design and implementation activity, project management tools need to be applied. The number and complexity of the tools depend on the size of the project.

CASE STUDY

Dow Corning's Communications Network Design

Dow Corning's communications network design was originally done much less formally than the method described in the text. Typical of smaller organizations, data network design was done by a data communications specialist, based on the requirements for new terminals and the desire to upgrade line speed or to take advantage of changing technologies. In those days, the network was small enough that a single specialist could keep track of the various parts of the network and how they were performing. He knew when it was time to look at a particular piece of the network and evaluate options for upgrading it. Because of his expertise and knowledge of product offerings, he was able to do the design "in his head," gain management approval, order the required equipment, and install the changes relatively rapidly. Data network design was handled more as a project than an ongoing process.

As the company grew and the data networks became more complicated, it became necessary to handle data network design in a more sophisticated way. LANs and manufacturing networks added new requirements and complexity to the task of planning for network growth and expansion. Also, network design has become so complicated because of the many products offered by vendors that involving a networking company in the design process became almost essential. However, Dow Corning maintains data and voice design and diagnostic expertise to assist in all network designs, even if the vendors are doing a majority of the work.

In 1997, Dow Corning signed a contract with WorldCom for a global WAN and in doing so contracted for WorldCom to provide much of the design expertise for the company's data network. WorldCom specialists provide network design assistance, while Dow Corning employees contribute information about traffic volume, growth rates, new application requirements, and service level needs.

For most projects other than the smallest, a project team is formed, responsibilities are assigned to the members of the project team, and target dates are set. Project leaders use the *Microsoft Project* project management software when warranted, but in almost all cases activity lists and Gantt charts are prepared so that progress can be tracked and measured. When new projects are about to be implemented, they are entered into the change management system, which automatically distributes the information to network people, the help desk, computer operations people, and others who could be affected.

Ameritech and AT&T have largely handled the voice network design. These vendors receive input from Dow Corning about actual or anticipated changes in facilities or personnel. In addition, the telephone companies perform periodic studies of actual traffic patterns and recommend changes to the network to Dow Corning's telecommunications management. Dow Corning's upgrade of its voice network to use AT&T's Software Defined Network (SDN) several years ago was partially based on the results of network design studies performed using AT&T's INOS tool.

QUESTIONS

1. When Dow Corning upgraded its worldwide network and all of its PC workstations in the mid-90s (as described in the case study at the end of Chapter 6), what project management techniques do you think they would have used? How would they have coordinated the project on a global basis considering the many languages that their employees speak?
2. Discuss the trade-offs of relying on the vendors to do voice network design and planning versus hiring or training a Dow Corning employee to do the work.

REVIEW QUESTIONS

1. Why is a phased approach to network analysis design used? What does it mean to say that network analysis and design are an iterative process?
2. Why is it necessary to have a formal request document for network analysis and design work to be done?
3. Compare and contrast the jobs of a computer systems analyst and a network analyst.
4. Discuss the criteria that might be used for prioritizing the list of network projects that could be undertaken, assuming that there are not enough people to work on them all.
5. Explain the concept of the creeping commitment of resources to a project. Why is it important?
6. What information can network simulation give the network designer?
7. Why is it so important to understand the requirements for a communications network before a designer begins the design work?
8. Why is it important to understand the geographic requirements for a network?
9. What are the components of the total cost of a network?
10. Why is it important to have users involved in the network design activity?
11. Why might communications software have to be written?
12. Explain the concepts of peak load and busy hour.
13. What do you think the hourly traffic pattern of data communications at the New York Stock Exchange would be? Draw a graph to illustrate your idea of the pattern.
14. Why is it important to investigate alternatives for a network design?
15. Referring to the list of potential benefits of a new network identified in the chapter, identify which are tangible and which are intangible. What other benefits could you add to the list?
16. Why is LAN management and administration often overlooked when LANs are designed?
17. Why is it important to investigate alternatives for a network design?
18. Why is it important to test error handling and error recovery procedures?
19. What is stress testing? Why is it important?
20. What is the purpose of the implementation cleanup phase?
21. What are some examples of telephone systems that are purposely designed so that some users experience a delay in getting their call answered?
22. What is a CCS? What is an Erlang?

23. Explain how telephone grade of service is calculated.

24. What is telephone call holding time?

25. Explain why the process of network analysis and design may be more or less formal and elaborate depending on the needs of the company and the size of the network design or upgrade.

PROBLEMS AND PROJECTS

1. A point-to-point circuit from San Diego to Miami is being designed. The projected traffic on the circuit is 4 million characters during the business day, with 15 percent of the traffic occurring during the busy hour. If less than a 2-second response time is required, what speed circuit would you recommend be installed? Why?

2. Determining the proper amount of documentation for a network is not an easy task. Describe several different levels of detail at which the documentation could be created. What are the advantages and disadvantages of the most detailed level? The least detailed level?

3. Management understands the economic implications of a new or expanded network better than the technical aspects. How would you quantify the benefits of a new network that would provide electronic mail capability to your company's 35 locations throughout the United States? Assume that the network would deliver any piece of electronic mail within 30 minutes after it is sent. What might be the benefits of tying your customers to the network and letting them use it to communicate more effectively?

4. The holding time for a voice call is typically longer than the holding time for the transmission of a data transaction and its response. Why?

5. Telephone companies design the public telephone equipment and networks to provide a P.01 grade of service, whereas business telephone systems often are designed to provide service in the P.01 to P.05 range. Is the design level for the business system better or worse than the public system? If better, how can private business afford the better grade of service? If worse, how can business afford to have inferior telephone service?

6. The impact of a 40 percent or 50 percent compound growth rate is often not realized by business people. Assume a network is handling 2,000 messages per day this year but that the traffic will grow by 50 percent for each of the next 5 years. How many messages will the network need to carry 5 years from now?

7. As a further exercise in compound growth, take the starting salary you expect to earn when you graduate from college and increase it 5 percent for every year you expect to work to see what your salary would be when you retire. Try it at 8 percent. (Hint: It's an easy program to write if you have access to a computer.)

Vocabulary

project team
creeping commitment of resources
project life cycle
busy hour
superhighway effect
request for proposal (RFP)
request for quotation (RFQ)
stress testing
workload generator
network simulator
driver
holding time
centa call seconds (CCS)
Erlang
Poisson capacity table
Erlang B capacity table
overdesigning
first-in-first-out (FIFO) queuing

References

Boardman, Bruce. "Making Sense of Network Chaos." *Network Computing* (April 17, 2000): 69–84.

Gurrie, Michael L., and Patrick J. O'Connor. *Voice/Data Telecommunications Systems.* Englewood Cliffs, NJ: Prentice-Hall, 1986.

Jewett, J., J. Shrago, and B. Yomtov. *Designing Optimal Voice Networks for Businesses, Government, and Telephone Companies.* Chicago: Telephony Publishing Corporation, 1980.

Martin, James. *Systems Analysis for Data Transmission.* Englewood Cliffs, NJ: Prentice-Hall, 1972.

Salinger, Anthony W., and Leigh Gerstenmaier. "How to Pin Down User Requirements." *Data Communications* (August 1985): 155.

Schaevitz, Alan Y. "Network Design Issues for the 1990s." *Business Communications Review* (November-December 1988): 49–51.

Thompson, H. Paul. "Increasing Data Communications Control Through System Simulation." *Information Systems Management,* vol. 4, no.1 (Winter 1987): 29.

PART SIX

Managing the Telecommunications Department

The purpose of Part Six is to describe how the telecommunications department is managed. Telecommunications management is often viewed as simply the operation of the communications network. In fact, the total picture is much broader. The telecommunications department can be seen as a company within a company. It has all of the same functions and responsibilities as the enterprise of which it is a part—marketing, product development, finance, and manufacturing. It is important to understand this part of telecommunications management, for without it, one's knowledge of the subject is incomplete.

Chapter 15 explains the broad scope of telecommunications department management. Alternative ways of organizing the department are discussed, management responsibilities are described, and several management issues unique to the telecommunications department are explored.

Chapter 16 looks at some of the newer telecommunications applications as well as issues that telecommunications management will have to deal with in the near future.

After studying Part Six, you will be knowledgeable about the way the telecommunications department fits into the organization and how it is managed. You will also understand what the department's responsibilities are and how they are carried out. Finally, you will gain some perspective on the future uses for telecommunications technology.

CHAPTER

15

Telecommunications Department Management

OBJECTIVES

After studying the material in this chapter, you should be able to

- explain why it is desirable to manage voice and data communications together;
- describe where the telecommunications department fits within the organization;
- describe the functions of the telecommunications department;
- describe the administrative activities of the telecommunications department;
- describe in detail the telecommunications management responsibilities;
- discuss several other issues that telecommunications management faces and with which it must be familiar.

INTRODUCTION

This chapter looks at the management of the telecommunications department, the organizational entity that is responsible for planning, designing, and operating telecommunications facilities for a business. We will look at how the department is organized and where it fits into the overall structure of the company. Several of the functions of the department were covered in detail in Chapters 13 and 14, so they are discussed only briefly here. The basic management responsibilities of staffing, planning, directing, and controlling, and how they are applied to the telecommunications department are examined in detail. Other issues important to the telecommunications manager and staff also are discussed in this chapter.

THE NEED FOR PROACTIVE TELECOMMUNICATIONS MANAGEMENT

As the strategic importance of telecommunications is realized, the telecommunications department moves into a more visible role in the company. Management's expectations are higher. Providing daily telecommunications service is as important as ever, but executive management is looking for opportunities to use telecommunications as a strategic weapon to obtain a competitive advantage. Executive management expects telecommunications management to lead the way and identify opportunities. The emphasis is on making money with telecommunications rather than on saving money. It's a subtle difference but one that has huge implications for the telecommunications department.

The telecommunications manager has to make a choice between being *proactive* and taking the lead in his or her organization in finding ways to use telecommunications to be successful, or being *reactive* and waiting until others complain or make requests for services. Clearly the former choice is much more desirable! A responsible telecommunications manager must be proactive in seeking out opportunities to apply available technologies for his or her organization's benefit. An added bonus is that such proactivity is likely to enhance the telecommunications manager's reputation and status. Telecommunications is a recognized profession, and the people who work in the field must demonstrate their professional expertise.

Companies like Cisco Systems, Inc., Federal Express, American Express, the airline companies, and banks simply could not do business the way they do it today if it weren't for the telecommunications capabilities they use. Keeping track of the millions of packages delivered overnight, allowing customers to do their own banking transactions using automatic teller machines, letting customers place their own orders for communications equipment—none of this would be possible without telecommunications. They are all examples of using telecommunications technology as a strategic weapon to obtain or maintain competitive advantage or competitive parity.

strategic asset

There are other reasons why companies have shown an increasing interest in telecommunications management. First, in general, there is a broader recognition that communications can be a strategic asset to a business. In certain industries, such as those that do heavy marketing by telephone or the transportation industry, this recognition came early. Companies in those industries have always been on the leading edge of communications development and usage. Now companies in other industries recognize that a good communications system, often coupled with a progressive web site on the Internet, adds value to the products and services they offer. Customers are being given access to their vendors' databases, and documentation is being exchanged electronically—all with the aim of providing better service and making it easier to do business. From the seller's prospective, the idea is to give customers more reasons to do business with them rather than with the competition. A proactive telecommunications management team will be on the leading edge of these developments within its company.

more oppportunities and choices

A second factor is that deregulation of the telecommunications industry has given businesses more choices and opportunities for communications services, but deregulation has brought with it the need for more analysis, judgment, and decision making. Before deregulation, companies had only one choice for their telephone and data communications equipment, circuits, and services—the telephone company. Since deregulation, there are many competitive communications products.

Third, technology is making the integration of voice and data communications feasible. In the past, there was little overlap of the two and few possibilities for sharing services or facilities, such as circuits. Therefore, there was little incentive to manage the voice and data together. Today, telephones are being used as computer terminals, and voice is being digitized so that electronically it looks like data. Therefore, for many organizations, it is more feasible and desirable than ever before to group all communications activity together under one management.

voice and data integration

Fourth, the increases in the number of online, realtime computer systems and the number of computer terminals using those systems are causing many companies to take a fresh look at what is being managed. In the past, most companies viewed their computer operations department as having the responsibility to run a computer with terminals attached via communications lines, as shown in Figure 15–1. With the rapid increase in the number of terminals and personal computers, many of

changing focus

Figure 15–1
Computer with a network of terminals attached.

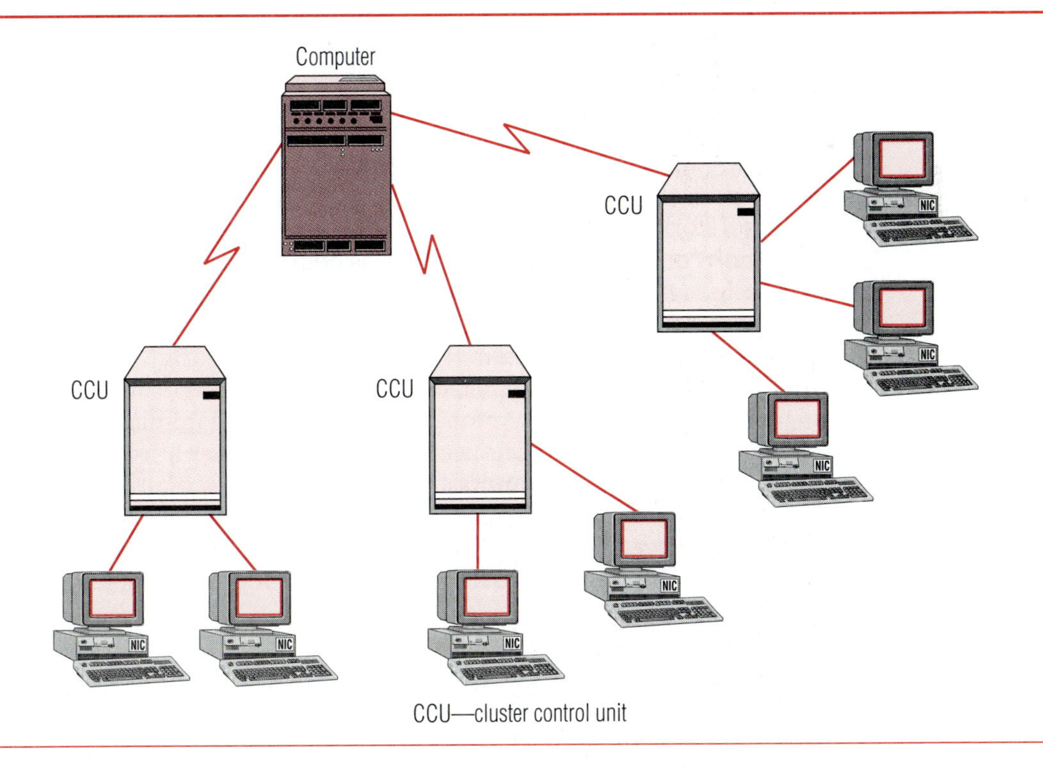

Figure 15–2
Communications network with terminals and computers attached.

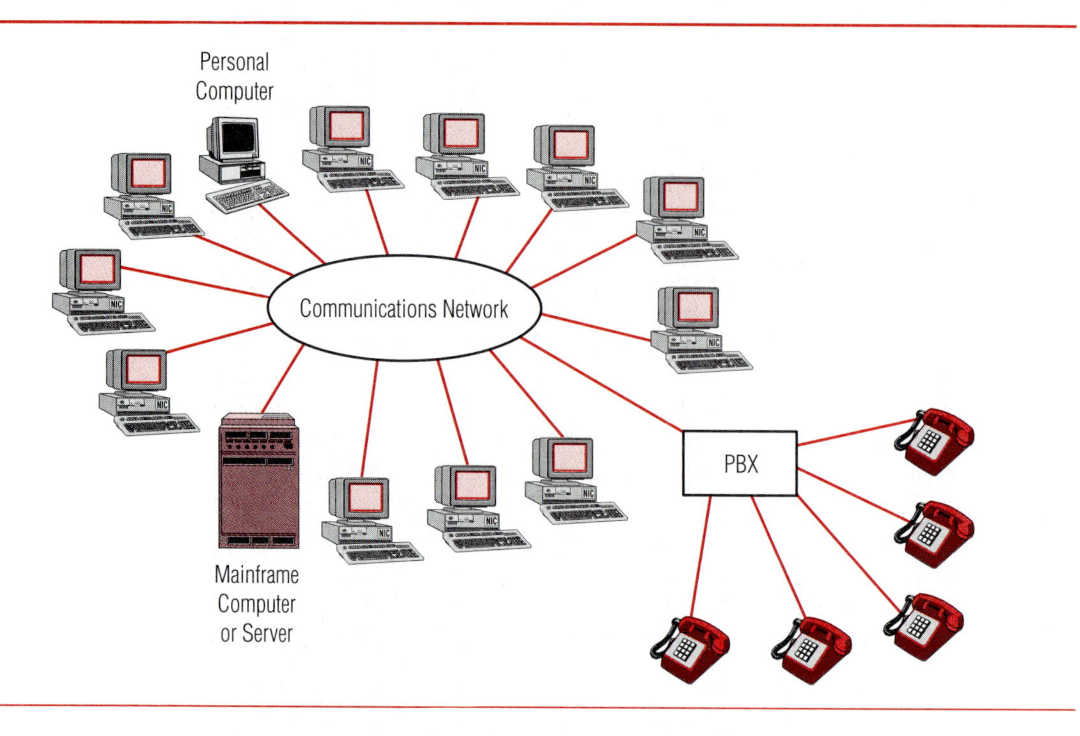

which are interconnected, companies are beginning to find that the operations department needs to focus primarily on running a communications network as seen in Figure 15–2. This view is further supported when one thinks of the modern digital PBX as being a specialized computer accessed by terminals called telephones. This change in emphasis from computer-centered management to communications network management is subtle but significant. It is adding new importance to the responsibilities of the communications managers within the firm.

higher cost

Finally, the amount of money being spent on communications is becoming a significant and noticeable part of the overall cost structure of many companies. Communications expenses were previously small enough that they could be, and often were, ignored. As more money is spent, however, the cost becomes more visible to senior management, who want to be reassured that the money is being well spent and that the growth is under control.

For many telecommunications managers, being proactive in external affairs, such as getting involved in industry groups or trade organizations, is also desirable. In some cases, companies with a large stake in telecommunications are getting involved at the legislative level, attempting to ensure that any telecommunications legislation is compatible with the company's aims and objectives.

WHERE THE TELECOMMUNICATIONS ORGANIZATION FITS WITHIN THE COMPANY

voice

Historically, the voice and data communications facilities of a company have been managed in two different parts of the organization. Voice communications management usually has been handled by the administration function or, in some cases, the facilities management department. The same person who handled the telephone system was often also responsible for other office services, such as copy machines, printing, facilities design, office furniture, and company cars. With so many responsibilities, the telephone system got scant attention unless there was a problem. Traditionally, there has been a great reliance on the salesperson from the telephone company to understand the business well enough to specify the right solution for a company's voice communications problems. It is fair to say that in the predivestiture days when the telephone company was a monopoly, voice communications in most companies was not really managed, only administered.

data

Data communications has grown up within the information systems department. It is a much newer discipline because computer-controlled data communications capabilities have only existed for about 35 years. As contrasted with the relatively slow, steady growth of telephones, data communications has experienced explosive growth as the use of terminals has blossomed. The management of data communications is generally regarded as being more complex than voice communications management. The technical capability of data communications people has often been higher than that of the voice communications people. Data communications management people have always operated in a less regulated environment than their voice management counterparts because they have been able to acquire their terminals from unregulated companies. They have grown accustomed to working with both unregulated and regulated suppliers.

combining voice and data functions

The voice communications management and data communications management in many companies have spent the last 35 years operating independently of one another. They have communicated with each other infrequently if at all, installed separate networks of lines, and, in some cases, offered competing capabilities or services to other departments in the company. Clearly, there is synergism to be achieved by combining the management of the voice and data communications facilities. The primary reasons cited by companies that have combined the two activities are efficiency, productivity, and cost advantage. With today's technology, communications circuits can be shared between voice and data, computerized PBXs require physical facilities like those that already exist in the mainframe computer room, redundant wiring in offices can be avoided, and the combining of staffs, equipment, and budget gives a communications manager more flexibility and opportunities for making trade-offs.

The reason that voice and data management have not been brought together in many companies is largely political. There is redundancy in the staffs of the two departments from the manager on down to the technician, and neither group wants to relinquish its responsibilities. However, as in any merger, there almost always is some pain and personnel dislocation when the two departments are brought together. The decision to merge the voice and data communications groups is a tough one, but it must be faced if the company is going to build an efficient, effective communications department and network for the future.

Another issue is where the combined communications organization will report. There are four basic choices:

1. to administration;
2. to information systems;
3. to some other function;
4. as a separate function reporting directly to a member of executive management.

Although there is no right answer, many companies have merged the voice and data communications groups and have the combined organization report within the information systems department. There are several reasons:

- The information systems department has a high profile and a history of coping with technological issues and problems.
- It is easier for data communications people to learn voice communications than for voice communications people to learn about data communications.
- The physical facilities needed for a PBX already exist in the computer room.
- Communications is viewed as being too technical to be set up as a separate function except in companies whose entire product line or service is communications-based.

Figure 15–3 shows how a company might have been organized when data communications and voice communications were separate entities managed in separate parts of the organization. Figure 15–4 depicts how the organization would look after voice and data communications were merged into one unit reporting to the Information Systems department.

In a vast majority of companies, the old data processing department has historically reported within the finance function, typically to the vice

Figure 15–3
Traditional placement of communications and word processing departments within the organization.

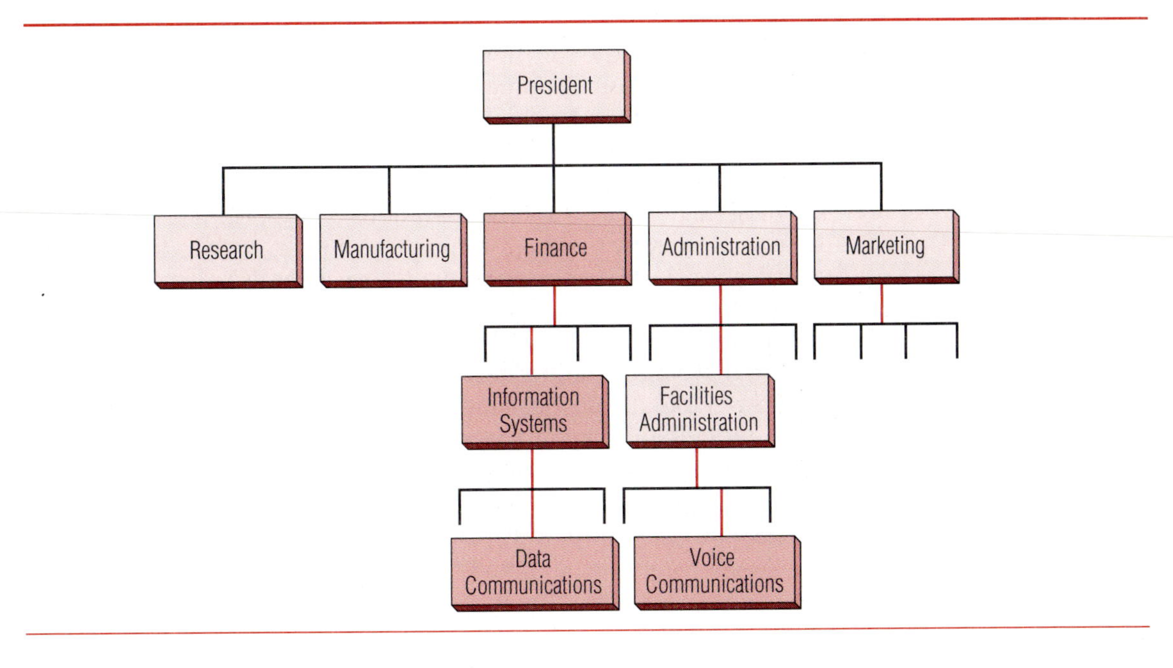

Figure 15–4
Organization after merger of data and voice communications.

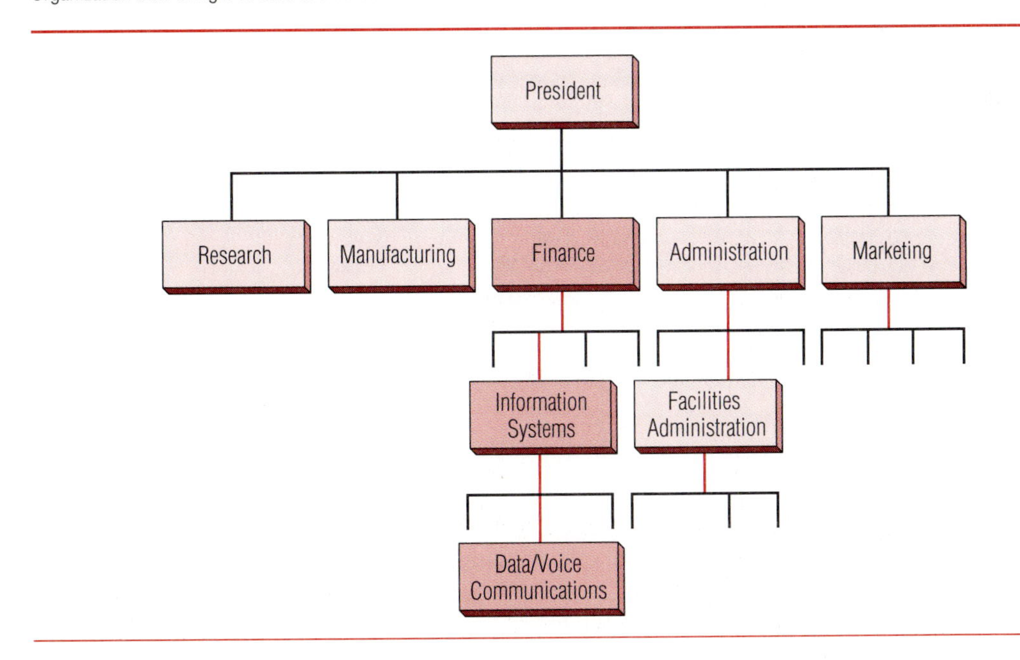

chief information officer

president of finance or the controller. The new Information Systems (IS) department may continue to report to finance as shown in Figure 15–4, or it could be set up as a separate function itself as shown in Figure 15–5. Setting IS up as a separate function usually raises the IS manager to a vice presidential level. In some companies, he or she is given the title of *chief information officer (CIO).* The chief information officer oversees all of the company's information handling technology, including data processing and telecommunications. This person concentrates on long-term strategy and planning, leaving day-to-day operations to subordinates. He or she usually reports directly to a high-ranking executive, such as the chief executive officer or the chairperson of the board. The CIO often is assigned the complicated job of figuring out how to integrate a confusing array of often incompatible computer and communications equipment into a single integrated system. Another responsibility is explaining to managers how they can make the best use of technology while making the technical people understand what management wants to do.

Today, the IS type of organization with a CIO exists in most companies in the information-intense industries of finance, banking, insurance, and transportation. Companies, such as Travelers Insurance, Bank of

Figure 15–5
The IS department as a separate function with the CIO at the vice presidential level.

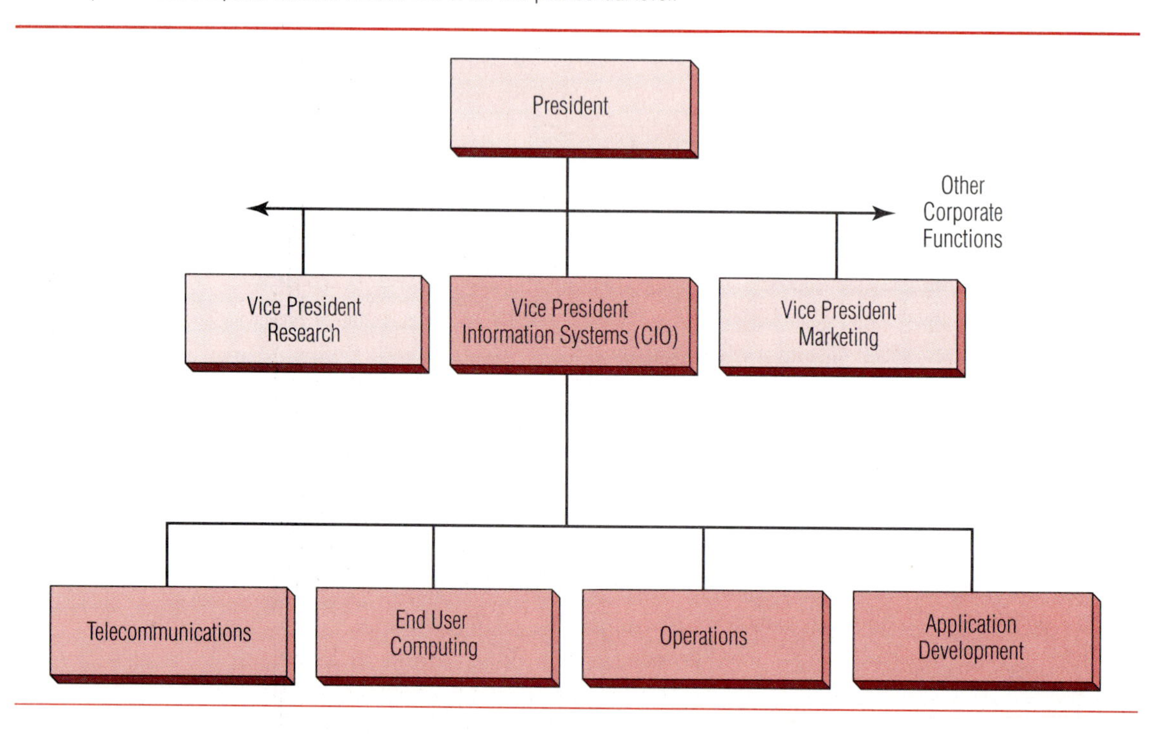

Figure 15–6
Alternative organization when telecommunications is of strategic importance to the company's success.

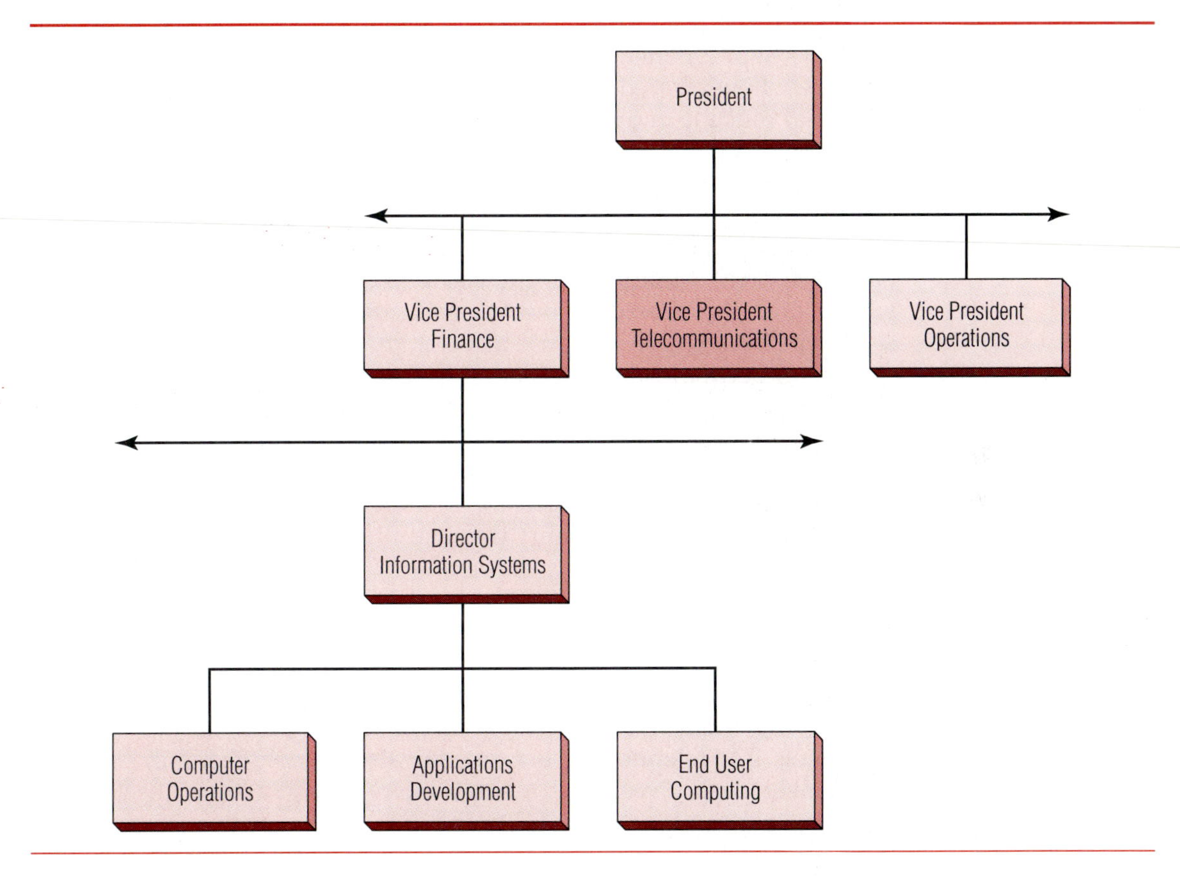

America, and United Airlines, have senior information officers with responsibilities like those just described.

Figure 15–6 shows an alternative organization that exists in a few companies where communications is of strategic importance to the company's success. Here, the telecommunications activity has been set up as a separate function with the chief communications officer at the vice presidential level, whereas information systems reports at a lower level to finance or another function.

THE FUNCTIONS OF THE TELECOMMUNICATIONS DEPARTMENT

Regardless of where the telecommunications organization reports, it has a certain set of responsibilities and activities to perform. Parts of the telecommunications organization are analogous to the major functions of the company, such as marketing, engineering, and production. The general principles of management that are taught in business management classes can and need to be applied to the management of the telecommunications department as well as to other departments in the company.

The following sections describe the basic business management functions that must be performed by the telecommunications department.

Design and Implementation of New Facilities and Services

All new communications facilities, including networks (both wide area and local area), building wiring, equipment, and services must be designed and engineered to meet specifications. Facilities and services should be designed in response to a statement of requirements based on the user's needs. Installation activities range from simple repetitive installation of new telephones and computer terminals to infrequent but complex tasks, such as wiring a new building or installing a new PBX or a new version of the software. Because most of the work in the design and engineering section is project-oriented, project management skills and techniques are important. This activity was discussed in detail in Chapter 14.

Network Operations and Technical Support

The network operations group is responsible for monitoring the status of the communications network and services, performing first-level problem solving when trouble occurs, and monitoring communications service levels. The telephone operators may also be included in this group.

A help desk gives users a single point of contact for questions or problems relating to the use of telephone or data communications services. Another activity, telecommunications maintenance, is divided into two parts: routine maintenance, which includes activities such as moves and changes of telephones or terminals, and installations of new versions of software, and nonroutine maintenance, which is a response to trouble when something doesn't work. Problem determination, problem solving, and repair are all part of this activity.

The technical support staff provides a high level of problem-solving capability for the most difficult problems. They also perform software maintenance and assist with other telecommunications activities.

Network operations and technical support were discussed in detail in Chapter 13.

Administrative Support

The administrative support group provides primarily clerical support to the telecommunications department. Familiarity with communications terminology and knowledge of at least the general concepts of the network and equipment are required, although detailed knowledge is not necessary.

The primary functions of the administrative support group are

- ordering/purchasing communications products and services;
- receiving and identifying products when they are delivered;
- inventorying all communications equipment;
- checking, approving, and paying communications bills;
- charging back for communications services rendered.

Many of these activities overlap with the work of other departments in the company, such as purchasing and accounting. Agreement must be reached with those departments about how much of the activity will be performed by the telecommunications administrative support people.

Other administrative activities are unique to the telecommunications department. These include

- arranging for adds, moves, and changes of telecommunications equipment;
- preparing and publishing the communications directory;
- registering new telecommunications system users;
- training users in basic telecommunications operations;
- maintaining telecommunications procedures;
- serving as telephone operators for the company's general telephone number.

Moves, Adds, and Changes (MAC) Communications systems are not static. The most frequently occurring activity affecting them is the addition or movement of telephones or terminals. As new people are hired, office layouts change, departments are moved, or a new building is built, communications network activity is required.

The move of a large computer center or a PBX telephone system is a very complex project that requires detailed planning and, in most cases, a full-time project leader for six months to a year. The complete requirements of such a move are beyond the scope of this book, but suffice it to say that numerous articles have been written on the subject, and consultants earn a living helping companies make a large move of this type.

moves, adds, and changes

Most moves, adds, and other changes, however, are relatively routine and can be coordinated by a person who is a good planner and communicator and who has some knowledge of the communications equipment. Despite the fact that moves can be handled routinely, they do require advanced notification so that the efforts of all of the parties involved in the move can be coordinated.

Preparing and Publishing the Communications Directory *Communication directories*, the modern version of the telephone book, provide a complete index of telephone numbers, e-mail addresses, regular mail addresses, voice mailbox identifications, and perhaps user identification codes for all employees in the organization. The directories range from simple typed lists in small organizations, to multipage books rivaling those of a small community, to online directories that are a part of an e-mail system. Although no general schedule can be defined that applies to all organizations, most companies that have an online directory update it in realtime, as changes are made. Whenever telephone numbers or e-mail addresses change, or employees join or leave the company, the online directory is updated. Organizations that print a directory usually update them at least quarterly or semiannually.

Online directories have the advantage of having the data stored in an easily updated central database that everyone can access from their workstations. This avoids the cumbersome and expensive process of producing and distributing paper telephone books and updates to them.

Registering Users Another administrative support function is the registration of new users to use the communications facilities. A part of the new employee orientation process may well be to assign the employee's telephone number and calling privileges. Registering people to use the data communications system and computers is another aspect of this responsibility. This often requires that the specific data transactions and files to which the employee will have access be identified and approved. Obviously there is a security implication to this process. Each company must develop procedures for authorizing an employee to access and use company data.

There may be a distinction between registering users to use the communications network and registering users to use a particular computer or computer application. It is most convenient if these two processes are combined into one, but for some organizations there is a clear distinction between them. In that case, the communications department registers users for the network, and the department that owns the computer uses a separate registration procedure to authorize access to it.

Training Users Sometimes the administrative support group has the responsibility for training people to use the communications system. Training can be simple, informal, and handled on a one-on-one basis, or it may be more highly structured and done in a classroom, depending on the size of the company and the complexity of the communications facilities. One good opportunity to do communications education is during new employee orientation. At that time, an introduction to the capabilities and features of the telephone system can be presented, and if appropriate, some basic terminal education can be given. With the proliferation of computer terminals and personal computers at all levels of our society, most people today have some familiarity with keyboards and personal computers, and therefore only a few comments about the unique characteristics of the company's terminals or computer systems may be required.

Education and training needs to be viewed as an *ongoing process*—not just a one-time event. Continuing training to follow up on the concepts taught in initial courses is very important to ensure that employees apply the new skills they learn and put them to work on the job. In order for companies to see year-to-year progress in their employees' use of new communications and other technological capabilities, the initial learning needs to be reinforced with follow-up training. Furthermore, people tend to slip back into old, bad habits, and follow-up training can remind them of techniques they may have forgotten about or stopped using.

continuing education

Unfortunately, with the pressures of business today, we often find that insufficient time or money is allocated to do a proper job of training. Furthermore, even when time and financial budgets are established, they are often the first to be cut if the telecommunications project is behind schedule or if budgets must be reduced. Telecommunications management should do their best to ensure that training does take place. It is never safe to assume that users will intuitively understand a new system or know how to use it properly.

Maintaining Procedures The administrative support group may be responsible for writing and maintaining general procedures for the communications system operations. Procedures, such as how to log on, how to make a long distance telephone call, and how to report trouble, need to be provided to everyone who will use the communications system. It goes without saying that procedures of this type should be prepared on a word processor so that they can be easily maintained because, like the network itself, they will change with time—usually faster than expected.

Internal operating procedures for the various groups in the communications department are best written by those groups. For example, operational procedures for the help desk, such as the questions to ask users when they have trouble, how to record problems, and how to escalate problems, are best written and maintained by the help desk personnel and their supervisors. Procedures and project management techniques for implementing new communications circuits should be developed by network analysts.

Telephone Operators Telephone operators may or may not report to the communications department. Many times, they have other responsibilities, such as doing secretarial work or acting as receptionists. When they perform the duties of a receptionist, they may report to an administrative or security function. In some companies, the telephone operators may be more customer service-oriented, primarily handling calls from customers. If this is the case, they should report to the marketing or customer service department. Having the telephone operators report to the telecommunications department is not critical to its success.

In small companies, the telephone operator almost always does other work besides answering the telephone. In addition to the possible duties just listed, the operator may also send e-mail messages and keep the telephone list or book up-to-date. In fact, the job may be primarily secretarial with only a secondary responsibility for general telephone answering.

Backup of Operational Personnel

Management must give attention to providing backups for people who are involved in day-to-day operational activities, such as telephone operators, maintenance technicians, and help desk personnel. When a company is dependent on its telecommunications facilities, those facilities need to be operational at all times. Yet employees will get sick and take vacations, and some means must be provided to cover for them when they are gone. Cross-training of two employees who have similar responsibilities is a frequently used option, but employees who are trained as backups must be given a chance to use their skills from time to time to ensure that they are able to step in and perform the job when they are needed.

TELECOMMUNICATIONS MANAGEMENT RESPONSIBILITIES

Since telecommunications can be viewed as a business within a business, the basic responsibilities of general management can be applied to telecommunications management. These functions are

- planning;
- staffing;
- organizing;
- directing;
- controlling.

What is unique to telecommunications management is that only recently has the need to apply these management techniques to the telecommunications activity been widely recognized. Some very large organizations have done a good job of managing their telecommunications for years, but most companies have gotten along without paying much attention to the management skills and functions in the telecommunications organization. Technicians or clerical people with few management skills and little management training were promoted to supervisory positions and charged with the responsibility to run the department. Before deregulation and divestiture, this approach sometimes worked because many of the issues that had to be dealt with were technical. But the environment is more complicated now. Competition exists, and the telecommunications department has more responsibility for selecting among expensive alternatives, making trade-offs, and deciding to what extent the company will provide communications support itself as opposed to delegating the duties to suppliers or other outsiders. The department must have a leader with management and decision-making skills in order to select the best communications options for the company. We will look at the management activities in detail and see how they apply to the management of the telecommunications department.

Planning

Of all of the managerial functions, planning is arguably the most important. It is known as the keystone of management, and an old saying is that "businesses don't plan to fail, they fail to plan." A lack of planning can lead to disorganization, which can hurt the organization's productivity and ability to stay competitive.

Planning is required in any organization. The exact format and frequency of the planning depend on the company and its culture. Some organizations have very formal, detailed, rigorously scheduled planning processes. Other organizations operate on a less formal basis, putting plans together only when required.

Before looking at specific types of plans, it is important to point out that the telecommunications plan cannot exist in a vacuum. It must relate to the strategic directions and plans of the company. In most cases, telecommunications projects are not undertaken for their own sake but are in support of some other activity within the company. For example, an expansion of the telephone system may be required to support a new telemarketing effort, or the data communications lines may need to be extended to support additional terminals for a new computer application, or the backbone connection to the company's Internet service provider (ISP) may need additional bandwidth to support a new e-business initiative on the Internet. All telecommunications plans should relate to the needs of the company and should show the benefit that will accrue to the business. Management needs to understand that telecommunications capabilities can help break down constraints of geography and time on the product or service that the company provides.

Staffing

A management function that is particularly challenging for communications managers is staffing the department to effectively accomplish the work to be done. Staffing is especially challenging because for a number of years there has been a shortage of trained, experienced communications people. Only recently have colleges and universities introduced programs focused on educating people for careers in telecommunications, and networking skills are still one of the most sought-after characteristics of communications and IT people. Historically, the three primary sources of communications people have been the telephone companies, the military, and internal people who learn telecommunications through on-the-job training.

sources of staff

Another complicating factor is the mix of skills required in the communications department. Building on the "company within a company" concept, it is apparent that planning, conceptualizing, operations, technical, and administrative skills are required by various members of the organization. When the department is small and has only a few people, it is difficult to obtain the right mix of skills. Larger departments have a better chance of obtaining people with the required skills or at least obtaining people with the aptitude to learn them. It may not be necessary for a telecommunications department to have all of the skills it needs. Outside consultants may be an effective supplement to the people within the telecommunications department, particularly when very specialized expertise is required. Outsourcing certain work is another option.

The Telecommunications Manager The telecommunications manager must be a conceptualizer and visionary of how communications capabilities can be applied to solve business problems. He or she must be articulate and persuasive, able to understand company problems and opportunities, and able to envision how the application of communications technology might address these problems and opportunities. Ideally, the manager will be included in high-level planning discussions with senior management to keep informed about the factors that are of strategic importance to the company and to ensure that the mission of the telecommunications department is lined up in support of those strategies. This is especially true today, as companies are developing e-business strategies and plans that have a heavy reliance on telecommunications networks. The manager needs to have an ability to grasp technical subject matter, although he or she need not personally be a technician or former technician.

qualities

The telecommunications manager must be able to plan and implement projects at various scopes and levels. He or she must be a decision maker who can deal as effectively with today's operational problems as with a 3-year plan. A college degree in telecommunications, management, or engineering usually is required for such a position.

credentials

Designers and Implementers Telecommunications network designers and implementers must have a good understanding of communications systems and product offerings. This knowledge usually is obtained through a combination of education and experience. Until recently there were no college courses to learn how to do this work.

qualities

Good designers are creative and innovative and have strong analytical skills. Engineering training is very desirable. Since most design and implementation work is done in project form, a knowledge of project management techniques is required. Network designers must also have good verbal and written communication skills. They work with eventual users of the system as well as with the vendors who provide the equipment or services. On large projects, it is feasible to take a strong project leader from another part of the company to manage the communications project if experienced people who have the specific communications systems knowledge are on the project team.

credentials

Generally, most large companies prefer that their network development people have a 4-year college degree, although this is not mandatory. There is a feeling, however, that people with a 4-year degree are more flexible in their thinking and more likely to advance and move into other responsible positions within the company. Experience would, in general, support this notion.

Network Operations Staff Network operations is the manufacturing facility of the telecommunications department. Like many manufacturing people, network operations people often have more contact with the customers or users of the communications network than might first be expected. Therefore, they must be service-oriented and strongly motivated to keep the network operational and performing properly at all times. As in the previously described jobs, good verbal communications skills are required in order to deal effectively with the network users, many of whom experience problems with the network.

qualities

In addition to having good people skills, network operations people must thoroughly understand the network hardware and software that they are operating. There is a good deal of interchangeability between computer operations and network operations personnel. In most organizations, the ability to trade people back and forth is desirable because it provides career path opportunities for people in both groups.

credentials

Network operations people normally have at least a high school diploma. Some companies are requiring a 2-year or 4-year degree for these jobs.

qualities

Technical Support Staff Technical support personnel must understand the complexities of both software and hardware and the complex interactions between them. They are usually described as self-starting, analytical problem solvers.

credentials

A 4-year college degree normally is required, and it is almost always supplemented by advanced technical education conducted by the vendors of the products being used. The telecommunications technical support people and the data processing technical support people often have similar backgrounds and career paths. In many organizations, they are combined into a single technical support group.

qualities

Administrative Support Staff The administrative support required in the telecommunications department is primarily clerical in nature. Experience with accounting and purchasing activities is desirable. A professional accountant may be required to ensure that proper accounting procedures are followed. The administrative people need to be thorough and oriented toward handling a mass of detail.

credentials

The administrative support people often attend some specialized telecommunications courses to have a better feel for the products and services they are handling.

Staff Selection When considering telecommunications department staffing, several questions need to be asked:

- Should experienced communications people be hired from the outside, or should current employees be trained in telecommunications skills and technologies?
- What career paths are available for communications employees? Will they be able to transfer to other functions within the company?
- Should some, or all, of the telecommunications work be contracted to outside firms? For what expertise should consultants be considered?

Management must recognize that because the telecommunications industry is growing rapidly, the market for good, experienced communications people is tight and salaries are high. But while communications professionals consider salary important, a wide variety of other factors contribute to whether they sign on with an organization and how long they stick around. Some of the other important attributes of a position are:

- a clearly defined career path;
- ongoing professional challenges to keep skills current;
- recognition of efforts;
- a sense of community within the organization.

Management can improve its chances of recruiting and keeping a strong telecommunications staff by paying attention to these factors and developing an environment in which communications professionals enjoy working.

When trade-offs must be made, it is normally better to select management people based on their management skills and knowledge of the company. This may well bias the decision in favor of a person who is already an employee. For technical people, the trade-off should be made in favor of specific telecommunications knowledge, experience, and education. Knowledge of the company is less important, and therefore going outside to hire a skilled individual is feasible.

Use of Consultants Every department of every company occasionally needs advice or assistance in areas where it has little or no expertise. In telecommunications, this is particularly true because of the rapid technological changes. Telecommunications managers may find that they need help evaluating a new technology, such as fiber optics, or the impact of a new regulation or tariff. Sometimes a vendor can help with this evaluation, but since the vendor has a product to sell, the information he or she provides may be biased. AT&T has good network evaluation and design tools that they make available to their customers, but obviously AT&T won't recommend a circuit provided by MCI or another carrier.

There are many telecommunications consultants in the field today. Some are independent, and some are associated with large consulting or public accounting firms. Before a consultant is engaged, it is important to determine the consultant's expertise in the particular field of interest. It is important to talk to other clients who have used the consultant to get their opinions about the consulting services provided. It is also important to find out exactly who will do the consulting work. The expertise and experience of the individual consultant is the primary determinant of the quality of the result and is ultimately more important than whether or not he or she happens to work for a "name" firm.

check references

It is important to define the exact scope of the work to be performed in writing and to ask the consultant for a written proposal and cost estimate as well as a list of all of the items or documents that will be delivered at the end of the consulting engagement. The proposal can be used as an indicator of how well the consultant understands the requirements of the job. Some consultants may suggest modifications to their proposal, either to expand the scope of the work to be done or to reduce the scope to hold costs down. In any case, it is important to reach an understanding of the scope before any work begins.

define the scope of the work

Organizing

Organizing is the grouping of people to accomplish the mission of the department. Obviously, telecommunications departments in large and in small companies are organized differently. In large companies, there are many specialists, whereas smaller organizations have fewer people, and they tend to be generalists. The structure of the department is also dependent on the sophistication and level of maturity of telecommunications in the company. As the company becomes more dependent on its telecommunications capability, the telecommunications department usually becomes larger and more structured in its operation. Different styles of organization are appropriate at different stages in the growth of telecommunications within the company.

One decision that management must make when organizing the telecommunications department is how much overlap there will be with other functions in the company. For example, the telecommunications department needs a purchasing procedure, and management must decide whether to set up its own purchasing group or use the company's general purchasing department. Another decision is to determine which telecommunications activities are going to be performed by employees and

Figure 15–7
Typical telecommunications department organization.

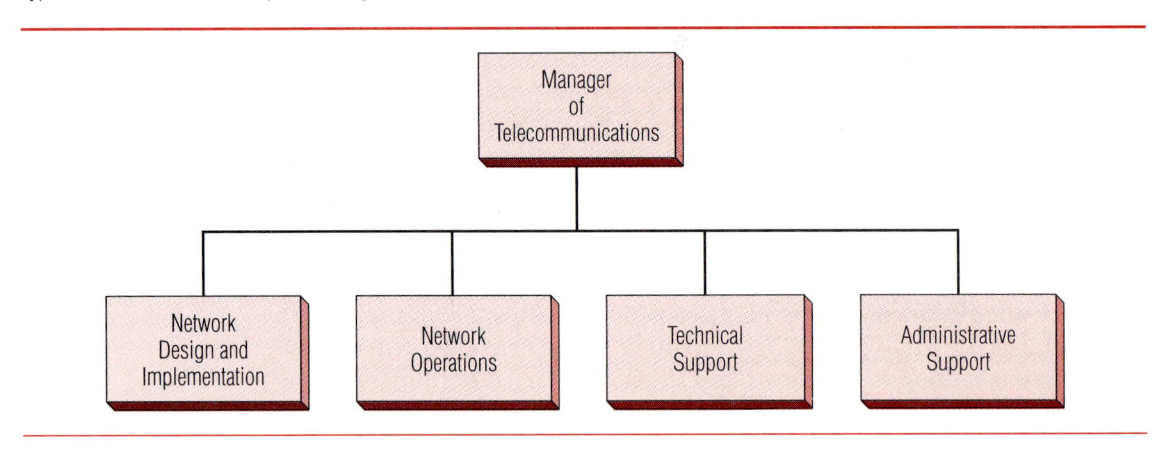

which will be contracted to outside services or performed by vendors. Ongoing activities, such as communications operations, are usually performed by company employees, whereas occasional work, such as laying wires and cables, is contracted to outside specialists.

As in any department, there are many ways to organize the people to carry out their responsibilities. The groupings of activities discussed in this book are typical of what is occurring in industry today, but they are not the only possibilities. Although each telecommunications department has certain basic activities to perform, there is always flexibility to tailor the organization to the talents of the individuals it comprises.

An organization chart of a typical telecommunications department is shown in Figure 15–7. This chart assumes that the voice- and data-related activities are handled by the same people.

Planning

In any company, planning is required. The exact format and the frequency of the planning depend on the company and its culture. Some organizations have very formal, detailed, rigorously scheduled planning processes. Other organizations operate on a less formal basis, putting plans together only when required.

Before looking at specific types of plans, it is important to point out that the telecommunications plan cannot exist in a vacuum. It must relate to the strategic directions and plans of the company. In most cases, telecommunications projects are not undertaken for their own sake but are in support of some other activity within the company. For example, an expansion of the telephone system may be required to support a new telemarketing effort, or the data communications lines may need to be extended to support additional terminals for a new computer application. All telecommunications plans should relate to the needs of the company and should show the benefit that will accrue to the business. Management needs to understand that telecommunications capabilities can help break down constraints of geography and time on the product or service that the company provides.

Long-Term Plan The long-term telecommunications plan should look 3 to 5 years into the future and give a vision of what telecommunications facilities and services will be required by the company and implemented in that time frame. Obviously, as a vision, it will never be 100 percent accurate, but it is important to put this type of plan together to give a sense of direction. The company may have a strong feeling, for example, that within 3 years it will begin operating in Europe. The 5-year plan may show the need to explore international communications options beginning in about 2 years in preparation for the anticipated expansion to Europe.

The telecommunications manager often puts this plan together. He or she has the most contact with senior management and others in the organization and may be more forward thinking and conceptual than others in the department. The long-term plan should be written, brief (10 to 15 pages at most), and as specific and pictorial as possible. It should be updated as appropriate (perhaps every 1 or 2 years) to reflect changing company strategies, new technological capabilities, and regulatory implications.

It is important to point out that in some industries, preparing a long-term plan may be an exercise in futility. Think of the many "dotcom" companies that exist today providing services on the Internet. Many of them didn't even exist 1 or 2 years ago, and for them to try to project what their organization will be like, and hence what their communications requirements will be in 5 or 10 years would be a waste of time. Some of these companies will disappear or be purchased and a few will be wildly successful, but in either case, a long-term plan would probably not have projected what needed to be done from a communications viewpoint.

Medium-Term Plan There should be a medium-term plan covering a 6- to 18-month time horizon and identifying telecommunications projects and their priorities for that time frame. This plan should identify the financial and people resources needed to complete the projects that make up the plan and the benefits that the implementation of the projects will bring to the company. The medium-term plan should be tied to the budgeting process to ensure that money will be available to implement the projects listed in the plan. It should show the starting and ending dates for each project but not the detailed tasks required for project implementation.

Like the long-term plan, the medium-term plan should be in writing. Each project listed in the plan should have information describing its scope, the reasons it is being done, its cost, benefits, starting and ending dates, and assigned people and equipment resources. One useful technique for medium-term planning is the *Gantt chart.* A Gantt chart, as

Gantt chart

Figure 15–8
A high-level Gantt chart of a medium-term telecommunications plan showing project start and end dates. A more detailed Gantt chart would show a finer breakdown of the tasks.

	1998												1999	
	J	F	M	A	M	J	J	A	S	O	N	D	J	F
Install N.Y. data circuit	1–15			4–30										
Install key system in Seattle					5–1		7–10							
Upgrade data circuit to 56 kbps			3–1						9–30					
Install new terminals in plant					5–15						11–30			
Install PBX in Atlanta								8–1					1–31	

shown in Figure 15–8, lists projects or activities on the left side and dates or other time periods across the top. A bar is drawn for each project showing when it starts and when it ends. The positions of the bars show how the projects or tasks overlap, and it is easy to see how they interrelate in time.

Project Plan There is also a need for a specific task-oriented plan for each project showing the tasks, worker days, cost, and people resources required to complete it. This plan should set targeted completion dates and identify responsibilities for each task. A suggested format is shown in Figure 15–9. This task list may be supplemented by a detailed Gantt chart for the project. This detailed chart would list the project tasks in the left-hand column and show how the tasks interrelate. The thought process that is required to put the task list and Gantt chart together is very worthwhile in itself because in the process of putting the chart together, problems, conflicts, and previously unrecognized activities are uncovered. Detailed project plans should be updated as often as required. Depending on the size of the project, this may mean that monthly, weekly, or even daily updates are required. In any case, it is important to communicate changes in the plan to all of the individuals involved in its implementation, as well as to users and management.

Regular reviews with the affected users and key management personnel are desirable. Depending on the size of the project and the level of management, reviews may be held weekly or monthly. Like the discussions within the project team, these reviews help ensure that the project team is designing the telecommunications system correctly. The reviews also provide an opportunity to check progress and discuss any difficulties or unexpected circumstances that have arisen. The level of detail discussed in these reviews depends on the level of management that is participating. User line management usually is interested in hearing more of

Figure 15–9
A telecommunications project plan.

Install New York Multipoint Data Circuit			
Task	**Responsibility**	**Target Date**	**Comment**
Get exact addresses of the four locations	Jones	1–15	
Order the circuit	Jones	1–20	90-day lead time
Order the modems	Smith	1–20	
Contact each location about delivery date	Smith	3–1	Marketing has name of contact person
Modems delivered to each location	Vendor	4–1	
Circuit installation	Vendor	4–20	
Test circuit and correct problems	Jones/Smith	4–25	With vendor help
Circuit operational	Jones	4–25	

the details than senior management, whose primary concerns are schedule and cost-related.

Directing

Management's responsibility for directing is to ensure that the mission and plans of the telecommunications organization are executed on a timely and accurate basis. This is best achieved by first specifying the overall mission for the department and then ensuring that it is understood by the members of the group.

principles

A similar approach is to develop a set of "telecommunications principles" that state how the organization intends to do telecommunications. Although there are always exceptions, the principles serve as a benchmark against which decisions can be measured. An example of one company's telecommunications principles is shown in Figure 15–10.

Each subgroup in the telecommunications department must know its responsibility for helping to achieve the overall objectives. Management must also ensure that each individual is motivated and can see how his or her individual efforts help the department to meet its objectives. To do this, specific objectives, goals, and standards of performance for each individual in the department are required. Objectives are frequently written annually, although there is significant benefit in reviewing and updating them more often.

Figure 15–10
Telecommunications principles.

Directional Statement: The telecommunications network will achieve a high level of connectivity, capacity, and compatibility among systems for the secure electronic transmission of information of any type (data, text, image, voice, video, fax, voice mail, etc.) to serve customers, employees, and others with whom we work.

Principles:

- ☐ As a target, our telecommunications capability will provide single workstation or terminal access to our computer and network functions from any company location.
- ☐ As a target, the company telecommunications network will make physical data processing locations transparent to the user.
- ☐ Our company will use a single wide area, multiprotocol, cost-effective, manageable telecommunications network that is capable of withstanding component failures without interrupting user service.
- ☐ The telecommunications network will be independent of a single vendor's technology set. As a target it will be compliant with the Open Systems Interconnection (OSI) telecommunications model as endorsed by the International Organization for Standardization.
- ☐ Planning for new buildings and other facilities, and major renovations will include provisions to satisfy telecommunications requirements.

Other common techniques used in directing the department are requiring each staff member to write a monthly report and holding project review meetings regularly. The purpose of both of these techniques is to improve communications within the department as well as to allow individuals to gain recognition for the results they are achieving.

Words that often are used to describe a successful telecommunications organization are *proactive, results-oriented,* and *service-oriented.* If these words are used by others in the company, it is a good sign that the department is being successful.

Controlling

Unfortunately, the word *controlling* often has a negative connotation, but it simply means ensuring that performance is taking place according to the plans. Control activities are very important to any management process. Even the best-made plans are subject to change. Part of managing is ensuring that changes in plans are made consciously and with a full understanding of the implications. In the telecommunications department, two major controls are financial controls and the controls applied to ensure the quality of the service being rendered.

Financial Controls Financial controls take several forms. First, a financial plan or budget for telecommunications expenses is put together. Then performance against the budget is measured. In some companies, telecommunications expenses are charged back to the users when services are rendered. If this is the case, chargeback rates must be established as a part of the budgeting process, based on the expenses that are expected to be incurred. By setting the chargeback rates higher than the expected expenses, the telecommunications department becomes a *profit center* generating more income than it spends. If the charge-out rates are lower than expenses or if no charge-outs are used, the telecommunications department is a *cost center.*

profit center

cost center

Having appropriate financial controls and overall financial management practices is important because the level of telecommunications expenditures continues to increase and become a more important part of a company's overall expenses.

Expense Budgeting Expense budgets usually are put together for the fiscal year used by the company; this may or may not correspond to the calendar year. First, the expenses for all existing people and communications facilities must be identified. Then all expected changes, such as the addition of new people, circuits, or terminals, must be identified and their cost added to the budget. Typically, the budget is broken into expense categories, as shown in Figure 15–11, and described in the following list:

- Salaries—the total amount of money, including overtime, to be paid to employees in the department during the year.
- Employee benefits—the amount of money to be paid for insurance, retirement, dental, and other benefit programs. It is often calculated as a percentage of the salaries and wages amount.
- Rental/lease—the amount of money to be paid to vendors periodically for the use of equipment that has not been purchased.
- Maintenance—the amount of money to be paid for the repair of equipment or for service contracts to cover the repair.
- Depreciation—money budgeted to cover the depreciation expense for equipment that has been purchased.
- Supplies—money for such items as ink cartridges and paper for printers, subscriptions to periodicals, general or special office supplies, and so forth.
- Education—money budgeted for vendor classes, general seminars, and other educational activities.
- Travel—money for the expenses of taking a trip, such as airline tickets, lodging, and meals.
- Utilities—money to pay the heating, air conditioning, and telephone bills.
- Building/corporate overhead (sometimes called "burden")—many companies assess a flat fee to each department for use of the office space and other corporate facilities. This is sometimes expressed as an overhead rate, which may be based on the number of square feet occupied, the number of people in the department, or the size of the rest of the budget.

The telecommunications manager may budget for literally hundreds of items. Sometimes it is desirable to break down further individual budget categories. Figure 15–12 shows the items that make up the rent/lease

Figure 15–11
The telecommunications department expense budget.

Telecommunications Budget	Jan	Feb	Mar	Apr	May	Jun	Jul	Aug	Sep	Oct	Nov	Dec	Total
Salaries	9,600	9,600	9,600	9,700	9,700	9,800	9,800	9,800	9,800	9,800	9,900	9,900	117,000
Benefits	2,688	2,688	2,688	2,716	2,716	2,744	2,744	2,744	2,744	2,744	2,772	2,772	32,760
Rental/Lease	12,150	12,300	12,300	12,350	12,500	12,500	12,700	12,900	12,900	12,900	12,900	13,500	151,900
Maintenance	5,250	5,250	5,400	5,400	5,400	5,400	5,700	5,700	5,700	5,800	5,800	5,900	66,700
Depreciation	2,200	2,250	2,300	2,350	2,400	2,400	2,600	2,600	2,650	2,700	2,700	2,700	29,850
Supplies	150	150	150	150	150	150	150	150	150	150	150	150	1,800
Education	400	400	400	400	400	400	400	400	400	400	400	400	4,800
Travel	500	500	900	600	600	400	700	200	1,500	1,000	400	0	7,300
Utilities	350	350	350	350	350	350	425	425	425	425	425	425	4,650
Overhead (Burden)	700	700	700	700	700	700	700	700	700	700	700	700	8,400
Monthly Total	**33,988**	**34,188**	**34,788**	**34,716**	**34,916**	**34,844**	**35,919**	**35,619**	**36,969**	**36,619**	**36,147**	**36,447**	**425,160**

Figure 15–12
Detailed budget for the rent/lease expenses category. Notice that some items are planned to be installed during the year; their budgeted amounts are 0 in the first few months.

Rent/Lease Budget	Jan	Feb	Mar	Apr	May	Jun	Jul	Aug	Sep	Oct	Nov	Dec	Total
Communic. Controller	875	875	875	875	875	875	875	875	875	875	875	875	10,500
Server 1	295	295	295	295	295	295	295	295	295	295	295	295	3,540
Server 2	315	315	315	315	315	315	315	315	315	315	315	315	3,780
Server 3	0	0	0	280	280	280	280	280	280	280	280	280	2,520
West Circuit	1,850	1,850	1,850	1,850	1,850	1,850	1,850	1,850	1,850	1,850	1,850	1,850	22,200
Plant Circuit A	470	470	470	470	470	470	470	470	470	470	470	470	5,640
Modem-West 1	165	165	165	165	165	165	165	165	165	165	165	165	1,980
Modem-West 2	165	165	165	165	165	165	165	165	165	165	165	165	1,980
Modem-Plant A1	225	225	225	225	225	225	225	225	225	225	225	225	2,700
Modem-Plant A2	225	225	225	225	225	225	225	225	225	225	225	225	2,700
Terminal 1	47	47	47	47	47	47	47	47	47	47	47	47	564
Terminal 2	68	68	68	68	68	68	68	68	68	68	68	68	816
Terminal 3	110	110	110	110	110	110	110	110	110	110	110	110	1,320
Terminal 4	0	0	68	68	68	68	68	68	68	68	68	68	680
Terminal 5	0	0	0	0	0	95	95	95	95	95	95	95	665
Monthly Total	**4,810**	**4,810**	**4,878**	**5,158**	**5,158**	**5,253**	**5,253**	**5,253**	**5,253**	**5,253**	**5,253**	**5,253**	**61,585**

item on a budget. Each of the items and its cost are listed. Items that will be added during the year show a zero cost until the month in which they are added. Totals are shown for each month and for each item.

A spreadsheet program is an invaluable tool to use during the budgeting process. As additions or other changes are made to the budget, items can be added or numbers corrected quickly with the spreadsheet program recalculating the monthly and yearly totals instantly. Breaking the budget apart by months is desirable to account for items that are added or deleted in the middle of the year, and it also serves as a monthly milestone for checking actual expenses against the budget as the year progresses.

Capital Budgeting In some companies, it also is necessary to prepare a capital budget. The capital budget is a plan for money to be spent on items that are purchased and become assets of the firm. Whereas conservative accounting practice dictates that as many items as possible are charged to current expenses, the Internal Revenue Service guidelines state that certain items with a long life must be capitalized (become assets) and their expense spread over the asset's life. This expense is recovered in the form of depreciation. When purchased (as opposed to rented or leased), most expensive telecommunications equipment must be capitalized and depreciated.

A capital budget is a list of the items to be purchased, their cost, expected useful life, and anticipated month of acquisition. The company may have special capital expenditure authorization procedures beyond the budgeting process so that specific approval for each major capital purchase is obtained even though the budget had been agreed to previously.

Cost Control During the month, invoices for installed equipment or new purchases come in from vendors. These invoices must be verified to ensure that they are correct before they are submitted to the accounting department for payment. As the bills are paid, they are charged against the telecommunications department budget.

verifying invoices

Periodically, it is necessary to check the total expenditures and compare them to the budget to ensure that costs are under control. Normally, reports are provided each month by the company's accounting department in a format similar to that in Figure 15–13. The left side of the report shows the actual expenditures for the current month, the budget for the month, and the variance from budget. The right side shows the same information on a year-to-date basis. With this type of report and the appropriate detailed information to back it up, the telecommunications manager can keep close track of the money his or her department is spending and, if necessary, take appropriate steps to keep the expenses in line with the budget.

checking expenditures

Chargeback It is very common for some or all of the telecommunications expenses to be directly charged back to the users with some sort of a recharge system. It is desirable to have all of the invoices for telecommunications equipment and services pass through the communications department, where they can be checked and approved by knowledgeable people. However, there is also benefit in passing the costs of the commu-

Figure 15–13
A monthly telecommunications department expense report.

Telecommunications Expenses
July 2001

Current Month				Year to Date		
Detail Actual Expenses	Budget	Variance	Type of Expense	Detail Actual Expenses	Budget	Variance
9,626.49	9,800	174	Salaries and Wages	66,841.99	67,400	558
2,695.42	2,744	49	Benefits	18,715.76	18,872	156
190.85	150	−41	Supplies	1,632.00	1,050	−582
5,219.38	5,700	481	Maintenance	36,266.32	39,900	3,634
1,119.21	300	−819	Outside Services	1,119.21	2,100	981
575.76	700	124	Travel	3,885.64	4,900	1,014
382.19	360	−22	Telephone	2,674.49	2,520	−154
12,519.38	12,700	181	Rental/Lease	82,184.58	84,300	2,115
830.00	830	0	Overhead	5,810.00	5,810	0
3,019.00	2,600	−419	Depreciation	20,874.00	19,130	−1,744
36,177.68	35,884	−292	Total Expenses	240,003.99	245,982	5,978
		−0.81%	Variance as a % of Budget			2.43%

nications system on to the users so that individuals realize that the service is not free. In most organizations, each department pays for its telephone equipment and long distance telephone charges through the internal chargeback system. It is also common for users of data terminals to purchase or otherwise pay for the terminal itself, and in many companies an internal bill for the use of the data network and other communications equipment, such as servers and routers also is generated. Many people believe that the only way to ensure that the costs of communications facilities are kept under control is to be sure that each user is paying his or her own way. If there is a perception that communications services are free, they are more likely to be abused.

communication services are not free

Quality Control Telecommunications departments should have standards of performance for the services they provide. The standards should be agreed on with the users and should meet their requirements. Typical standards set a level of expectation regarding network availability, response time, ability to obtain long distance lines, and so forth. Responsibilities must be assigned for monitoring and reporting performance against the standards and for making changes in the standards to address users' changing requirements. Typically, network operations is responsible for this activity.

The Telecommunications Audit Another type of control is the periodic audit of the telecommunications department and user departments by internal or external auditors. The purpose of the audit is to review the telecommunications activity to ensure that weaknesses in procedures or controls do not exist, and that telecommunications assets are being used properly. You can better understand the importance of this activity when you realize that in the United States today, toll fraud, the improper use of long distance service, is a 5.5 billion dollar problem!

Audits normally are conducted by individuals not involved in telecommunications and need not be limited to financial matters. Audits also can be used very effectively in the operations area to review existing methods and procedures for possible improvement. Typically, internal auditors report to a high-level executive in the company to maintain their independence from any department. Therefore, their reports usually require a formal response by the management of the department that has been audited. A partial list of the areas that are examined in a telecommunications audit is shown in Figure 15–14.

Figure 15–14
Checklist for a telecommunications audit.

Administrative:

- ☐ Policies and procedures are defined and communicated.
- ☐ An adequate management structure exists.
- ☐ Provision is made for the separation of duties and responsibilities.
- ☐ Purchases are only made with the approval of authorized personnel.

Personnel:

- ☐ Job descriptions exist and are up-to-date.
- ☐ People are properly trained for their jobs.
- ☐ Critical skills are backed up.

Network Operations:

- ☐ Standards of performance exist and are up-to-date.
- ☐ Service level agreements exist and are up-to-date, and service is monitored.
- ☐ The network is operated in accordance with established procedures.
- ☐ User support is provided to assure effective use of the network.
- ☐ A disaster backup plan exists and is tested at least once a year.

Asset Protection:

- ☐ Environmental controls are in place.
- ☐ Physical access to telecommunications equipment areas is limited to authorized personnel.
- ☐ Network hardware is acquired following standard procedures, is covered by maintenance contracts, and is inventoried.
- ☐ Network software is covered by appropriate licenses and maintenance contracts, and is inventoried.
- ☐ Changes to network software are properly authorized, thoroughly tested, and implemented as approved by management.
- ☐ Communications media are secure.

Access Control:

- ☐ User access to the network is restricted to authorized personnel.
- ☐ Access to systems software and related documentation is restricted to authorized personnel.

Properly used, the telecommunications audit can be a very effective tool for checking the performance and upgrading the telecommunications activity. From senior management's standpoint, the audit provides an independent perspective of the operations of the department. From the telecommunications manager's perspective, the audit may provide valuable insights. If problems surface, it may give the manager some additional clout to get the problems corrected.

COST EFFECTIVENESS

Up to this point, we have not emphasized the importance of cost effectiveness as an important influence on the way telecommunications is managed in the enterprise. Enlightened management wants the telecommunications system to be cost-effective but is not necessarily concerned that it operate at the least possible cost. Management is interested in the overall effectiveness of the money spent for a telecommunications system, in the effectiveness and efficiency of the network, and in its contribution to the overall success of the company.

The standard approach to telecommunications cost management has been to ensure that the monthly bills from the telecommunications vendors are accurate and paid on time and that the number of telecommunications lines and other equipment is kept at a minimum to provide the desired service. In the days when telecommunications was a monopoly in the United States, there were few other actions that the telecommunications manager could take.

competing services

With deregulation and competition, there are now many alternatives for most telecommunications services, and the telecommunications department has a responsibility to investigate competing alternatives. Long distance telephone service is available from several companies beside AT&T. Telephone systems may be provided by the telephone company or purchased from independent manufacturers such as Northern Telecommunications or Siemens-Rolm. When the manager compares alternatives, it is of course necessary to be sure that the services and products being compared are truly the same. This type of investigation places more of a burden on the telecommunications department; no longer can the manager just go to the telephone company and order the only product available.

renting, leasing, or buying equipment

Another cost management technique is the application of economic analysis tools to help make the financial decisions of the telecommunications department. In addition to selecting between competing alternatives, there are financial alternatives to evaluate—whether to rent, lease, or buy a particular product. The telecommunications manager and staff must know about and use techniques such as discounted cash flow and the present value of money. A study of these techniques is outside the scope of this book, but they are covered in all basic finance courses. If the company has finance department specialists using and applying these tools, it is wise to get them involved to ensure that the best financial decision is made for the company.

Another set of decisions the telecommunications staff must deal with is whether to perform certain services within the department or to outsource them. Examples include publishing the organization's telephone book, equipment installation and troubleshooting, and maintenance. Most companies today do not have the staff or expertise to perform all of the service or maintenance activities. Vendors offer maintenance contracts on their equipment and have technicians with special training. Independent service firms maintain many types of telecommunications equipment.

As usual, there are trade-offs. On the one hand, it may cost less to have certain services performed by outside organizations, but because they are not directly controlled, service may not be as timely as if it were done by company employees. On the other hand, the outside company may be able to do a more thorough job than inside technicians could do because of specialized training or test equipment.

All in all, the telecommunications managers must devote an adequate amount of attention to ensuring that their department is run in a cost-effective way. There are almost always opportunities to reduce cost without reducing service. Looking for improvements in cost efficiency and effectiveness is an important part of any manager's job.

OTHER MANAGEMENT ISSUES

Selling the Capabilities of the Telecommunications Department

In many companies, the telecommunications manager needs to work constantly to keep communications-based opportunities and capabilities in front of senior management and to look for opportunities for telecommunications to contribute to or improve the company's products, profits, or other capabilities. In some companies, telecommunications is still viewed as administrative overhead. The focus is on cost control, the emphasis being that less is better. Telecommunications department objectives are mainly related to saving money.

In many other companies, the potential value of telecommunications is well recognized. In these companies, management focuses first on the service required and then on the cost. Management in these companies is receptive to new ideas. Some ideas may, in fact, increase costs, but benefits can be achieved to offset the increased level of expenditure. The payoff of a new telecommunications project may occur over time, just as it does in other investments the company makes.

In a few companies, telecommunications is of such strategic importance that the companies could not survive without their communications facilities. In these companies, telecommunications is a money maker, not a cost center. To some extent, the more money spent on telecommunications, the more profit that is made by the firm. Financial institutions, transportation companies, and insurance companies are examples. In these companies, the telecommunications manager's job is one of prioritizing all of the possible projects that "could" be done rather than selling the idea that some telecommunications work "should" be done.

Project Justification Criteria

In most cases, telecommunications projects must go through some sort of justification process. This usually includes a look at the alternatives and a comparison of the anticipated costs of doing the project versus the benefits to be achieved. In most cases, the costs of a project are relatively easy to obtain and quantify. Benefits, however, often are less tangible and more difficult to quantify. While needs to be a continual effort made to attempt to quantify the benefits and put dollar values on them, the project sponsors should not be afraid to state what they believe the intangible or nonquantifiable benefits will be. Many times the ultimate decision to do a project is based on a perception or belief that particular benefits will accrue to the company even though their value cannot be specified accurately.

The criteria used for justifying telecommunications projects are closely related to management's perception of the strategic importance of telecommunications to the company's success. If telecommunications is viewed as a major contributor to the company's success, it is a lot easier to get new projects approved. If the telecommunications project can be characterized as being of strategic importance to the company, such as helping to beat the competition or as an investment in the future, it will be easier to gain approval for its implementation.

Transnational Data Flow

Organizations involved in international operations need to be aware of the changing guidelines and laws affecting the transmission of data across international boundaries. As was discussed in Chapter 2, companies planning to transmit data across international boundaries need to understand and monitor the changing laws and guidelines that govern such transmissions. Depending on the degree of international communications, the telecommunications manager or one of his or her staff may have the responsibility to monitor these issues. Several consulting services and magazines are available to help a company keep up with the changes in this area.

SURVEY OF TELECOMMUNICATIONS MANAGEMENT TRENDS

In late 1996, *Teleconnect* magazine conducted a survey of telecommunications managers working in various industries. Some of the results of the study were:

- Large companies and institutions still separate voice and data. Although voice and data communications managers talk to each other and know what they are doing and planning, actual work assignments are still segregated.
- Staffing levels are generally constant. Employers expect communication managers to find more productive ways of doing their work with existing staff, the same as they expect from other departments in the organization.
- Telephone call accounting is a necessity. These systems are used for calculating chargebacks and for tracking long distance charges, misuse, and telephone abuse, all judged to be very important activities.
- Companies are using equipment providers' technicians. The trend seems to be away from in-house service and maintenance and toward outsourcing.
- Managers are not happy with customer service from vendors. Vendors often want to sell the product they have and not what the customer needs. Vendor personnel change rapidly.
- Videoconferencing usage is increasing. Colleges and hospitals are on the leading edge; companies in industry seem to be lagging.

Most of these trends are consistent with the material you have studied in this chapter and other parts of the book. They should only be taken as guidelines, however, as there is no substitute for critical analysis of the particular situation in each organization. In any given case, there may be very good reasons to do things differently.

SUMMARY

Rapidly changing technology and regulatory environments combined with the growing use of telecommunications within business and government are forcing telecommunications managers to become more sophisticated in the way they run their organizations. In many ways, the telecommunications department can be viewed as a "business within a business," and traditional management techniques can be applied. The functions of the telecommunications department are analogous to the functions of the company itself. Telecommunications professionals must have skills in research development, product management, finance, marketing, administration, and performance measurement, and they should develop a good rapport with senior management.

CASE STUDY

Dow Corning's Telecommunications Management

Telecommunications management at Dow Corning has been an evolutionary process. Data communications has always been managed in the Information Technology department. Until 1985, the computer technical support or computer operations groups handled data communications work. Before 1982, voice communications were managed by the office services department, which also handled copying, printing, and other administrative functions. In 1982, the responsibility for planning the voice communications system was moved to IT and merged with the data communications responsibility in the computer technical support group. The technical support manager had voice communications planning added to his list of responsibilities. In 1985, a new telecommunications department within IT was formed. It was given the responsibility for planning, designing, and operating both voice and data communications networks. In 1995, the telecommunications department was merged back into the Global Information Technology Services group of the IT department because it was felt that close integration with the computer staff would be needed as the company upgraded its network to be LAN and server based.

High on management's list of concerns has been security. Internal users have had unique user identification codes and passwords for many years. In the last several years, there has been an increasing emphasis placed on making users aware of their security responsibilities. Most of the focus has been on ensuring that company data and computer applications are properly protected and that users are properly authorized. The company maintains logs of all terminal and computer usage, and data encryption equipment is installed where required.

A related concern and area of concentration has been disaster recovery planning. In 1986, Dow Corning signed a contract with Comdisco for disaster backup services for its mainframe computers, but now Dow Corning maintains its own offsite data center with backup servers and telecommunications hardware. The backup center is linked to the main computer center with optical fiber–based telecommunications lines using the FDDI protocol, and data is routinely moved back and forth between the centers. Testing of the disaster recovery plan is done regularly and focuses on ensuring that the company's network, computer, and applications software can be restored to operation at the disaster backup site and that data communications can be established between the backup site and all other Dow Corning locations. Testing of the backup capabilities is viewed as an ongoing process.

Telecommunications planning is done at several levels. A 3- to 5-year vision was developed in 1995. It hasn't been updated since, but management recognizes that an update is due because telecommunications technology has made significant advances. A 12-month project plan for the following year is put together late in the year to coincide with the company's budgeting process. Budgets are established for the fiscal year, which coincides with the calendar year. The list of planned telecommunications projects changes rapidly and must be updated about every 6 months. As specific projects are initiated, the project leader is responsible for developing a project plan that details the tasks to be performed and the timetable for completing them.

Virtually all vendor bills for telecommunications services are checked and approved by the telecommunications staff. The costs for telephones and long distance telephone charges are included in the IT budget and are not passed on to the users.

Dow Corning has five full-time people performing telecommunications administrative support activities for the Midland area and others around the world who have telecommunications administrative support activities as a part of their jobs. One person is responsible for working out the details of the data network moves, adds, and changes. She works closely with the users to ensure that when their personal computer workstation arrives it can be readily connected to the LAN. She also coordinates with the Dow Corning people who arrange offices to ensure that proper planning is done for the related personal computer moves.

Two people perform similar functions for telephones. They order new telephone lines from the telephone company when required and coordinate moves and changes when offices are rearranged. As in most large companies, this coordination takes a significant amount of time, since many people change jobs and offices each year. Dow Corning has adopted a general policy of changing telephone numbers when people change jobs: The telephone number stays with the position, not with the employee. However, there are always exceptions and special circumstances that must be accommodated.

Another person handles the administration of the voice mail system. She registers new users and trains them to use the voice mail system. She also checks the system's status and statistical reports to ensure that all of the circuits are working properly and being used equally. When required, she sends voice messages to all users informing them about new features or upcoming service activities.

The supervisor of the group says that a real challenge has been to seamlessly tie together the capabilities of the three Centrex systems, Rolm PBX, and Lucent voice messaging systems so that, to the customer or other person who calls Dow Corning, the voice system works as a unified whole, even though several pieces of equipment are involved. She says that proposed changes must be analyzed carefully and tested thoroughly, so that callers don't get directed to a voice mailbox when they want to talk to a person, or get put on hold for an inordinate amount of time.

Dow Corning's telephone operators are not a part of the telecommunications staff or the IT department. They report to the site security department and also serve as receptionists for the main entrance at the Dow Corning center. Since the Centrex telephone system has direct inward dialing (DID), most calls from the outside go directly to the person for whom they are intended. The only calls that the operators must handle are from people who call the general Dow Corning number because they don't know the specific person to whom they want to speak. Thus, the telephone operators are also information sources and must often question callers in order to direct their calls to the appropriate person or department.

QUESTIONS

1. Users at Dow Corning do not pay for telecommunications services they use. What are the pros and cons of the IT department absorbing all of the telecommunications costs in its budget?

2. Do you think that charging the users for telecommunications usage is a good idea, or does the time spent generating bills and other charge-out statements outweigh the benefits that accrue from doing the charging? Is the charge-out of these expenses an effective control mechanism?

3. When 3- to 5-year plans are developed for the telecommunications network, with whom should they be reviewed? Can technical people relate to such long-term thinking, or are they more concerned with what is going to happen in the next few months or the coming year?

4. Looking back through the Dow Corning case material, what controls can you find that are in place for the network's management and operation? Do you feel that more controls are needed?

5. Does it seem unusual that the telephone operators are not a part of the telecommunications staff but report to the security department instead? What are the pros and cons of this organization structure?

REVIEW QUESTIONS

1. Discuss why it is important to manage the telecommunications activity within a company. Has the need increased or decreased since 1990? Why or why not?
2. Explain why it is important for a telecommunications department to be proactive and forward thinking.
3. What are the factors that favor having the telecommunications department report to the chief information officer? If a chief information officer does not exist, to whom should the telecommunications department report? Why?
4. What is meant by the statement in the text that political reasons have kept many companies from merging their voice and data telecommunications departments? Why is it important to overcome these stumbling blocks?
5. Describe some situations in which it would be appropriate for the telecommunications department to use consultants.
6. The Nova Company needs a good technical person to install and maintain a new data communications network. How would you recommend Nova go about locating such a person? What attributes should the company be looking for when the managers interview candidates for the job?
7. An up-and-coming systems analyst in the Photometrics Corporation was just promoted to be the telecommunications supervisor. She has excellent verbal communications skills and gets along well with people at all levels in the company. She has 3 years of experience in systems analysis and programming and has performed extremely well, but she has no specific telecommunications or supervisory experience. What difficulties might this person encounter in her new job, and what kinds of people should she try to hire to complement her strengths and weaknesses?
8. In a small telecommunications department consisting of one full-time and one part-time person, how are all of the different tasks handled, such as planning, designing network enhancements, solving day-to-day problems, and operating the network?
9. What are the differences among a long-term, medium-term, and short-term telecommunications plan?
10. What are the consequences if a telecommunications department does not have a long-term, 3- to 5-year plan?
11. Why are financial management and controls important to the telecommunications department?
12. Explain the difference between expense budgeting and capital budgeting.
13. The text lists a number of types of telecommunications department expenses. Which of these do you think are most important to manage? Why? Which represent the largest dollar value for a typical telecommunications department?
14. The text gives several arguments in favor of charging out telecommunications costs directly to users. Why might some companies not charge the costs back to the users?
15. What is the purpose of having a telecommunications audit?
16. Explain why telecommunications management needs to look for opportunities to make the department more cost-effective.
17. What are some of the factors other than cost savings that can be used to help justify an expansion of the telecommunications network?
18. Why must the telecommunications manager be concerned about the physical security of the company's communications facilities?
19. Why is dial-in access an especially vulnerable point in a data communications network?

PROBLEMS AND PROJECTS

1. Use a spreadsheet program and the expense categories listed in the text to create a budget for a telecommunications department. Increase the personnel costs by 5 percent per year for five years and the rental expenses by 20 percent per year for the same time period. Observe what happens to the total budget for the department over the five-year period. How would you present this kind of projection to senior management to gain support for a long-term telecommunications plan? What arguments would you use to support why the telecommunications costs are going up at a rate that greatly exceeds the general rate of inflation and the 7 percent growth rate for other administrative expenses in the company?

2. Call back and handshake were two techniques described in the text that provide security against unauthorized access to a computer on dial-in lines. Invent another technique that would provide access control. Your technique could be implemented in hardware, software, or a combination of the two. Describe how it would work. See if your classmates can figure out a way to break your "security system."

3. Identify some of the different situations that would occur in a disaster caused by a tornado striking a computer and communications control center as compared to a fire striking the same location. How might these differences be reflected in the disaster recovery plan?

4. The Radix Supply company has a small but sophisticated online computer system providing most of the company's data processing capabilities. One of the people in the telecommunications group believes that the company telephone book should be put on the computer and made available online. In fact, she is advocating that Radix stop printing the telephone book every month and use only the online version. What are the factors the company must consider before the managers decide whether to eliminate the printed directory and have only an online version?

5. Visit a company with a telecommunications department. Find out where the department resides in the company's organization. To whom does the telecommunications manager report? How is the telecommunications department organized? How does the group recruit its people? Does the department use consultants and, if so, for what purposes? How often is an expense budget prepared? Are there regular reports of telecommunications expenses? What type of planning process does the department have? Are project leaders expected to prepare detailed project plans? How frequently is project progress reviewed? Who reviews projects?

6. The telecommunications principles shown in Figure 15–10 need to be tailored to each organization's unique circumstances. Suggest some upgrades to these principles that would be appropriate for the telecommunications department at your school or company.

Vocabulary

chief information officer (CIO)
communications directory
Gantt chart
profit center
cost center

References

Petersohn, Henry H. *Executive's Guide to Data Communications in the Corporate Environment.* Englewood Cliffs, NJ: Prentice-Hall, 1986.

Peterson, Kerstin. "Communication Managers' Survey." *Teleconnect* (October 1996): 90–91.

Schafer, Maria. "Good Networking Help Is Hard To Find." *Network Computing* (April 3, 2000): 100–103.

Sigler, Jerry L., and Gerard J. Cunningham. "Building a Financial Management Structure for Communications." *Business Communications Review* (August 1989): 18–21.

CHAPTER

16

Future Directions in Telecommunications

OBJECTIVES

After studying the material in this chapter, you should be able to

- describe many of the trends that are shaping the future capabilities of telecommunications systems;
- describe applications of telecommunications that are "just arriving" or are on the horizon for the near future;
- understand why these new telecommunications capabilities are likely to have a profound impact on our society.

INTRODUCTION

One aspect of telecommunications that makes the field so exciting is the constant and rapid technological change, a characteristic that is predicted to continue for the foreseeable future. When new communications technology becomes available, it brings with it the capability of applying communications to situations where it was previously not feasible, justifiable, or even possible in a way that meets user requirements. The constant technological change is one reason why some people have chosen to work in the field for many years.

As new technology becomes available, the potential uses for it are often not immediately evident. People must become familiar with the technology and figure out how it might be applied to solve a problem or to do something that has not been done before. Even when use of the technology seems evident, skeptics are likely to pooh-pooh the idea. When the telephone was first invented, one futurist in England declared that there would never be a market for the telephone in England because there was a perfectly adequate supply of messenger boys!

In this chapter, we will look at some of the discoveries, events, and ideas that are giving us insight into the uses of future communications capabilities. Some of the ideas will be great successes, whereas others will fail. Furthermore, for every idea discussed in this chapter, there are probably at least ten that are not covered, and more will emerge in the coming years.

TRENDS IN ELECTRONICS

The capabilities of electronic circuit chips have been increasing exponentially and are continuing to do so. The microprocessor chips that drive computers of all sizes allow the computing power of a room-sized mainframe computer of 10 years ago to reside in today's desktop personal computer. Microprocessor chips are ubiquitous. Today there are more than 15 billion microprocessors in existence—more than two for every person on earth! In the United States and most other developed countries, most homes—even those without personal computers—have a couple of dozen microprocessors. They're in TV sets, wristwatches, cameras, kitchen appliances, vacuum cleaners, and just about everything else that uses electricity. Cars typically have at least 10 microprocessors and one Mercedes Benz model has more than 50.

The price-performance of memory chips, microprocessors, and other related electronics has been coming down at a rate of 20 percent per year compounded. This trend has existed for the past 15 to 20 years and is expected to continue for the foreseeable future. Of course, this does not mean that every single chip decreases in price by 20 percent each year or that the decreases are uniform, but the trend is clear. As shown in Figure 16–1, at 20 percent per year, the price of a particular chip reduces to half its original cost in just over 4 years. If such chips represent a substantial part of the cost of a product sold in a highly competitive market, say a personal computer, the price of the end product is likely to drop at a similar rate.

digital signal processor

In communications equipment, microprocessors and memory chips find many uses. Two examples are adding intelligence to telephones and modems and allowing more memory buffering capability in PBXs and front-end processors. One type of microprocessor, called the *digital signal processor (DSP),* is having an especially significant impact. Digital signal processing is the technique of using microprocessors to analyze, enhance, or otherwise manipulate sounds, images, and other real world signals. A DSP chip processes a digitized signal through a series of mathematical algorithms over and over again to enhance or otherwise manipulate the signal. Before DSP chips became available, analog techniques that processed

Figure 16–1
Declining cost of circuit chips, assuming 20 percent per year cost improvement.

Year	$100 chip	$20 chip	$5 chip
1	$100.00	$20.00	$5.00
2	80.00	16.00	4.00
3	64.00	12.80	3.20
4	51.00	10.24	2.56
5	41.00	8.19	2.05

signals in their original wavelike form were used, but analog signal processing is much slower and less accurate than the digital method.

Applications of DSP that are of particular interest in communications are *voice compression* and *video signal compression*. The objective of compression is to reduce the number of bits required to carry a digitized signal while still maintaining its unique characteristics and quality. In voice signal compression, the objective is to strip out redundant bits of information, yet maintain tone and inflection so that the listener can recognize the speaker. Twenty to one compression of a voice signal is possible while still meeting the recognition objective.

compression

Video signal compression is even more interesting. A standard broadcast-quality, color television signal requires 92 million bits per second (Mbps) when transmitted digitally in uncompressed form. Today it is possible to compress that signal so that it can be transmitted at less than 36 Mbps. The challenge for cost-effective and universally available video conferencing services has been to compress the signal to the point that it can be transmitted over the public switched telecommunications network while providing good quality. Codecs used for this purpose are continuing to improve in quality and drop in price, making video conferencing an even more feasible business tool.

Another application for digital signal processing techniques is speech recognition. The idea of talking into a microphone attached to a device, such as a personal computer, and having the speech converted to text and a typewritten page emerge from the printer has long been the researcher's dream. Until recently, the speeds of the DSP microprocessors were simply not adequate to keep up with the rapid flow of normal human speech. Speech recognition systems, which have been available for several years, perform adequately if the speaker talks slowly, enunciates clearly, and limits his or her vocabulary to a predetermined set of words. Such devices are used in applications where simple instructions, such as "stop," "left," or "forward," are given to a machine.

speech recognition

One application of this technology is in the factory. Ford Motor Company installed such a system for its automobile paint inspectors. They speak into a headset and identify flaws in paint jobs of new cars as the cars roll down the assembly line. The inspectors use simple terms such as *right front fender, sag* or *right quarter roof, dirt*. The information is fed directly into a computer where it can be easily manipulated to stop aberrations in the painting process.

Other applications exist in telephone-based systems. You may be asked by the audio response unit to "Touch or say one to access your account balance, touch or say two to speak to a customer service representative," etc. When you call United Airlines to check on a flight, you are asked, "Flight 555 departs from Chicago and flies to Atlanta. Would you like to hear about the departure time from Chicago or the arrival time in Atlanta?" You can respond, "Departure," "Depart," "Arrive," "Arrival,"

"Chicago," "Atlanta," and probably several other similar words, and the system will understand what information you want to know.

Recognizing a few words, spoken slowly, is much simpler than recognizing continuous speech, however. In addition to the speed difference, other complications in continuous speech recognition include recognizing the difference between homonyms, such as *to, too,* and *two,* and recognizing the same word spoken with different regional accents, such as *drawer* versus *drawh.* The latter problem is solved in today's systems by "training" the speech recognition processor to understand a certain voice. The prospective user of the system begins by speaking a few predetermined words into the microphone. The speech recognition equipment analyzes the wave forms of the speaker's voice and compares them to standard wave forms for those words stored in its memory. It then generates conversion rules that allow a successful comparison between the wave form of the speaker's voice and the standard wave form for all words in its vocabulary.

Speech recognition equipment is available for personal computers that have sound cards and microphones. One package from IBM, sells for about $100. You can begin using the program right away with no training, if you pause briefly after each word. To train the system so that you can speak continuously, you need to spend about 30 minutes reading 50 to 280 sentences aloud to familiarize the system with your speech patterns.

TRENDS IN COMMUNICATIONS CIRCUITS

The single most universal trend with communications circuits is the rapid advance toward end-to-end digital capability. The advantages of digital transmission are clearly recognized, and most of the common carriers are moving as rapidly as possible to install the electronics that can handle digital signals in their central offices. The carriers are also converting microwave radios from analog to digital transmission and installing fiber-optic circuits to provide the high bandwidth that digital transmission requires.

As shown in Figure 16–2, in the United States today a high percentage of the public telephone network is operating in digital mode. The facilities of the three major long distance carriers, AT&T, WorldCom, and Sprint, are essentially all digital, and the long haul circuits use a high percentage of optical fiber cable. The equipment of the larger local telephone companies is highly digital, too, but many of the smaller companies still have older analog equipment and circuits.

digital telephone network

In the mid-1970s, data transmission on circuits leased from the telephone company topped out at about 9,600 bps. Dial-up circuits normally had a data rate of 300 bps or sometimes 1,200 bps. Today those same circuits routinely handle data at 64,000 bps, and, with special electronics at each end, at up to 8 Mbps as we saw in Chapter 9, where DSL circuits were described.

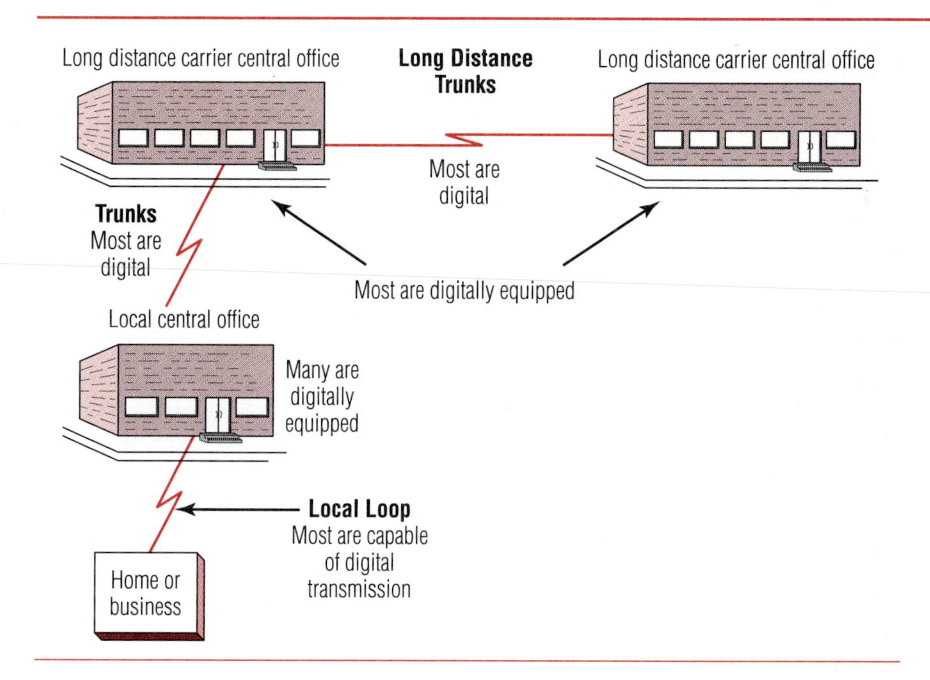

Figure 16–2 Amount of U.S. telephone system that is digitally equipped.

Because of the inherent delay associated with satellite transmission, since the early 1980s there has been a movement away from using satellite circuits for voice transmission. Satellite circuits are perfectly adequate for batch data transmission, television broadcasting, or the distribution of programs for entertainment—applications in which the time delay can be tolerated. They are less satisfactory when voice or interactive data transmission with rapid turnaround is required.

Three major trends in satellite technology will have an impact on the way satellites are used in the future:

1. higher bandwidth;
2. more powerful satellite transmitters;
3. low earth orbiting (non-geosynchronous) satellites, also known as LEOs.

The higher bandwidth is needed to help keep the cost of satellite transmission competitive with optical fibers. Higher powered transmitters on the satellites allow the receiving dishes to be smaller, as exemplified by the 18-inch satellite receiving dishes used with the direct broadcast satellites (DBS). The trade-off is that larger batteries and solar cells to charge them are required on the satellite.

Although commercial satellites have been around for years, advances in battery technology and digital compression techniques, as well as advances in the technology for manufacturing and launching satellites, have all dramatically slashed costs. That has made affordable mass-market products possible. More than 1,700 communications satellites will be launched before 2005, most of them into low-earth orbit, meaning that they are not geosynchronous. One of the difficult technologies to work out with these satellites is that, since they are not in continuous view of any point on earth, they must be able to seamlessly pass off the communications passing through them to another satellite as it comes into view of an earth station.

Optical fiber technology is making the biggest impact on communications circuits today. Common carriers and other communications companies are installing fiber cables at unprecedented rates. A single optical fiber of the newest type can carry 10 million phone conversations or upwards of 320 billion bits per second. In one laboratory experiment a fiber was driven at a terabit—a trillion bits per second, or enough capacity to handle 20 million phone calls at once! Although production systems lag behind laboratory experiments by several years, actual progress has exceeded even the most optimistic forecasts.

higher optical fiber capacity

The wider bandwidths and higher transmission speeds are needed by today's business and Internet applications. Figure 16–3 shows how long it would take to transmit several typical types of data and files that might be sent by an interactive application or downloaded from the Internet. You

Figure 16–3

Time required to transmit various types and sizes of data and files at varying transmission speeds, not including the effect of line turnaround time or retransmissions due to errors on the circuit.

		Time to Transmit (seconds)			
Transmission Speed (bits per second)	**Number of bytes Transmitted (per second)**	**Average Typewritten Page (4,000 bytes)**	**Online System Response—Graphics but No Pictures (20,000 bytes)**	**File Downloaded from the Internet (1 megabyte)**	**File Downloaded from the Internet (5 megabytes)**
1,200 bps	150	26.7 secs.	133.3 secs.	111.1 mins.	9.3 hrs.
2,400 bps	300	13.3	66.6	55.5 mins.	4.6 hrs.
4,800 bps	600	6.7	33.3	27.8 mins.	2.3 hrs.
9,600 bps	1,200	3.3	16.7	13.9 mins.	1.2 hrs.
56,000 bps	7,000	.6	2.9	2.4 mins.	11.9 mins.
1 Mbps	125,000	.03	.16	8.0 secs.	40.0 secs.
10 Mbps	1,250,000	.003	.016	.8 secs.	4.0 secs.
100 Mbps	12,500,000	.0003	.0016	.08 secs.	.4 secs.
1 Gbps	125,000,000	.00003	.00016	.008 secs.	.04 secs.

can see that at the speeds that were common a few years ago, the transmissions would take so long that no one would do them. In other words, many of the things we do with workstations and telecommunications today just weren't done a few years ago because they weren't practical. This trend will of course continue, and things that are impractical or very slow today will be commonplace in the not-too-distant future.

TRENDS IN STANDARDIZATION

In any new industry, standardization comes slowly. Initially, it is not clear which facets of the industry or its products should be standardized. Furthermore, each company producing products believes that it has better ideas than its competition. Since standards often represent compromises and, therefore, not necessarily the best technical solutions, companies must be convinced that they can still maintain their unique product identity and advantages even while meeting the standards.

In the communications industry, standards have been stronger and better developed in the telephone segment than in the data segment. This is because the telephone segment of the industry is older, and the need for any-to-any communications between telephones has been obvious from the start. Data communications standards are now evolving rapidly because users want the capability to connect any terminal to any computer and to have more flexibility in the connection than they have enjoyed in the past. Organizations of all types are realizing the benefits of having truly open computer and communications systems that allow wider choices of products from many vendors. The market is driving the vendors to provide standards-compliant products.

Because of the slow process of developing standards and the fact that standards usually represent a technical compromise, nonstandard and proprietary products and solutions to communications problems will continue to exist. Sometimes these products provide a unique ability for a company to solve a business problem, and the fact that the product falls outside of any standards may be irrelevant. At other times, the proprietary solution may provide better capability than the standard technique. For example, most vendors of video codecs include both the standard data compression method and a higher capacity proprietary technique in their products. If a conference is held between two parties with different types of video equipment, they use the standard compression algorithm. If the equipment at both ends is from the same vendor, Picturetel, for example, the Picturetel proprietary compression technique can be used to get better picture quality on the same speed circuit.

Sometimes, products evolve in areas where no standards work is under way. For example, each vendor of voice mail systems has its own proprietary method of coding the voice and storing it in the system's disk.

Lucent's voice mail products cannot exchange voice messages with Rolm Phonemail systems. In the future, the need to standardize the systems so that they can exchange voice messages may be identified.

A vendor that can solve an existing business communications problem can be highly successful, even if its product is nonstandard. By installing the proprietary solution, a user of the product is able to solve a problem months or years earlier than if he or she waited for the standardization process to occur and standardized products to become available. Both the vendor and user face the risk of potentially having to convert to standardized products in the future. This risk must be weighed against the potential benefit that the product/solution can bring in the meantime.

TRENDS IN REGULATION

In the United States, there are general trends toward continuing deregulation and encouraging competition. Competition has opened the communications industry and allowed many new companies to enter the market with innovative products. The cost of staying competitive over the long haul is high, however, and many companies are purchased or merged after a few years—and of course some fail.

The Telecommunications Act of 1996 opened more of the U.S. telecommunications market to competition, though, interestingly enough, some of the telephone companies are only now embracing their freedom because in order to compete in markets outside of their home territory, they must open their home market, the part of their business where they still have a monopoly, to competition.

In other parts of the world there is a broad trend toward deregulation, especially in developed countries. Less developed countries tend to be more protective of their telecommunications infrastructure, and thus tend to keep it highly regulated.

TRENDS IN THE APPLICATION OF COMMUNICATIONS TECHNOLOGY

The deregulated, competitive environment, coupled with rapid technological advances, acts as an enabler for new communications applications. The application of communications technology is the most exciting aspect of communications. This is where the benefits to business and individuals can be seen. This section discusses some of the things that can be done through the application of new communications technology and developments. Some of the applications are being performed today on a limited basis, whereas others will be available in the near future.

Telecommuting

Telecommuting is the use of an alternate work location for employees who are normally based in the office. Telecommuters typically work at home and use the telephone, modem, fax machine, and a personal computer connected via telephone lines to the mainframe computer at the office. Telecommuting works best for people who are used to having a certain amount of freedom and who work well alone. By working at home, they can have an unlimited amount of quiet time with no interruptions (assuming that the worker doesn't have three small children or a spouse who gives music lessons!). However, communications with associates may suffer, particularly the informal communications that are a vital part of any business environment. Too much telecommuting can be professionally risky if an employee gets out of the mainstream. Telecommuters usually schedule regular times to go into the office for face-to-face contact with their peers and superiors and for meetings.

Telecommuting can also work well for part-time employees who only want to work a few hours each day. They can work out a schedule with their employer so that they work at home a certain number of hours per week and go into the office 1 or 2 half-days to keep in touch with their peers and attend meetings. The employer usually covers the cost of the terminal and an additional telephone line in the employee's home.

Telecommunications technology allows many people to work at home and still be in touch with the activities and happenings at their office or work place. (Courtesy of Picture Tel Corp.)

Experienced telecommuters say that a few factors help determine if telecommuting will work in a specific situation:

- Having the right job. Positions that require a lot of independent work, such as writing, research, some kinds of consulting, and copyediting, are a natural fit. Many people feel that nearly all of today's white collar jobs have some component that can be done as well or better outside the office.
- Being the right kind of person. People who easily motivate themselves and who have been at their job long enough to have solid relationships with co-workers make better telecommuters.
- Having the right reason. The decision to telecommute should be based on whether a person can be more productive at the job by working out of the office at least a couple of days a week. It should not be made because a person wants to be at home when the kids get home from school.
- Having the right boss. Whether you are appropriately rewarded for the work you do at home depends on your boss. Unless your boss recognizes your work, telecommuting won't be an asset for your career.
- Working for the right company. Some companies aren't ready to have employees who telecommute. A technological commitment as well as a support commitment must be made to the telecommuter, and if the company isn't ready to provide those, the telecommuter is quite likely to be unsuccessful.

It should be emphasized that telecommuting is not a future application of communications technology but is in widespread use today. Many employees of the city of Fort Collins, Colorado, regularly work at home using a terminal. A lawyer who works for a Chicago law firm telecommutes from Colorado. He consults with his partners by telephone and sends reports to the office using his personal computer and a modem. A speechwriter works in his home and sends speeches to his clients electronically. A webmaster in Michigan works in the office 3 days a week and at home 2 days to be with her two small children. She has a PC workstation and second telephone line in her home, so she can be logged on to her company's network and still talk on the telephone. She is able to do most parts of her job from her home as easily as when she is in the office, and both she and her manager feel that the arrangement is satisfactory. Many more examples abound, and as our society gets used to the concept of having workers at home, telecommuting will become more widespread.

Electronic Document Interchange (EDI)

Electronic document interchange (EDI) is the use of communications techniques to transmit documents electronically. The term is used most frequently in a business context today, but there is no reason why EDI techniques cannot be used by consumers as well. EDI relies primarily on the development of standard formats for various business documents, such as invoices, purchase orders, and acknowledgments. Furthermore, there must be an agreement between businesses (or other parties) to use these formats and transmit the data electronically using a standard communications technique. The benefits of EDI are the more rapid transmission of the documentation supporting the business activities and transactions and the elimination of paper. These can lead to other benefits, such as reduced inventory levels or faster collection of money from creditors.

All of the technology required for EDI is in place today. The banking and grocery industries have been leaders in developing EDI capabilities, and the techniques are now spreading to other industries and across industry boundaries. The chemical industry is active in developing standards for the electronic exchange of business documents. The automobile industry has been active in promoting the use of EDI with its suppliers and has made it a requirement for the suppliers of certain commodities to use EDI to do business with the large auto companies.

Telediagnosis

Telediagnosis uses microprocessor chips and standard communications facilities to allow a device, such as an appliance or a computer, to automatically place a telephone call to a diagnostics center. IBM mainframe computers, for example, continually monitor their own operation. If they detect a problem, the computer places a telephone call to a diagnostics center and transmits relevant data about the problem it is experiencing. At the diagnostics center, engineers can examine the data, request other information, and essentially operate the mainframe from the remote location in diagnostic mode. Assuming that the problem is not disabling the computer, normal data processing can continue while the diagnostic work is under way. In a high percentage of cases, the remote diagnostic center can determine which part of the computer is failing and can dispatch a local technician with the correct replacement part.

This concept could just as well be implemented in home appliances, such as televisions, washing machines, or furnaces. Similarly, an automobile might be driven to the local dealer's garage, where it could be connected to a diagnostic computer that communicates with a central diagnostic computer at the automobile manufacturer's headquarters. The technology exists; economics will drive the general availability of such services.

Telemedicine

Telemedicine can be broadly defined as the use of telecommunications technologies to provide medical information and services. It involves the practice of delivering health care over a distance using telecommunications equipment as simple as telephones and fax machines, or as complex as the

This doctor is conferring with a patient at a remote location about her diagnosis and treatment plan. Her medical data had previously been sent to him using both data transmission and facsimile. (Courtesy of Medical College of Georgia)

use of PCs and full-motion interactive multimedia. Telemedicine involves the electronic transmission of medical information for the purposes of diagnosis and treatment of patients using standard and specialized telecommunications equipment. Telemedicine uses batch processing of patient data for diagnosis at a later time and realtime consulting and testing.

One of the first uses of telecommunications for medical purposes actually occurred back in the late 1950s in Nebraska. A microwave link was used for telepsychiatry consultations between a state mental hospital and the Nebraska Psychiatric Institute. About the same time, NASA and the U.S. Public Health Service developed a joint telemedicine program to serve the Papago Indian Reservation in Arizona. By 1997, there were about 72 telemedicine programs in the U.S. using high-end interactive video technology for teleconsultations.

There has always been a lack of health care providers, especially specialists, in rural areas. Telemedicine provides a means for specialists to electronically travel to these regions, thereby increasing the scope of their coverage without physically traveling. In addition, it improves the quality and availability of care to patients in remote areas.

What makes telemedicine exciting is the potential generated by the evolving digital technologies. With the costs of the technology coming down, the potential exists for telemedicine to make a substantial contribution to the reduction of the cost of health care. There are problems to be overcome, however. Will medical insurance cover the cost of telediagnosis and consultations? If a patient in one state has telemedical treatment by a doctor in another state, which state's laws apply? Is electronic medical record keeping sufficient? Suffice it to say that work is being done to answer these and many other questions concerning telemedicine, and it is highly likely that the use of telemedicine will expand in the future.

In Belgium not long ago, a surgeon using a joystick and video screen controlled a robotic arm with a scalpel made an incision into the abdomen of a man hospitalized in Utrecht, Netherlands, before a camera was inserted to make a diagnosis! Are you ready for this?

Automobile Navigation by Satellite

In this application, which is widely used in Japan and coming into use in the U.S., a car is equipped with a satellite receiver and a small television-like device that can display a map of the local area. The receiver picks up signals from three or four satellites, which are part of the *Global Positioning System (GPS)* originally developed by and for the U.S. military. In May 2000, the accuracy and resolution of the GPS was improved for non-military use so that even the least expensive GPS receivers can calculate their location within 20 feet. When a signal from a ground station is also used, the accuracy is even greater. Receivers in automobiles most commonly display a map showing the automobile's location. Some

This automobile is equipped with a GPS navigation system that assists the driver to reach his destination. (Courtesy of BMW of North America, Inc.)

SIDEBAR 16–1

TRACKING CONSUMERS BY SATELLITE

GET READY FOR, UMM, WIRELESS WINDOW SHOPPING. GeePS.com Inc. is testing technology that will create personal marketing zones around individuals strolling city streets or driving through the suburbs. Punch some keys on a handheld computer or a new-age cell phone, and you'll get information on nearby shops and restaurants. And merchants can try to turn passersby into customers by announcing special deals good for, say, the next hour. "This is how we're going to rescue the bricks-and-mortar stores," says Arshad Masood, president of the Cranbury (N.J.) startup.

If you've noticed how GeePS sounds just like GPS (short for the global positioning satellite system), that's no accident. The concept will exploit GPS signals to pinpoint the location of tomorrow's gee-whiz wireless gizmos to within 50 feet. Today, mobile computers and phones don't have GPS circuits, typically—but that's expected to change next year. GeePS isn't waiting. It has already launched tests in San Francisco and New York, temporarily locating people by ZIP code.

systems even give instructions to the driver on how to get to a destination. Oceangoing ships have used the GPS system for years, but until recently, the technology was too expensive for the average person to afford. Now GPS built into automobiles adds less than $1,000 to the car's price, and hand-held GPS receivers with mapping capabilities are available for less than $300, making them affordable for hikers, pleasure boaters, and dads who need a new toy!

Tracking Cargo Containers

In a similar application, cargo containers—20- to 40-foot-long boxes that carry products from manufacturers to customers all over the world—are tracked by satellite. Cargo containers are loaded with products ranging from tennis shoes to chemicals to appliances at the manufacturing plant, and are then trucked to a port where the containers are loaded onto ships for transport overseas. On any given day, some 5 million containers are on the move. Most shipments arrive safely, but some show up with the merchandise wet, smashed, or baked. Some don't show up at all.

It is hard to know where a container is at any moment during its journey, but one company, International Cargo Management Systems (ICMS), has a product that is changing that. ICMS has developed a briefcase-sized box full of electronics that is bolted inside the cargo container. The box can determine the container's location from existing satellites and also contains sensors to monitor temperature, humidity, and shocks or jolts. Using other satellites, the box radios the information to the manufacturer, the shipping company, the insurance company, or all three.

SIDEBAR 16–2

INTERNATIONAL CARGO MANAGEMENT SYSTEMS

The patented *ProfitMAX*™ cargo tracking and monitoring system provides "in-transit visibility" of cargo as it moves through the supply chain. It is comprised of a self-sufficient, microprocessor based mobile data collection unit interfacing with a central monitoring station, which in turn provides segmented data to a network of individual user computer stations.

The mobile unit resides with a truck trailer, cargo container or rail car, and travels anywhere the cargo goes on a global scale, recording location and status of the shipment. In addition to satellite tracking (GPS), its sensors monitor environmental conditions such as temperature, humidity, multi-axial shock, and door integrity (openings).

The *ProfitMAX*™ mobile unit communicates via wireless link (satellite or cellular) to a central computer monitoring station in San Antonio, Texas, which archives data to be retrieved by system users. The unit transmits status reports on a selected schedule and whenever an alarm condition occurs, and can be queried on a user-selected schedule. User needs and available power to the mobile unit will determine the potential frequency of transmissions. When the unit is not making scheduled or unscheduled reports, its communications platform returns to a "sleep" mode to conserve power. The mobile unit draws its operating power from self-contained batteries, from solar panel trickle-charge or from the host vehicle when a connection is available.

Because the mobile unit contains a microprocessor, it is capable of analyzing the data which it collects. If a shipment is subjected to an event which violates parameters set by the user, an "alarm" condition will immediately be noted. If an alarm message is received, the system can provide immediate notification to all designated parties by automated phone messaging, beeper, fax or electronic data transfer.

Data which is collected and routed through the system is encoded, partitioned and secure while stored in the mobile unit, during transmission and when received and archived in the central computer. It is available only to user-authorized parties. The ICMS monitoring station and customer support teams operate 24 hours a day, but most functions of this state-of-the-art system are fully automated.

The *ProfitMAX*™ system is uniquely designed to resolve the types of concerns commonly raised about international, intermodal, mission-critical, or high-value/high-theft shipments. *ProfitMAX*™ is an essential operating tool, providing a seamless information flow through the entire supply chain which can be used to manage resources as well as shipments; to analyze trends relating to routes, drivers, equipment performance and breakage and loss; and to settle insurance claims.

The central monitoring station not only collects and disburses reports, but also creates a library of data on cargo movement which becomes an invaluable tool for understanding and anticipating occurrences and trends in damage and loss, operating costs and other features of the supply chain.

A client/user will access data collected on his shipments through his computer running a standard Internet browser. The screen will display maps indicating precise locations of shipments, and will allow you to view current or historical measurements from internal sensors or GPS location. Data can be downloaded into familiar report formats and graphic

SIDEBAR 16–2 *(continued)*

presentations. The *ProfitMAX*™ server-side software enables the secure, electronic transfer of shipping documents linked with the cargo.

Rather than a data deluge, the user sets guidelines for the type and format of the information or events which he wants to view. Other data is simply archived for future use. The open architecture of the system allows for great flexibility to meet the needs of each company and to integrate with existing office systems. Through a process we call *Envisioneering*™, ICMS process engineers will examine with a user the implementation of a system which fits the needs of his operation as it relates to shipping, logistics, Business-to-Business (B2B) and supply chain management functions.

The *ProfitMAX*™ mobile unit is rugged, space-conscious and tamper-resistant. It can even be moved from one container or trailer to another in less than 15 minutes once the container has gone through the one-time preparation process. Since some shipments may not require monitoring, or containers may stand idle for periods of time, this mobility allows the user to maximize the service from the *ProfitMAX*™ units, and thereby maximize his investment in the product.

Electronic Journalism

Electronic journalism involves communications in several ways. Examples best illustrate the point:

- For a number of years, the *Wall Street Journal* and *USA Today* have been composed in a central location and transmitted via satellite circuits to several printing companies located around the world. The papers are printed and distributed on a regional basis, thereby giving simultaneous availability worldwide.
- Many newspapers allow letters to the editor and classified advertising to be sent to the paper via e-mail.
- Photographers use digital cameras that send images over telephone lines directly to computers at the publisher's office.
- When President Lincoln was shot, the news reached Europe via ship in about two weeks. When President Kennedy was shot, viewers in Europe watched the funeral live via satellite transmission. When President Reagan was shot, a reporter for a London newspaper, who was located in his office three blocks from the scene of the shooting, first heard about it via a telephone call from his editor in London, who had watched the shooting occur on live television!

Newspapers are now available on the Internet. Some are free, and some require a subscription. All of the network television news services have web sites. Technology allows us to identify topics that we are interested in and to have that information selectively distributed to us so that when we next connect to the World Wide Web the information is waiting. In the future, our television will be able to scan the programming on all channels and automatically record programs it knows are of interest to us.

Expansion of the Internet and the World Wide Web

The Internet really came into its own with the release of the first public browser, Netscape, in 1995. Research showed that in 1998, 37 percent of U.S. households were connected to the Internet, and that by 2003 that figure is projected to rise to 63 percent. A recent report published by the Yankee Group predicts that the number of wireless Internet users, those connecting mainly through wireless telephone networks, will grow from 3 million today to 50 million by 2004.

All of these users will require more bandwidth in the Internet backbone circuits. Peak volumes of traffic through the Internet have doubled every 6 to 8 months for the last 2 years. While there is no immediate danger of running out of capacity, thanks to the rapid expansion in the bandwidth of the nation's broadband backbone circuits, we could experience a slowdown as more audio- and video-rich data is transmitted. One observer estimates that we may need a 100 to 1,000-fold improvement in the network backbone, and servers may need to be 1,000 times faster than today.

A new project called Internet2™ is underway to conduct research into the technologies and techniques that will be needed as the Internet grows and expands. Internet2™ is a not-for-profit consortium led by over 17 U.S. universities developing and deploying advanced network applications to accelerate the creation of tomorrow's Internet. Internet is not a separate physical network and will not replace the Internet. The consortium is developing and testing new technologies that gradually will be implemented and that will enable revolutionary Internet applications. These applications will require performance that is not possible on today's Internet. Included in the research are projects such as digital libraries, virtual laboratories, and distant-independent learning. A key goal of the project is to accelerate the diffusion of advanced Internet technology, in particular into the commercial sector. For more information, visit the project's web site at www.internet2.edu.

Faster connections means more graphic-intensive web sites loaded with streaming video and audio clips. Net shopping will begin to resemble the realworld activity, since buyers will be able to look at products in detail, perhaps even in 3D. You may get most of your entertainment—movies, music, videos—and news on the Internet. Proactive applications may automatically retrieve data for you. Rather than requiring you to download data in batches, data may come to you in realtime. You might, for example get your bank transactions as they are processed by the bank, much as you receive e-mail today, soon after someone has sent it to you.

High-Speed Communications to the Home

As we have seen, many companies are working today to find ways to get high-speed communications capabilities into our homes. Telephone companies and cable television companies have been the most active in this work, but satellite companies with DBS capability are also in the marketplace. The four most likely methods to achieve the high-speed capability are:

1. using one of the xDSL techniques, discussed in Chapter 9, to carry a high-speed signal over standard telephone lines;
2. bringing the signal in on the television cable;
3. receiving the signal with a small dish antenna from a DBS-like satellite;
4. using wireless technology and having the signal broadcast directly to the house from a transmitter in the neighborhood.

Since the roll-out of these capabilities isn't occurring in all locations at the same time, it is also likely that some combination of these methods may be used, depending on the homeowner's needs and the ability of communications companies to deliver one or more of these services. Optical fibers weren't listed as a primary medium because of the cost to run them to individual homes; however, some neighborhoods, especially newly constructed ones, may have fiber cables installed, which would add a fifth alternative to the menu of choices.

A question some are asking is, "How much transmission speed will be enough for the Internet connections of the future?" No one knows the answer, but people who have high-speed connections to the Internet today say that when you don't have to wait for the net, you start using it for new activities—activities that once took ages. One person claims that he started checking the weather on the web every day before getting dressed and programming his PC to back up his hard disk to a site on the web every night. But high speed isn't everything. Another attribute that current users of high-speed connections like is being connected to the web all the time. There's no waiting for a PC to boot up, or for the computer to dial and connect to an ISP. This adds a level of convenience that is synergistic with the high-speed connection and reinforces the use of the Internet in new ways. Most people (and businesses) feel that there is tremendous advantage and productivity in being online all the time.

Smart Homes

A *smart home* is a house that has microprocessors to help control home functions, such as heating, lights, and security alarms. When coupled with telecommunications, the computers can be instructed from a remote location to perform certain functions. For example, one could call home just before leaving the office and tell the computer to turn the heat up or the oven on. A smart home might also contain smart appliances that can react to your needs and habits, or that have the built-in diagnostic capability discussed earlier in the chapter. For example, your water heater needs to know your travel schedule so that it can stop heating water when you are gone!

Another aspect of the smart home of the future is that it may have a common wiring system. Every piece of wiring will be connected to the same control center. This means that every plug in the house will be interchangeable. You will be able to plug in a hair dryer, a stereo speaker, a telephone, or a security system to any outlet, and each of those appliances or devices will send a coded signal to the control center to tell it what kind of electrical power or other signal it needs. The blender will send a signal requesting electrical power, and the audio speaker will send a signal requesting music from the entertainment center. A child who sticks a finger into an empty socket will not be shocked because fingers don't carry the coded signal asking for electricity. As houses with this type of wiring system become available, manufacturers will begin to offer smart appliances with the coded signals built in. In the meantime, there will be adapter plugs available so that the appliances that people own today can be used with the new wiring system.

common wiring system

With a common wiring system, it will be possible to control and monitor any appliance or any function from anyplace in the house or, via telecommunications, from anywhere else. Furthermore, since the appliances will all be interconnected through the central control system, they will be able to talk to each other. For example, when the clothes washer finishes a load, it will be able to flash a message on the television screen you are watching to tell you that it's time to put the clothes in the dryer. The alarm system will be able to call you by telephone and tell you if someone is in the house when you are away. A fire detector, in addition to calling an emergency number, could turn on televisions and radios and have them speak a fire warning message.

Pervasive and Embedded Computing and Networking

The continued and accelerating trend of embedding computing devices in almost every type of device will enable many of the capabilities described in the above sections. You may be surprised to know that 98 percent of the computing devices sold today are embedded in products that do not reveal the embedded computing devices to the users. Toasters, dishwashers, ovens, and other home appliances have all been automated.

Now, yet another revolution is about to take place allowing us to connect all of these computational components together through networking. When pervasive computers are coupled with wireless networking and GPS location systems, you have an amazing capability to do new things. A project at the University of Washington is seeking to develop a virtual neighbor for the elderly. Sensors would be used to collect data on traffic and resource-use patterns so that remote observers or automated agents can ensure that nothing is awry. Unusual situations, such as doors or windows left open overnight or a lack of motion around the home for an extended period, would trigger messages to those concerned. We might put a collar on a dog that has the capability to do wireless location via a GPS link. Sensors could be put in concrete to detect seismic activity. Communicating mobile robots could be sent into dangerous areas to detect the boundaries of a forest fire or an oil spill. All of the above applications and countless others are being researched, and some are very close to being practical.

SOCIOLOGICAL IMPACT OF TELECOMMUNICATIONS CAPABILITIES

Throughout this book, we have studied the marvelous new ways that advanced telecommunications capabilities will change and generally improve our lives. However, with the implementation of heretofore unimagined applications, brought about by the rapid advancement in technological capability, sociological impact and change is inevitable, and some of it will be troublesome. Do you really want to have a personal communication system that allows people to reach you anytime and anywhere? Do you want to have video on demand so that you can sit at home to watch new movies instead of going out and attending a theater with friends? What will happen to libraries as we know them today if all of the same information is available at home through the information highway?

Historically, when changes in technology have been intended for a single purpose (for example, to make something work better, faster, or more cheaply), they have sometimes had significant and often unanticipated effects in other areas. For example, the automobile was originally developed as nothing more than an automated horse and buggy, as implied by its original name, "horseless carriage." But the automobile transformed society in a way that the horse and carriage never could have done. Will developments in Internet technology or high-speed communications have an impact of similar magnitude?

Furthermore, there will be difficult new legal challenges. For example, electronic data interchange may raise the following questions. When is a contract made? What substitutes for a written signature? What if, despite intensive error checking, an amount or other crucial number in a contract or other legal document is changed as the document is electronically transmitted and the erroneous change is not noticed until days or years later?

With all of the trends pointing toward more information being available, faster, and at any location, an interesting question arises. One must ask whether individuals will be able to absorb and take advantage of all this data and information. Studies have shown that our society is more stressful than in years past because our senses receive so many stimuli every minute of every day. Are our human data acquisition and processing skills up to the challenge of dealing with the rapidly growing output of an ever-increasing number of microprocessors?

Clearly, one approach that will help us deal with all of the information available to us is the use of the computer itself. Using artificial intelligence techniques, the computer will digest, cull, and summarize data and present it to us in manageable forms and quantities. With proper instruction, the computer should be able to provide us with information that is meaningful and of interest while serving as a buffer to protect us from the millions of other information stimuli that will come our way.

It is also clear that the trends in electronics aren't changing. Microprocessors are getting faster and cheaper, bandwidth is getting wider and cheaper, and video resolution is getting higher and cheaper. The widespread use of the Internet and the World Wide Web leapt into our lives within a 2-year period in the mid-1990s. But also, thanks at least partly to those trends and others, people are getting more comfortable with the language and use of personal computers, e-mail, video conferencing, global satellite positioning, and countless other technological advances. People's ability to adapt is critical, because all of the technology in the world is useless if people don't or won't use it.

It must be recognized that the trends described in this chapter are likely to cause some new problems that will have to be solved. The invasion of privacy and the ability for someone to more easily monitor our daily activities are two simple examples that will have to be dealt with. It is a safe bet, however, that human intelligence and people's natural adaptability will ensure that ways will be found to deal with the problems. Furthermore, it is virtually certain that new communications technologies and capabilities will be put to use in ways we have never imagined.

SUMMARY

This chapter explored many emerging communications and computing trends and directions. The new capabilities that technology affords will shape our lives and lifestyles. As new technologies become available, there is always a transition period when the old and the new technologies are mixed. Organizations and individuals have large investments in installed equipment and capability, and they are only able to afford and take advantage of new technology at a certain rate. Therefore, compatibility of new with old is important, so that we can take advantage of new developments, while protecting the investment we have made in older equipment. We want to be able to use a new compact disc player but are not willing to throw out our record collection. Similarly, we want to add optical disks to our computers without making our magnetic disk obsolete. It is extremely important that vendors consider these compatibility issues when designing new products. Assuming that new technology can be implemented in an evolutionary fashion, the rapid growth of the computer and communications industries should continue indefinitely.

The implication for the telecommunications department in a company is that it will be working in a world of constant change. As new technologies emerge or standards come into widespread use, there will be good business reasons to take advantage of them. Taking advantage will mean changes to telecommunications hardware, software, and procedures—or all three simultaneously.

One of the key jobs of the telecommunications manager is to learn how to cope with constant change and to implement it smoothly and without disruption. Not implementing change is tantamount to putting the company at a competitive disadvantage or out of business. Implementing new products, services, or techniques carelessly is disruptive, may waste money, and can spur a loss of productivity or business effectiveness.

CASE STUDY

The Dow Corning Communications Network Vision

Dow Corning's vision of the future for telecommunications in the company has several facets. There is a strong belief that the personal computer workstations used by its employees will continue to become more powerful and that they will be used for an increasing variety of activities. Management also believes that the capacity and capability of the LANs and servers will continue to grow and expand. The company also believes that in order to take advantage of the increasing capability of the equipment and software that employees' have at their disposal, they must undergo continuous learning. To that end, the skill development efforts within the company have been increased.

To support the growth in workstation and application sophistication, higher transmission speeds to the desktops will be required, especially as it becomes possible to integrate voice and image transmission with data. LAN technology, especially gigabit Ethernet, and more extensive use of fiber optics and the FDDI standard, hold the promise of being able to deliver the needed speed, and the company is already using the technology.

Management recognizes that network and computer systems are going to have to be of higher reliability. They must approach the reliability of the public telephone system and essentially be operational all the time. In order to achieve this reliability, the network management procedures and practices must be improved so that the help desk knows of problems before the users. More redundancy must be added to the existing network to make it more fail-safe so that there is no single point of failure.

Finally, Dow Corning believes that the network can be used to help it communicate better with other companies with which it deals, most notably its customers and suppliers. Dow Corning has made its first forays into the world of electronic business using the Internet, and management believes that the company can improve its customer service and add value to its products by providing new types of information and new ways of doing business.

The implications of all of these changes for the telecommunications staff are significant because the capacity and capabilities of the network of the future are likely to far surpass today's capabilities. The department's people will need to acquire new technical and management skills in order to be ready to implement and operate new telecommunications systems. Having innovative and creative telecommunications people will continue to be an important success factor for the company.

QUESTIONS

1. Is the vision of more processing being done by personal computer workstations a fundamental change for Dow Corning, or simply a by-product of evolving hardware and software capabilities?
2. What kinds of value can Dow Corning, a specialty chemical company, add to the products it sells by using its information and telecommunications capabilities?

3. What would be the pros and cons of bringing in a nontechnical manager from another part of the company, someone with little telecommunications knowledge, to manage the telecommunications work at Dow Corning? What about bringing in a person with a similar background from outside the company?

4. What will be the most immediate effects on the telecommunications network if Dow Corning starts selling its products on the Internet?

PROBLEMS AND PROJECTS

1. Count the number of fractional horsepower motors you have in your home. Does the number surprise you? Can you count the number of microprocessors? Identify some novel places where those microprocessors will be used in the near future.

2. If sophisticated speech recognition equipment were available, it would theoretically be possible for you to dictate your college term papers into your personal computer and have the result printed immediately. Is your thought process organized well enough so that this would really work for you? What other obstacles will you have to overcome so that you can take advantage of the speech recognition capability when it is available to you?

3. Based on your knowledge of fiber-optic technology, what are some of the practical problems of running a fiber to every desk?

4. Thinking about the differences in the work of clerical workers and managers, identify some of the differing characteristics that would be required in a terminal designed for each group.

5. Engineers and other technical people tend to push for standardization in all parts of our society. What are some of the advantages of not having all communications standardized?

6. The banking industry has been using EDI for years to transmit money between financial institutions. If you had EDI capability available to you personally, can you think of any uses you could make of it?

7. If automobile navigation by satellite were widely available, as described in the text, how else could the communications capability be used by the driver or other occupants of the car?

8. The text mentioned that telecommuters must go into the office periodically for meetings. What would be required to allow them to participate in the meetings from their homes?

9. What are some sociological problems that may occur because of ubiquitous, high-speed data communications being readily available?

10. Describe how telemedicine can make a contribution to reducing the cost of health care.

Vocabulary

digital signal processor (DSP)
voice compression
video signal compression
telecommuting
electronic document interchange (EDI)
telediagnosis
telemedicine
Global Positioning System (GPS)
smart home

References

Alsop, Stewart. "May Your Net Connection Be as Fast as Mine." *Fortune* (October 11, 1999): 311.

Borriello, Gaettano, and Roy Want. "Embedded Computation Meets the World Wide Web." *Communications of the ACM* (May 2000): 59–66.

Estrin, Deborah, Ramish Govidan, and John Heidemann. "Embedding the Internet." *Communications of the ACM* (May 2000): 38–41.

Higgins, Kelly Jackson. "Diagnosis: Telemedicine Network Transports IP, ATM." *Network Computing* (December 13, 1999): 76.

Huston, Terry L., and Janis L. Huston. "Is Telemedicine a Practical Reality?" *Communications of the ACM* (June 2000): 91–95.

Lindquist, Christopher. "The Totally Digital House." *PC Computing* (January 2000): 144–164.

Pottie, G. J., and W. J. Kaiser. "Wireless Integrated Network Sensors." *Communications of the ACM* (June 2000): 91–95.

Stepanek, Marcia. "A Small Town Reveals America's Digital Divide." *Business Week* (October 4, 1999): 188–198.

Sukhatme, Gaurav S., and Maja J. Mataric. "Embedding Robots Into the Internet." *Communications of the ACM* (May 2000): 67–73.

Tennenhouse, David. "Proactive Computing." *Communications of the ACM* (May 2000): 43–50.

Vergano, Don. "Remote Care Can Come Home." *USA Today* (February 17, 2000): 10D.

Warner, Melanie. "Telesurgeons Operate with Video, Robots." *The Japan Times* (March 9, 1997).

_____. "Working at Home—The Right Way to Be a Star in Your Bunny Slippers." *Fortune* (March 3, 1997): 165–166.

APPENDICES

Appendix A

The ISO-OSI Model

The ISO-OSI model was introduced in Chapter 3 before you learned about the technical details of telecommunications; hence, the explanation about the purpose and functions of the layers of the OSI model was relatively simple. This appendix provides a somewhat more detailed and technical explanation of the functions of the seven layers.

Layer 1: Physical Link Layer Layer 1 defines the electrical standards and signaling required to make and break a connection on the physical link and to allow bit streams from terminals or computers to flow onto the network. It specifies the modem interface between the data terminal equipment and the line. For analog circuits, the most common interface is the V.24 or RS-232-C standard. The standard for the interface to digital circuits is X.21. As we saw in Chapter 8, the physical layer is concerned with voltage levels, current flows, and whether the transmission is simplex, half-duplex, or full-duplex. Besides the interface standards, the physical link layer defines how connections will be established and terminated. Other functions defined in this layer include monitoring channel signals, clocking on the channel, handling interrupts in the hardware, and informing the data link layer when a transmission has been completed.

Layer 1 is the only level of the architecture in which actual transmissions take place. Although we talk about layers 2 through 7 "communicating" with each other, it is important to remember that they must pass the information they wish to communicate down through the layers to layer 1 for transmission to the other correspondent in the communication.

Layer 2: Data Link Layer Layer 2 defines standards for structuring data into frames and sending the frames across the network. It is concerned with the following questions:

- How does a machine know where a frame of data starts and ends?
- How are transmission errors detected and corrected?
- How are polling and addressing handled?
- How are machines addressed?

The answers to these questions make up the data link protocol. The OSI protocol is High-Level Data Link Control (HDLC), which was discussed in Chapter 10.

For LANs, the data link layer is divided into two sublayers: the media access control (MAC) layer to define access to the network (the two alternatives are CSMA and token techniques, discussed in Chapter 10) and the logical link control (LLC), which defines the protocol, a subset of HDLC, for use on LANs.

It is important to know that the IEEE, which establishes the 802 standards for LANs, fully subscribes to the OSI architecture.

Layer 2 is normally located in the front-end processor (if one exists) or server at the host end of the circuit and in a remote cluster controller or intelligent terminal. Layer 2 must work closely with the modem to accomplish its work. Whereas modems ensure that bits are accurately sensed from the communications line at the receiving end, layer 2 groups the bits into characters for further processing.

Layer 2 is also responsible for checking the VRC, LRC, and CRC codes transmitted with the data, for acknowledging the receipt of good frames or blocks, and for requesting the retransmission of those that are in error. Another type of error that layer 2 handles can occur when data is received from the circuit faster than the receiver can handle it. When this happens, layer 2 signals the transmitter to either slow down or stop transmitting until further notice. This is called *pacing*.

Layer 3: Network Layer The primary functions of layer 3 are network addressing and routing. Layer 3 also generates acknowledgments that an entire message has been received correctly. When a message is received from layer 4 for transmission, layer 3 is responsible for breaking it into blocks or packets of a suitable size for transmission. In most packet switching networks, for example, the packet size might be 128 characters. Layer 3 assigns the correct destination address to the packet and determines how it should be routed through the network. Specifically, it decides on which communications circuit to send a packet or block.

At the receiving end, layer 3 reassembles the packets or blocks into messages before passing them up to layer 4. Layer 3 might receive a message from another computer that is not destined for its location. In that case, it turns the message around and sends it back down to level 2 for forwarding through the network to the ultimate destination. This situation occurs in multinode networks where some nodes relay data for others.

Over the past 30 years, countries and companies have established their own standards for layers 1 through 3 (without thinking of them as layers). Many networks have been implemented using these proprietary standards. Therefore, the changeover process to international standards is difficult and time-consuming, even if a country or company wants to make the change.

Layer 4: Transport Layer Layer 4 selects the route the transmission will take between two DTEs. A favorite analogy for layer 4 is to compare it to the postal system. A user mails a letter but does not know (or care) exactly how the letter is transported to its destination as long as the service is reliable. In a telecommunications network, layer 4 selects the network service to be used to transport the message when options exist. For example, if a computer has both leased lines and a packet switching network connected to it, layer 4 decides which of the two services should be used to transport a particular message. To a terminal or computer, layers 4, 3, 2, and 1 together provide the transportation service for the user's message.

Layer 4 also contains the capability to handle user addressing. At the transmitting station, this means that network addresses that are meaningful to the user, such as location codes, terminal names, or other mnemonic codes, are converted to addresses that are meaningful to the network software and hardware. These network addresses usually are binary numbers. On the receiving end, network addresses must be converted back to user addresses.

Another function of layer 4 is to control the flow of messages so that a fast computer cannot overrun a slow terminal. This flow control works in conjunction with the flow control at layer 2, but layer 4 is concerned with controlling entire messages, whereas layer 2 is concerned with controlling the flow of frames. Layer 4 could allow a message to be sent, but the individual frames that make up the message could be delayed by layer 2 because of slowdown signals that it had received from layer 2 at the receiving end.

Layer 4 also prevents the loss or duplication of entire messages. Whereas more detailed checks occur at lower levels to ensure that frames and blocks of data are received correctly, layer 4 must implement controls to ensure that entire messages are not lost and, if necessary, to request their retransmission.

Other functions of layer 4 include multiplexing several streams of messages from higher levels onto one physical circuit and adding appropriate headers to messages to be broadcast to many recipients.

The transport layer is sometimes implemented in the host computer and sometimes in the front-end processor when one exists, or in a higher-level protocol.

Layer 5: Session Layer A *session* is a temporary connection between machines or programs for an exchange of messages according to rules that have been agreed on for that exchange. The session is the first part of the communications process that is directly visible to the user. Users directly request the establishment of sessions between their terminals and computers when they begin the sign-on or log-on process.

Before a session can begin, the machines or programs must agree to the terms and conditions of the session, such as who transmits first, for how long, and so on. Clearly, there will be differences in these rules between interactive sessions and batch sessions and between terminals of different types. Given an appropriate terminal, there is no inherent reason why a user cannot have multiple sessions in progress simultaneously.

Layer 5 establishes, maintains, and breaks a session between two systems or users. If a session is unintentionally broken, session layer must reestablish it. Session layer also provides the ability for the user to abort a session. For example, the BREAK key or ESC key on the terminal may be used for this purpose.

The implementation of priorities for expediting some messages or traffic occurs in this layer, as do certain accounting functions concerning session duration. These are used to charge the user for network time.

The session layer is usually implemented in the host computer access method or server software.

Layer 6: Presentation Layer The presentation layer deals with the way data is formatted and presented to the user at the terminal at the receiving end of a connection. It also performs similar formatting at the transmitting end so that the data from a terminal is presented to the lower layers in a constant format for transmission. An application programmer on the host computer writes programs to talk to a standard virtual terminal, as was discussed in Chapter 11, and the layer 6 software performs a transformation to meet the specifications of the real terminal in use. Screen formatting, such as matching the message to the number of characters per line and the number of lines per screen, would also be done.

Other functions that occur in layer 6 are code conversion, data compaction, and data encryption. With the exception of data encryption, which is often implemented in hardware, the rest of layer 6 is almost always performed by software in the host computer.

Layer 7: Application or User Layer Layer 7 is the application program or user that is doing the communicating. This is the layer at which data editing, file updating, or user thinking occurs. This layer is the source or ultimate receiver of data transmitted through the network.

A great deal of effort has been expended to define and standardize some common elements of applications that operate in layer 7. The activity has six major thrusts.

1. Common application service elements (CASE)—this work is aimed at defining standards for such things as log on and password identification as well as checkpoint, restart, and backup processes.
2. Job transfer and manipulation (JTM)—this defines standards for the transfer of batch jobs from one computer to another.
3. File transfer, access, and management (FTAM)—this work is aimed at defining standards for the transfer of files between systems and for providing record-level access to a file on another computer.
4. Message oriented interchange systems (MOTIS)—concerned with defining standards for interconnecting the world's many message exchange systems, this work is also known as the ITU-T X.400 standard.
5. Office document architecture/office document interchange facility (ODA/ODIF)—this work is aimed at providing standards to allow the transfer, edit, and return of documents across systems from multiple vendors.
6. Virtual terminal services (VTS)—this work is concerned with precise definition of the virtual terminal concept, including character, graphics, and image terminals, as well as hard-copy devices.

Telecommunications Trade-Offs

Throughout this book there have been many references to the trade-offs made in planning, designing, or operating a communications system. Trade-offs occur because there are alternate solutions to many communications problems and different ways to design products and provide services. In some cases, the trade-off requires that an absolute *either/or* decision be made. For example, either vendor A or vendor B will be selected to provide the front-end processor. There is no way to buy half of the FEP from one vendor and half from another. In most situations, however, an absolute choice does not have to be made between the alternatives. For example, a company might well decide to have some leased circuits and some switched circuits; it doesn't have to select only one type or the other.

The important thing is to be aware of the options that exist so that you can make a proper selection to solve a problem optimally. Knowledge is the key, for if you don't know that options exist and there are trade-offs to be made, you can't choose between the alternatives.

Many trade-offs are interrelated. For example, the choice of an asynchronous versus a synchronous protocol may be directly related to the choice of terminals to be used for the application. The use of a vendor's proprietary protocol can occur only if that vendor has been selected to provide communications products. Some trade-off decisions are primarily made by the company that is using the communications products. Others are determined by the company in combination with the vendor. Still others are primarily made by the vendors, typically when they are designing their products or services.

The following list itemizes many communications trade-offs and alternatives. Depending on the level of detail, such a list could go on for pages. The trade-offs included here are the most important ones faced by network managers and designers.

CUSTOMER TRADE-OFFS

Network

- Mix of LAN and WAN technology, where appropriate
- Ethernet versus token ring versus another technology
- Type of wiring or cabling for a LAN
- Brand of network operating system
- Install and maintain the network with employees versus outsourcing the work

WAN Circuits

- Private network versus public network
- Private versus leased versus switched circuits
- Two-wire versus four-wire circuits
- Circuit speed versus cost
- Point-to-point circuits versus multipoint circuits
- Conditioned versus unconditioned circuits
- Compressing versus not compressing data transmissions
- Multiplexing versus concentrating versus inverse concentrating

Terminals

- Intelligent versus smart versus dumb terminals
- General-purpose terminal versus application-oriented terminal
- VDT versus printing terminal

Applications

- Voice mail versus text mail versus both
- Teleconferencing versus travel
- Encrypting versus not encrypting data
- The level of reliability, availability, and responsiveness that is required versus the cost of providing it

Other Trade-Offs

- Custom software versus off-the-shelf programs
- Leasing versus buying hardware
- Vendor A versus vendor B
- Performing certain telecommunications services inside the company versus contracting them to outsiders

CUSTOMER-VENDOR TRADE-OFFS

- Architected versus nonarchitected communications approach
- Terrestrial versus satellite circuits
- Analog versus digital transmission
- Asynchronous versus synchronous transmission

VENDOR TRADE-OFFS

- Front-end processor versus direct connection of WAN circuits to a computer
- Hardware versus software implementation of certain functions
- Entering versus not entering a certain geographic market
- Providing versus not providing a certain telecommunications product or service

Telecommunications Periodicals and Newsletters

The rapid growth of the telecommunications industry in recent years has brought with it an explosion in the number of trade journals and specialized newsletters aimed at providing information to telecommunications professionals. This list contains information about the most widely read telecommunications publications.

Communications News. Published monthly by Nelson Publishing, 2504 N. Tamiami Trail, Nokomis, FL 34275. (941) 966-9521.

Communications Today. Published weekly by Phillips Business Information, Inc., 1201 Seven Locks Road, Suite 300, Potomac, MD 20854. (301) 424-3338.

Datacomm Advisor: IDC's Newsletter Covering Network Management—Products, Services, Applications. Published monthly by International Data Corporation, 5 Speen Street, Framingham, MA 01701. (508) 872-8200.

Internet Week. Published weekly by CMP Publications, Inc., 600 Community Drive, Manhasset, NY 11030. (516) 562-5000.

Network Magazine. Published monthly by CMP Publications, Inc., P.O. Box 2013, Skokie, IL 60076. (800) 577-5356.

Network Strategy Report. Published by Forrester Research, Inc., 400 Technology Square, Cambridge, MA 02139. (617) 497-7090.

Network World. Published weekly by Network World, 118 Turnpike Road, Southborough, MA 01772. (508) 460-3333.

Satellite Communications. Published monthly by Intertec, 6151 Powers Ferry Road, N.W., Atlanta, GA 30339. (770) 955-2500.

Telecommunications. Published monthly by Horizon House, 685 Canton Street, Norwood, MA 02062. (781) 769-9750.

Teleconnect. Published monthly by CMP Publications, Inc., 12 West 21st Street, New York, NY 10010. (212) 691-8215.

Telephony. Published weekly by Intertech Publishing Corporation, One IBM Plaza, Suite 2300, Chicago, IL 60604. (312) 595-1080.

The Voice Report. Published bimonthly by United Communications Group, 11300 Rockville Pike, Suite 1100, Rockville, MD 20852. (301) 816-8950.

Telecommunications Professional and User Associations

This list contains the names, addresses, and telephone numbers of the major national telecommunications associations. Information about membership in these organizations can be obtained by contacting the association at the address listed. In addition to the groups listed here, there are numerous regional, state, and local telecommunications organizations throughout the country.

Association of College and University Telecommunication Administrators (ACUTA)
152 W. Zandale Dr., Suite 200
Lexington, KY 40503
(859) 278-3338

Canadian Business Telecommunications Alliance (CBTA)
161 Bay St., Suite 3650
P.O. Box 705
Trust Tower BCE Place
Toronto, Ontario
Canada M5J 2S1

Communications Managers' Association (CMA)
1201 Mt. Kemble Ave.
Morristown, NJ 07960
(800) 867-8008 or (973) 425-1700

Computer and Communications Industry Association (CCIA)
666 Eleventh Street N.W.
Suite 600
Washington, D.C. 20001
(202) 783-0070

Energy Telecommunications and Electrical Association (ENTELEC)
666 Eleventh Street N.W.
Suite 600
Washington, D.C. 20001
(281) 357-8700 or (888) 503-8700

Institute of Electrical and Electronics Engineers (IEEE)
Council on Communications
3 Park Avenue, 17th Floor
New York, NY 10016
(212) 419-7900

International Communications Association (ICA)
3530 Forest Lane
Suite 200
Dallas, TX 75234
(800) 422-4636

International Telecommunications Users' Group (INTUG)
INTUG Secretary
18 Westminster Palace Gardens
Artillery Row
London SW1P 1RR
England

North American Telecommunications Association (NATA)
P.O. Box 23015
Washington, D.C. 20036
(202) 479-0970

Organization for the Promotion and Advancement of Small Telecommunications Companies (OPASTCO)
21 Dupont Circle N.W., Suite 700
Washington, D.C. 20036
(202) 659-5990

Telecommunications Industry Association (TIA)
2001 Pennsylvania Ave. N.W., Suite 800
Washington, D.C. 20006
(202) 783-1338

The Information Technology & Telecommunications Association (TCA)
74 New Montgomery, Suite 230
San Francisco, CA 94105
(415) 777-4647

United States Telephone Association (USTA)
1401 H St. N.W., Suite 600
Washington, D.C. 20005
(202) 326-7300

Organizations That Conduct Telecommunications Seminars

Many public and private organizations conduct telecommunications education courses and seminars. In addition to those listed here, many colleges and universities have regularly scheduled classes and occasional seminars covering the gamut of telecommunications topics.

BCR Enterprises, Inc.
999 Oakmont Plaza Drive, Suite 100
Westmont, IL 60559
(800) 227-1234

C.M.P. Media, Inc.
12 W. 21st Street
New York, NY 10010
(212) 691-8215

International Communications Association (ICA)
2375 Villa Creek Dr., Suite 200
Dallas, TX 75234
(900) 422-4636

Peregrine Systems
616 Marriot Dr., Suite 500
Nashville, TN 37214
(615) 872-9000

Technology Transfer Institute
741 Tenth Street
Santa Monica, CA 90402-2899
(310) 394-8305

Tele-Strategies
1355 Beverly Road, Suite 110
McLean, VA 22101
(703) 734-7050

Tellabs, Inc.
4951 Indiana Avenue
Lisle, IL 60532
(630) 378-8800

United States Telecom Association (USTA)
1401 H St. N.W., Suite 600
Washington, D.C. 20005-2164
(202) 326-7300

Internet World Wide Web Addresses

There are many sites on the Internet that have valuable information about telecommunications. The following list could never be complete but lists some of the sites the author has found valuable, as well as the addresses of many of the major companies in the telecommunications marketplace. You may want to visit these web sites to gain additional or the most up-to-date information, or to do additional research. The Internet is dynamic, so web addresses do change, but hopefully those on this list will be quite stable.

Organization	Internet WWW Address (URL)
2wire	www.2wire.com
Alcatel	www.alcatel.com
Assistive Technology Industry Association	www.atia.com
AT&T Corporation	www.att.com
Bell South	www.bell-south.com
Cable Modem Information Network	www.cable-modem.net
Cable Modem University	www.catv.org
Canon	www.canon.com
Cisco Systems	www.cisco.com
Comdisco	www.comdisco.com
Competitive Local Exchange Carriers (CLEC)	www.clec.com
Digital Equipment Corporation	www.compaq.com
Dow Corning Corporation	www.dowcorning.com
Ericsson	www.ericsson.com
Federal Communications Commission (FCC)	www.fcc.gov
Fujitsu	www.fujitsu.com
Hewlett Packard	www.hp.com
IBM	www.ibm.com
Internet2 Project	www.internet2.edu
Lucent Technologies	www.lucent.com
Matsushita	www.matsushita.co.jp
Microsoft	www.microsoft.com
Motorola	www.motorola.com
N.V. Philips	www.philips.com

Organization	Internet WWW Address (URL)
Netscape	www.netscape.com
Nippon Electric Company (NEC)	www.nec.com
Nokia	www.nokia.com
Nortel Networks Corporation	www.nortelnetworks.com
North American Numbering Plan	www.nampa.com
Nippon Telegraph and Telephone Corporation (NTT)	www.ntt.com
QWEST Communications International Inc.	www.qwest.com
Ricoh	www.ricoh.com
RSA Security, Inc.	www.rsasecurity.com
SBC Communications, Inc.	www.sbc.com
Siemens	www.siemens.com
Sprint	www.sprint.com
Toshiba	www.toshiba.com
Verizon	www.verizon.com
Who is using the Internet	www.cyberatlas.com
WorldCom, Inc.	www.worldcom.com

Abbreviations and Acronyms in the Text

ACD automatic call distributor
ACK acknowledge
ACP Airline Control Program (IBM)
A/D analog to digital
ADPCM adaptive differential pulse code modulation
ADSL asymmetric digital subscriber line
AHS American Hospital Supply
AM amplitude modulation
ANI automatic number identification
ANSI American National Standards Institute
APA all points addressable
API application program interface
ARPA Advanced Research Projects Agency
ARQ automatic repeat request
ASCII American Standard Code for Information Interchange
ASR automatic-send-receive
ATM asynchronous transfer mode
ATM automatic teller machine
AT&T American Telephone and Telegraph Company
BALUN balanced-unbalanced
BCC block check character
BCD binary-coded decimal
BELLCORE Bell Communications Research
BISDN broadband ISDN
BISYNC binary synchronous communications
BNA Burroughs Network Architecture (Burroughs/Unisys)
BOC Bell Operating Company
bps bits per second
BSC binary synchronous communications
CAD computer-aided design or computer-assisted drafting
CAM communications access method
CASE common application service elements
CBX computerized branch exchange
CCIS common channel interoffice signaling
CCITT Consultative Committee on International Telegraphy and Telephony
CCS centa call seconds
CCU cluster control unit
CDMA code division multiple access
CDR call detail recording
CI–I Computer Inquiry I
CI–II Computer Inquiry II
CI–III Computer Inquiry III
CIO chief information officer
CIR committed information rate
CLEC competitive local exchange carrier
CMIP Common Management Information Protocol
CODEC coder/decoder
COE Council of Europe
CPE customer premise equipment
CR carriage return
CRC cyclic redundancy checking
CRT cathode ray tube
CRTC Canadian Radio-television and Telecommunications Commission
CSMA/CA carrier sense multiple access with collision avoidance
CSMA/CD carrier sense multiple access with collision detection
CSU channel service unit
D/A digital to analog
DAA data access arrangement
dB decibel
DBCS double byte character set
DBS direct broadcast satellite
DC direct current

DCE data circuit-terminating equipment
DDCMP Digital Data Communications Message Protocol (DEC)
DDD direct distance dialing
DDP distributed data processing
DEC Digital Equipment Corporation
DES data encryption standard
DID direct inward dialing
DLE data link escape
DNA Digital Network Architecture (DEC)
DOC Department of Communications (Australia)
DOD direct outward dialing
DP data processing
DPAM demand priority access method
DPSK differential phase shift keying
DSL digital subscriber line
DSP digital signal processor
DSS digital satellite service
DSU data service unit
DTE data terminal equipment
DTMF dual-tone-multifrequency
EBCDIC Extended Binary Coded Decimal Interchange Code
EDI electronic document interchange
EIA Electrical Industries Association
e-mail electronic mail
ENQ enquiry
EOA end of address
EOB end of block
EOM end of message
EOT end of transmission
EPSCS enhanced private switched communication service
ESC escape
ESS electronic switching system
ETB end of text block
ETN electronic tandem network
ETX end of text
EUC End User Computing (Dow Corning)
FAX facsimile
FCC Federal Communications Commission
FDDI fiber distributed data interface
FDM frequency division multiplexing
FDMA frequency division multiple access
FDX full-duplex
FEC forward error correction
FEP front-end processor
FIFO first in, first out
FM frequency modulation
FSK frequency shift keying
FTAM file transfer, access, and management
FTP file transfer protocol
FX foreign exchange
GE General Electric Company
GEIS General Electric Information Services
GHz gigahertz
GM General Motors Corporation
GOSIP Government Open Systems Interconnection Protocol
GPS global positioning system
GSM global system for mobile communications
GUI graphical user interface
HBO Home Box Office
HDLC high-level data link control
HDX half-duplex
HTML hypertext markup language
HTTP hypertext transfer protocol
Hz hertz
IBM IBM Corporation
ICC Interstate Commerce Commission
IDDD international direct distance dialing
IEEE Institute of Electrical and Electronic Engineers
ILEC incumbent local exchange carrier
IRM information resource management
ISDN Integrated Services Digital Network
ISO International Organization for Standardization
ISP Internet service provider
IP Internet protocol
IPX internetwork packet exchange (Novell)
ITB intermediate text block
ITT International Telephone and Telegraph Company
ITU International Telecommunications Union
ITU-T International Telecommunications Union-Telecommunications Standardization Sector
IXC interexchange carrier
JTM job transfer and manipulation
kbps thousands of bits per second
KDD Kokusai Denshin Denwa Co. (Japan)
kHz kilohertz
KSR keyboard-send-receive
LAN local area network
LAPB link access procedure, balanced
LAPD link access procedure, D-channel
LAPF link access procedure for frame mode bearer services
LATA local access and transport area
LCD liquid crystal display
LEC local exchange carrier (or company)
LED light-emitting diode
LF line feed
LLC logical link control
LPC linear predictive coding
LRC longitudinal redundancy checking
LU logical unit (IBM)
MAC media access control
MAC moves, adds, and changes
MAN metropolitan area network
MAP Manufacturing Automation Protocol
MAU multistation access unit
Mbps millions of bits per second
MCI MCI Communications Corporation

MFJ modified final judgment
MH modified Huffman encoding
MHz megahertz
MIB management information base
MIME multipurpose internet mail extensions
MMR modified modified read encoding
MODEM modulator/demodulator
MOTIS message-oriented interchange systems
MPT Ministry of Posts and Telecommunications (Japan, Russia)
MR modified read encoding
MTBF mean time between failures
MTSO mobile telephone switching office
MTTR mean time to repair
MUX multiplexer
NAK negative acknowledge
NANP North American numbering plan
NAU network addressable unit (IBM)
NCC network control center
NCP network control program
NEC Nippon Electric Company
NECA National Exchange Carriers Association
NetBIOS Network Basic Input Output System
NIC network interface card
NMS Netware Management System (Novell)
NOS network operating system
NRZ nonreturn to zero
NSA National Security Agency
NTS Windows NT server (Microsoft)
NTT Nippon Telegraph and Telephone Company
NUL null
OCC other common carrier
OCR optical character recognition
ODA office document architecture
ODIF office document interchange facility
OECD Organization for Economic Cooperation and Development
OFTEL Office of Telecommunications (United Kingdom)
ONA open network architecture
OSI Open Systems Interconnection
OV office vision (IBM)
PABX private automatic branch exchange
PACTEL Pacific Telesis Corporation
PAD packet assembly/disassembly
PAM phase amplitude modulation
PBX private branch exchange
PC personal computer
PCM pulse code modulation
PCN personal communications network
PCS personal communications service
PDN packet data network
PDU protocol data unit
PEL picture element
PERT program evaluation review technique
PFK program function key
PHS personal handyphone system
PIN personal identification number
Pixel picture element
PM phase modulation
POP point of presence
POS point of sale
POTS plain old telephone service
PPP point-to-point protocol
PSC public service commission
PSK phase shift keying
PSN public switched network
PSTN public switched telephone network
PTT post, telephone, and telegraph
PU physical unit (IBM)
PUC public utility commission
QAM quadrature amplitude modulation
RBOC Regional Bell Operating Company
RFP request for proposal
RFQ request for quotation
RJE remote job entry
RMON remote monitoring
RO receive-only
RVI reverse interrupt
SBT six-bit transcode
SCC specialized common carrier
SDLC synchronous data link control (IBM)
SDN software defined network
SDSL symmetric digital subscriber line
SLIP serial line internet protocol
SMDR station message detail recording
SMTP simple mail transfer protocol
SNA Systems Network Architecture (IBM)
SNMP Simple Network Management Protocol
SOH start of header
SONET synchronous optical network
SPX sequenced packet exchange (Novell)
SS7 signaling system 7
SSCP system service control point (IBM)
STDM statistical time division multiplexing
STP shielded twisted pair
STX start of text
SWIFT Society for Worldwide Interbank Financial Telecommunications
SYN synchronization
TASI time assignment speech interpolation
TCAM telecommunications access method
TCM telecommunications monitor
TCM trellis code modulation
TCP/IP transmission control protocol/internet protocol
TDM time division multiplexing
TDMA time division multiple access
TNDF transnational data flow
TOP Technical Office Protocol
TPF Transaction Processing Facility (IBM)
TPNS teleprocessing network simulator (IBM)
TTD temporary text delay
TWX teletypewriter exchange system
UDLC Universal Data Link Control (Sperry/Unisys)
UIFN universal international freephone numbering
UPC universal product code

URL	uniform resource locator
UTP	unshielded twisted pair
VAN	value-added network
VDSL	very-high-rate digital subscriber line
VDT	video display terminal
VDU	video display unit
VINES	Virtual Network Integrated Server (Banyan)
VoIP	Voice over IP
VRC	vertical redundancy checking
VSAT	very small aperture terminal
VTAM	Virtual Telecommunications Access Method (IBM)
VTS	virtual terminal services
WAIS	wide area information server
WAN	wide area network
WAP	wireless application protocol
WATS	Wide Area Telecommunications Service
WDM	wavelength division multiplexing
WWW	World Wide Web

GLOSSARY

access line. A telecommunications line that continuously connects a remote station to a switching exchange. A telephone number is associated with the access line.
access method. Computer software that moves data between main storage and input/output devices.
acknowledgment. The transmission, by a receiver, of acknowledgment characters as an affirmative response to a sender.
acknowledgment character (ACK). A transmission control character transmitted by a receiver as an affirmative response to a sender.
acoustic coupler. Telecommunications equipment that permits use of a telephone handset to connect a terminal to a telephone network.
acoustic coupling. A method of coupling data terminal equipment (DTE) or a similar device to a telephone line by means of transducers that use sound waves to or from a telephone handset or equivalent.
activation. In a network, the process by which a component of a node is made ready to perform the functions for which it was designed.
active line. A telecommunications line that is currently available for transmission of data.
adaptive differential pulse code modulation (ADPCM). A variation of pulse code modulation in which only the difference in signal samples is coded.
adaptive equalizer. An equalizer circuit in a modem that adjusts itself to the exact parameters of the incoming waveform based on the known characteristics of a standard training signal.
adaptive Huffman coding. A type of character compression in which the text is continuously scanned to ensure that the fewest bits are assigned to the characters appearing most frequently.
add-in circuit board protocol converter. An electronic circuit for converting one protocol to another. Contained on a circuit board, it can be added to a personal computer or other telecommunications device.
address. (1) A character or group of characters that identifies a data source or destination. (2) To refer to a device or an item of data by its address. (3) The part of the selection signals that indicates the destination of a call.
addressing. The means by which the originator or control station selects the unit to which it is going to send a message.
addressing characters. Identifying characters sent by a device on a telecommunications line that cause a particular station (or component) to accept a message sent to it.
aerial cable. A telecommunications cable connected to poles or similar overhead structures.
aeronautical telephone service. Telephone service provided in airplanes to communicate with telephones on the ground.
airline reservation system. An online application in which a computing system is used to keep track of seat inventories, flight schedules, passenger records, and other information. The reservation system is designed to maintain up-to-date data files and to respond, within seconds, to inquiries from ticket agents at locations remote from the computing system.
algorithm. A set of mathematical rules.
all-points-addressable (APA). An attribute of a VDT or printer that allows each individual dot on the screen or spot on a page to be individually addressed for output or input.
alphanumeric character. Pertaining to a character set that contains letters, digits, and usually other characters, such as punctuation marks.
American National Standards Institute (ANSI). An organization formed for the purpose of establishing voluntary industry standards.
American Standard Code for Information Interchange (ASCII). The standard code, using a coded character set consisting of 7-bit coded characters (8 bits, including the parity check), used for information interchange among data processing systems, data communications systems, and associated equipment. The ASCII character set consists of control characters and graphic characters.
amplifier. A device that, by enabling a received wave to control a local source of power, is capable of delivering an enlarged reproduction of the wave.
amplitude. The size or magnitude of a voltage or current analog waveform.
amplitude modulation (AM). (1) Modulation in which the amplitude of an alternating current is the characteristic varied. (2) The variation of a carrier signal's strength (amplitude) as a function of an information signal.
analog. Pertaining to data in the form of continuously variable physical quantities.

analog channel. A data communications channel on which the information transmitted can take any value between the limits defined by the channel. Voice-grade channels are analog channels.
analog signal. A signal that varies in a continuous manner. Examples are voice and music. *Contrast with* digital signal.
analog-to-digital (A/D) converter. A device that senses an analog signal and converts it to a proportional representation in digital form.
analysis. The methodical investigation of a problem and the separation of the problem into smaller related units for further detailed study.
analyst. A person who defines problems and develops algorithms and procedures for their solution.
answerback. The response of a terminal to remote control signals.
application level firewall. A firewall computer that examines and controls data at the application level. The server looks at entire messages and does a more detailed analysis of the appropriateness before making a decision as to whether to let the traffic pass.
application program interface (API). A mechanism for application programs to send data to and receive data from a LAN.
applications server. On a LAN, a computer that provides processing capacity for applications that are shared by many people.
architecture. A plan or direction that is oriented toward the needs of a user. An architecture describes "what"; it does not tell "how."
area code. A three-digit number identifying a geographic area of the U.S. and Canada to permit direct distance dialing on the telephone system.
asymmetric digital subscriber line (ADSL). A transmission technology that delivers high-speed signals over twisted pair telephone wires. Speeds vary, and are normally slower upstream than downstream, but are in the range of 16 to 640 Kbps upstream, and 1.5 to 9 Mbps downstream. *See also* digital subscriber line.
asymmetric key. In an encryption system, the key that is used for encryption and the key used for decryption are not the same. *See also* public key encryption.
asynchronous. Without a regular time relationship.
asynchronous transfer mode (ATM). A packet switching technique that uses fixed-length packets called cells and is designed to operate on high-speed lines. ATM effectively eliminates any delay in delivering the packets, making it suitable for voice or video.
asynchronous transmission (asynch). (1) Transmission in which the time of occurrence of the start of each character or block of characters is arbitrary. (2) Transmission in which each information character is individually synchronized (usually by the use of start elements and stop elements).
attention interruption. An I/O interruption caused by a terminal user pressing an attention key or its equivalent.
attenuation. A decrease in magnitude of current, voltage, or power of a signal in transmission between points. It is normally expressed in decibels.
attenuation distortion. The deformation of an analog signal that occurs when the signal does not attenuate evenly across its frequency range.
audio frequencies. Frequencies that can be heard by the human ear (approximately 15 Hertz to 20,000 Hertz).
audio response unit. An output device that provides a spoken response to digital inquiries from a telephone or other device. The response is composed from a prerecorded vocabulary of words and can be transmitted over telecommunications lines to the location from which the inquiry originated.
audiotex. A voice mail system that can access a database on a computer.
audit. To review and examine the activities of a system, mainly to test the adequacy and effectiveness of control procedures.
audit trail. A manual or computerized means for tracing the transactions affecting the contents of a record.
authorization code. A code, typically made up of the user's identification and password, used to protect against unauthorized access to data and system facilities.
auto answer. *See* automatic answering.
auto dial. *See* automatic dialing.
auto-poll. A feature that allows a piece of hardware to poll stations and accept a negative response without interrupting a higher level piece of hardware or software.
automated attendant. A type of telephone service in which a computer with a voice response unit is programmed to answer a telephone and direct the call to a person, based on input from the caller.
automatic answering. (1) Answering in which the called data terminal equipment (DTE) automatically responds to the calling signal; the call may be established whether or not the called DTE is attended. (2) A machine feature that allows a transmission control unit or a station to respond automatically to a call that it receives over a switched line.
automatic call distribution (ACD) unit. A device attached to a telephone system that routes the next incoming call to the next available agent.
automatic dialing. A capability that allows a computer program or an operator using a keyboard to send commands to a modem, causing it to dial a telephone number.
automatic number identification (ANI). A capability of the telephone system to provide the number of the calling party to the called party as the telephone is ringing. Usually the number is displayed on a small screen on the telephone or on a separate box with a screen that is connected between the telephone line and the telephone.
automatic repeat request (ARQ). An error correction technique. When the receiving DTE detects an error, it signals the transmitting DTE to resend the data.
automatic-send-receive (ASR). A teletypewriter unit with keyboard, printer, paper tape reader/transmitter, and paper tape punch. This combination of units may be used online or offline and, in some cases, online and offline concurrently.
automatic teller machine (ATM). A specialized computer terminal that enables consumers to conduct banking transactions without the assistance of a bank teller.

availability. Having a system or service operational when a user wants to use it.

backbone circuit. The main circuit in a network.
backbone network. The main network in a particular network system.
background noise. Phenomena in all electrical circuitry resulting from the movement of electrons. *Also known as* white noise *or* Gaussian noise.
balun. An acronym that stands for balanced-unbalanced; also the name of a device that connects two different types of wire or cable. The balun is a small transformer that converts the electrical and physical characteristics of one wire type to another, e.g., from coaxial cable to twisted-pair wire.
bandwidth. The difference, expressed in Hertz, between the two limiting frequencies of a band.
bandwidth-on-demand. A concept whereby a person or device can acquire (and presumably pay for) large amounts of bandwidth on very short notice to transmit data at very high speeds, and then relinquish the bandwidth for other uses when it is not needed.
bar code reader. A device that reads codes printed in the form of bars on merchandise or tags.
base group. Telephone company terminology for a 48 kHz signal that contains twelve 4 kHz voice signals.
baseband. A form of modulation in which signals are pulsed directly on the transmission medium. In local area networks, baseband also implies the digital transmission of data.
baseband transmission. Transmission using baseband techniques. The signal is transmitted in digital form using the entire bandwidth of a circuit or cable. Typically used in local area networks.
basic access. A method of accessing an ISDN network in which the user has two 64-kbps B channels and one 16-kbps D channel. This type of access is also known as 2B+D.
basic business transactions. Fundamental operational units of business activity.
basic services. Services performed by the common carriers to provide the transportation of information. Basic services are regulated. *Contrast with* enhanced services.
batch. A set of data accumulated over a period of time.
batch processing. (1) Processing data or performing jobs accumulated in advance so that each accumulation is processed or accomplished in the same run. (2) Processing data accumulated over a period of time.
batched communication. A large body of data sent from one station to another station in a network without intervening responses from the receiving unit. *Contrast with* inquiry/response communication.
baud. A unit of signaling speed equal to the number of discrete conditions or signal events per second. If the duration of a signal event is 20 milliseconds, the modulation rate is 1 second / 20 milliseconds = 50 baud.
Baudot code. A code for the transmission of data in which 5 equal-length bits represent one character. This code is used in some teletypewriter machines, where one start element and one stop element are added.
Bell Operating Companies (BOCs). The 22 telephone companies that were members of the Bell System before divestiture.
Bell System. The collection of companies headed by AT&T and consisting of the 22 Bell Operating Companies and the Western Electric Corporation. The Bell System was dismantled by divestiture on January 1, 1984.
bid. In the contention form of invitation or selection, an attempt by the computer or by a station to gain control of the line so that it can transmit data. A bid may be successful or unsuccessful in seizing a circuit in that group. *Contrast with* seize.
binary. (1) Pertaining to a selection, choice, or condition that has two possible values or states. (2) Pertaining to the base two numbering system.
binary code. A code that makes use of exactly two distinct characters, usually 0 and 1.
Binary-Coded Decimal (BCD) code. A binary-coded notation in which each of the decimal digits is represented by a binary numeral; for example, in binary-coded decimal notation that uses the weights 8-4-2-1, the number 23 is represented by 0010 0011. Compare this to its representation in the pure binary numeration system, which is 10111.
binary digit. (1) In binary notation, either the character 0 or 1. (2) *Synonym for* bit.
Binary Synchronous Communications (BISYNC). (1) Communications using binary synchronous protocol. (2) A uniform procedure, using a standardized set of control characters and control character sequences, for synchronous transmission of binary-coded data between stations.
bipolar, nonreturn-to-zero (NRZ). A signaling method whereby the voltage is constant during a bit time. Most commonly, a negative voltage represents one binary value and a positive voltage is used to represent the other.
bipolar, return-to-zero. Signals that have the 1 bits represented by a positive voltage and the 0 bits represented by a negative voltage. Between pulses, the voltage always returns to zero. *Contrast with* bipolar, nonreturn-to-zero.
bit. Synonym for binary digit.
bit-oriented protocol. A communications protocol that uses bits, singly or in combination, to control the communications.
bit rate. The speed at which bits are transmitted, usually expressed in bits per second (bps).
bit stream. A binary signal without regard to grouping by character.
bit stuffing. (1) The occasional insertion of a dummy bit in a bit stream. (2) In SDLC, a 0 bit inserted after all strings of five consecutive 1 bits in the header and data portion of the message. At the receiving end, the extra 0 bit is removed by the hardware.
bit synchronization. A method of ensuring that a communications circuit is sampled at the appropriate time to determine the presence or absence of a bit.
bits per second (bps). The basic unit of speed on a data communications circuit.
blank character. A graphic representation of the space character.

blink. Varying the intensity of one or more characters displayed on a VDT several times per second to catch the operator's attention.
block. (1) A string of records, a string of words, or a character string formed for technical or logic reasons to be treated as an entity. (2) A set of things, such as words, characters, or digits, handled as a unit. (3) A group of bits, or characters, transmitted as a unit. An encoding procedure generally is applied to the group of bits or characters for error control purposes. (4) That portion of a message terminated by an EOB or ETB line control character or, if it is the last block in the message, by an EOT or ETX line control character.
block check. That part of the error control procedure used for determining that a data block is structured according to given rules.
block check character (BCC). In longitudinal redundancy checking and cyclic redundancy checking, a character that is transmitted by the sender after each message block and is compared with a block check character computed by the receiver to determine if the transmission was successful.
block error rate. The ratio of the number of blocks incorrectly received to the total number of blocks sent.
block length. (1) The number of records, words, or characters in a block. (2) A measure of the size of a block, usually specified in units, such as records, words, computer words, or characters.
blocking. (1) The process of combining incoming messages into a single message. (2) In a telephone switching system, the inability to make a connection or obtain a service because the devices needed for the connection are in use.
Bluetooth. A technology for wireless connectivity within a 33-foot radius.
bridge. A device that allows data to be sent from one network to another so terminals on both networks can communicate as though a single network existed.
broadband. (1) A communications channel having a bandwidth greater than a voice-grade channel and therefore capable of higher speed data transmission. (2) In local area networks, an analog transmission with frequency division multiplexing.
broadband ISDN (BISDN). An enhanced ISDN service that provides full duplex data transmission at either 155.52 or 622.08 Mbps, or an asymmetrical circuit that provides 155.52 Mbps in one direction and 622.08 Mbps in the other.
broadband transmission. A transmission technique of a local area network in which the signal is transmitted in analog form with frequency division multiplexing.
broadcast. The simultaneous transmission to a number of stations.
broadcast routing. A type of routing in which messages are sent to all stations on a network. Stations for which the messages are not intended ignore them.
brouter. A device that provides the functions of a bridge and router.
browser. A program designed to allow easy access to the Internet's World Wide Web including its text, graphic, audio, and visual content. Popular browsers include Microsoft's Internet Explorer and Netscape's Navigator.
buffer. A portion of memory designated as a temporary storage place for data. Buffers are frequently used to hold data arriving from a telecommunications line until a complete unit of data is received, at which time the complete unit is passed to a computer for processing.
buffering. The storage of bits or characters until they are specifically released. For example, a buffered terminal is one in which the keyed characters are stored in an internal storage area or buffer until a special key, such as the CARRIAGE RETURN or ENTER key, is pressed. Then all of the characters stored in the buffer are transmitted to the host computer in one operation.
bus. (1) One or more conductors used for transmitting signals or power. (2) In a local area network, a physical facility from where data is transferred to all destinations, but from which only addressed destinations may read in accordance with appropriate conventions or protocols.
bus network. A network topology in which multiple nodes are attached to a single circuit of limited length. A bus network is typically a local area network that transmits data at high speed.
business machine. Customer-provided data terminal equipment (DTE) that connects to a communications common carrier's telecommunications equipment for the purpose of data movement.
business machine clocking. An oscillator supplied by the business machine for regulating the bit rate of transmission. *Contrast with* data set clocking.
busy hour. The hour of the day when the traffic carried on a network is the highest.
bypass. Installing private telecommunications circuits to avoid using those of a carrier.
byte. An 8-bit binary character operated on as a unit.
byte-count-oriented protocol. A protocol that uses a special character to mark the beginning of the header, followed by a count field that indicates how many characters are in the data portion of the message.

cable modem. A modem that links a DTE to a television system cable.
cabling plan. A document that describes how the wiring and/or cabling will be installed.
callback. A security technique used with dial-up lines. After a user calls and identifies himself or herself, the computer breaks the connection and calls the user back at a predetermined telephone number. In some systems, the number at which the user wishes to be called back can be specified when the initial connection is made and before the computer disconnects.
callback unit. A hardware device that performs the callback function.
call control procedure. The implementation of a set of protocols necessary to establish, maintain, and release a call.
call detail recording (CDR). *See* station message detail recording (SMDR).
call progress signal. A call control signal transmitted from the data circuit-terminating equipment (DCE) to the call-

ing data terminal equipment (DTE) to indicate the progress of the establishment of a call, the reason why the connection could not be established, or any other network condition.

call setup time. The time taken to connect a switched telephone call. The time between the end of dialing and answering by the receiving party.

camp-on. A method of holding a call for a line that is in use and of signaling when it becomes free.

carriage-return character (CR). A format effector that causes the print or display position to move to the first position on the same line. *Contrast with* line feed character (LF).

carrier. (1) A company that provides the telecommunications networks. *See* communications common carrier. (2) A communications signal. *See* carrier wave.

carrier sense multiple access with collision avoidance (CSMA/CA). A communications protocol used on local area networks in which a station listens to the circuit before transmitting in an attempt to avoid collisions.

carrier sense multiple access with collision detection (CSMA/CD). A communications protocol frequently used on local area networks in which stations, on detecting a collision of data caused by multiple simultaneous transmissions, wait a random period of time before retransmitting.

carrier system. A means of obtaining a number of channels over a single circuit by modulating each channel on a different carrier frequency and demodulating at the receiving point to restore the signals to their original form.

carrier wave. An analog signal that in itself contains no information.

cathode ray tube terminal (CRT). A particular type of video display terminal that uses a vacuum tube display in which a beam of electrons can be controlled to form alphanumeric characters or symbols on a luminescent screen, for example, by use of a dot matrix.

CCITT. *See* Consultative Committee on International Telegraphy and Telephony (CCITT).

cell. A fixed-length packet in an asynchronous transfer mode (ATM) system.

cell relay. *See* asynchronous transfer mode (ATM).

cellular telephone service. A system for handling telephone calls to and from moving automobiles. Cities are divided into small geographic areas called *cells.* Telephone calls are transmitted to and from low-power radio transmitters in each cell. Calls are passed from one transmitter to another as the automobile leaves one cell and enters another.

cellular telephone system. A telephone system in which the geographic area to be covered is divided into small sections called *cells.* A transmitter/receiver in each cell relays telephone calls to cellular telephones located within the cell.

centa. One hundred.

centa call seconds (CCS). A measure of equipment or circuit utilization. One centa call second is 100 seconds of utilization.

centi. One hundredth.

central office. In the United States, the place where communications common carriers terminate customer lines and locate the equipment that interconnects those lines.

central office switch. The equipment in a telephone company central office that allows any circuit to be connected to any other.

centralized network. *Synonym for* star network.

centralized routing. A routing system in which the destination of all messages is determined by a single piece of hardware or software.

Centrex. Central office telephone equipment serving subscribers at one location on a private automatic branch exchange basis. The system allows such services as direct inward dialing, direct distance dialing, and console switchboards.

change control. A disciplined approach to managing changes.

change coordination meeting. A meeting held to ensure that changes to a system are properly approved and communicated to all interested parties.

change management. The application of management principles to ensure that changes in a system are controlled to minimize the impact on system users.

channel. (1) A one-way communications path. (2) In information theory, that part of a communications system that connects the message source with the message sink.

channel group. *See* base group.

channel service unit (CSU). *See* data service unit (DSU).

character. A member of a set of elements upon which agreement has been reached and that is used for the organization, control or representation of data. Characters may be letters, digits, punctuation marks, or other symbols, often represented in the form of a spatial arrangement of adjacent or connected strokes or in the form of other physical conditions in data media.

character assignments. Unique groups of bits assigned to represent the various characters in a code.

character compression. A type of compression in which characters are represented by a shortened number of bits, depending on the frequency with which the character is used.

character-oriented protocol. A communications protocol that uses special characters to indicate the beginning and end of messages. BISYNC is a character-oriented protocol.

character set. A set of unique representations called *characters,* for example, the 26 letters of the English alphabet, 0 and 1 of the Boolean alphabet, the set of signals in the Morse code alphabet, and the 128 ASCII characters.

character stripping. A data compression technique in which leading and trailing control characters are removed from a message before it is sent through a telecommunications system.

character synchronization. A technique for ensuring that the proper sets of bits on a communications line are grouped to form characters.

Cheapernet. *See* thin Ethernet.

check bit. (1) A binary check digit, for example, a parity bit. (2) A bit associated with a character or block for the purpose of checking for the absence of error within the character or block.

checkpoint record. The contents of a computer's memory and other control information that are stored on disk or tape at predetermined intervals so that the computer may be restarted after a failure.

checkpoint/restart. The process of recording a checkpoint record on disk or other nonvolatile media and later using that information to restart a computer that has failed.

chief information officer (CIO). A title sometimes given to the highest ranking executive in charge of a company's information resources.

chip. (1) A minute piece of semiconductive material used in the manufacture of electronic components. (2) An integrated circuit on a piece of semiconductive material.

ciphertext. The character stream or text that is the output of an encryption algorithm. *See also* plaintext.

circuit. The path over which two-way communications take place.

circuit grade. The information-carrying capability of a circuit in speed or type of signal. The grades of circuits are broadband, voice, subvoice, and telegraph. For data use, these grades are identified with certain speed ranges.

circuit noise level. The ratio of the circuit noise to some arbitrary amount chosen as the reference. This ratio normally is indicated in decibels above the reference noise.

circuit speed. The number of bits that a circuit can carry per unit of time, typically 1 second. Circuit speed is normally measured in bits per second.

circuit-switched data transmission service. A service using circuit switching to establish and maintain a connection before data can be transferred between data terminal equipments (DTEs).

circuit switching. The temporary establishment of a connection between two pieces of equipment that permits the exclusive use until the connection is released. The connection is set up on demand and discontinued when the transmission is complete. An example is a dial-up telephone connection.

city code. In the telephone numbering system, a single- or multidigit code that is uniquely assigned to a city within a country.

cladding. The glass that surrounds the core of an optical fiber and acts as a mirror to the core.

client. In a client-server computing system, the user or using computer that takes advantage of the facilities or services of server computers.

client-server computing. A type of distributed processing in which certain computers, called servers, provide standardized capabilities, such as printing, database management, or communications, to other computers that are called clients.

clock. (1) A device that measures and indicates time. (2) A device that generates periodic signals used for synchronization. (3) Equipment that provides a time base used in a transmission system to control the timing of certain functions, such as sampling, and to control the duration of signal elements.

clock pulse. A synchronization signal provided by a clock.

clocking. The use of clock pulses to control synchronization of data and control characters.

cluster. A station that consists of a control unit (cluster controller) and the terminals attached to it.

cluster control unit (CCU). A device that can control the input/output operations of more than one device connected to it. A cluster control unit may be controlled by a program stored and executed in the unit, or it may be controlled entirely by hardware.

coaxial cable. A cable consisting of one conductor, usually a small copper tube or wire, within and insulated from another conductor of larger diameter, usually copper tubing or copper braid.

code. (1) A set of unambiguous rules specifying the manner in which data may be represented in a discrete form. (2) A predetermined set of symbols that have specific meanings.

code conversion. A process for changing the bit grouping for a character in one code into the corresponding bit grouping for a character in a second code.

code converter. A device that changes the representation of data, using one code in the place of another or one coded character set in the place of another.

code division multiple access (CDMA). A transmission technique used in digital radio technology that combines time division and frequency division multiple access techniques, yielding higher capacity and better security than either of them.

code efficiency. Using the least number of bits to convey the meaning of a character with accuracy.

code-independent data communications. A mode of data communications that uses a character-oriented link protocol that does not depend on the character set or code used by the data source.

code points. The number of possible combinations in a coding system.

code transparent data communication. A mode of data communications that uses a bit-oriented link protocol that does not depend on the bit sequence structure used by the data source.

codec. A device that converts analog signals to digital signals or vice versa.

coded character set. A set of unambiguous rules that establish a character set and the one-to-one relationships between the characters of the set and their coded representations. *Synonymous with* code.

coding scheme. *See* code (1).

collision. Two (or more) terminals trying to transmit a message at the same time, thereby causing both messages to be garbled and unintelligible at the receiving end.

committed information rate (CIR). The contracted transmission speed on a frame relay circuit. Data sent within the CIR is highly likely to get through the network unless extremely severe network congestion occurs. Data sent above the CIR is subject to discard if the network gets congested.

common carrier. *See* communications common carrier.

common channel interoffice signaling (CCIS). A system for sending signals between central offices in a telephone network.
Common Management Information Protocol (CMIP). An ISO standard protocol for exchanging network management commands and information between devices attached to a network.
communication. (1) A process that allows information to pass between a sender and one or more receivers. (2) The transfer of meaningful information from one location to a second location. (3) The art of expressing ideas, especially in speech and writing. (4) The science of transmitting information, especially in symbols.
communications access method (CAM). Computer software that reads and writes data from and to communications lines. *Synonym for* telecommunications access method (TCAM).
communications adapter. An optional hardware feature, available on certain processors, that permits telecommunications lines to be attached to the processors.
communications common carrier. In the USA and Canada, a public data transmission service that provides the general public with transmission service facilities, for example, a telephone or telegraph company.
communications controller. A hardware device that manages the details of line control and sometimes data routing through a network. *See also* front-end processor (FEP).
communications directory. An online or hard copy document that lists the names and telephone numbers of a company's employees and departments as well as other information, such as terminal names or user IDs, that is pertinent to communications.
communications facility. *See* telecommunications facility.
communications line. Deprecated term for telecommunications line or transmission line.
communications network. A collection of communications circuits managed as a single entity.
communications server. (1) A server on a LAN that provides connections to other computers or networks. (2) A server based telephone system that is attached to both the public telephone network and to a LAN. With time, the capabilities of these two types of communication server are coming closer together.
communications standards. Standards established to ensure compatibility among several communications services or several types of communications equipment.
communications theory. The mathematical discipline dealing with the probabilistic features of data transmission in the presence of noise.
compaction. *See* compression.
compandor (compressor-expandor). Equipment that compresses the outgoing speech volume range and expands the incoming speech volume range on a long distance telephone circuit. Such equipment can make more efficient use of voice telecommunications channels.
competitive local exchange carrier (CLEC). Carriers that have no ties to the old Bell system and are, in general, aggressively promoting new telecommunications services. In some cases these carriers are building their own new digital networks, and in other cases they lease network bandwidth from other LECs.
compression. The process of eliminating redundant characters or bits from a data stream before it is stored or transmitted.
computer-aided design (CAD). The use of a computer with special terminals and software for engineering design and drafting. *Synonym for* computer-assisted drafting (CAD).
computer-assisted drafting (CAD). *See* computer-aided design (CAD).
computer branch exchange (CBX). *See* private automatic branch exchange (PABX) *and* private branch exchange (PBX).
Computer Inquiry I (CI–I). A study conducted by the FCC, concluded in 1971, that examined the relationship between the telecommunications and data processing industries to determine which aspects of both industries should be regulated for the long term.
Computer Inquiry II (CI–II). A study conducted by the FCC, concluded in 1981, that accelerated the deregulation of the telecommunications industry.
Computer Inquiry III (CI–III). A study conducted by the FCC to determine to what extent AT&T and the BOCs are allowed to provide enhanced (data processing) services in the network.
computer network. A complex consisting of two or more interconnected computing units.
computer virus. An executable computer program that causes unwanted events in a computer such as the destruction of data. Most viruses attach themselves to other programs. Viruses are spread when programs to which they are attached are copied or downloaded.
concentration. The process of combining multiple messages into a single message for transmission. *Contrast with* deconcentration.
concentrator. (1) In data transmission, a functional unit that permits a common transmission medium to serve more data sources than there are channels currently available within the transmission medium. (2) Any device that combines incoming messages into a single message (concentration) or extracts individual messages from the data sent in a single transmission sequence (deconcentration).
conditioned line. A communications line on which the specifications for amplitude and distortion have been tightened. Signals traveling on a conditioned circuit are less likely to encounter errors than on an unconditioned circuit.
conditioning. The addition of equipment to a leased voice-grade circuit to provide minimum values of line characteristics required for data transmission.
conducted media. Any medium where the signal flows through a physical entity such as twisted pair wire, coaxial cable, or optical fiber.
configuration control. The maintenance of records that identify and keep track of all of the equipment in a system.

connection-oriented routing. A transmission through a network using a virtual or real circuit between the sender and receiver.

connectionless routing. A transmission through a network when no virtual or real circuit has been established between the sender and receiver. Packets are sent into the network by the sender, and each travels independently to the receiver, which must then reassemble them into a message.

Consultative Committee on International Telegraphy and Telephony (CCITT). The previous name for the International Telecommunications Union-Telecommunications Standardization Sector (ITU-T).

continuous ARQ. An error correction technique in which data blocks are continuously sent over the forward channel, while ACKs and NAKs are sent over the reverse channel. *Contrast with* stop and wait ARQ.

control character. A character whose occurrence in a particular context initiates, modifies, or stops a control operation. A control character may be recorded for use in a subsequent action, and it may have a graphic representation in some circumstances.

control station. A station on a network that assumes control of the network's operation. A typical control station exerts its control by polling and addressing. *Contrast with* slave station.

control terminal. Any active terminal on a network at which a user is authorized to enter commands affecting system operation.

control unit. A device that controls input/output operations at one or more devices. *See also* controller *and* cluster control unit (CCU).

controller. A device that directs the transmission of data over the data links of a network. Its operations may be controlled by a program executed in a processor to which the controller is connected, or they may be controlled by a program executed within the device. *See* cluster control unit (CCU) *and* communications controller.

conversational mode. A mode of operation of a data processing system in which a sequence of alternating entries and responses between a user and the system takes place in a manner similar to a dialogue between two persons. *Synonym for* interactive.

cordless telephones. Telephones in which the base and the handset contain small transceivers, which broadcast to each other and which allow the people using them to move away from the base unit during a conversation.

core. The glass or plastic center conductor of an optical fiber that provides the transmission carrying capability.

cost center. An accounting term used to designate a department or other entity where costs are accumulated. Departments that are cost centers do not make a profit by selling or recharging their services. *Contrast with* profit center.

creeping commitment of resources. A concept of project management that suggests that the resources dedicated to a project should only be increased as the scope of the project becomes better defined. The intention is to minimize the amount of resources spent in case the project is determined to be infeasible or is otherwise canceled.

crossbar switch. A device that makes a connection between one line in each of two sets of lines. The two sets are physically arranged along adjacent sides of a matrix of contacts or switch points.

crosstalk. The unwanted energy transferred from one circuit, called the *disturbing circuit*, to another circuit, called the *disturbed circuit*.

current beam position. On a CRT display device, the coordinates on the display surface at which the electron beam is aimed.

current loop. An interface between a terminal and a circuit that indicates 1 and 0 bits by the presence or absence of an electrical current.

cursor. (1) In computer graphics, a movable marker that indicates a position on a display space. (2) A displayed symbol that acts as a marker to help the user locate a point in text, in a system command, or in storage. (3) A movable spot of light on the screen of a display device that usually indicates where the next character is to be entered, replaced, or deleted.

cursor control keys. The keys that control the movement of the cursor.

customer premise equipment (CPE). Any communications equipment that is located on customer premises, such as telephones, personal computers, fax machines, telephone systems, PBXs, key systems, routers, and hubs.

cut-through switch. A type of switch that reads the destination address of a packet and immediately begins sending the packet to the destination before the entire packet has been received by the switch.

cycle. A complete wave of an analog signal. The frequency is the number of cycles that are completed in one second.

cycle stealing. Interrupting a computer to store each character coming from a telecommunications line in the computer's memory.

cyclic redundancy checking (CRC). (1) An error checking technique in which the check key is generated by a cyclic algorithm. (2) A system of error checking performed at both the sending and receiving stations after a block check character has been accumulated.

data. (1) A representation of facts, concepts, or instructions in a formalized manner suitable for communication, interpretation, or processing by human or automatic means. (2) Any representations, such as characters or analog quantities, to which meaning is, or might be, assigned.

data access arrangement (DAA). Equipment that permits attachment of privately owned data terminal equipment and telecommunications equipment to the public telephone network.

data circuit. Associated transmit and receive channels that provide a means of two-way data communications.

data circuit-terminating equipment (DCE). The equipment installed at the user's premises that provides all the functions required to establish, maintain, and terminate a connection and the signal conversion and coding between the data terminal equipment (DTE) and the line.

data communications. (1) The transmission and reception of data. (2) The transmission, reception, and validation

of data. (3) Data transfer between data source and data sink via one or more data links according to appropriate protocols.
data communications channel. A means of one-way transmission.
data encryption standard (DES). A cryptographic algorithm designed to encipher and decipher data using a 64-bit cryptographic key as specified in the Federal Information Processing Standard Publication 46, January 15, 1977.
data integrity. The quality of data that exists as long as accidental or malicious destruction, alteration, or loss of data is prevented.
data link. (1) The physical means of connecting one location to another to transmit and receive data. (2) The interconnecting data circuit between two or more pieces of equipment operating in accordance with a link protocol. It does not include the data source and the data sink.
data link control character. A control character intended to control or facilitate transmission of data over a network.
data link escape (DLE) character. A transmission control character that changes the meaning of a limited number of contiguous following characters or coded representations and that is used exclusively to provide supplementary transmission control characters.
data link protocol. The rules governing the operation of a data link. *See* protocol.
data network. The assembly of functional units that establishes data circuits between pieces of data terminal equipment (DTE).
data PBX. A switch especially designed for switching data calls. Data PBXs do not handle voice calls.
Data-Phone. Both a service mark and a trademark of AT&T and the Bell System. As a service mark, it indicates the transmission of data over the telephone network. As a trademark, it identifies the telecommunications equipment furnished by the Bell System for transmission services.
data processing (DP). The systematic performance of operations upon data, for example, handling, merging, sorting, and computing. *Synonym for* information processing.
data processing system. A system, including computer systems and associated personnel, that performs input, processing, storage, output, and control functions to accomplish a sequence of operations on data.
data security. The protection of data against unauthorized disclosure, transfer, modifications, or destruction, whether accidental or intentional.
data service unit/channel service unit (DSU/CSU). An interface device that ensures that the digital signal entering a communications line is properly shaped into square pulses and precisely timed.
data set clocking. A time-based oscillator supplied by the modem for regulating the bit rate of transmission.
data sink. (1) A functional unit that accepts data after transmission. It may originate error control signals. (2) The part of data terminal equipment (DTE) that receives data from a data link.
data terminal equipment (DTE). The part of a data station that serves as a data source, data sink, or both and provides for the data communications control function according to protocols.
data transfer rate. The average number of bits, characters, or blocks per unit of time transferred from a data source to a data sink. The rate is usually expressed as bits, characters, or blocks per second, minute, or hour.
datagram. In packet switching, a self-contained packet that is independent of other packets, that does not require acknowledgment, and that carries information sufficient for routing from the originating data terminal equipment (DTE) to the destination DTE without relying on earlier exchanges between DTEs and the network.
dB meter. A meter having a scale calibrated to read directly in decibel values at a reference level that must be specified (usually 1 milliwatt equals Φ dB). Used in audio-frequency amplifier circuits of broadcast stations, public address systems, and receiver output circuits to indicate volume level.
dBm. Decibel based on 1 milliwatt.
deactivation. In a network, the process of taking any element out of service, rendering it inoperable, or placing it in a state in which it cannot perform the functions for which it was designed.
decibel (dB). (1) A unit that expresses the ratio of two power levels on a logarithmic scale. (2) A unit for measuring relative power. The number of decibels is 10 times the logarithm (base 10) of the ratio of the measured power levels. If the measured levels are voltages (across the same or equal resistance), the number of decibels is 20 times the log of the ratio.
DECNET. A family of hardware and software that implement Digital Network Architecture (DNA).
deconcentration. The process of extracting individual messages from data sent in a single transmission sequence. *Contrast with* concentration.
decryption. Converting encrypted data into clear data. *Contrast with* encryption.
dedicated hardware protocol converter. Electronic circuitry that has as its sole purpose the conversion of one protocol to another.
delta modulation. A technique of digitizing an analog signal by comparing the values of two successive samples and assigning a 1 bit if the second sample has a greater value and a 0 bit if the second sample has a lesser value.
demand priority access method (DPAM). The media access technique used on 100VG-AnyLAN technology.
demarcation point. The physical and electrical boundary between the telephone company responsibility and the customer responsibility.
demodulation. The process of retrieving intelligence (data) from a modulated carrier wave. *Reverse of* modulation.
demodulator. A device that performs demodulation.
destination code. A code in a message header containing the name of a terminal or application program to which the message is directed.
destructive cursor. On a VDT device, a cursor that erases any character through which it passes as it is advanced, backspaced, or otherwise moved. *Contrast with* nondestructive cursor.

detector. Circuitry that separates a received signal into its component parts, typically the carrier and the modulation.
device control character. A control character used for the control of ancillary devices associated with a data processing system or data communications system, for example, for switching such devices on or off.
dial. To use a dial or push-button telephone to initiate a telephone call. In telecommunications, this action is taken to attempt to establish a connection between a terminal and a telecommunications device over a switched line.
dial backup. A technique for bypassing the failure of a private or leased circuit. When a failure occurs, a switched connection is made so that communications can be reinstated.
dial line. *Synonym for* switched connection.
dial pulse. An interruption in the DC loop of a calling telephone. The interruption is produced by breaking and making the dial pulse contacts of a calling telephone when a digit is dialed. The loop current is interrupted once for each unit of value of the digit.
dial pulsing. A signaling technique used to send a telephone number by generating electrical pulses on a telephone line.
dial tone. An audible signal indicating that a device is ready to be dialed.
dialback unit. A hardware device that performs the function of callback.
dialing. Deprecated term for calling.
dialogue. In an interactive system, a series of interrelated inquiries and responses analogous to a conversation between two people.
dial-up. The use of a dial or push-button telephone to initiate a station-to-station telephone call.
dial-up line. A line on which the connection is made by dialing. *See also* switched line.
dial-up terminal. A terminal on a switched line.
dibit. A group of two bits. In four-phase modulation, each possible dibit is encoded as one of four unique carrier phase shifts. The four possible states for a dibit are 00, 01, 10, and 11. *Contrast with* quadbit *and* tribit.
differential Manchester encoding. A digital signaling technique in which a 0 is represented by the presence of a transition at the beginning of the bit period and a 1 is represented by an absence of a transition at the beginning of the bit period. A mid-bit transition also exists to provide clocking.
differential phase shift keying (DPSK). A modulation technique in which the relative changes of the carrier signal phase are coded according to the data to be transmitted.
digital circuit. A circuit expressly designed to carry the pulses of digital signals.
Digital Data Communications Message Protocol (DDCMP). A byte-count-oriented protocol developed by Digital Equipment Corporation.
Digital Network Architecture (DNA). A communications architecture developed by Digital Equipment Corporation as a framework for all of the company's communications products.
digital satellite service (DSS). *See* direct broadcast satellite.
digital signal. A discrete or discontinuous signal; the various states of which are pulses that are discrete intervals apart. *Contrast with* analog signal.
digital signal processor (DSP). A microprocessor especially designed to analyze, enhance, or otherwise manipulate sounds, images, or other signals.
digital subscriber line (DSL). The generic name for a technology developed to enable telephone companies to deliver digitized signals to subscribers at about 1.5 Mbps over existing twisted pair copper telephone wire. *See also* asymmetric digital subscriber line, symmetric digital subscriber line, *and* very high rate digital subscriber line.
digital switching. A process in which connections are established by operations on digital signals without converting them to analog signals.
digital-to-analog (D/A) converter. A device that converts a digital value to a proportional analog signal.
digitize. To express or represent in a digital form data that is not discrete data, for example, to obtain a digital representation of the magnitude of a physical quantity from an analog representation of that magnitude.
digitizing distortion. *See* quantizing noise.
direct broadcast satellite (DBS). A satellite with the primary purpose of sending signals directly to small antennas in homes or businesses. Because the receiving antennas are small, the satellite normally has a relatively high-powered transmitter.
direct current (DC) signaling. Signaling caused by opening and closing a direct current electrical circuit.
direct distance dialing (DDD). A telephone exchange service that enables the telephone user to call subscribers outside of the user's local service area without operator assistance.
direct inward dialing (DID). A facility that allows a telephone call to pass through a telephone system directly to an extension without operator intervention.
direct outward dialing (DOD). A facility that allows an internal caller at an extension to dial an external number without operator assistance.
direct sequence. A spread spectrum transmission technique in which bits from the original signal are combined with bits generated by a pseudorandom bit stream generator using Boolean math. The receiver, using the same pseudorandom bit stream, can reverse the Boolean math process and recover the original bits. *See also* spread spectrum *and* frequency hopping.
disconnect. To disengage the apparatus used in a connection and to restore it to its ready condition.
disconnect signal. A signal transmitted from one end of a subscriber line or trunk to indicate at the other end that the established connection is to be disconnected.
discussion group. In the context of the Internet, a group of people that meet online to discuss topics of mutual interest and to exchange ideas.
disk server. A server on a LAN that provides simulated disks to other computers.
diskless workstation. A personal computer or other terminal that does not have a hard disk, making it dependent on a server or some other computer for disk storage.

dispersion. The difference in the arrival time between signals that travel straight through the core of a fiber-optic cable and those that reflect off the cladding and, therefore, travel a slightly longer path.
display device. (1) An output unit that gives a visual representation of data. (2) In computer graphics, a device capable of presenting display elements on a display surface, for example, a terminal screen, plotter, microfilm viewer, or printer.
distinctive ringing. A ringing cadence that indicates whether a call is internal or external.
distortion. The unwanted change in wave form that occurs between two points in a transmission system. The six major forms of distortion are (1) *bias:* a type of telegraph distortion resulting when the significant intervals of the modulation do not all have their exact theoretical durations, (2) *characteristic:* distortion caused by transients that, as a result of the modulation, are present in the transmission channel and depend on its transmission qualities, (3) *delay:* distortion that occurs when the envelope delay of a circuit or system is not constant over the frequency range required for transmission, (4) *end:* distortion of start-stop teletypewriter signals. The shifting of the end of all marking pulses from their proper positions in relation to the beginning of the start pulse, (5) *fortuitous ("jitter"):* a type of distortion that results in the intermittent shortening or lengthening of the signals. This distortion is entirely random in nature and can be caused by such things as battery fluctuations, hits on the line, and power induction, (6) *harmonic:* the resultant presence of harmonic frequencies (due to nonlinear characteristics of a transmission line) in the response when a sine wave is applied.
distributed data processing (DDP). *See* distributed processing.
distributed processing. Data processing in which some or all of the processing, storage, and control functions, in addition to input/output functions, are spread among several computers and connected by communications facilities.
distributed routing. A technique of routing messages in a network in which some or all of the nodes maintain tables that show how messages should be directed to destinations.
distribution cable. A subgrouping of individual telephone lines as they approach a central office.
distribution frame. A structure for terminating permanent wires of a telephone central office, private branch exchange, or private exchange and for permitting the easy change of connections between them by means of cross-connecting wires.
domain. In an SNA network, the resources that are under the control of one or more associated host processors.
dot matrix. A printing technique in which a matrix of wires push a ribbon against paper to leave an impression. In a variation of the dot matrix technique, the wires are heated and cause a chemical reaction on specially treated paper.
double byte character set (DBCS). A coding system that has 2^{16} code points that can be used to represent all characters in all languages. DBCSs are usually vendor-specific.
downlink. The rebroadcast of a microwave radio signal from a satellite back to earth.
downloading. The transmission of a file of data from a mainframe computer to a personal computer.
driver. *See* workload generator.
dropout. In data communications, a momentary loss in signal, usually due to the effect of noise or system malfunction.
drop wire. The wire running from a residence or business to a telephone pole or its underground equivalent.
dual-tone-multifrequency (DTMF). A method of signaling a desired telephone number by sending tones on the telephone line.
dumb terminal. A terminal that has little or no memory and is not programmable. A dumb terminal is totally dependent on the host computer for all processing capability.
duplex. *See* full-duplex (FDX).
duplex transmission. Data transmission in both directions at the same time.
dynamic routing. A technique used in data networks by which each node can determine the best way for a message to be sent to its destination.

E&M signaling. A type of signaling between a switch or PBX and a trunk in which the signaling information is transferred via two-state voltage conditions on two wires.
EBCDIC. *See* Extended Binary Coded Decimal Interchange Code.
echo. The reversal of a signal, bouncing it back to the sender, caused by an electrical wave bouncing back from an intermediate point or the distant end of a circuit.
echo check. A check to determine the correctness of the transmission of data in which the received data is returned to the source for comparison with the originally transmitted data.
echo suppressor. A device that permits transmission in only one direction at a time, thus eliminating the problems caused by the echo.
effective data transfer rate. The average number of bits, characters, or blocks per unit of time transferred from a data source to a data sink and accepted as valid.
800 service. A telephone service that lets subscribers make calls to certain zones at discounted rates. Inbound 800 service allows callers to make calls that are paid for by the called party.
electronic document interchange (EDI). The use of telecommunications to transmit documents electronically.
electronic mail (e-mail). The use of telecommunications for sending textual messages from one person to another. The capability of storing the messages in an electronic mailbox is normally a part of the electronic mail system.
electronic mailbox. Space on the disk of a computer to store electronic mail messages.
electronic switching system (ESS). Electronic switching computer for central office functions.
electronic tandem network (ETN). A private telephone network in which software in customer PBXs determines how calls should be routed over leased, private, or public telephone lines.

encryption. Transformation of data from the meaningful code that is normally transmitted, called *clear text,* to a meaningless sequence of digits and letters that must be decrypted before it becomes meaningful again. *Contrast with* decryption.
end of address (EOA). One or more transmission control characters transmitted on a line to indicate the end of nontext characters (for example, addressing characters).
end of block (EOB). A transmission control character that marks the end of a block of data.
end of message (EOM). The specific character or sequence of characters that indicates the end of a message or record.
end of text (ETX). A transmission control character sent to mark the end of the text of the message.
end of text block (ETB). A transmission control character sent to mark the end of a portion of the text of a message.
end of text character (ETX). A control character that marks the end of a message's text.
end of transmission block (ETB). A transmission control character used to indicate the end of a transmission block of data when data is divided into such blocks for transmission purposes.
end of transmission (EOT). A transmission control character used to indicate the conclusion of a transmission that may have included one or more messages.
end office. A local telephone company central office designed to serve consumers or businesses.
end system. In the context of an internet, a subnetwork that supports the users connected to it.
enhanced services. Communications services in which some processing of the information being transmitted takes place. *Contrast with* basic services.
enquiry (ENQ). A transmission control character used as a request for a response from the station with which the connection has been set up; the response may include station identification, the type of equipment in service, and the status of the remote station.
enter. To place a message on a circuit to be transmitted from a terminal to the computer.
envelope delay distortion. Distortion caused by the electrical phenomenon that not all frequencies propagate down a telecommunications circuit at exactly the same speed.
equal access. A part of the modified final judgment that specified that local telephone companies must provide all of the long distance companies access equal in type, quality, and price to that provided to AT&T.
equalization. Compensation for differences in attenuation (reduction or loss of signal) at different frequencies.
equalizer. Any combination of devices, such as coils, capacitors, or resistors, inserted in a transmission line or amplifier circuit to improve its frequency response.
equivalent four-wire system. A transmission system using frequency division to obtain full-duplex operation over only one pair of wires.
ergonomics. The study of the problems of people in adjusting to their environment, especially the science that seeks to adapt work or working conditions to suit the worker.
Erlang. A measure of communications equipment or circuit usage. One Erlang is 1 hour of equipment usage or 36 CCS.
Erlang B capacity table. A table for determining the number of circuits required to carry a certain level of telephone traffic. The Erlang B table assumes that the sources of traffic are infinite and that all unsuccessful call attempts are abandoned and not retried.
error. A discrepancy between a computed, observed, or measured value or condition and the true, specified, or theoretically correct value or condition.
error correcting code. A code in which each telegraph or data signal conforms to specific rules of construction so that departures from this construction in the receive signals can be automatically detected, permitting the automatic correction, at the receiving terminal, of some or all of the errors. Such codes require more signal elements than are necessary to convey the basic information.
error correction system. A system employing an error detecting code and so arranged that some or all of the signals detected as being in error are automatically corrected at the receiving terminal before delivery to the data sink. *Note:* In a packet switched data service, the error correcting system might result in the retransmission of at least one or more complete packets should an error be detected.
error detecting code. A code in which each element that is represented conforms to specific rules of construction so that if certain errors occur, the resulting representation will not conform to the rules, thereby indicating the presence of errors. Such codes require more signal elements than are necessary to convey the fundamental information.
error detection. The techniques employed to ensure that transmission and other errors are identified.
error message. An indication that an error has been detected.
error rate. A measure of the quality of a circuit or system; the number of erroneous bits or characters in a sample, frequently taken per 100,000 characters.
error ratio. The ratio of the number of data units in error to the total number of data units.
error recovery. The process of correcting or bypassing a fault to restore a system to a prescribed condition.
escape character (ESC). A code extension character used to indicate that the following character or group of characters is to be interpreted in a nonstandard way.
escape mechanism. A method of assigning an alternate meaning to characters in a coding system. *See* escape character (ESC).
Ethernet. A local area network that uses CSMA/CD protocol on a baseband bus.
exchange. A room or building equipped so that telecommunications lines terminated there may be interconnected as required. The equipment may include manual or automatic switching equipment.
exchange code. In the United States, the first 3 digits of a 7-digit telephone number. The exchange code designates the telephone exchange that serves the customer.
Extended Binary Coded Decimal Interchange Code (EBCDIC). A coding system consisting of 256 characters, each represented by 8 bits.
external modem. A modem that exists in its own box or cabinet.

facsimile (FAX) machine. A machine that scans a sheet of paper and converts the light and dark areas to electrical signals that can be transmitted over telephone lines.

facsimile (FAX) modem. A modem designed to follow the standards and algorithms required to send facsimiles from a computer to a facsimile machine or another computer that has a facsimile modem.

fast busy. A tone signal that indicates that a telephone call cannot be completed because all circuits are busy.

fast select polling. A polling technique in which a station without traffic to send does not need to return a character to the polling station.

fat client. An application designed such that almost all of the processing is done on the client computer, while little is done on the server. *See also* thin client.

Federal Communications Commission (FCC). A board of commissioners appointed by the president under the Communications Act of 1934. The commissioners regulate all interstate and foreign electrical telecommunications systems originating in the United States.

feeder cable. A grouping of several distribution cables as they approach a central office.

Fiber Distributed Data Interface (FDDI). A standard for transmitting data on an optical fiber.

figures shift. A physical shift in a teletypewriter that enables the printing of images, such as numbers, symbols, and uppercase characters.

file server. A server on a LAN that provides storage for data files.

file transfer protocol (FTP). An application layer protocol designed to efficiently transfer files between two computers. Frequently used on TCP/IP networks.

firewall. In the context of the Internet, a computer with special software installed between the Internet and a private network for the purpose of preventing unauthorized access to the private network.

first in, first out (FIFO) queuing. Queuing in the order that calls or transactions arrive. Calls that arrive first are serviced first.

five-level code. A telegraph code that uses five impulses for describing a character. Start and stop elements may be added for asynchronous transmission. A common five-level code is the Baudot code.

fixed equalizer. Electronic circuitry in a modem that shapes the transmitted wave using the assumption that the communications line has an average set of parameters.

flag. (1) Any of various types of indicators used for identification. (2) A bit sequence that signals the occurrence of some condition, such as the end of a word. (3) In high-level data link control (HDLC), the initial and final octets of a frame with the specific bit configuration of 01111110. A single flag may be used to denote the end of one frame and the start of another.

flat panel display. A technology for VDTs yielding a display that is much flatter and takes up less space on a desk than a CRT.

flat rate service. A method of charging for local calls that gives the user an unlimited number of calls for a flat monthly fee.

foreign exchange (FX) line. A service that connects a customer's telephone system to a telephone company central office that normally does not serve the customer's location.

format effector character. A character that controls the positioning of information on a terminal screen or paper.

formatted mode. A method of displaying output on a VDT in which the entire screen can be arranged in any desired configuration and transmitted to the terminal at one time. *Contrast with* line-by-line mode.

forward channel. The primary transmission channel in a data circuit. *Contrast with* reverse channel.

forward error correction (FEC). A technique of transmitting extra bits or characters with a block of data so that transmission errors can be corrected at the receiving end.

four-wire circuit. A path in which four wires (two for each direction of transmission) are presented to the station equipment. Leased circuits are four-wire circuits.

four-wire terminating set. An arrangement by which four-wire circuits are terminated on a two-wire basis for interconnection with two-wire circuits.

fractional T-1. The subdivision or multiplexing of T-1 circuits to provide circuit speeds that are a fraction of the T-1's capacity.

frame. (1) In SDLC, the vehicle for every command, every response, and all information that is transmitted using SDLC procedures. Each frame begins and ends with a flag. (2) In high-level data link control (HDLC), the sequence of contiguous bits bracketed by and including opening and closing flag (01111110) sequences.

frame relay. A low-overhead packet switching technique—designed to operate on high-speed lines—in which each packet keeps track of the destinations it has passed through.

framing bits. Noninformation-carrying bits used to make possible the separation of characters in a bit stream.

freeze-frame television. A television system in which the picture is only updated as needed, typically every 30 to 90 seconds.

frequency. An attribute of analog signals that describes the rate at which the current alternates. Frequency is measured in Hertz.

frequency division multiple access (FDMA). A transmission technique used in digital radio transmissions in which frequencies are shared among several users.

frequency division multiplexing (FDM). A technique of putting several analog signals on a circuit by shifting the frequencies of the signals to different ranges so that they do not interfere with one another.

frequency hopping. A spread spectrum transmission technique in which the signal is broadcast over a seemingly random series of radio frequencies, hopping from frequency to frequency at split second intervals. *See also* spread spectrum *and* direct sequence.

frequency modulation (FM). Modulation in which the frequency of an alternating current is the characteristic varied.

frequency shift keying (FSK). Frequency modulation of a carrier by a signal that varies between a fixed number of discrete values.

front-end processor (FEP). A processor that can relieve a host computer of certain processing tasks, such as line control, message handling, code conversion, error control, and application functions. *See also* communications controller.
full-duplex (FDX). A mode of operating a data link in which data may be transmitted simultaneously in both directions over two channels.
full-duplex transmission. Data transmission in both directions simultaneously on a circuit.
full-motion television. Television pictures in which 30 pictures are sent every second.
function key. On a terminal, a key, such as an ATTENTION or an ENTER key, that when pressed transmits a signal not associated with a printable or displayable character. Detection of the signal usually causes the system to perform some predefined function.

Gantt chart. A project management tool that shows projects, activities, or tasks (normally listed chronologically) on the left and dates across the top. Each activity is indicated by a bar on the chart that shows its starting and ending dates.
gateway. The connection between two networks that use different protocols. The gateway translates the protocols to allow terminals on the two networks to communicate.
Gaussian noise. *See* background noise.
general poll. A technique in which special invitation characters are sent to solicit transmission of data from all attached remote devices that are ready to send.
geosynchronous orbit. A satellite orbit that exactly matches the rotation speed of the earth. Thus, from the earth, the satellite appears to be stationary.
giga (G). One billion. For example, 1 gigaHertz equals 1,000,000,000 Hertz. One gigaHertz also equals 1,000 megaHertz and 1,000,000 kiloHertz.
global positioning system (GPS). A satellite-based system for precisely locating any point on earth. Receivers on earth pick up signals from three or more satellites and then through triangulation calculate the exact position of the receiver, usually measured in latitude and longitude.
global system for mobile communications (GSM). A cellular telephone technology based on TDMA technology but with a higher capacity and better call quality. It supports both voice and data transmission. GSM is the standard for cellular phone service in Europe and Asia. GSM service is available in the United States but only in limited areas.
Government Open Systems Interconnection Protocol (GOSIP). A U.S. government-specified subset of the OSI model that defines what parts of the OSI model the government will follow and, therefore, what products it will buy.
grade of service. A measure of the traffic-handling capability of a network from the point of view of sufficiency of equipment and trunking throughout a multiplicity of nodes.
graphic character. A character that can be displayed on a terminal screen or printed on paper.
graphical user interface (GUI). A way in which a user can interact with a computer using a pointing device, such as a mouse, to select icons displayed on the screen or indicate an action to be taken.
Gray code. A binary code in which sequential numbers are represented by binary expressions, each of which differs from the preceding expression in one place only.
group. Twelve 4-kHz voice signals multiplexed together into a 48-kHz signal.
groupware. A type of software that allows groups of people to talk to each other or to work together simultaneously.
guard band. *See* guard channel.
guard channel. The space between the primary signal and the edge of an analog channel.
guided media. *See* conducted media.

hacker. A term originally denoting a technically inclined individual who enjoyed pushing computers to their limits and making them perform tasks no one thought possible. Recently, a term describing a person with the mischievous, malevolent intent to access computers to change or destroy data or perform other unauthorized operations.
half-duplex (HDX). A mode of operation of a data link in which data may be transmitted in both directions but only in one direction at a time.
half-duplex transmission. Data transmission in either direction, one direction at a time.
hamming code. A data code that is capable of being corrected automatically.
handset. A telephone mouthpiece and receiver in a single unit that can be held in one hand.
handshake. A security technique, used on dial-up circuits, that requires that terminal hardware identify itself to the computer by automatically sending a predetermined identification code. The handshake technique is not controlled by the terminal operator.
handshaking. Exchange of predetermined signals when a connection is established between two dataset devices.
hardwired. Directly connected by wire or cable.
harmonic. The resultant presence of harmonic frequencies (due to nonlinear characteristics of a transmission line) in the response when a sine wave is applied.
header. The part of a data message containing information about the message, such as its destination, a sequence number, and perhaps a date or time.
help desk. The single point of contact for users when problems occur.
Hertz (Hz). A unit of frequency equal to one cycle per second.
hierarchical network. A network in which processing and control functions are performed at several levels by computers specially suited for the functions performed, for example, in factory or laboratory automation.
High-Level Data Link Control (HDLC). A bit-oriented data link protocol that exercises control of data links by the use of a specified series of bits rather than by the control characters. HDLC is the protocol standardized by ISO.
high-speed circuit. A circuit designed to carry data at speeds greater than voice-grade circuits. *Synonym for* wideband circuit.

hit. A transient disturbance to a data communications medium that could mutilate characters being transmitted.
holding time. The duration of a switched call. Most often applied in traffic studies to the duration of a telephone call.
home page. At a site on the WWW, the top page in a hierarchy. The home page normally indicates what information is available on other pages to which it is connected.
host computer. In a network, a computer that primarily provides services, such as computation, database access, or special programs or programming languages.
hot key. A key or combination of keys that allows the user to switch from one computer or session to view information from another computer or session.
hot standby. A standby computer or telecommunications line in place that is ready to take over automatically in case of failure.
hub. A device that serves as a connection point for all of the wires or cables in a LAN. Some hubs have intelligence and can perform error detection.
hub polling. A type of polling in which each station polls the next station in succession on the communications circuit. The last station polls the first station on the circuit.
Huffman coding. A type of character compression.
hybrid network. A network made up of a combination of various network topologies.
hybrid systems. A term applied to telephone systems that have some of the characteristics of key systems and some of PBXs.
hypertext. A form of text that has highlighted words that can be clicked with a mouse to connect the user with a source of additional information about the word or topic. A tool of the World Wide Web on the Internet.
hypertext markup language (HTML). A formatting tool used to format pages for the WWW.
hypertext transfer protocol (HTTP). The protocol used to carry WWW traffic between a WWW browser computer and the WWW server being accessed.

identification (ID) characters. Characters sent by a station to identify itself. TWX, BSC, and SDLC stations use ID characters.
idle character. (1) A control character that is sent when there is no information to be transmitted. (2) A character transmitted on a telecommunications line that does not print or punch at the output component of the accepting terminal.
idle line. *Synonym for* inactive line.
image. A faithful likeness of the subject matter of the original.
impulse noise. A sudden spike on the communications circuit when the received amplitude goes beyond a certain level, caused by transient electrical impulses, such as lightning, switching equipment, or a motor starting.
IMS/VS. (Information Management System/Virtual Storage). A database-data communications product developed and marketed by IBM. It allows users to access a computer-maintained database through remote terminals.
inactive line. A telecommunications line that is not currently available for transmitting data. *Contrast with* active line.
inactive node. In a network, a node that is neither connected to nor available for connection to another node.
in-band signals. Signal that occurs within the frequency range allowed for a voice signal. *Contrast with* out-of-band signals.
incumbent local exchange carrier (ILEC). Established carriers, many of which were originally part of the Bell system.
information. The meaning that is assigned to data.
information bearer channel. A channel provided for data transmission that is capable of carrying all the necessary information to permit communications, including such information as users' data synchronizing sequences and control signals. It may, therefore, operate at a greater signaling rate than that required solely for the users' data.
information bits. In data communications, those bits that are generated by the data source and that are not used for error control by the data transmission system.
information highway. A popular term for the Internet. However, when thinking of the information highway, people normally envision a network that delivers data at much higher speeds than today.
information interchange. The process of sending and receiving data in such a manner that the information content or meaning assigned to the data is not altered during the transmission.
information processing. *Synonym for* data processing (DP).
information resource management (IRM). An organization of the information-related resources of a company usually incorporating data processing, data communications, voice communications, office automation, and sometimes the company's libraries.
information security. The protection of information against unauthorized disclosure, transfer, modifications, or destruction, whether accidental or intentional.
infrared. Light waves below the visible spectrum. Infrared light can be used for limited distance, line of sight or near-line of sight transmission.
inquiry. A request for information from storage, for example, a request for the number of available airline seats or a machine statement to initiate a search of library documents.
inquiry and transaction processing. A type of application in which inquiries and records of transactions received from a number of terminals are used to interrogate or update one or more master files.
inquiry/response communication. In a network, the process of exchanging messages and responses, with one exchange usually involving a request for information (an inquiry) and a response that provides the information.
integrated circuit. A combination of interconnected circuit elements inseparably associated on or within a continuous substrate.
Integrated Services Digital Network (ISDN). An evolving set of standards for a digital, public telephone network.
intelligent terminal. A terminal that can be programmed.

intensifying. A method for highlighting characters on the screen of a VDT for easy identification by the user. A character, or any collection of dots on an all-points-addressable (APA) screen, that is made brighter than the other characters around it.
inter-LATA. Long distance calls between LATAs. Inter-LATA calls are handled by an interexchange carrier.
interactive. Pertaining to an application in which each entry calls forth a response from a system or program, as in an inquiry system or an airline reservation system. An interactive system may also be conversational, implying a continuous dialogue between the user and the system.
interactive voice response. A type of telephone service in which the caller can obtain varying responses or information from a computer by inputting digits from a Touchtone telephone.
interconnect industry. A segment of the communications industry that makes equipment for attachment to the telephone network that provides customers with alternatives such as decorative telephones and private telephone systems for business.
interexchange carrier (IXC). Long distance carriers. *Contrast with* local exchange carrier (LEC).
interface. A shared boundary. An interface might be a hardware component to link two devices, or it might be a portion of storage or registers accessed by two or more computer programs.
intermediate system. In the context of an internet, subnetworks that provide a communication path and provide necessary relaying and routing of messages.
intermediate text block (ITB). A character used to terminate an intermediate block of characters. The block check character is sent immediately following ITB, but no line turnaround occurs. The response following ETB or ETX also applies to all of the ITB checks immediately preceding the block terminated by ETB or ETX.
intermessage delay. The elapsed time between the receipt at a terminal of a system response and the time that a new transaction is entered. *Synonym for* think time.
internal modem. A modem contained on a single circuit card that can be inserted into a personal computer or other device.
international direct distance dialing (IDDD). A telephone exchange service that enables the telephone user to call subscribers in other countries without operator assistance.
International Organization for Standardization (ISO). An organization established to promote the development of standards to facilitate the international exchange of goods and services and to develop mutual cooperation in areas of intellectual, scientific, technological, and economic activity.
International Telecommunications Union (ITU). The specialized telecommunications agency of the United Nations, established to provide standardized communications procedures and practices, including frequency allocation and radio regulations, on a worldwide basis.
International Telecommunications Union-Telecommunications Standardization Sector (ITU-T). The part of the ITU that deals with global telecommunications standards.
internet. An interconnected set of networks.
Internet. A TCP/IP-based, interconnected set of government, research, education, commercial, and private networks.
Internet service provider (ISP). A company or organization that provides access to the Internet, typically for a fee.
internetwork address. A destination address that contains enough information to route a message to a node on a different network.
interoffice trunk. A direct trunk between local central offices in the same exchange.
intertoll trunk. A trunk between toll offices in different telephone exchanges.
intranet. A private network modeled after the WWW on which browsers and servers are used to provide access to information of use to the particular audience. Many organizations have implemented intranets as a way to disseminate information to employees.
inverse concentrator. Equipment that takes a high-speed data stream, for example from a computer, and breaks it apart for transmission over multiple slower speed circuits.
invitation. The process in which a processor contacts a station in order to allow the station to transmit a message if it has one ready. *See also* polling.
invitation list. A series of sets of polling characters or identification sequences associated with the stations on a line. The order in which sets of polling characters are specified determines the order in which polled stations are invited to enter messages on the line.
isochronous transmission. A data transmission process in which there is always an integral number of unit intervals between any two significant instants.

jack. A connecting device to which a wire or wires of a circuit may be attached and that is arranged for the insertion of a plug.
jitter. Small, rapid, unwanted amplitude or phase changes of an analog signal. Small variations of the pulses of a digital signal from their ideal positions in time.
job. A set of data that completely defines a unit of work for a computer. A job usually includes all necessary computer programs, linkages, files, and instructions to the operating system.
journaling. Recording transactions against a dataset so that the dataset can be reconstructed by applying transactions in the journal against a previous version of the dataset.
joystick. In computer graphics, a lever that can pivot in all directions and that is used as a locator device.
jumbo group. Six 600-channel master groups, giving a total of 3,600 channels. The bandwidth of a jumbo group is 1.440 Mhz.

Kanji. A character set of symbols used in Japanese ideographic alphabets.
key. (1) On a keyboard, a control or switch by means of which a specified function is performed. (2) To enter characters or data from a keyboard.

key-encrypting key. A key used in sessions with cryptography to encipher and decipher other keys.
key system. A small, private telephone system.
keyboard. (1) On a typewriter or terminal, an arrangement of typing and function keys laid out in a specified manner. (2) A systematic arrangement of keys by which a machine is operated or by which data is entered. (3) A device for the encoding of data by key depression, which causes the generation of the selected code element. (4) A group of numerical keys, alphabetical keys, and function keys used for entering information into a terminal and into the system.
keyboard-send-receive (KSR). A combination teletypewriter transmitter and receiver with transmission capability from a keyboard only.
kilo (k). One thousand. For example, 1 kilohertz equals 1,000 hertz.

LAN Manager. Network operating system software for a LAN produced by Microsoft.
laser. A device that transmits an extremely narrow beam of energy in the visible light spectrum.
layer. (1) In the open systems interconnection (OSI) architecture, a collection of related functions that comprise one level of a hierarchy of functions. Each layer specifies its own functions and assumes that lower level functions are provided. (2) In SNA, a grouping of related functions that are logically separate from the functions in other layers. The implementation of the functions in one layer can be changed without affecting functions in other layers.
leased circuit. A circuit that is owned by a common carrier but leased from them by another organization for full-time, exclusive use. *See also* private line.
leased line. *See* leased circuit.
least cost routing. Routing a telephone call so that the cost of the call is minimized.
letters shift. A physical shift in a teletypewriter that enables the printing of lowercase characters.
level. The amplitude of a signal.
levels of support. A concept related to a support organization that suggests that the minimum skills necessary to solve a problem should be used. *See also* problem escalation.
light-emitting diode (LED). A semiconductor chip that gives off visible or infrared light when activated.
lightpen. A specialized input device that is attached to a VDT by cable. It is held by the operator and pressed against the screen of the terminal to mark a spot or indicate a selection from several choices.
line. (1) On a terminal, one or more characters entered before a return to the first printing or display position. (2) A string of characters accepted by the system as a single block of input from a terminal, for example, all characters entered before a carriage return or all characters entered before the terminal user presses the ATTENTION key. (3) *See* circuit.
line-by-line mode. A mode of operation for terminals in which one line at a time is sent to or received from the computer.
line control. *Synonym for* protocol.
line feed (LF). A format effector that causes the print or display position to move to the corresponding position on the next line. *See also* carriage-return character (CR).
line group. One or more telecommunications lines of the same type that can be activated and deactivated as a unit.
line level. The signal level in decibels at a particular position on a telecommunications line.
line load. Usually a percentage of maximum circuit capability that reflects actual use during a span of time; for example, peak hour line load.
line noise. Noise originating in a telecommunications line.
line switching. *Synonym for* circuit switching.
line trace. In the network control program, an optional function that logs all activity on the line.
line turnaround. A process for half-duplex transmission in which one modem stops transmitting and becomes the receiver, and the receiving modem becomes the transmitter.
linear predictive coding (LPC). A technique of digitizing an analog signal by predicting which direction the analog signal will take. LPC samples the analog signal less often than other digitizing techniques, allowing the transmitted bit rate to be lower.
link. A segment of a circuit between two points.
link access procedure, balanced (LAPB). A subset of the HDLC protocol that operates in full-duplex, point-to-point mode. It is most commonly used between an X.25 DTE and a packet switching network.
link access procedure, D-channel (LAPD). A subset of HDLC that provides data link control on an ISDN D channel in AMB mode. LAPD always uses a 16-bit address, 7-bit sequence numbers, and a 16-bit CRC.
link access procedure for frame-mode bearer services (LAPF). A data link protocol for frame relay networks. LAPF is made up of a control protocol, which is similar to HDLC, and a core protocol, which is a subset of the control protocol. The control protocol uses 16- to 32-bit addresses, 7-bit sequence numbers, and a 16-bit CRC. The core protocol has no control field, which means that there is no mechanism for error control, hence streamlining the operation of the network.
liquid crystal display (LCD). A video display technology that uses two sheets of polarizing material with a liquid crystal solution between them. An electric current passed through the liquid causes the crystals to align so that light cannot pass through them.
local access and transport area (LATA). The local calling areas that were defined originally within the United States when divestiture occurred.
local area network (LAN). A limited distance network, usually existing within a building or several buildings in close proximity to one another. Transmission on a LAN normally occurs at speeds of 1 Mbps and up.
local calling. Telephone calling within a designated local service area.
local calls. Calls within a local service area.
local central office. A central office arranged for terminating subscriber lines and provided with trunks of establishing connections to and from other central offices.

local exchange carrier (LEC). The BOCs and the independent telephone companies.

local loop. A channel connecting the subscriber's equipment to the line-terminating equipment in the central office exchange.

local service area. Telephones served by a particular central office and (usually) several surrounding central offices.

Localtalk. A LAN cabling system for Apple computers.

lockout. In a telephone circuit controlled by an echo suppressor, the inability of one or both subscribers to get through because of either excessive local circuit noise or continuous speech from one subscriber.

logical circuit. In packet mode operation, a means of duplex transmission across a data link, comprising associated send and receive channels. A number of logical circuits may be derived from a data link by packet interleaving. Several logical circuits may exist on the same data link.

logical link control (LLC). The data link control protocol defined by the IEEE for use on LANs.

logical unit (LU). SNA's view of a communications user.

log off. The procedure by which a user ends a terminal session.

log on. The procedure by which a user begins a terminal session.

long distance calls. Calls outside of the local service area.

longitudinal parity check. (1) A parity check performed on a group of binary digits in a longitudinal direction for each track. (2) A system of error checking performed at the receiving station after a block check character has been accumulated.

longitudinal redundancy check (LRC). *Synonym for* longitudinal parity check.

loop back test. A procedure in which signals are looped from a test instrument through a modem or loopback switch and back to the test instrument for measurement.

loop network. A network configuration in which there is a single path between all nodes, and the path is a closed circuit.

low speed. Usually, a data transmission speed of 600 bps or less.

low-speed circuit. A circuit that is designed for telegraph and teletypewriter usage at speeds of from 45 to 600 bps and that cannot handle a voice transmission. Used by the public telex network. *Synonym for* subvoice-grade circuit.

main distribution frame. A frame that has one part on which the permanent outside lines entering the central office building terminate and another part on which cabling, such as the subscriber line cabling or trunk cabling, terminates. In a PBX, the main distribution frame is for similar purposes.

management information base (MIB). The database in which the SNMP protocol stores information about the operation of a network.

Manchester coding. A digital signaling technique in which there is a transition in the middle of each bit time. A 1 is encoded with a low level during the first half of the bit time and a high level during the second half. A 0 is encoded with a high level during the first half of the bit time and a low level during the second half.

Manufacturing Automation Protocol (MAP). A communications protocol, based on the OSI reference model, specifically oriented toward use in an automated manufacturing environment.

marine telephone service. Telephone service for boats and ships that uses radio or satellite links to connect from the ship to a shore station.

mark. The normal no-traffic line condition by which a steady signal is transmitted.

master group. Ten supergroups, each of which contains 60 voice channels. The bandwidth of a master group is 2.4 MHz.

master station. *See* control station.

mean time between failures (MTBF). For a stated period in the life of a function unit, the mean value of the lengths of time between consecutive failures under stated conditions.

mean time to repair (MTTR). The average time required for corrective maintenance.

measured rate service. A method of charging for local calls based on the number of calls, their duration, and the distance.

medium. *See* transmission medium.

medium access control (MAC). A technique for determining which of several stations on a local area network can use the network.

medium speed. Usually, a data transmission rate between 600 bps and the limit of a voice-grade facility.

mega (M). One million. For example, 1 megaHertz equals 1,000,000 Hertz. Also 1 megaHertz equals 1,000 kiloHertz.

mesh network. A network configuration in which there are one or more paths between any two nodes.

message. (1) An arbitrary amount of information whose beginning and end are defined or implied. (2) A group of characters and control bit sequences transferred as an entity.

message center. A location where messages are received from a communications network and either forwarded to another location or delivered to the intended recipient.

message queue. A line of messages that are awaiting processing or waiting to be sent to a terminal.

message routing. The process of selecting the correct circuit path for a message.

message switching. (1) In a data network, the process of routing messages by receiving, storing, and forwarding complete messages. (2) The technique of receiving a complete message, storing it, and then forwarding it to its destination unaltered.

message text. The part of a message that is of concern to the party ultimately receiving the message, that is, the message exclusive of the header or control information.

metropolitan area network (MAN). A network of limited geographic scope, generally defined as within a 50-mile radius. Standards for MANs are being defined by the IEEE.

micro. One millionth.

microcomputer. A computer system whose processing unit is a microprocessor. A basic microcomputer includes a microprocessor, storage, and an input/output facility, which may or may not be on one chip.

microprocessor. An integrated circuit that accepts coded instructions for execution. The instructions may be entered, integrated, or stored internally.

microwave radio. Radio transmissions in the 4 to 28 GHz range. Microwave radio transmissions require that the transmitting and receiving antennas be within sight of each other.

milli. One thousandth.

mobile telephone switching office (MTSO). The central office of a cellular telephone system. MTSOs are typically connected to cellular antenna towers and to the public switched telephone system's central offices by cable.

modem. A device that modulates and demodulates signals transmitted over data communications lines. One of the functions of a modem is to enable digital data to be transmitted over analog transmission facilities.

modified final judgment (MFJ). The stipulation that on January 1, 1984, AT&T would divest itself of all 22 of its associated operating companies in the Bell System.

modified Huffman (MH) encoding. A method of encoding facsimile data before it is transmitted.

modified modified read (MMR) encoding. A method of encoding facsimile data before it is transmitted.

modified read (MR) encoding. A method of encoding facsimile data before it is transmitted.

modulation. The process by which some characteristic of one wave is varied in accordance with another wave or signal. This technique is used in modems to make DTE signals compatible with communications facilities.

modulation rate. The reciprocal of the measure of the shortest nominal time interval between successive significant instants of the modulated signal. If this measure is expressed in seconds, the modulation rate is given in bauds.

modulator. A functional unit that converts a signal into a modulated signal suitable for transmission. *Contrast with* demodulator.

monitor. Software or hardware that observes, supervises, controls, or verifies the operations of a system.

mouse. In computer graphics, a locator device operated by moving it on a surface.

multidrop circuit. *See* multipoint circuit.

multimode. A type of optical fiber with a core approximately 50 microns (.050 millimeter) in diameter.

multiple sessions. Having several connections to different software applications at the same time. The capability is normally provided by hardware or software in the terminal or terminal control unit.

multiplexer (MUX). A device capable of interleaving the events of two or more activities or capable of distributing the events of an interleaved sequence to the respective activities.

multiplexing. (1) In data transmission, a function that permits two or more data sources to share a common transmission medium such that each data source has its own channel. (2) The division of a transmission facility into two or more channels either by splitting the frequency band transmitted by the channel into narrower bands, each of which is used to constitute a distinct channel (frequency division multiplexing), or by allotting this common channel to several different information channels, one at a time (time division multiplexing).

multipoint circuit. A circuit with several nodes connected to it.

multipoint line. A communication line that has several nodes attached.

multiport. *See* split stream operation.

multipurpose internet mail extensions (MIME). An extension to the SMTP mail transfer protocol that overcomes many of the SMTP's limitations, such as its inability to handle foreign characters.

multi-station access unit (MAU). A wiring hub in a token ring LAN.

multitasking. The capability of a computer operating system to appear to run two or more programs simultaneously by rapidly switching back and forth between them.

National Exchange Carriers Association (NECA). An organization of communications carriers that sets North American wide area network standards.

negative acknowledge character (NAK). A transmission control character transmitted by a station as a negative response to the station with which the connection has been set up.

negative polling limit. For a start-stop or BSC terminal, the maximum number of consecutive negative responses to polling that the communications controller accepts before suspending polling operations.

Netware. Network operating system software for a LAN produced by Novell.

network. (1) An interconnected group of nodes. (2) The assembly of equipment through which connections are made between data stations.

network addressable unit (NAU). In SNA, a logical unit, a physical unit, or a system services control point. It is the origin or the destination of information transmitted by the path control network.

network application. The use to which a network is put, such as data collection or inquiry/update.

network architecture. A set of design principles, including the organization of functions and the description of data formats and procedures, used as the basis for design and implementation of a telecommunications application network.

Network Basic Input-Output System (NetBIOS). The part of the DOS operating system that provides the interface between IBM and compatible personal computers and a network.

network computer. A limited-capability computer that allows connection to the Internet but does not have all of the capability of a regular personal computer.

network congestion. A network condition in which traffic is greater than the network can carry, for any reason.

network control center (NCC). A place from which a communications network is operated and monitored.

network control mode. The functions of a network control program that enable it to direct a communications controller to perform activities, such as polling, device addressing, dialing, and answering.
network control program (NCP). A program that controls the operation of a front-end processor or communications controller.
network interface card (NIC). A circuit card in a personal computer that provides the electrical interface to a network.
network management. The process of operating and controlling a telecommunications network so that it meets the requirements of its users.
network node. *Synonym for* node.
network operating system (NOS). The software that controls a LAN's operation.
network operations. The activities required to run a network on a daily basis and to keep it running when problems occur.
network operator. A person or program responsible for controlling the operation of all or part of a network.
network operator console. A system console or terminal in the network from which an operator controls the network.
network simulator. *See* workload generator.
network topology. The schematic arrangement of the links and nodes of a network.
networked society. A vision of the future in which most people have the ability to communicate with each other and with various computer-based systems by several means.
900 Service. A telephone service for which the caller pays the cost of the telephone call.
node. In a network, a point at which one or more functional units interconnect transmission lines. The term *node* derives from graph theory in which a node is a junction point of links, areas, or edges.
noise. (1) Random variations of one or more characteristics of any entity, such as voltage, current, or data. (2) A random signal of known statistical properties of amplitude, distribution, and spectral density. (3) Loosely, any disturbance tending to interfere with the normal operation of a device or system.
nondestructive cursor. On a VDT device, a cursor that does not erase characters through which it passes as it is advanced, backspaced, or otherwise moved. *Contrast with* destructive cursor.
noninformation bits. In data communications, those bits that are used for error control or other purposes which do not directly convey the meaning of the message.
nonswitched connection. A connection that does not have to be established by dialing. *Contrast with* switched connection.
nontransparent mode. A mode of transmission in which all control characters are treated as control characters (that is, not treated as text). *Contrast with* transparent mode.
N-out-of-M code. A coding system in which M bits are used to transmit a character and N of the bits must be 1s.
null character (NUL). A control character that is used to accomplish media-fill or time-fill and that may be inserted into or removed from a sequence of characters without affecting the meaning of the sequence. However, the control of equipment or the format may be affected by this character.
numbering plan. A uniform numbering system in which each telephone central office has a unique designation similar in form to that of all other offices connected to the nationwide dialing network. In one numbering plan, the first 3 of 10 dialed digits are the area code, the next 3 are the office code, and the remaining 4 are the station number.
numeric keypad. Extra keys on a keyboard that function like a 10-key calculator.

off-hook. Activated (in regard to a telephone set). By extension, a data set automatically answering on a public switched system is said to go off-hook. *Contrast with* on-hook.
offline. Pertaining to the operation of a functional unit without the continual control of a computer.
offline system. A system in which human operations are required between the original recording functions and the ultimate data processing function. This includes conversion operations as well as the necessary loading and unloading operations incident to the use of point-to-point or data-gathering systems. *Contrast with* online system.
on-hook. Deactivated (in reference to a telephone set). A telephone not in use is on-hook. *Contrast with* off-hook.
one-way communication. Communication in which information is always transferred in one preassigned direction.
one-way trunk. A trunk between central exchanges where traffic can originate on only one end.
online. (1) The state of being connected, usually to a computer. (2) Pertaining to the operation of a functional unit that is under the continual control of a computer.
online system. A system in which the input data enters the computer directly from the point of origin or in which output data is transmitted directly to where it is used.
open network architecture (ONA). A set of provisions imposed by the FCC on the BOCs and AT&T to ensure the competitive availability of and access to unregulated, enhanced network services.
Open Systems Interconnection (OSI) reference model. A telecommunications architecture proposed by the International Standards Organization (ISO).
open wire. (1) A conductor separately supported above the surface of the ground; that is, on insulators. (2) A broken wire.
open-wire line. A pole line in which the conductors are principally in the form of bare, uninsulated wire. Ceramic, glass, or plastic insulators are used to physically attach the bare wire to the telephone poles. Short circuits between the individual conductors are avoided by appropriate spacing.
operating system. The central control program that governs a computer hardware's operation.

optical character recognition (OCR). The process of scanning a document with a beam of light and detecting individual characters.
optical fiber. A communications medium made of very thin glass or plastic fiber that conducts light waves.
optical recognition. A device that can detect individual data items or characters and convert them into ASCII or another code for transmission to a computer.
OS/2 LAN Server. Network operating system software for a LAN produced by IBM.
oscilloscope. An instrument for displaying the changes in a varying current or voltage.
out-of-band signals. Signals outside of the frequency range allowed for a voice signal. *Contrast with* in-band signals.
out-pulsing. The pulses caused by a rotary dial opening and closing an electrical circuit when the dial is turned and released.
outsourcing. The transfer of some of the activities of an organization to another company, usually for the purpose of obtaining specialized service or lower cost.
overrun. Loss of data because a receiving device is unable to accept data at the rate it is transmitted.

pacing. A technique by which a receiving station controls the rate of transmission of a sending station to prevent overrun.
packet. A sequence of binary digits (including data and control signals) that is switched as a composite whole. The data, control signals, and possibly error control information are arranged in a specific format.
packet assembly/disassembly (PAD). The process of dividing a message into packets at the transmitting end and reassembling the message from the packets at the receiving end.
packet data network (PDN). A network that uses packet switching techniques for transmitting data.
Packet filtering. *See* packet level firewall.
packetizing. The process of dividing a message into packets.
packet level firewall. A firewall computer that examines all network traffic at the packet level and allows or denies packet passage from one network to the other based on the source and destination addresses. This technique is called *packet filtering.*
packet sequencing. A process of ensuring that packets are delivered to the receiving data terminal equipment (DTE) in the same sequence as they were transmitted by the sending DTE.
packet switched data transmission service. A user service involving the transmission and, if necessary, the assembly and disassembly of data in the form of packets.
packet switching. The technique of sending packets through a network, sometimes by diverse routes.
page mode. *See* formatted mode.
parallel transmission. (1) In data communications, the simultaneous transmission of a certain number of signal elements constituting the same telegraph or data signal. (2) The simultaneous transmission of the bits constituting an entity of data over a data circuit. *Contrast with* serial transmission.
parity bit. The binary digit appended to a group of binary digits to make the sum of all the digits either always odd (odd parity) or always even (even parity).
parity check. A redundancy check that uses a parity bit.
patch. (1) A temporary electrical connection. (2) To make an improvised modification.
patch cord. A cable with plugs at both ends that is used to connect two devices.
patch panel. Equipment that allows a piece of equipment to be temporarily connected to other equipment or a circuit using patch cords.
path. In a network, a route between any two nodes.
peer-to-peer. The ability of two computers to communicate directly without passing through or using the capability of a mainframe computer.
pel. *See* picture element (pixel, pel).
performance management. The application of management principles to ensure that the performance of a system meets the required parameters.
permanent virtual circuit. In packet switching networks, a full-time connection between two nodes.
personal communications network (PCN). The European name for personal communication service.
personal communications service (PCS). A type of communication in which a user carries a small telephone-transceiver, which allows him or her to be reached regardless of where he or she is located.
personal handyphone (PHS) system. The Japanese name for their personal communications system (PCS) network.
personal identification number (PIN). A secret code or password that allows a person access to certain facilities or capabilities on a computer system.
phase. An attribute of an analog signal that describes its relative position measured in degrees.
phase jitter. An unwanted change in the phase of the signal.
phase modulation. Modulation in which the phase angle of a carrier is the characteristic varied.
phase shift. The offset of an analog signal from its previous location. Phase shifts are measured in degrees.
phase shift keying (PSK). A modulation technique in which the phase of an analog signal is varied.
physical unit (PU). SNA's view of communications hardware.
picture element (pixel, pel). (1) The part of the area of the original document that coincides with the scanning spot at a given instant and that is of one intensity only with no distinction of the details that may be included. (2) In computer graphics, the smallest element of a display space that can be independently assigned color and intensity. (3) The area of the finest detail that can be effectively reproduced on the recording medium.
pixel. *See* picture element (pixel, pel).
plain old telephone service (POTS). Basic telephone service with no special features.
plaintext. Unencrypted information. *See also* ciphertext.
point of presence (POP). The location within a LATA at which customers are connected to an IXC.

point of sale (POS) terminal. A specialized terminal designed to be used by a clerk in a store and to enter sales transactions into a computer.
point-to-point line. A circuit connecting two nodes. *Contrast with* multipoint circuit.
point-to-point-protocol (PPP). An asynchronous protocol that includes capabilities for line testing, authentication, data compression and error correction. It is primarily used by personal computers to dial into a TCP/IP-based network though it may be used on leased lines as well.
Poisson capacity table. A table for determining the number of circuits required to carry a certain level of telephone traffic. The Poisson table assumes that the sources of traffic are infinite and that all blocked calls will be retried within a short period of time.
polling. (1) Interrogation of devices for purposes such as to avoid contention, to determine operational status, or to determine readiness to send or receive data. (2) The process in which stations are invited, one at a time, to transmit.
polling characters. A set of characters peculiar to a terminal and the polling operation. Response to these characters indicates to the computer whether the terminal has a message to enter.
polling delay. A user-specified delay between passes through an invitation list for either a line or a line group.
polling ID. The unique character or characters associated with a particular station.
polling list. A list that specifies the sequence in which stations are to be polled.
polynomial error checking. An error checking technique in which the bits of a block of data are processed by a mathematical algorithm using a polynomial function to calculate the block check character.
port. An access point for a circuit.
port speed. In the context of a frame relay circuit, the port speed is the maximum transmission speed of the line between the customer and the frame relay carrier.
Post, Telephone, and Telegraph (PTT). A generic term for the government-operated common carriers in countries other than the U.S. and Canada.
presentation services. The processing required to change the format of messages from that required or generated by an application program to or from the format required by a specific terminal.
preventive software maintenance. A philosophy of software maintenance that advocates the application of software corrections, whether the problem being corrected has been seen or not. The intention is to correct software problems before they are seen by users.
primary access. A method of accessing an ISDN network in which the user has twenty-three 64-kbps B channels and one 16-kbps D channel. This type of access is also known as 23B+D.
print server. A server on a LAN that provides the hardware and software to drive one or more printers.
private circuit. A telecommunications line that is owned by a company other than a communications carrier.
private automatic branch exchange (PABX). A private automatic telephone exchange that provides for the transmission of calls to and from the public telephone network. *See also* private branch exchange (PBX).
private branch exchange (PBX). A private telephone exchange connected to the public telephone network on the user's premises.
private key. In a public key encryption system, the key that is kept secret and never distributed.
private line. A communications circuit that is owned by a company other than a common carrier.
private network. A network built by a company for its exclusive use, which uses circuits available from a variety of sources.
problem escalation. The process of bringing a problem to the attention of higher levels of management and/or bringing more highly trained technical resources to work on the problem.
problem management. The application of management principles to ensure that problems are resolved as quickly as possible with the minimum resources.
problem tracking meeting. A periodic meeting to discuss the status of all open problems that have not yet been resolved.
profit center. A department or other entity that generates more revenue than it spends. Profit centers typically sell their services or recharge them to other departments in the company. *Contrast with* cost center.
program evaluation review technique (PERT). A project management technique that shows on a chart the interrelationships of activities.
program function keys. Special keys on a terminal keyboard that direct the computer to perform specific actions determined by the computer program.
project life cycle. A concept expressing the desirability of gradually adding people to a project in order to minimize the cost until the specifications and benefits of the project are known and management's commitment is assured.
project team. A group of people organized for the purpose of completing a project.
propagation delay. The time necessary for a signal to travel from one point to another.
protocol. (1) A specification for the format and relative timing of information exchanged between communicating parties. (2) The set of rules governing the operation of functional units of a communications system that must be followed if communications are to be achieved.
protocol analyzer. Test equipment that examines the bits on a communications circuit to determine whether the rules of a particular protocol are being followed.
protocol converter. Hardware or software that converts a data transmission from one protocol to another.
protocol data unit (PDU). A frame of logical link control protocol.
proxy server. A server that changes the network addresses from one form to another, so that users or computers on one network do not know the actual addresses of nodes on the other network.

public data transmission service. A data transmission service established and operated by an administration and provided by means of a public data network. Circuit-switched, packet-switched, and leased circuit data transmission services are feasible.
public key. In a public key encryption system, the key that is revealed (made public) to anyone who may want to use it.
public key encryption. An encryption technology that uses two keys. One is publicly known and the other is kept private and never distributed. A public-key cryptographic algorithm relies on one key for encryption and a different but related key for decryption.
public network. A network established and operated by communications common carriers or telecommunications administrations for the specific purpose of providing circuit-switched, packet-switched, and leased circuit services to the public.
public service commission (PSC). *See* public utility commission (PUC).
public switched network (PSN). A network that provides circuits switched to many customers. In the United States, there are three: telex, TWX, and telephone.
public switched telephone network (PSTN). *See* public switched network (PSN).
public telephone network. *See* public switched network (PSN).
public utility commission (PUC). An arm of state government that has jurisdiction over intrastate rates and services. *Also known as* public service commission (PSC).
pulse. A variation in the value of a quantity, short in relation to the time schedule of interest, the final value being the same as the initial value.
pulse code modulation (PCM). A process in which a signal is sampled, and the magnitude of each sample with respect to a fixed reference is quantized and converted by coding to a digital signal.
punchdown block. A connector for telephone wiring on which the connection is made by pushing (punching) the wire between two prongs with a special tool.
punched paper tape. A medium used on older teletypewriters. Characters were coded in the tape by punched holes.
push-button dialing. The use of keys or push buttons instead of a rotary dial to generate a sequence of digits to establish a circuit connection. The signal form is usually tones. *Contrast with* rotary dial.
push-button dialing pad. A 12-key device used to originate tone keying signals. It usually is attached to rotary dial telephones for use in originating data signals.

quadbit. A group of four bits. In 16-phase modulation, each possible quadbit is encoded as one of 16 unique carrier phase shifts. *Contrast with* dibit *and* tribit.
quadrature amplitude modulation (QAM). A combination of phase and amplitude modulation used to achieve high data rates while maintaining relatively low signaling rates.
quantization. The subdivision of the range of values of a variable into a finite number of nonoverlapping, and not necessarily equal, subranges or intervals, each of which is represented by an assigned value within the subrange. For example, a person's age is quantized for most purposes with a quantum of 1 year.
quantizing noise. The error introduced when an analog signal is digitized.
queue. A line or list formed by items in a system waiting for service, for example, tasks to be performed or messages to be transmitted in a message routing system.
queuing. The process of placing items that cannot be handled into a queue to await service.

radiated media. Transmission medium that propagates the signal through the air such as radio, infrared, or microwave.
radio paging. The broadcast of a special radio signal that activates a small portable receiver carried by the person being paged.
rate center. A specified geographic location used by telephone companies to determine mileage measurements for the application of interexchange mileage rates.
real enough time. Response time that is fast enough to meet the requirements of a particular application.
realtime. Pertaining to an application in which response to input is fast enough to affect subsequent input, such as a process control system or a computer-assisted instruction system.
receive-only (RO). A teletypewriter that has no keyboard. It is used where no input to the computer is desired or necessary.
reed relay. A switch that has contacts that open or close when an electrical current is applied.
regenerative repeater. *See* repeater.
Regional Bell Operating Company (RBOC). One of the seven corporations formed when divestiture occurred and that comprise the 22 Bell Operating Companies.
regional center. A control center connecting sectional centers of the telephone system together. Every pair of regional centers in the United States has a direct circuit group running from one center to the other.
relative transmission level. The ratio of the test-tone power at one point to the test-tone power at some other point in the system chosen as a reference point. The ratio is expressed in decibels. The transmission level at the transmitting switchboard is frequently taken as zero level reference point.
relay center. A central point at which message switching takes place; a message switching center.
reliability. Trouble-free operation.
remote batch. Data collected in a batch and then transmitted to the computer as a unit.
remote job entry (RJE). The process of submitting a job to a computer for processing using telecommunications lines. Normally the output of the processing is returned on the lines to the terminal from which the job was submitted.

remote terminal. A terminal attached to a computer via a telecommunications line.
repeater. A device that performs digital signal regeneration together with ancillary functions. Its function is to retime and retransmit the received signal impulses restored to their original shape and strength.
request for proposal (RFP). A letter or document sent to vendors asking them to show how a (communications) problem or situation can be addressed. Normally, the vendor's response to an RFP proposes a solution and quotes estimated prices.
request for quotation (RFQ). *See* request for proposal (RFP).
response. An answer to an inquiry.
response time. The elapsed time between the end of an inquiry or demand on a data processing system and the beginning of the response, for example, the length of time between an indication of the end of an inquiry and the display of the first character of the response at a user terminal.
reverse channel. A means of simultaneous communications from the receiver to the transmitter over half-duplex data transmission systems. The reverse channel is generally used only for the transmission of control information and operates at a much slower speed than the primary channel.
reverse video. A technique used with VDTs that reverses the character and background colors for highlighting purposes.
ring. (1) The signal made by a telephone to indicate an incoming call. (2) A part of a plug used to make circuit connections in a manual switchboard or patch panel. The ring is the connector attached to the negative side of the common battery that powers the station equipment. By extension, it is the negative battery side of a telecommunications line. *Contrast with* tip.
ring network. A network in which each node is connected to two adjacent nodes.
ringback tone. An audible signal indicating that the called party is being rung.
roll call polling. The most common implementation of a polling system, in which one station on a line is designated as the master and the others are slaves.
rotary dial. In a switched system, the conventional dialing method that creates a series of pulses to identify the called station. *Contrast with* push-button dialing *and* dual-tone multifrequency (DTMF).
router. A piece of hardware or software that directs messages toward their destination, often from one network to another.
routing code. The name given in some countries to the part of the telephone number known in the United States as the area code.
RS-232-C. A specification for the physical, mechanical, and electrical interface between data terminal equipment (DTE) and data circuit-terminating equipment (DCE). *See also* V.24.
RS-232-D. A 1987 revision and update to the RS-232-C specification that is exactly compatible with the V.24 standard.
RS-336. A specification for the interface between a modem (DCE) and a terminal or computer (DTE). This interface, unlike the RS-232-C, has a provision for the automatic dialing of calls under modem control.
RS-449. A specification for the interface between a modem (DCE) and a terminal or computer (DTE). This specification was designed to overcome some of the problems with the RS-232-C interface specification.
run length encoding. A type of compression in which the input text is scanned for repeating characters, which, when found, are reduced to shorter character strings.

satellite carrier. A company that offers communications services using satellites.
scrambler. A voice encryption device that makes the voice unintelligible to anyone without a descrambler, effectively rendering wiretapping useless.
screen. An illuminated display surface; for example, the display surface of a VDT or plasma panel.
search engine. A program that uses a keyword or keywords to search for information in a database. Most commonly used in the context of the Internet on which search engines allow users to search for information using multiple keywords and Boolean logic.
segment. A portion of a LAN that has been separated because of distance or traffic. Segments are connected to other segments by bridges or switches.
seize. To gain control of a line in order to transmit data. *Contrast with* bid.
serial. (1) Pertaining to the sequential performance of two or more activities in a single device. In English, the modifiers serial and parallel usually refer to devices, as opposed to sequential and consecutive, which refer to processes. (2) Pertaining to the sequential processing of the individual parts of a whole, such as the bits of a character or the characters of a word, using the same facilities for successive parts.
serial line internet protocol (SLIP). A protocol for carrying IP over dial-up or leased lines. SLIP contains little negotiation capability and does not support error detection or correction.
serial system. A system made up of a number of components connected in series. In telecommunications, a terminal, modem, and computer may be connected in a series forming a serial system.
serial transmission. (1) In data communications, transmission at successive intervals of signal elements constituting the same telegraph or data signal. The sequential elements may be transmitted with or without interruption, provided that they are not transmitted simultaneously. (2) The sequential transmission of the bits constituting an entity of data over a data circuit. *Contrast with* parallel transmission.
server. On a local area network, a computer with software that provides service to other devices on the LAN. Typical servers are file servers, print servers, and communications servers.

service level agreement. A set of performance objectives reached by consensus between the user and the provider of a service.
serving central office. A telephone subscriber's local central office.
session. (1) A connection between two stations that allows them to communicate. (2) The period of time during which a user of a terminal can communicate with an interactive system; usually, the elapsed time between log on and log off.
shielded twisted pair. Twisted pair wires surrounded by a metallic shield.
shielding. A metallic sheath that surrounds the center conductor of a cable. Coaxial cable has shielding around the center conductor.
sidetone. The small amount of signal fed back from the mouthpiece to the receiver of a telephone handset.
signal. A variation of a physical quantity, used to convey data.
signal processor. Electronic circuitry that is designed to manipulate a signal.
signal-to-noise ratio. The ratio of signal strength to noise strength.
signal transformation. The action of modifying one or more characteristics of a signal, such as its maximum value, shape, or timing.
signaling rate. The number of times per second that a signal changes. Signaling rate is measured in baud.
Signaling System No. 7 (SS7). A signaling system used among telephone company central offices to set up calls, indicate their status, and tear down the calls when they are completed.
simple mail transfer protocol (SMTP). An application layer protocol used by TCP/IP-based networks for the exchange of electronic mail.
Simple Network Management Protocol (SNMP). A protocol, which originated on the Internet, for exchanging network management commands and information between devices on a network.
simplex circuit. A circuit that carries communication in one direction only.
simplex communication. *Synonym for* one-way communication.
simplex transmission. Transmission on a telecommunications line in one direction only. Transmission in the other direction is not allowed.
simultaneous transmission. Transmission of control characters or data in one direction while information is being received in the other direction.
sine wave. The waveform of a single-frequency analog signal of constant amplitude and phase.
single-address message. A message that is to be delivered to only one destination.
single mode. A type of optical fiber that has a glass or plastic core approximately 5 microns (.005 millimeter) in diameter.
sink. In telecommunications, the receiver.
site license. An agreement, usually for software, that allows its use by an unlimited number of people at the site. The term *site* may be defined in different ways according to the terms of the agreement.
slave station. A data station that operates under the control of a master or control station. *Contrast with* control station.
smart home. A home that uses microprocessors to perform certain functions, such as controling heating, lighting, and security alarms.
smart terminal. A terminal that is not programmable but that has memory capable of being loaded with information.
software defined network (SDN). A bulk pricing offered by telephone companies designed for businesses or others who make large numbers of calls. Standard-switched telephone lines are used to carry the calls.
software metering. A method of keeping track of the number of simultaneous users of a piece of software installed on a server. The usual purpose of software metering is to prevent more people from simultaneously using the software than have been paid for in the software license agreement.
software protocol converter. A computer program that converts one protocol to another.
source. The transmitting station in a telecommunications system.
space. *See* space signal.
space signal. In asynchronous transmission, the space signal is the signal for a zero bit.
speaker cone. The paper-like membrane of a speaker that vibrates in response to the movement of a voice coil. The voice coil movement is caused by the interaction of an electrical signal in the coil with a magnet in the speaker.
specific polling. A polling technique that sends invitation characters to a device to find out whether the device is ready to enter data.
split stream operation. A modem feature that allows several slower speed data streams to be combined into one higher speed data stream for transmission, and split apart at the receiving end. The total data rate of the slower speed data streams must not exceed the capacity of the circuit.
spooling. A technique of queuing input or output between slow-speed and high-speed computer hardware. Print spooling is most common. Output from several computers is queued (spooled) to a disk until a printer is free.
spread spectrum. A radio transmission technique in which the frequency of the transmission is changed periodically.
standard test-tone power. One milliwatt (0 dBm) at 1,000 cycles per second.
star network. A network configuration in which there is only one path between a central or controlling node and each endpoint node.
start bit. *See* start signal.
start of header (SOH). A transmission control character used as the first character of a message heading.
start of text (STX). A transmission control character that precedes text and that may be used to terminate the message heading.

start signal. (1) A signal to a receiving mechanism to get ready to receive data or perform a function. (2) In a start/stop system, a signal preceding a character or block that prepares the receiving device for the reception of the code elements. A start signal is limited to one signal element generally having the duration of a unit interval. *Synonym for* start bit.
start/stop transmission. (1) Asynchronous transmission such that a group of signals representing a character is preceded by a start element and is followed by a stop element. (2) Asynchronous transmission in which a group of bits is preceded by a start bit that prepares the receiving mechanism for the reception and registration of a character and is followed by at least 1 stop bit that enables the receiving mechanism to come to an idle condition pending the reception of the next character.
station. One of the input or output points of a system that uses telecommunications facilities; for example, the telephone set in the telephone system or the point at which the business machine interfaces with the channel on a leased private line.
station extension. An extension telephone.
station features. Features of a telephone system that are activated by the user of the system.
station message detail recording (SMDR). A feature of a telephone system that records detailed information about telephone calls placed through the system.
station selection code. A Western Union term for an identifying call that is transmitted to an outlying telegraph receiver and automatically turns its printer on.
statistical time division multiplexing (STDM). A device that combines signals from several terminals. An STDM does not reserve specific time slots for each device but assigns time only when a device has data to send.
step-by-step switch. A switch that moves in synch with a pulse device, such as a rotary telephone dial. Each digit dialed moves successive selector switches to carry the connection forward until the desired line is reached.
stop and wait ARQ. An error checking technique in which each block of data must be acknowledged before the next block can be sent. *Contrast with* continuous ARQ.
stop bit. In start/stop transmission, the bit that indicates the end of a character. *See also* start/stop transmission.
stop signal. (1) A signal to a receiving mechanism to wait for the next signal. (2) In a start/stop system, a signal following a character or block that prepares the receiving device for the reception of a subsequent character or block.
store-and-forward. An application in which input is transmitted, usually to a computer, stored, and then later delivered to the recipient.
store-and-forward switch. A type of switch that reads entire packets and stores them, if necessary, before sending them to the destination.
Street Talk. A directory service for networks produced by Banyan.
stress testing. Placing a heavy load on a system to see if it performs properly.
Strowger switch. A step-by-step switch named after its inventor, Almon B. Strowger. *See also* step-by-step switch.
subnetwork. (1) In the OSI reference model, layers 1, 2, and 3 together constitute the subnetwork. (2) A portion of the network. (3) In the context of the Internet, one of the interconnected networks that also continues to operate on its own and maintain its own identity.
subordinate station. *See* slave station.
subscriber's loop. *See* local loop.
subvoice-grade circuit. A circuit of bandwidth narrower than that of voice-grade circuits. Such circuits are usually subchannels of a voice-grade line.
supergroup. Five 48 kHz groups of 12 voice channels each. The total bandwidth of a supergroup is 240 kHz.
superhighway effect. A concept that suggests that the capacity of new facilities is often exceeded faster than anticipated because users, finding the capability better than expected, use it to a greater extent or for purposes beyond those for which it was originally designed.
switch. (1) In the context of a LAN, a device that connects two or more LAN segments together and allows all of the connections to operate simultaneously. (2) In the context of the telephone system, a device in the central office that makes the connection for telephone calls. (3) Sometimes a PBX is referred to as a switch.
switched connection. (1) A mode of operating a data link in which a circuit or channel is established to switching facilities, as, for example, in a public switched network. (2) A connection that is established by dialing. *Contrast with* nonswitched connection.
switched line. A telecommunications line in which the connection is established by dialing. *See also* dial-up line *and* leased circuit.
switched telecommunications network. A switched network furnished by communications common carriers or telecommunications administrations.
switched virtual circuit. In packet switching networks, a temporary connection between two nodes established only for the duration of a session.
switchhook. A switch on a telephone set, associated with the structure supporting the receiver or handset. It is operated by the removal or replacement of the receiver or handset on the support.
switching center. A location that terminates multiple circuits. It is capable of interconnecting circuits or transferring traffic between circuits.
switching office. A telephone company location that contains switching equipment.
symmetric digital subscriber line (SDSL). One of the DSL family of services that provides equal speed channels in both directions. SDSL is capable of speeds up to 768 Kbps in each direction and is targeted at business customers. *See also* digital subscriber line.
synchronization character (SYN). In a data message, a character that is inserted in the data stream from time to time by the transmitting station to ensure that the receiver is maintaining character synchronization and properly grouping the bits into characters. The synchronization

characters are removed by the receiver and do not remain in the received message.
synchronous. (1) Pertaining to two or more processes that depend on the occurrences of specific events, such as common timing signals. (2) Occurring with a regular or predictable time relationship.
Synchronous Data Link Control (SDLC). A bit-oriented data link protocol developed by IBM. SDLC is a proper subset of HDLC.
synchronous line control. A scheme of operating procedures and control signals by which telecommunications lines are controlled.
synchronous optical network (SONET). A standard for transmitting data on optical fibers.
synchronous transmission. (1) Data transmission in which the time of occurrence of each signal representing a bit is related to a fixed time frame. (2) Data transmission in which the sending and receiving instruments are operating continuously at substantially the same frequency and are maintained, by means of correction, in a desired phase relationship.
system features. Features of a telephone system that are available to all users and that may be automatically activated on behalf of the user.
Systems Network Architecture (SNA). A seven-layer communications architecture developed by IBM to serve as a basis for future telecommunications products.

T-carrier system. A family of high-speed, digital transmission systems, designated according to their transmission capacity.
tariff. The published rate for a specific unit of equipment, facility, or type of service provided by a telecommunications carrier. Also, the vehicle by which the regulating agencies approve or disapprove of such facilities or services. Thus, the tariff becomes a contract between the customer and the telecommunications facility.
Technical Office Protocol (TOP). A communications architecture, based on the OSI reference model, specifically oriented toward office automation.
telecommunications. (1) Any transmission, emission, or reception of signs, signals, writing, images, and sounds or intelligence of any nature by wire, radio, optical, or other electromagnetic systems. (2) Communication, as by telegraph or telephone.
telecommunications access method (TCAM). Communications software that controls communications lines. *See also* communications access method (CAM).
telecommunications control unit. *See* front-end processor (FEP).
telecommunications facility. Transmission capabilities or the means for providing such capabilities.
telecommunications monitor (TCM). Computer software that governs the overall operation of a network and may provide transaction processing or other services.
telecommuting. Using telecommunications to work from home or other locations instead of on the business's premises.
telediagnosis. Using telecommunications to diagnose a problem from a remote location.
telegraph. A system employing the interruption or change in polarity of direct current for the transmission of signals.
telegraph-grade circuit. A circuit suitable for transmission by teletypewriter equipment. Normally, the circuit is considered to employ DC signaling at a maximum speed of 75 baud.
telephone company. Any common carrier providing public telephone system service.
telephony. Transmission of speech or other sounds.
teleprinter. Equipment used in a printing telegraph system.
teleprocessing. Remote access data processing.
teletex. A standardized communications messaging technology allowing automatic, error-free transmissions between terminals at speeds 48 times greater than telex. Teletex is the logical successor to telex and TWX.
Teletype. Trademark of AT&T, usually referring to a series of teleprinter equipment, such as tape punches, reperforators, and page printers, that is used for telecommunications.
teletypewriter. A slow-speed terminal with a keyboard for input and paper for receiving printed output.
teletypewriter exchange (TWX). Teletypewriter service provided by Western Union in which suitably arranged teletypewriter stations are provided with lines to a central office for access to other such stations throughout the United States and Canada. Both Baudot and ASCII coded machines are used. Business machines may also be used with certain restrictions.
telex. An international message-switching service that uses teleprinters to produce hardcopy of the messages.
telex network. An international public messaging service using slow-speed teletypewriter equipment and Baudot code to exchange messages between subscribers. In the United States, telex service is provided by Western Union.
telnet. A capability of the Internet that allows a person or computer to log on to another computer on the Internet.
terminal. (1) A device, usually equipped with a keyboard and a display device, capable of sending and receiving information over a link. (2) A point in a system or network at which data can either enter or leave.
terminal component. A separately addressable part of a terminal that performs an input or output function, such as the display component of a keyboard-display device or the printer component of a keyboard-printer device.
terminal emulation program. A program that makes a personal computer act like a terminal that is recognized by the host.
terminal session. *See* session.
test tone. A tone used in identifying circuits for trouble location or for circuit adjustment.
text. The part of a data message containing the subject matter of interest to the user.

thick Ethernet. An Ethernet operating on traditional 0.4-inch coaxial cable.
thin client. An application designed such that little or no processing is done on the client computer; most processing is done on the server. *See also* fat client.
thin Ethernet. An Ethernet operating on more flexible 0.25-inch coaxial cable.
think time. The time between the receipt of a message at a terminal until the next message is entered by the user. *Synonym for* intermessage delay *and for* user response time.
tie line. *See* tie trunk.
tie trunk. A telephone line or channel directly connecting two branch exchanges.
time assignment speech interpolation (TASI). A technique of multiplexing telephone calls by taking advantage of the pauses in normal speech and assigning the channel to another call during the pause.
time division multiple access (TDMA). A transmission technique used in digital radio transmission in which the use of a frequency is divided into time slots that are shared among several users.
time division multiplexing (TDM). A technique that divides a circuit's capacity into time slots, each of which is used by a different voice or data signal.
timesharing. (1) Pertaining to the interleaved use of time on a computer system that enables two or more users to execute computer programs concurrently. (2) A mode of operation of a data processing system that provides for the interleaving in time of two or more processes in one processor. (3) A method of using a computing system that allows a number of users to execute programs concurrently and to interact with the programs during execution.
tip. The end of the plug used to make circuit connections in a manual switchboard or patch panel. The tip is the connector attached to the positive side of the common battery that powers the station equipment. By extension, it is the positive battery side of a telecommunications line. *Contrast with* ring.
tip and ring. Telephone company jargon for the two wires that carry a telephone signal. The term is a carry-over from earlier days when the plug at the telephone operator's console had two connection points, the tip and the ring.
token. A particular character in a token-oriented protocol. The terminal that has the token has the right to use the communications circuit.
toll. In public switched systems, a charge for a connection beyond an exchange boundary that is based on time and distance.
toll calls. Calls outside a local service area.
toll office. A central office at which channels and toll circuits terminate. Whereas there is usually one particular central office in a city, larger cities may have several central offices where toll message circuits terminate.
toll trunk. A communications circuit between telephone company toll offices.
tone dialing. *See* dual-tone-multifrequency (DTMF).
tone signaling. Signaling performed by sending tones on a circuit.
topology. The way in which a network's circuits are configured.
torn tape message system. Deprecated name for a teletypewriter-based message switching system in which paper tape was torn off one teletypewriter and read into another.
torn tape switching center. A location where operators tear off the incoming printed and punched paper tape and transfer it manually to the proper outgoing circuit.
total cost of ownership. A concept that tries to identify all of the costs associated with the acquisition and ownership of a device or piece of equipment during its lifetime. The concept of the total cost of ownership was first applied to personal computers by the Gartner Group in the early 1990s.
touch-sensitive. A VDT screen that can detect the location of the user's finger using either a photosensitive or resistive technique.
Touchtone. AT&T's tradename for dual-tone-multifrequency dialing.
trackball. In computer graphics, a ball, movable about its center, that is used as a locator device.
traffic. Transmitted and received messages.
trailer. The part of a data message following the text.
transaction. An exchange between a terminal and another device, such as a computer, that accomplishes a particular action or result; for example, the entry of a customer's deposit and the updating of the customer's balance.
transaction processing system. A system in which the users run prewritten programs to perform business transactions, generally of a somewhat repetitive nature.
transceiver. A terminal that can transmit and receive traffic.
transient error. An error that occurs once or at unpredictable intervals.
transition. The switching from one state (for example, positive voltage) to another (negative voltage).
transmission. (1) The process of sending data from one place for reception elsewhere. (2) In data communications, a series of characters including headings and texts. (3) The process of dispatching a signal, message, or other form of intelligence by wire, radio, telegraphy, telephony, facsimile, or other means. (4) One or more blocks or messages. *Note:* Transmission implies only the sending of data; the data may or may not be received.
transmission code. A code for sending information over telecommunications lines.
transmission control character. (1) Any control character used to control or facilitate transmission of data between data terminal equipment (DTE). (2) Characters transmitted over a line that are not message data but that cause certain control operations to be performed when encountered. Among such operations are addressing, polling, message delimiting and blocking, transmission error checking, and carriage return.
Transmission Control Protocol/Internet Protocol (TCP/IP). A set of transmission rules for interconnecting communications networks. TCP/IP is heavily supported by the U.S. government.
transmission control unit. *See* front-end processor (FEP).

transmission efficiency. The ratio of information bits to total bits transmitted.
transmission medium. Any material substance that can be, or is, used for the propagation of signals, usually in the form of modulated radio, light, or acoustic waves, from one point to another, such as an optical fiber, cable, bundle, wire, dielectric slab, water, or air. *Note:* Free space can also be considered a transmission medium for electromagnetic waves.
transmit. (1) To send data from one place for reception elsewhere. (2) To move an entity from one place to another; for example, to broadcast radio waves, to dispatch data via a transmission medium, or to transfer data from one data station to another via a line.
transnational data flow (TNDF). The transmission of data across national borders.
transparent data. Data that is not recognized as containing transmission control characters. Transparent data is sometimes preceded by a control byte and a count of the amount of data following.
transparent mode. A mode of binary synchronous text transmission in which data, including normally restricted data link control characters, is transmitted only as specific bit patterns. Control characters that are intended to be effective are preceded by a DLE character. *Contrast with* nontransparent mode.
trellis code modulation (TCM). A specialized form of quadrature amplitude modulation that codes the data so that many bit combinations are invalid. TCM is used for high-speed data communications.
tribit. A group of 3 bits. In eight-phase modulation, each possible tribit is encoded as 1 of 8 unique carrier phase shifts. *Contrast with* dibit *and* quadbit.
triple DES. An enhancement to the standard data encryption technique, which doubles the encryption key length to 112 bits.
trouble ticket. An online record or paper form that is filled out to document the symptoms of a problem and the action being taken to correct it.
trunk. A telephone channel between two central offices or switching devices that is used to provide a telephone connection between subscribers.
trunk group. The trunks between two switching centers, individual message distribution points, or both, that use the same multiplex terminal equipment.
tuning. The process of making adjustments to a system to improve its performance.
turnaround time. The actual time required to reverse the direction of transmission from send to receive or vice versa when a half-duplex circuit is used. For most telecommunications facilities, there will be time required by line propagation and line effects, modem timing, and machine reaction. A typical time is 200 milliseconds on a half-duplex telephone connection.
twisted pair wires. A pair of wires insulated with a plastic coating and twisted together that is used as a medium for telecommunications circuits.
two-wire circuit. A metallic circuit formed by two conductors insulated from each other. It is possible to use the two conductors as a one-way transmission path, a half-duplex, or a duplex path.

unbuffered. A terminal in which a character is transmitted to the computer as soon as a key on the keyboard is pressed.
unguided media. *See* radiated media.
Unicode. A standardized coding system that has 2^{16} code points that can be used to represent all the characters of all languages.
unified messaging. A system that allows the handling of e-mail, voice mail, and faxes through one electronic mailbox, so that a person can connect to it and receive all of his or her messages, regardless of how they were originally sent.
uniform resource locator (URL). The address that is used to specify a server and home page on the World Wide Web (WWW).
unipolar. A digital signaling technique in which a 1 bit is represented by a positive voltage pulse and a 0 bit by no voltage.
universal international freephone numbering (UIFN). A worldwide toll-free calling service under which the calls are paid for by the called party. Similar to 800 service in the United States.
universal service. The attribute of the telephone system that allows any station to connect to any other.
unshielded twisted pair (UTP). *See* twisted pair wires.
uplink. The microwave radio signal beamed up to a satellite.
uploading. The transmission of a file of data from a personal computer to a mainframe.
uptime. The time that a telecommunications network is operating and available to be used.
user. (1) The ultimate source or destination of information flowing through a system. (2) A person, process, program, device, or system that employs a user application network for data processing and information exchange.
user friendly. A terminal or system that is easy to learn and easy to use.
user group. In the context of the Internet, groups of people who get together online to discuss particular topics and exchange ideas.
user response time. The time it takes the user to see what the computer displayed, interpret it, type the next transaction, and press the ENTER key. *See also* think time.
user written commands. Control sequences written by users to perform predefined functions.
usenet newsgroup. A type of discussion group on the Internet that notifies registered members of new messages since the last log on.

V.24. An ITU-T specification for the interface between a modem (DCE) and a terminal or computer (DTE). The interface is identical to the RS-232-D. *See also* RS-232-D *and* RS-232-C.
V.32. An ITU-T standard for transmitting data at 9,600 bps, full duplex on a switched circuit.

V.32bis. An ITU-T standard for transmitting data at 14,400 bps, full duplex on a switched circuit.
V.34. An ITU-T standard for transmitting data at 28,800 bps, full duplex on a switched circuit. V.34 assumes that most of the transmission will occur on a relatively error-free digital circuit.
V.34bis. An ITU-T standard for transmitting data at 33,600 bps, full duplex on a switched circuit. V.34bis assumes that most of the transmission will occur on a relatively error-free digital circuit.
V.42. An ITU-T standard for error detection and error correction.
V.42bis. An ITU-T standard that specifies how modems will compress data before transmitting.
V.90. An ITU-T standard for transmitting data at 56,000 bps, full duplex on a switched circuit. V.90 assumes that at least one end of the communications line has a pure digital connection to the telephone network. V.90 transmission is asymmetric, in that the 56 kbps data rate is only achieved on the half of the transmission from the all-digital end of the connection. Transmissions from the analog end follow the V.34bis standard and occur at a maximum rate of 33.6 kbps.
value-added carrier. A carrier that provides enhanced communications services. Normally, some type of computation is provided in addition to the basic communications service.
value-added network (VAN). A public data network that contains intelligence that provides enhanced communications services.
vertical redundancy check (VRC). A parity check performed on each character of data as the block is received.
very-high-rate digital subscriber line (VDSL). One of the DSL family of services that transmits data at speeds of 51 to 55 Mbps over short twisted pair telephone lines of up to 1,000 feet, and as low as 13 Mbps at 4,000 feet. Early versions of this technology are asymmetric, like ADSL, and have an upstream channel of 1.6 to 2.3 Mbps. *See also* digital subscriber line *and* asymmetric digital subscriber line.
very small aperture terminal (VSAT). A satellite system using the Ku band of microwave frequencies, which require small receiving antennas.
video conferencing. Meetings conducted in rooms equipped with television cameras and receivers for remote users' participation.
video display terminal (VDT). A computer terminal with a screen on which characters or graphics are displayed and (normally) a keyboard that is used to enter data.
video display unit (VDU). *See* video display terminal (VDT).
video signal compression. The process of reducing the number of bits required to carry a digitized video signal while maintaining adequate quality.
videotex. An application in which the computer is able to store text and images in digital form and transmit them to remote terminals for display or interaction.
VINES. Network operating system software for a LAN produced by Banyan.
virtual circuit. A temporary circuit built between the sender and receiver of a telecommunications transmission (e.g., for a telephone call).
virtual network. As contrasted with a leased or private network, a virtual network appears to the customer as though it is dedicated for his exclusive use, but in reality it uses the public switched telephone network to provide service.
virtual telecommunications access method (VTAM). IBM's primary telecommunications access method.
virtual terminal. A concept that allows an application program to send or receive data to or from a generic terminal. Other software transforms the input and output to correspond to the actual characteristics of the real terminal being used.
virus. *See* computer virus.
voice-band. The 300 Hz to 3,300 Hz band used on telephone equipment for the transmission of voice and data.
voice compression. The process of reducing the number of bits required to carry a digitized voice signal while maintaining the essential characteristics of speech.
voice-grade circuit. A circuit suitable for transmission of speech, digital or analog data, or facsimile, generally with a frequency range of about 300 Hz to 3,300 Hz. Voice-grade circuits can transmit data at speeds up to 19,200 bps.
voice mail. A messaging service that people use to leave voice messages for others. The system provides a voice mailbox on a computer in which the voice messages are digitized and stored. The voice equivalent of an electronic mail system for textual messages.
voice over IP (VoIP). The ability to send voice signals over a network that uses the IP protocol. This is notable because networks based on the IP protocol were originally designed to handle data transmissions.
voice response unit. Hardware designed to respond to input signals with a spoken voice.

WATS. *See* 800 Service.
wavelength division multiplexing (WDM). A technique used on optical fibers in which many light beams of different wavelengths are transmitted along a single fiber simultaneously without interfering with one another. Each light beam can carry many individually modulated data streams, allowing very high data rates to be achieved.
Web browser. Software that provides access to the World Wide Web and sometimes other parts of the Internet.
white noise. *See* background noise.
wide area network (WAN). A network that covers a large geographic area, requiring the crossing of public right-of-ways and the use of circuits provided by a common carrier.
Wide Area Telecommunications Services (WATS). An older name for 800 service. *See* 800 service.
wideband circuit. A circuit designed to carry data at speeds greater than voice-grade circuits. *Synonym for* high-speed circuit.
Windows NT. Network operating system software for a LAN produced by Microsoft.
wireless application protocol (WAP). A protocol for transmitting WWW pages to cellular telephones. WAP downsizes fat, graphics-rich web pages so that they are usable on small cell phone displays.

wireless communications. Communications in which the media is not wire or cable but the signal is broadcast by radio.
workload generator. Computer software designed to generate transactions or other work for a computer or network for testing purposes.
workstation. (1) The place where a terminal operator sits or stands to do work. It contains a working surface, terminal, chair, and other equipment or supplies needed by the person to do his or her job. (2) A powerful microcomputer that contains specialized software to assist a person in doing his or her job.
World Wide Web (WWW). A collection of servers on the Internet that provide home pages that use hypertext markup language (HTML). The servers are identified by uniform resource locators (URLs).

X.21. A specification for an interface between data terminal equipment (DTE) and a digital public telephone network.
X.21 BIS. A specification for an interface between data terminal equipment (DTE) and an analog telephone network. X.21 BIS is electrically virtually identical to the RS-232-C and V.24 interface specifications.
X.25 standard for data transmission. The first three layers of the OSI reference model. A standard for data transmission using a packet switching network.
X.400 standard for electronic mail. A standard for the transmission of electronic mail.
X.500 standard for network directories. A standard that specifies how to create and maintain a directory of e-mail users and their network addresses.
Xmodem protocol. An asynchronous protocol developed for use between microcomputers, especially for transfers of data files between them.

Ymodem protocol. An asynchronous protocol developed for use between microcomputers, especially for transfers of data files between them.

zero slot LAN. A LAN that connects to personal computers or workstations using the computer's serial or parallel port.
Zmodem protocol. An asynchronous protocol developed for use between microcomputers, especially for transfers of data files between them.

INDEX

Linux

CAPACITY PLANNING

CIRCUIT UTILIZATION STUDY + OUTAGES

NETWORK SECURITY P564/565

P616

P504

SWITCHES TRAFFIC ANALYSIS

NETWORK — AUDIT

NW WORKLOAD MANAGEMENT

P593 — TRAFFIC LOADS

P574. COMMUNICATION TECHNICAL SUPPORT

P570 MODEM CABINET

DISASTER RECOVERY PLANING

P558/557 CHANGE MANAGEMENT P569

PROTOCOL MANAGEMENT P.630 180-183

P640

P.SODC

TELECOM DGT P.628

AN P185

180 Frequency 177-180

P640 TELECOM MANAGEMENT P.

DR PLANNING

EDI - page 677

679

GPS -

(ICMS)

569

TRAINING

CARGO MANAGEMENT

(ARCHI-

RADIO 753

P.557

REQUIREMENTS P592

STANDARDS

TRAFFIC FLOW P593

LAN MANAGEMENT

TRAFFIC LOAD P.492

MAIN TRAFFIC

NETWORK P.825

"RF TECHNOLOG

COSTS page 608

P192

(RADIO SIGNAL